Molecular Modelling

PRINCIPLES AND
APPLICATIONS

Andrew R. Leach

Glaxo Wellcome Research and Development
and the University of Southampton

To Christine

Addison Wesley Longman Limited
Edinburgh Gate, Harlow
Essex CM20 2JE, England
and Associated Companies throughout the world

First published 1996

British Library Cataloguing in Publication Data
A catalogue entry for this title is available from the British Library.

ISBN 0-582-23933-8

Produced by Longman Singapore Publishers (Pte) Ltd.
Printed in Singapore

Contents

Preface

Molecular modelling used to be restricted to a small number of scientists who had access to the necessary computer hardware and software. Its practitioners wrote their own programs, managed their own computer systems and mended them when they broke down. Today's computer workstations are much more powerful than the mainframe computers of even a few years ago and can be purchased relatively cheaply. It is no longer necessary for the modeller to write computer programs as software can be obtained from commercial software companies and academic laboratories. Molecular modelling can now be performed in any laboratory or classroom.

This book is intended to provide an introduction to some of the techniques used in molecular modelling and computational chemistry, and to illustrate how these techniques can be used to study physical, chemical and biological phenomena. A major objective is to provide, in one volume, some of the theoretical background to the vast array of methods available to the molecular modeller. I also hope that the book will help the reader to select the most appropriate method for a problem and so make the most of his or her modelling hardware and software. Many modelling programs are extremely simple to use and are often supplied with seductive graphical interfaces which obviously helps to make modelling techniques more accessible, but it can also be very easy to select a wholly inappropriate technique or method.

Most molecular modelling studies involve three stages. In the first stage a model is selected to describe the intra- and inter-molecular interactions in the system. The two most common models that are used in molecular modelling are quantum mechanics and molecular mechanics. These models enable the energy of any arrangement of the atoms and molecules in the system to be calculated, and allow the modeller to determine how the energy of the system varies as the positions of the atoms and molecules change. The second stage of a molecular modelling study is the calculation itself, such as an energy minimisation, a molecular dynamics or Monte Carlo simulation, or a conformational search. Finally, the calculation must be analysed, not only to calculate properties but also to check that it has been performed properly.

The book is organised so that some of the techniques discussed in later chapters refer to material discussed earlier, though I have tried to make each chapter as independent of the others as possible. Some readers may therefore be pleased to know that it is not essential to completely digest the chapters on quantum mechanics and molecular mechanics in order to read about methods for searching conformational space! Readers with experience in one or more areas may of course wish to be more selective.

I have tried to provide as much of the underlying theory as seems appropriate to enable the reader to understand the fundamentals of each method. In doing so I have assumed some background knowledge of quantum mechanics, statistical mechanics, conformational analysis and mathematics. A reader with an undergraduate degree in chemistry should have covered this material, which should also be familiar to many undergraduates in the final year of their degree course. Full discussion can be found in the suggestions for further reading at the end of each chapter. I have also attempted to provide a reasonable selection of original references, though in a book of this scope it is obviously impossible to provide a comprehensive coverage of the literature. In this context, I apologise in advance if any technique is inappropriately attributed.

In Chapter 1 we consider some of the historical background to molecular modelling and discuss a number of important general principles that are common to many modelling methods. We also examine the use of computer graphics, the Internet and the World-Wide Web and the molecular modelling literature. Chapter 1 concludes with a brief summary of some relevant mathematical concepts. Chapters 2 and 3 describe quantum mechanics and molecular mechanics, which are the two major methods used to model the interactions within a molecular system. These methods can be used to calculate the energy of a given arrangement of the atoms as well as certain other properties. In Chapters 4–8 we examine energy minimisation, molecular dynamics, Monte Carlo simulations and conformational analysis. These techniques use an appropriate energy model to determine a wide range of structural and thermodynamic properties. The final two chapters describe various techniques that combine concepts from previous chapters. In Chapter 9 we discuss the calculation of free energies using computer simulation, continuum solvent models, and methods for simulating chemical reactions. Chapter 10 is concerned with computational methods for discovering and designing new molecules, such as database searching, *de novo* design and quantitative structure–activity relationships.

The range of systems that can be considered in molecular modelling is extremely broad, from isolated molecules through simple atomic and molecular liquids to polymers, biological macromolecules such as proteins and DNA and solids. Many of the techniques are illustrated with examples chosen to reflect the breadth of applications. It is inevitable that for reasons of space some techniques must be dealt with in a rudimentary fashion (or not at all), and that many interesting and important applications cannot be described. Molecular modelling is a rapidly developing discipline, and has benefited from the dramatic improvements in computer hardware and software of recent years. Calculations that were major undertakings only a few years ago can now be performed using personal computing facilities. Thus, examples used to indicate the 'state of the art' at the time of writing will invariably be routine within a short time.

Symbols and physical constants

This list contains the most frequently used symbols and physical constants ordered according to approximate appearance in the text.

λ	Lagrange multiplier
r, θ, ϕ	spherical polar coordinates
i, **j**, **k**	orthogonal unit vectors along x, y, z axes
ϕ, θ, ψ	Euler angles
$\langle x\rangle$ or $\bar{x}$	arithmetic mean value of x
I	unit matrix
i	square root of -1
$\hat{\mathbf{r}}$	unit vector
α	exponent in Gaussian function (normal distribution)
σ	standard deviation
σ^2	variance
h	Planck's constant (6.62618×10^{-34} Js)
$\hbar$	$h/2\pi$ (1.05459×10^{-34} Js)
m	particle mass
Ψ	molecular wavefunction
∇^2	$\partial^2/\partial x^2 + \partial^2/\partial y^2 + \partial^2/\partial z^2$ ('del-squared')
$\mathscr{H}$	Hamiltonian
ψ	spatial orbital
α, β	spin functions ('spin up' and 'spin down')
χ	spin orbital (product of spatial orbital and a spin function)
ϕ	basis function/atomic orbital (usually labelled ϕ_μ, ϕ_ν, ϕ_λ, ϕ_σ)
dv or d**r**	indicates an integral over all spatial coordinates
dσ	indicates an integral over all spin coordinates
dτ	indicates an integral over all spatial and spin coordinates
r_{ij}	distances between two particles i and j (usually electrons in quantum mechanics)
R_{AB}	distance between two nuclei A and B
δ_{ij}	Kronecker delta ($\delta_{ij} = 1$ if i = j; $\delta_{ij} = 0$ if i $\neq$ j)
$\mathscr{K}$	exchange operator
$\mathscr{J}$	Coulomb operator
$\mathscr{H}^{core}$	core Hamiltonian operator
F	Fock matrix
S	overlap matrix
S_{ij}	overlap integral between orbitals i and j

f	Fock operator
$\mathbf{C}$	matrix of basis function coefficients
$\mathbf{E}$	matrix of orbital energies
$\mathbf{P}$	density matrix
ζ	Slater exponent
K	number of basis functions
N	number of electrons
M	number of nuclei
α	Coulomb integral in Hückel theory
β	resonance integral in Hückel theory
α	atomic or molecular polarisability
$\rho(\mathbf{r})$	electron density at $\mathbf{r}$
$\phi(\mathbf{r})$	electrostatic potential at $\mathbf{r}$
q_i	partial charge on atom i
Z_A	nuclear charge on atom A
μ	dipole moment
Θ	quadruple moment
l	bond length
k	force constant
θ	bond angle
τ, ω	torsion angle
$\mathbf{E}$	electric field
ε	well depth parameter in Lennard-Jones pairwise potential function
σ	collision diameter used in Lennard-Jones function
r^*, r_m	separation corresponding to minimum in Lennard-Jones function
A	coefficient of r^{-12} in Lennard-Jones function
C	coefficient of r^{-6} in Lennard-Jones function
ε_0	dielectric constant *in vacuo*
ε_r	relative permittivity
ε	dielectric constant ($\varepsilon = \varepsilon_0\varepsilon_r$)
k_B	Boltmann's constant (1.38066×10^{-23} JK^{-1})
N_A	Avogadro's number (6.02205×10^{23} mol^{-1})
N	number of particles in system
$\mathbf{r}$	atomic or molecular position
$\mathbf{r}^N$	denotes positions of the N particles in the system
$\mathbf{p}$	linear momentum
$\mathbf{p}^N$	denotes momenta of the N particles in the system
$\mathbf{P}$	total linear momentum of system
$\mathcal{V}$	total potential energy of system (often written as a function of $\mathbf{r}^N$)
v	pairwise potential energy
$\mathbf{x}_k$	3N Cartesian vector at point k
$\mathbf{g}_k$	gradient at point k (1st derivative of $\mathcal{V}$ with respect to the coordinates)
$\boldsymbol{\mathcal{V}}''$	Hessian matrix of second derivatives of $\boldsymbol{\mathcal{V}}'$ with respect to the coordinates
ν	normal mode vibrational frequency
$\mathbf{H}$	inverse Hessian matrix
Q	canonical ensemble partition function
Z	configurational integral

$\mathscr{K}$	kinetic energy
ρ	density
T	temperature
Λ	de Broglie wavelength $(=\sqrt{[h^2/2\pi m k_B T]})$
t	time
P	pressure
V	volume
E	instantaneous energy
U	internal energy
A	Helmholtz free energy
G	Gibbs free energy
C_v	heat capacity at constant volume
$g(r)$	pair radial distribution function
S	switching function
L	length of box in periodic simulations
s	statistical inefficiency
ω	angular velocity
μ	reduced mass
W	virial
$\mathbf{v}$	velocity
$\mathbf{a}$	acceleration
$\mathbf{b}$	third derivative of position with respect to time $(d^3\mathbf{r}/dt^3)$
δt	time step in molecular dynamics simulations
$\mathbf{f}_{ij}$	force between particles i and j
$\langle A \rangle$	Ensemble average value of property A
q	generalised coordinate
C_{xy}	un-normalised correlation function
c_{xy}	normalised correlation function
τ	coupling parameter
κ	isothermal compressibility
γ	collision frequency (in stochastic dynamics)
S	order parameter
ρ	probability density function
$\boldsymbol{\pi}$	transition matrix
$\boldsymbol{\alpha}$	stochastic matrix
ξ	random number (usually in range 0 to 1)
z	activity
μ	chemical potential
W	Rosenbluth weight
$\mathbf{G}$	metric matrix (in distance geometry)
p_i	ith principal component
$\mathbf{Z}$	variance–covariance matrix
λ	coupling parameter (used in free energy calculations)
$W(\mathbf{r}^N)$	weighting function used in umbrella sampling
$\mathscr{N}$	number density $(= N/V)$
σ	Hammett substitution constant
P	partition coefficient of solute between two solvents
π	$\log(P_X/P_H)$ for a substituent X relative to a hydrogen substituent

Acknowledgements

The idea of a book covering the techniques of molecular modelling and computational chemistry arose while I was preparing material for a third-year undergraduate course in the Department of Chemistry at the University of Southampton given jointly with Dominic Tildesley. Kathy Hick from Longman was visiting the Department at about the same time and expressed an interest in publishing such a textbook. Had I known that the subsequent two and a half years would see not only the birth of our first child but also a change of job and a house move I am not sure that I would have been very keen on the project. However, thanks largely to the support and patience of my family, the end is in sight and I am finally able to compose these words of acknowledgement.

While writing this book I have benefited enormously from discussions with many colleagues, friends and collaborators. I am particularly grateful to those who gave their time to read and comment upon draft copies of various chapters. I would therefore like to thank the following (in alphabetical order): Dr D B Adolf, Dr J M Blaney, Professor A V Chadwick, Dr P S Charifson, Dr C-W Chung, Dr A Cleasby, Dr A Emerson, Dr J W Essex, Dr D V S Green, Dr I R Gould, Dr M M Hann, Dr C A Leach, Dr M Pass, Dr D A Pearlman, Dr C A Reynolds, Dr D W Salt, Dr M Saqi, Professor J I Siepmann, Dr W C Swope, Dr N R Taylor, Dr P J Thomas, Professor D J Tildesley, Mr O Warschkow. My discussions with Jonathan Essex and Dominic Tildesley have been particularly fruitful in shaping the sections involving computer simulations. Teri Klein very kindly helped with the laborious task of proofreading. Any errors that remain are of course entirely my own responsibility. If you do find any, I would like to know! I will also be pleased to receive any constructive suggestions, comments or criticisms. I would also like to thank Prof. T E Ferrin, Dr S E Greasley, Dr M M Hann, Dr H Jhoti, Dr S N Jordan, Professor G R Luckhurst, Dr P M McMeekin, Dr A Nicholls, Dr P Popelier and Dr T E Klein for help with the production of the figures.

I would also like to acknowledge those who have guided my path through the field, including Dr J G Vinter, Dr C K Prout, Dr D P Dolata, Professor P A Kollman, Professor I D Kuntz and Professor R Langridge. Many others too numerous to mention have also helped to make my research career intellectually stimulating and above all, fun!

I have been very fortunate in my publishing team at Longman. Alexandra Seabrook took over from Kathy Hick as senior publisher and Christopher Leeding and Bill Jenkins performed their onerous editorial tasks with true commitment. I would also like to thank everyone else involved in the production and marketing of the book.

This book is dedicated to my wife Christine who has not only assisted in its preparation by checking several sections but has also provided much needed moral support. She has endured without complaint that common trait among authors which elevates the need to finish the book above all other tasks (such as painting the house or fixing the garden fence). I hope that the final result compensates to some extent for the many evenings and weekends that she has spent on her own while I was working.

Molecular modelling would not be what it is today without the efforts of those who develop computer hardware and software and I would like to acknowledge the authors of the following computer programs which were used to generate figures and/or data described in the text. All calculations were performed using Silicon Graphics computers.

InsightII: Biosym Technologies of San Diego, California, USA (Figures 1.4, 1.5, 1.7, 6.21, 8.19, 8.39, 8.40, 8.41, 10.1, 10.21)

Netscape: Netscape Communications Corp (Figure 1.8)

Gaussian 92: Frisch M J, G W Trucks, M Head-Gordon, P M W Gill, M W Wong, J B Foresman, B G Johnson, H B Schlegel, M A Robb, E S Replogle, R Gomperts, J L Andres, K Raghavachari, J S Binkley, C Gonzalez, R L Martin, D J Fox, D J DeFrees, J Baker, J J P Stewart and J A Pople. Gaussian Inc, Pittsburgh, Pennsylvania, USA (Data for Figures 2.23, 3.19, 4.15, 4.22)

Spartan: Wavefunction Inc., Irvine, California, USA (Figures 2.17, 2.18, 2.19, 2.24)

SPASMS (San Francisco Package of Applications for the Simulation of Molecular Systems). Spellmeyer D A, W C Swope, E-R Evensen, T Cheatham, D M Ferguson and P A Kollman. University of California San Francisco, USA (data for Figures 5.2, 6.3, 6.10)

GRASP (Graphical Representation and Analysis of Surface Properties): A Nicholls, Columbia University, New York, USA (Figures 9.22, 9.23)

Micromol: Colwell S M, A R Marshall, R D Amos and N C Handy 1985. Quantum Chemistry on Microcomputers, *Chemistry in Britain* **21**:655–659 (some of the data for the calculation described in section 2.5.5)

Dials and Windows: Ravishanker G, S Swaminathan, D L Beveridge, R Lavery and H Sklenar 1989. *Journal of Biomolecular Structure and Dynamics* **6**:669–699. Wesleyan University, USA (Figure 6.15)

PROCHECK: Laskowski R, M W MacArthur, D S Moss and J M Thornton 1993. Procheck—A program to check the stereochemical quality of protein structures. *Journal of Applied Crystallography* **26**:283–291 (Figure 8.3)

HIPPO: Zsoldos Z 1995. New methods for *de novo* 3D Structure Design. PhD thesis, University of Leeds (Figure 10.22). Also at http://chem.leeds.ac.uk/ICAMS/eccc/hippo.html (WWW)

Molscript: Kraulis P J 1991. Molscript—A program to produce both detailed and schematic plots of protein structures. *Journal of Applied Crystallography* **24**: 946–950 (Figure 8.40)

GRID: Goodford P J 1985. A Computational Procedure for Determining

Energetically Favorable Binding Sites on Biologically Important Macromolecules. *Journal of Medicinal Chemistry* **28**:849–857. Molecular Discovery Ltd., Oxford, United Kingdom (data for Figure 10.21)

GCG: Genetics Computer Group, Inc. University Research Park, 575 Science Drive, Suite B, Madison, Wisconsin 53711 USA. (data for Figure 8.42)

The Regents of the University of California (Figure 1.8)

Cambridge Structural Database: Allen F H, S A Bellard, M D Brice, B A Cartwright, A Doubleday, H Higgs, T Hummelink, B G Hummelink-Peters, O Kennard, W D S Motherwell, J R Rodgers and D G Watson 1979. The Cambridge Crystallographic Data Centre: Computer-Based Search, Retrieval, Analysis and Display of Information. *Acta Crystallographica* **B35**:2331–2339. Cambridge Crystallographic Data Centre, Cambridge, United Kingdom (data for Figures 8.28, 8.31)

Quanta: Molecular Simulations Inc., 200 Fifth Avenue, Waltham, MA 02154 USA. (Figure 8.22)

The following programs were used to produce draft copies of the manuscript and diagrams: Microsoft Word, version 5.1 (Microsoft Corp.); Gnuplot (T Williams and C Kelley); Kaleidagraph (Abelbeck Software); and Chem3D (CambridgeSoft Corp.).

We are grateful to the following for permission to reproduce copyright material:

American Chemical Society for Fig. 3.48 (Pranata and Jorgensen, 1991), Fig. 4.23 (Doubleday *et al.*, 1985), Fig. 6.13 (Jorgensen, Binning and Bigot, 1981), Fig. 9.28 (Chandrasekhar and Jorgensen, 1985) and Fig. 9.30 (Aqvist, Fothergill and Warshel, 1993); American Institute of Physics for Fig. 4.29 (Gonzalez and Schlegel, 1988), Fig. 6.2 (Alder and Wainwright, 1959), Fig. 6.12 (Guillot, 1991); Gordon and Breach Publishers for Fig. 7.15 (Cracknell, 1994).

Whilst every effort has been made to trace the owners of copyright material, in a few cases this has proved impossible, and we take this opportunity to offer our apologies to any copyright holders whose rights we may have unwittingly infringed.

1 Useful concepts in molecular modelling

1.1 Introduction

What is molecular modelling? 'Molecular' clearly implies some connection with molecules. The Oxford English Dictionary defines 'model' as 'a simplified or idealised description of a system or process, often in mathematical terms, devised to facilitate calculations and predictions'. Molecular modelling would therefore appear to be concerned with ways to mimic the behaviour of molecules and molecular systems. Today, molecular modelling is invariably associated with computer modelling, but it is quite feasible to perform some simple molecular modelling studies using mechanical models or a pencil, paper and hand calculator. Nevertheless, computational techniques have revolutionised molecular modelling to the extent that most calculations could not be performed without the use of a computer. This is not to imply that a more sophisticated model is necessarily any better than a simple one, but computers have certainly extended the range of models that can be considered and the systems to which they can be applied.

The 'models' that most chemists first encounter are molecular models such as the 'stick' models devised by Dreiding or the 'space filling' models of Corey, Pauling and Koltun (commonly referred to as CPK models). These models enable three-dimensional representations of the structures of molecules to be constructed. An important advantage of these models is that they are interactive, enabling the user to pose 'what if...' or 'is it possible to...' questions. These structural models continue to play an important role both in teaching and in research, but molecular modelling is also concerned with more abstract models, many of which have a distinguished history. An obvious example is quantum mechanics, the foundations of which were laid many years before the first computers were constructed.

There is a lot of confusion over the meaning of the terms 'theoretical chemistry', 'computational chemistry' and 'molecular modelling'. Indeed, many practitioners use all three labels to describe aspects of their research, as the occasion demands! 'Theoretical chemistry' is often considered synonymous with quantum mechanics, whereas computational chemistry encompasses not only quantum mechanics but also molecular mechanics, minimisation, simulations, conformational analysis and other computer-based methods for understanding and predicting the behaviour of molecular systems. Molecular modellers use all of these methods and so we shall not concern ourselves with semantics but rather shall consider any theoretical or computational technique that provides insight into the behaviour of molecular systems to be an example of molecular modelling. If a distinction has to be made, it is in the emphasis that molecular modelling places on the representation and manipulation of the structures of molecules, and properties that are dependent upon those three-dimensional structures. The prominent part that computer graphics has played in molecular modelling has led some scientists to consider molecular modelling as little more than a method for generating 'pretty pictures' but the technique is now firmly established, widely used and accepted as a discipline in its own right.

In the rest of this chapter we shall discuss a number of concepts and techniques that are relevant to many areas of molecular modelling and so do not sit comfortably in any individual chapter. We will also define some of the terms that will be used throughout the book.

1.2 Coordinate systems

It is obviously important to be able to specify the positions of the atoms and/or molecules in the system to a modelling program.* There are two common ways in which this can be done. The most straightforward approach is to specify the Cartesian (x, y, z) coordinates of all the atoms present. The alternative is to use *internal coordinates*, in which the position of each atom is described relative to other atoms in the system. Internal coordinates are usually written as a Z-matrix. The Z-matrix contains one line for each atom in the system. A sample Z-matrix for the staggered conformation of ethane (see Figure 1.1) is as follows:

1	C						
2	C	1.54	1				
3	H	1.0	1	109.5	2		
4	H	1.0	2	109.5	1	180.0	3
5	H	1.0	1	109.5	2	60.0	4
6	H	1.0	2	109.5	1	−60.0	5
7	H	1.0	1	109.5	2	180.0	6
8	H	1.0	2	109.5	1	60.0	7

In the first line of the Z-matrix we define atom 1, which is a carbon atom. Atom number 2 is also a carbon atom that is a distance of 1.54 Å from atom 1 (columns 3 and 4). Atom 3 is a hydrogen atom that is bonded to atom 1 with a bond length of 1.0 Å. The angle formed by atoms 2–1–3 is 109.5°, information that is speci-

* For a system containing a large number of independent molecules it is common to use the term 'configuration' to refer to each arrangement; this use of the word configuration is not to be confused with its standard chemical meaning as a different bonding arrangement of the atoms in a molecule.

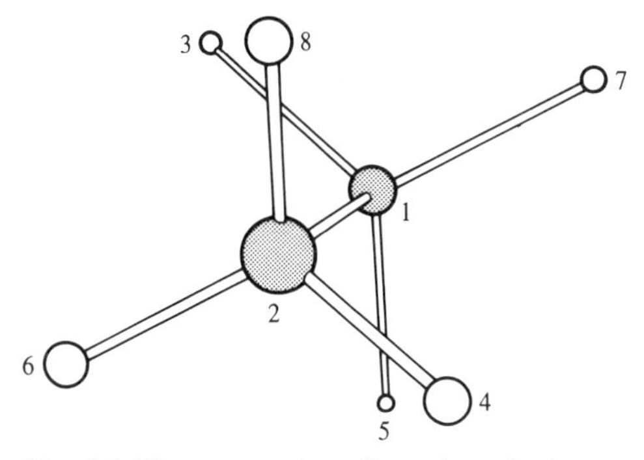

Fig. 1.1 The staggered conformation of ethane.

fied in columns 5 and 6. The fourth atom is a hydrogen, a distance of 1.0 Å from atom 2, the angle 4–2–1 is 109.5°, and the torsion angle* for atoms 4–2–1–3 is 180°. Thus for all except the first three atoms, each atom has three internal coordinates: the distance of the atom from one of the atoms previously defined, the angle formed by the atom and two of the previous atoms, and the torsion angle defined by the atom and three of the previous atoms. Fewer internal coordinates are required for the first three atoms because the first atom can be placed anywhere in space (and so it has no internal coordinates); for the second atom it is only necessary to specify its distance from the first atom and then for the third atom only a distance and an angle are required.

It is always possible to convert internal to Cartesian coordinates and vice versa. However, one coordinate system is usually preferred for a given application. Internal coordinates can usefully describe the relationship between the atoms in a single molecule, but Cartesian coordinates may be more appropriate when describing a collection of discrete molecules. Internal coordinates are commonly used as input to quantum mechanics programs whereas calculations using molecular mechanics are usually done in Cartesian coordinates. The total number of coordinates that must be specified in the internal coordinate system is six fewer than the number of Cartesian coordinates for a non-linear molecule. This is because we are at liberty to arbitrarily translate and rotate the system within Cartesian space without changing the relative positions of the atoms.

* The torsion angle (also referred to as a dihedral) of four atoms A–B–C–D is defined as the angle between the two planes, one containing atoms A, B and C and the other containing atoms B, C and D, as shown in Figure 1.2. A torsion angle can vary through 360° although the range $-180°$ to $+180°$ is most commonly used. We shall adopt the IUPAC definition of a torsion angle in which an eclipsed conformation corresponds to a torsion angle of 0° and a *trans* or *anti* conformation to a torsion angle of 180°. The reader should note that this may not correspond to some of the definitions used in the literature, where the *trans* arrangement is defined as a torsion angle of 0°. If one looks along the bond B–C, then the torsion angle is the angle through which it is necessary to rotate the bond AB in a clockwise sense in order to superimpose the two planes, Figure 1.2.

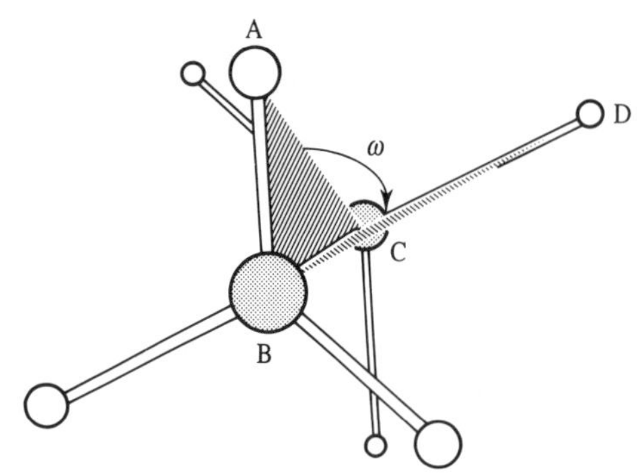

Fig. 1.2 A torsion angle A–B–C–D is defined as the angle between the planes A, B, C and B, C, D.

1.3 Potential energy surfaces

In molecular modelling the Born–Oppenheimer approximation is invariably assumed to operate. This enables the electronic and nuclear motions to be separated; the much smaller mass of the electrons means that they can rapidly adjust to any change in the nuclear positions. Consequently, the energy of a molecule in its ground electronic state can be considered a function of the nuclear coordinates only. If some or all of the nuclei move then the energy will usually change. The new nuclear positions could be the result of a simple process such as a single bond rotation or it could arise from the concerted movement of a large number of atoms. The magnitude of the accompanying rise or fall in the energy will depend upon the type of change involved. For example, about 3 kcal/mol is required to change the covalent carbon–carbon bond length in ethane by 0.1 Å away from its equilibrium value, but only about 0.1 kcal/mol is required to increase the non-covalent separation between two argon atoms by 1 Å from their minimum energy separation. For small isolated molecules, rotation about single bonds usually involves the smallest changes in energy. For example, if we rotate the carbon–carbon bond in ethane, keeping all of the bond lengths and angles fixed in value, then the energy varies in an approximately sinusoidal fashion as shown in Figure 1.3, with minima at the three staggered conformations. The energy in this case can be considered a function of a single coordinate only (i.e. the torsion angle of the carbon–carbon bond), and as such can be displayed graphically, with energy along one axis and the value of the coordinate along the other.

Changes in the energy of a system can be considered as movements on a multi-dimensional 'surface' called the *energy surface*. We shall be particularly interested in stationary points on the energy surface, where the first derivative of the energy is zero with respect to the internal or Cartesian coordinates. At a stationary point the forces on all the atoms are zero. Minimum points are one type of stationary point; these correspond to stable structures. Methods for locating stationary points will be discussed in more detail in Chapter 4, together with a more detailed consideration of the concept of the energy surface.

1.4 Molecular graphics

Computer graphics has had a dramatic impact upon molecular modelling. It should always be remembered, however, that there is much more to molecular modelling than computer graphics. It is the interaction between molecular graphics and the underlying theoretical methods that has enhanced the accessibility of molecular modelling methods and assisted the analysis and interpretation of such calculations.

Molecular graphics systems have evolved from delicate and temperamental pieces of equipment that cost hundreds of thousands of pounds and occupied entire rooms, to today's inexpensive workstations that fit on or under a desk and yet are hundreds of times more powerful. Over the years, two different types of molecular graphics display have been used in molecular modelling. First to be developed were vector devices, which construct pictures using an electron gun to draw lines (or dots) on the screen, in a manner similar to an oscilloscope. Vector devices were the mainstay of molecular modelling for almost two decades, but have now been largely superseded by raster devices. These divide the screen into a large number of small 'dots', called pixels. Each pixel can be set to any of a large number of colours, and so by setting each pixel to the appropriate colour it is possible to generate the desired image.

Molecules are most commonly represented on a computer graphics screen using 'stick' or 'space-filling' representations, which are analogous to the Dreiding and Corey–Pauling–Koltun (CPK) mechanical models. Sophisticated variations on these two basic types have been developed, such as the ability to colour molecules by atomic number and the inclusion of shading and lighting effects which give 'solid' models a more realistic appearance. Some of the commonly used molecular representations are shown in Figure 1.4 (colour plate section). Computer-generated models do have some advantages when compared

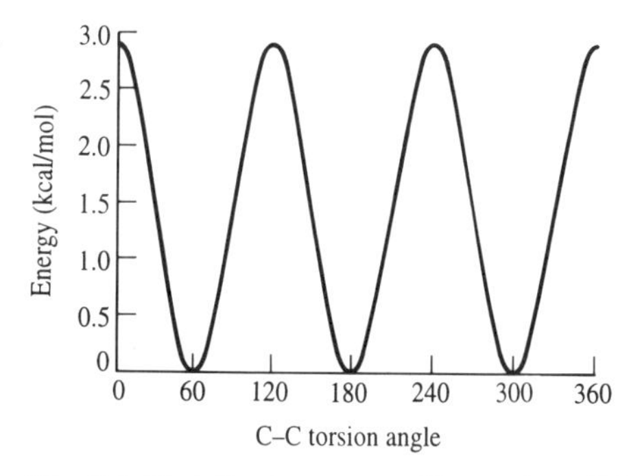

Fig. 1.3 Variation in energy with rotation of the carbon–carbon bond in ethane.

to their mechanical counterparts. Of particular importance is the fact that a computer model can be very easily interrogated to provide quantitative information, from simple geometrical measures such as the distance between two atoms to more complex quantities such as the energy or surface area. Quantitative information such as this can be very difficult if not impossible to obtain from a mechanical model. Nevertheless, mechanical models may still be preferred in certain types of situations, due to the ease with which they can be manipulated and viewed in three dimensions. A computer screen is inherently two-dimensional whereas molecules are three-dimensional objects. Nevertheless, some impression of the three-dimensional nature of an object can be represented on a computer screen using techniques such as depth cueing (in which those parts of the object that are further away from the viewer are made less bright) and through the use of perspective. Specialised hardware enables more realistic three-dimensional stereo images to be viewed. In the near future 'virtual reality' systems may enable a scientist to interact with a computer generated molecular model in much the same way that a mechanical model can be manipulated.

Even the most basic computer graphics program provides some standard facilities for the manipulation of models, including the ability to translate, rotate and 'zoom' the model towards and away from the viewer. More sophisticated packages can provide the scientist with quantitative feedback on the effect of altering the structure. For example, as a bond is rotated then the energy of each structure could be calculated and displayed interactively.

For large molecular systems it may not always be desirable to include every single atom in the computer image; the sheer number of atoms can result in a very confusing and cluttered picture. A clearer picture may be achieved by omitting certain atoms (e.g. hydrogen atoms) or by representing groups of atoms as single 'pseudo-atoms'. The techniques that have been developed for displaying protein structures nicely illustrate the range of computer graphics representation possible (the use of computational techniques to investigate the structures of proteins is considered in Chapter 8). Proteins are polymers constructed from amino acids and even a small protein may contain several thousand atoms. One way to produce a clearer picture is to dispense with the explicit representation of any atoms and to represent the protein using a 'ribbon'. Proteins are also commonly represented using the cartoon drawings developed by J Richardson, an example of which is shown in Figure 1.5 (colour plate section). The cylinders in this figure represent an arrangement of amino acids called an α-helix, and the flat arrows an alternative type of regular structure called a β-strand. The regions between the cylinders and the strands have no such regular structure and are represented as 'tubes'.

1.5 Surfaces

Many of the problems that are studied using molecular modelling involve the non-covalent interaction between two or more molecules. The study of such interactions is often facilitated by examining the van der Waals, molecular or accessible surfaces of the molecule. The *van der Waals surface* is simply constructed from the overlapping van der Waals spheres of the atoms, Figure 1.6. It corresponds to a CPK or space-filling model. Let us now consider the approach of a small 'probe' molecule, represented as a single van der Waals sphere, up to the van der Waals surface of a larger molecule. The finite size of the probe sphere means that there will be regions of 'dead space', crevices that are not accessible to the probe as it rolls about on the larger molecule. This is illustrated in Figure 1.6. The amount of dead space increases with the size of the probe; conversely, a probe of zero size would be able to access all of the crevices. The *molecular surface* [Richards 1977] is traced out by the inward-facing part of the probe sphere as it rolls on the van der Waals surface of the molecule. The molecular surface contains two different types of surface element. The *contact surface* corresponds to those regions where the probe is actually in contact with the van der Waals surface of the 'target'. The *re-entrant* surface regions occur where there are crevices that are too narrow for the probe molecule to penetrate. The molecular surface is usually defined using a water molecule as the probe, represented as a sphere of radius 1.4 Å.

The *accessible surface* is also widely used. As originally defined by Lee and Richards [Lee and Richards 1971] this is the surface that is traced by the centre of the probe molecule as it rolls on the van der Waals surface of the molecule (Figure 1.6). The centre of the probe molecule can thus be placed at any point on the accessible surface and not penetrate the van der Waals spheres of any of the atoms in the molecule.

Widely used algorithms for calculating the molecular and accessible surfaces were developed by Connolly [Connolly 1983a; Connolly 1983b], and others [e.g. Richmond 1984] have described formulae for the calculation of exact or approximate values of the surface area. There are many ways to represent surfaces, some of which are illustrated in Figure 1.7 (colour plate section). As shown, it may also be possible to endow a surface with a translucent quality which enables the molecule inside the surface to be displayed. Clipping can also be used to cut through the surface to enable the 'inside' to be viewed. In addition, properties such as the electrostatic potential can be calculated on the surface and represented using an appropriate colour scheme.

1.6 Computer hardware and software

One cannot fail to be amazed at the pace of development in the computer industry, where the ratio of performance-to-price has increased by an order of magnitude every five years or so. The workstations that are commonplace in many laboratories now offer a real alternative to centrally maintained 'supercomputers' for molecular modelling calculations, especially as a workstation can be dedicated to a single task whereas the supercomputer has to be shared with many other users. Nevertheless, in the immediate future there will always be some

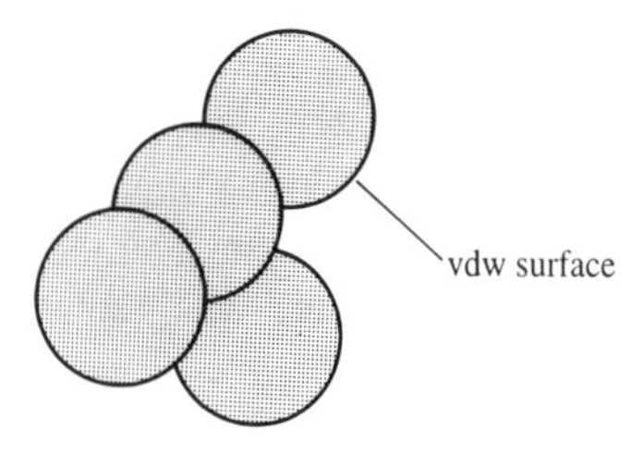

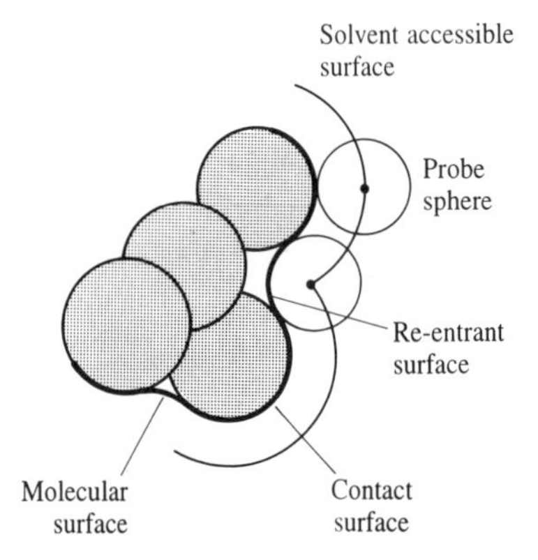

Fig. 1.6 The van der Waals (vdw) surface of a molecule corresponds to the outward facing surfaces of the van der Waals spheres of the atoms. The molecular surface is generated by rolling a spherical probe (usually of radius 1.4 Å to represent a water molecule) on the van der Waals surface. The molecular surface is constructed from contact and re-entrant surface elements. The centre of the probe traces out the accessible surface.

calculations that require the power that only a supercomputer can offer. One of the most important hardware developments for the molecular modelling community was the vector processor. These dramatically reduce the time required to perform calculations where the same arithmetical operations are performed on a large amount of data (such as on all atoms in the system). The speed of any computer system is ultimately constrained by the speed at which electrical signals can be transmitted. This means that there will come a time when no further enhancements can be made using machines with 'traditional' single-processor serial architectures and parallel computers will play an ever more important role.

A parallel computer couples processors together in such a way that a calculation is divided into small pieces with the results being combined at the end. Certain types of calculation are more amenable to parallel processing than others, but a significant amount of effort is being spent converting existing algorithms to run efficiently on parallel architectures. Completely new methods are also being developed to take maximum advantage of the opportunities of parallel processing.

To perform molecular modelling calculations one also requires appropriate programs (the software). The software used by molecular modellers ranges from simple programs that perform just a single task to highly complex packages that integrate many different methods. There is also an extremely wide variation in the price of software! Some programs have been so widely used and tested that they can be considered to have reached the status of a 'gold standard' against which similar programs are compared. One hesitates to specify such programs in print, but three items of software have been so widely used and cited that they can safely be afforded the accolade. These are the Gaussian series of programs for performing *ab initio* quantum mechanics, the MOPAC/AMPAC programs for semi-empirical quantum mechanics and the MM2 program for molecular mechanics.

Various pieces of software were used to generate the data for the examples and illustrations throughout this book. Some of these were written specifically for the task; some were freely available programs; others were commercial packages. I have decided not to describe specific programs in any detail, as such descriptions rapidly become outdated. Nevertheless, all items of software are accredited where appropriate. Please note that the use of any particular piece of software does not imply any recommendation!

1.7 Units of length and energy

It will be noted that our Z-matrix for ethane has been defined using the angstrom as the unit of length ($1\ \text{Å} \equiv 10^{-10}\ \text{m} \equiv 100\ \text{pm}$). The angstrom is a non-SI unit, but is a very convenient one to use, as most bond lengths are of the order of 1–2 Å. One other very common non-SI unit found in the molecular modelling literature is the kilocalorie ($1\ \text{kcal} \equiv 4.1840\ \text{kJ}$). Other systems of units are employed in other types of calculation, such as the atomic units used in quantum mechanics (discussed in Chapter 3). It is important to be aware of, and familiar with, these non-standard units as they are widely used in the literature and throughout this book.

1.8 The molecular modelling literature

The number of scientific papers concerned with molecular modelling methods is rising rapidly, as is the number of journals in which such papers are published. This reflects the tremendous diversity of problems to which molecular modelling can be applied and the ever-increasing availability of molecular modelling methods. It does, however, mean that it can be very difficult to remain up-to-date with the field. A number of specialist journals are devoted to theoretical chemistry, computational chemistry and molecular modelling, each with their own particular emphasis. Relevant papers are also published in the more 'general' journals. Many scientists are now fortunate to have access to electronic catalogues of publications which can be searched to find relevant papers. One particularly valuable source of information on molecular modelling methods are the *Reviews in Computational Chemistry*, edited by Boyd and Lipkowitz [Lipkowitz and Boyd 1990–1996]. Each of these volumes contains chapters on a variety of subjects, each written by an appropriate expert.

1.9 The Internet

The Internet is a collection of local and national computer networks that enable scientists to communicate with each other around the world. The Internet is still largely used for electronic mail, but extremely rapid growth is being observed in other areas, particularly the 'World-Wide Web' (WWW), an information system originally developed at CERN that enables pieces of information to be linked together, even though they may be held on different computers throughout the world [Winter *et al.* 1995]. For example, a document held on a computer in London may contain a link (called a hyperlink) to another document stored on a computer in California. A user reading the London document would activate the link and the Californian document would be automatically downloaded. At the time of writing, the Mosaic and Netscape programs are the most popular way to access the WWW; both of these are very intuitive and easy to use. A sample WWW 'page' is shown in Figure 1.8 (colour plate section). Highlighted items are the links; clicking such an item activates the link and downloads the appropriate data. The link need not necessarily be to a textual document, but could be to a picture or even a movie clip. Within the molecular modelling context, a link may be established to a file containing the coordinates of a molecule. Making the connection activates a graphics program which loads the coordinates, and so provides the user with an interactive representation of the information. Computer programs can also be downloaded from the Internet in this way. Some sites have even set up computer servers to which modelling calculations can be submitted with the results being returned via electronic mails and more than one 'electronic conference' on molecular modelling has been held with participants from many different countries. A limited selection of useful WWW sites is given in Appendix 1.1.

1.10 Mathematical concepts

A full appreciation of all of the techniques of molecular modelling would require a mathematical treatment beyond that appropriate to a book of this size. However, a proper understanding does benefit from some knowledge of mathematical concepts such as vectors, matrices, differential equations, complex numbers, series expansions and Lagrangian multipliers and some very elementary statistical concepts. There is only space in this book for a cursory introduction to these mathematical concepts and ideas, with very brief descriptions and some key results. The suggestions for further reading provide detailed background information on all of the mathematical topics required.

1.10.1 Series expansions

There are various series expansions that are useful for approximating functions. Particularly important is the *Taylor series*: if $f(x)$ is a continuous, single-valued function of x with continuous derivatives $f'(x), f''(x)\ldots$ then we can expand the function about a point x_0 as follows:

$$f(x_0+x)=f(x_0)+\frac{x}{1!}f'(x_0)+\frac{x^2}{2!}f''(x_0)+\frac{x^3}{3!}f'''(x_0)+\cdots+\frac{x^n}{n!}f^{(n)}(x_0) \tag{1.1}$$

Taylor series are often truncated after the term involving the second derivative, which makes the function vary in a quadratic fashion. This is a common assumption in many of the minimisation algorithms that we will discuss in Chapter 4.

A *Maclaurin series* is a specific form of the Taylor series for which $x_0=0$. Some standard expansions in Taylor series form are:

$$e^x = 1+x+\frac{x^2}{2!}+\frac{x^3}{3!}+\frac{x^4}{4!}+\cdots \tag{1.2}$$

$$\sin x = x-\frac{x^3}{3!}+\frac{x^5}{5!}-\cdots \tag{1.3}$$

$$\ln(1+x) = x-\frac{x^2}{2}+\frac{x^3}{3}-\frac{x^4}{4}+\cdots \tag{1.4}$$

The *binomial expansion* is used for functions of the form $(1+x)^\alpha$:

$$(1+x)^\alpha = 1+\alpha x+\alpha(\alpha-1)\frac{x^2}{2!}+\alpha(\alpha-1)(\alpha-2)\frac{x^3}{3!}+\cdots \tag{1.5}$$

All these series must have $|x|<1$ to be convergent.

1.10.2 Vectors

A vector is a quantity with both magnitude and direction. For example, the velocity of a moving body is a vector quantity as it defines both the direction in which the body is travelling and the speed at which it is moving. In Cartesian coordinates a vector such as the velocity will have three components, indicating the contribution to the overall motion from the component motions along the x, y and z directions. The addition and subtraction of vectors can be understood using geometrical constructions, as shown in Figure 1.9. Thus if we want to calculate the force on an atom due to its interactions with all other atoms in the system (as

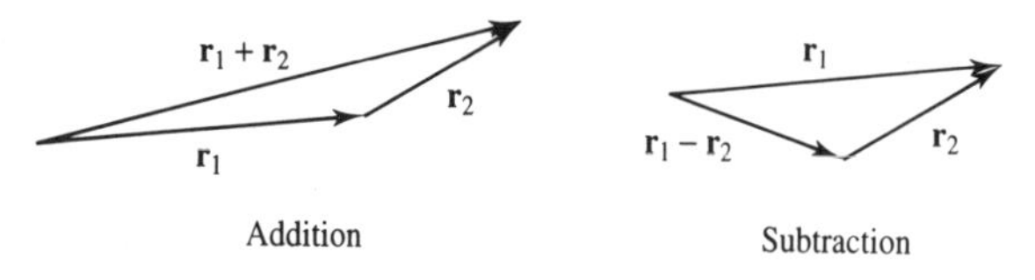

Fig. 1.9 The addition and subtraction of vectors.

required in molecular dynamics calculations, see Chapter 6), we would perform a vector sum of all the individual forces.

Some of the common manipulations that are performed with vectors include the scalar product, vector product and scalar triple product, which we will illustrate using vectors $\mathbf{r}_1$, $\mathbf{r}_2$ and $\mathbf{r}_3$ that are defined in a rectangular Cartesian coordinate system:

$$\begin{aligned}\mathbf{r}_1 &= x_1\mathbf{i} + y_1\mathbf{j} + z_1\mathbf{k}\\ \mathbf{r}_2 &= x_2\mathbf{i} + y_2\mathbf{j} + z_2\mathbf{k}\\ \mathbf{r}_3 &= x_3\mathbf{i} + y_3\mathbf{j} + z_3\mathbf{k}\end{aligned} \tag{1.6}$$

i, **j** and **k** are orthogonal unit vectors along the x, y and z axes. The *scalar product* is defined as:

$$\mathbf{r}_1.\mathbf{r}_2 = |\mathbf{r}_1|\,|\mathbf{r}_2|\cos\theta \tag{1.7}$$

$|\mathbf{r}_1|$ and $|\mathbf{r}_2|$ are the magnitudes of the two vectors ($|\mathbf{r}_1| = \sqrt{[x_1^2 + y_1^2 + z_1^2]}$) and θ is the angle between them (Figure 1.10). The angle can be calculated as follows:

$$\cos\theta = \frac{x_1x_2 + y_1y_2 + z_1z_2}{|\mathbf{r}_1|\,|\mathbf{r}_2|} \tag{1.8}$$

The scalar product of two vectors is thus a scalar.

The *vector product* of two vectors $\mathbf{r}_1 \times \mathbf{r}_2$ (sometimes written $\mathbf{r}_1 \wedge \mathbf{r}_2$) is a new vector ($\mathbf{v}$), in a direction perpendicular to the plane containing the two original vectors (Figure 1.10). The direction of this new vector is such that $\mathbf{r}_1$, $\mathbf{r}_2$ and the new vector form a right-handed system. If $\mathbf{r}_1$ and $\mathbf{r}_2$ are three-component vectors then the components of $\mathbf{v}$ are given by:

$$\mathbf{v} = (y_1z_2 - z_1y_2)\mathbf{i} + (z_1x_2 - x_1z_2)\mathbf{j} + (x_1y_2 - y_1x_2)\mathbf{k} \tag{1.9}$$

Note that the vector product $\mathbf{r}_2 \times \mathbf{r}_1$ is not the same as the vector product $\mathbf{r}_1 \times \mathbf{r}_2$ as it corresponds to a vector in the opposite direction. The vector product is thus not commutative.

The *scalar triple product* $\mathbf{r}_1.(\mathbf{r}_2 \times \mathbf{r}_3)$ equals the scalar product of $\mathbf{r}_1$ with the vector product of $\mathbf{r}_2$ and $\mathbf{r}_3$. The result is a scalar. The scalar triple product has a useful geometrical interpretation; it is the volume of the parallelepiped whose sides correspond to the three vectors (Figure 1.10).

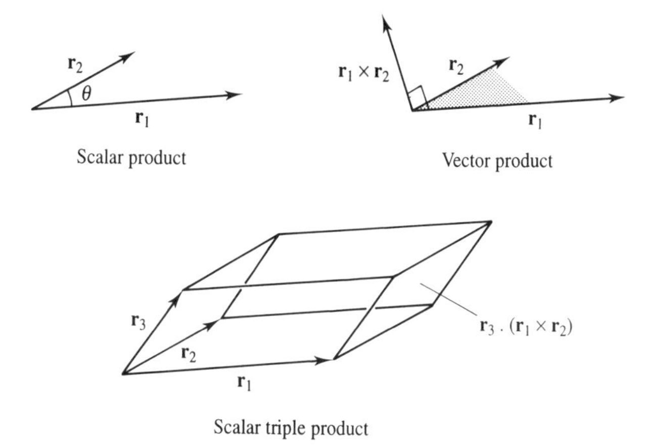

Fig. 1.10 The scalar product, vector product and scalar triple product.

1.10.3 Matrices, eigenvectors and eigenvalues

A matrix is a set of quantities arranged in a rectangular array. An $m \times n$ matrix has m rows and n columns. A vector can thus be considered to be a one column matrix. Matrix addition and subtraction can only be performed with matrices of the same order. For example:

$$\text{If} \quad \mathbf{A} = \begin{pmatrix} 4 & 7 \\ -3 & 5 \\ 8 & -2 \end{pmatrix} \quad \text{and} \quad \mathbf{B} = \begin{pmatrix} -4 & 3 \\ 5 & 2 \\ -5 & 3 \end{pmatrix}$$

$$\text{Then} \quad \mathbf{A} + \mathbf{B} = \begin{pmatrix} 0 & 10 \\ 2 & 7 \\ 3 & 1 \end{pmatrix}; \quad \mathbf{A} - \mathbf{B} = \begin{pmatrix} 8 & 4 \\ -8 & 3 \\ 12 & -5 \end{pmatrix} \tag{1.10}$$

Multiplication of two matrices (**AB**) is only possible if the number of columns in **A** is equal to the number of rows in **B**. If **A** is an $m \times n$ matrix and **B** is an $n \times o$ matrix then the product **AB** is an $m \times o$ matrix. Each element (i, j) in the matrix **AB** is obtained by taking each of the n values in the ith row of **A** and multiplying by the corresponding value in the jth column of **B**. To illustrate with a simple example:

$$\text{If} \quad \mathbf{A} = \begin{pmatrix} 3 & -2 & 5 \\ -3 & 4 & 1 \end{pmatrix} \quad \text{and} \quad \mathbf{B} = \begin{pmatrix} 0 & 3 \\ -2 & 4 \\ 1 & 6 \end{pmatrix}$$

Then

$$\mathbf{AB} = \begin{pmatrix} (3\times 0)+(-2\times -2)+(5\times 1) & (3\times 3)+(-2\times 4)+(5\times 6) \\ (-3\times 0)+(4\times -2)+(1\times 1) & (-3\times 3)+(4\times 4)+(1\times 6) \end{pmatrix}$$
$$= \begin{pmatrix} 9 & 31 \\ -7 & 13 \end{pmatrix} \tag{1.11}$$

We shall often encounter square matrices, which have the same number of rows and columns. A diagonal matrix is a square matrix in which all the elements are zero except for those on the diagonal. The *unit* or *identity* matrix **I** is a special type of diagonal matrix in which all the non-zero elements are 1; thus the 3 × 3 unit matrix is:

$$\mathbf{I} = \begin{pmatrix} 1 & 0 & 0 \\ 0 & 1 & 0 \\ 0 & 0 & 1 \end{pmatrix} \tag{1.12}$$

A matrix is *symmetric* if it is a square matrix with elements such that the elements above and below the diagonal are mirror images; $A_{ij} = A_{ji}$.

Multiplication of a matrix by its inverse gives the unit matrix:

$$\mathbf{A}^{-1}\mathbf{A} = \mathbf{I}$$

To compute the inverse of a square matrix it is necessary to first calculate its *determinant*, $|\mathbf{A}|$. The determinants of 2 × 2 and 3 × 3 matrices are calculated as follows:

$$\begin{vmatrix} a & b \\ c & d \end{vmatrix} = ad - bc \tag{1.13}$$

$$\begin{vmatrix} a & b & c \\ d & e & f \\ g & h & i \end{vmatrix} = a\begin{vmatrix} e & f \\ h & i \end{vmatrix} - b\begin{vmatrix} d & f \\ g & i \end{vmatrix} + c\begin{vmatrix} d & e \\ g & h \end{vmatrix}$$
$$= a(ei - hf) - b(di - fg) + c(dh - eg) \tag{1.14}$$

For example:

$$\begin{vmatrix} 3 & 6 \\ -2 & 3 \end{vmatrix} = 21; \quad \begin{vmatrix} 4 & 2 & -2 \\ 2 & 5 & 0 \\ -2 & 0 & 3 \end{vmatrix} = 28 \tag{1.15}$$

As can be seen, the determinant of a 3 × 3 matrix can be written as a sum of determinants of 2 × 2 matrices, obtained by first selecting one of the rows or columns in the matrix (the top row was chosen in our example). For each element A_{ij} in this row, the row and column in which that number appears are deleted (i.e. the ith row and the jth column). This leaves a 2 × 2 matrix whose determinant is calculated and then multiplied by $(-1)^{i+j}$. The result of this calculation is called the *cofactor* of the element A_{ij}. For example, the cofactor of the element A_{12} in the 3×3 matrix

$$\mathbf{A} = \begin{pmatrix} 4 & 2 & -2 \\ 2 & 5 & 0 \\ -2 & 0 & 3 \end{pmatrix}$$

is -6. When calculating the determinant the cofactor is multiplied by the element A_{ij}. The determinant of larger matrices can be obtained by extensions of the scheme illustrated above; thus the determinant of a 4×4 matrix is initially written in terms of 3×3 matrices, which in turn can be expressed in terms of 2×2 matrices.

Determinants have many useful and interesting properties. The determinant is zero if any two of its rows and columns are identical. The sign of the determinant is reversed by exchanging any pair of rows or any pair of columns. If all elements of a row (or column) are multiplied by the same number, then the value of the determinant is multiplied by that number. The value of a determinant is unaffected if equal multiples of the values in any row (or column) are added to another row (or column).

The vector product and the scalar triple product can be conveniently written as matrix determinants. Thus:

$$\mathbf{r}_1 \times \mathbf{r}_2 = \begin{vmatrix} \mathbf{i} & \mathbf{j} & \mathbf{k} \\ x_1 & y_1 & z_1 \\ x_2 & y_2 & z_2 \end{vmatrix} \tag{1.16}$$

$$\mathbf{r}_1 . (\mathbf{r}_2 \times \mathbf{r}_3) = \begin{vmatrix} x_1 & y_1 & z_1 \\ x_2 & y_2 & z_2 \\ x_3 & y_3 & z_3 \end{vmatrix} \tag{1.17}$$

The *transpose* of a matrix, $\mathbf{A}^{\mathrm{T}}$, is the matrix obtained by exchanging its rows and columns. Thus the transpose of an $m \times n$ matrix is an $n \times m$ matrix:

$$\text{If} \quad \mathbf{A} = \begin{pmatrix} 4 & 7 \\ -3 & 5 \\ 8 & -2 \end{pmatrix} \quad \mathbf{A}^{\mathrm{T}} = \begin{pmatrix} 4 & -3 & 8 \\ 7 & 5 & -2 \end{pmatrix} \tag{1.18}$$

The transpose of a square matrix is of course another square matrix. The transpose of a symmetric matrix is itself. One particularly important transpose matrix is the *adjoint* matrix, adj$\mathbf{A}$, which is the transpose matrix of cofactors. For example, the matrix of cofactors of the 3×3 matrix

$$\mathbf{A} = \begin{pmatrix} 4 & 2 & -2 \\ 2 & 5 & 0 \\ -2 & 0 & 3 \end{pmatrix} \quad \text{is} \quad \begin{pmatrix} 15 & -6 & 10 \\ -6 & 8 & -4 \\ 10 & -4 & 16 \end{pmatrix} \tag{1.19}$$

In this case the adjoint matrix is the same as the matrix of cofactors (as **A** is a symmetric matrix). The *inverse* of a matrix is obtained by dividing the elements of the adjoint matrix by the determinant:

$$\mathbf{A}^{-1} = \frac{\text{adj}\mathbf{A}}{|\mathbf{A}|} \tag{1.20}$$

Thus the inverse of our 3 × 3 matrix is:

$$\mathbf{A}^{-1} = \begin{pmatrix} 15/28 & -3/14 & 5/14 \\ -3/14 & 2/7 & -1/7 \\ 5/14 & -4 & 4/7 \end{pmatrix} \tag{1.21}$$

One of the most common matrix calculations involves finding its *eigenvalues* and *eigenvectors*. An eigenvector is a column matrix **x** such that:

$$\mathbf{Ax} = \lambda\mathbf{x} \tag{1.22}$$

λ is the associated eigenvalue. The eigenvector problem can be reformulated as follows:

$$\mathbf{Ax} = \lambda\mathbf{xI} \Rightarrow \mathbf{Ax} - \lambda\mathbf{xI} = 0 \Rightarrow (\mathbf{A} - \lambda\mathbf{I})\mathbf{x} = 0 \tag{1.23}$$

A trivial solution to this equation is $\mathbf{x} = 0$. For a non-trivial solution, we require that the determinant $|\mathbf{A} - \lambda\mathbf{I}|$ equals zero. One way to determine the eigenvalues and their associated eigenvectors is thus to expand the determinant to give a polynomial equation in λ. For our 3 × 3 symmetric matrix this gives:

$$\begin{pmatrix} 4-\lambda & 2 & -2 \\ 2 & 5-\lambda & 0 \\ -2 & 0 & 3-\lambda \end{pmatrix} \tag{1.24}$$

or:

$$(4-\lambda)(5-\lambda)(3-\lambda) - 2[2(3-\lambda)] - 2[2(5-\lambda)] = 0 \tag{1.25}$$

This can be factorised to give:

$$(1-\lambda)(7-\lambda)(4-\lambda) = 0 \tag{1.26}$$

The eigenvalues are thus $\lambda_1 = 1$, $\lambda_2 = 4$, $\lambda_3 = 7$. The corresponding eigenvectors are:

$$\lambda_1 = 1 : \mathbf{x}_1 = \begin{pmatrix} 2/3 \\ -1/3 \\ 2/3 \end{pmatrix} \quad \lambda_2 = 4 : \mathbf{x}_2 = \begin{pmatrix} -1/3 \\ 2/3 \\ 2/3 \end{pmatrix}$$

$$\lambda_3 = 7 : \mathbf{x}_3 = \begin{pmatrix} 2/3 \\ 2/3 \\ -1/3 \end{pmatrix} \tag{1.27}$$

Here, we have expressed the eigenvectors as vectors of unit length; any multiple of each eigenvector would also be a solution. **A** is a real, symmetric matrix. The eigenvalues of such matrices are always real and orthogonal (i.e. the scalar products of all pairs of eigenvectors are zero). This can be easily seen in our example.

As can be readily envisaged, expanding the determinant and solving a polynomial in λ is not the most efficient way to determine the eigenvalues and eigenvectors of larger matrices. Matrix diagonalisation methods are much more common. *Diagonalisation* of a matrix **A** involves finding a matrix **U** such that:

$$\mathbf{U}^{-1}\mathbf{A}\mathbf{U} = \mathbf{D} \tag{1.28}$$

D is the diagonal matrix of eigenvalues. When **A** is a real symmetric matrix, then **U** is the matrix of eigenvectors and $\mathbf{U}^{-1}$ is the inverse matrix of eigenvectors. Thus, for our example:

$$\begin{pmatrix} 2/3 & -1/3 & 2/3 \\ -1/3 & 2/3 & 2/3 \\ 2/3 & 2/3 & -1/3 \end{pmatrix} \begin{pmatrix} 4 & 2 & -2 \\ 2 & 5 & 0 \\ -2 & 0 & 3 \end{pmatrix} \begin{pmatrix} 2/3 & -1/3 & 2/3 \\ -1/3 & 2/3 & 2/3 \\ 2/3 & 2/3 & -1/3 \end{pmatrix} = \begin{pmatrix} 1 & 0 & 0 \\ 0 & 4 & 0 \\ 0 & 0 & 7 \end{pmatrix} \tag{1.29}$$

Note that for a real symmetric matrix **A**, the inverse $\mathbf{U}^{-1}$ is the same as the transpose, $\mathbf{U}^{\mathrm{T}}$.

Many methods have been devised for diagonalising matrices; some of these are specific to certain classes of matrices such as the class of real symmetric matrices. Many modelling techniques require us to calculate the eigenvalues and eigenvectors of a matrix, including self-consistent field quantum mechanics (section 2.5), the distance geometry method for exploring conformational space (section 8.6) and principal components analysis (section 8.12.1). The class of *positive definite* matrices is important in energy minimisation and finding transition structures; the eigenvalues of a positive definite matrix are all positive. A *positive semidefinite* matrix of rank m has m positive eigenvalues.

1.10.4 Complex numbers

A complex number has two components: a real part (a) and an imaginary part (b), as follows:

$$x = a + bi \tag{1.30}$$

i is the square root of -1 ($i = \sqrt{-1}$). Complex numbers enable certain types of equations that have no real solutions to be solved. For example, the roots of the equation $x^2 - 2x + 3 = 0$ are $x = 1 + \sqrt{2}i$ and $x = 1 - \sqrt{2}i$. A complex number can be considered as a vector in a two-dimensional coordinate system. Complex numbers are commonly represented using an *Argand diagram*, in which the x coordinate corresponds to the real part of the complex number and the y coordinate to the imaginary part (Figure 1.11).

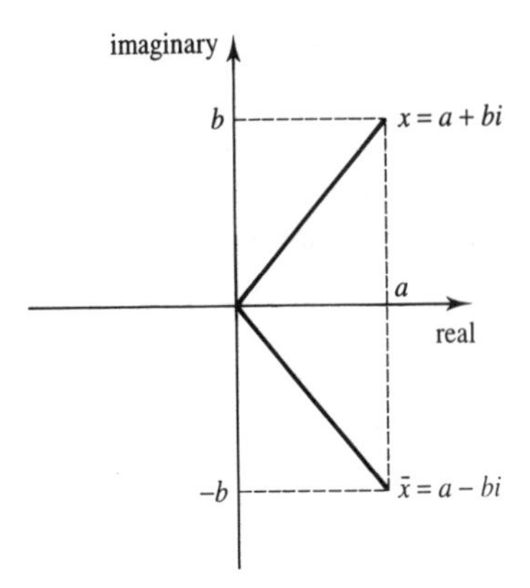

Fig. 1.11 The Argand diagram used to represent complex numbers.

Arithmetical operations on complex numbers are performed much as for vectors. Thus, if $x = a + bi$ and $y = c + di$, then:

$$x + y = (a + c) + (b + d)i \tag{1.31}$$

$$x - y = (a - c) + (b - d)i \tag{1.32}$$

$$xy = (ac - bd) + (ad + bc)i \tag{1.33}$$

The *complex conjugate*, $\bar{x}$ equals $a - bi$ and is obtained by reflecting x in the real axis in the Argand diagram.

A commonly used relationship involving complex numbers is:

$$\mathrm{e}^{i\theta} = \cos\theta + i\sin\theta \tag{1.34}$$

where θ is any real number. This relationship is used in Fourier analysis (section 6.9) and can be derived from the expansions of the exponential, cosine and sine functions:

$$\mathrm{e}^{i\theta} = 1 + i\theta - \frac{\theta^2}{2!} - \frac{i\theta^3}{3!} + \frac{\theta^4}{4!} + \cdots \tag{1.35}$$

$$\sin\theta = \theta - \frac{\theta^3}{3!} + \frac{\theta^5}{5!} - \cdots \tag{1.36}$$

$$\cos\theta = 1 - \frac{\theta^2}{2!} + \frac{\theta^4}{4!} - \cdots \tag{1.37}$$

Various other relationships can be defined. For example:

$$\cos\theta = \frac{\mathrm{e}^{i\theta} + \mathrm{e}^{-i\theta}}{2} \qquad \sin\theta = \frac{\mathrm{e}^{i\theta} - \mathrm{e}^{-i\theta}}{2i} \tag{1.38}$$

1.10.5 Lagrange multipliers

Lagrange multipliers can be used to find the stationary points of functions, subject to a set of constraints. Suppose we wish to find the stationary points of a function $f(x, y) = 4x^2 + 3x + 2y^2 + 6y$ subject to the constraint $y = 4x + 2$. In the Lagrange method the constraint is written in the form $g(x, y) = 0$:

$$g(x, y) = y - 4x - 2 = 0 \tag{1.39}$$

To find stationary points $f(x, y)$ subject to $g(x, y) = 0$ we first determine the total derivative df, which is set equal to zero:

$$\mathrm{d}f = \frac{\partial f}{\partial x}\mathrm{d}x + \frac{\partial f}{\partial y}\mathrm{d}y = (8x + 3)\mathrm{d}x + (4y + 6)\mathrm{d}y = 0 \tag{1.40}$$

Without the constraint the stationary points would be determined by setting the two partial derivatives $\partial f/\partial x$ and $\partial f/\partial y$ equal to zero, as x and y are independent. With the constraint, x and y are no longer independent but are related via the derivative of the constraint function g:

$$\mathrm{d}g = \frac{\partial g}{\partial x}\mathrm{d}x + \frac{\partial g}{\partial y}\mathrm{d}y = -4\mathrm{d}x + \mathrm{d}y = 0 \tag{1.41}$$

The derivative of the constraint function, dg, is multiplied by a parameter λ (the *Lagrange multiplier*) and added to the total derivative df:

$$\left(\frac{\partial f}{\partial x} + \lambda\frac{\partial g}{\partial x}\right)\mathrm{d}x + \left(\frac{\partial f}{\partial y} + \lambda\frac{\partial g}{\partial y}\right)\mathrm{d}y = 0 \tag{1.42}$$

The value of the Lagrange multiplier is obtained by setting each of the terms in brackets to zero. Thus for our example we have:

$$8x + 3 - 4\lambda = 0 \tag{1.43}$$

$$4y + 6 + \lambda = 0 \tag{1.44}$$

From these two equations we can obtain a further equation linking x and y:

$$\lambda = 2x + 3/4 = -6 - 4y \quad \text{or} \quad x = -27/8 - 2y \tag{1.45}$$

Combining this with the constraint equation enables us to identify the stationary point, which is at $(-59/72, -23/18)$.

This simple example could of course have been solved by simply substituting the constraint equation into the original function, to give a function of just one of the variables. However, in many cases this is not possible. The Lagrange multiplier method provides a powerful approach which is widely applicable to problems involving constraints such as in constraint dynamics (section 6.5) and in quantum mechanics.

1.10.6 Multiple integrals

Many of the theories used in molecular modelling involve multiple integrals. Examples include the two-electron integrals found in Hartree–Fock theory, and the integral over the positions and momenta used to define the partition function, Q. In fact, most of the multiple integrals that have to be evaluated are double integrals.

A 'traditional' or one-dimensional integral corresponds to the area under the curve between the imposed limit, as illustrated in Figure 1.12. Multiple integrals are simply extensions of these ideas to more dimensions. We shall illustrate the principles using a function of two variables, $f(x, y)$. The double integral

$$\iint_A \mathrm{d}x\mathrm{d}y\, f(x,y) \equiv \iint_A f(x,y)\mathrm{d}x\mathrm{d}y \qquad (1.46)$$

is the sum of the volume elements $f(x, y)\delta x\delta y$ (see Figure 1.12) over the area A as δx and δy tend to zero. Note that the 'dxdy' can be put either immediately after the integral sign or at the end; in this book we often use the first method for multiple integrals.

Some multiple integrals can be written as a product of single integrals. This occurs when $f(x, y)$ is itself a product of functions $g(x)h(y)$, in which case the integral can be separated:

$$\iint_A \mathrm{d}x\mathrm{d}y g(x)h(y) = \int \mathrm{d}x g(x) \int \mathrm{d}y h(y) \qquad (1.47)$$

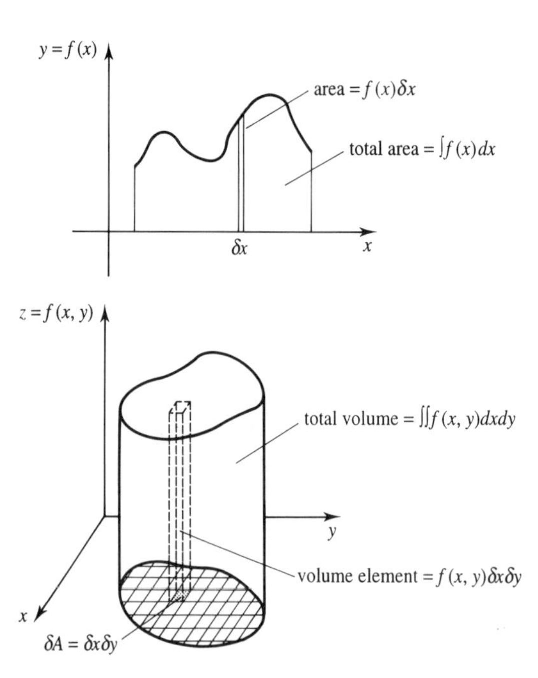

Fig. 1.12 Single and double integrals. Figure adapted in part from Boas M L, 1983, *Mathematical Methods in the Physical Sciences*. 2nd Edition. New York, Wiley.

For example:

$$\int_{-1}^{1} \mathrm{d}x \int_{-\pi/2}^{+\pi/2} \mathrm{d}y\, x^2 \cos y = \int_{-1}^{1} \mathrm{d}x\, x^2 (\sin y)_{-\pi/2}^{+\pi/2} = 2\left(\frac{x^3}{3}\right)_{-1}^{+1} = \frac{4}{3} \tag{1.48}$$

We will use the separation of multiple integrals throughout our discussion of quantum mechanics and computer simulation methods (Chapters 3, 6, 7 and 8).

1.10.7 Some basic elements of statistics

Statistics is concerned with the collection and interpretation of numerical data. The subject is a vast and complex one, and all we shall do here is to state some of the definitions commonly used and to explain some of the terminology.

The *arithmetic mean* of a set of observations is the sum of the observations divided by the number of observations:

$$\bar{x} = \frac{1}{N} \sum_{i=1}^{N} x_i \tag{1.49}$$

N is the number of observations. The mean may also be written $\langle x \rangle$. The *variance*, σ^2, indicates the extent to which the set of observations cluster around the mean value and equals the average of the squared deviations from the mean:

$$\sigma^2 = \frac{1}{N} \sum_{i=1}^{N} (x_i - \bar{x})^2 \tag{1.50}$$

The variance can also be calculated using the following formula which may be more convenient:

$$\sigma^2 = \frac{1}{N}\left[\sum_{i=1}^{N} (x_i^2) - \frac{1}{N}\left(\sum_{i=1}^{N} x_i\right)^2\right] \tag{1.51}$$

The *standard deviation*, σ, equals the (positive) square root of the variance:

$$\sigma = \sqrt{\frac{1}{N} \sum_{i=1}^{N} (x_i - \bar{x})^2} \tag{1.52}$$

It is often desired to compare the distribution of observations in a population with a theoretical distribution. The *normal distribution* (also called the Gaussian distribution) is a particularly important theoretical distribution in molecular modelling. The probability density function for a general normal distribution is:

$$f(x) = \frac{1}{\sigma\sqrt{2\pi}} \exp[-(x - \bar{x})^2 / 2\sigma^2] \tag{1.53}$$

The factor before the exponential ensures that the integral of the function $f(x)$ from $-\infty$ to $+\infty$ equals 1.

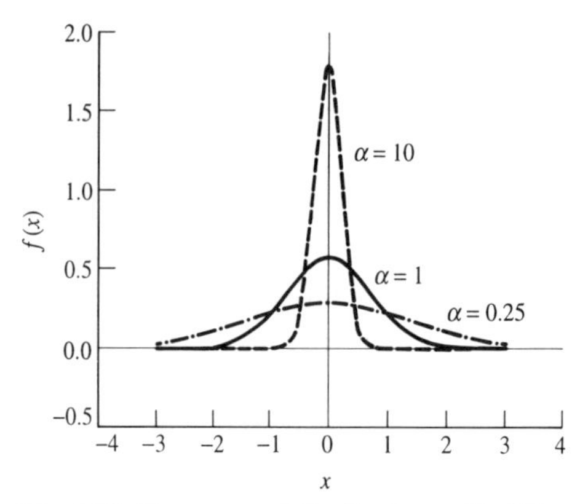

Fig. 1.13 Three normal distributions with different values of α (equation (1.54)). The functions are normalised so that the area under each curve is the same.

The distribution is often written in terms of a parameter α:

$$f(x) = \sqrt{\frac{\alpha}{\pi}}\,\mathrm{e}^{-\alpha(x-\bar{x})^2} \tag{1.54}$$

In Figure 1.13 we show three normal distributions that all have zero mean but different values of the variance (σ^2). A variance larger than 1 (small α) gives a flatter function and a variance less than 1 (larger α) gives a sharper function.

1.11 Appendix 1.1 A selection of websites

The addresses provided here were active and correct at the time of writing, but are liable to change. Simply type the address shown into your WWW browser; if you have problems contact your systems administrator.

http://www.osc.edu/chemistry.html
Computational Chemistry List (a discussion forum for all those interested in computational chemistry)
http://bionmr1.rug.ac.be/chemistry/overview.html
An overview of chemistry mailing lists
http://www.ch.ic.ac.uk/
Imperial College Chemistry home page (contains links to a host of useful modelling-related pages)
http://molbio.info.nih.gov/modeling/
National Institutes of Health molecular modelling home page
http://expasy.hcuge.ch/swissmod/SWISS-MODEL.html
Swiss-Model (automated protein modelling server)
http://www.pdb.bnl.gov
Brookhaven databank (protein structures)
http://csdvx2.ccdc.cam.ac.uk/
Cambridge Crystallographic Data Centre
http://edisto.awod.com/netsci/
Network Science: An on-line journal of science and computers
http://www.acs.org/
American Chemical Society
http://chemistry.rsc.org/rsc/
Royal Society of Chemistry (U.K.)
http://www.syp.toppan.co.jp:8082/bcsjstart.html
Chemical Society of Japan
http://ecs.tu-bs.de/ecs
European Chemical Society

Further reading

Bachrach S M 1996. *The Internet: A Guide for Chemists.* Washington, D.C., American Chemical Society.

Boas M L 1983. *Mathematical Methods in the Physical Sciences.* New York, Wiley.

December J and N Randall 1994. *The World Wide Web Unleashed.* Indianapolis, Sams Publishing.

Murray-Rust P M 1996. *On first looking into the World Wide Web.* Englewood Cliffs, NJ, Prentice Hall.

Press W H, B P Flannery, S A Teukolsky, W T Vetterling 1992. *Numerical Recipes in Fortran.* Cambridge, UK, Cambridge University Press.

Stephenson G 1973. *Mathematical Methods for Science Students.* London, Longman.

References

Bolin J T, D J Filman, D A Matthews, R C Hamlin and J Kraut 1982. Crystal Structures of *Escherichia coli* and *Lactobacillus casei* Dihydrofolate Reductase Refined at 1.7 Ångstroms Resolution. I. Features and Binding of Methotrexate. *Journal of Biological Chemistry* **257**:13650–13662.

Connolly M L 1983a. Solvent-accessible Surfaces of Proteins and Nucleic Acids. *Science* **221**:709–713.

Connolly M L 1983b. Analytical Molecular Surface Calculation. *Journal of Applied Crystallography* **16**:548–558.

Lee B, F M Richards 1971. The Interpretation of Protein Structures: Estimation of Static Accessibility. *Journal of Molecular Biology* **55**:379–400.

Lipkowitz K B and D B Boyd 1990–1996. *Reviews in Computational Chemistry,* vols 1–7. New York, VCH.

Richards F M 1977. Areas, Volumes, Packing and Protein Structure. *Annual Review in Biophysics and Bioengineering* **6**:151–176.

Richmond T J 1984. Solvent Accessible Surface Area and Excluded Volume in Proteins. *Journal of Molecular Biology* **178**:63–88.

Winter M J, H S Rzepa and B J Whitaker 1995. Surfing the Chemical Net. *Chemistry in Britain* September 1995:685–689 and http://www.ch.ic.ac.uk/rzepa/cib/.

2 Quantum mechanical models

2.1 Introduction

Our aim in this chapter will be to establish the basic elements of those quantum mechanical methods that are most widely used in molecular modelling. We shall assume some familiarity with the elementary concepts of quantum mechanics as found in most 'general' physical chemistry textbooks, but little else other than some basic mathematics (see section 1.10). There are also many excellent introductory texts to quantum mechanics. Quantum mechanics does of course predate the first computers by many years, and it is a tribute to the pioneers in the field that so many of the methods in common use today are based upon their efforts. The early applications were restricted to atomic, diatomic or highly symmetrical systems which could be solved by hand. The development of quantum mechanical techniques that are more generally applicable and that can be implemented on a computer (thereby eliminating the need for much laborious hand-calculation) means that quantum mechanics can now be used to perform calculations on molecular systems of real, practical interest. Quantum mechanics explicitly represents the electrons in a calculation, and so it is possible to derive properties that depend upon the electronic distribution and in particular to investigate chemical reactions in which bonds are broken and formed. These qualities, which differentiate quantum mechanics from the empirical force field methods described in Chapter 3 will be emphasised in our discussion of typical applications.

There are a number of quantum theories for treating molecular systems. The first we shall examine, and the one which has been most widely used is *molecular orbital theory*. However, alternative approaches have been developed, some of which we shall also describe, albeit briefly. We will be primarily concerned with the *ab initio* and semi-empirical approaches to quantum mechanics but will also mention techniques such as Hückel theory and valence bond theory. An alternative approach to quantum mechanics, density functional theory, is considered in section 9.13.

Quantum mechanics is often considered to be a difficult subject and a cursory glance at the following pages in this chapter may simply serve to reinforce that view! However, if followed carefully it is possible to see how models that are developed for very simple systems can be applied to much more complex systems. As a consequence our treatment does require some consideration of the mathematical background to the simplest and most common types of calculation.

Our strategy in developing the underlying theory of molecular orbital quantum mechanical calculations is as follows. First we revise some key features of quantum mechanics including the hydrogen atom. We then discuss the functional form of an acceptable wavefunction for a molecular system and show how to calculate the energy of such a system from the wavefunction. This leads to the problem of determining the wavefunction itself and how this can be done using routine mathematical methods. We will then be in a position to understand how quantum mechanical calculations can be performed for 'real' systems and will have the background necessary to consider more advanced topics.

The starting point for any discussion of quantum mechanics is, of course, the Schrödinger equation. The full, time-dependent form of this equation is

$$\left\{-\frac{\hbar^2}{2m}\left(\frac{\partial^2}{\partial x^2}+\frac{\partial^2}{\partial y^2}+\frac{\partial^2}{\partial z^2}\right)+\mathscr{V}\right\}\Psi(\mathbf{r},t)=i\hbar\frac{\partial\Psi(\mathbf{r},t)}{\partial t} \tag{2.1}$$

Equation (2.1) refers to a single particle (e.g. an electron) of mass m which is moving through space (given by a position vector $\mathbf{r}=x\mathbf{i}+y\mathbf{j}+z\mathbf{k}$) and time ($t$) under the influence of an external field $\mathscr{V}$ (which might be the electrostatic potential due to the nuclei of a molecule). $\hbar$ is Planck's constant divided by 2π and i is the square root of -1. Ψ is the *wavefunction* which characterises the particle's motion; it is from the wavefunction that we can derive various properties of the particle. When the external potential $\mathscr{V}$ is independent of time then the wavefunction can be written as the product of a spatial part and a time part: $\Psi(\mathbf{r},t)=\psi(\mathbf{r})T(t)$. We shall only consider situations where the potential is independent of time, which enables the time-dependent Schrödinger equation to be written in the more familiar, time-independent form:

$$\left\{-\frac{\hbar^2}{2m}\nabla^2+\mathscr{V}\right\}\Psi(\mathbf{r})=E\Psi(\mathbf{r}) \tag{2.2}$$

Here, E is the energy of the particle and we have used the abbreviation ∇^2 (pronounced 'del-squared'):

$$\nabla^2=\frac{\partial^2}{\partial x^2}+\frac{\partial^2}{\partial y^2}+\frac{\partial^2}{\partial z^2} \tag{2.3}$$

It is usual to abbreviate the left-hand side of equation (2.1) to $\mathscr{H}\Psi$, where $\mathscr{H}$ is the *Hamiltonian operator*:

$$\mathscr{H}=-\frac{\hbar^2}{2m}\nabla^2+\mathscr{V} \tag{2.4}$$

This reduces the Schrödinger equation to $\mathscr{H}\Psi=E\Psi$. To solve the Schrödinger equation it is necessary to find values of E and functions Ψ such that when the wavefunction is operated upon by the Hamiltonian, it returns the wavefunction multiplied by the energy. The Schrödinger equation falls into the category of equations known as partial differential eigenvalue equations in which an operator acts on a function (the eigenfunction) and returns the function multiplied by a scalar (the eigenvalue). A simple example of an eigenvalue equation is:

$$\frac{d}{dx}(y) = ry \tag{2.5}$$

The operator here is d/dx. One eigenfunction of this equation is $y = e^{ax}$ with the eigenvalue r being equal to a. Equation (2.5) is a first-order differential equation. The Schrödinger equation is a second-order differential equation as it involves the second derivative of Ψ. A simple example of an equation of this type is

$$\frac{d^2 y}{dx^2} = ry \tag{2.6}$$

The solutions of equation (2.6) have the form $y = A\cos kx + B\sin kx$, where A, B and k are constants. In the Schrödinger equation Ψ is the eigenfunction and E the eigenvalue.

2.1.1 Operators

The concept of an operator is an important one in quantum mechanics. The *expectation value* (which we can consider to be the average value) of a quantity such as the energy, position or linear momentum can be determined using an appropriate operator. The most commonly used operator is that for the energy, which is the Hamiltonian operator itself, $\mathcal{H}$. The energy can be determined by calculating the following integral:

$$E = \frac{\int \Psi^* \mathcal{H} \Psi d\tau}{\int \Psi^* \Psi d\tau} \tag{2.7}$$

The two integrals in equation (2.7) are performed over all space (i.e. from $-\infty$ to $+\infty$ in the x, y and z directions). Note the use of the complex conjugate notation (Ψ^*) which reminds us that the wavefunction may be a complex number. This equation can be derived by premultiplying both sides of the Schrödinger equation, $\mathcal{H}\Psi = E\Psi$, by the complex conjugate of the wavefunction, Ψ^*, and integrating both sides over all space. Thus:

$$\int \Psi^* \mathcal{H} \Psi d\tau = \int \Psi^* E \Psi d\tau \tag{2.8}$$

E is a scalar and so can be taken outside the integral, thus leading to equation (2.7). If the wavefunction is normalised then the denominator in equation (2.7) will equal one.

The Hamiltonian operator is composed of two parts that reflect the contributions of kinetic and potential energies to the total energy. The kinetic energy operator is

$$-\frac{\hbar^2}{2m}\nabla^2$$

and the operator for the potential energy simply involves multiplication by the appropriate expression for the potential energy. For an electron in an isolated atom or molecule the potential energy operator comprises the electrostatic interactions between the electron and the nuclei and the interactions between the electron and the other electrons. For a single electron and a single nucleus with Z protons the potential energy operator is thus:

$$\mathscr{V} = -\frac{Ze^2}{4\pi\varepsilon_0 r} \tag{2.9}$$

Another operator is that for linear momentum along the x direction, which is

$$\frac{\hbar}{i}\frac{\partial}{\partial x}.$$

The expectation value of this quantity can thus be obtained by evaluating the following integral:

$$p_x = \frac{\int \Psi^* \frac{\hbar}{i}\frac{\partial}{\partial x}\Psi \mathrm{d}\tau}{\int \Psi^* \Psi \mathrm{d}\tau} \tag{2.10}$$

2.1.2 Atomic units

Quantum mechanics is primarily concerned with atomic particles; electrons, protons and neutrons. When the properties of such particles (e.g. mass, charge etc.) are expressed in 'macroscopic' units then the value must usually be multiplied or divided by several powers of 10. It is preferable to use a set of units that enables the results of a calculation to be reported as 'easily manageable' values. One way to achieve this would be to multiply each number by an appropriate power of 10. However, further simplification can be achieved by recognising that it is often necessary to carry quantities such as the mass of the electron or electronic charge all the way through a calculation. These quantities are thus also incorporated into the atomic units. The atomic units of length, mass and energy are as follows.

1 unit of charge equals the absolute charge on an electron, $|e| = 1.60219 \times 10^{-19}$C
1 mass unit equals the mass of the electron, $m_e = 9.10593 \times 10^{-31}$kg
1 unit of length (1 Bohr) is given by $a_0 = h^2/4\pi^2 m_e e^2 = 5.29177 \times 10^{-11}$ m
1 unit of energy (1 Hartree) is given by $E_a = e^2/4\pi\varepsilon_0 a_0 = 4.35981 \times 10^{-18}$ J

The atomic unit of length is the radius of the first orbit in Bohr's treatment of the hydrogen atom. It also turns out to be the most probable distance of a 1s electron from the nucleus in the hydrogen atom. The atomic unit of energy corresponds to the interaction between two electronic charges separated by the Bohr radius. The total energy of the 1s electron in the hydrogen atom equals −0.5 Hartrees. In atomic units Planck's constant $h = 2\pi$ and so $\hbar \equiv 1$.

2.1.3 *Exact solutions to the Schrödinger equation*

The Schrödinger equation can be solved exactly for only a few problems, such as the particle in a box, the harmonic oscillator, the particle on a ring, the particle on a sphere and the hydrogen atom, all of which are dealt with in introductory textbooks. A common feature of these problems is that it is necessary to impose certain requirements (often called *boundary conditions*) on possible solutions to the equation. Thus, for a particle in a box with infinitely high walls, the wavefunction is required to go to zero at the boundaries. For a particle on a ring the wavefunction must have a periodicity of 2π because it must repeat every traversal of the ring. An additional requirement on solutions to the Schrödinger equation is that the wavefunction at a point $\mathbf{r}$ when multiplied by its complex conjugate is the probability of finding the particle at the point (this is the Born interpretation of the wavefunction). The square of an electronic wavefunction thus gives the electron density at any given point. If we integrate the probability of finding the particle over all space, then the result must be one as the particle must be somewhere:

$$\int \Psi^* \Psi \mathrm{d}\tau = 1 \tag{2.11}$$

$\mathrm{d}\tau$ indicates that the integration is over all space. Wavefunctions which satisfy this condition are said to be *normalised*. It is usual to require the solutions to the Schrödinger equation to be orthogonal:

$$\int \Psi_m^* \Psi_n \mathrm{d}\tau = 0 \; (m \neq n) \tag{2.12}$$

A convenient way to express both the orthogonality of different wavefunctions and the normalisation conditions uses the *Kronecker delta*:

$$\int \Psi_m^* \Psi_n \mathrm{d}\tau = \delta_{mn} \tag{2.13}$$

When used in this context, the Kronecker delta can be taken to have a value of one if m equals n and zero otherwise. Wavefunctions that are both orthogonal and normalised are said to be *orthonormal*.

2.2 One-electron atoms

In an atom that contains a single electron, the potential energy depends upon the distance between the electron and the nucleus as given by the Coulomb equation. The Hamiltonian thus takes the following form:

$$\mathscr{H} = -\frac{\hbar^2}{2m}\nabla^2 - \frac{Ze^2}{4\pi\varepsilon_0 r} \tag{2.14}$$

In atomic units the Hamiltonian is:

$$\mathscr{H} = -\frac{1}{2}\nabla^2 - \frac{Z}{r} \tag{2.15}$$

For the hydrogen atom, the nuclear charge, Z equals $+1$. r is the distance of the electron from the nucleus. The helium cation, He^+, is also a one-electron atom but has a nuclear charge of $+2$. As atoms have spherical symmetry it is more convenient to transform the Schrödinger equation to polar coordinates r, θ and ϕ where r is the distance from the nucleus (located at the origin), θ is the angle to the z axis and ϕ is the angle from the x axis in the xy plane (Figure 2.1). The solutions can be written as the product of a radial function $R(r)$ that depends only on r and an angular function $Y(\theta, \phi)$ called a *spherical harmonic* that depends on θ and ϕ:

$$\Psi_{nlm} = R_{nl}(r)Y_{lm}(\theta, \phi) \tag{2.16}$$

The wavefunctions are commonly referred to as *orbitals* and are characterised by three quantum numbers: n, m and l. The quantum numbers can adopt values as follows:

n: principal quantum number: 0, 1, 2, ...
l: azimuthal quantum number: 0, 1, ... $(n-1)$
m: magnetic quantum number: $-l$, $-(l-1)$, ... 0 ... $(l-1)$, l.

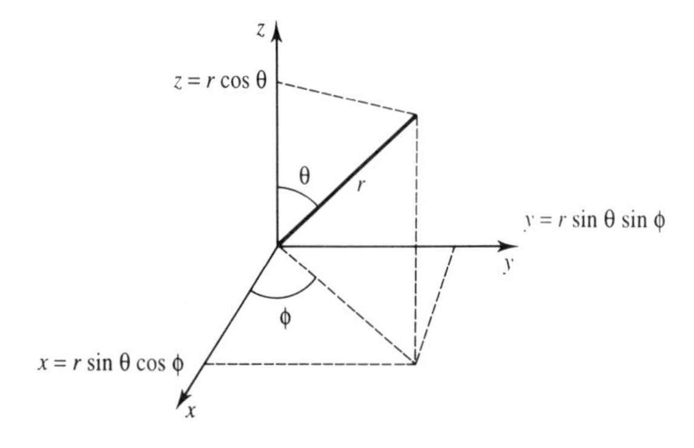

Fig. 2.1 The relationship between spherical polar and Cartesian coordinates.

Table 2.1 Radial function for one-electron atoms.

n	l	$R_{nl}(r)$
1	0	$2\zeta^{3/2}\ \exp(-\zeta r)$
2	0	$2\zeta^{3/2}(1-\zeta r)\exp(-\zeta r)$
2	1	$(4/3)^{1/2}\zeta^{5/2}r\ \exp(-\zeta r)$
3	0	$(2/3)^{1/2}\zeta^{3/2}(3-6\zeta r+2\zeta^2 r^2)\exp(-\zeta r)$
3	1	$(8/9)^{1/2}\zeta^{5/2}(2-\zeta r)r\ \exp(-\zeta r)$
3	2	$(8/45)^{1/2}\zeta^{7/2}r^2\ \exp(-\zeta r)$

The full radial function is:

$$R_{nl}(r) = -\left[\left(\frac{2Z}{na_0}\right)^3 \frac{(n-l-1)!}{2n[(n+l)!]^3}\right]^{\frac{1}{2}} \exp\left(-\frac{\rho}{2}\right)\rho^l L_{n+1}^{2l+1}(\rho) \tag{2.17}$$

$\rho = 2Zr/na_0$ where a_0 is the Bohr radius.* The term in square brackets is a normalising factor. $L_{n+1}^{2l+1}(\rho)$ is a special type of function called a Laguerre polynomial. We shall rarely be interested in any other than the first few members of the series; moreover, they simplify considerably if atomic units are used and if we write them in terms of the *orbital exponent* $\zeta = Z/n$. The first few members of the series for low values of n are given in Table 2.1 and are illustrated graphically in Figure 2.2. As can be seen, the radial part of the wavefunction is a polynomial multiplied by a decaying exponential.

The angular part of the wavefunction is the product of a function of θ and a function of ϕ:

$$Y_{lm}(\theta, \phi) = \Theta_{lm}(\theta)\Phi_m(\phi) \tag{2.18}$$

These functions are:

$$\Phi_m(\phi) = \frac{1}{\sqrt{2\pi}}\exp(im\phi) \tag{2.19}$$

$$\Theta_{lm}(\theta) = \left[\frac{(2l+1)}{2}\frac{(l-|m|)!}{(l+|m|)!}\right]^{\frac{1}{2}} P_l^{|m|}(\cos\ \theta) \tag{2.20}$$

The functions $\Phi_m(\phi)$ are just the solutions to the Schrödinger equation for a particle on a ring. The term in square brackets for the function $\Theta_{lm}(\theta)$ is a normalising factor. $P_l^{|m|}(\cos\theta)$ is a member of a series of functions called the associated Legendre polynomials (the 'Legendre polynomials' are functions for which $|m| = 0$). The total orbital angular momentum of an electron in the orbital is given by $l(l+1)\hbar$ and the component of the angular momentum along the $\theta = 0$ axis is given by $l\hbar$. The energy of each solution is a function of the principal quantum number only; thus orbitals with the same value of n but different l and m are degenerate. The orbitals are often represented as shown in Figure 2.3. These graphical representations are not necessarily the same as the solutions given

* Strictly, a_0 in this case is given by $a_0 = h^2/\pi^2\mu e$ where μ is the reduced mass, $\mu = m_e M/(m_e + M)$; M is the mass of the nucleus.

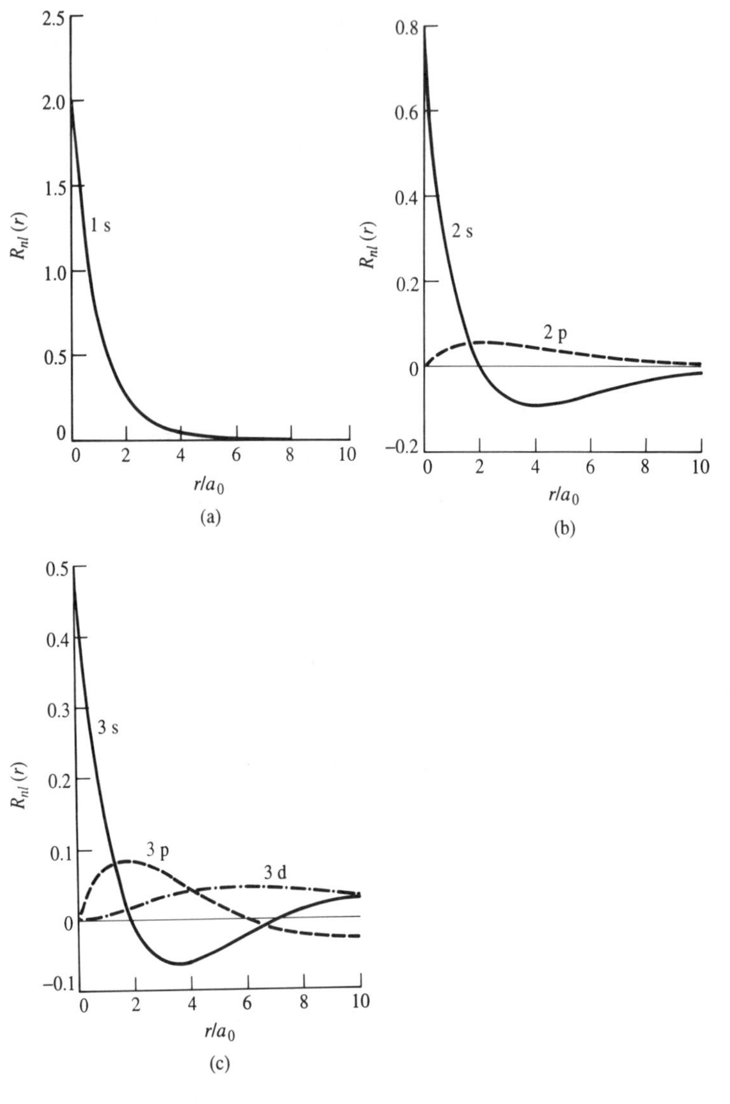

Fig. 2.2 The functions $R_{nl}(r)$ for the first three values of the principal quantum number. (a) 1s; (b) 2s and 2p; (c) 3s, 3p and 3d.

above. For example, the 'correct' solutions for the 2p orbitals comprise one real and two complex functions:

$$2p(+1) = \sqrt{(3/4\pi)}\, R(r) \sin\theta\, e^{i\phi} \tag{2.21}$$

$$2p(0) = \sqrt{(3/4\pi)}\, R(r) \cos\theta \tag{2.22}$$

$$2p(-1) = \sqrt{(3/4\pi)}\, R(r) \sin\theta\, e^{-i\phi} \tag{2.23}$$

$R(r)$ is the radial part of the wavefunction and $\sqrt{(3/4\pi)}$ is a normalisation factor for the angular part. The 2p(0) function is real and corresponds to the $2p_z$ orbital that is pictured in Figure 2.3. A linear combination of the two remaining

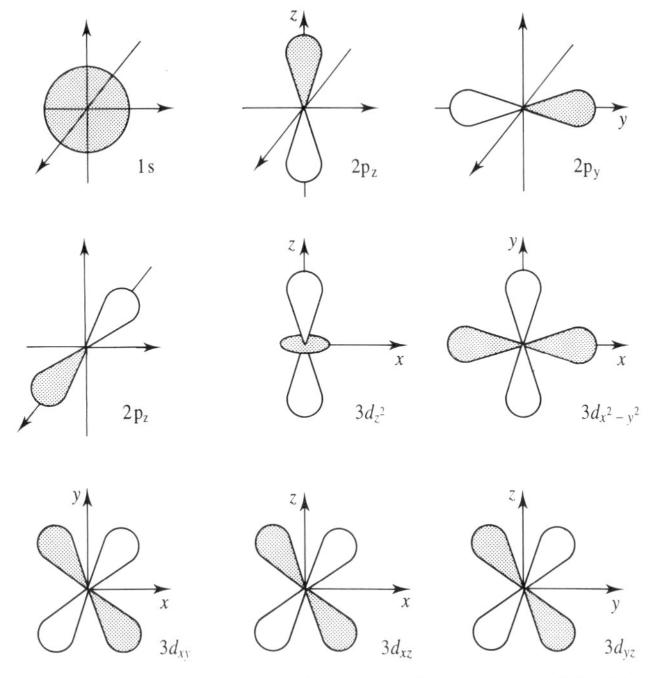

Fig. 2.3 The common graphical representations of s, p and d orbitals.

2p solutions is used to generate two 'real' 2p wavefunctions, making use of the relationship $\exp(i\phi) = \cos\phi + i\sin\phi$ (section 1.10.4). These linear combinations are the $2p_x$ and $2p_y$ orbitals shown in Figure 2.3.

$$2p_x = 1/2[2p(+1) + 2p(-1)] = \sqrt{(3/4\pi)}\, R(r)\sin\theta\cos\phi \qquad (2.24)$$

$$2p_y = -1/2[2p(+1) - 2p(-1)] = \sqrt{(3/4\pi)}\, R(r)\sin\theta\sin\phi \qquad (2.25)$$

These linear combinations still have the same energy as the original complex wavefunctions. This is a general property of degenerate solutions of the Hamiltonian operator. The reason why they are labelled $2p_x$ and $2p_y$ is that in polar coordinates the Cartesian coordinates x, y and z have the same angular dependence as the orbitals in Figure 2.3:

$$x = r\sin\theta\cos\phi$$
$$y = r\sin\theta\sin\phi$$
$$z = r\cos\theta \qquad (2.26)$$

The solutions of the Schrödinger equation are either real or occur in degenerate pairs. These pairs are complex conjugates that can then be combined to give energetically equivalent real solutions. It is only when dealing with certain types of operators that it is necessary to retain a complex wavefunction (for the 2p functions, the operator that corresponds to angular momentum about the z axis falls into this category). In fact, to simplify matters we will almost always ignore the complex notation from now on and will deal with real orbitals.

Finally, we should note that the solutions are all orthogonal to each other; if the product of any pair of orbitals is integrated over all space, the result is zero unless the two orbitals are the same. Orthonormality is achieved by multiplying by an appropriate normalisation constant.

The orbital picture has proved invaluable for providing insight and qualitative interpretations into the nature of the bonding in and reactivity of chemical systems. It is one which we would like to retain for polyelectronic systems to provide a unifying theme that links the simplest systems with much more complicated ones.

2.3 Polyelectronic atoms and molecules

Solving the Schrödinger equation for atoms with more than one electron is complicated by a number of factors. The first complication is that the Schrödinger equation for such systems cannot be solved exactly, even for the helium atom. The helium atom has three particles (two electrons and one nucleus) and is an example of a *three-body problem*. No exact solutions can be found for systems that involve three (or more) interacting particles. Thus, any solutions we might find for polyelectronic atoms or molecules can only be approximations to the real, true solutions of the Schrödinger equation. One consequence of there being no exact solution is that the wavefunction may adopt more than one functional form; no form is necessarily more 'correct' than another. In fact, the most general form of the wavefunction will be an infinite series of functions.

A second complication with multi-electron species is that we must account for electron spin. Spin is characterised by the quantum number s, which for an electron can only take the value $1/2$. The spin angular momentum is quantised such that its projection on the z axis is either $+\hbar$ or $-\hbar$. These two states are characterised by the quantum number m_s which can have values of $+1/2$ or $-1/2$ and are often referred to as 'up spin' and 'down spin' respectively. Electron spin is incorporated into the solutions to the Schrödinger equation by writing each one-electron wavefunction as the product of a spatial function that depends on the coordinates of the electron and a spin function that depends on its spin. Such solutions are called *spin orbitals* which we will represent using the symbol χ. The spatial part (which will be referred to as an orbital and represented using ϕ for atomic orbitals and ψ for molecular orbitals) describes the distribution of electron density in space and is analogous to the orbital diagrams in Figure 2.3. The spin part defines the electron spin and is labelled α or β. These spin functions have the value 0 or 1 depending on the quantum number m_s of the electron. Thus $\alpha(1/2)=1$, $\alpha(-1/2)=0$, $\beta(+1/2)=0$, $\beta(-1/2)=1$. Each spatial orbital can accommodate two electrons, with paired spins. In order to predict the electronic structure of a polyelectronic atom or a molecule, the *Aufbau principle* is employed, in which electrons are assigned to the orbitals, two electrons per orbital. We need to remember that electrons occupy degenerate states with a maximum number of unpaired electrons (Hund's rules), and that there are certain situations where it is energetically more favourable to place an unpaired electron in a higher energy spatial orbital rather than pair it with another electron. However, such situations are rare, particularly for molecular systems and for most of the systems that we shall be interested in the number of electrons, N, will be an even number that occupy the $N/2$ lowest energy orbitals.

Electrons are indistinguishable. If we exchange any pair of electrons, then the distribution of electron density remains the same. According to the Born interpretation, the electron density is equal to the square of the wavefunction. It therefore follows that the wavefunction must either remain unchanged when two electrons are exchanged, or else it must change sign. In fact, for electrons the wavefunction is required to change sign: this is the *antisymmetry principle.*

2.3.1 The Born–Oppenheimer approximation

It was stated above that the Schrödinger equation cannot be solved exactly for any molecular systems. However, it is possible to solve the equation exactly for the simplest molecular species, $H_2{}^+$ (and isotopically equivalent species such as HD^+), when the motion of the electrons is decoupled from the motion of the nuclei in accordance with the Born–Oppenheimer approximation. The masses of the nuclei are much greater than the masses of the electrons (the resting mass of the lightest nucleus, the proton, is 1836 times heavier than the resting mass of the electron). This means that the electrons can adjust almost instantaneously to any changes in the positions of the nuclei. The electronic wavefunction thus depends only on the positions of the nuclei and not on their momenta. Under the Born–Oppenheimer approximation the total wavefunction for the molecule can be written in the following form:

$$\Psi_{\text{tot}}(\text{nuclei, electrons}) = \Psi(\text{electrons})\Psi(\text{nuclei}) \tag{2.27}$$

The total energy equals the sum of the nuclear energy (the electrostatic repulsion between the positively charged nuclei) and the electronic energy. The electronic energy comprises the kinetic and potential energy of the electrons moving in the electrostatic field of the nuclei, together with electron–electron repulsion: $E_{\text{tot}} = E(\text{electrons}) + E(\text{nuclei})$.

When the Born–Oppenheimer approximation is used we concentrate on the electronic motions; the nuclei are considered to be fixed. For each arrangement of the nuclei the Schrödinger equation is solved for the electrons alone in the field of the nuclei. If it is desired to change the nuclear positions then it is necessary to add the nuclear repulsion to the electronic energy in order to calculate the total energy of the configuration.

2.3.2 The helium atom

We now return to the helium atom, with our objective being to find a wavefunction that describes the behaviour of the electrons. The Born–Oppenheimer approximation is not of course relevant to systems with just one nucleus and the wavefunction will be a function of the two electrons (which we shall label 1 and 2 with positions in space $\mathbf{r}_1$ and $\mathbf{r}_2$). As noted above, for polyelectronic systems any solution we find can only ever be an approximation to the true solution. There are a number of ways in which approximate solutions to the Schrödinger equation can be found. One approach is to find a simpler but related problem that can be more easily solved, and then consider how the differences between the two problems change the Hamiltonian and thereby affect the solutions. This is called *perturbation theory*, and is most appropriate when the differences between the real and simple problems are small. For example, a perturbation approach to tackling the helium atom might choose as the related system a 'pseudo atom', containing two electrons that interact with the nucleus but not with each other. Although this is a 'three body' problem, the lack of any interaction between the electrons means that it can be solved exactly using the method of the separation of variables. The separation of variables technique can be applied whenever the Hamiltonian can be divided into parts that are themselves dependent solely upon subsets of the coordinates. The equation to be solved in this case is:

$$\left\{-\frac{\hbar^2}{2m}\nabla_1^2 - \frac{Ze^2}{4\pi\varepsilon_0 r_1} - \frac{\hbar^2}{2m}\nabla_2^2 - \frac{Ze^2}{4\pi\varepsilon_0 r_2}\right\}\Psi(\mathbf{r}_1, \mathbf{r}_2) = E\Psi(\mathbf{r}_1, \mathbf{r}_2) \quad (2.28)$$

Or, in atomic units:

$$\left\{-\frac{1}{2}\nabla_1^2 - \frac{Z}{r_1} - \frac{1}{2}\nabla_2^2 - \frac{Z}{r_2}\right\}\Psi(\mathbf{r}_1, \mathbf{r}_2) = E\Psi(\mathbf{r}_1, \mathbf{r}_2) \quad (2.29)$$

We can abbreviate this equation to

$$\{\mathscr{H}_1 + \mathscr{H}_2\}\Psi(\mathbf{r}_1, \mathbf{r}_2) = E\Psi(\mathbf{r}_1, \mathbf{r}_2) \quad (2.30)$$

$\mathscr{H}_1$ and $\mathscr{H}_2$ are the individual Hamiltonians for electrons 1 and 2. Let us assume that the wavefunction can be written as a product of individual one-electron wavefunctions, $\phi_1(\mathbf{r}_1)$ and $\phi_2(\mathbf{r}_2)$: $\Psi(\mathbf{r}_1, \mathbf{r}_2) = \phi_1(\mathbf{r}_1)\phi_2(\mathbf{r}_2)$. Then we can write:

$$[\mathscr{H}_1 + \mathscr{H}_2]\phi_1(\mathbf{r}_1)\phi_2(\mathbf{r}_2) = E\phi_1(\mathbf{r}_1)\phi_2(\mathbf{r}_2) \quad (2.31)$$

Premultiplying by $\phi_1(\mathbf{r}_1)\phi_2(\mathbf{r}_2)$ and integrating over all space gives:

$$\iint \mathrm{d}\tau_1 \mathrm{d}\tau_2 \phi_1(\mathbf{r}_1)\phi_2(\mathbf{r}_2)[\mathscr{H}_1 + \mathscr{H}_2]\phi_1(\mathbf{r}_1)\phi_2(\mathbf{r}_2)$$
$$= \iint \mathrm{d}\tau_1 \mathrm{d}\tau_2 \phi_1(\mathbf{r}_1)\phi_2(\mathbf{r}_2)\phi_1(\mathbf{r}_1)\phi_2(\mathbf{r}_2) \quad (2.32)$$

or

$$\int d\tau_1 \phi_1(\mathbf{r}_1)\mathscr{H}_1\phi_1(\mathbf{r}_1)\int d\tau_2\phi_2(\mathbf{r}_2)\phi_2(\mathbf{r}_2) + \int d\tau_1\phi_1(\mathbf{r}_1)\phi_1(\mathbf{r}_1)\int d\tau_2\phi_2(\mathbf{r}_2)\mathscr{H}_2\phi_2(\mathbf{r}_2) = E\int d\tau_1\phi_1(\mathbf{r}_1)\phi_1(\mathbf{r}_1)\int d\tau_2\phi_2(\mathbf{r}_2)\phi_2(\mathbf{r}_2) \quad (2.33)$$

If we assume that the wavefunctions are normalised then it can easily be seen that the total energy E is the sum of the individual orbital energies E_1 and E_2 ($E_1 = \int d\tau_1\phi_1(\mathbf{r}_1)\mathscr{H}_1\phi_1(\mathbf{r}_1)$ and $E_2 = \int d\tau_2\phi_2(\mathbf{r}_2)\mathscr{H}_2\phi_2(\mathbf{r}_2)$). When the separation of variables method is used the solutions for each electron are just those of the hydrogen atom (1s, 2s, etc.) in equation (2.17) with $Z=2$.

We now wish to establish the general functional form of possible wavefunctions for the two electrons in this pseudo helium atom. We will do so by considering first the spatial part of the wavefunction. We will show how to derive functional forms for the wavefunction in which the exchange of electrons is independent of the electron labels and does not affect the electron density. The simplest approach is to assume that each wavefunction for the helium atom is the product of the individual one-electron solutions. As we have just seen, this implies that the total energy is equal to the sum of the one-electron orbital energies, which is not correct as it ignores electron–electron repulsion. Nevertheless, it is a useful illustrative model. The wavefunction of the lowest energy state then has each of the two electrons in a 1s orbital:

$$1s(1)1s(2) \quad (2.34)$$

'1s(1)' indicates a 1s function that depends on the coordinates of electron 1 ($\mathbf{r}_1$) and '1s(2)' indicates a 1s function that depends upon the coordinates of electron 2 ($\mathbf{r}_2$). This wavefunction satisfies the indistinguishability criterion, for we obtain the same function when we exchange the electrons – 1s(1)1s(2) is the same as 1s(2)1s(1). Its energy is twice that of a single electron in a 1s orbital. What of the first excited state, in which one electron is promoted to the 2s orbital? Two possible wavefunctions for this state are:

$$1s(1)2s(2) \quad (2.35)$$

$$1s(2)2s(1) \quad (2.36)$$

Do these wavefunctions satisfy the indistinguishability criterion? In other words, do we get the same function (or its negative) when we exchange the electrons? We do not, for when the two electrons (1 and 2) are exchanged then a different wavefunction is obtained: '1s(1)2s(2)' and '1s(2)2s(1)' are not the same, nor is one simply minus the other. However, linear combinations of these two wavefunctions do not suffer from the labelling problem and so we might anticipate that functional forms such as the following might constitute acceptable solutions to the Schrödinger equation for the pseudo-helium atom:

$$(1/\sqrt{2})[1s(1)2s(2) + 1s(2)2s(1)] \tag{2.37}$$

$$(1/\sqrt{2})[1s(1)2s(2) - 1s(2)2s(1)] \tag{2.38}$$

The factor $(1/\sqrt{2})$ ensures that the wavefunction is normalised. Of the three acceptable spatial forms that we have described so far, two are symmetric (i.e. do not change sign when the electron labels are exchanged) and one is antisymmetric (the sign changes when the electrons are exchanged):

$$1s(1)1s(2) \qquad \text{symmetric} \tag{2.39}$$

$$(1/\sqrt{2})[1s(1)2s(2) + 1s(2)2s(1)] \qquad \text{symmetric} \tag{2.40}$$

$$(1/\sqrt{2})[1s(1)2s(2) - 1s(2)2s(1)] \qquad \text{antisymmetric} \tag{2.41}$$

We now need to consider the effects of electron spin. For two electrons 1 and 2 there are four spin states; $\alpha(1)$, $\beta(1)$, $\alpha(2)$, $\beta(2)$. The indistinguishability criterion holds for the spin components as well, and so the following combinations of spin wavefunctions are possible:

$$\alpha(1)\alpha(2) \qquad \text{symmetric} \tag{2.42}$$

$$\beta(1)\beta(2) \qquad \text{symmetric} \tag{2.43}$$

$$(1/\sqrt{2})[\alpha(1)\beta(2) + \alpha(2)\beta(1)] \qquad \text{symmetric} \tag{2.44}$$

$$(1/\sqrt{2})[\alpha(1)\beta(2) - \alpha(2)\beta(1)] \qquad \text{antisymmetric} \tag{2.45}$$

When we combine the spatial and spin wavefunctions, the overall wavefunction must be antisymmetric with respect to exchange of electrons. It is therefore only admissible to combine a symmetric spatial part with an antisymmetric spin part, or an antisymmetric spatial part with a symmetric spin part. The following functional forms are therefore permissible functional forms for the wavefunctions of the ground and first few excited states of the helium atom:

$$(1/\sqrt{2})1s(1)1s(2)[\alpha(1)\beta(2) - \alpha(2)\beta(1)] \tag{2.46}$$

$$(1/2)[1s(1)2s(2) + 1s(2)2s(1)][\alpha(1)\beta(2) - \alpha(2)\beta(1)] \tag{2.47}$$

$$(1/\sqrt{2})[1s(1)2s(2) - 1s(2)2s(1)]\alpha(1)\alpha(2) \tag{2.48}$$

$$(1/\sqrt{2})[1s(1)2s(2) - 1s(2)2s(1)]\beta(1)\beta(2) \tag{2.49}$$

$$(1/2)[1s(1)2s(2) - 1s(2)2s(1)][\alpha(1)\beta(2) + \alpha(2)\beta(1)] \tag{2.50}$$

2.3.3 *General polyelectronic systems and Slater determinants*

We now turn to the general case. What is an appropriate functional form of the wavefunction for a polyelectronic system (not necessarily an atom) with N electrons that satisfies the antisymmetry principle? First, we note that the following functional form of the wavefunction is inappropriate:

$$\Psi(1, 2, \ldots N) = \chi_1(1)\chi_2(2) \;\ldots\; \chi_N(N) \tag{2.51}$$

This product of spin orbitals is unacceptable because it does not satisfy the antisymmetry principle; exchanging pairs of electrons does not give the negative of the wavefunction. This formulation of the wavefunction is known as a *Hartree product*. The energy of a system described by a Hartree product equals the sum of the one-electron spin orbitals. A key conclusion of the Hartree product description is that the probability of finding an electron at a particular point in space is independent of the probability of finding any other electron at that point in space. In fact, it turns out that the motions of the electrons are correlated. In addition, the Hartree product assumes that specific electrons have been assigned to specific orbitals, whereas the antisymmetry principle requires that the electrons are indistinguishable. Recall that for the helium atom, an acceptable functional form for the lowest-energy state is:

$$\begin{aligned} \Psi &= 1s(1)1s(2)[\alpha(1)\beta(2) - \alpha(2)\beta(1) \\ &\equiv 1s(1)1s(2)\alpha(1)\beta(2) - 1s(1)1s(2)\alpha(2)\beta(1) \end{aligned} \tag{2.52}$$

This can be written in the form of a 2 × 2 determinant:

$$\begin{vmatrix} 1s(1)\alpha(1) & 1s(1)\beta(1) \\ 1s(2)\alpha(2) & 1s(2)\beta(2) \end{vmatrix} \tag{2.53}$$

The two spin orbitals are

$$\chi_1 = 1s(1)\alpha(1) \quad \text{and} \quad \chi_2 = 1s(1)\beta(1). \tag{2.54}$$

A determinant is the most convenient way to write down the permitted functional forms of a polyelectronic wavefunction that satisfies the antisymmetry principle. In general, if we have N electrons in spin orbitals $\chi_1, \chi_2, \ldots, \chi_N$ (where each spin orbital is the product of a spatial function and a spin function) then an acceptable form of the wavefunction is:

$$\Psi = \frac{1}{\sqrt{N!}} \begin{vmatrix} \chi_1(1) & \chi_2(1) & \cdots & \chi_N(1) \\ \chi_1(2) & \chi_2(2) & \cdots & \chi_N(2) \\ \vdots & \vdots & & \vdots \\ \chi_1(N) & \chi_2(N) & \cdots & \chi_N(N) \end{vmatrix} \tag{2.55}$$

As before, $\chi_1(1)$ is used to indicate a function that depends on the space and spin coordinates of the electron labelled '1'. The factor $1/\sqrt{N!}$ ensures that the wavefunction is normalised; we shall see later why the normalisation factor has this particular value. This functional form of the wavefunction is called a *Slater determinant* and is the simplest form of an orbital wavefunction that satisfies the antisymmetry principle. The Slater determinant is a particularly convenient and concise way to represent the wavefunction due to the special properties of determinants. Exchanging any two rows of a determinant, a process which corresponds to exchanging two electrons, changes the sign of the determinant and therefore directly leads to the antisymmetry property. If any two rows of a determinant are identical, which would correspond to two electrons being assigned to the same spin orbital, then the determinant vanishes. This can be considered a manifestation of the Pauli principle which states that no two electrons can have the same set of quantum numbers. The Pauli principle also leads to the notion that each spatial orbital can accommodate two electrons of opposite spins.

When the Slater determinant is expanded, a total of $N!$ terms result. This is because there are $N!$ different permutations of N electrons. For example, for a three-electron system with spin orbitals χ_1, χ_2 and χ_3 the determinant is

$$\Psi = \frac{1}{\sqrt{12}} \begin{vmatrix} \chi_1(1) & \chi_2(1) & \chi_3(1) \\ \chi_1(2) & \chi_2(2) & \chi_3(2) \\ \chi_1(3) & \chi_2(3) & \chi_3(3) \end{vmatrix} \tag{2.56}$$

Expansion of the determinant gives the following expression (ignoring the normalisation constant):

$$\begin{aligned} &\chi_1(1)\chi_2(2)\chi_3(3) - \chi_1(1)\chi_3(2)\chi_2(3) + \chi_2(1)\chi_3(2)\chi_1(3) \\ &\quad - \chi_2(1)\chi_1(2)\chi_3(3) + \chi_3(1)\chi_1(2)\chi_2(3) - \chi_3(1)\chi_2(2)\chi_1(3) \end{aligned} \tag{2.57}$$

This expansion contains 6 terms ($\equiv 3!$). The six possible permutations of three electrons are: 123, 132, 213, 231, 312, 321. Some of these permutations involve single exchanges of electrons; others involve the exchange of two electrons. For example, the permutation 132 can be generated from the initial permutation by exchanging electrons 2 and 3. If we do so then the following wavefunction is obtained:

$$\begin{aligned} &\chi_1(1)\chi_2(3)\chi_3(2) - \chi_1(1)\chi_3(3)\chi_2(2) + \chi_2(1)\chi_3(3)\chi_1(2) \\ &\qquad - \chi_2(1)\chi_1(3)\chi_3(2) + \chi_3(1)\chi_1(3)\chi_2(2) - \chi_3(1)\chi_2(3)\chi_1(2) \\ &\quad = -\chi_1(1)\chi_2(2)\chi_3(3) + \chi_1(1)\chi_3(2)\chi_2(3) \\ &\qquad\quad - \chi_2(1)\chi_3(2)\chi_1(3) + \chi_2(1)\chi_1(2)\chi_3(3) \\ &\qquad\quad - \chi_3(1)\chi_1(2)\chi_2(3) + \chi_3(1)\chi_2(2)\chi_1(3) \\ &\quad = -\Psi \end{aligned} \tag{2.58}$$

By contrast, the permutation 312 requires that electrons 1 and 3 are exchanged and then electrons 1 and 2 are exchanged. This gives rise to an unchanged wavefunction. In general, an odd permutation involves an odd number of electron exchanges and leads to a wavefunction with a changed sign; an even permutation involves an even number of electron exchanges and returns the wavefunction unchanged.

For any sizeable system the Slater determinant can be tedious to write out, let alone the equivalent full orbital expansion, and so it is common to use a shorthand notation. Various notation systems have been devised. In one system the terms along the diagonal of the matrix are written as a single row determinant. For the 3×3 determinant we therefore have:

$$\begin{vmatrix} \chi_1(1) & \chi_2(1) & \chi_3(1) \\ \chi_1(2) & \chi_2(2) & \chi_3(2) \\ \chi_1(3) & \chi_2(3) & \chi_3(3) \end{vmatrix} \equiv |\chi_1 \quad \chi_2 \quad \chi_3| \tag{2.59}$$

The normalisation factor is assumed. It is often convenient to indicate the spin of each electron in the determinant; this is done by writing a bar when the spin part is β (spin down); a function without a bar indicates an α spin (spin up). Thus, the following are all commonly used ways to write the Slater determinantal wavefunction for the beryllium atom (which has the electronic configuration $1s^2 2s^2$):

$$\begin{aligned} \Psi &= \frac{1}{\sqrt{24}} \begin{vmatrix} \phi_{1s}(1) & \bar{\phi}_{1s}(1) & \phi_{2s}(1) & \bar{\phi}_{2s}(1) \\ \phi_{1s}(2) & \bar{\phi}_{1s}(2) & \phi_{2s}(2) & \bar{\phi}_{2s}(2) \\ \phi_{1s}(3) & \bar{\phi}_{1s}(3) & \phi_{2s}(3) & \bar{\phi}_{2s}(3) \\ \phi_{1s}(4) & \bar{\phi}_{1s}(4) & \phi_{2s}(4) & \bar{\phi}_{2s}(4) \end{vmatrix} \\ &\equiv |\phi_{1s} \quad \bar{\phi}_{1s} \quad \phi_{2s} \quad \bar{\phi}_{2s}| \\ &\equiv |1s \quad \bar{1s} \quad 2s \quad \bar{2s}| \end{aligned} \tag{2.60}$$

An important property of determinants is that a multiple of any column can be added to another column without altering the value of the determinant. This means that the spin orbitals are not unique; other linear combinations give the same energy. To illustrate this, consider the first excited state configuration of the helium atom ($1s^1 2s^1$), which can be written as the following 2×2 determinant:

$$\begin{vmatrix} 1s(1)\alpha(1) & 2s(1)\alpha(1) \\ 1s(2)\alpha(2) & 2s(2)\alpha(2) \end{vmatrix} = 1s(1)\alpha(1)2s(2)\alpha(2) - 1s(2)\alpha(2)2s(1)\alpha(1) \tag{2.61}$$

We now introduce two new 'spin orbitals':

$$\chi_1' = \frac{1s + 2s}{\sqrt{2}}\alpha; \quad \chi_2' = \frac{1s - 2s}{\sqrt{2}}\alpha \tag{2.62}$$

With these new orbitals the value of the determinant is as follows:

$$\begin{aligned} &\begin{vmatrix} \chi_1'(1) & \chi_2'(1) \\ \chi_1'(2) & \chi_2'(2) \end{vmatrix} \\ &\quad = \frac{[1s(1) + 2s(1)][1s(2) - 2s(2)]\alpha(1)\alpha(2)}{2} \\ &\qquad - \frac{[1s(1) - 2s(1)][1s(2) + 2s(2)]\alpha(1)\alpha(2)}{2} \\ &\quad \equiv -\Psi \end{aligned} \tag{2.63}$$

This can be helpful because it may enable more meaningful sets of orbitals to be generated from the original solutions. Molecular orbital calculations may give solutions that are 'smeared out' throughout the entire molecule whereas we may find orbitals that are localised in specific regions (e.g. in the bonds between atoms) to be more useful.

2.4 Molecular orbital calculations

2.4.1 Calculating the energy from the wavefunction: the hydrogen molecule

In our treatment of molecular systems we first show how to determine the energy for a given wavefunction, and then demonstrate how to calculate the wave function for a specific nuclear geometry. In the most popular kind of quantum mechanical calculations performed on molecules each molecular spin orbital is expressed as a linear combination of atomic orbitals (the LCAO approach). Thus each molecular orbital can be written as a summation of the following form:

$$\psi_{\mathrm{i}} = \sum_{\mu=1}^{\mathrm{K}} c_{\mu \mathrm{i}} \phi_{\mu} \tag{2.64}$$

ψ_{i} is a (spatial) molecular orbital, ϕ_{μ} is one of K atomic orbitals and $c_{\mu\mathrm{i}}$ is a coefficient. In a simple LCAO picture of the lowest energy state of molecular hydrogen, H_2, there are two electrons with opposite spins in the lowest energy spatial orbital (labelled $1\sigma_{\mathrm{g}}$) which is formed from a linear combination of two hydrogen-atom 1s orbitals:

$$1\sigma_{\mathrm{g}} = A(1\mathrm{s}_{\mathrm{A}} + 1\mathrm{s}_{\mathrm{B}}) \tag{2.65}$$

A is the normalisation factor whose value is not important in our present discussion. To calculate the energy of the ground state of the hydrogen molecule for a fixed internuclear distance we first write the wavefunction as a 2×2 determinant:

$$\Psi = \begin{vmatrix} \chi_1(1) & \chi_2(1) \\ \chi_1(2) & \chi_2(2) \end{vmatrix} = \chi_1(1)\chi_2(2) - \chi_1(2)\chi_2(1) \tag{2.66}$$

where

$$\begin{aligned} \chi_1(1) &= 1\sigma_{\mathrm{g}}(1)\alpha(1) \\ \chi_2(1) &= 1\sigma_{\mathrm{g}}(1)\beta(1) \\ \chi_1(2) &= 1\sigma_{\mathrm{g}}(2)\alpha(2) \\ \chi_2(2) &= 1\sigma_{\mathrm{g}}(2)\beta(2) \end{aligned}$$

For the hydrogen molecule, the Hamiltonian comprises the kinetic energy operator for each electron plus the potential energy operator due to the Coulomb attraction between the two electrons and the two nuclei, and the repulsion between the two electrons. In atomic units the Hamiltonian is thus

$$\mathscr{H} = -\frac{1}{2}\nabla_1^2 - \frac{1}{2}\nabla_2^2 - \frac{Z_{\mathrm{A}}}{r_{1\mathrm{A}}} - \frac{Z_{\mathrm{B}}}{r_{1\mathrm{B}}} - \frac{Z_{\mathrm{A}}}{r_{2\mathrm{A}}} - \frac{Z_{\mathrm{B}}}{r_{2\mathrm{B}}} + \frac{1}{r_{12}} \tag{2.67}$$

The electrons have been labelled 1 and 2 and the nuclei have been labelled A and B. For H_2 the nuclear charges Z_A and Z_B are both equal to 1. First we need to consider how to calculate the energy of this state of the hydrogen molecule. This is obtained using equation (2.7):

$$E = \frac{\int \Psi \mathscr{H} \Psi \mathrm{d}\tau}{\int \Psi \Psi \mathrm{d}\tau} \tag{2.68}$$

In general, a quantum mechanical calculation provides molecular orbitals that are normalised but the total wavefunction is not. The normalisation constant for the wavefunction of the two-electron hydrogen molecule is $1/\sqrt{2}$ and so the denominator in equation (2.68) is equal to 2.

We now substitute the hydrogen molecule wavefunction into equation (2.68), to provide the following:

$$\begin{aligned} E = \tfrac{1}{2}\int\int \mathrm{d}\tau_1 \mathrm{d}\tau_2 \Big\{ & [\chi_1(1)\chi_2(2) - \chi_2(1)\chi_1(2)] \\ & \times [-\tfrac{1}{2}\nabla_1^2 - \tfrac{1}{2}\nabla_2^2 - (1/r_{1A}) - (1/r_{1B}) - (1/r_{2A}) \\ & - (1/r_{2B}) + (1/r_{12})] \\ & \times [\chi_1(1)\chi_2(2) - \chi_2(1)\chi_1(2)] \Big\} \end{aligned} \tag{2.69}$$

$\mathrm{d}\tau_i$ indicates that the integration is over the spatial and spin coordinates of electron i. It is useful to separate the Hamiltonian operator into two H_2^+ Hamiltonians plus the interelectronic repulsion term:

$$\begin{aligned} E = \frac{1}{2}\int\int \mathrm{d}\tau_1 \mathrm{d}\tau_2 \Big\{ & [\chi_1(1)\chi_2(2) - \chi_2(1)\chi_1(2)] \\ & \times [\mathscr{H}_1 + \mathscr{H}_2 + (1/r_{12})] \\ & \times [\chi_1(1)\chi_2(2) - \chi_2(1)\chi_1(2)] \Big\} \end{aligned} \tag{2.70}$$

where

$$\mathscr{H}_1 = -\frac{1}{2}\nabla_1^2 - \frac{1}{r_{1A}} - \frac{1}{r_{1B}} \text{ and } \mathscr{H}_2 = -\frac{1}{2}\nabla_2^2 - \frac{1}{r_{2A}} - \frac{1}{r_{2B}} \tag{2.71}$$

We can now start to separate the integral in equation (2.70) into individual terms and identify the various contributions to the electronic energy:

$$\begin{aligned} E = & \iint d\tau_1 d\tau_2 \chi_1(1)\chi_2(2)(\mathscr{H}_1)\chi_1(1)\chi_2(2) \\ & - \iint d\tau_1 d\tau_2 \chi_1(1)\chi_2(2)(\mathscr{H}_1)\chi_2(1)\chi_1(2) + \cdots \\ & + \iint d\tau_1 d\tau_2 \chi_1(1)\chi_2(2)(\mathscr{H}_2)\chi_1(1)\chi_2(2) \\ & - \iint d\tau_1 d\tau_2 \chi_1(1)\chi_2(2)(\mathscr{H}_2)\chi_2(1)\chi_1(2) + \cdots \\ & + \iint d\tau_1 d\tau_2 \chi_1(1)\chi_2(2)\left(\frac{1}{r_{12}}\right)\chi_1(1)\chi_2(2) \\ & - \iint d\tau_1 d\tau_2 \chi_1(1)\chi_2(2)\left(\frac{1}{r_{12}}\right)\chi_2(1)\chi_1(2) + \cdots \end{aligned} \tag{2.72}$$

Each of these individual terms can be simplified if we recognise that terms dependent upon electrons other than those in the operator can be separated out. For example, the first term in the expansion, equation (2.72) is:

$$\iint d\tau_1 d\tau_2 \chi_1(1)\chi_2(2)(\mathscr{H}_1)\chi_1(1)\chi_2(2) \tag{2.73}$$

The operator $\mathscr{H}_1$ is a function of the coordinates of electron 1 only, and so terms involving electron 2 can be separated out as follows:

$$\begin{aligned} & \iint d\tau_1 d\tau_2 \chi_1(1)\chi_2(2)(\mathscr{H}_1)\chi_1(1)\chi_2(2) \\ & \quad = \int d\tau_2 \chi_2(2)\chi_2(2) \int d\tau_1 \chi_1(1)\left(-\tfrac{1}{2}\nabla_1^2 - \frac{1}{r_{1A}} - \frac{1}{r_{1B}}\right)\chi_1(1) \end{aligned} \tag{2.74}$$

If the molecular orbitals are normalised, the integral $\int d\tau_2 \chi_2(2)\chi_2(2)$ equals 1. Further simplification can be achieved by splitting the integral involving electron 1 into separate integrals over the spatial and spin parts; the integral over spin orbitals is equal to the product of an integral over the spatial coordinates and an integral over the spin coordinates:

$$\begin{aligned} & \int d\tau_1 \chi_1(1)\left(-\nabla_1^2 - \frac{1}{r_{1A}} - \frac{1}{r_{1B}}\right)\chi_1(1) \\ & \quad = \int dv_1 1\sigma_g(1)\left(\tfrac{1}{2} - \nabla_1^2 - \frac{1}{\mathrm{r}_{1A}} - \frac{1}{\mathrm{r}_{1B}}\right)1\sigma_g(1) \int d\sigma_1 \alpha(1)\alpha(1) \end{aligned}$$

dv indicates integration over spatial coordinates and $d\sigma$ indicates integration over the spin coordinates. The integral over the spin coordinates equals 1. This expression corresponds to the sum of the kinetic and potential energy of an electron in the orbital $1\sigma_g$ in the electrostatic field of the two bare nuclei. This integral can in turn be expanded by substituting the atomic orbital combination for $1\sigma_g$:

$$\int dv_1 1\sigma_g(1)\left(-\tfrac{1}{2}\nabla_1^2 - \frac{1}{r_{1A}} - \frac{1}{r_{1B}}\right)1\sigma_g(1)$$
$$= A^2 \int dv_1\{1s_A(1) + 1s_B(1)\}\left(-\tfrac{1}{2}\nabla_1^2 - \frac{1}{r_{1A}} - \frac{1}{r_{1B}}\right)\{1s_A(1)$$
$$+ 1s_B(1)\} \tag{2.75}$$

A is the normalisation constant. The integral in equation (2.75) can in turn be factorised to give a sum of integrals, each of which involves a pair of atomic orbitals:

$$\int dv_1\{1s_A(1) + 1s_B(1)\}\left(-\tfrac{1}{2}\nabla_1^2 - \frac{1}{r_{1A}} - \frac{1}{r_{1B}}\right)\{1s_A(1) + 1s_B(1)\}$$
$$= \int dv_1 1s_A(1)\left(-\tfrac{1}{2}\nabla_1^2 - \frac{1}{r_{1A}} - \frac{1}{r_{1B}}\right)1s_A(1)$$
$$+ \int dv_1 1s_A(1)\left(-\tfrac{1}{2}\nabla_1^2 - \frac{1}{r_{1A}} - \frac{1}{r_{1B}}\right)1s_B(1) + \cdots \tag{2.76}$$

Let us now apply the same procedure to the second term in equation (2.72):

$$\iint d\tau_1 d\tau_2 \chi_1(1)\chi_2(2)(\mathcal{H}_1)\chi_1(1)\chi_2(2)$$
$$= \int d\tau_1 \chi_1(1)(\mathcal{H}_1)\chi_2(1) \int d\tau_2 \chi_2(2)\chi_1(2) \tag{2.77}$$

This particular integral is zero because the molecular orbitals are orthogonal and so the integral over the coordinates of electron 2 equals zero:

$$\int d\tau_2 \chi_2(2)\chi_1(2) = 0 \tag{2.78}$$

A similar procedure can be applied to the other integrals involving electron–nuclear interactions; it turns out that there are four non-zero integrals, each of which is equal to the energy of a single electron in the field of the two hydrogen nuclei.

There remain four integrals arising from electron–electron interactions. These are:

$$\iint d\tau_1 d\tau_2 \chi_1(1)\chi_2(2)\left(\frac{1}{r_{12}}\right)\chi_1(1)\chi_2(2)$$
$$+ \iint d\tau_1 d\tau_2 \chi_2(1)\chi_1(2)\left(\frac{1}{r_{12}}\right)\chi_2(1)\chi_1(2)$$
$$- \iint d\tau_1 d\tau_2 \chi_1(1)\chi_2(2)\left(\frac{1}{r_{12}}\right)\chi_2(1)\chi_1(2)$$
$$- \iint d\tau_1 d\tau_2 \chi_2(1)\chi_1(2)\left(\frac{1}{r_{12}}\right)\chi_1(1)\chi_2(2) \tag{2.79}$$

The first two of these can be simplified as follows:

$$\iint d\tau_1 d\tau_2 \chi_1(1)\chi_2(2)\left(\frac{1}{r_{12}}\right)\chi_1(1)\chi_2(2)$$
$$= \iint dv_1 dv_2 1\sigma_g(1)1\sigma_g(2)\left(\frac{1}{r_{12}}\right)1\sigma_g(1)1\sigma_g(2)$$
$$\times \int d\sigma_1 \alpha(1)\alpha(1) \int d\sigma_2 \beta(2)\beta(2)$$
$$= \iint dv_1 dv_2 1\sigma_g(1)1\sigma_g(1)\left(\frac{1}{r_{12}}\right)1\sigma_g(2)1\sigma_g(2) \qquad (2.80)$$

According to the Born interpretation of the wavefunction, $1\sigma_g(\mathbf{r}_1)1\sigma_g(\mathbf{r}_1)$ equals the electron density of electron 1 in orbital $1\sigma_g$ at a position $\mathbf{r}_1$. Similarly, $1\sigma_g(\mathbf{r}_2)1\sigma_g(\mathbf{r}_2)$ is the electron density of electron 2. The electrostatic repulsion between these regions of electron density thus equals $1\sigma_g(\mathbf{r}_1)1\sigma_g(\mathbf{r}_1)\times(1/r_{12}) \times 1\sigma_g(\mathbf{r}_2)1\sigma_g(\mathbf{r}_2)$ where r_{12} is the distance between the two electrons. The integral of this function over all space thus corresponds to the electrostatic (Coulomb) repulsion between the two orbitals.

If we substitute the atomic orbital expansion, we obtain a series of two-electron integrals, each of which involves four atomic orbitals:

$$\iint dv_1 dv_2 1\sigma_g(1)1\sigma_g(2)\left(\frac{1}{r_{12}}\right)1\sigma_g(1)1\sigma_g(2)$$
$$= \iint dv_1 dv_2 1s_A(1)1s_A(2)\left(\frac{1}{r_{12}}\right)1s_A(1)1s_A(2)$$
$$+ \iint dv_1 dv_2 1s_A(1)1s_A(2)\left(\frac{1}{r_{12}}\right)1s_A(1)1s_B(2) + \cdots \qquad (2.81)$$

The remaining two integrals from equation (2.79) are:

$$\iint d\tau_1 d\tau_2 \chi_1(1)\chi_2(2)\left(\frac{1}{r_{12}}\right)\chi_2(1)\chi_1(2)$$
$$= \iint dv_1 dv_2 1\sigma_g(1)1\sigma_g(2)\left(\frac{1}{r_{12}}\right)1\sigma_g(1)1\sigma_g(2)$$
$$\times \int d\sigma_1 \alpha(1)\beta(1) \int d\sigma_2 \beta(2)\alpha(2) \qquad (2.82)$$

$$\iint d\tau_1 d\tau_2 \chi_2(1)\chi_1(2)\left(\frac{1}{r_{12}}\right)\chi_1(1)\chi_2(2)$$
$$= \iint dv_1 dv_2 1\sigma_g(1)1\sigma_g(2)\left(\frac{1}{r_{12}}\right)1\sigma_g(1)1\sigma_g(2)$$
$$\times \int d\sigma_1 \beta(1)\alpha(1) \int d\sigma_2 \alpha(2)\beta(2) \qquad (2.83)$$

Both of these integrals are zero due to the orthogonality of the electron spin states α and β.

The triplet excited state of H_2 is obtained by promoting an electron to a higher energy molecular orbital. This higher energy (antibonding) orbital is written $1\sigma_u$ and can be considered to arise from two 1s orbitals as follows:

$$1\sigma_u = A(1s_A - 1s_B) \tag{2.84}$$

The triplet state has two unpaired electrons with the same spin (α) and so the wavefunction state is:

$$\begin{vmatrix} 1\sigma_g\alpha(1) & 1\sigma_u\alpha(1) \\ 1\sigma_g\alpha(2) & 1\sigma_u\alpha(2) \end{vmatrix} \tag{2.85}$$

If we now expand the expression for the energy as for the ground state, terms analogous to the electron–nucleus and electron–electron interactions can again be obtained. However, the cross terms are no longer equal to zero as was the case for the ground state because the electron spins are now the same (both α). For example, compare with equation (2.82):

$$\begin{aligned} \iint d\tau_1 d\tau_2 \chi_1(1)\chi_2(2)\left(\frac{1}{r_{12}}\right)\chi_2(1)\chi_1(2) \\ = \iint dv_1 dv_2 1\sigma_g(1)1\sigma_u(2)\left(\frac{1}{r_{12}}\right)1\sigma_g(2)1\sigma_u(1) \\ \times \int d\sigma_1\alpha(1)\alpha(1)\int d\sigma_2\alpha(2)\alpha(2) \end{aligned} \tag{2.86}$$

This contribution is called the *exchange interaction*. This appears with a minus sign in the expression for the total energy and so acts to stabilise the triplet $1s^1 2s^1$ state over the analogous singlet state.

2.4.2 The energy of a general polyelectronic system

The hydrogen molecule is such a small problem that all of the integrals can be written out in full. This is rarely the case in molecular orbital calculations. Nevertheless, the same principles are used to determine the energy of a polyelectronic molecular system. For an N-electron system, the Hamiltonian takes the following general form:

$$\mathcal{H} = \left(-\frac{1}{2}\sum_{i=1}^{N}\nabla_i^2 - \frac{1}{r_{1A}} - \frac{1}{r_{1B}}\cdots + \frac{1}{r_{12}} + \frac{1}{r_{13}} + \cdots\right) \tag{2.87}$$

As with the hydrogen molecule, we have adopted the convention that the nuclei are labelled using capital letters A, B, C etc. and the electrons are labelled 1, 2, 3 ...

Recall that the Slater determinant for a system of N electrons in N spin orbitals can be written:

$$\begin{vmatrix} \chi_1(1) & \chi_2(1) & \chi_3(1) & \cdots & \chi_N(1) \\ \chi_1(2) & \chi_2(2) & \chi_3(2) & \cdots & \chi_N(2) \\ \chi_1(3) & \chi_2(3) & \chi_3(3) & \cdots & \chi_N(3) \\ \vdots & \vdots & \vdots & & \vdots \\ \chi_1(N) & \chi_2(N) & \chi_3(N) & \cdots & \chi_N(N) \end{vmatrix} \tag{2.88}$$

Each term in the determinant can thus be written $\chi_i(1)\chi_j(2)\chi_k(3)\} \ldots \chi_u\,(N-1)\chi_v(N)$ where i, j, k ... u, v is a series of N integers.

As usual, the energy can be calculated from $E = \int \Psi \mathscr{H} \Psi / \int \Psi \Psi$:

$$\begin{aligned} \int \Psi \mathscr{H} \Psi = \int \cdots \int & \mathrm{d}\tau_1 \mathrm{d}\tau_2 \cdots \mathrm{d}\tau_N \\ & \times \Bigg\{ [\chi_i(1)\chi_j(2)\chi_k(3)\cdots] \\ & \quad \times \Bigg(-\tfrac{1}{2}\sum_i \nabla_i^2 - (1/r_{1A}) - (1/r_{1B}) \cdots + (1/r_{12}) \\ & \qquad + (1/r_{13}) + \cdots \Bigg) [\chi_i(1)\chi_j(2)\chi_k(3)\cdots] \Bigg\} \end{aligned} \tag{2.89}$$

$$\begin{aligned} \int \Psi \Psi = \int \cdots \int & \mathrm{d}\tau_1 \mathrm{d}\tau_2 \cdots \mathrm{d}\tau_N \{ [\chi_i(1)\chi_j(2)\chi_k(3)\cdots] \\ & \times [\chi_i(1)\chi_j(2)\chi_k(3)\cdots] \} \end{aligned} \tag{2.90}$$

We can now see why the normalisation factor of the Slater determinantal wavefunction is $1/\sqrt{N!}$ If each determinant contains $N!$ terms then the product of two Slater determinants, [determinant][determinant] contains $(N!)^2$ terms. However, if the spin orbitals form an orthonormal set then only products of identical terms from the determinant will be non-zero when integrated over all space. We can illustrate this with the three-electron example. Considering just the first two terms in the expansion we obtain the following:

$$\begin{aligned} \iiint & \mathrm{d}\tau_1 \mathrm{d}\tau_2 \mathrm{d}\tau_3 [\chi_1(1)\chi_2(2)\chi_3(3) - \chi_1(1)\chi_3(2)\chi_2(3) + \cdots] \\ & \times [\chi_1(1)\chi_2(2)\chi_3(3) - \chi_1(1)\chi_3(2)\chi_2(3) + \cdots] \end{aligned} \tag{2.91}$$

When multiplied out this gives:

$$\begin{aligned}&\iiint d\tau_1 d\tau_2 d\tau_3[\chi_1(1)\chi_2(2)\chi_3(3)][\chi_1(1)\chi_2(2)\chi_3(3)]\\&\quad-\iiint d\tau_1 d\tau_2 d\tau_3[\chi_1(1)\chi_2(2)\chi_3(3)][\chi_1(1)\chi_3(2)\chi_2(3)]\\&\quad+\cdots\\&\quad+\iiint d\tau_1 d\tau_2 d\tau_3[\chi_1(1)\chi_3(2)\chi_2(3)][\chi_1(1)\chi_3(2)\chi_2(3)]\\&\quad+\cdots\end{aligned} \tag{2.92}$$

The first of the integrals in equation (2.92) equals 1 (if the spin orbitals are normalised). The second term is zero because the terms involving both electrons 2 and 3 are different (for example, the integral $\int d\tau_2\chi_2(2)\chi_3(2)$ will be zero due to the orthogonality of the spin orbitals χ_2 and χ_3). The third term in equation (2.92) will be equal to 1, and so on. It turns out that there are $N!$ such non-zero terms. Thus if each individual term in the determinant is normalised, then:

$$\int \Psi\Psi = N! \tag{2.93}$$

Hence the normalisation factor for the determinantal wavefunction is $1/\sqrt{N!}$

Turning now to the numerator in the energy expression (equation (2.8)), this can be broken down into a series of one-electron and two-electron integrals, as for the hydrogen molecule. Each of these individual integrals has the general form:

$$\int\cdots\int d\tau_1 d\tau_2\ldots[term1]operator[term2] \tag{2.94}$$

[*term*1] and [*term*2] each represent one of the $N!$ terms in the Slater determinant. To simplify this integral, we first recognise that all spin orbitals involving an electron that does not appear in the operator can be taken outside the integral. For example, if the operator is $1/r_{1\text{A}}$, then all spin orbitals other than those that depend on the coordinates of electron 1 can be separated from the integral. The orthogonality of the spin orbitals means that the integral will be zero unless all indices involving these other electrons are the same in [*term*1] and [*term*2]. Again, to use our three-electron system as an example:

$$\begin{aligned}&\iiint d\tau_1 d\tau_2 d\tau_3[\chi_1(1)\chi_2(2)\chi_3(3)]\left(-\frac{1}{r_{1\text{A}}}\right)[\chi_1(1)\chi_2(2)\chi_3(3)]\\&\quad=\iint d\tau_2 d\tau_3[\chi_2(2)\chi_3(3)][\chi_2(2)\chi_3(3)]\int d\tau_1\chi_1(1)\left(-\frac{1}{r_{1\text{A}}}\right)\chi_1(1)\\&\quad=\int d\tau_1\chi_1(1)\left(-\frac{1}{r_{1\text{A}}}\right)\chi_1(1)\end{aligned} \tag{2.95}$$

But:

$$\iiint d\tau_1 d\tau_2 d\tau_3 [\chi_1(1)\chi_2(2)\chi_3(3)]\left(-\frac{1}{r_{1A}}\right)[\chi_1(1)\chi_3(2)\chi_2(3)]$$
$$= \iint d\tau_2 d\tau_3 [\chi_2(2)\chi_3(3)][\chi_3(2)\chi_2(3)] \int d\tau_1 \chi_1(1)\left(-\frac{1}{r_{1A}}\right)\chi_1(1)$$
$$= 0 \tag{2.96}$$

For integrals that involve two-electron operators (i.e. $1/r_{ij}$), only those terms that do not involve the coordinates of the two electrons can be taken outside the integral. For example:

$$\iiint d\tau_1 d\tau_2 d\tau_3 [\chi_1(1)\chi_2(2)\chi_3(3)]\left(\frac{1}{r_{12}}\right)[\chi_1(1)\chi_2(2)\chi_3(3)]$$
$$= \iint d\tau_1 d\tau_2 [\chi_1(1)\chi_2(2)]\left(\frac{1}{r_{12}}\right)[\chi_1(2)\chi_2(2)] \int d\tau_3 \chi_3(3)\chi_3(3)$$
$$= \iint d\tau_1 d\tau_2 [\chi_1(1)\chi_2(2)]\left(\frac{1}{r_{12}}\right)[\chi_1(2)\chi_2(2)] \tag{2.97}$$

But

$$\iiint d\tau_1 d\tau_2 d\tau_3 [\chi_1(1)\chi_2(2)\chi_3(3)]\left(\frac{1}{r_{12}}\right)[\chi_1(1)\chi_3(2)\chi_2(3)]$$
$$= \iint d\tau_1 d\tau_2 [\chi_1(1)\chi_2(2)]\left(\frac{1}{r_{12}}\right)[\chi_1(2)\chi_3(2)] \int d\tau_3 \chi_3(3)\chi_2(3)$$
$$= 0 \tag{2.98}$$

As a consequence of these results, most of the individual integrals in the expansion will be zero. Nevertheless, it can be readily envisaged that there will still be an extremely large number of integrals to consider for all except the smallest problems. It is thus more convenient to write the energy expression in a concise form that recognises the three types of interaction that contribute to the total electronic energy of the system.

First, there is the kinetic and potential energy of each electron moving in the field of the nuclei. The energy associated with this contribution for the molecular orbital χ_i is often written H_{ii}^{core} and for M nuclei is given by:

$$H_{ii}^{core} = \int d\tau_1 \chi_i(1)\left(-\frac{1}{2}\nabla_i^2 - \sum_{A=1}^{M}\frac{Z_A}{r_{iA}}\right)\chi_i(1) \tag{2.99}$$

For N electrons in N molecular orbitals this contribution to the total energy is:

$$E_{total}^{core} = \sum_{i=1}^{N}\int d\tau_1 \chi_i(1)\left(-\frac{1}{2}\nabla_i^2 - \sum_{A=1}^{M}\frac{Z_A}{r_{iA}}\right)\chi_i(1) = \sum_{i=1}^{N} H_{ii}^{core} \tag{2.100}$$

Here we have followed convention and have used the label '1' wherever there is an integral involving the coordinates of a single electron, even though the actual electron may not be 'electron 1'. Similarly, when it is necessary to consider two electrons then the labels 1 and 2 are conventionally employed. H_{ii}^{core} makes a favourable (i.e. negative) contribution to the electronic energy.

The second contribution to the energy arises from the electrostatic repulsion between pairs of electrons. This interaction depends on the electron–electron distance and, as we have seen, is calculated from integrals such as:

$$J_{ij}\int\int d\tau_1 d\tau_2 \chi_i(1)\chi_j(2)\left(\frac{1}{r_{12}}\right)\chi_i(1)\chi_j(2) \tag{2.101}$$

The symbol J_{ij} is often used to represent this Coulomb interaction between electrons in spin orbitals i and j, and is unfavourable (i.e. positive). The total electrostatic interaction between the electron in orbital χ_i and the other $N-1$ electrons is a sum of all such integrals, where the summation index j runs from 1 to N, excluding i:

$$\begin{aligned} E_i^{coulomb} &= \sum_{j\neq i}^{N}\int d\tau_1 d\tau_2 \chi_i(1)\chi_j(2)\frac{1}{r_{12}}\chi_j(2)\chi_i(1) \\ &\equiv \sum_{j\neq i}^{N}\int d\tau_1 d\tau_2 \chi_i(1)\chi_i(1)\frac{1}{r_{12}}\chi_j(2)\chi_j(2) \end{aligned} \tag{2.102}$$

The total Coulomb contribution to the electronic energy of the system is obtained as a double summation over all electrons, taking care to count each interaction just once:

$$E_{total}^{coulomb} = \sum_{i=1}^{N}\sum_{j=i+1}^{N}\int d\tau_1 d\tau_2 \chi_i(1)\chi_i(1)\frac{1}{r_{12}}\chi_j(2)\chi_j(2) = \sum_{i=1}^{N}\sum_{j=i+1}^{N} J_{ij} \tag{2.103}$$

The third contribution to the energy is the exchange 'interaction'. This has no classical counterpart and arises because the motions of electrons with parallel spins are correlated: whereas there is a finite probability of finding two electrons with opposite (i.e. paired) spins at the same point in space, where the spins are the same then the probability is zero. This can be considered a manifestation of the Pauli principle, for if two electrons occupied the same region of space and had parallel spins then they could be considered to have the same set of quantum numbers. Electrons with the same spin thus tend to 'avoid' each other, and they experience a lower Coulombic repulsion giving a lower (i.e. more favourable) energy. The exchange interaction involves integrals of the form:

$$K_{ij}\int\int d\tau_1 d\tau_2 \chi_i(1)\chi_j(2)\left(\frac{1}{r_{12}}\right)\chi_i(2)\chi_j(1) \tag{2.104}$$

This integral is only non-zero if the spins of the electrons in the spin orbitals χ_i and χ_j are the same. The energy due to exchange is often represented as K_{ij}. The exchange energy between the electron in spin orbital χ_i and the other $N-1$ electrons is:

$$E_i^{\text{exchange}} = \sum_{j \neq i}^{N} \iint d\tau_1 d\tau_2 \chi_i(1)\chi_j(2)\left(\frac{1}{r_{12}}\right)\chi_i(2)\chi_j(1) \tag{2.105}$$

The total exchange energy is calculated thus:

$$E_{\text{total}}^{\text{exchange}} = \sum_{i=1}^{N} \sum_{j'=i+1}^{N} \iint d\tau_1 d\tau_2 \chi_i(1)\chi_j(2)\left(\frac{1}{r_{12}}\right)\chi_i(2)\chi_j(1) = \sum_{j=1}^{N} \sum_{j'=i+1}^{N} K_{ij} \tag{2.106}$$

The prime on the counter j′ indicates that the summation is only over electrons with the same spin as electron i.

2.4.3 *Shorthand representations of the one- and two-electron integrals*

Various shorthand ways have been devised to represent the integrals involved in an electronic structure calculation. The two-electron integrals J_{ij} and K_{ij} are particularly long-winded to write out. In one scheme the Coulomb interaction J_{ij} is written as:

$$\langle \chi_i^* \chi_j^* \left| \frac{1}{r_{12}} \right| \chi_i \chi_j \rangle \tag{2.107}$$

In this notation the complex parts are written on the left-hand side and the real parts on the right. Sometimes the χ symbol is eliminated:

$$\left\langle \text{ij} \left| \frac{1}{r_{12}} \right| \text{ij} \right\rangle \tag{2.108}$$

The exchange integrals would be written

$$\left\langle \text{ij} \left| \frac{1}{r_{12}} \right| \text{ji} \right\rangle$$

in this notation.

A notation that is widely used in the chemical literature writes the orbitals that are functions of electron 1 on the left-hand side (with the complex conjugate orbital first if appropriate) and the orbitals that are functions of electron 2 on the right-hand side (again with the complex conjugate orbital first). In this notation, which is the one that we will adopt, the Coulomb integral is written (ii | jj) and the exchange integral (ij | ij). The one-electron integrals such as equation (2.99) are written as follows:

$$\left(\text{i} \left| -\frac{1}{2}\nabla_i^2 - \sum_{A=1}^{M} \frac{Z_A}{r_{iA}} \right| \text{j} \right) \equiv \int d\tau_1 \chi_i(1) \left(-\frac{1}{2}\nabla_i^2 - \sum_{A=1}^{M} \frac{Z_A}{r_{iA}} \right) \chi_j(1) \tag{2.109}$$

When calculating the total energy of the system, we should not forget the Coulomb interaction between the nuclei; this is constant within the Born–Oppenheimer approximation for a given spatial arrangement of nuclei. When it is desired to change the nuclear positions, it is of course necessary to take the internuclear repulsion energy into account, which is calculated using the Coulomb equation:

$$\sum_{A=1}^{M}\sum_{B=A+1}^{M}\frac{Z_A Z_B}{R_{AB}} \tag{2.110}$$

2.4.4 The energy of a closed-shell system

In molecular modelling we are usually concerned with the ground states of molecules, most of which have closed-shell configurations. In a closed-shell system containing N electrons in $N/2$ orbitals, there are two spin orbitals associated with each spatial orbital ψ_i: $\psi_i\alpha$ and $\psi_i\beta$. The electronic energy of such a system can be calculated in a manner analogous to that for the hydrogen molecule. First, there is the energy of each electron moving in the field of the bare nuclei. For an electron in a molecular orbital χ_i, this contributes an energy H_{ii}^{core}. If there are two electrons in the orbital then the energy is $2H_{ii}^{core}$ and for $N/2$ orbitals the total contribution to the energy will be

$$\sum_{i=1}^{N/2} 2H_{ii}{}^{core}$$

If we consider the electron–electron terms, the interaction between each pair of orbitals ψ_i and ψ_j involves a total of four electrons. There are four ways in which two electrons in one orbital can interact in a Coulomb sense with two electrons in a second orbital, thus giving $4J_{ij}$. However, there are just two ways to obtain paired electrons from this arrangement, giving a total exchange contribution of $-2K_{ij}$. Finally, the Coulomb interaction between each pair of electrons in the same orbital must be included; there is no exchange interaction because the electrons have paired spins. The total energy is thus given as:

$$E = 2\sum_{i=1}^{N/2} H_{ii}^{core} + \sum_{i=1}^{N/2}\sum_{j=i+1}^{N/2}(4J_{ij} - 2K_{ij}) + \sum_{i=1}^{N/2} J_{ii} \tag{2.111}$$

A more concise form of this equation can be obtained if we recognise that $J_{ii} = K_{ii}$:

$$E = 2\sum_{i=1}^{N/2} H_{ii}^{core} + \sum_{i=1}^{N/2}\sum_{j=1}^{N/2}(2J_{ij} - K_{ij}) \tag{2.112}$$

2.5 The Hartree–Fock equations

In our hydrogen molecule calculation in section 2.4.1 the molecular orbitals were provided as input, but in most electronic structure calculations we are usually trying to calculate the molecular orbitals. How do we go about this? We must remember that for many-body problems there is no 'correct' solution; we therefore require some means to decide whether one proposed wavefunction is 'better' than another. Fortunately, the *variation theorem* provides us with a mechanism for answering this question. The theorem states that the energy calculated from an approximation to the true wavefunction will always be greater than the true energy. Consequently, the better the wavefunction, the lower the energy. The 'best' wavefunction is obtained when the energy is a minimum. At a minimum, the first derivative of the energy, δE will be zero. The Hartree–Fock equations are obtained by imposing this condition on the expression for the energy, subject to the constraint that the molecular orbitals remain orthonormal. The orthonormality condition is written in terms of the *overlap integral*, S_{ij}, between two orbitals i and j. Thus

$$S_{\mathrm{ij}} = \int \chi_{\mathrm{i}}\chi_{\mathrm{j}}\mathrm{d}\tau = \delta_{\mathrm{ij}} \text{ } (\delta_{\mathrm{ij}} \text{ is the Kronecker delta}) \tag{2.113}$$

This type of constrained minimisation problem can be tackled using the method of Lagrange multipliers. In this approach (see section 1.10.5 for a brief introduction to Lagrange multipliers) the derivative of the function to be minimised is added to the derivatives of the constraint(s) multiplied by a constant called a Lagrange multiplier. The sum is then set equal to zero. If the Lagrange multiplier for each of the orthonormality conditions is written λ_{ij}, then:

$$\delta E + \delta \sum_{\mathrm{i}} \sum_{\mathrm{j}} \lambda_{\mathrm{ij}} S_{\mathrm{ij}} = 0 \tag{2.114}$$

In the Hartree–Fock equations the Lagrange multipliers are actually written $-2\varepsilon_{\mathrm{ij}}$ to reflect the fact that they are related to the molecular orbital energies. The equation to be solved is thus:

$$\delta E - 2\delta \sum_{\mathrm{i}} \sum_{\mathrm{j}} \varepsilon_{\mathrm{ij}} S_{\mathrm{ij}} = 0 \tag{2.115}$$

We will not describe in detail how this equation is solved, as it is rather complicated. However, a qualitative picture is possible. The major difference between polyelectronic systems and systems with single electrons is the presence of interactions between the electrons, which as we have seen are expressed as Coulomb and exchange integrals. Suppose we are given the task of finding the 'best' (i.e. lowest energy) wavefunction for a polyelectronic system. We wish to retain the orbital picture of the system, in which single electrons are assigned to individual spin orbitals. The problem is to find a solution which simultaneously enables all the electronic motions to be taken into account, as a change in the spin orbital for one electron will influence the behaviour of an electron in another spin orbital due to the coupling of the electronic motions. We concentrate on a single electron in a spin orbital χ_i in the field of the nuclei and the other electrons in their (fixed) spin orbitals χ_j. The Hamiltonian operator for the electron in χ_i contains three terms appropriate to the three different contributions to the energy that were identified above (core, Coulomb, exchange). The result can be written as an integro-differential equation for χ_i that has the following form:

$$\left[-\frac{1}{2}\nabla_i^2 - \sum_{A=1}^{M}\frac{Z_A}{r_{iA}}\right]\chi_i(1) + \sum_{j\neq i}\left[\int d\tau_2 \chi_j(2)\chi_j(2)\frac{1}{r_{12}}\right]\chi_i(1) - \sum_{j\neq i}\left[\int d\tau_2 \chi_j(2)\chi_i(2)\frac{1}{r_{12}}\right]\chi_i(1) = \sum_j \varepsilon_{ij}\chi_j(1) \tag{2.116}$$

This expression can be tidied up by introducing three operators that represent the contributions to the energy of the spin orbital χ_i in the 'frozen' system:

The core Hamiltonian operator, $\mathscr{H}^{core}(1)$:

$$\mathscr{H}^{core}(1) = -\frac{1}{2}\nabla_1^2 - \sum_{A=1}^{M}\frac{Z_A}{r_{1A}} \tag{2.117}$$

In the absence of any interelectronic interactions this would be the only operator present, corresponding to the motion of a single electron moving in the field of the bare nuclei.

The Coulomb operator, $\mathscr{J}_j(1)$:

$$\mathscr{J}_j(1) = \int d\tau_2 \chi(2)\frac{1}{r_{12}}\chi_j(2) \tag{2.118}$$

This operator corresponds to the average potential due to an electron in χ_j.

The exchange operator $\mathscr{K}_j(1)$:

$$\mathscr{K}_j(1)\chi_i(1) = \left[\int d\tau_2 \chi_j(2)\frac{1}{r_{12}}\chi_i(2)\right]\chi_j(1) \tag{2.119}$$

The form of this operator is rather unusual, insofar as it must be defined in terms of its effect when acting on the spin orbital χ_i.

Equation (2.116) can thus be written:

$$\mathscr{H}^{core}(1)\chi_i(1) + \sum_{j\neq i}^{N}\mathscr{J}_j(1)\chi_i(1) - \sum_{j\neq i}^{N}\mathscr{K}_j(1)\chi_i(1) = \sum_j \varepsilon_{ij}\chi_j(1) \tag{2.120}$$

Making use of the fact that $\{\mathscr{J}_i(1) - \mathscr{K}_i(1)\}\chi_i(1) = 0$ leads to the following form:

$$\left[\mathscr{H}^{core}(1) + \sum_{j=1}^{N}\{\mathscr{J}_j(1) - \mathscr{K}_j(1)\}\right]\chi_i(1) = \sum_{j=1}^{N}\varepsilon_{ij}\chi_j(1) \tag{2.121}$$

Or, more simply:

$$\mathscr{f}_i \chi_i = \sum_j \varepsilon_{ij} \chi_j \tag{2.122}$$

$\mathscr{f}_i$ is called the *Fock operator*:

$$\mathscr{f}_i(1) = \mathscr{H}^{\text{core}}(1) + \sum_{j=1}^{N} \{\mathscr{J}_j(1) - \mathscr{K}_j(1)\} \tag{2.123}$$

For a closed-shell system, the Fock operator has the following form:

$$\mathscr{f}_i(1) = \mathscr{H}^{\text{core}}(1) + \sum_{j=1}^{N/2} \{2\mathscr{J}_j(1) - \mathscr{K}_j(1)\} \tag{2.124}$$

The Fock operator is an effective one-electron Hamiltonian for the electron in the polyelectronic system. However, written in this form of equation (2.122), the Hartree–Fock equations do not seem to be particularly useful: on the left-hand side we have the Fock operator acting on the molecular orbital χ_i, but this returns not the molecular orbital multiplied by a constant as in a normal eigenvalue equation, but rather a series of orbitals χ_j' multiplied by some unknown constants ε_{ij}. This is because the solutions to the Hartree–Fock equations are not unique. We have already seen that the value of a determinant is unaffected when the multiple of any column is added to another column. If such a transformation is performed on the Slater determinant, then a different set of constants ε_{ij}' would be obtained with the spin orbitals χ_i being linear combinations of the first set. Certain transformations give rise to localised orbitals which are particularly useful for understanding the chemical nature of the system. These localised orbitals are no more 'correct' than a delocalised set. Fortunately, it is possible to manipulate the equations (2.122) mathematically so that the Lagrangian multipliers are zero unless the indices i and j are the same. The Hartree–Fock equations then take on the standard eigenvalue form:

$$\mathscr{f}_i \chi_i = \varepsilon_i \chi_i \tag{2.125}$$

Recall that in setting up these equations, each electron has been assumed to move in a 'fixed' field comprising the nuclei and the other electrons. This has important implications for the way in which we attempt to find a solution, for any solution that we might find by solving the equation for one electron will naturally affect the solutions for the other electrons in the system. The general strategy is called a *self-consistent field* (SCF) approach. One way to solve these equations is as follows. First, a set of trial solutions χ_i to the Hartree–Fock eigenvalue equations are obtained. These are used to calculate the Coulomb and exchange operators. The Hartree–Fock equations are solved, giving a second set of solutions χ_i, which are used in the next iteration. The SCF method thus gradually refines the individual electronic solutions that correspond to lower and lower total energies until the point is reached at which the results for all the electrons are unchanged, when they are said to be *self-consistent*.

2.5.1 Hartree–Fock calculations for atoms and Slater's rules

The Hartree–Fock equations are usually solved in different ways for atoms and for molecules. For atoms, the equations can be solved numerically if it is assumed that the electron distribution is spherically symmetrical. However, these numerical solutions are not particularly useful. Fortunately, analytical approximations to these solutions, which are very similar to those obtained for the hydrogen atom, can be used with considerable success. These approximate analytical functions thus have the form:

$$\psi = R_{nl}(r)Y_{lm}(\theta, \phi) \tag{2.126}$$

Y is a spherical harmonic (as for the hydrogen atom) and R is a radial function. The radial functions obtained for the hydrogen atom cannot be used directly for polyelectronic atoms due to the screening of the nuclear charge by the inner shell electrons, but the hydrogen atom functions are acceptable if the orbital exponent is adjusted to account for the screening effect. Even so, the hydrogen atom functions are not particularly convenient to use in molecular orbital calculations due to their complicated functional form. Slater [Slater 1930] suggested a simpler analytical form for the radial functions:

$$R_{nl}(r) = (2\zeta)^{n+\frac{1}{2}}[(2n)!]^{-1/2}r^{n-1}\mathrm{e}^{-\zeta r} \tag{2.127}$$

These functions are universally known as *Slater-type orbitals* (STOs) and are just the leading term in the appropriate Laguerre polynomials. The first three Slater functions are as follows:

$$R_{1s}(r) = 2\zeta^{3/2}\mathrm{e}^{-\zeta r} \tag{2.128}$$

$$R_{2s}(r) = R_{2p}(r) = \left(\frac{4\zeta^5}{3}\right)^{1/2} r\mathrm{e}^{-\zeta r} \tag{2.129}$$

$$R_{3s}(r) = R_{3p}(r) = R_{3d}(r) = \left(\frac{8\zeta^7}{45}\right)^{1/2} r^2\mathrm{e}^{-\zeta r} \tag{2.130}$$

To obtain the whole orbital we must multiply $R(r)$ by the appropriate angular part. For example, we would use the following expressions for the 1s, 2s and $2p_z$ orbitals:

$$\phi_{1s}(\mathbf{r}) = \sqrt{(\zeta^3/\pi)}\exp(-\zeta r) \tag{2.131}$$

$$\phi_{2s}(\mathbf{r}) = \sqrt{(\zeta^5/3\pi)}\mathbf{r}\exp(-\zeta r) \tag{2.132}$$

$$\phi_{2p_z}(\mathbf{r}) = \sqrt{(\zeta^5/\pi)}\exp(-\zeta r)\cos\theta \tag{2.133}$$

Slater provided a series of empirical rules for choosing the orbital exponents ζ, which are given by:

$$\zeta = \frac{Z - \sigma}{n^*} \tag{2.134}$$

Z is the atomic number and σ is a *shielding constant*, determined as below. n^* is an effective principal quantum number which takes the same value as the true principal quantum number for $n = 1$, 2 or 3, but for $n = 4$, 5, 6 has the values 3.7, 4.0, 4.2 respectively. The shielding constant is obtained as follows:

First divide the orbitals into the following groups:

(1s); (2s, 2p); (3s, 3p); (3d); (4s, 4p); (4d); (4f); (5s, 5p); (5d)

For a given orbital, σ is obtained by adding together the following contributions:

(a) zero from an orbital further from the nucleus than those in the group;

(b) 0.35 from each other electron in the same group, but if the other orbital is the 1s then the contribution is 0.3;

(c) 1.0 for each electron in a group with a principal quantum number 2 or more fewer than the current orbital;

(d) for each electron with a principal quantum number 1 fewer than the current orbital: 1.0 if the current orbital is d or f; 0.85 if the current orbital is s or p.

The shielding constant for the valence electrons of silicon is obtained using Slater's rules as follows. The electronic configuration of Si is $(1s^2)(2s^22p^6)(3s^23p^2)$. We therefore count 3×0.35 under rule (b), 2.0 under rule (c), and 8×0.85 under rule (d), giving a total of 9.85. When subtracted from the atomic number (14) this gives 4.15 for the value of $Z - \sigma$.

2.5.2 *Linear combination of atomic orbitals (LCAO) in Hartree–Fock theory*

Direct solution of the Hartree–Fock equations is not a practical proposition for molecules and so it is necessary to adopt an alternative approach. The most popular strategy is to write each spin orbital as a linear combination of single electron orbitals:

$$\psi_{\mathrm{i}} = \sum_{\nu=1}^{K} c_{\nu\mathrm{i}} \phi_{\nu} \tag{2.135}$$

The one-electron orbitals ϕ_ν are commonly called *basis functions* and often correspond to the atomic orbitals. We will label the basis functions with the Greek letters μ, ν, λ and σ. In the case of equation (2.135) there are K basis functions and we should therefore expect to derive a total of K molecular orbitals (although not all of these will necessarily be occupied by electrons). The smallest number of basis functions for a molecular system will be that which can just accommodate all the electrons in the molecule. More sophisticated calculations use more basis functions than a minimal set. At the *Hartree–Fock limit* the energy of the system can be reduced no further by the addition of any more basis functions; however, it may be possible to lower the energy below the Hartree–Fock limit by using a functional form of the wavefunction that is more extensive than the single Slater determinant.

In accordance with the variation theorem we require the set of coefficients $c_{\nu\mathrm{i}}$ that gives the lowest energy wavefunction, and some scheme for changing the coefficients to derive that wavefunction. For a given basis set and a given functional form of the wavefunction (i.e. a Slater determinant) the best set of coefficients is that for which the energy is a minimum, at which point

$$\frac{\partial E}{\partial c_{\mathrm{vi}}} = 0$$

for all coefficients c_{vi}. The objective is thus to determine the set of coefficients that gives the lowest energy for the system.

2.5.3 Closed-shell systems and the Roothaan–Hall equations

We shall initially consider a closed-shell system with N electrons in $N/2$ orbitals. The derivation of the Hartree–Fock equations for such a system was first proposed by Roothaan [Roothaan 1951] and (independently) by Hall [Hall 1951]. The resulting equations are known as the Roothaan equations or the Roothaan–Hall equations. Unlike the integro-differential form of the Hartree–Fock equations, equation (2.116), Roothaan and Hall recast the equations in matrix form which can be solved using standard techniques and can be applied to systems of any geometry. We shall identify the major steps in the Roothaan approach, starting with the expression for the Hartree–Fock energy for our closed-shell system, equation (2.112):

$$E = 2\sum_{\mathrm{i}=1}^{N/2} H_{\mathrm{ii}}^{\mathrm{core}} + \sum_{\mathrm{i}=1}^{N/2}\sum_{\mathrm{j}=1}^{N/2}(2J_{\mathrm{ij}} - K_{\mathrm{ij}}) \tag{2.136}$$

The corresponding Fock operator is (equation 2.124):

$$f_{\mathrm{i}}(1) = \mathscr{H}^{\mathrm{core}}(1) + \sum_{\mathrm{j}=1}^{N/2}\{2\mathscr{J}_{\mathrm{j}}(1) - \mathscr{K}_{\mathrm{j}}(1)\} \tag{2.137}$$

We now introduce the atomic orbital expansion for the orbitals ψ_{i} and substitute for the corresponding spin orbital χ_{i} into the Hartree–Fock equation, $f_{\mathrm{i}}(1)\chi_{\mathrm{i}}(1) = \varepsilon_{\mathrm{i}}\chi_{\mathrm{i}}(1)$:

$$f_{\mathrm{i}}(1)\sum_{\mathrm{v}=1}^{K} c_{\mathrm{vi}}\phi_{\mathrm{v}}(1) = \varepsilon_{\mathrm{i}}\sum_{\mathrm{v}=1}^{K} c_{\mathrm{vi}}\phi_{\mathrm{v}}(1) \tag{2.138}$$

Premultiplying each side by $\phi_{\mu}(1)$ (where ϕ_{μ} is also a basis function) and integrating gives the following matrix equation:

$$\sum_{\mathrm{v}=1}^{K} c_{\mathrm{vi}}\int \mathrm{d}v_1\phi_{\mu}(1)f_{\mathrm{i}}(1)\phi_{\mathrm{v}}(1) = \varepsilon_{\mathrm{i}}\sum_{\mathrm{v}=1}^{K} c_{\mathrm{vi}}\int \mathrm{d}v_1\phi_{\mu}(1)\phi_{\mathrm{v}}(1) \tag{2.139}$$

$\int dv_1\phi_{\mu}(1)\phi_{\mathrm{v}}(1)$ is the overlap integral between the basis functions μ and v, written $S_{\mu\mathrm{v}}$. Unlike the molecular orbitals which will be required to be orthonormal, the overlap between two basis functions is not necessarily zero (for example, they may be located on different atoms).

The elements of the *Fock matrix* are given by

$$F_{\mu\nu} = \int dv_1 \phi_\mu(1) f_i(1) \phi_\nu(1) \tag{2.140}$$

The Fock matrix elements for a closed-shell system can be expanded as follows by substituting the expression for the Fock operator:

$$\begin{aligned} F_{\mu\nu} = & \int dv_1 \phi_\mu(1) \mathcal{H}^{\text{core}}(1) \phi_\nu(1) \\ & + \sum_{j=1}^{N/2} \int dv_1 \phi_\mu(1)[2\mathcal{J}_j(1) - \mathcal{K}_j(1)] \phi_\nu(1) \end{aligned} \tag{2.141}$$

The elements of the Fock matrix can thus be written as the sum of core, Coulomb and exchange contributions. The core contribution is:

$$\begin{aligned} & \int dv_1 \phi_\mu(1) \mathcal{H}^{\text{core}}(1) \phi_\nu(1) \\ & = \int dv_1 \phi_\mu(1) \left[-\frac{1}{2}\nabla^2 - \sum_{A=1}^{M} \frac{Z_A}{|r_1 - R_A|} \right] \phi_\nu(1) \equiv H_{\mu\nu}^{\text{core}} \end{aligned} \tag{2.142}$$

The core contributions thus require the calculation of integrals that involve basis functions on up to two centres (depending upon whether ϕ_μ and ϕ_ν are centred on the same nucleus or not). Each element $H_{\mu\nu}^{\text{core}}$ can in turn be obtained as the sum of a kinetic energy integral and a potential energy integral corresponding to the two terms in the one-electron Hamiltonian.

The Coulomb and exchange contributions to the Fock matrix element $F_{\mu\nu}$ are together given by:

$$\sum_{j=1}^{N/2} \int dv_1 \phi_\mu(1)[2\mathcal{J}_j(1) - \mathcal{K}_j(1)] \phi_\nu(1) \tag{2.143}$$

Recall that the Coulomb operator $\mathcal{J}_j(1)$ due to interaction with a spin orbital χ_j is given by

$$\mathcal{J}_j(1) = \int d\tau_2 \chi_j(2) \frac{1}{r_{12}} \chi_j(2).$$

We need to write each of the two occurrences of the spin orbital χ_j in this integral in terms of the appropriate linear combination of basis functions:

$$\mathcal{J}_j(1) = \int d\tau_2 \sum_{\sigma=1}^{K} c_{\sigma j} \phi_\sigma(2) \frac{1}{r_{12}} \sum_{\lambda=1}^{K} c_{\lambda j} \phi_\lambda(2) \tag{2.144}$$

We have used the indices σ and λ for the basis functions here. Similarly, the exchange contribution can be written:

$$\mathcal{K}_j(1)\chi_i(1) = \left[\int d\tau_2 \sum_{\sigma=1}^{K} c_{\sigma j} \phi_\sigma(2) \frac{1}{r_{12}} \chi_i(2) \right] \sum_{\lambda=1}^{K} c_{\lambda j} \phi_\lambda(2) \tag{2.145}$$

When the Coulomb and exchange operators are expressed in terms of the basis functions and the orbital expansion is substituted for χ_i, then their contributions to the Fock matrix element $F_{\mu\nu}$take the following form:

$$\sum_{j=1}^{N/2} \int dv_1 \phi_\mu(1)[2\mathscr{J}_i(1) - \mathscr{K}_j(1)]\phi_\nu(1)$$

$$= \sum_{j=1}^{N/2} \sum_{\lambda=1}^{K} \sum_{\sigma=1}^{K} c_{\lambda j} c_{\sigma j} \left[\begin{array}{l} 2\int dv_1 dv_2 \phi_\mu(1)\phi_\nu(1)\frac{1}{r_{12}}\phi_\lambda(2)\phi_\sigma(2) \\ -\int dv_1 dv_2 \phi_\mu(1)\phi_\lambda(1)\frac{1}{r_{12}}\phi_\nu(2)\phi_\sigma(2) \end{array} \right]$$

$$\equiv \sum_{j=1}^{N/2} \sum_{\lambda=1}^{K} \sum_{\sigma=1}^{K} c_{\lambda j} c_{\sigma j}[2(\mu\nu|\lambda\sigma) - (\mu\lambda|\nu\sigma)] \tag{2.146}$$

We have used the shorthand notation for the integrals in the final expression. Note that the two-electron integrals may involve up to four different basis functions (μ, ν, λ, σ), which may in turn be located at four different centres. This has important consequences for the way in which we try to solve the equations.

It is helpful to simplify equation (2.146) by introducing the *charge density matrix*, **P**, whose elements are defined as:

$$P_{\mu\nu} = 2\sum_{i=1}^{N/2} c_{\mu i} c_{\nu i} \quad \text{and} \quad P_{\lambda\sigma} = 2\sum_{i=1}^{N/2} c_{\lambda i} c_{\sigma i} \tag{2.147}$$

Note that the summations are over the $N/2$ occupied orbitals. Other properties can be calculated from the density matrix; for example, the electronic energy is:

$$E = \frac{1}{2}\sum_{\mu=1}^{K}\sum_{\nu=1}^{K} P_{\mu\nu}(H_{\mu\nu}^{\text{core}} + F_{\mu\nu}) \tag{2.148}$$

The electron density at a point **r** can also be expressed in terms of the density matrix:

$$\rho(\mathbf{r}) = \sum_{\mu=1}^{K}\sum_{\nu=1}^{K} P_{\mu\nu}\phi_\mu(\mathbf{r})\phi_\nu(\mathbf{r}) \tag{2.149}$$

The expression for each element $F_{\mu\nu}$ of the Fock matrix elements for a closed-shell system of N electrons then becomes:

$$F_{\mu\nu} = H_{\mu\nu}^{\text{core}} + \sum_{\lambda=1}^{K}\sum_{\sigma=1}^{K} P_{\lambda\sigma}\left[(\mu\nu|\lambda\sigma) - \tfrac{1}{2}(\mu\lambda|\nu\sigma)\right] \tag{2.150}$$

This is the standard form for the expression for the Fock matrix in the Roothaan–Hall equations.

2.5.4 Solving the Roothaan–Hall equations

The Fock matrix is a $K \times K$ square matrix that is symmetric if real basis functions are used. The Roothaan–Hall equations (2.139) can be conveniently written as a matrix equation:

$$\mathbf{FC} = \mathbf{SCE} \tag{2.151}$$

The elements of the $K \times K$ matrix $\mathbf{C}$ are the coefficients $c_{\nu i}$:

$$\mathbf{C} = \begin{pmatrix} c_{1,1} & c_{1,2} & \cdots & c_{1,K} \\ c_{2,1} & c_{2,2} & \cdots & c_{2,K} \\ \vdots & \vdots & & \vdots \\ c_{K,1} & c_{K,2} & \cdots & c_{K,K} \end{pmatrix} \tag{2.152}$$

$\mathbf{E}$ is a diagonal matrix whose elements are the orbital energies:

$$\mathbf{E} = \begin{pmatrix} \varepsilon_1 & 0 & \cdots & 0 \\ 0 & \varepsilon_2 & \cdots & 0 \\ \vdots & \vdots & \ddots & \vdots \\ 0 & 0 & \cdots & \varepsilon_K \end{pmatrix} \tag{2.153}$$

Let us consider how we might solve the Roothaan–Hall equations and thereby obtain the molecular orbitals. The first point we must note is that the elements of the Fock matrix, which appear on the left hand side of equation (2.151), depend on the molecular orbital coefficients $c_{\nu i}$, which also appear on the right-hand side of the equation. Thus an iterative procedure is required to find a solution.

The one-electron contributions $H_{\mu\nu}^{\text{core}}$ due to the electrons moving in the field of the bare nuclei do not depend on the basis set coefficients and remain unchanged throughout the calculation. However, the Coulomb and exchange contributions do depend on the coefficients and we would expect these to vary throughout the calculation. The individual two-electron integrals $(\mu\nu \mid \lambda\sigma)$ are, however, constant throughout the calculation. An obvious strategy is thus to calculate and store these integrals for later use.

Having written the Roothaan–Hall equations in matrix form we would obviously like to solve them using standard matrix eigenvalue methods (discussed in section 1.10.3). However, standard eigenvalue methods would require an equation of the form $\mathbf{FC} = \mathbf{CE}$. The Roothaan–Hall equations only adopt such a form if the overlap matrix, $\mathbf{S}$, is equal to the unit matrix, $\mathbf{I}$ (in which all diagonal elements are equal to 1 and all off-diagonal elements are zero). The functions ϕ are usually normalised but they are not necessarily orthogonal (for example, because they are located on different atoms) and so there will invariably be non-zero off-diagonal elements of the overlap matrix. To solve the Roothaan–Hall equations using standard methods they must be transformed. This corresponds to transforming the basis functions so that they form an orthonormal set. We seek a matrix $\mathbf{X}$, such that $\mathbf{X}^{\mathrm{T}}\mathbf{SX} = \mathbf{I}$. $\mathbf{X}^{\mathrm{T}}$ is the transpose of $\mathbf{X}$, obtained by interchanging rows and columns. There are various ways in which $\mathbf{X}$ can be calculated; in *symmetric orthogonalisation*, the overlap matrix is diagonalised. Diagonalisation involves finding the matrix **U** such that

$$\mathbf{U}^{\mathrm{T}}\mathbf{S}\mathbf{U} = \mathbf{D} = \mathrm{diag}(\lambda_1 \ldots \lambda_K) \tag{2.154}$$

D is the diagonal matrix containing the eigenvalues λ_i of **S**, and **U** contains the eigenvectors of **S**. $\mathbf{U}^{\mathrm{T}}$ is the transpose of the matrix **U**. (This expression is often written $\mathbf{U}^{-1}\mathbf{S}\mathbf{U} = \mathbf{D}$ since for real basis functions $\mathbf{U}^{-1} = \mathbf{U}^{\mathrm{T}}$.) Then the matrix **X** is given by $\mathbf{X} = \mathbf{U}\mathbf{D}^{-1/2}\mathbf{U}^{\mathrm{T}}$ where $\mathbf{D}^{-1/2}$ is formed from the inverse square roots of **D**. We shall write **X** as $\mathbf{S}^{-1/2}$, as it can be considered to be the inverse square root of the overlap matrix: $\mathbf{S}^{-1/2}\mathbf{S}\mathbf{S}^{-1/2} = \mathbf{I}$.

The Roothaan–Hall equations can now be manipulated as follows. Both sides of equation (2.151) are pre-multiplied by the matrix $\mathbf{S}^{-1/2}$:

$$\mathbf{S}^{-1/2}\mathbf{F}\mathbf{C} = \mathbf{S}^{-1/2}\mathbf{S}\mathbf{C}\mathbf{E} = \mathbf{S}^{1/2}\mathbf{C}\mathbf{E} \tag{2.155}$$

Inserting the unit matrix, in the form $\mathbf{S}^{-1/2}\mathbf{S}^{1/2}$ into the left-hand side gives:

$$\mathbf{S}^{-1/2}\mathbf{F}(\mathbf{S}^{-1/2}\mathbf{S}^{1/2})\mathbf{C} = \mathbf{S}^{1/2}\mathbf{C}\mathbf{E} \tag{2.156}$$

or

$$\mathbf{S}^{-1/2}\mathbf{F}\mathbf{S}^{-1/2}(\mathbf{S}^{1/2}\mathbf{C}) = (\mathbf{S}^{1/2}\mathbf{C})\mathbf{E} \tag{2.157}$$

Equation (2.157) can be written $\mathbf{F}'\mathbf{C}' = \mathbf{C}'\mathbf{E}$, where $\mathbf{F}' = \mathbf{S}^{-1/2}\mathbf{F}\mathbf{S}^{-1/2}$ and $\mathbf{C}' = \mathbf{S}^{1/2}\mathbf{C}$.

The matrix equation $\mathbf{F}'\mathbf{C}' = \mathbf{C}'\mathbf{E}$ can be solved using standard methods; a solution only exists if the determinant $|\mathbf{F}' - \mathbf{E}\mathbf{I}|$ equals zero. In simple cases this can be done by multiplying out the determinant to give a polynomial (the secular equation) whose roots are the eigenvalues ε_i, but for large matrices a much more practical approach involves the diagonalisation of $\mathbf{F}'$. The matrix of coefficients, $\mathbf{C}'$, are the eigenvectors of $\mathbf{F}'$. The basis function coefficients **C** can then be obtained from $\mathbf{C}'$ using $\mathbf{C} = \mathbf{S}^{-1/2}\mathbf{C}'$. A common scheme for solving the Roothaan–Hall equations is thus as follows:

1. Calculate the integrals to form the Fock matrix, **F**.
2. Calculate the overlap matrix, **S**.
3. Diagonalise **S**.
4. Form $\mathbf{S}^{-1/2}$.
5. Guess, or otherwise calculate an initial density matrix, **P**.
6. Form the Fock matrix using the integrals and the density matrix **P**.
7. Form $\mathbf{F}' = \mathbf{S}^{-1/2}\mathbf{F}\mathbf{S}^{-1/2}$.
8. Solve the secular equation $|\mathbf{F}' - \mathbf{E}\mathbf{I}| = 0$ to give the eigenvalues **E** and the eigenvectors $\mathbf{C}'$ by diagonalising $\mathbf{F}'$.
9. Calculate the molecular orbital coefficients, **C** from $\mathbf{C} = \mathbf{S}^{-1/2}\mathbf{C}'$.
10. Calculate a new density matrix, **P**, from the matrix **C**.
11. Check for convergence. If the calculation has converged, stop. Otherwise repeat from step 6 using the new density matrix **P**.

This procedure requires an initial guess of the density matrix, **P**. The simplest approach is to use the null matrix, which corresponds to ignoring all the electron–electron terms so that the electrons just experience the bare nuclei. This can sometimes lead to convergence problems which may be prevented if a lower level of theory (such as semi-empirical or extended Hückel) is used to provide the initial guess. Moreover, a better guess may enable the calculation to be performed more quickly. A variety of criteria can be used to establish whether the calculation has converged or not. For example, the density matrix can be compared with that from the previous iteration, and/or the change in energy can be monitored together with the basis set coefficients.

The result of a Hartree–Fock calculation is a set of K molecular orbitals where K is the number of basis functions in the calculation. The N electrons are then fed into these orbitals in accordance with the Aufbau principle, two electrons per orbital, starting with the lowest energy orbitals. The remaining orbitals do not contain any electrons; these are known as the *virtual orbitals*. Alternative electronic configurations can be generated by exciting electrons from the occupied orbitals to the virtual orbitals; these excited configurations are used in more advanced calculations that will be discussed in section 2.8.

A Hartree–Fock calculation provides a set of orbital energies, ε_i. What is the significance of these? The energy of an electron in a spin orbital is calculated by adding the core interaction $H_{\mathrm{ii}}^{\mathrm{core}}$ to the Coulomb and exchange interactions with the other electrons in the system:

$$\varepsilon_{\mathrm{i}} = H_{\mathrm{ii}}^{\mathrm{core}} + \sum_{\mathrm{j}=1}^{N/2} (2J_{\mathrm{ij}} - K_{\mathrm{ij}}) \tag{2.158}$$

The total electronic energy of the ground state is given by equation (2.112):

$$E = 2\sum_{\mathrm{i}=1}^{N/2} H_{\mathrm{ii}}^{\mathrm{core}} + \sum_{\mathrm{i}=1}^{N/2}\sum_{\mathrm{j}=1}^{N/2} (2J_{\mathrm{ij}} - K_{\mathrm{ij}}) \tag{2.159}$$

The total energy is therefore not equal to the sum of the individual orbital energies, but is related as follows:

$$E = \sum_{\mathrm{i}=1}^{N} \varepsilon_i - \sum_{\mathrm{i}=1}^{N/2}\sum_{\mathrm{j}=1}^{N/2} (2J_{\mathrm{ij}} - K_{\mathrm{ij}}) \tag{2.160}$$

The reason for the discrepancy is that the individual orbital energies include contributions from the interaction between that electron and all the nuclei and all other electrons in the system. The Coulomb and exchange interactions between pairs of electrons are therefore counted twice when summing the individual orbital energies.

2.5.5 A simple illustration of the Roothaan–Hall approach

We will illustrate the stages involved in the Roothaan–Hall approach using the helium hydrogen molecular ion, HeH^+ as an example. This is a two-electron system. Our objective here is to show how the Roothaan–Hall method can be used to derive the wavefunction, for a fixed internuclear distance of 1 Å. We use HeH^+ rather than H_2 as our system as the lack of symmetry in HeH^+ makes the procedure more informative. There are two basis functions, $1s_A$ (centred on the helium atom) and $1s_B$ (on the hydrogen). The numerical values of the integrals that we shall use in our calculation were obtained using a Gaussian series approximation to the Slater orbitals (the STO-3G basis set, which is described in section 2.6). This detail need not concern us here. Each wavefunction is expressed as a linear combination of the two 1s atomic orbitals centred on the nuclei A and B:

$$\begin{aligned} \psi_1 &= c_{1A} 1s_A + c_{1B} 1s_B \\ \psi_2 &= c_{2A} 1s_A + c_{2B} 1s_B \end{aligned} \tag{2.161}$$

First, it is necessary to calculate the various one- and two-electron integrals and to formulate the Fock and overlap matrices, each of which will be a 2×2 symmetric matrix (as there are two orbitals in the basis set). The diagonal elements of the overlap matrix, **S**, are equal to 1.0 as each basis function is normalised; the off-diagonal elements have smaller, but non-zero values that are equal to the overlap between $1s_A$ and $1s_B$ for the internuclear distance chosen. The matrix **S** is:

$$\mathbf{S} = \begin{pmatrix} 1.0 & 0.392 \\ 0.392 & 1.0 \end{pmatrix} \tag{2.162}$$

The core contributions $H_{\mu\nu}^{\text{core}}$ can be calculated as the sum of three 2×2 matrices comprising the kinetic energy (**T**) and nuclear attraction terms for the two nuclei A and B ($\mathbf{V}_A$ and $\mathbf{V}_B$). The elements of these three matrices are obtained by evaluating the following integrals:

$$\begin{aligned} \mathbf{T}_{\mu\nu} &= \int \mathrm{d}v_1 \phi_\mu(1)\left(-\tfrac{1}{2}\nabla^2\right)\phi_\nu(1) \\ \mathbf{V}_{A,\mu\nu} &= \int \mathrm{d}v_1 \phi_\mu(1)\left(-\frac{Z_A}{r_{1A}}\right)\phi_\nu(1) \\ \mathbf{V}_{B,\mu\nu} &= \int \mathrm{d}v_1 \phi_\mu(1)\left(-\frac{Z_B}{r_{1B}}\right)\phi_\nu(1) \end{aligned} \tag{2.163}$$

The matrices are:

$$\mathbf{T} = \begin{pmatrix} 1.412 & 0.081 \\ 0.081 & 0.760 \end{pmatrix} \quad \mathbf{V}_{\mathrm{A}} = \begin{pmatrix} -3.344 & -0.758 \\ -0.758 & -1.026 \end{pmatrix}$$

$$\mathbf{V}_{\mathrm{B}} = \begin{pmatrix} -0.525 & -0.308 \\ -0.308 & -1.227 \end{pmatrix} \tag{2.164}$$

$\mathbf{H}^{\mathrm{core}}$ is the sum of these three:

$$\mathbf{H}^{\mathrm{core}} = \begin{pmatrix} -2.457 & -0.985 \\ -0.985 & -1.493 \end{pmatrix} \tag{2.165}$$

As far as the two-electron integrals are concerned, with two basis functions there are a total of 16 possible two-electron integrals. There are however only six unique two-electron integrals as the indices can be permuted as follows:

(i) $(1s_A1s_A \mid 1s_A1s_A) = 1.056$
(ii) $(1s_A1s_A \mid 1s_A1s_B) = (1s_A1s_A \mid 1s_B1s_A) = (1s_A1s_B \mid 1s_A1s_A)$
$= (1s_B1s_A \mid 1s_A1s_A) = 0.303$
(iii) $(1s_A1s_B \mid 1s_A1s_B) = (1s_A1s_B \mid 1s_B1s_A) = (1s_B1s_A \mid 1s_A1s_B)$
$= (1s_B1s_A \mid 1s_B1s_A) = 0.112$
(iv) $(1s_A1s_A \mid 1s_B1s_B) = (1s_B1s_B \mid 1s_A1s_A) = 0.496$
(v) $(1s_A1s_B \mid 1s_B1s_B) = (1s_B1s_A \mid 1s_B1s_B) = (1s_B1s_B \mid 1s_A1s_B)$
$= (1s_B1s_B \mid 1s_B1s_A) = 0.244$
(vi) $(1s_B1s_B \mid 1s_B1s_B) = 0.775$

To reiterate, these integrals are calculated as follows:

$$(\mu\nu|\lambda\sigma) = \iint \mathrm{d}v_1 \mathrm{d}v_2 \phi_\mu(1)\phi_\nu(1)\frac{1}{r_{12}}\phi_\lambda(2)\phi_\sigma(2) \tag{2.166}$$

Having calculated the integrals, we are now ready to start the SCF calculation. To formulate the Fock matrix it is necessary to have an initial guess of the density matrix, **P**. The simplest approach is to use the null matrix in which all elements are zero. In this initial step the Fock matrix **F** is therefore equal to $\mathbf{H}^{\mathrm{core}}$.

The Fock matrix must next be transformed to $\mathbf{F}'$ by pre- and post-multiplying by $\mathbf{S}^{-1/2}$:

$$\mathbf{S}^{-1/2} = \begin{pmatrix} -1.065 & -0.217 \\ -0.217 & 1.065 \end{pmatrix} \tag{2.167}$$

$\mathbf{F}'$ for this first iteration is thus:

$$\mathbf{F}' = \begin{pmatrix} -2.401 & -0.249 \\ -0.249 & -1.353 \end{pmatrix} \tag{2.168}$$

Diagonalisation of $\mathbf{F}'$ gives its eigenvalues and eigenvectors which are:

$$\mathbf{E} = \begin{pmatrix} -2.458 & 0.0 \\ 0.0 & -1.292 \end{pmatrix} \quad \mathbf{C}' = \begin{pmatrix} 0.975 & -0.220 \\ 0.220 & 0.975 \end{pmatrix} \tag{2.169}$$

The coefficients **C** are obtained from $\mathbf{C} = \mathbf{S}^{-1/2}\mathbf{C}'$ and are thus:

$$\mathbf{C} = \begin{pmatrix} 0.991 & -0.446 \\ 0.022 & 1.087 \end{pmatrix} \tag{2.170}$$

To formulate the density matrix **P** we need to identify the occupied orbital(s). With a two-electron system both electrons occupy the orbital with the lowest energy (i.e. the orbital with the lowest eigenvalue). At this stage the lowest energy orbital is:

$$\psi = 0.991\ 1s_A + 0.022\ 1s_B \tag{2.171}$$

The orbital is composed largely of the s orbital on the helium nucleus; in the absence of any electron–electron repulsion the electrons tend to congregate near the nucleus with the larger charge. The density matrix corresponding to this initial wavefunction is:

$$\mathbf{P} = \begin{pmatrix} 1.964 & 0.044 \\ 0.044 & 0.001 \end{pmatrix} \tag{2.172}$$

The new Fock matrix is formed using **P** and the two-electron integrals together with $\mathbf{H}^{\text{core}}$. For example, the element F_{11} is given by:

$$\begin{aligned} F_{11} = H_{11}^{\text{core}} &+ P_{11}\left[(1s_A 1s_A|1s_A 1s_A) - \tfrac{1}{2}(1s_A 1s_A|1s_A 1s_A)\right] \\ &+ P_{12}\left[(1s_A 1s_A|1s_A 1s_B) - \tfrac{1}{2}(1s_A 1s_A|1s_A 1s_B)\right] \\ &+ P_{21}\left[(1s_A 1s_A|1s_B 1s_B) - \tfrac{1}{2}(1s_A 1s_B|1s_A 1s_B)\right] \\ &+ P_{12}\left[(1s_A 1s_A|1s_B 1s_B) - \tfrac{1}{2}(1s_A 1s_B|1s_A 1s_B)\right] \end{aligned} \tag{2.173}$$

The complete Fock matrix is:

$$\mathbf{F} = \begin{pmatrix} -1.406 & -0.690 \\ -0.690 & -0.618 \end{pmatrix} \tag{2.174}$$

The energy that corresponds to this Fock matrix (calculated using equation (2.148)) is −3.870 Hartrees. In the next iteration, the various matrices are as follows:

$$\begin{aligned} &\mathbf{F}' = \begin{pmatrix} -1.305 & -0.347 \\ -0.347 & -0.448 \end{pmatrix} \quad \mathbf{E} = \begin{pmatrix} -1.427 & 0.0 \\ 0.0 & -0.325 \end{pmatrix} \\ &\mathbf{C}' = \begin{pmatrix} 0.943 & -0.334 \\ 0.334 & 0.943 \end{pmatrix} \quad \mathbf{C} = \begin{pmatrix} 0.931 & -0.560 \\ 0.150 & 1.076 \end{pmatrix} \\ &\mathbf{P} = \begin{pmatrix} 1.735 & 0.280 \\ 0.280 & 0.045 \end{pmatrix} \quad \mathbf{F} = \begin{pmatrix} -1.436 & -0.738 \\ -0.738 & -0.644 \end{pmatrix} \\ &\text{Energy} = -3.909 \text{ Hartrees} \end{aligned} \tag{2.175}$$

The calculation proceeds as illustrated in Table 2.2, which shows the variation in the coefficients of the atomic orbitals in the lowest energy wavefunction and the energy for the first four SCF iterations. The energy is converged to six decimal places after six iterations and the charge density matrix after nine iterations.

Table 2.2 Variation in basis set coefficients and electronic energy for the HeH^+ molecule.

Iteration	$c(1s_A)$	$c(1s_B)$	Energy
1	0.991	0.022	−3.870
2	0.931	0.150	−3.909
3	0.915	0.181	−3.911
4	0.912	0.187	−3.911

The final wavefunction still contains a large proportion of the 1s orbital on the helium atom, but less than was obtained without the two-electron integrals.

2.5.6 Application of the Hartree–Fock equations to molecular systems

We are now in a position to consider how the Hartree–Fock theory we have developed can be used to perform practical quantum mechanical calculations on molecular systems. This is an appropriate place in our discussion to distinguish the two major categories of quantum mechanical molecular orbital calculations: the *ab initio* and the semi-empirical methods. *Ab initio* strictly means 'from the beginning', or 'from first principles', which would imply that a calculation using such an approach would require as input only physical constants such as the speed of light, Planck's constant, the masses of elementary particles and so on. *Ab initio* in fact usually refers to a calculation which uses the full Hartree–Fock/ Roothaan–Hall equations, without ignoring or approximating any of the integrals or any of the terms in the Hamiltonian. The *ab initio* methods do rely upon calibration calculations and this has led some quantum chemists, notably Dewar (who has played a large part in the development of semi-empirical methods), to claim that any real difference between the *ab initio* and the semi-empirical methods is entirely pedagogical. By contrast, semi-empirical methods simplify the calculations, using parameters for some of the integrals and/or ignoring some of the terms in the Hamiltonian. First we shall consider *ab initio* methods.

2.6 Basis sets

The basis sets most commonly used in quantum mechanical calculations are composed of atomic functions. An obvious choice would be the Slater type orbitals for many electron atoms. Slater functions are unfortunately not particularly amenable to implementation in molecular orbital calculations. This is because some of the integrals are difficult, if not impossible, to evaluate, particularly when the atomic orbitals are centred on different nuclei. It is relatively straightforward to calculate integrals involving one or two centres, such as $(\mu\mu \mid \mu\mu)$, $(\mu\mu \mid \nu\nu)$, $(\mu\nu \mid \nu\nu)$ and $(\mu\nu \mid \mu\nu)$. Three- and four-centre integrals are also feasible with Slater functions if the atomic orbitals are located on the same atom. However, three- and four-centre integrals are very difficult if the atomic orbitals are based on different atoms. It is common in *ab initio* calculations to replace the Slater orbitals by functions based upon Gaussians. A Gaussian function has the form $\exp(-\alpha r^2)$, and *ab initio* calculations use basis functions comprising integral powers of x, y and z multiplied by $\exp(-\alpha r^2)$:

$$x^a y^b z^c \exp(-\alpha r^2). \tag{2.176}$$

α determines the radial extent (or 'spread') of a Gaussian function; a function with a large value of α does not spread very far whereas a small value of α gives a large spread. The *order* of these Gaussian-type functions is determined by the powers of the Cartesian variables; a zeroth-order function has $a+b+c=0$; a first-order function has $a+b+c=1$, and so on. There is thus one zeroth-order function, three first-order functions and six second-order functions. The idea of using Gaussian functions in quantum mechanical calculations is often ascribed to S F Boys [Boys 1950]. A major advantage of Gaussian functions is that the product of two Gaussians can be expressed as a single Gaussian, located along the line joining the centres of the two Gaussians m and n (Figure 2.4):

$$\exp(-\alpha_m r_m^2)\exp(-\alpha_n r_n^2) = \exp\left(-\frac{\alpha_m \alpha_n}{\alpha_m + \alpha_n} r_{mn}^2\right)\exp(-\alpha r_c^2) \tag{2.177}$$

r_{mn} is the distance between the centres m and n and the orbital exponent α of the combined function is related to the exponents α_m and α_n by:

$$\alpha = \alpha_m + \alpha_n \tag{2.178}$$

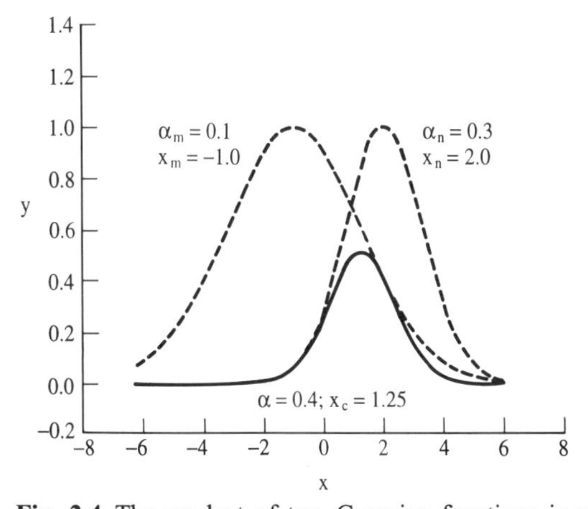

Fig. 2.4 The product of two Gaussian functions is another Gaussian centred along the line joining their centres. In this case the equations of the two functions are $y = \exp[-0.1(x+1.0)^2]$ and $y = \exp[-0.3(x-2.0)^2]$ and the equation of the product is $y = \exp(-27/40)[-0.4(x-1.25)^3]$ (equation (2.177)).

r_C is the distance from the point C which has coordinates:

$$x_c = \frac{\alpha_m x_m + \alpha_n x_n}{\alpha_m + \alpha_n}; \; y_c = \frac{\alpha_m y_m + \alpha_n y_n}{\alpha_m + \alpha_n}; \; z_c = \frac{\alpha_m z_m + \alpha_n z_n}{\alpha_m + \alpha_n} \tag{2.179}$$

x_m, y_m, z_m and x_n, y_n, z_n are the centres of the two original Gaussians m and n respectively.

Thus, in a two-electron integral of the form $(\mu\nu \mid \lambda\sigma)$, the product $\phi_\mu(1)\phi_\nu(1)$ (where ϕ_μ and ϕ_ν may be on different centres) can be replaced by a single Gaussian function that is centred at the appropriate point C. For Cartesian Gaussian functions the calculation is more complicated than for the example we have stated above due to the presence of the Cartesian functions, but even so, efficient methods for performing the integrals have been devised.

The zeroth-order Gaussian function g_s has the same angular symmetry as s atomic orbitals and the first-order Gaussian functions g_x, g_y and g_z have the same symmetries as the $2p_x$, $2p_y$ and $2p_z$ atomic orbitals. In normalised form these are:

$$g_s(\alpha, r) = \left(\frac{2\alpha}{\pi}\right)^{3/4} e^{-\alpha r^2} \tag{2.180}$$

$$g_x(\alpha, r) = \left(\frac{128\alpha^5}{\pi^3}\right)^{1/4} x e^{-\alpha r^2} \tag{2.181}$$

$$g_y(\alpha, r) = \left(\frac{128\alpha^5}{\pi^3}\right)^{1/4} y e^{-\alpha r^2} \tag{2.182}$$

$$g_z(\alpha, r) = \left(\frac{128\alpha^5}{\pi^3}\right)^{1/4} z e^{-\alpha r^2} \tag{2.183}$$

The six second-order functions have the following form, exemplified by two of the functions:

$$g_{xx}(\alpha, r) = \left(\frac{2048\alpha^7}{9\pi^3}\right)^{1/4} x^2 e^{-\alpha r^2} \tag{2.184}$$

$$g_{xy}(\alpha, r) = \left(\frac{2048\alpha^7}{9\pi^3}\right)^{1/4} xy e^{-\alpha r^2} \tag{2.185}$$

These second-order functions do not all have the same angular symmetry as the 3d atomic orbitals, but a set comprising g_{xy}, g_{xz} and g_{yz} together with two linear combinations of the g_{xx}, g_{yy} and g_{zz} do give the desired result:

$$g_{3zz-rr} = \tfrac{1}{2}(2g_{zz} - g_{xx} - g_{yy}) \tag{2.186}$$

$$g_{xx-yy} = \sqrt{\left(\frac{3}{4}\right)}(g_{xx} - g_{yy}) \tag{2.187}$$

The remaining sixth linear combination has the symmetry properties of an s function:

$$g_{rr} = \sqrt{5}(g_{xx} + g_{yy} + g_{zz}) \tag{2.188}$$

The advantages of Gaussian functions are countered by some serious shortcomings. This can be readily seen from a graphical comparison of the 1s Slater function and its 'best' Gaussian approximation, Figure 2.5. Unlike the Slater functions the Gaussian functions do not have a cusp at the origin and they also decay towards zero more quickly. It is found that replacing a Slater type orbital by a single Gaussian function leads to unacceptable errors. However, this problem can be overcome if each atomic orbital is represented as a linear combination of Gaussian functions. Each linear combination has the following form:

$$\phi_\mu = \sum_{\mathrm{i}=1}^{L} d_{\mathrm{i}\mu}\phi_\mathrm{i}(\alpha_{\mathrm{i}\mu}) \tag{2.189}$$

$d_{\mathrm{i}\mu}$ is the coefficient of the primitive Gaussian function ϕ_i, which has exponent $\alpha_{\mathrm{i}\mu}$. L is the number of functions in the expansion. For example, the linear combinations of Gaussian 1s functions that can be used to represent a 1s Slater type orbital with exponent $\xi = 1$ are given in Table 2.3.

The coefficients and the exponents are found by least-squares fitting, in which the overlap between the Slater type function and the Gaussian expansion is maximised. Thus, for the 1s Slater type orbital we seek to maximise the following integral:

$$S = \frac{1}{\sqrt{\pi}}\left(\frac{2\alpha}{\pi}\right)^{3/4}\int \mathrm{d}r\mathrm{e}^{-r}\mathrm{e}^{-\alpha r^2} \tag{2.190}$$

A graphical comparison of the 1s Slater type orbital and the four Gaussian expansions in Table 2.3 is shown in Figure 2.6. It is clear that the fit improves as the number of Gaussian functions increases, but even so, the addition of many more Gaussian functions cannot properly describe the exponential tail in the 'true' function and the cusp at the nucleus. This means that Gaussian functions underestimate the long-range overlap between atoms and the charge and spin density at the nucleus.

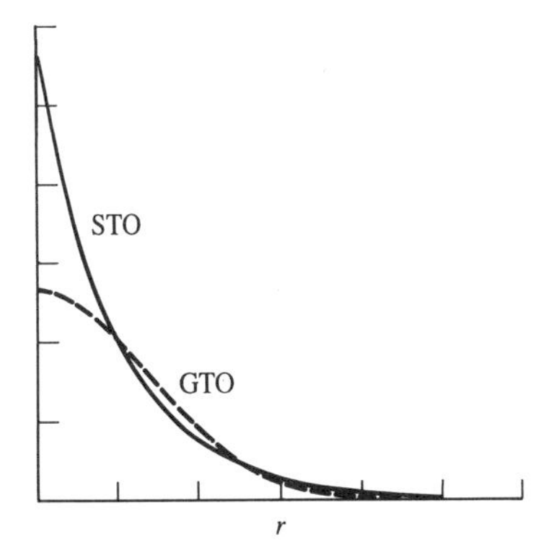

Fig. 2.5 The 1s Slater-type orbital and the best Gaussian equivalent.

Table 2.3 Coefficients and exponents for best-fit Gaussian expansions for the 1s Slater-type orbital [Hehre *et al.* 1969].

Number of Gaussians	Exponent, α	Expansion coefficient, d
1	0.270 950	1.00
2	0.151 623	0.678 914
	0.851 819	0.430 129
3	0.109 818	0.444 635
	0.405 771	0.535 28
	2.227 66	0.154 329
4	0.088 0187	0.291 626
	0.265 204	0.532 846
	0.954 620	0.260 141
	5.216 86	0.056 7523

A Gaussian expansion contains two parameters, the coefficient and the exponent. The most flexible way to use Gaussian functions in *ab initio* molecular orbital calculations permits both of these parameters to vary during the calculation. Such a calculation is said to use *uncontracted* or *primitive* Gaussians. However, calculations with primitive Gaussians require a significant computational effort and so basis sets that consist of *contracted* Gaussian functions are most commonly employed. In a contracted function the contraction coefficients and exponents are pre-determined and remain constant during the calculation. The series of Gaussian functions in such cases is commonly referred to as a *contraction* with the *contraction length* being the number of terms in the expansion. A further approximation that is often employed for the sake of computational efficiency is to use the same Gaussian exponents for the s and p

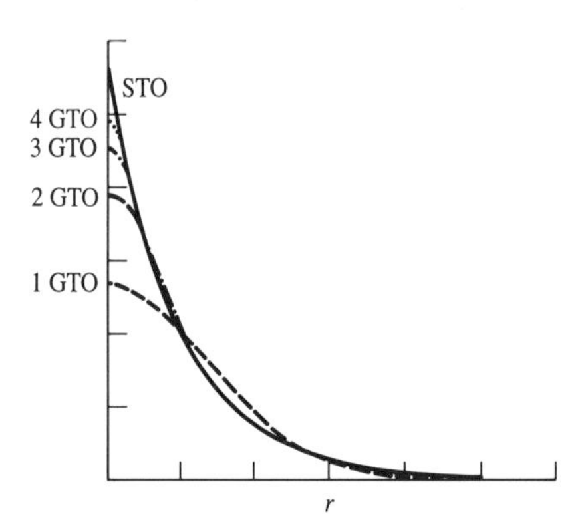

Fig. 2.6 Comparison of 1s Slater-type orbital and Gaussian expansions with up to four terms.

orbitals in a given shell. This clearly restricts the flexibility of the basis set, but it does have the advantage of significantly reducing the number of numerically different integrals that need to be calculated.

Quantum chemists have devised efficient short-hand notation schemes to denote the basis set used in an *ab initio* calculation, although this does mean that a proliferation of abbreviations and acronyms are introduced. However, the codes are usually quite simple to understand. We shall concentrate on the notation used by Pople and co-workers in their Gaussian series of programs.

A *minimal basis set* is a representation that, strictly speaking, contains just the number of functions that are required to accommodate all the filled orbitals in each atom. In practice, a minimal basis set normally includes all of the atomic orbitals in the shell. Thus, for hydrogen and helium a single s-type function would be required; for the elements from lithium to neon the 1s, 2s and 2p functions are used, and so on. The basis sets STO-3G, STO-4G, etc. (in general, STO-*n*G) are all minimal basis sets in which *n* Gaussian functions are used to represent each orbital. It is found that at least three Gaussian functions are required to properly represent each Slater type orbital and so the STO-3G basis set is the 'absolute minimum' that should be used in an *ab initio* molecular orbital calculation. In fact, it is found that there is often little difference between the results obtained with the STO-3G basis set and the larger minimal basis sets with more Gaussian functions, although for hydrogen-bonded complexes STO-4G can perform significantly better. The STO-3G basis set does perform remarkably well in predicting molecular geometries, though this is due in part to a fortuitous cancellation of errors. Of course, the computational effort increases with the number of functions in the Gaussian expansion.

The minimal basis sets are well-known to have several deficiencies. There are particular problems with compounds containing atoms at the end of a period, such as oxygen or fluorine. Such atoms are described using the same number of basis functions as the atoms at the beginning of the period despite the fact that they have more electrons. A minimal basis set only contains one contraction per atomic orbital and as the radial exponents are not allowed to vary during the calculation the functions cannot expand or contract in size in accordance with the molecular environment. The third drawback is that a minimal basis set cannot describe non-spherical aspects of the electronic distribution. For example, for a second row element such as carbon the only functions that incorporate any anisotropy are the $2p_x$, $2p_y$ and $2p_z$ functions. As the radial components of these functions are required to be the same, no one component (x, y or z) can differ from another.

These problems with minimal basis sets can be addressed if more than one function is used for each orbital. A basis set which doubles the number of functions in the minimal basis set is described as a *double zeta* basis. Thus, a linear combination of a 'contracted' function and a 'diffuse' function gives an overall result that is intermediate between the two. The basis set coefficients of the contracted and the diffuse functions are automatically calculated by the SCF procedure which thus automatically determines whether a more contracted or a more diffuse representation of that particular orbital is required. Such an approach can provide a solution to the anisotropy problem because it is then possible to have different linear combinations for the p_x, p_y and p_z orbitals.

An alternative to the double zeta basis approach is to double the number of functions used to describe the valence electrons but to keep a single function for the inner shells. The rationale for this approach is that the core orbitals, unlike the valence orbitals, do not affect chemical properties very much and vary only slightly from one molecule to another. The notation used for such *split valence* double zeta basis sets is exemplified by 3-21G. In this basis set three Gaussian functions are used to describe the core orbitals. The valence electrons are also represented by three Gaussians: the contracted part by two Gaussians and the diffuse part by one Gaussian. The most commonly used split valence basis sets are 3-21G, 4-31G and 6-31G.

Simply increasing the number of basis functions (triple zeta, quadruple zeta, etc.) does not necessarily improve the model. In fact, it can give rise to wholly erroneous results, particularly for molecules with a strongly anisotropic charge distribution. All of the basis sets we have encountered so far use functions that are centred on atomic nuclei. The use of split valence basis sets can help to surmount the problems with non-isotropic charge distribution but not completely. The charge distribution about an atom in a molecule is usually perturbed in comparison with the isolated atom. For example, the electron cloud in an isolated hydrogen atom is symmetrical, but when the hydrogen atom is present in a molecule the electrons are attracted towards the other nuclei. The distortion can be considered to correspond to mixing p-type character into the 1s orbital of the isolated atom, to give a form of sp hybrid. In a similar manner, the unoccupied d orbitals introduce asymmetry into p orbitals (Figure 2.7). The most common solution to this problem is to introduce *polarisation functions* into the basis set. The polarisation functions have a higher angular quantum number and so correspond to p orbitals for hydrogen and d orbitals for the first and second row elements.

The use of polarisation basis functions is indicated by an asterisk (*). Thus, 6-31G* refers to a 6-31G basis set with polarisation functions on the heavy (i.e. non-hydrogen) atoms. Two asterisks (e.g. 6-31G**) indicate the use of polarisation (i.e. p) functions on hydrogen and helium. The 6-31G** basis set is particularly useful where hydrogen acts as a bridging atom. Partial polarisation

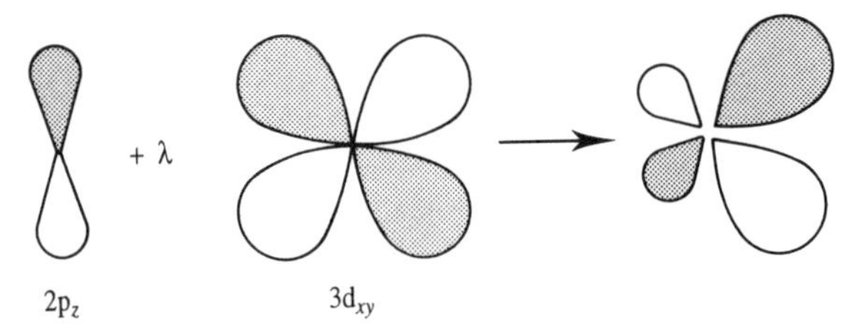

Fig. 2.7 The addition of a $3d_{xy}$ orbital to $2p_z$ gives a distorted orbital. Figure adapted from Hehre W J, L Radom, P v R Schleyer, J A Hehre 1986. *Ab Initio Molecular Orbital Theory.* New York, Wiley.

basis sets have also been developed. For example, the 3-21G$^{(*)}$ basis set has the same set of Gaussians as the 3-21G basis set (i.e. three functions for the inner shell, two contracted functions and one diffuse function for the valence shell) supplemented by six d-type Gaussians for the second-row elements. This basis set therefore attempts to account for d-orbital effects in molecules containing second-row elements. There are no special polarisation functions on first-row elements, which are described by the 3-21G basis set.

A deficiency of the basis sets described so far is their inability to deal with species which have a significant amount of electron density away from the nuclear centres such as anions and molecules containing lone pairs. This failure arises because the amplitudes of the Gaussian basis functions are rather low far from the nuclei. To remedy this deficiency highly diffuse functions can be added to the basis set. These basis sets are denoted using a '+'; thus the 3-21+G basis set contains an additional single set of diffuse s- and p-type Gaussian functions. '++' indicates that the diffuse functions are included for hydrogen as well as for heavy atoms. At these levels the terminology starts to become a little unwieldy. For example, the 6-311++G(3df, 3pd) basis set uses a single zeta core and triple zeta valence representation with additional diffuse functions on all atoms. The '(3df, 3pd)' indicates three sets of d functions and one set of f functions for first-row atoms and three sets of p functions and one set of d functions for hydrogen.

The basis sets that we have considered thus far are sufficient for most calculations. However, for some high-level calculations a basis set that effectively enables the basis set limit to be achieved is required. The *even-tempered* basis set is designed to achieve this; each function in this basis set is the product of a spherical harmonic and a Gaussian function multiplied by a power of the distance from the origin:

$$\chi_{klm}(\rho, \theta, \phi) = \exp(-\zeta_k^2)r^l Y_{lm}(\theta, \phi) \tag{2.191}$$

The orbital exponent ζ_k, is expressed as a function of two parameters α and β as follows:

$$\zeta_k = \alpha\beta^k \qquad k = 1, 2, 3, \ldots N \tag{2.192}$$

The even-tempered basis set consists of the following sequence of functions; 1s, 2p, 3d, 4f, ... that correspond to increasing values of k. The advantage of this basis set is that it is relatively easy to optimise the exponents for a large sequence of basis functions.

2.6.1 *Creating a basis set*

There is no definitive method for generating basis sets, and the construction of a new basis set is very much an art. Nevertheless, there are a number of well-established approaches that have resulted in widely used basis sets. We have already seen how linear combinations of Gaussian functions can be fitted to Slater-type orbitals by minimising the overlap (see Figure 2.6 and Table 2.3). The Gaussian exponents and coefficients are derived by least-squares fitting to the desired functions, such as Slater-type orbitals. When using basis sets that have been fitted to Slater orbitals it is often advantageous to use Slater exponents that are different to those obtained from Slater's rules. In general, better results for molecular calculations are obtained if larger Slater exponents are used for the valence electrons; this has the effect of giving a 'smaller', less diffuse orbital. For example, a value of 1.24 is widely used for the Slater exponent of hydrogen rather than the 1.0 that would be suggested by Slater's rules. It is straightforward to derive a basis set for a different Slater exponent if the Gaussian expansion has been fitted to a Slater-type orbital with $\zeta = 1.0$. If the Slater exponent ζ is replaced by a new value, ζ', then the respective Gaussian exponents α and α' are related by:

$$\frac{\alpha'}{\alpha} = \frac{\zeta'^2}{\zeta^2} \tag{2.193}$$

A doubling of the Slater exponent thus corresponds to a quadrupling of the Gaussian exponent. The expansion coefficients remain the same. For example, to obtain the exponents of the Gaussian functions for hydrogen in the STO-3G basis set we need to multiply the appropriate values in Table 2.3 by 1.24^2, giving exponents of 0.168 856, 0.623 913 and 3.425 25. This strategy can be quite powerful; the STO-*n*G basis sets were originally defined with exponents that reproduce 'best atom' values for the core orbitals but the exponents for the valence electrons were values that give optimal performance for a selected set of small molecules. For example, the suggested exponent for the valence orbitals in carbon was 1.72 rather than the 1.625 predicted by Slater's rules. The core orbitals have a Slater exponent of 5.67.

Basis sets can be constructed using an optimisation procedure in which the coefficients and the exponents are varied to give the lowest atomic energies. Some complications can arise when this approach is applied to larger basis sets. For example, in an atomic calculation the diffuse functions can move towards the nucleus, especially if the core region is described by only a few basis functions. This is contrary to the role of diffuse functions, which is to enhance the description in the internuclear region. It may therefore be necessary to construct the basis set in stages, first determining the diffuse functions, using many basis functions for the core, and then optimising the basis functions for the core region keeping the diffuse functions fixed. In many of the popular Gaussian basis sets the coefficients and exponents of the core orbitals are designed to reproduce calculations on atoms whereas the valence basis functions are parametrised to reproduce the properties of a carefully selected set of molecular data.

The basis sets of Dunning [Dunning 1970] are obtained in a rather different way to those of Pople and co-workers. The first step is to perform an atomic SCF calculation using a set of primitive Gaussian functions, in which the exponents are optimised to give the lowest energy for the atom. This set of primitive Gaussian functions (usually far too many for general use in molecular calculations) is then contracted to a smaller number of Gaussian functions, so drastically reducing the number of integrals that need be calculated. For example, Huzinga optimised the exponents of an uncontracted basis set that contained nine functions of s symmetry and five functions of p symmetry for the first-row elements [Huzinga 1965]. This (9s5p) basis set represents the 1s, 2s and three 2p orbitals and in fact corresponds to 24 basis functions per atom $(9+3 \times 5)$. The primitive Gaussians in this uncontracted basis set are then apportioned to the basis functions in the new, contracted basis set, which contains three s functions and two p functions and is written [3s2p]. No primitive is assigned to more than one of the contracted basis functions. The 1s orbital is constructed from six primitives, the 2s orbital from one set of two primitives and one set containing just one primitive, and the 2p orbitals are represented by one contracted function containing five primitives and one contracted function that contains the remaining primitive. The final basis set, which is illustrated in Table 2.4 for nitrogen, contains a total of nine basis functions rather than the original 24. Each of the primitive functions appears in just one basis function with its original exponent. The ratios of the coefficients of the primitives in the contracted basis set are equal to the ratios of the coefficients determined in the atomic SCF calculation. The major advantage of this approach is that calculations with the smaller basis set give results that are almost as good as calculations using the full basis set but with much less computational effort.

2.7 Open-shell systems

The Roothaan–Hall equations are not applicable to open-shell systems, which contain one or more unpaired electrons. Radicals are by definition open-shell systems as are some ground-state molecules such as NO and O_2. Two approaches have been devised to treat open-shell systems. The first of these is *spin-restricted*

Table 2.4 Exponents and contraction coefficients for the three s-type and the two p-type Gaussian functions in the basis set of Dunning for nitrogen [Dunning 1970].

Exponent	Coefficient	Exponent	Coefficient	Exponent	Coefficient
1s		2s		2s	
5900	0.001 190	7.193	−0.160 405	0.2133	1.000 000
887.5	0.009 099	1.707	1.058 215		
204.7	0.044 145				
59.84	0.150 464				
20.00	0.356 741				
7.193	0.446 533				
2.686	0.145 603				
2p		2p			
26.79	0.018 254	0.1654	1.000 000		
5.956	0.116 461				
1.707	0.390 178				
0.5314	0.637 102				

Hartree–Fock (RHF) theory, which uses combinations of singly and doubly occupied molecular orbitals. The closed-shell approach that we have developed thus far is a special case of RHF theory. The doubly occupied orbitals use the same spatial functions for electrons of both α and β spin. The orbital expansion equation (2.135) is employed together with the variational method to derive the optimal values of the coefficients. The alternative approach is the *spin-unrestricted Hartree–Fock* (UHF) theory of Pople and Nesbet [Pople and Nesbet 1954], which uses two distinct sets of molecular orbitals: one for electrons of α spin and the other for electrons of β spin. Two Fock matrices are involved, one for each type of spin, with elements as follows:

$$F^{\alpha}_{\mu\nu} = H^{\text{core}}_{\mu\nu} + \sum_{\lambda=1}^{K}\sum_{\sigma=1}^{K} [[P^{\alpha}_{\lambda\sigma} + P^{\beta}_{\lambda\sigma}](\mu\nu|\lambda\sigma) - P^{\alpha}_{\lambda\sigma}(\mu\lambda|\nu\sigma)] \tag{2.194}$$

$$F^{\beta}_{\mu\nu} = H^{\text{core}}_{\mu\nu} + \sum_{\lambda=1}^{K}\sum_{\sigma=1}^{K} [[P^{\alpha}_{\lambda\sigma} + P^{\beta}_{\lambda\sigma}](\mu\nu|\lambda\sigma) - P^{\beta}_{\lambda\sigma}(\mu\lambda|\nu\sigma)] \tag{2.195}$$

UHF theory also uses two density matrices, the full density matrix being the sum of these two:

$$P^{\alpha}_{\mu\nu} = \sum_{\text{i}=1}^{\alpha_{\text{occ}}} c^{\alpha}_{\mu\text{i}} c^{\alpha}_{\nu\text{i}} P^{\beta}_{\mu\nu} = \sum_{\text{i}=1}^{\beta_{\text{occ}}} c^{\beta}_{\mu\text{i}} c^{\beta}_{\nu\text{i}} \tag{2.196}$$

$$P_{\mu\nu} = P^{\alpha}_{\mu\nu} + P^{\beta}_{\mu\nu} \tag{2.197}$$

The summations in equations (2.196) and (2.197) are over the occupied orbitals with α and β spin as appropriate. Thus $\alpha_{\text{occ}} + \beta_{\text{occ}}$ equals the total number of electrons in the system. In a closed-shell Hartree–Fock wavefunction the distribution of electron spin is zero everywhere because the electrons are paired. In an open-shell system, however, there is an excess of electron spin that can be expressed as the spin density, analogous to the electron density. The spin density $\rho^{\text{spin}}(\mathbf{r})$ at a point $\mathbf{r}$ is given by:

$$\rho^{\text{spin}}(\mathbf{r}) = \rho^{\alpha}(\mathbf{r}) - \rho^{\beta}(\mathbf{r}) = \sum_{\mu=1}^{K}\sum_{\nu=1}^{K} [P^{\alpha}_{\mu\nu} - P^{\beta}_{\mu\nu}]\phi_{\mu}(\mathbf{r})\phi_{\nu}(\mathbf{r}) \tag{2.198}$$

Clearly, the UHF approach is more general and indeed the restricted Hartree–Fock approach is a special case of unrestricted Hartree–Fock. Figure 2.8 illustrates the conceptual difference between the RHF and the UHF models. Unrestricted wavefunctions are also the most appropriate way to deal with other problems such as molecules near the dissociation limit. The simplest example of this type of behaviour is the H_2 molecule, the ground state of which is a singlet with a bond length of approximately 0.75 Å. The restricted wavefunction is the appropriate Hartree–Fock wavefunction, with two paired electrons in a single spatial orbital. As the bond length increases towards the dissociation limit, this description is clearly inappropriate, for hydrogen is experimentally observed to dissociate to two hydrogen atoms. This behaviour cannot be achieved using a restricted Hartree–Fock wavefunction, which requires the two electrons to occupy

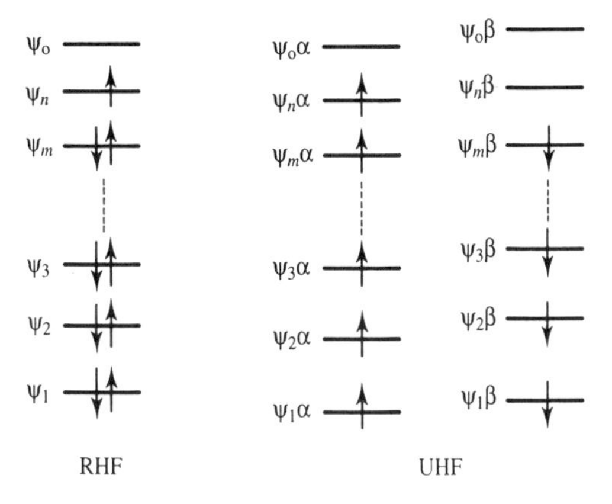

Fig. 2.8 The conceptual difference between the RHF and UHF models.

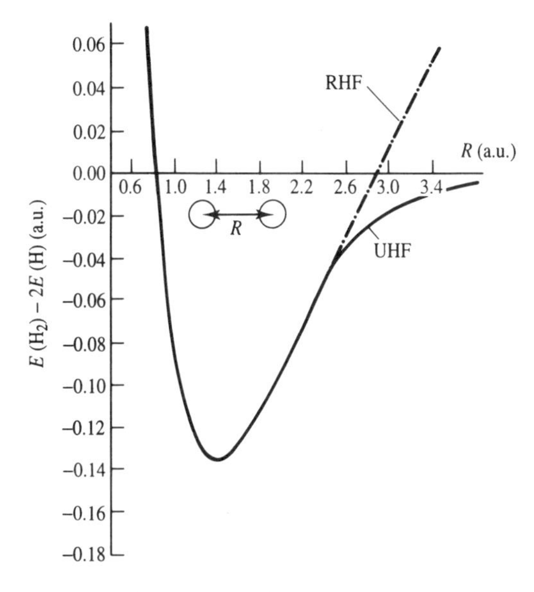

Fig. 2.9 UHF and RHF dissociation curves for H_2. After Szabo A, N S Ostlund 1982. *Modern Quantum Chemistry. Introduction to Advanced Electronic Structure Theory.* New York, McGraw-Hill.

the same spatial orbital, but it is appropriately described by a UHF wavefunction. Beyond about 1.2 Å the 'correct' wavefunction for hydrogen must thus be obtained using UHF theory. The results obtained by calculating the potential energy curves of the hydrogen molecule using the RHF and UHF theories are shown in Figure 2.9. As can be seen, RHF theory gives a dissociation energy that is much too large, whereas the UHF theory shows the correct dissociation behaviour.

2.8 Electron correlation

The most significant drawback of Hartree–Fock theory is that it fails to adequately represent electron correlation. In the self-consistent field method the electrons are assumed to be moving in an average potential of the other electrons, and so the instantaneous position of an electron is not influenced by the presence of a neighbouring electron. In fact, the motions of electrons are correlated and they tend to 'avoid' each other more than Hartree–Fock theory would suggest, giving rise to a lower energy. The correlation energy is defined as the difference between the Hartree–Fock energy and the exact energy. Neglecting electron correlation can lead to some clearly anomalous results, especially as the dissociation limit is approached. For example, an uncorrelated calculation would predict that the electrons in H_2 spend equal time on both nuclei, even when they are infinitely separated. Hartree–Fock geometries and relative energies for equilibrium structures are often in good agreement with experiment and as many molecular modelling applications are concerned with species at equilibrium it might be considered that correlation effects are not so important. Nevertheless, there is increasing evidence that the inclusion of correlation effects is warranted, especially when quantitative information is required. Moreover, electron correlation is crucial in the study of dispersive effects (which we shall consider in section 3.9.1) which play a major role in intermolecular interactions.

2.8.1 Configuration interaction

There are a number of ways in which correlation effects can be incorporated into an *ab initio* molecular orbital calculation. A popular approach is configuration interaction (CI), in which excited states are included in the description of an electronic state. To illustrate the principle, let us consider a lithium atom. The ground state of lithium can be written $1s^2 2s^1$ (although we have used the conventional nomenclature here, we should remember that the wavefunction is really a Slater determinant). Excitation of the outer valence electron gives states such as $1s^2 3s^1$. A better description of the overall wavefunction is a linear combination of the ground and excited state wavefunctions. If a Hartree–Fock calculation is performed with K basis functions then $2K$ spin orbitals are obtained. If these $2K$ spin orbitals are filled with N electrons ($N < 2K$) there will be $2K - N$ unoccupied, virtual orbitals. The wavefunction obtained from the single determinant approach that we have considered thus far is expressed only in terms of the occupied orbitals. For example, a very simple calculation on H_2 using as a basis set just the 1s orbitals on each hydrogen results in two molecular orbitals ($1\sigma_g$ and $1\sigma_u$). In the ground state, the $1\sigma_g$ orbital is filled with two electrons. An excited state can be generated by replacing one or more of the occupied spin orbitals with a virtual spin orbital. Possible excited states for the hydrogen molecule might thus include $1\sigma_g^1\sigma_u^1$ and $1\sigma_u^2$ (in fact, the first of these two configurations cannot be combined with the ground state, as we shall see). In addition to the replacement of single spin orbitals by single virtual orbitals, two spin orbitals can be replaced by two virtual orbitals, three spin orbitals by three virtual orbitals, and so on. In general, the CI wavefunction can be written

$$\Psi = c_0\Psi_0 + c_1\Psi_1 + c_2\Psi_2 + \cdots \tag{2.199}$$

Ψ_0 is the single determinant wavefunction obtained by solving the Hartree–Fock equations. Ψ_1, Ψ_2, etc., are wavefunctions (expressed as determinants) that represent configurations derived by replacing one or more of the occupied spin orbitals by a virtual spin orbital. The energy of the system is then minimised in order to determine the coefficients c_0, c_1, etc., using a linear variational approach, just as for a single-determinant calculation. A CI calculation thus involves an additional level of complexity; each configuration is written in terms of molecular orbitals, which in turn are expressed as a linear combination of basis functions. The number of integrals can become extremely large. The total number of ways to permute N electrons and K orbitals is $(2K!)/[N!(2K-N)!]$. This is a very large number for all except small values of K and N, which explains why it is not usual to consider all possibilities (termed full configuration interaction) except for very small systems. However, full CI is important because it is the most complete treatment possible within the limitations imposed by the basis set. It is common practice to limit the excited states considered. For example, in configuration interaction singles (CIS) only wavefunctions that differ from the Hartree–Fock wavefunction by a single spin orbital are included. Higher levels of theory involve double substitutions (configuration interaction doubles, CID) or both singles and double substitutions (configuration interaction singles and doubles, CISD). Even at the CIS or CID levels, the number of excited states to be included can be very large, and it may be desirable (or necessary) to restrict the spin orbitals that are involved in the substitutions. For example, only excitations involving the highest occupied molecular orbital (HOMO) and the lowest unoccupied molecular orbital (LUMO) may be permitted. Alternatively, the orbitals corresponding to the inner electron core may be neglected (the 'frozen core' approximation). Some of these options are illustrated in Figure 2.10.

Not all excitations necessarily help to lower the energy; some determinants do not mix with the ground state. A consequence of *Brillouin's theorem* is that single excitations do not mix directly with the single determinant, ground-state wavefunction Ψ_0. It would therefore be anticipated that double excitations would be most important and that single excitations would have no effect on the energy of the ground state. However, the single excitations can interact with the double excitations, which in turn interact with Ψ_0, and so single excitations do have a small indirect effect on the energy. The determinants of triple and higher excitations also do not interact directly with Ψ_0 (though they may do indirectly via other levels of excitation). This is because the Hamiltonian contains elements involving at most interactions between pairs of electrons, and so if the Slater determinants differ by more than two electron functions, their integral over all space will be zero.

In a 'traditional' CI calculation the determinants in the expansion, equation (2.199), are those obtained from a Hartree–Fock calculation; only the coefficients c_0, c_1 etc. are permitted to vary. Clearly, a better (i.e. lower-energy) wavefunction

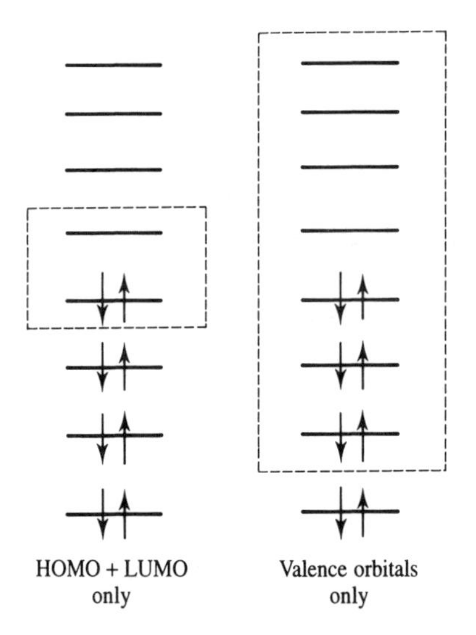

Fig. 2.10 Some of the ways in which excited state wavefunctions can be included in a configuration interaction calculation. Figure adapted from Hehre W J, L Radom, P v R Schleyer, J A Hehre 1986. *Ab Initio Molecular Orbital Theory.* New York, Wiley.

should be obtained if the coefficients of the basis functions themselves can also vary as well as the coefficients of the determinants. This approach is known as the multiconfiguration self-consistent field method (MCSCF). MCSCF theory is considerably more complicated than the Roothaan–Hall equations and well beyond the scope of our discussion. One MCSCF technique that has attracted considerable attention is the complete active-space SCF method (CASSCF) of Roos [Roos *et al.* 1980]. CASSCF enables very large numbers of configurations to be included in the calculation by dividing the molecular orbitals into three sets; those which are doubly occupied in all configurations, those which are unoccupied in all configurations, and then all the remaining 'active' orbitals. The list of configurations is generated by considering all possible arrangements of the active electrons among the active orbitals.

A CI calculation is variational: the energy obtained is guaranteed to be greater than the 'true' energy. A drawback of CI calculations other than those performed at the full CI level is that they are not size consistent. Simply put, this means that the energy of a number N of non-interacting atoms or molecules is not equal to N times the energy of a single atom or molecule. Another consequence of size consistency is that as the bond length in a diatomic molecule increases to infinity so the energy of the system should become equal to the sum of the energies of the respective atoms. To illustrate why this lack of size consistency arises, consider CID calculations on Be_2 and on two beryllium atoms. The electronic configuration of Be is $1s^2 2s^2$ and so if we label the two atoms A and B, then the wavefunction for each of the two separated atoms will include the configuration $1s_A^2\, 2p_A^2\, 1s_B^2\, 2p_B^2$ ($\equiv 1s_A^2\, 1s_B^2\, 2p_A^2\, 2p_B^2$), in which two electrons have been promoted in each beryllium atom from the 2s to the 2p orbitals. This configuration represents a *quadruple* excitation for the beryllium dimer, which has the electronic configuration $1s_A^2\, 1s_B^2\, 2s_A^2\, 2s_B^2$. This quadruply excited configuration is not included in the CID wavefunction for the dimer which is restricted to double excitations. In fact, the energy of a CI calculation including only doubly excited states is expected to scale in proportion to $\sqrt{N}$, where N is the number of non-interacting species present, rather than N.

2.8.2 Many body perturbation theory

Møller and Plesset proposed an alternative way to tackle the problem of electron correlation [Møller and Plesset 1934]. Their method is based upon Rayleigh–Schrödinger perturbation theory, in which the 'true' Hamiltonian operator $\mathscr{H}$ is expressed as the sum of a 'zeroth-order' Hamiltonian $\mathscr{H}_0$ (for which a set of molecular orbitals can be obtained) and a perturbation, $\mathscr{V}$:

$$\mathscr{H} = \mathscr{H}_0 + \mathscr{V} \tag{2.200}$$

The eigenfunctions of the true Hamiltonian operator are Ψ_i with corresponding energies E_i. The eigenfunctions of the zeroth-order Hamiltonian are written $\Psi_i^{(0)}$ with energies $E_i^{(0)}$. The ground state wavefunction is thus $\Psi_0^{(0)}$ with energy $E_0^{(0)}$. To devise a scheme by which it is possible to gradually improve the eigenfunctions and eigenvalues of $\mathscr{H}_0$ we can write the true Hamiltonian as follows:

$$\mathscr{H} = \mathscr{H}_0 + \lambda\mathscr{V} \tag{2.201}$$

λ is a parameter that can vary between 0 and 1; when λ is zero then $\mathscr{H}$ is equal to the zeroth-order Hamiltonian but when λ is one then $\mathscr{H}$ equals its true value. The eigenfunctions Ψ_i and eigenvalues E_i of $\mathscr{H}$ are then expressed in powers of λ:

$$\Psi_i = \Psi_i^{(0)} + \lambda\Psi_i^{(1)} + \lambda^2\Psi_i^{(2)} \cdots = \sum_{n=0} \lambda^n \Psi_i^{(n)} \tag{2.202}$$

$$E_i = E_i^{(0)} + \lambda E_i^{(1)} + \lambda^2 E_i^{(2)} + \cdots = \sum_{n=0} \lambda^n E_i^{(n)} \tag{2.203}$$

$E_i^{(1)}$ is the first-order correction to the energy, $E_i^{(2)}$ is the second-order correction and so on. These energies can be calculated from the eigenfunctions as follows:

$$E_i^{(0)} = \int \Psi_i^{(0)} \mathscr{H}_0 \Psi_i^{(0)} d\tau \tag{2.204}$$

$$E_i^{(1)} = \int \Psi_i^{(0)} \mathscr{V} \Psi_i^{(0)} d\tau \tag{2.205}$$

$$E_i^{(2)} = \int \Psi_i^{(0)} \mathscr{V} \Psi_i^{(1)} d\tau \tag{2.206}$$

$$E_i^{(3)} = \int \Psi_i^{(0)} \mathscr{V} \Psi_i^{(2)} d\tau \tag{2.207}$$

To determine the corrections to the energy it is therefore necessary to determine the wavefunctions to a given order. In Møller–Plesset perturbation theory the unperturbed Hamiltonian $\mathcal{H}_0$ is the sum of the one-electron Fock operators for the N electrons:

$$\mathcal{H}_0 = \sum_{i=1}^{N} f_i = \sum_{i=1}^{N} \left(\mathcal{H}^{\text{core}} + \sum_{j=1}^{N} (\mathcal{J}_i + \mathcal{K}_i) \right) \tag{2.208}$$

The Hartree–Fock wavefunction, $\Psi_0^{(0)}$ is an eigenfunction of $\mathcal{H}_0$ and the corresponding zeroth-order energy $E_0^{(0)}$ is equal to the sum or orbital energies for the occupied molecular orbitals:

$$E_0^{(0)} = \sum_{i=1}^{\text{occupied}} \varepsilon_i \tag{2.209}$$

In order to calculate higher order wavefunctions we need to establish the form of the perturbation, $\mathcal{V}$. This is the difference between the 'real' Hamiltonian $\mathcal{H}$ and the zeroth-order Hamiltonian, $\mathcal{H}_0$. Remember that the Slater determinant description, based on an orbital picture of the molecule, is only an approximation. The true Hamiltonian is equal to the sum of the nuclear attraction terms and electron repulsion terms:

$$\mathcal{H} = \sum_{i=1}^{N} (\mathcal{H}^{\text{core}}) + \sum_{i=1}^{N} \sum_{j=i+1}^{N} \frac{1}{r_{ij}} \tag{2.210}$$

Hence the perturbation $\mathcal{V}$ is given by:

$$\mathcal{V} = \sum_{i=1}^{N} \sum_{j=i+1}^{N} \frac{1}{r_{ij}} - \sum_{j=1}^{N} (\mathcal{J}_j + \mathcal{K}_j) \tag{2.211}$$

The first-order energy $E_0^{(1)}$ is given by:

$$E_0^{(1)} = -\frac{1}{2} \sum_{i=1}^{N} \sum_{j=1}^{N} [(\text{ii}|\text{jj}) - (\text{ij}|\text{ij})] \tag{2.212}$$

The sum of the zeroth-order and first-order energies thus corresponds to the Hartree–Fock energy (compare with equation (2.112) which gives the equivalent result for a closed-shell system):

$$E_0^{(0)} + E_0^{(1)} = \sum_{i=1}^{N} \varepsilon_i - \frac{1}{2} \sum_{i=1}^{N} \sum_{j=1}^{N} [(\text{ii}|\text{jj}) - (\text{ij}|\text{ij})] \tag{2.213}$$

To obtain an improvement on the Hartree–Fock energy it is therefore necessary to use Møller–Plesset perturbation theory to at least second order. This level of theory is referred to as MP2 and involves the integral $\int \Psi_0^{(0)} \mathcal{V} \Psi_0^{(1)} d\tau$. The higher-order wavefunction $\Psi_0^{(1)}$ is expressed as linear combinations of solutions to the zeroth-order Hamiltonian:

$$\Psi_0^{(1)} = \sum_j c_j^{(1)} \Psi_j^{(0)} \tag{2.214}$$

The $\Psi_j^{(0)}$ in equation (2.214) will include single, double, etc., excitations obtained by promoting electrons into the virtual orbitals obtained from a Hartree–Fock calculation. The second-order energy is given by:

$$E_0^{(2)} = \sum_{i}^{\text{occupied}} \sum_{j>i} \sum_{a}^{\text{virtual}} \sum_{b>a} \frac{\iint d\tau_1 d\tau_2 \chi_i(1)\chi_j(2)\left(\frac{1}{r_{12}}\right)[\chi_a(1)\chi_b(2) - \chi_b(1)\chi_a(2)]}{\varepsilon_a + \varepsilon_b - \varepsilon_i - \varepsilon_j} \quad (2.215)$$

These integrals will be non-zero only for double excitations, according to the Brillouin theorem. Third- and fourth-order Møller–Plesset calculations (MP3 and MP4) are also available as standard options in many *ab initio* packages.

The advantage of many-body perturbation theory is that it is size independent, unlike configuration interaction—even when a truncated expansion is used. However, Møller–Plesset perturbation theory is not variational and can sometimes give energies that are lower than the 'true' energy. Møller–Plesset calculations are computationally intensive and so their use is often restricted to 'single-point' calculations at a geometry obtained using a lower level of theory. They are at present the most popular way to incorporate electron correlation in molecular quantum mechanical calculations, especially at the MP2 level. A Møller–Plesset calculation is specified using the level of theory used (e.g. MP2, MP3) together with the basis set. Thus MP2/6-31G* indicates a second-order Møller–Plesset calculation with the 6-31G* basis set.

2.9 Practical considerations when performing *ab initio* calculations

Ab initio calculations can be extremely time-consuming, especially when using the higher levels of theory or when the nuclei are free to move, as in a minimisation calculation (see Chapter 4). Various 'tricks' have been developed which can significantly reduce the computational effort involved. Many of these options are routinely available in the major software packages and are invoked by the specification of simple keywords. One common tactic is to combine different levels of theory for the various stages of a calculation. For example, a lower level of theory can be used to provide the initial guess for the density matrix prior to the first SCF iteration. Lower levels of theory can also be used in other ways. Suppose we wish to determine some of the electronic properties of a molecule in a minimum energy structure. Energy minimisation requires that the nuclei move, and is typically performed in a series of steps, at each of which the energy (and frequently the gradient of the energy) must be calculated. Minimisation is therefore a computationally expensive procedure, particularly when performed at the high level of theory. To reduce this computational burden a lower level of theory can be employed for the geometry optimisation. A 'single point' calculation using a high level of theory is then performed at the geometry so obtained to give a wavefunction from which the properties are determined. The assumption here of course is that the geometry does not change much between the two levels of theory. Such calculations are denoted by slashes (/). For example, a calculation that is described as '6-31G*/STO-3G' indicates that the geometry was determined using the STO-3G basis set and the wavefunction was obtained using the 6-31G* basis set. Two slashes are used when each calculation is itself described using a slash, such as when electron correlation methods are used. For example, 'MP2/6-31G*//HF/6-31G*' indicates a geometry optimisation using a Hartree–Fock calculation with a 6-31G* basis set followed by a single-point calculation using the MP2 method for incorporating electron correlation, again using a 6-31G* basis set.

2.9.1 Convergence of self-consistent field calculations

In an SCF calculation the wavefunction is gradually refined until self-consistency is achieved. For closed-shell ground-state molecules this is usually quite straightforward and the energy converges after a few cycles. However, in some cases convergence is a problem, and the energy may oscillate from one iteration to the next or even diverge rapidly. Various methods have been proposed to deal with such situations. A simple strategy is to use an average set of orbital coefficients rather than the set obtained from the immediately preceding iteration. The coefficients in this average set can be weighted according to the energies of zeach iteration. This tends to weed out those coefficients that give rise to higher energies.

The initial guess of the density matrix may influence the convergence of the SCF calculation; a null matrix is the simplest approach, but better results may be obtained by using a density matrix from a calculation performed at a lower level of theory. For example, the density matrix from a semi-empirical calculation may be used as the starting point for an *ab initio* calculation. Conversely, such an approach may itself lead to problems if there is a significant difference between the density matrices for the lower and the higher levels of theory.

A more sophisticated method that has often been very successful is Pulay's direct inversion of the iterative subspace (DIIS) [Pulay 1980]. Here, the energy is assumed to vary as a quadratic function of the basis set coefficients. In DIIS the coefficients for the next iteration are calculated from their values in the previous steps. In essence, one is predicting where the minimum in the energy will lie from a knowledge of the points that have been visited and by assuming that the energy surface adopts a parabolic shape.

2.9.2 The direct SCF method

An *ab initio* calculation can be logically considered to involve two separate stages. First, the various one- and two-electron integrals are calculated. This is a computationally intensive task and considerable effort has been expended finding ways to make the calculation of the integrals as efficient as possible. In the second stage, the wavefunction is determined using the variation theorem. In a 'traditional' SCF calculation all of the integrals are first calculated and stored on disk, to be retrieved later during the SCF calculation as required. The number of integrals to be stored may run into millions and this inevitably leads to delays in accessing the data, particularly as the retrieval of information from a disk requires physical movement of the read head and so is slow. Modern computers (both workstations and supercomputers) have much faster (and cheaper) processing units, and many of these machines also have a substantial amount of internal memory that can be accessed in a fraction of the time it takes to read data from the disk. In a direct SCF calculation, the integrals are not stored on the disk but are kept in memory or recalculated when required [Almlöf *et al.* 1982].

A much-quoted 'fact' is that *ab initio* calculations scale as the fourth power of the number of basis functions for ground-state, closed-shell systems. This scaling factor arises because each two-electron integral $(\mu\nu \mid \lambda\sigma)$ involves four basis functions, so the number of two-electron integrals would be expected to increase in proportion to the fourth power of the number of basis functions. In fact, the number of such integrals is not exactly equal to the fourth power of the number of basis functions because many of the integrals are related by symmetry. We can calculate exactly the number of two-electron integrals that are required in a Hartree–Fock *ab initio* calculation as follows. There are seven different types of two-electron integrals:

1. $(ab \mid cd) \equiv (ab \mid dc) \equiv (ba \mid cd) \equiv (ba \mid dc) \equiv (cd \mid ab) \equiv (cd \mid ba) \equiv (dc \mid ab) \equiv (dc \mid ba)$
2. $(aa \mid bc) \equiv (aa \mid cb) \equiv (bc \mid aa) \equiv (cb \mid aa)$
3. $(ab \mid ac) \equiv (ab \mid ca) \equiv (ba \mid ac) \equiv (ba \mid ca) \equiv (ac \mid ab) \equiv (ac \mid ba) \equiv (ca \mid ab) \equiv (ca \mid ba)$
4. $(aa \mid bb) \equiv (bb \mid aa)$
5. $(ab \mid ab) \equiv (ab \mid ba) \equiv (ba \mid \mathrm{ab}) \equiv (ba \mid ba)$
6. $(aa \mid ab) \equiv (aa \mid ba) \equiv (ab \mid aa) \equiv (ba \mid aa)$
7. $(aa \mid aa)$

For a basis set with K basis functions, there are $K(K-1)(K-2)(K-3)$ integrals of type $(ab \mid cd)$, but due to symmetry only one-eighth of these are unique as shown. Similarly, there are $2K(K-1)(K-2)$ of type (2); $4K(K-1)(K-2)$ of type (3); $K(K-1)$ of type (4); $2K(K-1)$ of type (5); $4K(K-1)$ of type (6) and K of type 7. Thus, a basis set with 200 functions has a total of 202 015 050 unique two-electron integrals. For all but the smallest of basis sets most integrals are of type (1) which is why an *ab initio* problem is often considered to scale as $\mathrm{K}^4/8$ ($200^4/8 = 200\ 000\ 000$). Including electron correlation adds significantly to the computational cost; for example, MP2 calculations scale as the fifth power of the number of basis functions. Electron correlation methods may also require significantly more memory and disk than the comparable SCF calculation.

In practice, *ab initio* calculations often scale as a significantly smaller power than four. It is found that in favourable cases the computational cost of a direct SCF calculation on a large molecule scales as approximately the square of the number of basis functions used. This significant reduction (from four to two) is due to several factors. We have already noted some of the ways in which a carefully chosen basis set can reduce the computational effort, for example by making many of the integrals (particularly the two-electron integrals) identical by using the same Gaussian exponents for s and p orbitals in the same shell. Symmetry in the molecule may also be exploited, sometimes to great effect. The most effective way to reduce the computational effort is to identify integrals which are so small that ignoring them (i.e. setting them to zero) will not affect the results. The number of 'important' integrals is believed to scale as $K^2 \ln K$. The negligible integrals are determined by calculating an upper limit for each integral. This can be done rapidly and so those integrals that are guaranteed to be negligible can be identified and so ignored. The cutoff value which determines whether an integral is explicitly calculated or is set to zero can vary from one program to another, so it is always useful to check its value if different programs give different results for a given calculation.

2.9.3 Setting up the calculation and the choice of coordinates

The traditional way to provide the nuclear coordinates to a quantum mechanical program is via a Z-matrix, in which the positions of the nuclei are defined in terms of a set of internal coordinates (see section 1.2). Some programs also accept coordinates in Cartesian format, which can be more convenient for large systems. It can sometimes be important to choose an appropriate set of internal coordinates, especially when locating minima or transition points or when following reaction pathways. This is discussed in more detail in section 4.7.

2.9.4 Calculating derivatives of the energy

Considerable effort has been spent devising efficient ways of calculating the first and second derivatives of the energy with respect to the nuclear coordinates. Derivatives are primarily used during minimisation procedures for finding equilibrium structures and are also used by methods which locate transition structures and determine reaction pathways. To calculate derivatives of the energy it is necessary to calculate the derivatives of the various electron integrals. For Gaussian basis sets the derivatives can be obtained analytically, and it is relatively straightforward to obtain first derivatives for many levels of theory. The time taken to calculate the derivatives is comparable to that required for the calculation of the total energy. Second derivatives are more difficult and expensive to calculate, even at the lower levels of theory.

2.9.5 Basis set superposition error

Suppose we wish to calculate the energy of formation of a bimolecular complex, such as the energy of formation of a hydrogen-bonded water dimer. Such complexes are sometimes referred to as 'supermolecules'. One might expect that this energy value could be obtained by first calculating the energy of a single water molecule, then calculating the energy of the dimer, and finally subtracting the energy of the two isolated water molecules (the 'reactants') from that of the dimer (the 'products'). However, the energy difference obtained by such an approach will invariably be an overestimate of the true value. The discrepancy arises from a phenomenon known as *basis set superposition error* (BSSE). As the two water molecules approach, the energy of the system falls not only because of the favourable intermolecular interactions but also because the basis functions on each molecule provide a better description of the electronic structure around the other molecule. It is clear that the BSSE would be expected to be particularly significant when small, inadequate basis sets are used (e.g. the minimal basis STO-*n*G basis sets) which do not provide for an adequate representation of the electron distribution far from the nuclei, particularly in the region where non-covalent interactions are strongest. One way to estimate the basis set superposition error is via the counterpoise correction method of Boys and Bernardi in which the entire basis set is included in all calculations [Boys and Bernardi 1970]. Thus, in the general case:

$$A + B \rightarrow AB \tag{2.216}$$

$$\Delta E = E(AB) - [E(A) + E(B)] \tag{2.217}$$

The calculation of the energy of the individual species A is performed in the presence of 'ghost' orbitals of B; that is, without the nuclei or electrons of B. A similar calculation is performed for B using ghost orbitals on A. An alternative approach is to use a basis set in which the orbital exponents and contraction coefficients have been optimised for molecular calculations rather than for atoms. The relevance of the basis set superposition error and its dependence upon the basis set and the level of theory employed (i.e. SCF or with electron correlation) remains a subject of much research.

2.10 Approximate molecular orbital theories

Ab initio calculations can be extremely expensive in terms of the computer resources required. Nevertheless, improvements in computer hardware and the availability of easy-to-use programs have helped to make *ab initio* methods a widely used computational tool. The approximate quantum mechanical methods require significantly less computational resources. Indeed, the earliest approximate methods such as Hückel theory predate computers by many years. Moreover, by their incorporation of parameters derived from experimental data some approximate methods can calculate some properties more accurately then even the highest level of *ab initio* methods.

Many approximate molecular orbital theories have been devised. Most of these methods are not in widespread use today in their original form. Nevertheless, the more widely used methods of today are derived from earlier formalisms which we will therefore consider where appropriate. We will concentrate on the semi-empirical methods developed in the research groups of Pople and Dewar. The former pioneered the CNDO, INDO and NDDO methods which are now relatively little used in their original form, but provided the basis for subsequent work by the Dewar group, whose research resulted in the popular MINDO/3, MNDO and AM1 methods. Our aim will be to show how the theory can be applied in a practical way; not only to highlight their successes but also to show where problems were encountered and how these problems were overcome. We will also consider the Hückel molecular orbital approach and the extended Hückel method. Our discussion of the underlying theoretical background of the approximate molecular orbital methods will be based on the Roothaan–Hall framework we have already developed. This will help us to establish the similarities and the differences with the *ab initio* approach.

2.11 Semi-empirical methods

A discussion of semi-empirical methods starts most appropriately with the key components of the Roothaan–Hall equations which for a closed-shell system are:

$$\mathbf{FC} = \mathbf{SCE} \tag{2.218}$$

$$F_{\mu\nu} = H_{\mu\nu}^{\text{core}} + \sum_{\lambda=1}^{K}\sum_{\sigma=1}^{K} P_{\lambda\sigma}\left[(\mu\nu|\lambda\sigma) - \tfrac{1}{2}(\mu\lambda|\nu\sigma)\right] \tag{2.219}$$

$$P_{\lambda\sigma} = 2\sum_{\mathrm{i}=1}^{N/2} c_{\lambda\mathrm{i}}c_{\sigma\mathrm{i}} \tag{2.220}$$

$$H_{\mu\nu}^{\text{core}} = \int \mathrm{d}v_1\phi_\mu(1)\left[-\frac{1}{2}\nabla^2 - \sum_{\mathrm{A}=1}^{M}\frac{Z_\mathrm{A}}{|r_1 - R_\mathrm{A}|}\right]\phi_\nu(1) \tag{2.221}$$

In *ab initio* calculations all elements of the Fock matrix are calculated using equation (2.219), irrespective of whether the basis functions ϕ_μ, ϕ_ν, ϕ_λ and ϕ_σ are on the same atom, on atoms that are bonded or on atoms that are not formally bonded. To discuss the semi-empirical methods it is useful to consider the Fock matrix elements in three groups: $F_{\mu\mu}$ (the diagonal elements); $F_{\mu\nu}$ (where ϕ_μ and ϕ_ν are on the same atom) and $F_{\mu\nu}$ (where ϕ_μ and ϕ_ν are on different atoms).

We have mentioned several times that the greatest proportion of the time requird to perform an *ab initio* Hartree–Fock SCF calculation is invariably spent calculating and manipulating integrals. The most obvious way to reduce the computational effort is therefore to neglect or approximate some of these integrals. Semi-empirical methods achieve this in part by explicitly considering only the valence electrons of the system; the core electrons are subsumed into the nuclear core. The rationale behind this approximation is that the electrons involved in chemical bonding and other phenomena that we might wish to investigate are those in the valence shell. By considering all the valence electrons the semi-empirical methods differ from those theories (e.g. Hückel theory) that explicitly consider only the π-electrons of a conjugated system and which are therefore limited to specific classes of molecules. The semi-empirical calculations invariably use basis sets comprising Slater-type s, p and sometimes d orbitals. The orthogonality of such orbitals enables further simplifications to be made to the equations.

A feature common to the semi-empirical methods is that the overlap matrix, **S** (in equation 2.218), is set equal to the identity matrix, **I**. Thus all diagonal elements of the overlap matrix are equal to 1 and all off-diagonal elements are zero. Some of the off-diagonal elements would naturally be zero due to the use of orthogonal basis sets on each atom, but in addition the elements that correspond to the overlap between two atomic orbitals on different atoms are also set to zero. The main implication of this is that the Roothaan–Hall equations are simplified: $\mathbf{FC} = \mathbf{SCE}$ becomes $\mathbf{FC} = \mathbf{CE}$, and so is immediately in standard matrix form. It is important to note that setting **S** equal to the identify matrix does not mean that all overlap integrals are set to zero in the calculation of Fock matrix elements. Indeed, it is important specifically to include some of the overlaps in even the simplest of the semi-empirical models.

2.11.1 *Zero-differential overlap*

Many semi-empirical theories are based upon the zero-differential overlap approximation (ZDO). In this approximation, the overlap between pairs of different orbitals is set to zero for all volume elements dv:

$$\phi_\mu \phi_\nu \mathrm{d}v = 0 \tag{2.222}$$

This directly leads to the following result for the overlap integrals:

$$S_{\mu\nu} = \delta_{\mu\nu} \tag{2.223}$$

If the two atomic orbitals ϕ_μ and ϕ_ν are located on different atoms then the differential overlap is referred to as diatomic differential overlap; if ϕ_μ and ϕ_ν are on the same atom then we have monatomic differential overlap. If the ZDO approximation is applied to the two-electron repulsion integral $(\mu\nu \mid \lambda\sigma)$ then the integral will equal zero if $\mu \neq \nu$ and/or if $\lambda \neq \sigma$. This can be concisely written using the Kronecker delta:

$$(\mu\nu|\lambda\sigma) = (\mu\mu|\lambda\lambda)\delta_{\mu\nu}\delta_{\lambda\sigma} \tag{2.224}$$

It can immediately be seen that all three- and four-centre integrals are set to zero under the ZDO approximation. If the ZDO approximation is applied to all orbital pairs then the Roothaan–Hall equations for a closed-shell molecule (equation (2.219)) simplify considerably to give the following for $\mu \equiv \nu$:

$$F_{\mu\mu} = H_{\mu\mu}^{\text{core}} + \sum_{\lambda=1}^{K} P_{\lambda\lambda}(\mu\mu|\lambda\lambda) - \tfrac{1}{2}P_{\mu\mu}(\mu\mu|\mu\mu) \tag{2.225}$$

The summation over λ includes $\lambda = \mu$ and the terms in $(\mu\mu|\mu\mu)$ can be separated to give:

$$F_{\mu\mu} = H_{\mu\mu}^{\text{core}} + \tfrac{1}{2}P_{\mu\mu}(\mu\mu|\mu\mu) + \sum_{\lambda=1;\lambda\neq\mu}^{K} P_{\lambda\lambda}(\mu\mu|\lambda\lambda) \tag{2.226}$$

For $\nu \neq \mu$ we have:

$$F_{\mu\nu} = H_{\mu\nu}^{\text{core}} - \tfrac{1}{2}P_{\mu\nu}(\mu\mu|\nu\nu) \tag{2.227}$$

Sensible results cannot be obtained by simply applying the ZDO approximation to all pairs of orbitals *carte blanche*. There are two major reasons for this.

The first consideration is that the total wavefunction and the molecular properties calculated from it should be the same when a transformed basis set is used. We have already encountered this requirement in our discussion of the transformation of the Roothaan–Hall equations to an orthogonal set. To reiterate: suppose a molecular orbital is written as a linear combination of atomic orbitals:

$$\psi_{\text{i}} = \sum_{\mu} c_{\mu\text{i}}\phi_{\mu} \tag{2.228}$$

If an alternative basis set is used in which the basis functions are just linear combinations of the original basis functions, then the same wavefunction can be written as a linear combination of these new transformed functions:

$$\psi_{\text{i}} = \sum_{\alpha} c_{\alpha\text{i}}\phi'_{\alpha} \tag{2.229}$$

$$\phi'_{\alpha} = \sum_{\mu_{\alpha}} t_{\mu_{\alpha}}\phi_{\mu} \tag{2.230}$$

$t_{\mu_{\alpha}}$ are the coefficients of the original basis functions in the linear expansion of the transformed basis set. Different types of transformation are possible; for example, some transformations mix orbitals with the same principal and azimuthal quantum numbers (e.g. mixing $2p_x$, $2p_y$ and $2p_z$); others mix orbitals with the same principal quantum number but different azimuthal quantum numbers (e.g. mixing 2s, $2p_x$, $2p_y$ and $2p_z$ orbitals to give sp^3 hybrid orbitals); yet other transformations mix orbitals located on different atoms. Suppose we mix $2p_x$ and $2p_y$ atomic orbitals on the same atom. The differential overlap between these two orbitals is

$2p_x 2p_y$. We now introduce the following two new coordinates which correspond to a rotation in the xy plane:

$$x' = \frac{1}{\sqrt{2}}(x + y) \tag{2.231}$$

$$y' = \frac{1}{\sqrt{2}}(-x + y) \tag{2.232}$$

The overlap between the $2p'_x$ and $2p'_y$ orbitals in this new coordinate system is $\frac{1}{2}(2p_y^2 - 2p_x^2)$. If the zero differential overlap approximation were applied, then different results would be obtained for the two coordinate systems unless the overlap in the new, transformed system was also ignored.

The second reason why the ZDO approximation is not applied to all pairs of orbitals is that the major contributors to bond formation are the electron–core interactions between pairs of orbitals and the nuclear cores (i.e. $H_{\mu\nu}^{\text{core}}$). These interactions are therefore not subjected to the ZDO approximation (and so do not suffer from any transformation problems).

2.11.2 CNDO

The complete neglect of differential overlap (CNDO) approach of Pople, Santry and Segal was the first method to implement the zero-differential overlap approximation in a practical fashion [Pople *et al.* 1965]. To overcome the problems of rotational invariance, the two-electron integrals $(\mu\mu|\lambda\lambda)$ where μ and λ are on different atoms A and B were set equal to a parameter γ_{AB} which only depends on the nature of the atoms A and B and the internuclear distance, and not on the type of orbital. The parameter γ_{AB} can be considered to be the average electrostatic repulsion between an electron on atom A and an electron on atom B. When both atomic orbitals are on the same atom the parameter is written γ_{AA} and represents the average electron–electron repulsion between two electrons on an atom A.

With this approximation we can divide the elements of the Fock matrix into three groups: $F_{\mu\mu}$ (the diagonal elements), $F_{\mu\nu}$ (where μ and ν are on different atoms) and $F_{\mu\nu}$ (where μ and ν are on the same atom). To obtain $F_{\mu\mu}$ we substitute γ_{AB} for the two-electron integrals $(\mu\mu \mid \lambda\lambda)$ where μ and λ are on different atoms, and γ_{AA} where μ and λ are on the same atom into the Fock matrix equations, equations (2.233)–(2.235):

$$F_{\mu\mu} = H_{\mu\mu}^{\text{core}} + \sum_{\lambda=1;\,\lambda \text{ on A}}^{K} P_{\lambda\lambda}\gamma_{AA} - \tfrac{1}{2}P_{\mu\mu}\gamma_{AA} + \sum_{\lambda=1;\,\lambda \text{ not on A}}^{K} P_{\lambda\lambda}\gamma_{AB} \tag{2.233}$$

$$F_{\mu\nu} = H_{\mu\nu}^{\text{core}} - \tfrac{1}{2}P_{\mu\nu}\gamma_{AA};\quad \mu \text{ and } \nu \text{ both on atom A} \tag{2.234}$$

$$F_{\mu\nu} = H_{\mu\nu}^{\text{core}} - \tfrac{1}{2}P_{\mu\nu}\gamma_{AB};\quad \mu \text{ and } \nu \text{ on different atoms, A and B} \tag{2.235}$$

Equation (2.233) is rather untidy, involving summations over basis functions on atom A and basis functions not on atom A. It is often simplified by writing P_{AA} as the total electron density on atom A where

$$P_{AA} = \sum_{\lambda \text{ on A}}^{A} P_{\lambda\lambda}.$$

A similar expression can also be introduced for P_{BB}. With this notation $F_{\mu\mu}$ simplifies to:

$$F_{\mu\mu} = H_{\mu\mu}^{core} + (P_{AA} - \tfrac{1}{2}P_{\mu\mu})\gamma_{AA} + \sum_{B \neq A} P_{BB}\gamma_{AB} \tag{2.236}$$

The core Hamiltonian expressions, $H_{\mu\mu}^{core}$ and $H_{\mu\nu}^{core}$ correspond to electrons moving in the field of the parent nucleus and the other nuclei. In semi-empirical methods the core electrons are subsumed into the nucleus and so the nuclear charges are altered accordingly (for example, carbon has a nuclear 'charge' of $+4$).

In CNDO $H_{\mu\mu}^{core}$ is separated into an integral involving the atom on which ϕ_μ is situated (labelled A), and all the others (labelled B). Thus:

$$H_{\mu\mu}^{core} = U_{\mu\mu} - \sum_{B \neq A} V_{AB} \tag{2.237}$$

where

$$U_{\mu\mu} = \left(\mu \left| -\frac{1}{2}\nabla^2 - \frac{Z_A}{|\mathbf{r}_1 - \mathbf{R}_A|} \right| \mu\right) \quad \text{and} \quad V_{AB} = \left(\mu \left| \frac{Z_B}{|\mathbf{r}_1 - \mathbf{R}_B|} \right| \mu\right). \tag{2.238}$$

$U_{\mu\mu}$ is thus the energy of the orbital ϕ_μ in the field of its own nucleus (A) and core electrons; $-V_{AB}$ is the energy of the electron in the field of another nucleus (B). To maintain consistency with the way in which the two-electron integrals are treated, terms

$$\left(\mu \left| \frac{Z_B}{|\mathbf{r}_1 - \mathbf{R}_B|} \right| \mu\right)$$

must be the same for all orbitals ϕ_μ on atom A (i.e. the interaction energy between any electron in an orbital on atom A with the core of atom B is equal to V_{AB}).

We next consider $H_{\mu\nu}^{core}$, where ϕ_μ and ϕ_ν are both on the same atom, A. In this case the core Hamiltonian has the following form:

$$\begin{aligned} H_{\mu\nu}^{core} &= \left(\mu \left| -\frac{1}{2}\nabla^2 - \frac{Z_A}{|\mathbf{r}_1 - \mathbf{R}_A|} \right| \nu\right) - \sum_{B \neq A} \left(\mu \left| \frac{Z_B}{|\mathbf{r}_1 - \mathbf{R}_B|} \right| \nu\right) \\ &= U_{\mu\nu} - \sum_{B \neq A} \left(\mu \left| \frac{Z_B}{|\mathbf{r}_1 - \mathbf{R}_B|} \right| \nu\right) \end{aligned} \tag{2.239}$$

As ϕ_μ and ϕ_ν are on the same atom then $U_{\mu\nu}$ is zero due to the orthogonality of atomic orbitals. The term

$$\left(\mu\left|\frac{Z_\mathrm{B}}{|\mathbf{r}_1-\mathbf{R}_\mathrm{B}|}\right|\nu\right)$$

is zero in accordance with the zero-differential overlap approximation. Thus $H_{\mu\nu}^{\mathrm{core}}$ is zero in CNDO.

Finally, if ϕ_μ and ϕ_ν are on two different atoms A and B, then we can write:

$$H_{\mu\nu}^{\mathrm{core}}=\left(\mu\left|-\frac{1}{2}\nabla^2-\frac{Z_\mathrm{A}}{|\mathbf{r}_1-\mathbf{R}_\mathrm{A}|}-\frac{Z_\mathrm{B}}{|\mathbf{r}_1-\mathbf{R}_\mathrm{B}|}\right|\nu\right)$$
$$-\sum_{\mathrm{C}\neq\mathrm{A,B}}\left(\mu\left|-\frac{Z_\mathrm{C}}{|\mathbf{r}_1-\mathbf{R}_\mathrm{C}|}\right|\nu\right) \tag{2.240}$$

The second term corresponds to the interaction of the distribution $\phi_\mu\phi_\nu$ with the atoms C ($\neq$A, B). These interactions are ignored. The first part (known as the *resonance integral* and commonly written $\beta_{\mu\nu}$) is not subject to the ZDO approximation because it is the main cause of bonding. In CNDO the resonance integral is made proportional to the overlap integral, $S_{\mu\nu}$:

$$H_{\mu\nu}^{\mathrm{core}}=\beta_{\mathrm{AB}}^0 S_{\mu\nu} \tag{2.241}$$

β_{AB}^0 is a parameter which depends on the nature of the atoms A and B.

With these approximations the Fock matrix elements for CNDO become:

$$F_{\mu\mu}=U_{\mu\mu}+\sum_{\mathrm{B}\neq\mathrm{A}}V_{\mathrm{AB}}+(P_{\mathrm{AA}}-\tfrac{1}{2}P_{\mu\mu})\gamma_{\mathrm{AA}}+\sum_{\mathrm{B}\neq\mathrm{A}}P_{\mathrm{BB}}\gamma_{\mathrm{AB}} \tag{2.242}$$

$$F_{\mu\nu}=-\tfrac{1}{2}P_{\mu\nu}\gamma_{\mathrm{AA}} \qquad \mu \text{ and } \nu \text{ on the same atom, A} \tag{2.243}$$

$$F_{\mu\nu}=\beta_{\mathrm{AB}}^0 S_{\mu\nu}-\tfrac{1}{2}P_{\mu\nu}\gamma_{\mathrm{AB}} \qquad \mu \text{ on A and and } \nu \text{ on B} \tag{2.244}$$

To perform a CNDO calculation requires the following to be calculated or specified: the overlap integrals, $S_{\mu\nu}$, the core Hamiltonians $U_{\mu\mu}$, the electron–core interactions V_{AB}, the electron repulsion integrals γ_{AB} and γ_{AA} and the bonding parameters β_{AB}^0. The CNDO basis set comprises Slater-type orbitals for the valence shell with the exponents being chosen using Slater's rules (except for hydrogen where an exponent of 1.2 is used as this value is more appropriate to hydrogen atoms in molecules). Thus the basis set comprises 1s for hydrogen and 2s, $2p_x$, $2p_y$ and $2p_z$ for the first-row elements. The overlap integrals are calculated explicitly (the overlap between two basis functions on the same atom is of course zero with an s, p basis set). The electron repulsion integral parameter γ_{AB} is calculated using valence s functions on the two atoms A and B:

$$\gamma_{\mathrm{AB}}=\iint \mathrm{d}v_1\mathrm{d}v_2\,\phi_{\mathrm{s,A}}(1)\phi_{\mathrm{s,A}}(1)\left(\frac{1}{r_{12}}\right)\phi_{\mathrm{s,B}}(2)\phi_{\mathrm{s,B}}(2) \tag{2.245}$$

The use of spherically symmetric s orbitals avoids the problems associated with transformations of the axes. The core Hamiltonians ($U_{\mu\mu}$) are not calculated, but are obtained from experimental ionisation energies. This is because it is important to distinguish between s and p orbitals in the valence shell (i.e. the 2s and 2p orbitals for the first row elements), and without explicit core electrons this is difficult to achieve. The resonance integrals, β^0_{AB}, are written in terms of empirical single atom values as follows:

$$\beta^0_{AB} = 1/2(\beta^0_A + \beta^0_B)$$

The β^0 values are chosen to fit the results of minimal basis set *ab initio* calculations on diatomic molecules.

The electron–core interaction, V_{AB}, is calculated as the interaction between an electron in a valence s orbital on atom A with the nuclear core of atom B:

$$V_{AB} = \int dv_1 \phi_{s,A}(1) \frac{Z_B}{|\mathbf{r}_1 - \mathbf{R}_B|} \phi_{s,A}(1) \tag{2.246}$$

CNDO is rightly recognised as the first in a long line of important semi-empirical models. However, there were some important limitations with the model. One especially serious deficiency of the first version of CNDO (first introduced in 1965 [Pople and Segal 1965, Pople *et al.* 1965] and now known as CNDO/1) is that two neutral atoms show a significant (and incorrect) attraction even when separated by several angstroms. The predicted equilibrium distances for diatomic molecules are also too short and the dissociation energies too large. These effects are due to electrons on one atom penetrating the valence shell of another atom and so experiencing a nuclear attraction. This penetration effect can be quantified more explicitly as follows. The net charge on an atom B equals the difference between its nuclear charge and the total electron density: $Q_B = Z_B - P_{BB}$. If we now substitute for P_{BB} $(= Z_B - Q_B)$ in the diagonal elements of the Fock matrix, equation (2.242) we obtain:

$$F_{\mu\mu} = U_{\mu\mu} + (P_{AA} - \tfrac{1}{2}P_{\mu\mu})\gamma_{AA} + \sum_{B \neq A} [-Q_B\gamma_{AB} + (Z_B\gamma_{AB} - V_{AB})] \tag{2.247}$$

$-Q_B\gamma_{AB}$ is the contribution from the total charge on atom B; this is zero if the atomic charge is exactly balanced by the electron density. $Z_B\gamma_{AB} - V_{AB}$ is called the *penetration integral*. It was this contribution that caused the anomalous results for two neutral atoms at large separation. In the second version of CNDO (CNDO/2 [Pople and Segal 1966]) the penetration integral effect was eliminated by putting $V_{AB} = Z_B\gamma_{AB}$. The core Hamiltonian $U_{\mu\mu}$ was also defined differently in CNDO/2, using both ionisation energies and electron affinities.

2.11.3 INDO

CNDO makes no allowance for the fact that the interaction between two electrons depends upon their relative spins. This effect can be particularly severe for electrons on the same atom. Thus, in CNDO all two-electron integrals $(\mu\nu \mid \lambda\nu)$ are set to zero, and integrals $(\mu\mu \mid \nu\nu)$ and $(\mu\mu \mid \mu\mu)$ are forced to be equal (to γ_{AA}). The next development was the intermediate neglect of differential overlap model

(INDO [Pople *et al.* 1967]) which includes monatomic differential overlap for one-centre integrals (i.e. for integrals involving basis functions centred on the same atom). This enables the interaction between two electrons on the same atom with parallel spins to have a lower energy than the comparable interaction between electrons with paired spins. For this reason the Fock matrix elements are usually written with the spin (α or β) explicitly specified. The elements $F_{\mu\mu}$ and $F_{\mu\nu}$ (where μ and ν are located on atom A) then change from their CNDO/2 values as follows:

$$F^{\alpha}_{\mu\mu} = U_{\mu\mu} + \sum_{\lambda \text{ on A}} \sum_{\sigma \text{ on A}} [P_{\lambda\sigma}(\mu\mu|\lambda\sigma) - P^{\alpha}_{\lambda\sigma}(\mu\lambda|\mu\sigma)] + \sum_{B \neq A} (P_{BB} - Z_B)\gamma_{AB} \tag{2.248}$$

$$F^{\alpha}_{\mu\nu} = U_{\mu\nu} + \sum_{\lambda \text{ on A}} \sum_{\sigma \text{ on A}} [P_{\lambda\sigma}(\mu\nu|\lambda\sigma) - P^{\alpha}_{\lambda\sigma}(\mu\lambda|\nu\sigma)] \quad \mu \text{ and } \nu \text{ both on atom A} \tag{2.249}$$

In equation (2.248) we have included the CNDO/2 approximation $V_{AB} = Z_B\gamma_{AB}$. The matrix element $F_{\mu\nu}$ where μ and ν are on different atoms is the same as in CNDO/2:

$$F^{\alpha}_{\mu\nu} = \tfrac{1}{2}(\beta^0_A + \beta^0_B)S_{\mu\nu} - P^{\alpha}_{\mu\nu}\gamma_{AB} \tag{2.250}$$

In a closed-shell system, $P^{\alpha}_{\mu\nu} = P^{\beta}_{\mu\nu} = \frac{1}{2}P_{\mu\nu}$ and the Fock matrix elements can be obtained by making this substitution. If a basis set containing s, p orbitals is used, then many of the one-centre integrals nominally included in INDO are equal to zero, as are the core elements $U_{\mu\nu}$. Specifically, only the following one-centre, two-electron integrals are non-zero: $(\mu\mu\,|\,\mu\mu)$, $(\mu\mu\,|\,\nu\nu)$ and $(\mu\nu\,|\,\mu\nu)$. The elements of the Fock matrix that are affected can then be written as follows:

$$F_{\mu\mu} = U_{\mu\mu} + \sum_{\nu \text{ on A}} \left[P_{\nu\nu}(\mu\mu|\nu\nu - \tfrac{1}{2}P_{\nu\nu}(\mu\nu|\mu\nu)\right] + \sum_{B \neq A} (P_{BB} - Z_B)\gamma_{AB} \tag{2.251}$$

$$F_{\mu\nu} = \tfrac{3}{2}P_{\mu\nu}(\mu\nu|\mu\nu) - \tfrac{1}{2}P_{\mu\nu}(\mu\mu|\nu\nu) \qquad \mu, \nu \text{ on the same atom} \tag{2.252}$$

Some of the one-centre two-electron integrals in INDO are semi-empirical parameters, obtained by fitting to atomic spectroscopic data. The core integrals $U_{\mu\mu}$ are obtained in a slightly different fashion to that of CNDO/2, to take into account the new electronic configurations under the INDO model for atoms and their cations and anions. An INDO calculation requires little additional computational effort compared with the corresponding CNDO calculation and has the key advantage that states of different multiplicities can be distinguished. For example, in CNDO the singlet and triplet configurations $1s^22s^22p^2$ of carbon have the same energy whereas these can be distinguished using INDO. Two of the systems considered in the original INDO publication were the methyl and ethyl radicals, the unpaired electron density being compared with experimentally determined hyperfine coupling constants. INDO gave a much more favourable result for these systems than CNDO.

2.11.4 NDDO

The next level of approximation is the neglect of diatomic differential overlap model (NDDO [Pople *et al.* 1965]); this theory only neglects differential overlap between atomic orbitals on different atoms. Thus all of the two-electron, two-centre integrals of the form $(\mu\nu \mid \lambda\sigma)$ where μ and ν are on the same atom and λ and σ are also on the same atom are retained. The Fock matrix elements become:

$$F_{\mu\mu} = H_{\mu\mu}^{\text{core}} + \sum_{\lambda \text{ on A}} \sum_{\sigma \text{ on A}} \left[P_{\lambda\sigma}(\mu\mu|\lambda\sigma) - \tfrac{1}{2}P_{\lambda\sigma}(\mu\lambda|\mu\sigma)\right] + \sum_{B \neq A} \sum_{\lambda \text{ on B}} \sum_{\sigma \text{ on B}} P_{\lambda\sigma}(\mu\mu|\lambda\sigma) \tag{2.253}$$

$$F_{\mu\nu} = H_{\mu\nu}^{\text{core}} + \sum_{\lambda \text{ on A}} \sum_{\sigma \text{ on A}} [P_{\lambda\sigma}(\mu\nu|\lambda\sigma) - \tfrac{1}{2}P_{\lambda\sigma}(\mu\lambda|\nu\sigma)] + \sum_{B \neq A} \sum_{\lambda \text{ on B}} \sum_{\sigma \text{ on B}} P_{\lambda\sigma}(\mu\nu|\lambda\sigma); \quad \mu \text{ and } \nu \text{ both on A} \tag{2.254}$$

$$F_{\mu\nu} = H_{\mu\nu}^{\text{core}} - \frac{1}{2} \sum_{\lambda \text{ on B}} \sum_{\sigma \text{ on A}} P_{\lambda\sigma}(\mu\sigma|\nu\lambda); \quad \mu \text{ on A and } \nu \text{ on B} \tag{2.255}$$

It is again possible to tidy up equations (2.253) and (2.254) when an s, p basis set is used:

$$F_{\mu\mu} = H_{\mu\mu}^{\text{core}} + \sum_{\nu \text{ on A}} \left[P_{\nu\nu}(\mu\mu|\nu\nu) - \tfrac{1}{2}P_{\nu\nu}(\mu\nu|\mu\nu)\right] + \sum_{B \neq A} \sum_{\lambda \text{ on B}} \sum_{\sigma \text{ on B}} P_{\lambda\sigma}(\mu\mu|\lambda\sigma) \tag{2.256}$$

$$F_{\mu\nu} = H_{\mu\nu}^{\text{core}} + \tfrac{3}{2}P_{\mu\nu}(\mu\nu|\mu\nu) - \tfrac{1}{2}P_{\mu\nu}(\mu\mu|\nu\nu) + \sum_{B \neq A} \sum_{\lambda \text{ on B}} \sum_{\sigma \text{ on B}} P_{\lambda\sigma}(\mu\nu|\lambda\sigma) \tag{2.257}$$

Whereas the computation required for an INDO calculation is little more than for the analogous CNDO calculation, in NDDO the number of two-electron two-centre integrals is increased by a factor of approximately 100 for each pair of heavy atoms in the system.

2.11.5 MINDO/3

The CNDO, INDO and NDDO methods as originally devised and implemented are now little used, in comparison with the methods subsequently developed by Dewar and colleagues, but they were of considerable importance in showing how a systematic series of approximations could be used to develop methods of real practical value. Moreover, the calculations could be performed in a fraction of the time required to solve the full Roothaan–Hall equations. However, they did not produce very accurate results, largely because they were parametrised upon the results from relatively low-level *ab initio* calculations, which themselves agreed poorly with experiment. They were also limited to small classes of molecules and they often required a good experimental geometry to be supplied as input because the geometry optimisation algorithms were not very sophisticated.

It was through the introduction of the MINDO/3 method by Bingham, Dewar and Lo [Bingham *et al.* 1975a–1975d] that a wider audience was able to apply semi-empirical methods in their own research. MINDO/3 was not so much a significant change in the theory, being based upon INDO (MINDO stands for modified INDO), but it did differ significantly in the way in which the method was parametrised, making much more use of experimental data. It also incorporated a geometry optimisation routine (the Davidon–Fletcher–Powell method; see Chapter 4) which enabled the program to accept crude initial geometries as input and derive the associated minimum energy structures.

MINDO/3 uses an s, p basis set and its Fock matrix elements are:

$$F_{\mu\mu} = U_{\mu\mu} + \sum_{\nu \text{ on A}} \left(P_{\nu\nu}(\mu\mu|\nu\nu) - \tfrac{1}{2}P_{\nu\nu}(\mu\nu|\mu\nu)\right) + \sum_{\mathrm{B}\neq\mathrm{A}} (P_{\mathrm{BB}} - Z_{\mathrm{B}})\gamma_{\mathrm{AB}} \qquad (2.258)$$

$$F_{\mu\nu} = -\tfrac{1}{2}P_{\mu\nu}(\mu\nu|\mu\nu) \quad \mu \text{ and } \nu \text{ both on the same atom A} \qquad (2.259)$$

$$F_{\mu\nu} = H_{\mu\nu}^{\text{core}} - \tfrac{1}{2}P_{\mu\nu}(\mu\nu|\mu\nu) = H_{\mu\nu}^{\text{core}} - \tfrac{1}{2}P_{\mu\nu}\gamma_{\mathrm{AB}} \quad \mu \text{ on A and } \nu \text{ on B} \qquad (2.260)$$

The two-centre repulsion integrals γ_{AB} in MINDO/3 are calculated using the following function:

$$\gamma_{\mathrm{AB}} = \frac{e^2}{\left[R_{\mathrm{AB}}^2 + \dfrac{1}{4}\left(\dfrac{e^2}{\bar{g}_{\mathrm{A}}} + \dfrac{e^2}{\bar{g}_{\mathrm{B}}}\right)^2\right]^{\frac{1}{2}}} \qquad (2.261)$$

$\bar{g}_{\mathrm{A}}$ is the average of the one-centre two-electron integrals $g_{\mu\nu}$ on atom A (i.e. $g_{\mu\nu} \equiv (\mu\mu\,|\,\nu\nu)$) and $\bar{g}_{\mathrm{B}}$ is the equivalent average for atom B. This seemingly complex function for γ_{AB} is in fact quite simple; at large R_{AB} it tends towards the Coulomb's law expression e^2/R_{AB} and as R_{AB} tends to zero it approaches the average of the one-centre integrals on the two atoms. The two-centre, one-electron integrals $H_{\mu\nu}^{\text{core}}$ are given in MINDO/3 by:

$$H_{\mu\nu}^{\text{core}} = S_{\mu\nu}\beta_{\text{AB}}(I_\nu + I_\nu) \tag{2.262}$$

$S_{\mu\nu}$ is the overlap integral; I_μ and I_ν are ionisation potentials for the appropriate orbitals and β_{AB} is a parameter dependent upon both of the two atoms A and B.

The core–core interaction between pairs of nuclei was also changed in MINDO/3 from the form used in CNDO/2. One way to correct the fundamental problems with CNDO/2 such as the repulsion between two hydrogen atoms (or indeed any neutral molecules) at all distances is to change the core–core repulsion term from a simple Coulombic expression ($E_{\text{AB}} = Z_{\text{A}}Z_{\text{B}}/R_{\text{AB}}$) to:

$$E_{\text{AB}} = Z_{\text{A}}Z_{\text{B}}\gamma_{\text{AB}} \tag{2.263}$$

In fact, while this correction gives the desired behaviour at relatively long separations, it does not account for the fact that as two nuclei approach each other the screening by the core electrons decreases. As the separation approaches zero the core–core repulsion should be described by Coulomb's law. In MINDO/3 this is achieved by making the core–core interaction a function of the electron–electron repulsion integrals as follows:

$$E_{\text{AB}} = Z_{\text{A}}Z_{\text{B}}\{\gamma_{\text{AB}} + [(e^2/R_{\text{AB}}) - \gamma_{\text{AB}}]\exp(-\alpha_{\text{AB}}R_{\text{AB}})\} \tag{2.264}$$

α_{AB} is a parameter dependent upon the nature of the atoms A and B. For OH and NH bonds a slightly different core–core interaction was found to be more appropriate:

$$E_{\text{XH}} = Z_{\text{X}}Z_{\text{H}}\{\gamma_{\text{XH}} + [(e^2/R_{\text{XH}}) - \gamma_{\text{XH}}]\alpha_{\text{XH}}\exp(-R_{\text{XH}})\} \tag{2.265}$$

The parameters for MINDO/3 were obtained in an entirely different way than for previous semi-empirical methods. Some of the values that were fixed in CNDO, INDO and NNDO were permitted to vary during the MINDO/3 parametrisation procedure. For example, the exponents of the Slater atomic orbitals were allowed to vary from the values given by Slater's rules and indeed the exponents for s and p orbitals were not required to be the same. $U_{\mu\mu}$ and β_{AB} were also regarded as variable parameters. Another key difference was that the MINDO/3 parametrisation used experimental data such as molecular geometries and heats of formation, rather than theoretical values from *ab initio* calculations or data from atomic spectra. The parametrisation effort was a considerable undertaking and it was only at the fourth attempt that an acceptable model was obtained (as is implicit in the appearance of the '3' in the name). For example, just to parametrise two atoms such as carbon and hydrogen using a set of 20 molecules required between 30 000 and 50 000 SCF calculations for each parametrisation scheme that was investigated.

2.11.6 MNDO

MINDO/3 proved to be very successful when it was introduced; it is important to realise that even simple *ab initio* calculations were beyond the computational resources of all but a few research groups in the 1970s. However, there were some significant limitations. For example, heats of formation of unsaturated molecules were consistently too positive; the errors in calculated bond angles were often quite large; and the heats of formation for molecules containing adjacent atoms with lone pairs were too negative. Some of these limitations were due to the use of the INDO approximation, and in particular the inability of INDO to deal with systems containing lone pairs. Dewar and Thiel therefore introduced the modified neglect of diatomic overlap (MNDO) method which was based on NDDO [Dewar and Thiel 1977a]. The Fock matrix elements in MNDO were as follows:

$$F_{\mu\mu} = H_{\mu\mu}^{\text{core}} + \sum_{\text{v on A}} \left[P_{\text{vv}}(\mu\mu|\text{vv}) - \tfrac{1}{2}P_{\text{vv}}(\mu\text{v}|\mu\text{v})\right] + \sum_{\text{B}\neq\text{A}} \sum_{\lambda\,\text{on B}} \sum_{\sigma\,\text{on B}} P_{\lambda\sigma}(\mu\mu|\lambda\sigma) \tag{2.266}$$

where $$H_{\mu\mu}^{\text{core}} = U_{\mu\mu} - \sum_{\text{B}\neq\text{A}} V_{\mu\mu\text{B}} \tag{2.267}$$

$$F_{\mu\text{v}} = H_{\mu\text{v}}^{\text{core}} + \tfrac{3}{2}P_{\mu\text{v}}(\mu\text{v}|\mu\text{v}) - \tfrac{1}{2}P_{\mu\text{v}}(\mu\mu|\text{vv}) + \sum_{\text{B}\neq\text{A}} \sum_{\lambda\,\text{on B}} \sum_{\sigma\,\text{on B}} P_{\lambda\sigma}(\mu\text{v}|\lambda\sigma) \quad \mu \text{ and v both on A} \tag{2.268}$$

where $$H_{\mu\text{v}}^{\text{core}} = -\sum_{\text{B}\neq\text{A}} V_{\mu\text{vB}} \tag{2.269}$$

$$F_{\mu\text{v}} = H_{\mu\text{v}}^{\text{core}} - \tfrac{1}{2}\sum_{\lambda\,\text{on B}} \sum_{\sigma\,\text{on A}} P_{\lambda\sigma}(\mu\sigma|\text{v}\lambda) \quad \mu \text{ on A and v on B} \tag{2.270}$$

where $$H_{\mu\text{v}}^{\text{core}} = \tfrac{1}{2}S_{\mu\text{v}}(\beta_{\text{v}} + \beta_{\text{v}}) \tag{2.271}$$

The similarity with the NDDO expressions, equations (2.253)–(2.257) can clearly be seen; the major new features are the appearance of terms $V_{\mu\mu\text{B}}$ and $V_{\mu\text{vB}}$ and a new form for the two-centre, one-electron core resonance integrals which depend upon the overlap $S_{\mu\text{v}}$ and parameters β_μ and β_v as shown in equation (2.271). $V_{\mu\mu\text{B}}$ and $V_{\mu\text{vB}}$ are two-centre, one-electron attractions between an electron distribution $\phi_\mu\phi_\mu$ or $\phi_\mu\phi_\text{v}$ respectively on atom A and the core of atom B. These are expressed as follows:

$$V_{\mu\mu\text{B}} = -Z_\text{B}(\mu_\text{A}\mu_\text{A}|\text{s}_\text{B}\text{s}_\text{B}) \tag{2.272}$$

$$V_{\mu\text{vB}} = -Z_\text{B}(\mu_\text{A}\text{v}_\text{A}|\text{s}_\text{B}\text{s}_\text{B}) \tag{2.273}$$

The core–core repulsion terms are also different in MNDO from those in MINDO/3 with OH and NH bonds again being treated separately:

$$E_{AB} = Z_A Z_B (s_A s_A | s_B s_B)\{1 + \exp(-\alpha_A R_{AB}) + \exp(-\alpha_B R_{AB})\} \quad (2.274)$$

$$E_{XH} = Z_X Z_H (s_X s_X | s_H s_H)\{1 + R_{XH} \exp(-\alpha_X R_{XH})/R_{AB}) + \exp(-\alpha_H R_{XH})\} \quad (2.275)$$

Perhaps the most significant advantage of MNDO over MINDO/3 is the use throughout of monatomic parameters; MINDO/3 requires diatomic parameters in the resonance integral (β_{AB}) and the core–core repulsion (α_{AB}). It has been possible to expand MNDO to cover a much wider variety of elements such as aluminium, silicon, germanium, tin, bromine and lead. However, the use of an s,p basis set in the original MNDO method did mean that the method could not be applied to most transition metals, which require a basis set containing d orbitals. In addition, hypervalent compounds of sulphur and phosphorus are not modelled well. In more recent versions of the MNDO method d orbitals have been explicitly included for the heavier elements [Thiel and Voityuk 1994]. Another serious limitation of MNDO is its inability to accurately model intermolecular systems involving hydrogen bonds (for example, the heat of formation of the water dimer is far too low in MNDO). This is because of a tendency to overestimate the repulsion between atoms when they are separated by a distance approximately equal to the sum of their van der Waals radii. Conjugated systems can also present difficulties for MNDO. An extreme example of this occurs with compounds such as nitrobenzene in which the nitro group is predicted to be orthogonal to the aromatic ring rather than conjugated with it. In addition, MNDO energies are too positive for sterically crowded molecules and too negative for molecules containing four-membered rings.

2.11.7 AM1

The Austin Model 1 (AM1) model was the next semi-empirical theory produced by Dewar's group [Dewar *et al.* 1985]. AM1 was designed to eliminate the problems with MNDO which were considered to arise from a tendency to overestimate repulsions between atoms separated by distances approximately equal to the sum of their van der Waals radii. The strategy adopted was to modify the core–core term using Gaussian functions. Both attractive and repulsive Gaussian functions were used; the attractive Gaussians were designed to overcome the repulsion directly and were centred in the region where the repulsions were too large. Repulsive Gaussian functions were centred at smaller internuclear separ-

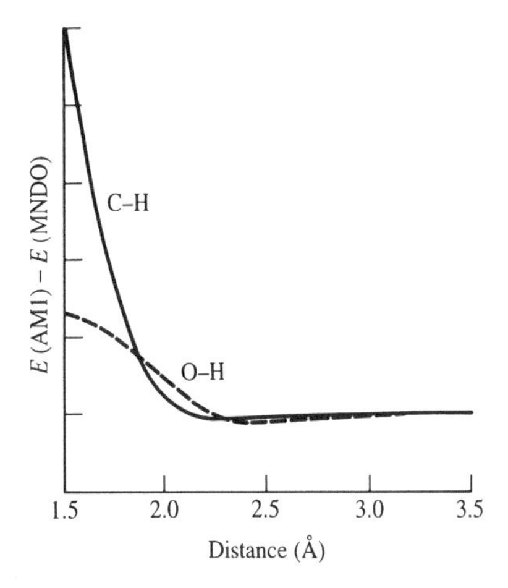

Fig. 2.11 The difference in the core–core energy for AM1 and MNDO for carbon–hydrogen and oxygen–hydrogen interactions.

ations. With this modification the expression for the core–core term was related to the MNDO expression by:

$$E_{AB} = E_{MNDO} + \frac{Z_A Z_B}{R_{AB}} \left\{ \sum_i K_{A_i} \exp[-L_{A_i}(R_{AB} - M_{A_i})^2] + \sum_j K_{B_j} \exp\left[-L_{B_j}(R_{AB} - M_{B_j})^2\right] \right\} \quad (2.276)$$

The additional terms are spherical Gaussian functions with a width determined by the parameter L. It was found that the values of these parameters were not critical and many were set to the same value. The M and K parameters were optimised for each atom, together with the α parameters in the exponential terms in equations (2.274) and (2.275). In the original parametrisation of AM1 there are four terms in the Gaussian expansion for carbon, three for hydrogen and nitrogen and two for oxygen (both attractive and repulsive Gaussians were used for carbon, hydrogen and nitrogen but only repulsive Gaussians for oxygen). The effect of including these Gaussian functions can be seen in Figure 2.11, which plots the difference in the MNDO and AM1 core–core terms for the carbon–hydrogen and oxygen–hydrogen interactions. The inclusion of these Gaussians significantly increased the number of parameters per atom, from 7 in the MNDO to between 13 and 16 per atom in AM1. This, of course, made the parametrisation process considerably more difficult. Overall, AM1 was a significant improvement over MNDO and many of the deficiencies associated with the core repulsion were corrected.

2.11.8 PM3

PM3 is also based on MNDO (the name derives from the fact that it is the third parametrisation of MNDO, AM1 being considered the second) [Stewart 1989a;

Stewart 1989b]. The PM3 Hamiltonian contains essentially the same elements as that for AM1 but the parameters for the PM3 model were derived using an automated parametrisation procedure devised by J J P Stewart. By contrast, many of the parameters in AM1 were obtained by applying chemical knowledge and 'intuition'. As a consequence, some of the parameters have significantly different values in AM1 and PM3 even though both methods use the same functional form and they both predict various thermodynamic and structural properties to approximately the same level of accuracy. Some problems do remain with PM3. One of the most important of these is the rotational barrier of the amide bond which is much too low and in some cases almost non-existent. This problem can be corrected through the use of an empirical torsional potential (see section 3.5). There is still considerable debate over the relative merits of the AM1 and PM3 approaches to parametrisation.

2.11.9 SAM1

The most recent offering from the Dewar group is SAM1, which stands for 'Semi-Ab-initio Model 1' [Dewar *et al.* 1993]. The name was chosen to reflect Dewar's belief that methods like AM1 offer such a significant enhancement over the earlier semi-empirical methods like CNDO/2 that they should be given a different generic name. In SAM1 a standard STO-3G Gaussian basis set is used to evaluate the electron repulsion integrals; close inspection of the results from AM1 and MNDO suggested that steric effects were overestimated because of the way in which the electron repulsion integrals were calculated. The resulting integrals were then scaled, partly to enable some of the effects of electron correlation to be included and partly to compensate for the use of a minimal basis set. The Gaussian terms in the core–core repulsion were retained to fine-tune the model. The number of parameters in SAM1 is no greater than in AM1 and fewer than in PM3. It does take longer to run (by up to two orders of magnitude) though it was felt that with the improvements in computer hardware such an increase was acceptable.

2.11.10 Programs for semi-empirical quantum mechanical calculations

The popularity of the MNDO, AM1 and PM3 methods is due in large part to their implementation in the MOPAC and AMPAC programs. The starting geometry is usually supplied as a Z-matrix, but Cartesian coordinates are also accepted. The programs are able to perform many kinds of calculations, including minimisation of the initial geometry, location of transition points and calculation of vibrational frequencies. Electron correlation is implicitly included in the semi-empirical methods because of the way in which they are parametrised, but specific electron correlation methods have also been developed for use with the various levels of approximation; this in turn necessitates the modification of some parameters.

The contributions of the Dewar group are rightly recognised as particularly significant in the development of semi-empirical methods, but other research groups have also made important contributions. The SINDO1 and ZINDO programs have been developed in the groups of Jug and Zerner respectively and both contain novel features. The ZINDO program of Zerner and co-workers can perform a wide variety of semi-empirical calculations and has been particularly useful for calculations on transition metal and lanthanide compounds and for predicting molecular electronic spectra.

2.12 Hückel theory

Hückel theory can be considered the 'grandfather' of approximate molecular orbital methods, having been formulated in the early 1930s [Hückel 1931], Hückel theory is limited to conjugated π-systems and was originally devised to explain the non-additive nature of certain properties of aromatic compounds. For example, the properties of benzene are much different from those of the hypothetical 'cyclohexatriene' molecule. Although Hückel theory as originally formulated is relatively little used in research today, extensions to it such as extended Hückel theory are still employed and can provide qualitative insights into the electronic structure of important classes of molecules. Hückel theory is also widely used for teaching purposes to introduce a 'real' theory that can be applied to relatively complex systems with little more than pencil and paper or a simple computer program.

Hückel theory separates the π-system from the underlying σ-framework and constructs molecular orbitals into which the π-electrons are then fed in the usual way according to the Aufbau principle. The π-electrons are thus considered to be moving in a field created by the nuclei and the 'core' of σ-electrons. The molecular orbitals are constructed from linear combinations of atomic orbitals and so the theory is an LCAO method. For our purposes it is most appropriate to consider Hückel theory in terms of the CNDO approximation (in fact, Hückel theory was the first ZDO molecular orbital theory to be developed). Let us examine the three types of Fock matrix elements in equations (2.242)–(2.244). First, $F_{\mu\mu}$. In a neutral species, the net charge on each atom will be approximately zero, and so if we take equation (2.247) in which penetration effects have been eliminated then we are left with $U_{\mu\mu} + (P_{AA} - 0.5P_{\mu\mu})\gamma_{AA}$. Now if each nucleus (A) in the π-system is the same (i.e. carbon) then this expression will be approximately constant for all nuclei being considered. The matrix elements $F_{\mu\mu}$ are often (confusingly) called Coulomb integrals in Hückel theory and are assigned the symbol α. All off-diagonal elements of the Fock matrix are assumed to be zero with the exception of elements $F_{\mu\nu}$, where μ and ν are π-orbitals on two bonded atoms. These $F_{\mu\nu}$ are assumed to be constant, are assigned the symbol β, and are known as resonance integrals. The Fock matrix in Hückel theory thus has as many rows and columns as the number of atoms in the π-system with diagonal elements that are all set to α. All off-diagonal elements F_{ij}

are zero unless there is a bond between the atoms i and j in which case the element is β. For benzene the Fock matrix is of the following form (atom labelling as in Figure 2.12):

$$\begin{pmatrix} \alpha & \beta & 0 & 0 & 0 & \beta \\ \beta & \alpha & \beta & 0 & 0 & 0 \\ 0 & \beta & \alpha & \beta & 0 & 0 \\ 0 & 0 & \beta & \alpha & \beta & 0 \\ 0 & 0 & 0 & \beta & \alpha & \beta \\ \beta & 0 & 0 & 0 & \beta & \alpha \end{pmatrix} \tag{2.277}$$

As with the other semi-empirical methods that we have considered so far, the overlap matrix is equal to the identity matrix. The following simple matrix equation must then be solved:

$$\mathbf{FC} = \mathbf{CE} \tag{2.278}$$

The equation can be solved by standard methods to give the basis set coefficients and the molecular orbital energies **E**. The orbital energies for benzene are $E_1 = \alpha + 2\beta$: E_2, $E_3 = \alpha + \beta$; E_4, $E_5 = \alpha - \beta$; $E_6 = \alpha - 2\beta$, and so the ground state places two electrons in ψ_1 and two each in the two degenerate orbitals ψ_2 and ψ_3. The lowest energy orbital ψ_1 is a linear combination of the six carbon p orbitals.

Hückel theory was extended to cover various other systems, including those with heteroatoms, but it was not particularly successful and has largely been superseded by other semi-empirical methods. Nevertheless, for appropriate problems Hückel theory can be very useful. One example is the calculations of P W Fowler and colleagues who studied the relationship between geometry and electronic structure for a range of buckminsterfullerenes (of which the parent molecule, C_{60}, was discovered in 1985) [Fowler 1993]. The fullerenes (or 'buckyballs') are excellent candidates for Hückel theory as they are composed of carbon and have extensive π-systems; three examples are shown in Figure 2.13.

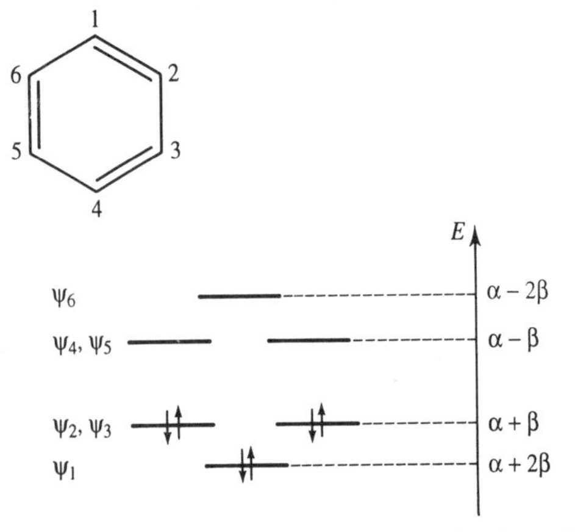

Fig. 2.12 Benzene and its Hückel molecular orbitals.

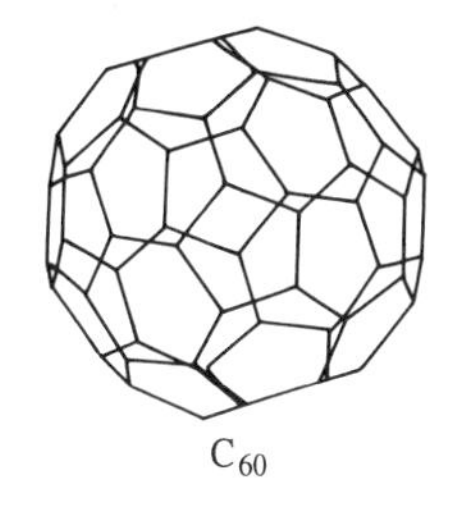

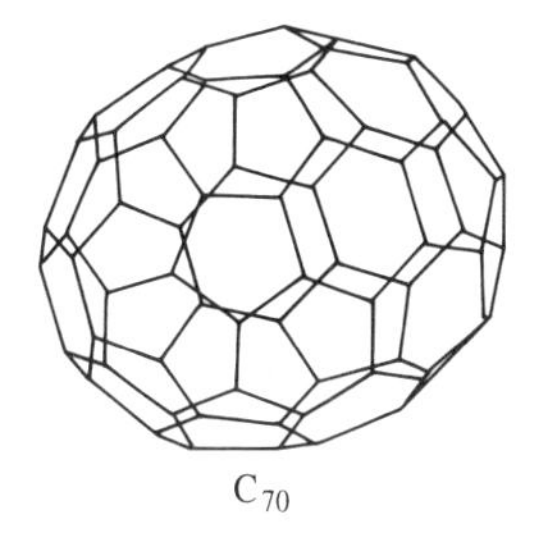

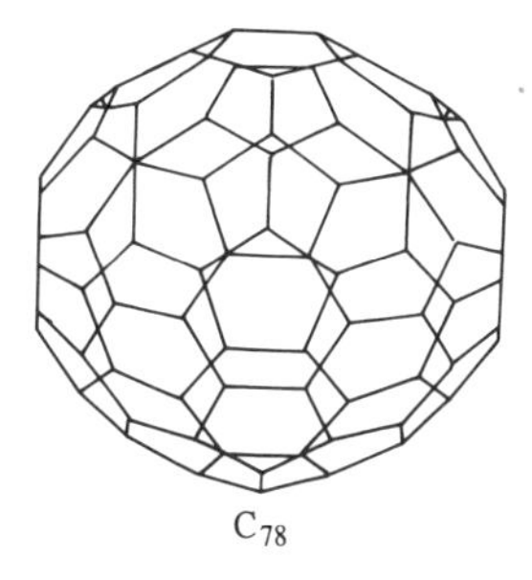

Fig. 2.13 Three fullerenes C_{60}, C_{70} and C_{78}.

The results of their calculations were summarised in two rules. The first rule states that at least one isomer C_n with a properly closed p shell (i.e. bonding HOMO, antibonding LUMO) exists for all $n = 60 + 6k$ ($k = 0, 2, 3 \ldots$ but not 1). Thus C_{60}, C_{72}, C_{78}, etc., are in this group. The second rule is for carbon cylinders and states that a closed shell structure is found for $n = 2p(7 + 3k)$ (for all k). C_{70} is the parent of this family. The calculations were extended to cover different types of structures and fullerenes doped with metals.

2.12.1 Extended Hückel theory

The Hückel theory method is clearly limited, in part because it is restricted to π-systems. The extended Hückel method is a molecular orbital theory that takes account of all the valence electrons in the molecule [Hoffmann 1963]. It is largely associated with R Hoffmann, who received the Nobel prize for his contributions.

The equation to be solved is **FC** = **SCE**, with the Fock matrix elements taking the following simple forms:

$$F_{\mu\mu}^{AA} = H_{\mu\mu} = -I_\mu \tag{2.279}$$

$$F_{\mu\nu}^{AB} = H_{\mu\nu} = -\tfrac{1}{2}K(I_\mu + I_\nu)S_{\mu\nu} \tag{2.280}$$

In these equations, μ and ν are two atomic orbitals (e.g. Slater-type orbitals), I_μ is the ionisation potential of the orbital and K is a constant, which was originally set to 1.75. The formula for the off-diagonal elements $H_\mu\nu$ (where μ and ν are on different atoms) was originally suggested by R S Mulliken. These off-diagonal matrix elements are calculated between all pairs of valence orbitals and so extended Hückel theory is not limited to π-systems.

The extended Hückel approach has proved to be rather successful for such a simple theory; for example, the famous Woodward–Hoffmann rules (see section 4.9.4) were based upon calculations using this model. Extended Hückel theory has found particular application in those areas where alternative theories cannot be used. This is largely due to the fact that the basis set requires no more than experimentally determined ionisation potentials. It is particularly useful for studying systems containing metals; these systems are problematic for many other methods due to the lack of suitable basis sets.

2.13 Valence bond theories

An entirely different way to treat the electronic structure of molecules is provided by valence bond theory, which was developed at about the same time as the molecular orbital approach. However, valence bond theory was not so amenable to calculations on large molecules, and molecular orbital theory came to dominate electronic structure theory for such systems. Nevertheless, valence bond theories are often considered to be more appropriate for certain types of problems than molecular orbital theory, especially when dealing with processes that involve bonds being broken and/or formed.

Valence bond theory is usually introduced using the the famous Heitler–London model of the hydrogen molecule [Heitler and London 1927]. This model considers two non-interacting hydrogen atoms (a and b) in their ground states that are separated by a long distance. The wavefunction for this system is:

$$\Psi = \phi_{1sa}(1)\phi_{1sb}(2) \tag{2.281}$$

As the two hydrogen atoms approach to form a hydrogen molecule, such a wavefunction is inappropriate as it implies that electron 1 remains confined to orbital 1sa and electron 2 to orbital 1sb. This clearly violates the indistinguishability principle, and so a linear combination is used

$$\Psi_{vb} \propto \phi_{1sa}(1)\phi_{1sb}(2) + \phi_{1sa}(2)\phi_{1sb}(1) \tag{2.282}$$

The corresponding molecular orbital function for this system is:

$$\Psi_{\mathrm{mo}} \propto \phi_{1\mathrm{sa}}(1)\phi_{1\mathrm{sb}}(2) + \phi_{1\mathrm{sa}}(2)\phi_{1\mathrm{sb}}(1) + \phi_{1\mathrm{sa}}(1)\phi_{1\mathrm{sa}}(2) + \phi_{1\mathrm{sb}}(1)\phi_{1\mathrm{sb}}(2) \quad (2.283)$$

The additional terms in the molecular orbital wavefunction correspond to states with the two electrons in the same orbital, which endows ionic character to the bond (H^+H^-). The valence bond wavefunction does not include any ionic character and in fact it correctly describes the dissociation into two hydrogen atoms. The simple valence bond and molecular orbital pictures in equations (2.282) and (2.283) are extremes, with the 'true' wavefunction being somewhere in the middle. The valence bond representation can be improved by including a degree of ionic character as follows:

$$\Psi_{\mathrm{vb}} \propto \phi_{1\mathrm{sa}}(1)\phi_{1\mathrm{sb}}(2) + \phi_{1\mathrm{sa}}(2)\phi_{1\mathrm{sb}}(1) + \lambda[\phi_{1\mathrm{sa}}(1)\phi_{1\mathrm{sa}}(2) + \phi_{1\mathrm{sb}}(1)\phi_{1\mathrm{sb}}(2)] \quad (2.284)$$

λ is a parameter that can be varied to give the 'correct' amount of ionic character. Another way to view the valence bond picture is that the incorporation of ionic character corrects the overemphasis that the valence bond treatment places on electron correlation. The molecular orbital wavefunction underestimates electron correlation and requires methods such as configuration interaction to correct for it.

One widely used valence bond theory is the generalised valence bond (GVB) method of Goddard and co-workers [Bobrowicz and Goddard 1977]. In the simple Heitler–London treatment of the hydrogen molecule the two orbitals are the non-orthogonal atomic orbitals on the two hydrogen atoms. In the GVB theory the analogous wavefunction is written:

$$\Psi_{\mathrm{GVB}} \propto u(1)v(2) + u(2)v(1) \quad (2.285)$$

u and v are non-orthogonal orbitals that are each expressed as a basis set expansion with the coefficients being variationally optimised to minimise the energy. The construction of the wavefunction from orbitals that are not necessarily orthogonal orbitals is characteristic of many valence bond theories and complicates the computational problem. The GVB approach is particularly successful for describing the electronic nature of systems as they approach dissociation.

2.14 Calculating molecular properties using quantum mechanics

Quantum mechanics can be used to calculate a wide range of properties. In addition to thermodynamic and structural values, quantum mechanics can be used to derive properties dependent upon the electronic distribution. Such properties often cannot be determined by any other method. In the rest of this chapter we shall provide a flavour of the ways in which quantum mechanics is used in molecular modelling. Some applications, such as the location of transition structures and the use of quantum mechanics in deriving force field parameters will be discussed in the appropriate sections in later chapters.

2.14.1 *Energies, Koopman's theorem and ionisation potentials*

The energy of an electron in an orbital (equation (2.158)) is often equated with the energy required to remove the electron to give the corresponding ion. This is *Koopman's theorem*. Two important caveats must be remembered when applying Koopman's theorem and comparing the results with experimentally determined ionisation potentials. The first of these is that the orbitals in the ionised state are assumed to be the same as in the unionised state; they are 'frozen'. This neglects the fact that the orbitals in the ionised state will be different from those in the unionised state. The energy of the ionised state will thus tend to be higher than it 'should' be, giving too large an ionisation potential. The second caveat is that the Hartree–Fock method does not include the effects of electron correlation. The correction due to electron correlation would be expected to be greater for the unionised state than for the ionised state as the former has more electrons. Fortunately, therefore, the effect of electron correlation often opposes the effect of the frozen orbitals, resulting in many cases in good agreement between experimentally determined ionisation potentials and calculated values.

A Hartree–Fock SCF calculation with K basis functions provides K molecular orbitals, but many of these will not be occupied by any electrons; they are the 'virtual' spin orbitals. If we were to add an electron to one of these virtual orbitals then this should provide a means of calculating the electron affinity of the system. Electron affinities predicted by Koopman's theorem are always positive when Hartree–Fock calculations are used because the virtual orbitals always have a positive energy. However, it is observed experimentally that many neutral molecules will accept an electron to form a stable anion and so have negative electron affinities. This can be understood if one realises that electron correlation would be expected to add to the error due to the 'frozen' orbital approximation, rather than to counteract it as for ionisation potentials.

2.14.2 *Calculation of electric multipoles*

Some of the most important properties that a quantum mechanical calculation provides are the electric multipole moments of the molecule. The electric multipoles reflect the distribution of charge in a molecule. The simplest electric moment (apart from the total net charge on the molecule) is the dipole. The dipole moment of a distribution of charges q_i located as positions $\mathbf{r}_i$ is given by $\sum q_i \mathbf{r}_i$. If there are just two charges $+q$ and $-q$ separated by a distance r then the dipole moment is qr. A dipole moment of 4.8 Debye corresponds to two charges equal in magnitude to the electronic charge e separated by 1 Å. The dipole moment is a vector quantity, with components along the three Cartesian axes. The dipole moment of a molecule has contributions from both the nuclei and the electrons.

The nuclear contributions can be calculated using the formula for a system of discrete charges:

$$\boldsymbol{\mu}_{\text{nuclear}} = \sum_{A=1}^{M} Z_A \mathbf{R}_A \tag{2.286}$$

The electronic contribution arises from a continuous function of electron density and must be calculated using the appropriate operator:

$$\boldsymbol{\mu}_{\text{electronic}} = \int d\tau \Psi_0 \left(\sum_{i=1}^{N} -\mathbf{r}_i \right) \Psi_0 \tag{2.287}$$

The dipole moment operator is a sum of one-electron operators $\mathbf{r}_i$, and as such the electronic contribution to the dipole moment can be written as a sum of one-electron contributions. The electronic contribution can also be written in terms of the density matrix, **P** as follows:

$$\boldsymbol{\mu}_{\text{electronic}} = \sum_{\mu=1}^{K} \sum_{\nu=1}^{K} P_{\mu\nu} \int d\tau \phi_\mu (-\mathbf{r}) \phi_\nu \tag{2.288}$$

The electronic contribution to the dipole moment is thus determined from the density matrix and a series of one-electron integrals $\int d\tau \phi_\mu (-\mathbf{r}) \phi_\nu$. The dipole moment operator, **r**, has components in the x, y and z directions, and so these one-electron integrals are divided into their appropriate components; for example, the x component of the electronic contribution to the dipole moment would be determined using:

$$\mu_x = \sum_{\mu=1}^{K} \sum_{\nu=1}^{K} P_{\mu\nu} \int d\tau \phi_\mu (-x) \phi_\nu \tag{2.289}$$

The quadrupole is the next electric moment. A molecule has a non-zero electric quadrupole moment when there is a non-spherically symmetrical distribution of charge. A quadrupole can be considered to arise from four charges that sum to zero which are arranged so that they do not lead to a net dipole. Three such arrangements are shown in Figure 2.14. Whereas the dipole moment has components in the x, y and z directions, the quadrupole has nine components from all pairwise combinations of x and y and is represented by a 3×3 matrix as follows:

$$\Theta = \begin{pmatrix} \sum q_i x_i^2 & \sum q_i x_i y_i & \sum q_i x_i z_i \\ \sum q_i y_i x_i & \sum q_i y_i^2 & \sum q_i y_i z_i \\ \sum q_i z_i x_i & \sum q_i z_i y_i & \sum q_i z_i^2 \end{pmatrix} \tag{2.290}$$

The three moments higher than the quadrupole are the octopole, hexapole and decapole. Methane is an example of a molecule whose lowest non-zero multipole moment is the octopole. The entire set of electric moments is required to completely and exactly describe the distribution of charge in a molecule. However, the series expansion is often truncated after the dipole or quadrupole as these are often the most significant.

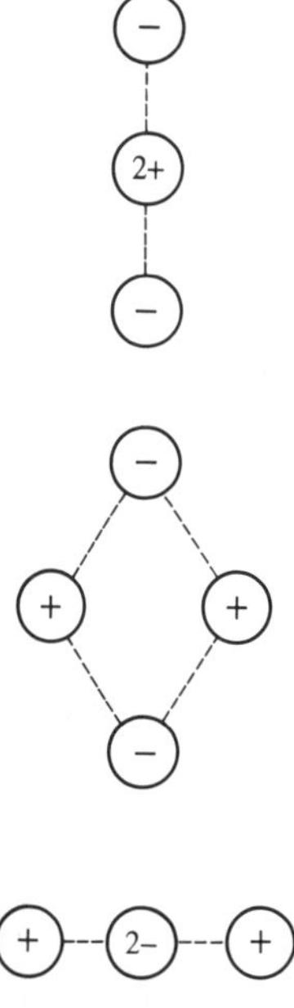

Fig. 2.14 A quadrupole moment can be obtained from various arrangements of two positive and two negative charges.

Extensive comparisons have been made of experimental and calculated dipole moments (and in some cases the higher moments, though these are difficult to determine accurately by experiment). Factors such as the basis set and electron correlation can have a significant impact on the accuracy of the results but it is found in many cases that the errors are systematic and that a simple scaling factor can be used to convert the results of a calculation with a small basis set to those obtained from experiment or with a much larger basis set. To illustrate how calculated dipole moments can vary, Table 2.5 provides the dipole moments for formaldehyde calculated at the experimental geometry using a variety of basis sets. It is also important to note that the dipole moment can be very sensitive to the geometry from which it is calculated. Thus errors in the dipole moment may reflect not only deficiencies in the dipole calculation but also the use of an inappropriate basis set to calculate the geometry.

Table 2.5 Dipole moments calculated for formaldehyde using various basis sets at the experimental geometry.

STO-3G	1.5258	3-21G	2.2903	4-31G	3.0041
6-31G*	2.7600	6-31G**	2.7576	6-311G**	2.7807
Expt.	2.34				

2.14.3 The total electron density distribution and molecular orbitals

The electron density $\rho(\mathbf{r})$ at a point $\mathbf{r}$ can be calculated from the Born interpretation of the wavefunction as a sum of squares of the spin orbitals at the point $\mathbf{r}$ for all occupied molecular orbitals. For a system of N electrons occupying $N/2$ real orbitals, we can write:

$$\rho(\mathbf{r}) = 2\sum_{\mathrm{i}=1}^{N/2} |\psi_{\mathrm{i}}(\mathbf{r})|^2 \tag{2.291}$$

If we express the molecular orbital ψ_{i} as a linear combination of basis functions, then the electron density at a point $\mathbf{r}$ is given as:

$$\begin{aligned}\rho(\mathbf{r}) &= 2\sum_{\mathrm{i}=1}^{N/2}\left(\sum_{\mu=1}^{K} c_{\mu\mathrm{i}}\phi_{\mu}(\mathbf{r})\right)\left(\sum_{\nu=1}^{K} c_{\nu\mathrm{i}}\phi_{\nu}(\mathbf{r})\right)\\ &= 2\sum_{\mathrm{i}=1}^{N/2}\sum_{\mu=1}^{K} c_{\mu\mathrm{i}}c_{\mu\mathrm{i}}\phi_{\mu}(\mathbf{r})\phi_{\mu}(\mathbf{r})\\ &\quad + 2\sum_{\mathrm{i}=1}^{N/2}\sum_{\mu=1}^{K}\sum_{\nu=\mu+1}^{K} 2c_{\mu\mathrm{i}}c_{\nu\mathrm{i}}\phi_{\mu}(\mathbf{r})\phi_{\nu}(\mathbf{r})\end{aligned} \tag{2.292}$$

Equation (2.292) can be tidied up considerably if it is written in terms of the elements of the density matrix

$$\left(P_{\mu\nu} = 2\sum_{\mathrm{i}=1}^{N/2} c_{\mu\mathrm{i}}c_{\nu\mathrm{i}}\right)$$

$$\begin{aligned}\rho(\mathbf{r}) &= \sum_{\mu=1}^{K}\sum_{\nu=1}^{K} P_{\mu\nu}\phi_{\mu}(\mathbf{r})\phi_{\nu}(\mathbf{r})\\ &= \sum_{\mu=1}^{K} P_{\mu\mu}\phi_{\mu}(\mathbf{r})\phi_{\mu}(\mathbf{r}) + 2\sum_{\mu=1}^{K}\sum_{\nu=\mu+1}^{K} P_{\mu\nu}\phi_{\mu}(\mathbf{r})\phi_{\nu}(\mathbf{r})\end{aligned} \tag{2.293}$$

The integral of $\rho(\mathbf{r})$ over all space equals the number of electrons in the system, N:

$$N = \int d\mathbf{r}\rho(\mathbf{r}) = 2\sum_{\mathrm{i}=1}^{N/2}\int d\mathbf{r}|\psi_{\mathrm{i}}(\mathbf{r})|^2 \tag{2.294}$$

If the overlap between two orbitals ϕ_{μ} and ϕ_{ν} is written $S_{\mu\nu}$ and if the basis functions are assumed to be normalised ($S_{\mu\mu} = 1$) then:

$$N = \sum_{\mu=1}^{K} P_{\mu\mu} + 2\sum_{\mu=1}^{K}\sum_{\nu=\mu+1}^{K} P_{\mu\nu}S_{\mu\nu} \tag{2.295}$$

The electron density can be visualised in several ways. One approach is to construct contours on slices through the molecule, such that each contour connects points of equal density, as shown in Figure 2.15 for formamide. The electron density can also be represented as an isometric projection (or a 'relief map', Figure 2.16), in which the height above the plane represents the magnitude of the electron density. These diagrams show that the electron density tends to be greatest near the nuclei, as would be expected. The electron density can also be represented as a solid object, whose surface connects points of equal density. The surface shown in Figure 2.17 (colour plate section) corresponds to an electron density of 0.0001 a.u. around formamide. Other properties such as the electrostatic potential can be mapped onto this surface, as we shall see in section 2.14.8.

The electron density distribution of individual molecular orbitals may also be determined and plotted. The highest occupied molecular orbital (HOMO) and the lowest unoccupied molecular orbital (LUMO) are often of particular interest as these are the orbitals most commonly involved in chemical reactions. As an illustration, the HOMO and LUMO for formamide are displayed in Figures 2.18 and 2.19 (colour plate section) as surface pictures.

2.14.4 Population analysis

Population analysis methods partition the electron density among the nuclei, so that each nucleus has a 'number' (not necessarily an integral number) of electrons associated with it. Such a partitioning provides a way to calculate the atomic charge on each nucleus. It should be noted that there is no quantum mechanical operator for the atomic charge and so any partitioning scheme must be arbitrary. Hence many methods have been devised. Here we will consider Mulliken and

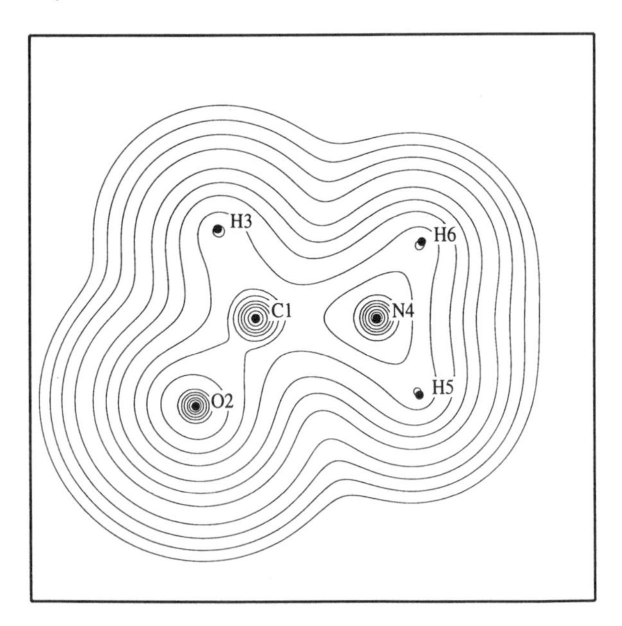

Fig. 2.15 Contour map showing the variation in electron density around formamide.

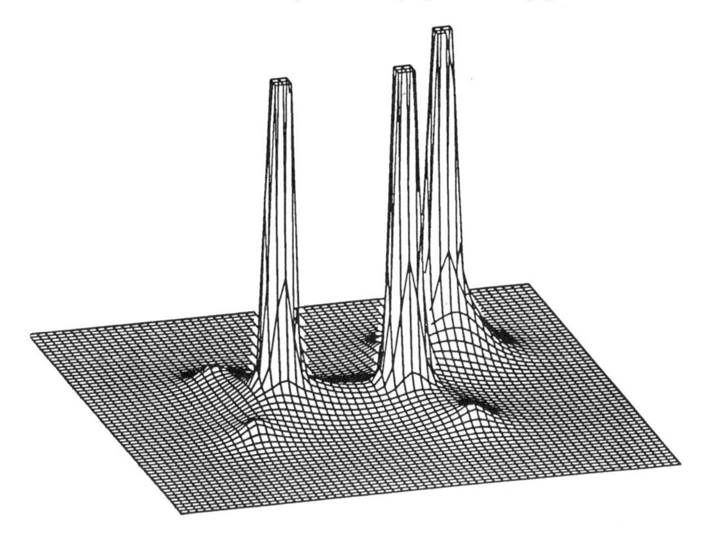

Fig. 2.16 Isometric projection of the electron density around formamide.

Löwdin analysis and Bader's theory of atoms in molecules. The alternatives include natural population analysis [Reed *et al.* 1985; Bachrach 1994]. Wiberg and Rablen have compared a number of methods for calculating atomic charges and we refer to some of their results in the following discussion [Wiberg and Rablen 1993]. To illustrate the variation that can be obtained in the results, for methane they found that the charge on the carbon atom varied from −0.473 to +0.244, depending upon the method chosen! We will also consider the problem of calculating atomic charges in more detail in Chapter 3 on molecular mechanics.

2.14.5 Mulliken and Löwdin population analysis

R S Mulliken suggested a widely used method for performing population analysis [Mulliken 1955]. The starting point is equation (2.295) which relates the total number of electrons to the density matrix and to the overlap integrals. In the Mulliken method, all of the electron density ($P_{\mu\mu}$) in an orbital is allocated to the atom on which ϕ_μ is located. The remaining electron density is associated with the overlap population, $\phi_\mu\phi_\nu$. For each element $\phi_\mu\phi_\nu$ of the density matrix, half of the density is assigned to the atom on which ϕ_μ is located and half to the atom on which ϕ_ν is located. The net charge on an atom A is then calculated by subtracting the number of electrons from the nuclear charge, Z_A:

$$q_A = Z_A - \sum_{\mu=1;\mu \text{ on A}}^{K} P_{\mu\mu} - \sum_{\mu=1;\mu \text{ on A}}^{K} \sum_{\nu=1;\nu\neq\mu}^{K} P_{\mu\nu}S_{\mu\nu} \tag{2.296}$$

Mulliken population analysis is a trivial calculation to perform once a self-consistent field has been established and the elements of the density matrix have been determined. However, there are some serious shortcomings to the method, as Mulliken himself pointed out.

A Mulliken analysis depends upon the use of a balanced basis set, in which an equivalent number of basis functions is present on each atom in the molecule. For example, it is possible to calculate a wavefunction for a molecule such as water in which all of the basis functions reside on the oxygen atom; if a large enough basis set is used then a quite reasonable wavefunction for the whole molecule can be obtained. However, the Mulliken analysis would put all of the charge on the oxygen. This is an extreme example of a general problem; p, d and f orbitals are spread quite far from the nucleus with which they are associated, and so may be very close to other atoms, yet the charge associated with electron occupation of such orbitals is assigned to the atom on which the orbital is centred. The equal apportioning of electrons between pairs of atoms, even if their electronegativities are very different, can lead in some cases to quite unrealistic values for the net atomic charge. *In extremis*, some orbitals may 'contain' a negative number of electrons and others more than two electrons, in clear contradiction of the Pauli principle. A Mulliken analysis assumes that each basis function can be associated with an atomic centre and so is not applicable if basis functions not centred on the nuclei are used. The atomic charges can be very dependent upon the basis set; for example, Wiberg and Rablen found that the charge on the central carbon in isobutene changed from +0.1 with a 6-31G* basis set to +1.0 for a 6-311++G** basis set.

In the Löwdin approach to population analysis [Löwdin 1970; Cusachs and Politzer 1968] the atomic orbitals are transformed to an orthogonal set, along with the molecular orbital coefficients. The transformed orbitals ϕ'_μ in the orthogonal set are given by:

$$\phi'_\mu = \sum_{\nu=1}^{K} (\mathbf{S}^{-1/2})_{\nu\mu} \phi_\nu \tag{2.297}$$

The electron population associated with an atom becomes:

$$q_A = Z_A - \sum_{\mu=1;\,\mu \text{ on } A}^{K} (\mathbf{S}^{1/2}\mathbf{P}\mathbf{S}^{1/2})_{\mu\mu} \tag{2.298}$$

Löwdin population analysis avoids the problem of negative populations or populations greater than two. Some quantum chemists prefer the Löwdin approach to that of Mulliken as the charges are often closer to chemically intuitive values and are less sensitive to basis set.

2.14.6 *Partitioning electron density: the theory of atoms in molecules*

R F W Bader's theory of 'atoms in molecules' [Bader 1985] provides an alternative way to partition the electrons among the atoms in a molecule. Bader's theory has been applied to many different problems, but for the purposes of our present discussion we will concentrate on its use in partitioning electron density. The Bader approach is based upon the concept of a *gradient vector path*, which is a curve around the molecule such that it is always perpendicular to the electron density contours. A set of gradient paths is drawn in Figure 2.20 for formamide. As can be seen, some of the gradient paths terminate at the atomic nuclei. Other gradient paths are attracted to points (called critical points) that are not located at the nuclei; particularly common are the bond critical points which are located between bonded atoms. Other types of critical point can occur; for example, a *ring critical point* is found in the centre of a benzene ring.

The bond critical points are points of minimum electron charge density between two bonded atoms. If we follow the contour in three-dimensional space from such a point down the gradient path along which the density decreases most rapidly then this gives a means to partition the density. This is shown in Figure 2.21 for hydrogen fluoride and in Figure 2.22 for formamide. This procedure can be performed for each bond, resulting in a three-dimensional partitioning of the electron density. The electron population that is assigned to each atom is then calculated by numerically integrating the charge density within the region surrounding that atom.

Wiberg and Rablen found that the charges obtained with the atoms in molecules method were relatively invariant to the basis set. The charges from this method were also consistent with the experimentally determined C–H bond dipoles in methane (in which the carbon is positive) and ethyne (in which the carbon is negative) unlike most of the other methods they examined.

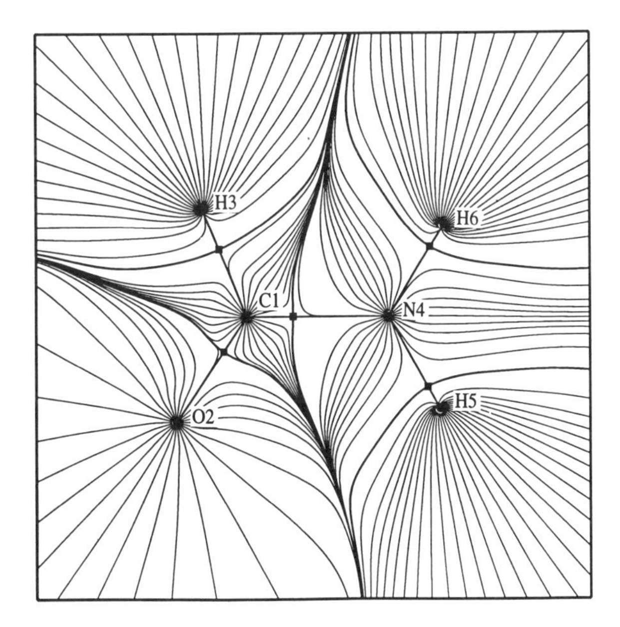

Fig. 2.20 Gradient vector paths around formamide. The paths terminate at atoms or at bond critical points (indicated with squares).

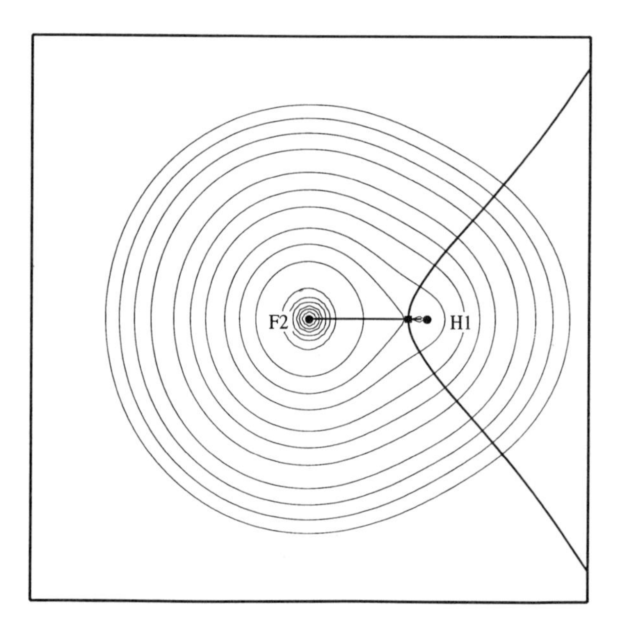

Fig. 2.21 Partitioning the electron density in hydrogen fluoride.

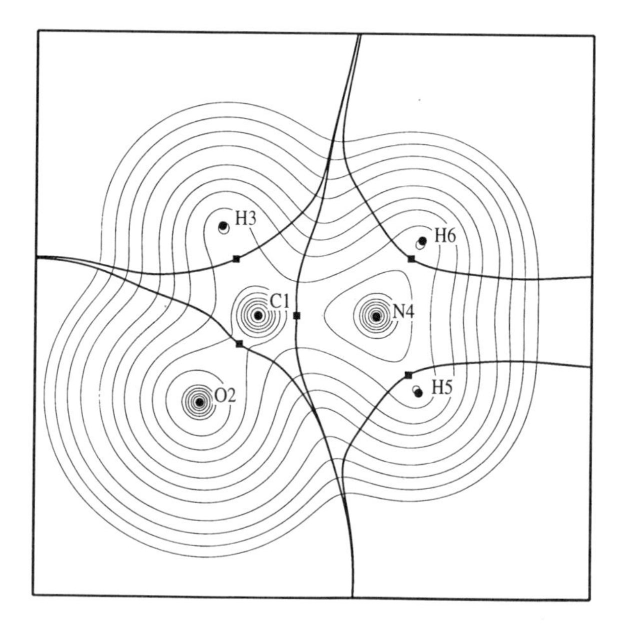

Fig. 2.22 Partitioning the electron density in formamide.

2.14.7 **Bond orders**

As with atomic charges, the bond order is not a quantum mechanical observable and so various methods have been proposed for calculating the bond orders in a molecule. K B Wiberg proposed a 'bond index' for use with an orthonormalised CNDO basis set [Wiberg 1966]. For two atoms A and B this index has the form:

$$W_{AB} = \sum_{\mu \text{ on A}} \sum_{\nu \text{ on B}} |P_{\mu\nu}|^2 \tag{2.299}$$

$P_{\mu\nu}$ is the appropriate element of the density matrix.

Mayer defined the bond order between two atoms as follows [Mayer 1983]:

$$B_{AB} = \sum_{\mu \text{ on A}} \sum_{\nu \text{ on B}} \left[(\mathbf{PS})_{\mu\nu}(\mathbf{PS})_{\nu\mu} + (\mathbf{P}^s\mathbf{S})_{\mu\nu}(\mathbf{P}^s\mathbf{S})_{\nu\mu} \right] \tag{2.300}$$

$\mathbf{P}$ is the total spinless density matrix ($\mathbf{P} = \mathbf{P}^\alpha + \mathbf{P}^\beta$) and $\mathbf{P}^s$ is the spin density matrix ($\mathbf{P}^s = \mathbf{P}^\alpha - \mathbf{P}^\beta$). For a closed-shell system Mayer's definition of the bond order reduces to:

$$B_{AB} = \sum_{\mu \text{ on A}} \sum_{\nu \text{ on B}} (\mathbf{PS})_{\mu\nu}(\mathbf{PS})_{\nu\mu} \tag{2.301}$$

Mayer's index is equivalent to Wiberg's formula if the overlap matrix $\mathbf{S}$ is a unit matrix. The bond orders obtained from Mayer's formula often seem intuitively reasonable, as illustrated in Table 2.6 for some simple molecules. The method has also been used to compute the bond orders for intermediate structures in reactions of the form $H + XH \rightarrow HX + H$ and $X + H_2 \rightarrow XH + H$ (X = F, Cl, Br). The results suggested that bond orders were a useful way to describe the similarity of the transition structure to the reactants or to the products. Moreover, the bond orders were approximately conserved along the reaction pathway.

As with methods for allocating electron density to atoms, neither the Wiberg nor the Mayer method is necessarily 'correct', though each appears to be a useful measure of the bond order that conforms to accepted pictures of bonding in molecules.

Table 2.6 Bond order obtained from the Mayer bond order scheme [Mayer 1983].

Molecule	Bond	STO-3G	4-31G
H_2	H–H	1.0	1.0
Methane	C–H	0.99	0.96
Ethene	C=C	2.01	1.96
	C–H	0.98	0.96
Ethyne	C≡C	3.00	3.27
	C–H	0.98	0.86
Water	O–H	0.95	0.80
N_2	N≡N	3.0	2.67

2.14.8 Electrostatic potentials

The electrostatic potential at a point $\mathbf{r}$, $\phi(\mathbf{r})$, is defined as the work done to bring unit positive charge from infinity to the point. The electrostatic interaction energy between a point charge q located at $\mathbf{r}$ and the molecule equals $q\phi(\mathbf{r})$. The electrostatic potential has contributions from both the nuclei and from the electrons, unlike the electron density which only reflects the electronic distribution. The electrostatic potential due to the M nuclei is:

$$\phi_{\text{nucl}}(\mathbf{r}) = \sum_{A=1}^{M} \frac{Z_A}{|\mathbf{r} - \mathbf{R}_A|} \tag{2.302}$$

The potential due to the electrons is obtained from the appropriate integral of the electron density:

$$\phi_{\text{elec}}(\mathbf{r}) = -\int \frac{d\mathbf{r}'\rho(\mathbf{r})}{|\mathbf{r}' - \mathbf{r}|} \tag{2.303}$$

The total electrostatic potential equals the sum of the nuclear and the electronic contributions:

$$\phi(\mathbf{r}) = \phi_{\text{nucl}}(\mathbf{r}) + \phi_{\text{elec}}(\mathbf{r}) \tag{2.304}$$

The electrostatic potential has proved to be particularly useful in rationalising the interactions between molecules and molecular recognition processes. This is because electrostatic forces are primarily responsible for long-range interactions between molecules. The electrostatic potential varies through space, and so it can be calculated and visualised in the same way as the electron density. Electrostatic potential contours can be used to propose where electrophilic attack might occur; electrophiles are often attracted to regions where the electrostatic potential is most negative. For example, the experimentally determined position of electrophilic attack at the nucleic acid cytosine is at N3 (Figure 2.23). This atom is next to a minimum in the electrostatic potential (also shown in Figure 2.23), as pointed out by Politzer and Murray [Politzer and Murray 1991].

Non-covalent interactions between molecules often occur at separations where the van der Waals radii of the atoms are just touching and so it is often most useful to examine the electrostatic potential in this region. For this reason, the electrostatic potential is often calculated at the molecular surface (defined in section 1.5) or the equivalent isodensity surface as shown in Figure 2.24 (colour plate section). Such pictorial representations can be used to qualitatively assess the degree of electrostatic similarity between two molecules.

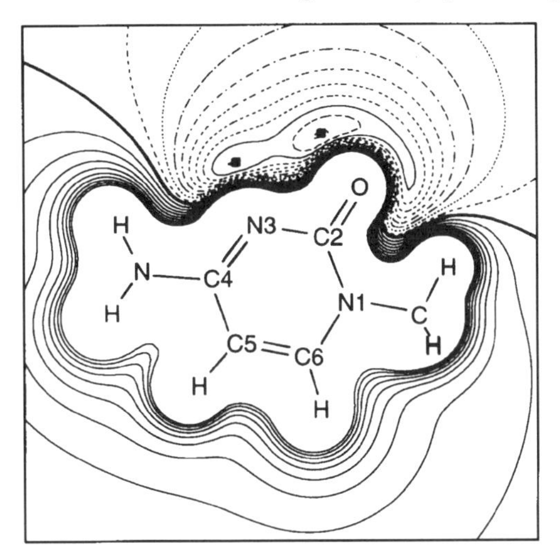

Fig. 2.23 Electrostatic potential contours around cytosine. Negative contours are dashed, the zero contour is bold. The minima near N3 and O are marked.

2.14.9 Thermodynamic and structural properties

The total energy of a system is equal to the sum of the electronic energy and the Coulombic nuclear repulsion energy:

$$E_{\text{tot}} = E_{\text{elec}} + \sum_{A=1}^{M} \sum_{B=A+1}^{M} \frac{Z_A Z_B}{R_{AB}} \tag{2.305}$$

A more useful quantity for comparison with experiment is the heat of formation, which is defined as the enthalpy change when one mole of a compound is formed from its constituent elements in their standard states. The heat of formation can thus be calculated by subtracting the heats of atomisation of the elements and the atomic ionisation energies from the total energy. Unfortunately, *ab initio* calculations that do not include electron correlation provide uniformly poor estimates of heats of formation; a single determinant wavefunction will give errors in bond dissociation energies of 25–40 kcal/mol even at the Hartree–Fock limit for diatomic molecules. Including electron correlation significantly reduces the errors. Unfortunately, the amount of time required for such calculations means that they are restricted to a small number of rather simple molecules.

When combined with an energy minimisation algorithm, quantum mechanics can be used to calculate equilibrium geometries of molecules. The results of such calculations can be compared with the structures obtained from gas-phase experiments using microwave spectroscopy, electronic spectroscopy and electron diffraction. Extensive tables listing comparisons between calculations and experiment for many molecules have been produced by several authors. Not surprisingly, the agreement between theory and experiment for *ab initio* calculations generally improves as one increases the size of the basis set and includes electron correlation. Hehre *et al.* suggest that the 3-21G basis set offers a good compromise between performance and applicability [Hehre *et al.* 1986]. It is often found that errors in structural predictions are systematic rather than random. For example, STO-3G bond lengths are generally too long whilst 6-31G* bond lengths tend to be too short. By analysing the trends in such calculations it can be possible to derive scaling factors which enable more accurate predictions to be made for each level of theory.

Quantum mechanics can be used to calculate the relative energies of conformations and the energy barriers between them. Experimental data is available for both relative stabilities and barrier heights in some cases, though this tends to be limited to relatively simple molecules. Butane is one molecule that has been investigated in great detail, with its *gauche* and *anti* conformations and the barriers that separate them. The energy difference between the *syn* and *anti* conformations of butane (Figure 2.25) was found to fall significantly with increasing basis set size, particularly when correlated levels of theory were employed [Allinger *et al.* 1990; Wiberg and Murcko 1988]. However, the smaller energy difference between the minimum energy *anti* and *gauche* conformations can be calculated quite accurately even with a relatively small basis set. Quantum mechanics calculations of the change in energy as a bond is rotated are often used to parametrise the torsional terms in molecular mechanics force fields, as will be discussed in section 3.17.

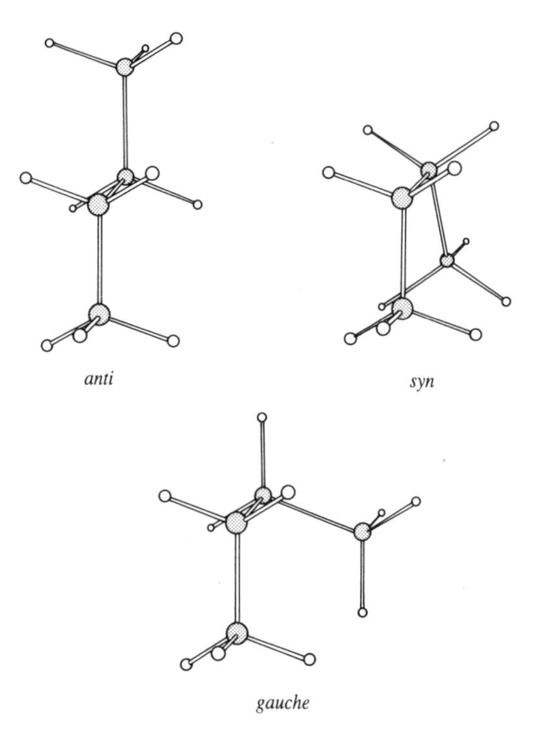

Fig. 2.25 *syn*, *anti* and *gauche* conformations of butane (C–C–C–C torsion angles 0°, 180° and ±60° respectively).

2.15 Performance of semi-empirical methods

Our discussion of the application of quantum mechanics calculations has not been explicitly directed towards any particular quantum mechanical theory, but has – implicitly at least – been written with *ab initio* methods in mind. All of the properties we have considered so far can also be determined using semi-empirical methods. Extensive tables detailing the performance of the popular semi-empirical methods have been published, both in the original papers and also in review articles, some of which are listed at the end of this chapter. The parametrisation of the semi-empirical approaches typically includes geometrical variables, dipole moments, ionisation energies and heats of formation. In Table 2.7 we provide a summary of the performance of the MINDO/3, MNDO, AM1, PM3 and SAM1 semi-empirical methods from data supplied in the original publications. The performance of successive semi-empirical methods has gradually improved from one method to another, though one should always remember that anomalous results may be obtained for certain types of system. Some of these limitations were outlined in the discussion of the various semi-empirical methods in section 2.11. It is worth emphasising that some of the major drawbacks with the semi-empirical methods arise simply because one is trying to calculate properties that were not given a major consideration in the parametrisation process. For example, many of the molecules used for the parametrisation of the MNDO, AM1 and PM3 methods had little or no conformational flexibility and so it is therefore not so surprising that some rotational barriers are not calculated with the same accuracy as (say) heats of formation. In addition, to achieve optimal performance for specific classes of molecules (e.g. the amino acids) or specific properties (e.g. conformational barriers) then it would be appropriate to include representative systems during the parametrisation procedure.

2.16 Energy component analysis

The interaction between atoms and molecules can vary from the weak attraction between a pair of closed-shell atoms (e.g. two rare gas atoms in a molecular beam) to the large energy associated with the formation of a chemical bond. Intermediate between these two extremes are interactions due to hydrogen bonding or electron donor–acceptor processes. In these intermediate cases it is often difficult to determine what factors are important in contributing to the interaction. For example, what can a hydrogen bond be ascribed to?

Morokuma analysis is a method for decomposing the energy change on formation of an intermolecular complex into five components: electrostatic, polarisation, exchange repulsion, charge transfer and mixing [Morokuma 1977]. Suppose we have performed *ab initio* SCF calculations on two molecules, X and Y, and on the intermolecular complex (or 'supermolecule') XY. The wavefunctions obtained can be written $A\Psi_X^0$, $A\Psi_Y^0$ and $A\Psi_{XY}^0$. 'A' indicates the use of an antisymmetrized wavefunction (e.g. a Slater determinant). The sum of

Table 2.7 Comparison of quantities calculated with various semi-empirical methods.

	MINDO/3	MNDO	AM1	PM3	SAM1	Reference
138 Heats of formation (kcal/mol)	11.0	6.3				Dewar and Thiel 1977b
228 Bond lengths	0.022 Å	0.014 Å				
91 Angles	5.6°	2.8°				
57 Dipole moments	0.49 D	0.38 D				
58 Heats of formation for hydrocarbons (kcal/mol)	9.7	5.87	5.07			Dewar *et al.* 1985
80 Heats of formation for species with N and/or O (kcal/mol)	11.69	6.64	5.88			
46 Dipole moments	0.54 D	0.32 D	0.26 D			
29 Ionisation energies	0.31 eV	0.39 eV	0.29 eV			
406 Heats of formation (kcal/mol)			8.82	7.12	5.21	Dewar *et al.* 1993
196 Dipole moments			0.35 D	0.40 D	0.32 D	

the energies of the isolated molecules is E_0 and the energy of the supermolecule is E_4 (we follow the original notation of Morokuma). The interaction energy ΔE is thus given by $E_4 - E_0$. The five components are calculated as follows.

The electrostatic contribution equals the interaction between the unperturbed electron distributions of the two isolated species, A and B. It is identical to the classical Coulomb interaction and equals the difference $E_1 - E_0$, where E_1 is the energy associated with the product of the two individual wavefunctions, Ψ_1:

$$\Psi_1 = A\Psi_A^0 A\Psi_B^0 \tag{2.306}$$

The electronic distributions of both X and Y will be changed by the presence of the other molecule. These polarisation effects cause a dipole to be induced in (say) molecule Y due to the charge distribution in molecule X and vice versa. Polarisation also affects the higher-order multipoles. To calculate the polarisation contribution we first calculate molecular wavefunctions Ψ_A and Ψ_B in the presence of the other molecule. The energy of the wavefunction Ψ_2 is determined as E_2, where Ψ_2 is:

$$\Psi_2 = A\Psi_X A\Psi_Y \tag{2.307}$$

The polarisation contribution equals $E_2 - E_1$ and is always attractive.

In determining Ψ_1 and Ψ_2, no electron exchange interactions are considered. The overlap between the electron distributions of X and Y at short range causes a repulsion because to bring together electrons with the same spin into the same region of space ultimately leads to a violation of the Pauli principle.

The exchange repulsion is calculated as $E_3 - E_1$, where E_3 is the energy of the wavefunction Ψ_3:

$$\Psi_3 = A(\Psi_X^0.\Psi_Y^0) \tag{2.308}$$

Ψ_3 is derived from the undistorted wavefunctions of X and Y but the exchange of electrons is permitted. The exchange term is always repulsive.

The charge transfer term arises from the transfer of charge (i.e. electrons) from occupied molecular orbitals on one molecule to unoccupied orbitals on the other molecule. This contribution is calculated as the difference between the energy of the supermolecule XY when this charge transfer is specifically allowed to occur, and an analogous calculation in which it is not.

The Morokuma formalism also requires an additional, 'mixing' or 'coupling' term to be included. This equals the difference between the total SCF difference, ΔE, and the sum of the four contributions (electrostatic, polarisation, exchange repulsion and charge transfer). The mixing term has little physical significance and is used because the four components do not completely account for the entire interaction energy (it is a fudge factor!). Fortunately, it is often relatively small.

Morokuma studied a number of hydrogen-bonded complexes using this scheme in order to assess the contribution from each component. The systems studied were typically of intermolecular complexes involving small molecules such as H_2O, HF and NH_3. In addition, Morokuma and his colleagues also examined a series of electron donor–acceptor complexes such as H_3N–BF_3, OC–BH_3, HF–ClF and benzene–$OC(CN)_2$. He also studied the basis-set dependence of the results and observed that the energy components were more sensitive than the energy differences. For example, a minimal STO-3G basis set overestimates the charge transfer contribution whereas double zeta basis sets tend to exaggerate the electrostatic interaction.

2.16.1 The water dimer

The water dimer $(H_2O)_2$ has been subject to perhaps the closest scrutiny of all hydrogen-bonded complexes. A variety of stable geometries are available to the water dimer, in which one or more hydrogen bonds are present. There has been considerable debate over the relative energies of these structures and even some dispute over which structures are actually at minimum points on the energy surface [Smith *et al.* 1990]. As might be expected, the results depend upon the basis set used. A linear geometry is observed experimentally and is also predicted to be the most stable structure by *ab initio* calculations with a wide variety of basis sets (see Figure 2.26). Using a 6-31G** basis set, Umeyama and Morokuma calculated that the −5.6 kcal/mol stabilisation energy was composed of −7.5 kcal/mol electrostatic stabilisation, 4.3 kcal/mol exchange repulsion, −0.5 kcal/mol polarisation and −1.8 kcal/mol charge transfer [Umeyama and Morokuma 1977]. The 'mixing term' contributed −0.1 kcal/mol. Thus the hydrogen bond in the water dimer was considered to arise primarily from electrostatic effects with a smaller charge transfer contribution. Morokuma and

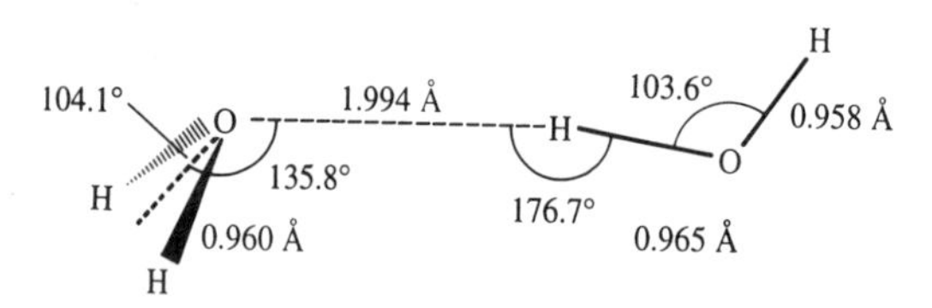

Fig. 2.26 The linear structure of the water dimer [Smith *et al.* 1990].

Umeyama also extended their analysis of charge transfer to investigate whether this was due to transfer from the proton donor to the acceptor, or from acceptor to donor. The results showed that approximately 90% of the charge transfer resulted from proton acceptor to proton donor transfer.

Morokuma analysis was widely used in the years after its introduction; it is less popular now as some problems have been encountered when trying to interpret the results with the larger basis sets that are feasible with today's faster computers and improved algorithms. In particular, when diffuse basis sets are used then there is a substantial amount of intermolecular overlap even at relatively large distances which can make it difficult to factor out the different components. Nevertheless, the approach is certainly a useful way to assess the major causes of a particular type of intermolecular interaction, if only to provide a qualitative picture.

Further reading

Atkins P W 1974. *Quanta: a Handbook of Concepts*. Oxford, Oxford University Press.

Atkins P W 1983. *Molecular Quantum Mechanics*. Oxford, Oxford University Press.

Atkins P W 1994. *Physical Chemistry*. 5th Edition. Oxford, Oxford University Press.

Clark T 1985. *A Handbook of Computational Chemistry: A Practical Guide to Chemical Structure and Energy Calculations*. New York, Wiley-Interscience.

Dewar M J S 1969. *The Molecular Orbital Theory of Organic Chemistry*. New York, McGraw-Hill.

Hehre W J, L Radom, P v R Schleyer and J A Pople 1986. *Ab Initio Molecular Orbital Theory*. New York, Wiley.

Hinchliffe A 1988. *Computational Quantum Chemistry*. Chichester, Wiley.

Hinchliffe A 1995. *Modelling Molecular Structures*. Chichester, Wiley.

Hirst D M *A Computational Approach to Chemistry*. Oxford, Blackwell Scientific.

Pople J A and D L Beveridge 1970. *Approximate Molecular Orbital Theory*. New York, McGraw-Hill.

Richards W G and D L Cooper 1983. *Ab Initio Molecular Orbital Calculations for Chemists*. 2nd Edition. Oxford, Clarendon Press.

Schaeffer H F III (Editor) 1977. *Applications of Electronic Structure Theory*. New York, Plenum Press.

Schaeffer H F III (Editor) 1977. *Methods of Electronic Structure Theory.* New York, Plenum Press.
Stewart J J P 1990. MOPAC: A Semi-Empirical Molecular Orbital Program. *Journal of Computer-Aided Molecular Design* **4**:1–45.
Stewart J J P 1990. Semi-empirical Molecular Orbital Methods. In Lipkowitz K B and D B Boyd (Editors). *Reviews in Computational Chemistry.* New York, VCH Publishers: 45–82.
Szabo A and N S Ostlund 1982. *Modern Quantum Chemistry. Introduction to Advanced Electronic Structure Theory.* New York, McGraw-Hill.
Zerner M C 1991. Semi-empirical Molecular Orbital Methods. In Lipkowitz K B, D B Boyd (Editors). *Reviews in Computational Chemistry.* Volume 2. New York, VCH Publishers: 313–366.

References

Allinger N L, R S Grev, B F Yates and H F Schaeffer III 1990. The Syn Rotational Barrier in Butane. *The Journal of the American Chemical Society* **112**:114–118.
Almlöf J, K Faegri Jr and K Korsell 1982. Principles for a Direct SCF Approach to LCAO-MO *Ab Initio* Calculations. *The Journal of Computational Chemistry* **3**:385–399.
Bachrach S M 1994. Population Analysis and Electron Densities from Quantum Mechanics. In Lipkowitz K B and D B Boyd (Editors). *Reviews in Computational Chemistry.* Volume 5. New York, VCH Publishers: 171–227.
Bader R F W 1985. Atoms in Molecules. *Accounts of Chemistry Research* **18**:9–15.
Bingham R C, M J S Dewar and D H Lo 1975a. Ground States of Molecules. XXV. MINDO/3. An Improved Version of the MINDO Semi-empirical SCF-MO Method. *The Journal of the American Chemical Society* **97**:1285–1293.
Bingham R C, M J S Dewar and D H Lo 1975b. Ground States of Molecules. XXVI. MINDO/3. Calculations for Hydrocarbons. *The Journal of the American Chemical Society* **97**:1294–1301.
Bingham R C, M J S Dewar and D H Lo 1975c. Ground States of Molecules. XXVII. MINDO/3. Calculations for CHON Species. *The Journal of the American Chemical Society* **97**:1302–1306.
Bingham R C, M J S Dewar and D H Lo 1975d. Ground States of Molecules. XXVIII. MINDO/3. Calculations for Compounds Containing Carbon, Hydrogen, Fluorine and Chlorine. *The Journal of the American Chemical Society* **97**:1307–1310.
Bobrowicz F W and W A Goddard III 1977. The Self-Consistent Field Equations for Generalized Valence Bond and Open-Shell Hartree–Fock Wave Functions. In *Modern Theoretical Chemistry III*; Schaeffer H F III (Editor). New York, Plenum: 79–127.

Boys S F 1950. Electronic Wave Functions. I. A General Method of Calculation for the Stationary States of Any Molecular System. *Proceedings of the Royal Society (London)* **A200**:542–554.

Boys S F and F Bernardi 1970. The Calculation of Small Molecular Interactions by the Differences of Separate Total Energies. Some Procedures with Reduced Errors. *Molecular Physics* **19**:553–566.

Cusachs L C and Politzer 1968. On the Problem of Defining the Charge on an Atom in a Molecule. *Chemical Physics Letters* **1**:529–531.

Dewar M J S and Thiel W 1977a. Ground States of Molecules. 38. The MNDO Method. Approximations and Parameters. *The Journal of the American Chemical Society* **99**:4899–4907.

Dewar M J S and Thiel W 1977b. Ground States of Molecules. 39. MNDO Results for Molecules Containing Hydrogen, Carbon, Nitrogen and Oxygen. *The Journal of the American Chemical Society* **99**:4907–4917.

Dewar M J S, E G Zoebisch, E F Healy and J J P Stewart 1985. AM1: A New General Purpose Quantum Mechanical Model. *Journal of the American Chemical Society* **107**:3902–3909.

Dewar M J S, C Jie and J Yu 1993. SAM1; The First of a New Series of General Purpose Quantum Mechanical Molecular Models. *Tetrahedron* **49**:5003–5038.

Dunning T H Jr 1970. Gaussian Basis Functions for Use in Molecular Calculations. I. Contraction of (9s5p) Atomic Basis Sets for First-Row Atoms. *The Journal of Chemical Physics* **53**:2823–2883.

Fowler P W 1993. Systematics of Fullerenes and Related Clusters. *Philosophical Transactions of the Royal Society (London)* **A343**:39–52.

Hall G G 1951. The Molecular Orbital Theory of Chemical Valency VIII. A Method for Calculating Ionisation Potentials. *Proceedings of the Royal Society (London)* **A205**:541–552.

Hehre W J, R F Stewart and J A Pople 1969. Self-Consistent Molecular-Orbital Methods. I. Use of Gaussian Expansions of Slater-Type Atomic Orbitals. *The Journal of Chemical Physics* **51**:2657–2664.

Heitler W and F London 1927. Wechselwirkung neutraler Atome und Homöopolare Bindung nach der Quantenmechanik. *Zeitschrift für Physik* **44**:455–472.

Hoffmann R 1963. An Extended Hückel Theory. I. Hydrocarbons. *The Journal of Chemical Physics* **39**:1397–1412.

Hückel Z 1931. Quanten theoretische Beiträge zum Benzolproblem. I. Die Electron enkonfiguration des Benzols. *Zeitschrift für Physik.* **70**:203–286.

Huzinga S 1965. Gaussian-Type Functions for Polyatomic Systems I. *The Journal of Chemical Physics* **42**:1293–1302.

Löwdin P-Q 1970. On the Orthogonality Problem. *Advances in Quantum Chemistry* **5**:185–199.

Mayer I, 1983. Charge, Bond Order and Valence in the *Ab Initio* SCF Theory. *Chemical Physics Letters* **97**:270–274.

Møller C and M S Plesset 1934. Note on an Approximate Treatment for Many-Electron Systems. *Physical Review* **46**:618–622.

Morokuma K 1977. Why Do Molecules Interact? The Origin of Electron Donor–Acceptor Complexes, Hydrogen Bonding, and Proton Affinity. *Accounts of Chemical Research* **10**:294–300.

Mulliken R S 1955. Electronic Population Analysis on LCAO-MO Molecular Wave Functions I. *The Journal of Chemical Physics* **23**:1833–1846.

Politzer P and J S Murray 1991. Molecular Electrostatic Potentials and Chemical Reactivity. In Lipkowitz K B, D B Boyd (Editors). *Reviews in Computational Chemistry.* Volume 2. New York, VCH Publishers: 273–312.

Pople J A and R K Nesbet 1954. Self-consistent orbitals for radicals. *The Journal of Chemical Physics* **22**:571–572.

Pople J A and G A Segal 1965. Approximate Self-Consistent Molecular Orbital Theory. II. Calculations with Complete Neglect of Differential Overlap. *The Journal of Chemical Physics* **43**:S136–S149.

Pople J A and G A Segal 1966. Approximate Self-Consistent Molecular Orbital Theory. III. CNDO Results for AB_2 and AB_3 systems. *The Journal of Chemical Physics* **44**:3289–3296.

Pople J A, D P Santry and G A Segal 1965. Approximate Self-Consistent Molecular Orbital Theory. I. Invariant Procedures. *The Journal of Chemical Physics* **43**:S129–S135.

Pople J A, D L Beveridge and P A Dobosh 1967. Approximate Self-Consistent Molecular–Orbital Theory. V. Intermediate Neglect of Differential Overlap. *The Journal of Chemical Physics* **47**:2026–2033.

Pulay P 1980. Convergence Acceleration of Iterative Sequences. The Case of SCF Iteration. *Chemical Physics Letters* **73**:393–398.

Reed A E, R B Weinstock and F Weinhold 1985. Natural Population Analysis. *The Journal of Chemical Physics* **83**:735–746.

Roos B O, P R Taylor and E M Siegbahm 1980. A complete active space SCF method (CASSCF) using a density matrix formulated super-CI approach. *Chemical Physics* **48**:157–173.

Roothaan C C J 1951. New Developments in Molecular Orbital Theory. *Reviews of Modern Physics* **23**:69–89.

Slater J C 1930. Atomic Shielding Constants. *Physical Review* **36**:57–64.

Smith B J, D J Swanton, J A Pople, H F Schaeffer III and L Radom 1990. Transition Structures for the Interchange of Hydrogen Atoms within the Water Dimer. *The Journal of Chemical Physics* **92**:1240–1247.

Stewart J J P 1989a. Optimisation of Parameters for Semi-empirical Methods I. Method. *The Journal of Computational Chemistry* **10**:209–220.

Stewart J J P 1989b. Optimisation of Parameters for Semi-empirical Methods II. Applications. *The Journal of Computational Chemistry* **10**:221– 264.

Thiel W and A A Voityuk 1994. Extension of MNDO to d Orbitals: Parameters and Results for Silicon. *Journal of Molecular Structure (Theochem)* **313**:141–154.

Umeyama H and K Morokuma 1977. The Origin of Hydrogen Bonding. An Energy Decomposition Study. *The Journal of the American Chemical Society* **99**:1316–1332.

Wiberg K B 1966. Application of the Pople–Santry–Segal CNDO Method to the

Cyclopropylcarbinyl and Cyclobutyl Cation and to Bicyclobutane. *Tetrahedron* **24**:1083–1096.

Wiberg K B and M A Murcko 1988. Rotational Barriers. 2. Energies of Alkane Rotamers. An Examination of Gauche Interactions. *The Journal of the American Chemical Society* **110**:8029–8038.

Wiberg K B and P R Rablen 1993. Comparison of Atomic Charges Derived via Different Procedures. *Journal of Computational Chemistry* **14**:1504–1518.

3 Empirical force field models: molecular mechanics

3.1 Introduction

Many of the problems that we would like to tackle in molecular modelling are unfortunately too large to be considered by quantum mechanical methods. Quantum mechanics deals with the electrons in a system, so that even if some of the electrons are ignored (as in the semi-empirical schemes) a large number of particles must still be considered and the calculations are time-consuming. Force field methods (also known as molecular mechanics) ignore the electronic motions and calculate the energy of a system as a function of the nuclear positions only. Molecular mechanics is thus invariably used to perform calculations on systems containing significant numbers of atoms. In some cases force fields can provide answers that are as accurate as even the highest level quantum mechanical calculations, in a fraction of the computer time. Molecular mechanics cannot of course provide properties that depend upon the electronic distribution in a molecule.

That molecular mechanics works at all is due to the validity of several assumptions. The first of these is the Born–Oppenheimer approximation, without which it would be impossible to contemplate writing the energy as a function of the nuclear coordinates at all. Molecular mechanics is based upon a rather simple model of the interactions within a system with contributions from processes such as the stretching of bonds, the opening and closing of angles and the rotations about single bonds. Even when simple functions (e.g. Hooke's law) are used to describe these contributions the force field can perform quite acceptably. Transferability is a key attribute of a force field, for it enables a set of parameters developed and tested on a relatively small number of cases to be applied to a much wider range of problems. Moreover, parameters developed from data on small molecules can be used to study much larger molecules such as polymers.

3.1.1 A simple molecular mechanics force field

Many of the molecular modelling force fields in use today can be interpreted in terms of a relatively simple four-component picture of the intra- and intermolecular forces within the system. Energetic penalties are associated with the deviation of bonds and angles away from their 'reference' or 'equilibrium' values, there is a function that describes how the energy changes as bonds are rotated, and finally the force field contains terms that describe the interaction between non-bonded parts of the system. More sophisticated force fields may have additional terms, but invariably contain these four components. An attractive feature of this representation is that the various terms can be ascribed to changes in specific internal coordinates such as bond lengths, angles, the rotation of bonds or movements of atoms relative to each other. This makes it easier to understand how changes in the force field parameters affect its performance, and also helps in the parametrisation process. One functional form for such a force field that can be used to model single molecules or assemblies of atoms and/or molecules is:

$$\begin{aligned}\mathscr{V}(\mathbf{r}^N) &= \sum_{\text{bonds}} \frac{k_i}{2}(l_i - l_{i,0})^2 + \sum_{\text{angles}} \frac{k_i}{2}(\theta_i - \theta_{i,0})^2 \\ &+ \sum_{\text{torsions}} \frac{V_n}{2}(1 + \cos(n\omega - \gamma)) \\ &+ \sum_{i=1}^{N} \sum_{j=i+1}^{N} \left(4\varepsilon_{ij} \left[\left(\frac{\sigma_{ij}}{r_{ij}} \right)^{12} - \left(\frac{\sigma_{ij}}{r_{ij}} \right)^{6} \right] + \frac{q_i q_j}{4\pi\varepsilon_0 r_{ij}} \right)\end{aligned} \tag{3.1}$$

$\mathscr{V}(\mathbf{r}^N)$ denotes the potential energy which is a function of the positions ($\mathbf{r}$) of N particles (usually atoms). The various contributions are schematically represented in Figure 3.1. The first term in equation (3.1) models the interaction between pairs of bonded atoms, modelled here by a harmonic potential that gives the increase in energy as the bond length l_i deviates from the reference value $l_{i,0}$. The second term is a summation over all valence angles in the molecule, again modelled using a harmonic potential (a valence angle is the angle formed between three atoms A–B–C in which A and C are both bonded to B). The third term in equation (3.1) is a torsional potential that models how the energy changes as a bond rotates. The fourth contribution is the non-bonded term. This is calculated between all pairs of atoms (i and j) that are in different molecules or that are in the same molecule but separated by at least three bonds (i.e. have a 1, n relationship where $n \geq 4$). In a simple force field the non-bonded term is usually modelled using a Coulomb potential term for electrostatic interactions and a Lennard–Jones potential for van der Waals interactions.

We shall discuss the nature of these different contributions in more detail in sections 3.3–3.9, but here we consider how the simple force field of equation (3.1) would be used to calculate the energy of a conformation of propane (Figure 3.2). Propane has ten bonds; two C–C bonds and eight C–H bonds. The C–C bonds are symmetrically equivalent but the C–H bonds fall into two classes; one group corresponding to the two hydrogens bonded to the central methylene (CH_2) carbon, and one group corresponding to the six hydrogens bonded to the methyl carbons. In some sophisticated force fields different parameters would be used for

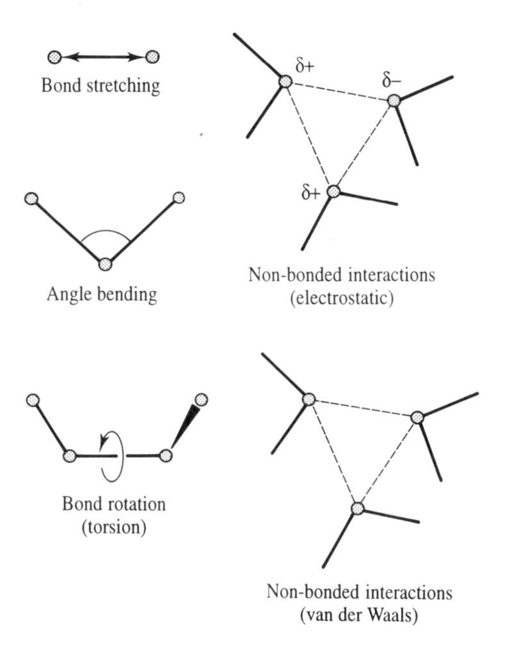

Figure 3.1 Schematic representation of the four key contributions to a molecular mechanics force field: bond stretching, angle bending, torsional terms and non-bonded interactions.

these two different types of C–H bond, but in most force fields the same bonding parameters (i.e. k_i and $l_{i,0}$) would be used for each of the eight C–H bonds. This is an example of the way in which the same parameters can be used for a wide variety of molecules. There are 18 different valence angles in propane, comprising one C–C–C angle, 10 C–C–H angles, and 7 H–C–H angles. Note that all angles are included in the force field model even though some of them may not be independent of the others. There are 18 torsional terms: 12 H–C–C–H torsions and 6 H–C–C–C torsions. Each of these is modelled with a cosine series expansion that has minima at the *trans* and *gauche* conformations. Finally, there are 27 non-bonded terms to calculate, comprising 21 H–H interactions and 6 H–C interactions. The electrostatic contribution would be calculated using Coulomb's law from partial atomic charges associated with each atom and the van der Waals contribution as a Lennard–Jones potential with appropriate ε_{ij} and σ_{ij} parameters. A sizeable number of terms are thus included in the force field model, even for a molecule as simple as propane. Even so, the number of terms (73) is many fewer than the number of integrals that would be involved in an equivalent *ab initio* quantum mechanical calculation.

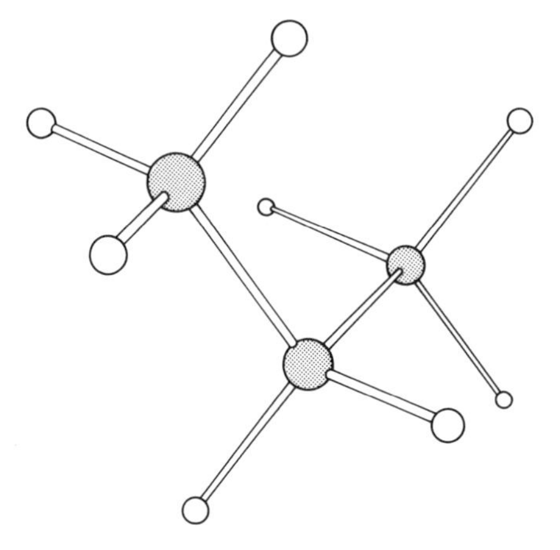

Figure 3.2 A typical force field model for propane contains 10 bond stretching terms, 18 angle bending terms, 18 torsional terms and 27 non-bonded interactions.

3.2 Some general features of molecular mechanics force fields

To define a force field one must specify not only the functional form but also the parameters (i.e. the various constants such as k_i, V_n and σ_{ij} in equation (3.1)); two force fields may use an identical functional form yet have very different parameters. Moreover, force fields with the same functional form but different parameters, and force fields with different functional forms, may give results of comparable accuracy. A force field should be considered as a single entity; it is not strictly correct to divide the energy into its individual components, let alone to take some of the parameters from one force field and mix them with parameters from another force field. Nevertheless, some of the terms in a force field are sufficiently independent of the others (particularly the bond and angle terms) to make this an acceptable approximation in certain cases.

The force fields used in molecular modelling are primarily designed to reproduce structural properties but they can also be used to predict other properties, such as molecular spectra. However, molecular mechanics force fields can rarely predict spectra with great accuracy (although the more recent molecular mechanics force fields are much better in this regard). A force field is generally designed to predict certain properties and will be parametrised accordingly. While it is useful to try to predict other quantities which have not been included in the parametrisation process it is not necessarily a failing if a force field is unable to do so.

Transferability of the functional form and parameters is an important feature of a force field. Transferability means that the same set of parameters can be used to model a series of related molecules, rather than having to define a new set of parameters for each individual molecule. For example, we would expect to be able to use the same set of parameters for all *n*-alkanes. Transferability is clearly important if we want to use the force field to make predictions. Only for some small systems where particularly accurate work is required may it be desirable to develop a model specific to that molecule.

One important point that we should bear in mind as we undertake a deeper analysis of molecular mechanics is that force fields are *empirical*; there is no 'correct' form for a force field. Of course, if one functional form is shown to perform better than another it is likely that form will be favoured. Most of the force fields in common use do have a very similar form, and it is tempting to assume that this must therefore be the optimal functional form. Certainly such models tend to conform to a useful picture of the interactions present in a system, but it should always be borne in mind that there may be better forms, particularly when developing a force field for new classes of molecules. The functional forms employed in molecular mechanics force fields are often a compromise between accuracy and computational efficiency; the most accurate functional form may often be unsatisfactory for efficient computation. As the performance of computers increases so it becomes possible to incorporate more sophisticated models. An additional consideration is that in order to use techniques such as energy minimisation and molecular dynamics, it is usually desirable to be able to calculate the first and second derivatives of the energy with respect to the atomic coordinates.

A concept that is common to most force fields is that of an *atom type*. When preparing the input for a quantum mechanics calculation it is usually necessary to specify the atomic numbers of the nuclei present, together with the geometry of the system and the overall charge and spin multiplicity. For a force field the overall charge and spin multiplicity are not explicitly required, but it is usually necessary to assign an atom type to each atom in the system. The atom type is more than just the atomic number of an atom; it usually contains information about its hybridisation state and sometimes the local environment. For example, it is necessary in most force fields to distinguish between sp^3-hybridised carbon atoms (which adopt a tetrahedral geometry), sp^2-hybridised carbons (which are trigonal) and sp-hybridised carbons (which are linear). Each force field parameter is expressed in terms of these atom types, so that the reference angle θ_0 for a tetrahedral carbon atom would be near 109.5° and that for a trigonal carbon would be near 120°. The atom types in some force fields reflect the neighbouring environment as well as the hybridisation and can be quite extensive for some atoms. For example, the MM2/MM3 force fields of Allinger and co-workers that are widely used for calculations on 'small' molecules [Allinger 1977; Allinger *et al.* 1989] distinguish the following types of carbon atom: sp^3, sp^2, sp, carbonyl, cyclopropane, radical, cyclopropene and carbonium ion. In the AMBER force field of Kollman and co-workers [Weiner *et al.* 1984; Cornell *et al.* 1995] the carbon atom at the junction between a six and a five-membered ring (e.g. in the amino acid tryptophan) is assigned an atom type that is different from the carbon atom in an isolated five-membered ring such as histidine, which in turn is different from the atom type of a carbon atom in a benzene ring. Indeed, the AMBER force field uses different atom types for a histidine amino acid depending upon its protonation state (Figure 3.3). Other, more general force fields would assign these atoms to the same generic 'sp^2 carbon' atom type. It is often found that force fields which are designed for modelling specific classes of molecules (such as proteins and nucleic acids, in the case of AMBER) use more specific atom types than force fields designed for general purpose use.

We now discuss in some detail the individual contributions to a molecular mechanics force field, giving a selection of the various functional forms that are in common use. We shall then consider the important task of parametrisation, in which values for the many force constants are derived. Our discussion will be illuminated by examples chosen from contemporary force fields in widespread use and the MM2/MM3 and AMBER force fields in particular.

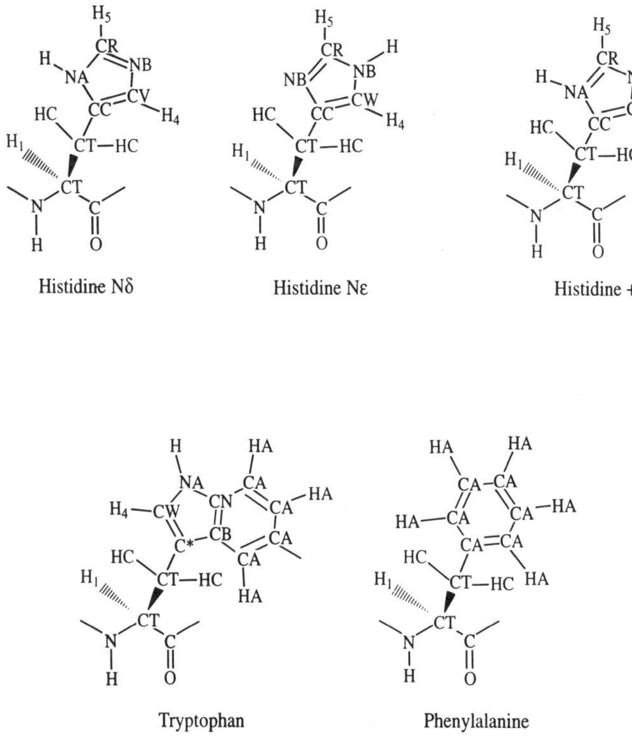

Histidine Nδ

Histidine Nε

Histidine +

Tryptophan

Phenylalanine

Figure 3.3 AMBER atom types for the amino acids histidine, tryptophan and phenylalanine. There are three possible protonation states of histidine.

3.3 Bond stretching

The potential energy curve for a typical bond has the form shown in Figure 3.4. Of the many functional forms used to model this curve, that suggested by Morse is particularly useful. The Morse potential has the form:

$$\mathscr{V}(l) = D_e\{1 - \exp[-a(l - l_0)]\}^2 \tag{3.2}$$

D_e is the depth of the potential energy minimum and $a = \omega\sqrt{(\mu/2D_e)}$ where μ is the reduced mass and ω is the frequency of the bond vibration. ω is related to the stretching constant of the bond, k, by $\omega = \sqrt{(k/\mu)}$. l_0 is the reference value of the bond. The Morse potential is not usually used in molecular mechanics force fields. In part this is because it is not particularly amenable to efficient computation but also because it requires three parameters to be specified for each bond. Moreover it is rare in molecular mechanics calculations for bonds to deviate significantly from their equilibrium values; the Morse curve describes a wide range of behaviour from the strong equilibrium behaviour to dissociation. Consequently, simpler expressions are often used. The most elementary approach is to use a Hooke's law formula in which the energy varies with the square of the displacement from the reference bond length l_0:

$$\mathscr{V}(l) = \frac{k}{2}(l - l_0)^2 \tag{3.3}$$

The astute reader will have noticed our use of the term 'reference bond length' (sometimes called the 'natural bond length') for the parameter l_0. This parameter is commonly called the 'equilibrium' bond length, but to do so can be misleading. The reference bond length is the value that the bond adopts when all other terms in the force field are set to zero. The equilibrium bond length, by contrast, is the value that is adopted in a minimum energy structure, when all other terms in the force field contribute. The complex interplay between the various components in the force field means that the bond may well deviate slightly from its reference value, in order to compensate for other contributions to the energy. It is also

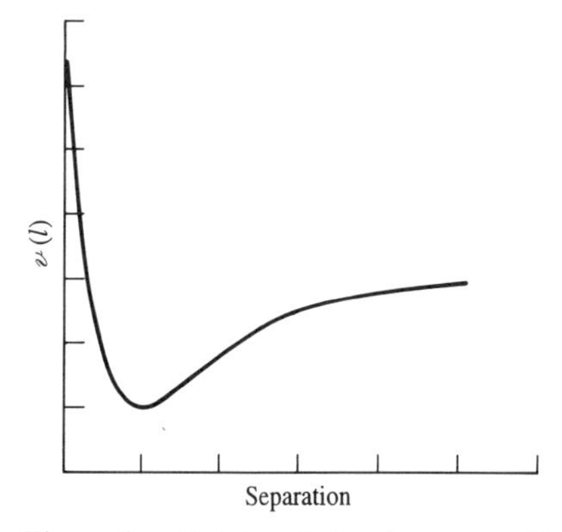

Figure 3.4 Variation in bond energy with interatomic separation.

important to recognise that 'real' molecules undergo vibrational motion (even at absolute zero, there is a zero-point energy due to vibrational motion. A true bond stretching potential is not harmonic but has a shape similar to that in Figure 3.4, which means that the 'average' length of the bond in a vibrating molecule will deviate from the equilibrium value for the hypothetical motionless state. The effects are usually small, but they are significant if one wishes to predict bond lengths to thousandths of an angstrom. When comparing the results of calculations with experimental data, one must also remember that different experimental techniques measure different 'equilibrium' values, especially when the experiments are performed at different temperatures. The errors in experimentally determined bond lengths can be quite large; for example, libration of a molecule in a crystal means that the bond lengths determined by X-ray methods at room temperature may have errors as large as 0.015 Å. MM2 is parametrised to fit the values obtained by electron diffraction, which give the mean distances between atoms averaged over the vibrational motion at room temperature.

The forces between bonded atoms are very strong and considerable energy is required to cause a bond to deviate significantly from its equilibrium value. This is reflected in the magnitude of the force constants for bond stretching; some typical values from the MM2 force field are shown in Table 3.1, where it can be seen that those bonds one would intuitively expect to be stronger have large force constants (contrast C–C with C=C and N≡N). A deviation of just 0.2 Å from the reference value l_0 with a force constant of 300 kcal mol^{-1} Å^{-2} would cause the energy of the system to rise by 12 kcal/mol.

The Hooke's law functional form is a reasonable approximation to the shape of the potential energy curve at the bottom of the potential well, at distances that correspond to bonding in ground-state molecules. It is less accurate away from equilibrium (Figure 3.5). To more accurately model the Morse curve, cubic and higher terms can be included and the bond stretching potential can be written as follows:

$$\mathcal{V}(l) = \frac{k}{2}(l - l_0)^2[1 - k'(l - l_0) - k''(l - l_0)^2 - k'''(l - l_0)^3 \ldots] \qquad (3.4)$$

An undesirable side-effect of an expansion that includes just a quadratic and a cubic term (as is employed in MM2) is that, far from the reference value, the

Table 3.1 Force constants and reference bond lengths for selected bonds [Allinger 1977].

Bond	l_0 (Å)	k (kcal mol^{-1} Å^{-2})
Csp^3–Csp^3	1.523	317
Csp^3–Csp^2	1.497	317
Csp^2=Csp^2	1.337	690
Csp^2=O	1.208	777
Csp^3–Nsp^3	1.438	367
C–N (amide)	1.345	719

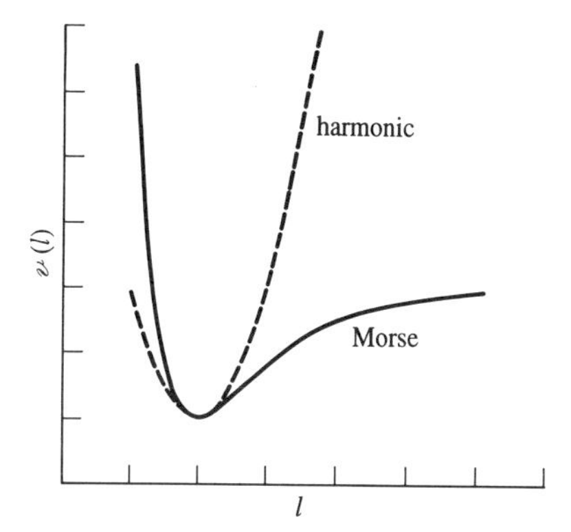

Figure 3.5 Comparison of the simple harmonic potential (Hooke's law) with the Morse curve.

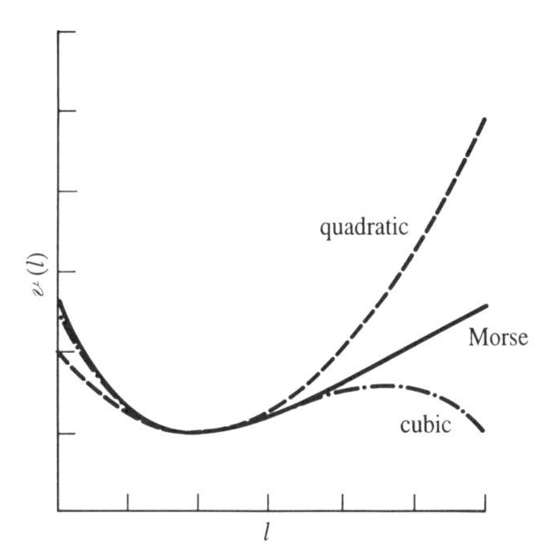

Figure 3.6 A cubic bond stretching potential passes through a maximum but gives a better approximation to the Morse curve close to the equilibrium structure than the quadratic form.

cubic function passes through a maximum. This can lead to a catastrophic lengthening of bonds (Figure 3.6). One way to accommodate this problem is to use the cubic contribution only when the structure is sufficiently close to its equilibrium geometry and is well inside the 'true' potential well. MM3 also includes a quartic term; this eliminates the inversion problem and leads to an even better description of the Morse curve.

3.4 Angle bending

The deviation of angles from their reference values is also frequently described using a Hooke's law or harmonic potential:

$$\mathscr{v}(\theta) = \frac{k}{2}(\theta - \theta_0)^2 \tag{3.5}$$

The contribution of each angle is characterised by a force constant and a reference value. Rather less energy is required to distort an angle away from equilibrium than to stretch or compress a bond, and the force constants are proportionately smaller, as can be observed in Table 3.2.

As with the bond stretching terms, the accuracy of the force field can be improved by the incorporation of higher order terms. MM2 contains a quartic term in addition to the quadratic term. Higher order terms have also been included to treat certain pathological cases such as very highly strained molecules. The general form of the angle bending term then becomes:

$$\mathscr{v}(\theta) = \frac{k}{2}(\theta - \theta_0)^2[1 - k'(\theta - \theta_0) - k''(\theta - \theta_0)^2 - k'''(\theta - \theta_0)^3 \ldots] \tag{3.6}$$

3.5 Torsional terms

The bond stretching and angle bending terms are often regarded as 'hard' degrees of freedom, in that quite substantial energies are required to cause significant deformations from their reference values. Most of the variation in structure and relative energies is due to the complex interplay between the torsional and non-bonded contributions.

The existence of barriers to rotation about chemical bonds is fundamental to understanding the structural properties of molecules and conformational analysis. The three minimum energy staggered conformations and three maximum energy eclipsed structures of ethane are a classic example of the way in which the energy changes with a bond rotation. Quantum mechanical calculations suggest that this

Table 3.2 Force constants and reference angles for selected angles [Allinger 1977].

Angle	θ_0	k (kcal mol^{-1}deg^{-1})
Csp^3–Csp^3–Csp^3	109.47	0.0099
Csp^3–Csp^3–H	109.47	0.0079
H–Csp^3–H	109.47	0.0070
Csp^3–Csp^2–Csp^3	117.2	0.0099
Csp^3–Csp^2=Csp^2	121.4	0.0121
Csp^3–Csp^2=O	122.5	0.0101

barrier to rotation can be considered to arise from antibonding interactions between the hydrogen atoms on opposite ends of the molecule; the antibonding interactions are minimised when the conformation is staggered and are at a maximum when the conformation is eclipsed. Many force fields are used for modelling flexible molecules where the major changes in conformation are due to rotations about bonds; in order to simulate this it is essential that the force field properly represents the energy profiles of such changes.

Not all molecular mechanics force fields use torsional potentials; it may be possible to rely upon non-bonded interactions between the atoms at the end of each torsion angle (the 1,4 atoms) to achieve the desired energy profile. However, most force fields for 'organic' molecules do use explicit torsional potentials with a contribution from each bonded quartet of atoms A–B–C–D in the system. Thus there would be nine individual torsional terms for ethane and 24 for benzene (6 × C–C–C–C, 12 × C–C–C–H and 6 × H–C–C–H). Torsional potentials are almost always expressed as a cosine series expansion. One functional form is:

$$\mathcal{V}(\omega) = \sum_{n=0}^{N} \frac{V_n}{2}[1 + \cos(n\omega - \gamma)] \tag{3.7}$$

ω is the torsion angle.

An alternative but equivalent expression is:

$$\mathcal{V}(\omega) = \sum_{n=0}^{N} C_n \cos(\omega)^n \tag{3.8}$$

V_n in equation (3.7) is often referred to as the 'barrier' height, but to do so is misleading, obviously so when more than one term is present in the expansion. Moreover, other terms in the force field equation contribute to the barrier height as a bond is rotated, especially the non-bonded interactions between the 1,4 atoms. The value of V_n does, however, give a qualitative indication of the relative barriers to rotation; for example, V_n for an amide bond will be larger than for a bond between two sp^3 carbon atoms. n in equation (3.7) is the *multiplicity*; its value gives the number of minimum points in the function as the bond is rotated through 360°. γ (the phase factor) determines where the torsion angle passes through its minimum value. For example, the energy profile for rotation about the single bond between two sp^3 carbon atoms could be represented by a single torsional term with $n = 3$ and $\gamma = 0°$. This would give a threefold rotational profile with minima at torsion angles of +60°, −60° and 180° and maxima at ±120° and 0°. A double bond between two sp^2 carbon atoms would have $n = 2$ and $\gamma = 180°$, giving minima at 0° and 180°. The value of V_n would also be significantly larger for the double bond than for the single bond. The effects of varying V_n, n and γ are illustrated in Figure 3.7 for commonly occurring torsional potentials.

Many of the torsional terms in the AMBER force field contain just one term from the cosine series expansion, but for some bonds it was found necessary to include more than one term. For example, to correctly model the tendency of

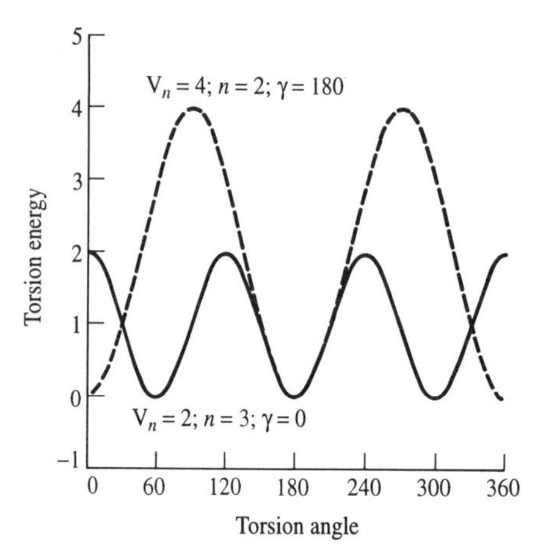

Figure 3.7 Torsional potential varies as shown for different values of V_n, n and γ.

O–C–C–O bonds to adopt a *gauche* conformation, a torsional potential with two terms was used for the O–C–C–O contribution:

$$\mathscr{V}(\omega_{\text{C-O-O-C}}) = 0.25(1 + \cos 3\omega) + 0.25(1 + \cos 2\omega) \tag{3.9}$$

The torsional energy for a OCH_2–CH_2O fragment (found in the sugars in DNA) varies with the torsion angle ω as shown in Figure 3.8. Another feature of the AMBER force field is its use of general torsional parameters. The energy profile for rotation about a bond that is described by a general torsional potential depends solely upon the atom types of the two atoms that comprise the central bond, and not upon the atom types of the terminal atoms. For example, all torsion angles in which the central bond is between two sp^3-hybridised carbon atoms (e.g. H–C–C–H, C–C–C–C, H–C–C–C) are assigned the same torsional parameters, unless the torsion is a special case such as O–C–C–O. In its treatment of the torsional contribution, AMBER takes a position intermediate between those force fields which only ever use a single term in the torsional expansion and those which consistently use more terms for all torsions. MM2 falls into the latter category; it uses three terms in the expansion:

$$\mathscr{V}(\omega) = \frac{V_1}{2}(1 + \cos \omega) + \frac{V_2}{2}(1 - \cos 2\omega) + \frac{V_3}{2}(1 + \cos 3\omega) \tag{3.10}$$

A physical interpretation has been ascribed to each of the three terms in the MM2 torsional expansion from an analysis of *ab initio* calculations on simple fluorinated hydrocarbons. The first, onefold term corresponds to interactions between bond dipoles, which are due to differences in electronegativity between bonded atoms. The twofold term is due to the effects of hyperconjugation (in alkanes) and conjugation effects (in alkenes) which provide 'double bond' char-

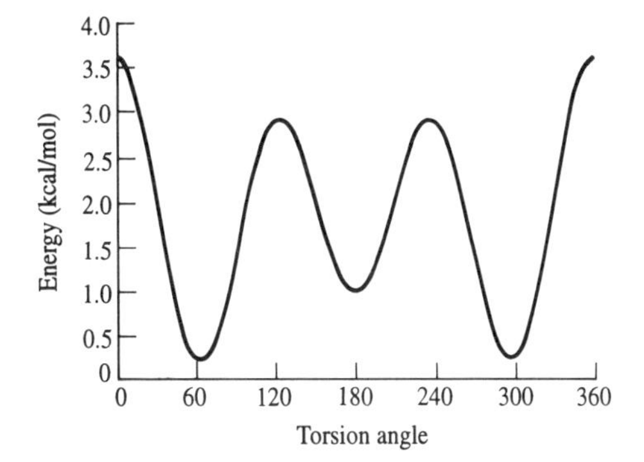

Figure 3.8 Variation in torsional energy (AMBER force field) with O–C–C–O torsion angle (ω) for OCH_2–CH_2O fragment. The minimum energy conformations arise for $\omega = 60°$ and $300°$.

acter to the bond. The threefold term corresponds to steric interactions between the 1,4 atoms. It was found that the additional terms in the torsional potential were especially important for systems containing heteroatoms, such as the halogenated hydrocarbons and molecules containing CCOC and CCNC fragments.

With careful parametrisation a force field which uses more than one term in the torsional expansion will be more successful than a force field that only uses a single term (and this is borne out by the MM2 force field). The major drawback is that many parameters are required to model even a modest range of molecules.

3.6 Improper torsions and out-of-plane bending motions

Let us consider how cyclobutanone would be modelled using a force field containing just standard bond stretching and angle bending terms of the type in equation (3.1). The equilibrium structure obtained with such a force field would have the oxygen atom located out of the plane formed by the adjoining carbon atom and the two carbon atoms bonded to it, as shown in Figure 3.9. In this structure, the angles to the oxygen adopt values close to the reference value of 120°. Experimentally, it is found that the oxygen atom remains in the plane of the cyclobutane ring, even though the C–C=O angles are large (133°). This is because the π-bonding energy, which is maximised in the coplanar arrangement, would be much reduced if the oxygen were bent out of the plane. To achieve the desired geometry it is necessary to incorporate an additional term (or terms) in the force field that keeps the sp^2 carbon and the three atoms bonded to it in the same plane. The simplest way to achieve this is to use an *out-of-plane* bending term.

There are several ways in which out-of-plane bending terms can be incorporated into a force field. One approach is to treat the four atoms as an 'improper' torsion angle (i.e. a torsion angle in which the four atoms are not

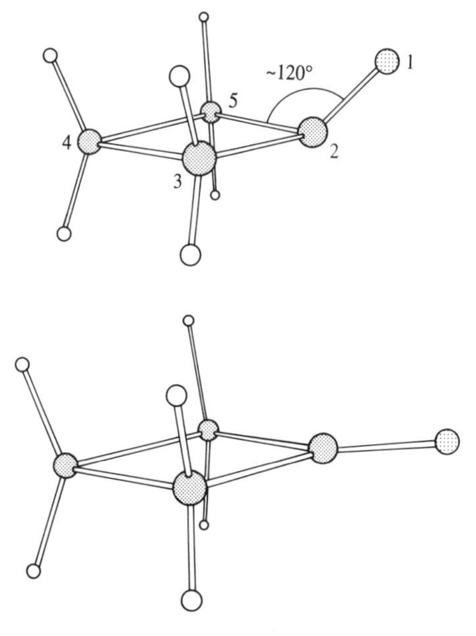

Figure 3.9 Without an out-of-plane term, the oxygen atom in cyclobutane is predicted to lie out of the plane of the ring (top) rather than in the plane.

bonded in the sequence 1–2–3–4). One way to define an improper torsion for cyclobutane would involve the atoms 1–5–3–2 in Figure 3.7. A torsional potential of the following form is then used to maintain the improper torsion angle at 0° or 180°:

$$\mathcal{V}(\omega) = k(1 - \cos 2\omega) \tag{3.11}$$

Various other ways to incorporate the out-of-plane bending contribution are possible. For example, one definition that is closer to an 'out of plane bend' involves a calculation of the angle between a bond from the central atom and the plane defined by the central atom and the other two atoms, Figure 3.10. A value of 0° corresponds to all four atoms being coplanar. A third approach is to calculate the height of the central atom above a plane defined by the other three atoms, Figure 3.10. With these two definitions the deviation of the out-of-plane coordinate (be it an angle or a distance) can be modelled using a harmonic potential of the form

$$\mathcal{V}(\theta) = \frac{k}{2}\theta^2; \qquad \mathcal{V}(h) = \frac{k}{2}h^2 \tag{3.12}$$

Of these three functional forms, the improper torsion definition is most widely used as it can then be easily included with the 'proper' torsional terms in the force field. However, it has been suggested that the other two functional forms are better ways to implement out-of-plane bending in the force field. Out-of-plane terms may also be used to achieve a particular geometry. For example, if it is

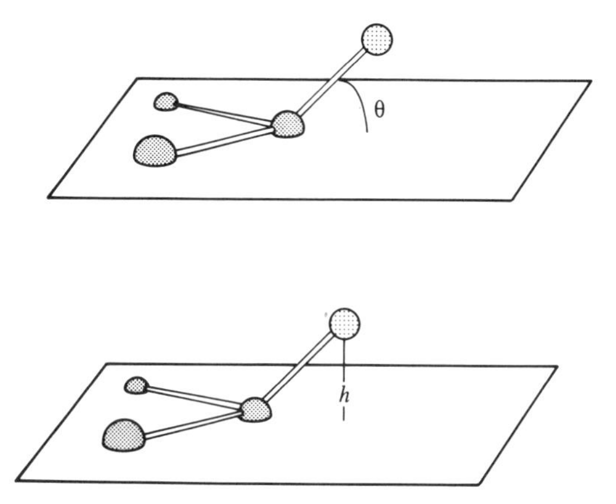

Figure 3.10 Two ways to model the out-of-plane bending contributions.

desired to ensure that an aromatic ring such as benzene maintains an approximately planar structure then this can be achieved using a suitable set of out-of-plane bending terms involving atoms on opposite sides of the ring, Figure 3.11. Improper torsional terms are commonly used in the so-called united atom force fields to maintain stereochemistry at chiral centres (see section 3.14). It is important to remember that out-of-plane terms may not always be necessary, and that to include such terms may have a deleterious effect on the performance of the force field. Vibrational frequencies in particular are often rather sensitive to the presence of out-of-plane terms.

3.7 Cross terms and non-bonded interactions

The presence of cross terms in a force field reflects coupling between the internal coordinates. For example, as a bond angle is decreased it is found that the adjacent bonds stretch to reduce the interaction between the 1,3 atoms, as illustrated in Figure 3.12. Cross terms were found to be important in force fields designed to predict vibrational spectra that were the forerunners of molecular

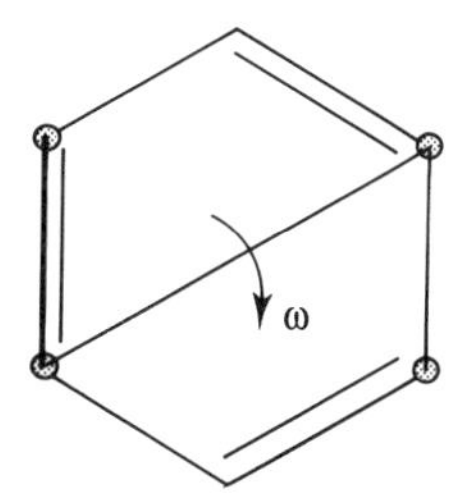

Figure 3.11 Improper torsional terms can be used to keep a benzene ring planar.

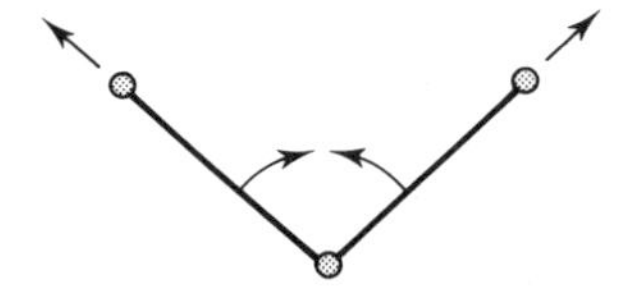

Figure 3.12 Coupling between the stretching of the bonds as an angle closes.

mechanics force fields, and so it is not surprising that cross terms must often be included in a molecular mechanics force field to achieve optimal performance. One should in principle include cross terms between all contributions to a force field. However, only a few cross terms are generally found to be necessary in order to accurately reproduce structural properties; more may be needed to reproduce other properties such as vibrational frequencies, which are more sensitive to the presence of such terms. In general, any interactions involving motions that are far apart in a molecule can usually be set to zero. Most cross terms are functions of two internal coordinates, such as stretch–stretch, stretch–bend and stretch–torsion terms, but cross terms involving more than two internal coordinates such as the bend–bend–torsion have also been used. Various functional forms are possible for the cross terms. For example, the stretch–stretch cross term between two bonds 1 and 2 can be modelled as:

$$\mathcal{V}(l_1, l_2) = \frac{k_{l_1,l_2}}{2}[(l_1 - l_{1,0})(l_2 - l_{2,0})] \tag{3.13}$$

The stretching of the two bonds adjoining an angle could be modelled using an equation of the following form (as in MM2 and MM3):

$$\mathcal{V}(l_1, l_2, \theta) = \frac{k_{l_1,l_2,\theta}}{2}[(l_1 - l_{1,0}) + (l_2 - l_{2,0})](\theta - \theta_0) \tag{3.14}$$

In a *Urey–Bradley* force field, angle bending is achieved using 1,3 non-bonded interactions rather than an explicit angle bending potential. The stretch–bond term in such a force field would be modelled by a harmonic function of the distance between the 1,3 atoms:

$$\mathcal{V}(r_{1,3}) = \frac{k_{r_{1,3}}}{2}(r_{1,3} - r^0_{1,3})^2 \tag{3.15}$$

A stretch–torsion cross term can be used to model the stretching of a bond that occurs in an eclipsed conformation. Two possible functional forms are:

$$\mathcal{V}(l, \omega) = k(l - l_0)\cos n\omega \tag{3.16a}$$

$$\mathcal{V}(l, \omega) = k(l - l_0)[1 + \cos n\omega] \tag{3.16b}$$

n is the periodicity of the rotation about the bond ($n = 3$ for sp^3–sp^3 bonds).

Torsion–bend and torsion–bend–bend terms may also be included; the latter, for example, would couple two angles A–B–C and B–C–D to a torsion angle A–B–C–D. Maple, Dinur and Hagler used quantum mechanics calculations to investigate which of the cross terms are most important and suggested that the stretch–stretch, stretch–bend, bend–bend, stretch–torsion and bend–bend–torsion were most important [Dinur and Hagler 1991] (schematically illustrated in Figure 3.13).

Independent molecules and atoms interact through non-bonded forces, which also play an important role in determining the structure of individual molecular species. The non-bonded interactions do not depend upon a specific bonding relationship between atoms. They are 'through-space' interactions, and are usually modelled as a function of some inverse power of the distance. The non-

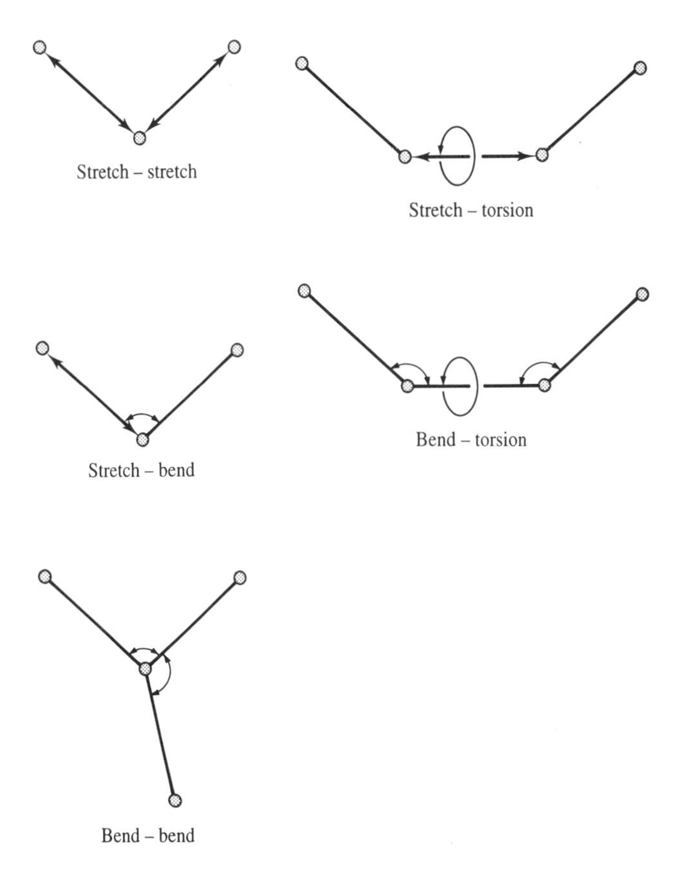

Figure 3.13 Schematic illustration of the cross terms that are believed to be most important in force fields. Adapted from Dinur U, A T Hagler 1991. New Approaches to Empirical Force Fields in Lipkowitz K B, D B Boyd (Editors). *Reviews in Computational Chemistry.* New York, VCH Publishers: 99–164.

bonded terms in a force field are usually considered in two groups, one comprising electrostatic interactions and the other van der Waals interactions.

3.8 Electrostatic interactions

3.8.1 *The central multipole expansion*

Electronegative elements attract electrons more than less electronegative elements, giving rise to an unequal distribution of charge in a molecule. This charge distribution can be represented in a number of ways, one common approach being an arrangement of fractional point charges throughout the molecule. These charges are designed to reproduce the electrostatic properties of the molecule. If the charges are restricted to the nuclear centres they are often referred to as *partial atomic charges* or *net atomic charges*. The electrostatic interaction between two molecules (or between different parts of the same molecule) is then calculated as a sum of interactions between pairs of point charges, using Coulomb's law:

$$\mathscr{V} = \sum_{i=1}^{N_A} \sum_{j=1}^{N_B} \frac{q_i q_j}{4\pi\varepsilon_0 r_{ij}} \tag{3.17}$$

N_A and N_B are the numbers of point charges in the two molecules. This approach to the representation and calculation of electrostatic interactions will be considered in more detail in section 3.8.2. First, we shall consider an alternative approach to the calculation of electrostatic interactions which treats a molecule as a single entity and is (in principle at least) capable of providing a very efficient way to calculate electrostatic intermolecular interactions. This is the *central multipole expansion*, which is based upon the electric moments or multipoles: the charge, dipole, quadrupole, octopole and so on introduced in section 2.14.2. These moments are usually represented by the following symbols: q (charge), μ (dipole), Θ (quadrupole) and Φ (octopole). We are often interested in the lowest non-zero electric moment. Thus species such as Na^+, Cl^-, NH_4^+, $CH_3CO_2^-$ have the charge as their lowest non-zero moment. For many uncharged molecules the dipole is the lowest non-zero moment. Molecules such as N_2 and CO_2 have the quadrupole as their lowest non-zero moment. The lowest non-zero moment for methane and tetrafluoromethane is the octopole. Each of these multipole moments can be represented by an appropriate distribution of charges. Thus a dipole can be represented using two charges placed an appropriate distance apart. A quadrupole can be represented using four charges and an octopole by eight charges. A complete description of the charge distribution around a molecule requires all of the non-zero electric moments to be specified. For some molecules, the lowest non-zero moment may not be the most significant and it may therefore be unwise to ignore the higher-order terms in the expansion without first checking their values.

To illustrate how the multipolar expansion is related to a distribution of charges in a system, let us consider the simple case of a molecule with two charges q_1 and q_2, positioned at $-z_1$ and z_2, respectively, Figure 3.14. The electrostatic potential at point P (a distance r from the origin, r_1 from charge q_1 and r_2 from charge q_2) is then given by:

$$\phi(r) = \frac{1}{4\pi\varepsilon_0}\left(\frac{q_1}{r_1} + \frac{q_2}{r_2}\right) \tag{3.18}$$

By applying the cosine rule this can be written as follows (see Figure 3.14):

$$\phi(r) = \frac{1}{4\pi\varepsilon_0}\left(\frac{q_1}{\sqrt{r^2 + z_1^2 + 2rz_1\cos\theta}} + \frac{q_2}{\sqrt{r^2 + z_1^2 - 2rz_1\cos\theta}}\right) \tag{3.19}$$

If $r \gg z_1$ and $r \gg z_2$ then this expression can be expanded as follows:

$$\phi(r) = \frac{1}{4\pi\varepsilon_0}\left(\frac{q_1 + q_2}{r} + \frac{(q_2 z_2 - q_1 z_1)\cos\theta}{r^2} + \frac{(q_1 z_1^2 + q_2 z_2^2)(3\cos^2\theta - 1)}{2r^3} + \ldots\right) \tag{3.20}$$

We can now associate the appropriate terms in the expansion with the various electric moments:

$$\phi(r) = \frac{1}{4\pi\varepsilon_0}\left(\frac{q}{r} + \frac{\mu\cos\theta}{r^2} + \frac{\Theta(3\cos^2\theta - 1)}{2r^3} + \cdots\right) \tag{3.21}$$

Thus $(q_1 + q_2)$ is the charge; $(q_2 z_2 - q_1 z_1)$ is the dipole; $(q_1 z_1^2 + q_2 z_2^2)$ is the quadrupole, and so on. One interesting feature about a charge distribution is that only the first non-zero moment is independent of the choice of origin. Thus, if a molecule is electrically neutral (i.e. $q_1 + q_2 = 0$) then its dipole moment is independent of the choice of origin. This can be demonstrated for our two charge

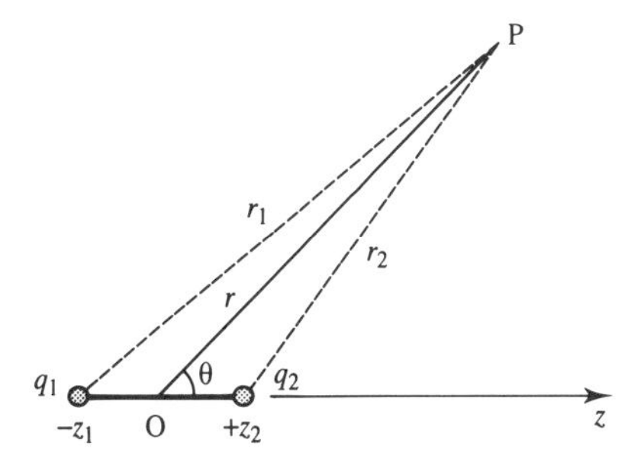

Figure 3.14 The electrostatic potential due to two point charges.

system as follows. If the position of the origin is now moved to a point $-z'$, then the dipole moment relative to this new origin is given by:

$$\mu' = q_2(z_2 + z') - q_1(z_1 - z') = \mu + qz' \tag{3.22}$$

Only if the total charge on the system (q) equals zero will the dipole moment be unchanged. Similar arguments can be used to show that if both the charge and the dipole moment are zero then the quadrupole moment is independent of the choice of origin. For convenience, the origin is often taken to be the centre of mass of the charge distribution.

The electric moments are examples of *tensor properties*: the charge is a rank 0 tensor (which is the same as a scalar quantity); the dipole is a rank 1 tensor (which is the same as a vector, with three components along the x, y and z axes); and the quadrupole is a rank 2 tensor and so has nine components that can be represented as a 3×3 matrix. In general, a tensor of rank n has 3^n components.

For a distribution of charges (one not restricted to lie along one of the Cartesian axes), the dipole moment is given by:

$$\boldsymbol{\mu} = \sum q_\mathrm{i}\mathbf{r}_\mathrm{i} \tag{3.23}$$

The components of the dipole moment along the x, y and z axes are $\sum q_\mathrm{i}x_\mathrm{i}$, $\sum q_\mathrm{i}y_\mathrm{i}$ and $\sum q_\mathrm{i}z_\mathrm{i}$. The analogous way to define the quadrupole moment is as follows:

$$\Theta = \begin{pmatrix} \sum q_\mathrm{i}x_\mathrm{i}^2 & \sum q_\mathrm{i}x_\mathrm{i}y_\mathrm{i} & \sum q_\mathrm{i}x_\mathrm{i}z_\mathrm{i} \\ \sum q_\mathrm{i}y_\mathrm{i}x_\mathrm{i} & \sum q_\mathrm{i}y_\mathrm{i}^2 & \sum q_\mathrm{i}y_\mathrm{i}z_\mathrm{i} \\ \sum q_\mathrm{i}z_\mathrm{i}x_\mathrm{i} & \sum q_\mathrm{i}z_\mathrm{i}y_\mathrm{i} & \sum q_\mathrm{i}z_\mathrm{i}^2 \end{pmatrix} \tag{3.24}$$

This definition of the quadrupole is obviously dependent upon the orientation of the charge distribution within the coordinate frame. Transformation of the axes can lead to alternative definitions that may be more informative. Thus the quadrupole moment is commonly defined as follows:

$$\Theta = \frac{1}{2}\begin{pmatrix} \sum_\mathrm{i} q_\mathrm{i}(3x_\mathrm{i}^2 - r_\mathrm{i}^2) & 3\sum_\mathrm{i} q_\mathrm{i}x_\mathrm{i}y_\mathrm{i} & 3\sum_\mathrm{i} q_\mathrm{i}x_\mathrm{i}z_\mathrm{i} \\ 3\sum_\mathrm{i} q_\mathrm{i}x_\mathrm{i}z_\mathrm{i} & \sum_\mathrm{i} q_\mathrm{i}(3y_\mathrm{i}^2 - r_\mathrm{i}^2) & 3\sum_\mathrm{i} q_\mathrm{i}y_\mathrm{i}z_\mathrm{i} \\ 3\sum_\mathrm{i} q_\mathrm{i}x_\mathrm{i}z_\mathrm{i} & 3\sum_\mathrm{i} q_\mathrm{i}y_\mathrm{i}z_\mathrm{i} & \sum_\mathrm{i} q_\mathrm{i}(3z_\mathrm{i}^2 - r_\mathrm{i}^2) \end{pmatrix} \tag{3.25}$$

In equation (3.25) $r_\mathrm{i}^2 = x_\mathrm{i}^2 + y_\mathrm{i}^2 + z_\mathrm{i}^2$. This definition enables one to assess the deviation from spherical symmetry as a spherically symmetric charge distribution will have

$$\sum_\mathrm{i} q_\mathrm{i}x_\mathrm{i}^2 = \sum_\mathrm{i} q_\mathrm{i}y_\mathrm{i}^2 = \sum_\mathrm{i} q_\mathrm{i}z_\mathrm{i}^2 = \frac{1}{3}\sum_\mathrm{i} q_\mathrm{i}r_\mathrm{i}^2$$

and so the diagonal elements of the tensor will be zero. Quadrupoles are also reported in terms of the *principal axes*; these are three mutually perpendicular axes α, β and γ that are linear combinations of x, y and z such that the quadrupole tensor is diagonal (i.e. off-diagonal elements are zero):

$$\Theta = \begin{pmatrix} \Theta_{\alpha\alpha} & 0 & 0 \\ 0 & \Theta_{\beta\beta} & 0 \\ 0 & 0 & \Theta_{\gamma\gamma} \end{pmatrix} \tag{3.26}$$

Let us now consider the effect of placing another molecule with a linear charge distribution (charges q_1' and q_2') with its centre of mass at the point P. The relative orientation of the two molecules can be described in terms of four parameters (the distance joining their centres of mass and three angles as shown in Figure 3.15). The electrostatic interaction between the two molecules is calculated by multiplying each charge by the potential at that point and adding the result for each charge. The following expression is the result [Buckingham 1959]:

$$\mathscr{V}(q,q') = \frac{1}{4\pi\varepsilon_0}\left\{\begin{array}{l} \dfrac{qq'}{r} \\ +\dfrac{1}{r^2}(q\mu'\cos\theta + q'\mu\cos\theta') \\ +\dfrac{\mu\mu'}{r^3}(2\cos\theta\cos\theta' + \sin\theta\sin\theta'\cos\zeta) \\ +\dfrac{1}{2r^3}[q\Theta'(3\cos^2\theta' - 1) + q'\Theta(3\cos^2\theta - 1)] \\ +\dfrac{3}{2r^4}[\mu\Theta'\{\cos\theta(3\cos^2\theta' - 1) \\ \qquad +2\sin\theta\sin\theta'\cos\theta'\cos\zeta\} \\ \qquad +\mu'\Theta\{\cos\theta'(3\cos^2\theta - 1) \\ \qquad +2\sin\theta'\sin\theta\cos\theta\cos\zeta\}] \\ +\dfrac{3\Theta\Theta'}{4r^5}[1 - 5\cos^2\theta - 5\cos^2\theta' \\ \qquad +17\cos^2\theta\cos^2\theta' + 2\sin^2\theta\sin^2\theta'\cos^2\zeta \\ \qquad +16\sin\theta\sin\theta'\cos\theta\cos\theta'\cos\zeta] \\ +\cdots \end{array}\right\} \tag{3.27}$$

The energy of interaction between two charge distributions is thus an infinite series that includes charge–charge, charge–dipole, dipole–dipole, charge–quadrupole, dipole–quadrupole interactions, quadrupole–quadrupole terms and so on. These terms depend on different inverse powers of the separation r. If the molecules are neutral (i.e. $q = q' = 0$) then the leading term in the expansion is that due to the dipole–dipole interaction which varies as r^{-3}. This is a key result, for the range of the dipole–dipole interaction (r^{-3}) is much less than that of the Coulomb interaction (r^{-1}), Figure 3.16. This will be important in later chapters, where we shall collect atoms together into neutral groups. The electrostatic inter-

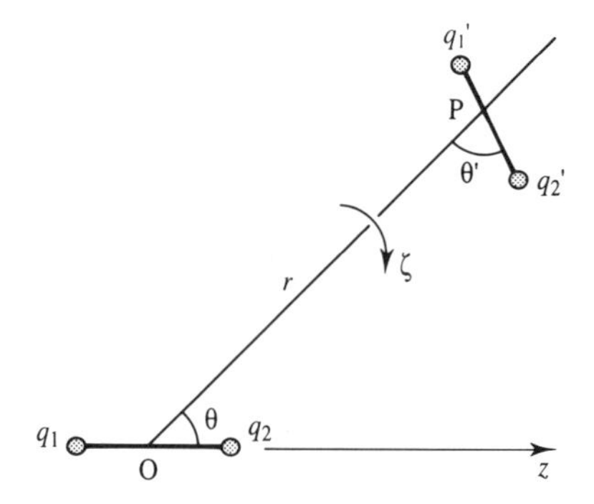

Figure 3.15 The relative orientation of two dipoles.

action between these groups then decays as r^{-3} rather than the r^{-1} dependence of each individual charge–charge interaction. This can be seen from Figure 3.16, in which the functions r^{-1} and r^{-3} have been plotted as a function of distance. Even when the dipole–dipole interaction energy has fallen off almost to zero the charge–charge interaction energy is still significant. In general, the interaction energy between two multipoles of order n and m decreases as $r^{-(n+m+1)}$. It should be emphasised again that these expressions are only valid when the separation of the two molecules r, is much larger than the internal dimensions of the molecules. The favourable arrangements for the various multipoles are shown in Figure 3.17.

A central multipole expansion therefore provides a way to calculate the electrostatic interaction between two molecules. The multipole moments can be obtained from the wavefunction and can therefore be calculated using quantum mechanics (see section 2.14.2) or can be determined from experiment. One example of the use of a multipole expansion is the benzene model of Claessens, Ferrario and Ryckaert [Claessens *et al.* 1983]. Benzene has no charge and no

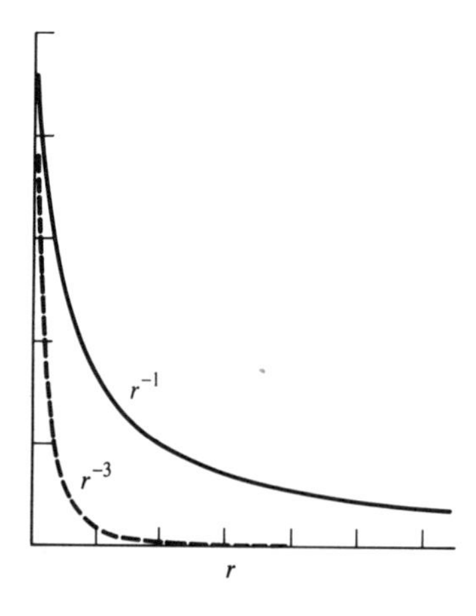

Figure 3.16 The charge–charge energy decays much more slowly ($\propto r^{-1}$) than the dipole–dipole energy ($\propto r^{-3}$).

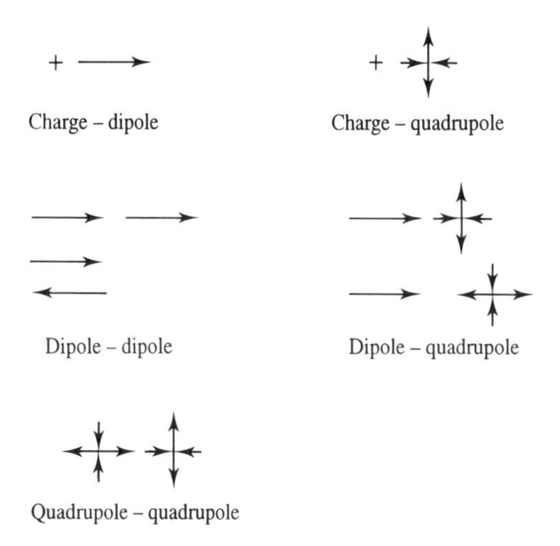

Figure 3.17 The most favourable orientations of various multipoles. Figure adapted from Buckingham A D 1959. Molecular Quadrupole Moments. *Quarterly Reviews of the Chemical Society* **13**:183–214.

dipole moment, but it does have a sizeable quadrupole. The inclusion of the quadrupole was found to give clearly superior results in molecular dynamics simulations of the liquid state over models that lacked any electronic contribution.

The main advantage of the multipolar description in calculating the electrostatic interactions between molecules is its efficiency. For example, the charge–charge interaction energy between two benzene molecules would require 144 charge–charge interactions with a partial atomic charge model rather than the single quadrupole–quadrupole term. Unfortunately, the multipole expansion is not applicable when the molecules are separated by distances comparable to the molecular dimensions. The formal condition for convergence of the multipolar interaction energy is that the distance between two interacting molecules should be larger than the sum of the distances from the centre of each molecule to the furthest part of its charge distribution. If a sphere is constructed around each molecule, positioned on its centre of mass, with a radius that encompasses all of the charge distribution, then the multipole expansion for the interaction between two molecules will converge if these spheres do not intersect. Even if one requires the sphere to encompass just the nuclei in a molecule (i.e. ignoring the fact that the charge distribution around a molecule extends to infinity) there may still be problems. For example, the convergence sphere for a molecule such as butane would extend beyond the van der Waals radii in some directions, enabling other molecules to penetrate the convergence sphere, as illustrated in Figure 3.18. Another problem is that the multipolar expansion may be slow to converge. The multipolar expansion is often located at the centre of mass, but this may not be the best choice to achieve the most rapid convergence.

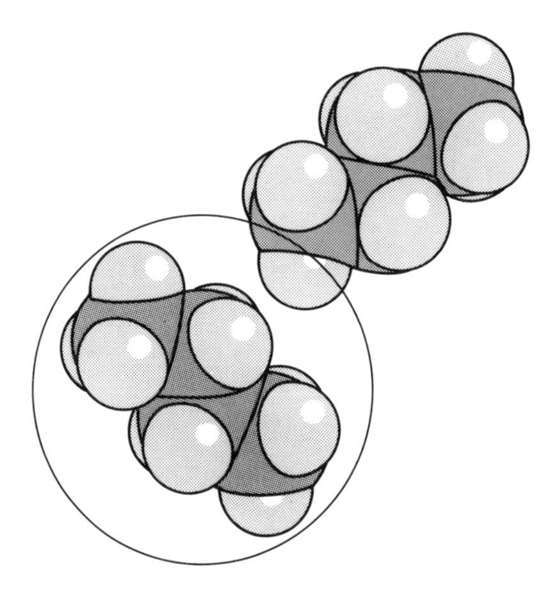

Figure 3.18 The convergence sphere of the multipole expansion for a molecule such as butane may be penetrated by another molecule.

There are other difficulties with the central multipole expansion. The multipole moments are properties of the entire molecule and so cannot be used to determine intramolecular interactions. The central multipole model thus tends to be restricted to calculations involving small molecules that are kept fixed in conformation during the calculation, and where the interactions between molecules act at their centres of mass. It can be a complicated procedure to calculate the forces acting on a molecule with a multipole model. The interaction between multipoles of zero order (i.e. charges) gives rise to a simple translational force. Multipoles of a higher order have directionality and interactions between these produce a torque, or twisting force. Moreover, whereas the charge–charge forces are equal and opposite, the torque acting on molecule i due to another molecule j is not necessarily equal and opposite to the torque on molecule j due to molecule i.

3.8.2 *Point-charge electrostatic models*

We therefore return to the point-charge model for calculating the electrostatic interactions. If sufficient point charges are used then all of the electric moments can be reproduced and the multipole interaction energy, equation (3.27), is exactly equal to that calculated from the Coulomb summation, equation (3.17).

An accurate representation of a molecule's electrostatic properties may require charges to be placed at locations other than at the atomic nuclei. A simple example of this is molecular nitrogen, which has a dipole moment of zero. The total charge on nitrogen is zero, and so an atomic partial charge model would put zero charge on each nucleus. However, nitrogen does have a quadrupole moment that significantly affects its properties. The simplest way to model this is to place three partial charges along the bond: a charge of $-q$ at each nucleus and $+2q$ at the centre of mass. The quadrupole–quadrupole interaction between two nitrogen molecules can then be calculated by summing nine pairs of charge–charge interactions. The value of q can be calculated using the following relationship between the quadrupole moment and the partial charge:

$$\Theta = 2q(l/2)^2 \tag{3.28}$$

l is the bond length. The experimental quadrupole moment is consistent with a charge, q, of approximately $0.5e$. In fact, a better representation of the electrostatic potential around the nitrogen molecule is obtained using the five-charge model shown in Figure 3.19.

An alternative to the point charge model is to assign dipoles to the bonds in the molecule. The electrostatic energy is then given as a sum of dipole–dipole interaction energies. This approach (which is adopted in MM2) can be unwieldy for molecules that have a formal charge and which require charge–charge and charge–dipole terms to be included in the energy expression. Charged species are dealt with naturally using the point charge model.

3.8.3 Calculating partial atomic charges

Given the widespread use of the partial atomic charge model, it is important to consider how the charges are obtained. For simple species the atomic charges required to reproduce the electric moments can be calculated exactly if the geometry is known. For example, the experimentally determined dipole moment of HF (1.82 D) can be reproduced by placing equal but opposite charges of $0.413e$ on the two atomic nuclei (assuming a bond length of 0.917 Å). The tetrahedral arrangement of the hydrogens about the carbon in methane means that each hydrogen atom has an identical charge equal to one quarter the charge on the carbon. The molecule is electrically neutral with zero dipole and quadrupole moments but a non-zero octopole moment which can be reproduced using a hydrogen charge of approximately $0.14e$.

In some cases the atomic charges are chosen to reproduce thermodynamic properties calculated using a molecular dynamics or Monte Carlo simulation. A series of simulations is performed and the charge model is modified until satisfactory agreement with experiment is obtained. This approach can be quite powerful despite its apparent simplicity, but it is only really practical for small molecules or simple models.

The electrostatic properties of a molecule are a consequence of the distribution of the electrons and the nuclei and thus it is reasonable to assume that one should be able to obtain a set of partial atomic charges using quantum mechanics. Unfortunately, the partial atomic charge is not an experimentally observable quantity and cannot be unambiguously calculated from the wavefunction. This explains why numerous ways to determine partial atomic charges have been

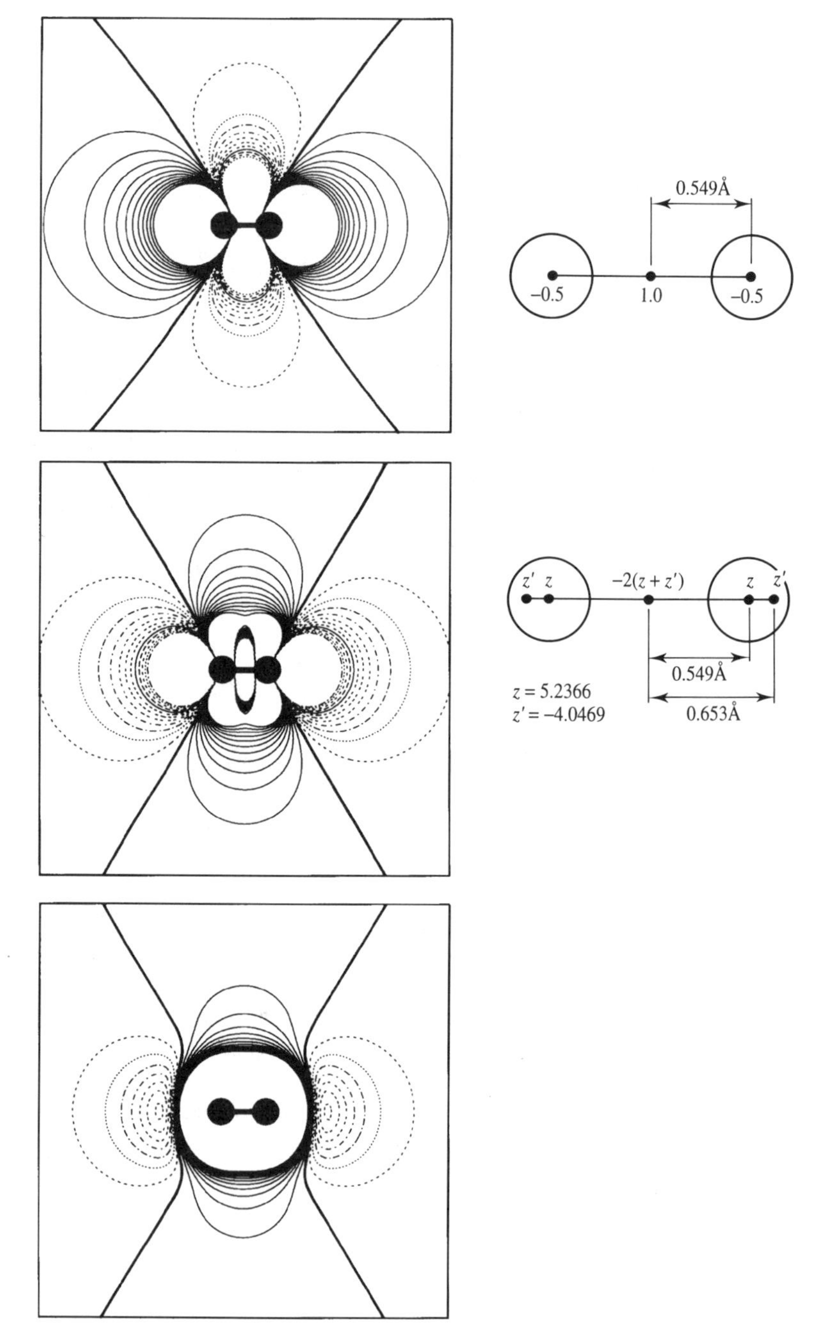

Figure 3.19 Two charge models for N_2 with the electrostatic potentials that they generate. Also shown is the electrostatic potential calculated using *ab initio* quantum mechanics (6-31G* basis set). Negative contours are dashed and the zero contour is bold.

proposed, and why there is still considerable debate as to the 'best' method to derive them. Indirect comparisons of the various methods are possible, usually by calculating appropriate quantities from the charge model and then comparing the results with either experiment or quantum mechanics. For example, one might examine how well the charge model reproduces the experimental or quantum mechanical multipole moments or the electrostatic potential around the molecule.

We have already encountered in section 2.14.4 the population analysis method for calculating partial atomic charges. Such sets of charges (commonly referred to as *Mulliken charges* when obtained from that particular partitioning scheme) are often considered to be inappropriate for accurately representing the interactions between molecules. This is because Mulliken charges are primarily dependent upon the constitution of the molecule – how the atoms are bonded together – rather than being designed to reproduce the properties that determine how molecules interact with each other, such as the electrostatic potential. The importance of the electrostatic potential in intermolecular interactions has resulted in much interest in schemes that calculate charges consistent with this particular property.

3.8.4 Charges derived from the molecular electrostatic potential

The electrostatic potential at a point is the force acting on a unit positive charge placed at that point. The nuclei give rise to a positive (i.e. repulsive) force whereas the electrons give rise to a negative potential. The electrostatic potential is an observable quantity that can be determined from a wavefunction using equations (2.302) and (2.303):

$$\phi(\mathbf{r}) = \phi_{\text{nucl}}(\mathbf{r}) + \phi_{\text{elec}}(\mathbf{r}) = \sum_{\text{A}=1}^{M} \frac{Z_{\text{A}}}{|\mathbf{r} - \mathbf{R}_{\text{A}}|} - \int \frac{\text{d}\mathbf{r}'\rho(\mathbf{r})}{|\mathbf{r}' - \mathbf{r}|} \tag{3.29}$$

The electrostatic potential is a continuous property, and is not easily represented by an analytical function. Consequently, it is necessary to derive a discrete representation for use in numerical analysis. The objective is to derive the set of partial charges (usually partial atomic charges) that best reproduces the quantum mechanical electrostatic potential at a series of points surrounding the molecule. A solution to this problem was suggested by Cox and Williams [Cox and Williams 1981]. The electrostatic potential at each of the chosen points is calculated from the wavefunction. A least-squares fitting procedure is then employed to determine the set of partial atomic charges that best reproduces the electrostatic potential at the points, subject to the constraint that the sum of the charges should be equal to the net charge on the molecule. Symmetry conditions may also be imposed to ensure that the charges on symmetrically equivalent atoms are equal. It is also possible to require the atomic charges to reproduce other electrostatic properties of the molecules such as the dipole moment. The fitting procedure minimises the sum of squares of the differences in the electrostatic potential. Thus, if the electrostatic potential at a point is ϕ_{i}^{0} and if the value from the charge model is $\phi_{\mathrm{i}}^{\mathrm{calc}}$, then the objective is to minimise the following function:

$$R = \sum_{\mathrm{i}=1}^{N_{\mathrm{points}}} w_{\mathrm{i}}(\phi_{\mathrm{i}}^{0} - \phi_{\mathrm{i}}^{\mathrm{calc}})^2 \tag{3.30}$$

N_{points} is the number of points and w_{i} is a weighting factor that enables different points to be given different degrees of 'importance' in the fitting process. One of the charges is dependent on the values of the others (because the sum must equal Z, the molecular charge). This Nth charge has a value given by:

$$q_N = Z - \sum_{\mathrm{j}=1}^{N-1} q_{\mathrm{j}} \tag{3.31}$$

The electrostatic potential due to the charges q_{j} at the point i is given by Coulomb's law:

$$\phi_{\mathrm{i}}^{\mathrm{calc}} = \sum_{\mathrm{j}=1}^{N-1} \frac{q_{\mathrm{j}}}{4\pi\varepsilon_0 r_{\mathrm{ij}}} + \frac{Z - \sum_{\mathrm{j}=1}^{N-1} q_{\mathrm{j}}}{4\pi\varepsilon_0 r_{\mathrm{i}N}} \tag{3.32}$$

r_{ij} is the distance from the charge j to the point i. At a minimum value of the error function, R, the first derivative is equal to zero with respect to all charges q_k:

$$\frac{\partial R}{\partial q_k} = -2 \sum_{\mathrm{i}=1}^{N_{\mathrm{points}}} w_{\mathrm{i}}(\phi_{\mathrm{i}}^{0} - \phi_{\mathrm{i}}^{\mathrm{calc}}) \left(\frac{\partial \phi_{\mathrm{i}}^{\mathrm{calc}}}{\partial q_k} \right) = 0 \tag{3.33}$$

This equation can be written in the following form:

$$\sum_{\mathrm{i}=1}^{N_{\mathrm{points}}} w_{\mathrm{i}} \left(\phi_{\mathrm{i}}^{0} - \frac{Z}{r_{\mathrm{i}N}} \right) \left(\frac{1}{r_{\mathrm{i}k}} - \frac{1}{r_{\mathrm{i}N}} \right) = \sum_{\mathrm{j}=1}^{N-1} \left[\sum_{\mathrm{i}=1}^{N_{\mathrm{points}}} w_{\mathrm{i}} \left(\frac{1}{r_{\mathrm{i}k}} - \frac{1}{r_{\mathrm{i}N}} \right) \left(\frac{1}{r_{\mathrm{ij}}} - \frac{1}{r_{\mathrm{i}N}} \right) \right] \frac{q_{\mathrm{j}}}{4\pi\varepsilon_0} \tag{3.34}$$

When expressed in this way, then the set of equations can be recast as a matrix equation of the form $\mathbf{Aq} = \mathbf{a}$. The charges $\mathbf{q}$ are then determined using standard matrix methods via $\mathbf{q} = \mathbf{A}^{-1}\mathbf{a}$.

The points i (1, 2, ... N_{points}) where the potential is fitted can be chosen in a variety of ways, but should be taken from the region where it is most important to correctly model intermolecular interactions. This region is just beyond the van der Waals radii of the atoms involved. Cox and Williams selected points from a regular grid in a shell defined by two surfaces, one corresponding to the union of the van der Waals radii plus 1.2 Å and the others approximately 1 Å beyond that. The CHELP procedure of Chirlian and Francl [Chirlian and Francl 1987] uses spherical shells, 1 Å apart, centred on each atom with points symmetrically distributed on the surface. Any points within the van der Waals radius of any atom in the system are discarded and the shells extend to 3 Å from the van der Waals surface of the molecule. The CHELP method employs a Lagrange multiplier method to find the atomic charges, rather than an iterative least-squares procedure. This minimises the error function R (equation (3.30)) subject to the constraint that the charges sum to the total molecular charge. Such an analysis yields a set of $N+1$ equations in $N+1$ unknowns and can be solved using standard matrix methods. The CHELPG algorithm of Breneman and Wiberg [Breneman and Wiberg 1990] combines the regular grid of points of Cox and Williams with the Lagrange multiplier method of Chirlian and Francl as the results from CHELP were found to change if the molecule was reorientated in the coordinate system. In CHELPG a cubic grid of points (spaced 0.3–0.8 Å apart) is used and all grid points that lie within the van der Waals radius of any atom are discarded, together with all points that lie further than 2.8 Å away from any atom.

The algorithm of Singh and Kollman used to derive the charges in the 1984 AMBER force field uses points on a series of molecular surfaces, constructed using gradually increasing van der Waals radii for the atoms [Singh and Kollman 1984]. The points at which the potential was fitted were located on these shells. For the 1995 AMBER force field a modified version of this electrostatic potential method was employed (termed 'restrained electrostatic potential fit', or RESP [Bayly *et al.* 1993]). The RESP algorithm uses hyperbolic restraints on non-hydrogen atoms. These restraints have the effect of reducing the charges on some atoms, particularly buried carbon atoms, which can be assigned artificially high charges in standard electrostatic potential fitting methods. The RESP charges also vary less with the molecular conformation.

3.8.5 Deriving charge models for large systems

Molecular mechanics is used to model systems containing thousands of atoms such as polymers. How then can charges be derived for such species? Clearly one cannot routinely perform quantum mechanical calculations on a molecule with so many atoms and so it must be broken into fragments of a suitable size. In some cases the fragments might appear relatively easy to define; for example, many polymeric systems are constructed by connecting together chemically defined monomeric units. The atomic charges for each monomer should be obtained from calculations on suitable fragments that recreate the immediate local environment of the fragment in the larger molecule. For example, partial atomic charges for amino acids are often obtained from calculations on a 'dipeptide' fragment (see Figure 3.20) which is more akin to the environment within a protein than in an isolated amino acid.

The charge sets obtained from electrostatic potential fitting can be highly dependent upon the basis set used to derive the wavefunction. Moreover, the charges do not always improve if a larger basis set is used. It is generally considered that the 6-31G* basis set gives reasonable results for calculations relevant to condensed phases. In many cases it is possible to scale the results of a calculation using a small basis set or even a lower level of theory (such as a semi-

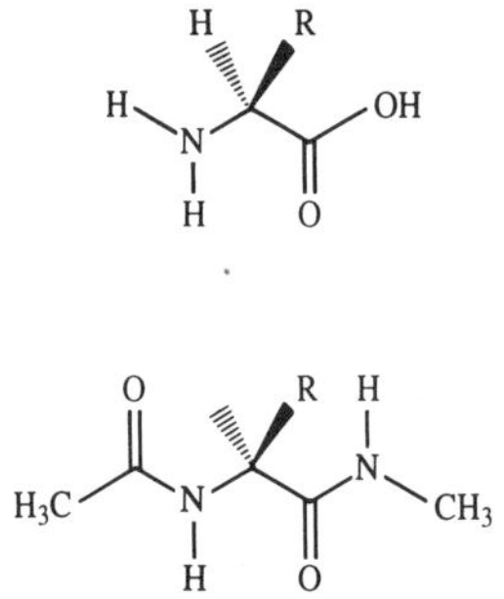

Figure 3.20 The charges used for calculations on proteins are best derived using a suitable fragment for each amino acid that reflects the environment within the protein (bottom), rather than the isolated amino acid (top).

empirical calculation) to obtain results comparable to those of a high-level calculation. Of the various semi-empirical methods available, MNDO appears to give the best correspondence with the charges derived from *ab initio* calculations and scaling factors have been determined by several research groups [Ferenczy *et al.* 1990; Luque *et al.* 1990; Bezler *et al.* 1990]. An additional complicating factor is that the charges obtained from electrostatic potential fitting will often depend upon the conformation for which the quantum mechanical calculation was performed [Williams 1990]. One solution is to perform a series of charge calculations for different conformations and then use a charge model in which each charge is weighted according to the relative population of that particular conformation as calculated from the Boltzmann distribution [Reynolds *et al.* 1992]. In a few charge models the charges vary continuously with the conformation [Rappé and Goddard 1991; Dinur and Hagler 1995].

3.8.6 Rapid methods for calculating atomic charges

Some methods calculate atomic charges solely from information about the atoms present in the molecule and the way in which the atoms are connected. The great advantage of such methods is that they are very fast, and can be used to calculate the charge distributions for large numbers of molecules (e.g. in a database). We will consider the Gasteiger and Marsili method [Gasteiger and Marsili 1980] as an example.

The Gasteiger–Marsili approach uses the concept of the *partial equalisation of orbital electronegativity.* Electronegativity is a concept well known to chemists, being defined by Pauling as 'the power of an atom to attract electrons to itself'. Mulliken subsequently defined the electronegativity of an atom A as the average of its ionisation potential I_A and its electron affinity E_A:

$$\chi_A = \frac{1}{2}(I_A + E_A) \tag{3.35}$$

As Mulliken pointed out, the ionisation potential and electron affinity are specific to a given valence state of an atom, and therefore the electronegativities of an atom's valence states would not be expected to be the same. This idea can be extended to the concept of orbital electronegativity, which is the electronegativity of a specific orbital in a given valence state. For example, an sp orbital has a higher electronegativity than an sp^3 orbital. The orbital electronegativity will also depend on the occupancy of the orbital; an empty orbital will be better able to attract an electron than an orbital with a single electron, which in turn will be better than an orbital with two electrons. The electronegativity of an orbital will also be affected by the charges in other orbitals. Gasteiger and Marsili assumed a polynomial relationship between the orbital electronegativity $\chi_{\mu A}$ of an orbital ϕ_μ in atom A and the charge Q_A on the atom A:

$$\chi_{\mu A} = a_\mu + b_{\mu A} Q_A + \chi_{\mu A} Q_A^2 \tag{3.36}$$

Values of the coefficients a, b and c were derived for common elements in their usual valence states (for example, for carbon there are different values for sp^3, $sp^2\pi$ and $sp\pi^2$ valence states).

Electrons flow from the less electronegative elements to the more electronegative ones. This flow of electrons results in a positive charge on the less electronegative atoms and a negative charge on the more electronegative atoms, and as such the flow acts to equalise the electronegativities. Total equalisation of electronegativity does not, however, lead to chemically sensible results. This effect is modelled in the Gasteiger and Marsili approach by an iterative procedure, in which less and less charge is transferred between bonded atoms at each step. The electron charge transferred from an atom A to an atom B (where B is more electronegative than A) in iteration k is given by:

$$Q^{\langle k \rangle} = \frac{\chi_B^{\langle k \rangle} - \chi_A^{\langle k \rangle}}{\chi_A^{+}} \alpha^k \tag{3.37}$$

In equation (3.37), $Q^{\langle k \rangle}$ is the charge (in electrons) transferred; $\chi_A^{\langle k \rangle}$ and $\chi_B^{\langle k \rangle}$ are the electronegativities of the atoms A and B; χ_A^{+} is the electronegativity of the cation of the less electronegative atom and α is a damping factor which is raised to the power k. Gasteiger and Marsili set α to $1/2$. The charge on each atom is initially assigned its formal charge. In each iteration, the electronegativities are calculated using equation (3.36) and hence the charge to be transferred. The total charge on an atom at the end of each iteration is thus obtained by adding the charge transferred from all bonds to the atom to the value of the charge from the previous iteration. The damping factor α^k reduces the influence of the more electronegative atoms. This influence decreases with each iteration. With a damping factor of $1/2$ rapid convergence is achieved, usually within four or five steps.

3.8.7 Beyond partial atomic charge models

Most of the charge models that we have considered so far place the charge on the nuclear centres. Atom-centred charges have many advantages. For example, the electrostatic forces due to charge–charge interactions then act directly on the nuclei. This is important if one wishes to calculate the forces on the nuclei as is required for energy minimisation or a molecular dynamics simulation. Nuclear-centred charges do nevertheless suffer from some drawbacks. In particular, they assume that the charge density about each atom is spherically symmetrical. However, an atom's valence electrons are often distributed in a far from spherical manner, especially in molecules that contain features such as lone pairs and π-electron clouds above aromatic ring systems.

3.8.8 Distributed multipole models

One way to represent the anisotropy of a molecular charge distribution is to use *distributed multipoles*. In this model, point charges, dipoles, quadrupoles and higher multipoles are distributed throughout the molecule. These distributed multipoles can be determined in various ways but the distributed multipole analysis (DMA) model of A J Stone [Stone 1981; Stone and Alderton 1985] is probably the best-known example. The DMA method calculates the multipoles from a quantum mechanics wavefunction defined in terms of Gaussian basis functions. As we saw in section 2.6, the overlap between two Gaussian functions can be represented by another Gaussian located at a point (P) along the line that connects them. Each product of basis functions $\phi_\mu \phi_\nu$ thus corresponds to a charge density at P. This density can be expressed as a multipole expansion about P. The highest multipole moment in the local expansion depends upon the basis set used; no multipole moment higher than the sum of the angular quantum numbers of the basis set is possible. Thus, when using a basis set that contains just s and p functions there will be local multipoles no higher than the quadrupole. The crucial feature is that the local multipole expansion about P can be represented as a multipole expansion about another nearby point S. In the distributed multipole approach, a set of site points is chosen and then the local multipole expansion for each pair of basis functions is 'moved' from the relevant point P to one of the sites S.

There are no limitations on the number or location of the multipole sites S; a natural set to use is obtained by placing a site point on each atomic nucleus. In some applications (especially for small molecules) additional sites are defined at the centres of bonds. For example, Stone derived a distributed multipole model for nitrogen from a Dunning [5s4p2d] basis set with two polarisation functions. This model contains charges of $+0.60$ on the nuclei and a charge of -1.20 at the centre of the bond, together with a dipole on each of the two nuclei and a quadrupole located at the centre of the bond (see Figure 3.21). Four HF charges are placed on the two nuclei and at the centre of the bond with a dipole and a quadrupole on the fluorine and a small dipole at the centre of the bond (Figure

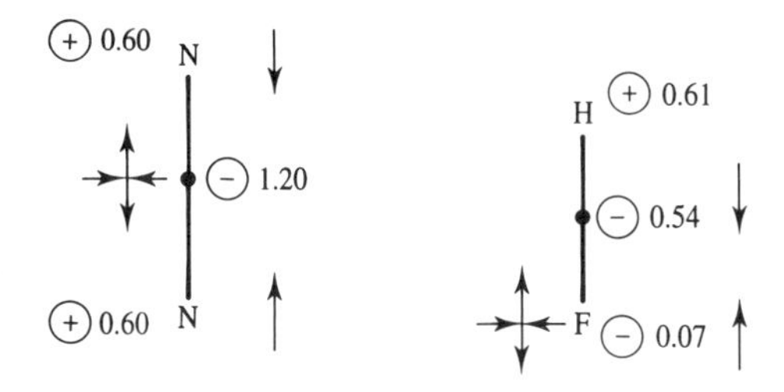

Figure 3.21 Distributed multipole models for N_2 and HF. Figure adapted from Stone A J and M Alderton 1985. Distributed Multipole Analysis Methods and Applications. *Molecular Physics* **56**:1047–1064.

3.21). In larger molecules not every atom may be given a site, such as hydrogen atoms bonded to apolar atoms. It is also possible to restrict the order of the multipole expansion at a given atom so that, for example, only a charge component would be present on a polar hydrogen with the higher moments being represented by multipoles on the atom to which it is bonded. An important consideration when choosing the multipole sites is that when a local multipole expansion is moved, the resulting multipole expansion is no longer a truncated series. However, the smaller the distance between P and the corresponding site point S, the quicker the series converges. In practice, therefore, each local multipole moment expansion is either moved to the nearest site point or is divided between the two nearest site points when they are equally close. With a basis set that contains just s and p functions and multipole sites at the atomic nuclei, it is usually found that the distributed multipole series converges rapidly after the quadrupole term. The multipoles themselves can vary considerably with the basis set used to perform the *ab initio* calculation, but the various electronic properties derived from them usually do not change much.

The distributed multipole model automatically includes non-spherical, anisotropic effects due to features such as lone pairs or π-electrons. The original applications of the DMA approach were to small molecules such as diatomics and triatomics. The method has since been used to develop models for nuclei acids and for peptides and has even been applied to the undecapeptide cyclosporin [Price *et al.* 1989], which contains 199 atoms (the quantum mechanical calculation on this molecule used 1000 basis functions). However, the distributed multipole models have not yet been widely incorporated into force fields, not least because of the additional computational effort required with this type of model. It can be complicated to calculate the atomic forces with the distributed multipole model; in particular, multipoles that are not located on atoms generate torques that must be further analysed to determine the forces on the nuclei.

3.8.9 Using charge schemes to study aromatic–aromatic interactions

The attractive interactions between molecules containing π-systems have long been studied by theoreticians and experimentalists. Such systems are involved in a variety of phenomena, including the stacking of the nucleic acid bases in DNA, the packing of aromatic molecules in crystals and interactions between amino acid side chains in proteins. A variety of orientations are observed for aromatic dimers, ranging from edge-on, T-shaped structures to face-to-face structures (Figure 3.22). Within these two families the molecules can move relative to each other, so that, for example, in a face-to-face arrangement the atoms are overlaid or are staggered. In the T-shaped structure the large quadrupole moments of the benzene molecules adopt their most favourable orientations.

One very simple model of the interactions in such systems was devised by Hunter and Saunders [Hunter and Saunders 1990] who wanted to explain the stacking behaviour of aromatic systems such as the porphyrins shown in Figure 3.23. It is experimentally observed that these molecules adopt a cofacial arrangement with their centres offset as shown. Hunter and Saunders placed point charges not only at the nuclei, but also at locations above and below each atom, perpendicular to the plane of the ring. Thus in benzene each carbon atom was given a charge of $+1$ and also had two associated charges of $-1/2$ above and below the ring (Figure 3.24). The electrostatic interaction between two ring systems is calculated in the usual way by summing the charge–charge interactions using Coulomb's law. A major advantage of the Hunter–Saunders approach is its computational simplicity. Moreover, it can be extended to cover a wide range of atom types and so applied to many systems [Vinter 1994]. Hunter and Saunders summarised the results of their investigations on porphyrins in three rules:

1. π–π repulsion dominates in a face-to-face geometry;
2. π–σ attraction dominates in an edge-on geometry;
3. π–σ attraction dominates in an offset π-stacked geometry.

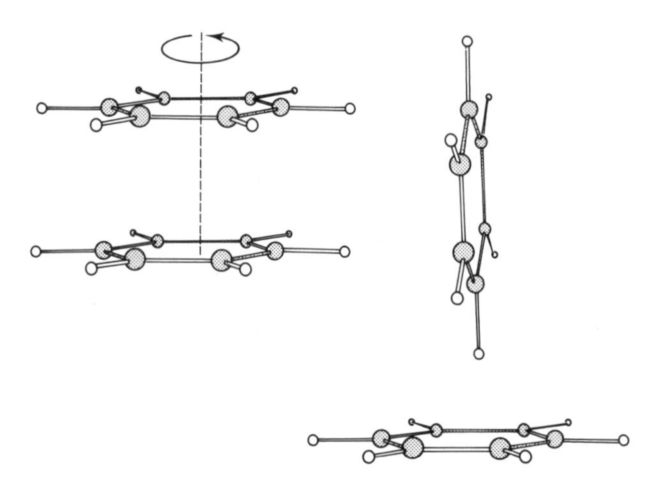

Figure 3.22 Face-to-face (left) and T-shaped (right) orientations of the benzene dimer.

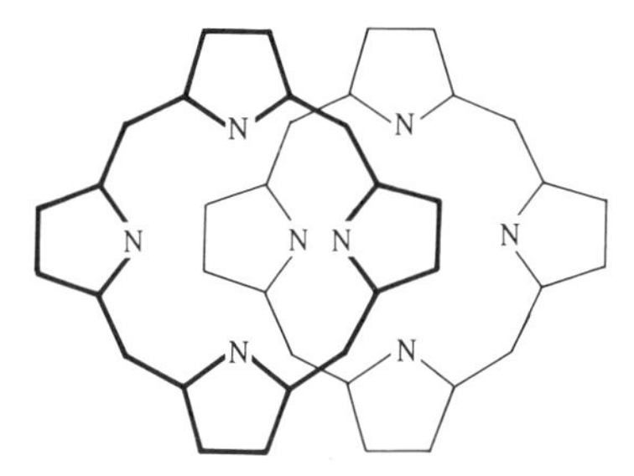

Figure 3.23 Porphyrin system typical of those studied by Hunter and Saunders [Hunter and Saunders 1990].

The interactions between aromatic systems have also been studied using point charge models, central multipoles and distributed multipoles. Fowler and Buckingham examined homodimers of *sym*-triazine and 1,3,5-trifluorobenzene (Figure 3.25) [Fowler and Buckingham 1991]. They were particularly keen to calculate how the electrostatic energy changed as the rings were twisted in the face-to-face geometry. All but one of the energy models suggested that the staggered orientations were the arrangements of minimum energy, but the energy difference between the eclipsed and staggered structures varied widely depending upon the model. The central multipole model was found to be ineffective due to convergence problems. Three different point-charge models were considered, all of which gave acceptable energy curves. The distributed multipole model also performed well, being comparable to the most accurate of the point-charge models.

3.8.10 Polarisation

Our discussion of electronic effects has concentrated so far on 'permanent' features of the charge distribution. Electrostatic interactions also arise from

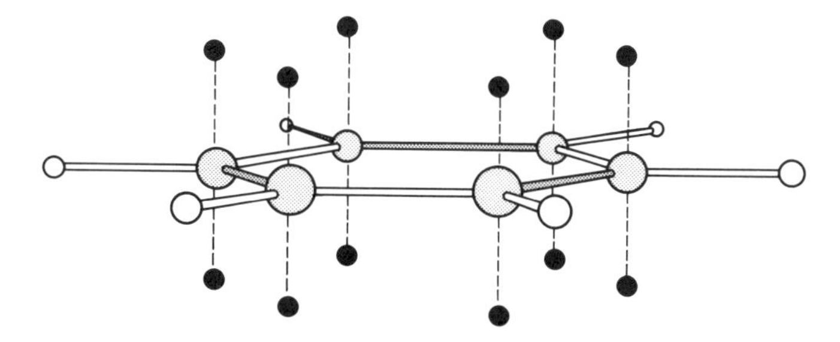

Figure 3.24 Anisotropic model of benzene developed by Hunter and Saunders [Hunter and Saunders 1990].

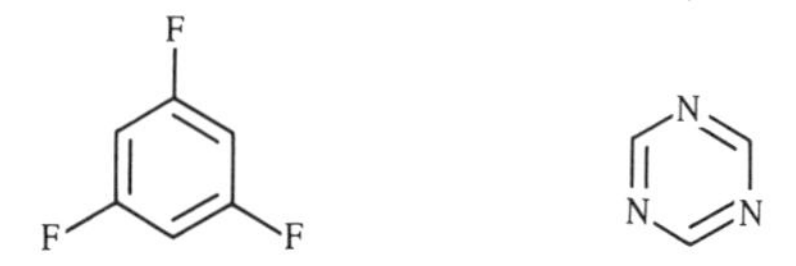

Figure 3.25 *Sym*-triazine and 1,3,5-trifluorobenzene.

changes in the charge distribution of a molecule or atom caused by an external field, a process called *polarisation*. The primary effect of the external electric field (which in our case will be caused by neighbouring molecules) is to induce a dipole in the molecule. The magnitude of the induced dipole moment $\boldsymbol{\mu}_{\text{ind}}$ is proportional to the electric field $\mathbf{E}$ with the constant of proportionality being the polarisability α:

$$\boldsymbol{\mu}_{\text{ind}} = \alpha \mathbf{E} \tag{3.38}$$

The energy of interaction between a dipole $\boldsymbol{\mu}_{\text{ind}}$ and an electric field $\mathbf{E}$ (the induction energy) is determined by calculating the work done in charging the field from zero to E, calculated from the following integral:

$$\mathscr{V}(\alpha, E) = -\int_0^E \mathrm{d}\mathbf{E}\boldsymbol{\mu}_{\text{ind}} = -\int_0^E \mathrm{d}\mathbf{E}\alpha\mathbf{E} = -\tfrac{1}{2}\alpha E^2 \tag{3.39}$$

In strong electric fields contributions to the induced dipole moment that are proportional to E^2 or E^3 can also be important, and higher-order moments such as quadrupoles can also be induced. We will not be concerned with such contributions.

For isolated atoms, the polarisability is isotropic – it does not depend on the orientation of the atom with respect to the applied field and the induced dipole is in the direction of the electric field, as in equation (3.38). However, the polarisability of a molecule is often anisotropic. This means that the direction of the induced dipole is not necessarily in the same direction as the electric field. The polarisability of a molecule is often modelled as a collection of isotropically polarisable atoms. A small molecule may alternatively be modelled as a single isotropic polarisable centre.

Let us consider the electric field due to a dipole μ aligned along the z axis. The magnitude of the electric field at a point P due to the dipole (see Figure 3.26) is:

$$E(r, \theta) = \frac{\mu\sqrt{1 + 3\cos^2\theta}}{4\pi\varepsilon_0 r^3} \tag{3.40}$$

The induction energy with another molecule of polarisability α placed at P is therefore

$$\mathscr{V}(r, \theta) = -\alpha\mu^2 \frac{1 + 3\cos^2\theta}{(4\pi\varepsilon_0 r^3)^2} \tag{3.41}$$

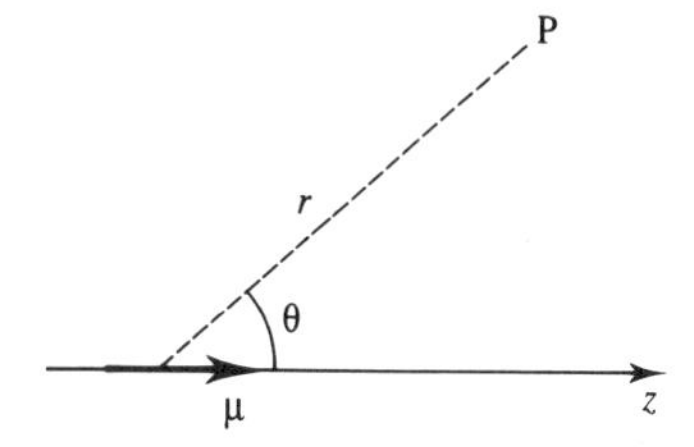

Figure 3.26 Electric field at point P due to dipole at the origin.

The interaction between a dipole and an induced dipole is independent of the disorientating effect of thermal motion whereas the dipole–dipole interaction between two permanent dipoles does vary with the relative orientation of the two dipoles. This is because the induced dipole follows the direction of the permanent dipole even as the molecules change their orientations as a consequence of molecular collisions.

An important consideration when modelling polarisation effects is that the dipole induced on a molecule (A) will affect the charge distribution of another molecule (B). The electric field at A due to the dipole(s) on B will in turn be affected. The presence of other molecules can also influence the interaction. Consider the polarisation interaction between a polar molecule and a neighbour (Figure 3.27). A third molecule may reduce the size of the electric field on the second molecule and so lower the induction energy. This type of three-body effect would be expected to be particularly significant when polarisable atoms are close to polar groups. Polarisation is a cooperative effect and as such is modelled using a set of coupled equations which are typically solved iteratively. Initially, the induced dipoles are set to zero. An initial approximation to each induced dipole is then calculated from the permanent charges (i.e. partial atomic charges). The electric field due to these induced dipoles is then added to the electric field from the permanent charges. This gives a refined value of the electric field from which a new induced dipole can be determined. The calculation continues until the induced dipoles do not change significantly between iterations.

A variety of schemes for including polarisation into molecular mechanics force fields have been devised. One approach is to model the polarisation effects at the atomic level, with dipoles being induced on each atom [Dang *et al.* 1991]. The magnitude of the dipole induced on an atom i is given by:

$$\boldsymbol{\mu}_{\mathrm{ind,i}} = \alpha_{\mathrm{i}} \mathbf{E}_{\mathrm{i}} \tag{3.42}$$

α_i is the atomic polarisability, assumed to be isotropic. Appropriate values of α_i have been determined for various systems. The electric field, $\mathbf{E}_i$ at atom i is the vector sum of the field due to the permanent and induced dipoles of the other atoms in the system:

$$\mathbf{E}_{\mathrm{i}} = \sum_{\mathrm{j} \neq \mathrm{i}} \frac{q_{\mathrm{j}} \mathbf{r}_{\mathrm{ij}}}{r_{\mathrm{ij}}^3} + \sum_{\mathrm{j} \neq \mathrm{i}} \frac{\boldsymbol{\mu}_{\mathrm{j}}}{r_{\mathrm{ij}}^3} \left(3\mathbf{r}_{\mathrm{ij}} \frac{\mathbf{r}_{\mathrm{ij}}}{r_{\mathrm{ij}}^2} - 1 \right) \tag{3.43}$$

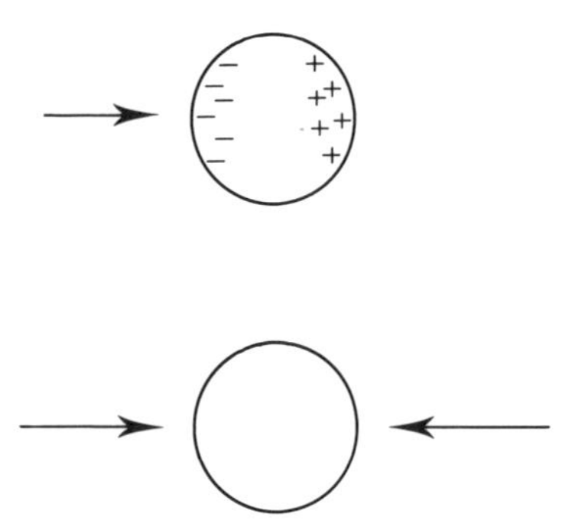

Figure 3.27 The polarisation interaction between a dipole and a polarisable molecule can be affected by the presence of a second dipole and is therefore a many-body effect.

$\mathbf{r}_i$ and $\mathbf{r}_j$ are the position vectors of the atoms i and j. Convergence of these equations in procedures such as molecular dynamics, where successive configurations are generated, can be accelerated if the induced dipoles obtained at each current step are used as the starting points for the next configuration.

An alternative way to model polarisation effects is exemplified by the water model of Sprik and Klein [Sprik and Klein 1988], where the polarisation centre is represented as a collection of closely spaced charges whose values are permitted to vary but whose total sums to zero. In the water model, shown in Figure 3.28, four tetrahedrally arranged charges are used to model the polarisation centre. These charges endow the molecule with an induced dipole moment of any magnitude and direction. The charges are determined iteratively for each configuration of the system. The isotropic polarisability of a simple ion can similarly be treated using two charges of equal magnitude but opposite sign placed either side of the ion. The direction of the 'bond' linking the two polarisation charges and the ion can reorient to change the direction of the induced dipole. In a subsequent refinement of this model Sprik and Klein replaced the point charges by Gaussian charge distributions at the polarisation sites; these were better at modelling features such as hydrogen bonding.

Due to the computational expense, polarisation effects are often included in a calculation only when their effect is likely to be significant, such as simulations of ionic solutions. These systems usually contain atoms or ions and small molecules only. It is important to be aware of the following problem when using atomic polarisabilities. Consider a diatomic molecule. The application of an external field will induce dipoles on both atoms. The dipole on one atom will also contribute to the electric field at the other atom, and thereby influence its induced dipole, but the model takes no account of the fact that the charge distributions on the two atoms are inherently linked. For this reason (and for reasons of computational efficiency) it is common to treat small molecules such as water as single polarisable centres when calculating polarisation effects.

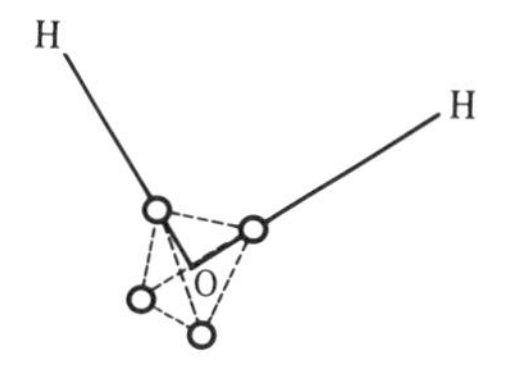

Figure 3.28 Polarisable models of water and ions developed by Sprik and Klein. Figure adapted from Sprik M 1993. Effective Pair Potentials and Beyond in Allen M P, D J Tildesley, Editors. *Computer Simulation in Chemical Physics.* Dordrecht, Kluwer Academic Publishers.

3.8.11 Solvent dielectric models

All of the formulae that we have written for electrostatic energies, potentials and forces have included the permittivity of free space, ε_0. This is as one would expect for species acting in a vacuum. However, under some circumstances a different dielectric model is used in the equations for the electrostatic interactions. This is often done when it is desired to mimic solvent effects, without actually including any explicit solvent molecules. One effect of a solvent is to dampen the electrostatic interactions. A very simple way to model this damping effect is to increase the permittivity, most easily by using an appropriate value for the relative permittivity in the Coulomb's law equation (i.e. $\varepsilon = \varepsilon_0\varepsilon_r$). An alternative approach is to make the dielectric dependent upon the separation of the charged species; this gives rise to the so-called distance-dependent dielectric models. The simplest implementation of a distance-dependent dielectric is to make the relative permittivity proportional to the distance. The interaction energy between two charges q_i and q_j then becomes:

$$\mathscr{V}(r) = \frac{1}{4\pi\varepsilon_0}\frac{q_i q_j}{r^2} \tag{3.44}$$

The simple distance-dependent dielectric has no physical basis and so it is not generally recommended, except when no alternative is possible. More sophisticated distance-dependent functions can also be employed. Many of these have an approximately sigmoidal shape in which the relative permittivity is low at short distances and then rises towards the bulk value at long distances. One example of such a function is [Smith and Pettit 1994]:

$$\varepsilon_{\text{eff}}(r) = \varepsilon_r - \frac{\varepsilon_r - 1}{2}[(rS)^2 + 2rS + 2]\text{e}^{-rS} \tag{3.45}$$

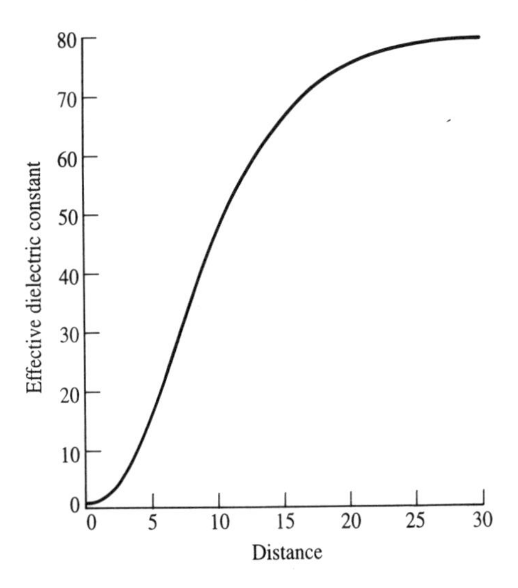

Figure 3.29 A sigmoidal dielectric model smoothly varies the effective permittivity from 80 to 1 as shown.

The value of ε_{eff} varies from a value of 1 at zero separation to ε_r (the bulk permittivity of the solvent) at large distances, in a manner determined by the parameter S (which is typically given a value between 0.15 Å^{-1} and 0.3 Å^{-1}), Figure 3.29. Sigmoidal functions give better behaviour than the simple distance-dependent dielectric model. However, it may be difficult to choose the appropriate value for the bulk dielectric ε_r when performing calculations on large solutes as the shortest distance between two charges may be through the solute molecule rather than through the solvent, Figure 3.30.

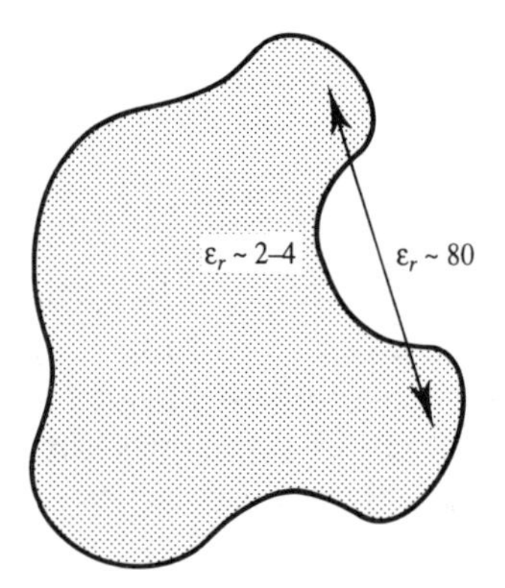

Figure 3.30 A line joining two points may pass through regions of different permittivity.

The polarisation term can be a major contributor to the free energy of solvation of a solute and a variety of schemes have been devised to incorporate such effects where the solvent is modelled as a continuum. We shall discuss these methods in more detail in sections 9.8–9.11.

3.9 van der Waals interactions

Electrostatic interactions cannot account for all of the non-bonded interactions in a system. The rare gas atoms are an obvious example; all of the multipole moments of a rare gas atom are zero and so there can be no dipole–dipole or dipole–induced dipole interactions. But there clearly must be interactions between the atoms; how else could the rare gases have liquid and solid phases or show deviations from ideal gas behaviour? Deviations from ideal gas behaviour were famously quantitated by van der Waals, thus the forces that give rise to such deviations are often referred to as van der Waals forces.

If we were to study the interaction between two isolated argon atoms using a molecular beam experiment then we would find that the interaction energy varies with the separation in a manner as shown in Figure 3.31. The other rare gases show a similar behaviour. The essential features of this curve are as follows. The interaction energy is zero at infinite distance (and indeed is negligible even at relatively short distances). As the separation is reduced, the energy decreases, passing through a minimum at a distance of approximately 3.8 Å for argon. The energy then rapidly increases as the separation decreases further. The force between the atoms, which equals minus the first derivative of the potential energy with respect to distance, is also shown in Figure 3.31. A variety of experiments

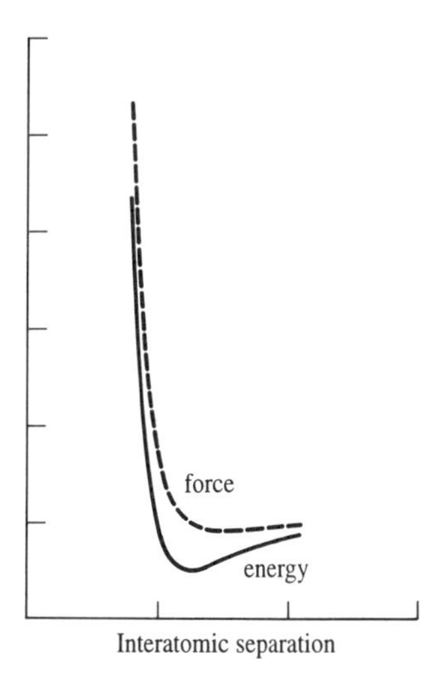

Figure 3.31 The interaction energy and the force between two argon atoms.

have been used to provide evidence for the nature of the van der Waals interactions, including gas imperfections, molecular beams, spectroscopic studies and measurements of transport properties.

3.9.1 Dispersive interactions

The curve in Figure 3.31 is usually considered to arise from a balance between attractive and repulsive forces. The attractive forces are long-range whereas the repulsive forces act at short distances. The attractive contribution is due to *dispersive forces*. London first showed how the dispersive force could be explained using quantum mechanics [London 1930] and so this interaction is sometimes referred to as the London force. The dispersive force is due to instantaneous dipoles which arise during the fluctuations in the electron clouds. An instantaneous dipole in a molecule can in turn induce a dipole in neighbouring atoms, giving rise to an attractive inductive effect.

A simple model to explain the dispersive interaction was proposed by Drude. This model consists of 'molecules' with two charges, $+q$ and $-q$ separated by a distance r. The negative charge performs simple harmonic motion with angular frequency ω along the z axis about the stationary positive charge (Figure 3.32). If the force constant for the oscillator is k and if the mass of the oscillating charge is m, then the potential energy of an isolated Drude molecule is $\frac{1}{2}kz^2$, where z is the separation of the two charges. ω is related to the force constant by $\omega = \sqrt{(k/m)}$. The Schrödinger equation for a Drude molecule is:

$$-\frac{\hbar^2}{2m}\frac{\partial^2\psi}{\partial z^2} + \frac{1}{2}kz^2\psi = E\psi \tag{3.46}$$

This is the Schrödinger equation for a simple harmonic oscillator. The energies of the system are given by $E_v = (v + \frac{1}{2})\hbar\omega$ and the zero-point energy is $\frac{1}{2}\hbar\omega$.

We now introduce a second Drude molecule, identical to the first, with the positive charge also located on the z axis and an oscillating negative charge (Figure 3.32). When the two molecules are infinitely separated, they do not interact and the total ground-state energy of the system is just twice the zero-point energy of a single molecule, $\hbar\omega/2\pi$. As the molecules approach (along the z axis) there are interactions between the two dipoles and the interaction energy between

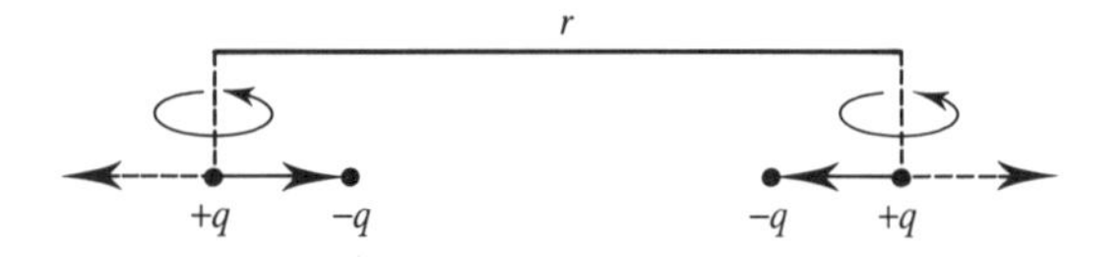

Figure 3.32 The Drude model for dispersive interactions. Figure adapted from Rigby M, E B Smith, W A Wakeham and G C Maitland 1986. *The Forces Between Molecules*. Oxford, Clarendon Press.

the two 'molecules' can be shown to be approximately given by (see Appendix 3.1):

$$\mathscr{V}(r) = -\frac{\alpha^4 \hbar\omega}{2(4\pi\varepsilon_0)^2 r^6} \tag{3.47}$$

The Drude model thus predicts that the dispersion interaction varies as $1/r^6$.

The two-dimensional Drude model can be extended to three dimensions, the result being:

$$\mathscr{V}(r) = -\frac{3\alpha^4 \hbar\omega}{4(4\pi\varepsilon_0)^2 r^6} \tag{3.48}$$

The Drude model only considers the dipole–dipole interaction; if higher order terms, due to dipole–quadrupole, quadrupole–quadrupole, etc. interactions are included as well as other terms in the binomial expansion, then the energy of the Drude model is more properly written as a series expansion:

$$\mathscr{V}(r) = \frac{C_6}{r^6} + \frac{C_8}{r^8} + \frac{C_{10}}{r^{10}} + \cdots \tag{3.49}$$

All of the coefficients C_n are negative, implying an attractive interaction. Despite its simplicity, the Drude model gives quite reasonable results; if just the C_6 term is included then for argon the resulting dispersion energy is only about 25% too small.

3.9.2 The repulsive contribution

Below about 3 Å, even a small decrease in the separation between a pair of argon atoms causes a large increase in the energy. This increase has a quantum mechanical origin and can be understood in terms of the Pauli principle, which formally prohibits any two electrons in a system from having the same set of quantum numbers. The interaction is due to electrons with the same spin, therefore the short-range repulsive forces are often referred to as *exchange forces*. They are also known as overlap forces. The effect of exchange is to reduce the electrostatic repulsion between pairs of electrons by forbidding them to occupy the same region of space (i.e. the internuclear region). The reduced electron density in the internuclear region leads to repulsion between the incompletely shielded nuclei. At very short internuclear separations, the interaction energy varies as $1/r$ due to this nuclear repulsion, but at larger separations the energy decays exponentially, as $\exp(-2r/a_0)$, where a_0 is the Bohr radius.

3.9.3 Modelling van der Waals interactions

The dispersive and exchange–repulsive interactions between atoms and molecules can be calculated using quantum mechanics, though such calculations are far from trivial, requiring electron correlation and large basis sets. For a force field we require a means to accurately model the interatomic potential curve (Figure 3.31), using a simple empirical expression that can be rapidly calculated. The need for a function that can be rapidly evaluated is a consequence of the large number of van der Waals interactions that must be determined in many of the systems that we would like to model. The best known of the van der Waals potential functions is the *Lennard–Jones 12–6 function*, which takes the following form for the interaction between two atoms:

$$\mathcal{V}(r) = 4\varepsilon\left[\left(\frac{\sigma}{r}\right)^{12} - \left(\frac{\sigma}{r}\right)^{6}\right] \tag{3.50}$$

The Lennard–Jones 12–6 potential contains just two adjustable parameters: the collision diameter σ (the separation for which the energy is zero) and the well depth ε. These parameters are graphically illustrated in Figure 3.33. The Lennard–Jones equation may also be expressed in terms of the separation at which the energy passes through a minimum, r_m (also written r^*). At this separation, the first derivative of the energy with respect to the internuclear distance is zero (i.e. $\partial\mathcal{V}/\partial r = 0$) from which it can easily be shown that $r_m = 2^{1/6}\sigma$. We can thus also write the Lennard–Jones 12–6 potential function as follows:

$$\mathcal{V}(r) = \varepsilon\{(r_m/r)^{12} - 2(r_m/r)^6\} \tag{3.51}$$

or

$$\mathcal{V}(r) = A/r^{12} - C/r^6 \tag{3.52}$$

A is equal to εr_m^{12} (or $4\varepsilon\sigma^{12}$) and C is equal to $2\varepsilon r_m^6$ (or $4\varepsilon\sigma^6$).

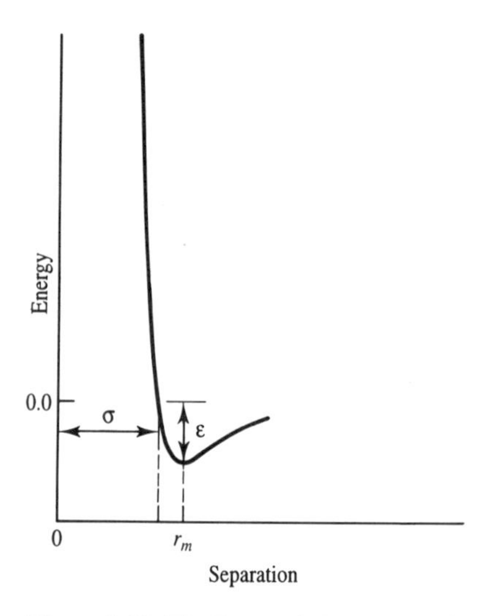

Figure 3.33 The Lennard–Jones potential.

The Lennard–Jones potential is characterised by an attractive part that varies as r^{-6} and a repulsive part that varies as r^{-12}. These two components are drawn in Figure 3.34. The r^{-6} variation is of course the same power-law relationship found for the leading term in theoretical treatments of the dispersion energy such as the Drude model. There are no strong theoretical arguments in favour of the repulsive r^{-12}, especially as quantum mechanics calculations suggest an exponential form. The twelfth power term is found to be quite reasonable for rare gases but is rather too steep for other systems such as hydrocarbons. However, the 6–12 potential is widely used, particularly for calculations on large systems as r^{-12} can be rapidly calculated by squaring the r^{-6} term. The r^{-6} term can also be calculated from the square of the distance without having to perform a computationally expensive square root calculation. Different powers have also been used for the repulsive part of the potential; values of 9 or 10 give a less steep curve and are used in some force fields. These alternative expressions are often derived from a van der Waals potential of the following general form:

$$\mathcal{V}(r) = \varepsilon\left[\frac{m}{n-m}\left(\frac{r_m}{r}\right)^n - \frac{n}{n-m}\left(\frac{r_m}{r}\right)^m\right] \tag{3.53}$$

Equation (3.53) returns the Lennard–Jones potential for $n = 12$ and $m = 6$.

Several formulations in which the r^{-12} term is replaced by a theoretically more realistic exponential expression have been proposed. These include the *Buckingham potential*:

$$\mathcal{V}(r) = \varepsilon\left[\frac{6}{\alpha-6}\mathrm{e}^{-\alpha\left(\frac{r}{r_m}-1\right)} - \frac{\alpha}{\alpha-6}\left(\frac{r_m}{r}\right)^6\right] \tag{3.54}$$

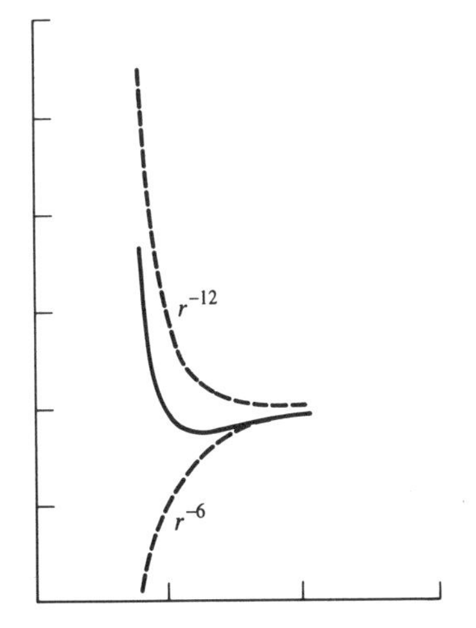

Figure 3.34 The Lennard–Jones potential is constructed from a repulsive component ($\propto r^{-12}$) and an attractive component ($\propto r^{-6}$).

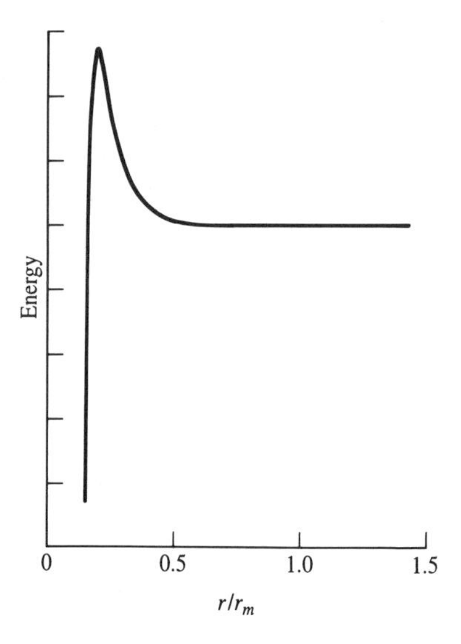

Figure 3.35 A drawback of the Buckingham potential is that it becomes steeply attractive at short distances.

There are three adjustable parameters in the Buckingham potential (ε, r_m and α). A value of α between approximately 14 and 15 gives a potential that closely corresponds to the Lennard–Jones 12–6 potential in the minimum energy region. When using the Buckingham potential it is important to remember that at very short distances the potential becomes strongly attractive, as shown in Figure 3.35. This could lead to nuclei being fused together during a calculation and so the program must check that atoms are not becoming too close. The *Hill potential* is an exponential–6 potential with just two parameters: the minimum energy radius r_m and the well depth ε [Hill 1948]:

$$\mathcal{V}(r) = -2.25\varepsilon(r_m/r)^6 + 8.28 \times 10^5 \varepsilon \exp(-r/0.0736 r_m) \tag{3.55}$$

The Hill potential was originally developed to enable the more realistic exponential term to be written in terms of Lennard–Jones parameters. The coefficients 2.25, 8.25×10^5 and 0.0736 in equation (3.35) were determined by fitting to data for the rare gases, and were assumed to be applicable to other non-polar gases. A Morse potential may also be used to model the van der Waals interactions in a force field, with appropriate parameters.

3.9.4 van der Waals interactions in polyatomic systems

The interaction energy between molecules depends not only upon their separation but also their relative orientations and, where appropriate, their conformations. It is usual to calculate the van der Waals interaction energy between two molecules using a site model in which the interaction is determined as the sum of the interactions between all pairs of sites on the two molecules. The sites are often identified with the nuclear positions, but this need not necessarily be the case.

Polyatomic systems invariably involve the calculation of van der Waals interactions between different types of atoms. For example, to calculate the Lennard–Jones interaction energy between two carbon monoxide molecules using a two-site model would require not only van der Waals parameters for the carbon–carbon interactions and the oxygen–oxygen interactions but also for the carbon–oxygen interactions. A system containing N different types of atom would require $N(N-1)/2$ sets of parameters for the interaction between unlike atoms. The determination of van der Waals parameters can be a difficult and time-consuming process and so it is common to assume that parameters for the cross interactions can be obtained from the parameters of the pure atoms using *mixing rules*. In the commonly used Lorentz–Berthelot mixing rules, the collision diameter σ_{AB} for the A–B interaction equals the arithmetic mean of the values for the two pure species and the well depth ε_{AB} is given as the geometric mean:

$$\sigma_{AB} = \tfrac{1}{2}(\sigma_{AA} + \sigma_{BB}) \tag{3.56}$$

$$\varepsilon_{AB} = \sqrt{(\varepsilon_{AA}\varepsilon_{BB})} \tag{3.57}$$

When written in terms of the separation of minimum energy (r^* or r_m), the following notation may be encountered:

$$r^*_{AB} = R^*_{AA} + R^*_{BB} \tag{3.58}$$

R^*_{AA} and R^*_{BB} are atomic parameters, equal to one half of r^*_{AA} and r^*_{BB} respectively.

The Lorentz–Berthelot combining rules are most successful when applied to similar species. Their major failing is that the well depth can be overestimated by the geometric mean rule. Some force fields calculate the collision diameter for mixed interactions as the geometric mean of the values for the two component atoms. Jorgensen's OPLS force field falls into this category [Jorgensen and Tirado-Reeves 1988].

In some force fields the interaction sites are not all situated on the atomic nuclei. For example, in the MM2 program, the van der Waals centres of hydrogen atoms bonded to carbon are placed not at the nuclei but are approximately 10% along the bond towards the attached atom. The rationale for this is that the electron distribution about small atoms such as oxygen, fluorine, but particularly hydrogen, is distinctly non-spherical. The single electron from the hydrogen is involved in the bond to the adjacent atom and there are no other electrons that can contribute to the van der Waals interactions. Some force fields also require lone pairs to be defined on particular atoms; these have their own van der Waals and electrostatic parameters.

The van der Waals and electrostatic interactions between atoms separated by three bonds (i.e. the 1,4 atoms) are often treated differently from other non-bonded interactions. The interaction between such atoms contributes to the rotational barrier about the central bond, in conjunction with the torsional potential. These 1,4 non-bonded interactions are often scaled down by an empirical factor; for example, a factor of 2.0 is suggested for both the electrostatic and van der Waals terms in the 1984 AMBER force field (a scale factor of 1/1.2 is used for the electrostatic terms in the 1995 AMBER force field). There are several reasons why one would wish to scale the 1,4 interactions. The error associated with the use of an r^{-12} repulsion term (which is too steep compared with the more correct exponential term) would be most significant for 1,4 atoms. In addition, when two 1,4 atoms come close together some redistribution of the charge along the connecting bonds would be expected that would act to reduce the interaction. Such a charge redistribution would not be possible for two atoms at a similar distance apart if they were in different molecules.

The parameters for the van der Waals interactions can be obtained in a variety of ways. In the early force fields, such parameters were often determined from an analysis of crystal packing. The objective of such studies was to produce a set of van der Waals parameters which enabled the experimental geometries and thermodynamic properties such as the heat of sublimation to be reproduced as accurately as possible. More recent force fields derive their van der Waals parameters using liquid simulations in which the parameters are optimised to reproduce a range of thermodynamic properties such as the densities and enthalpies of vaporisation for appropriate liquids.

3.9.5 *Reduced units*

The Lennard–Jones potential is completely specified by the two parameters ε and σ. This means that the results of a calculation performed on (say) liquid argon can be easily converted to give equivalent results for another noble gas. For this reason is it common to simulate the rare gases in terms of reduced units with ε and σ both set to 1. The results can then be converted to any system as appropriate. For example, the reduced density ρ^* is related to the real density by $\rho^* = \rho\sigma^3$; the reduced energy E^* is given by $E^* = E/\varepsilon$, and so on. Electrostatic interactions given by Coulomb's law are also often written in terms of a reduced unit of charge, which corresponds to each charge being divided by $\sqrt{(4\pi\varepsilon_0)}$. This means that Coulomb's law takes the less cumbersome form:

$$\mathcal{V}(q_1, q_2) = q_1 q_2 / r_{12} \quad \text{or} \quad \mathcal{V}(q_1, q_2) = q_1 q_2 / \varepsilon_r r_{12} \tag{3.59}$$

3.10 Many-body effects in empirical potentials

The electrostatic and van der Waals energies that we have considered so far are calculated between pairs of interaction sites. The total non-bonded interaction energy is thus determined by adding together the interactions between all pairs of sites in the system. However, the interaction between two molecules can be affected by the presence of a third, fourth or more molecules. For example, the interaction energy between three molecules A, B and C is not in general given by the sum of the pairwise interaction energies: $\mathscr{v}(\mathrm{A,B,C}) \neq \mathscr{v}(\mathrm{A,B}) + \mathscr{v}(\mathrm{A,C}) + \mathscr{v}(\mathrm{B,C})$. We have already seen an example of a non-pairwise contribution, namely the polarisation interaction which is determined using a self-consistent procedure.

Three-body effects can significantly affect the dispersion interaction. For example, it is believed that three-body interactions account for approximately 10% of the lattice energy of crystalline argon. For very precise work, interactions involving more than three atoms may have to be taken into account, but they are usually small enough to be ignored. A potential that includes both two- and three-body interactions would be written in the following general form:

$$\mathscr{V}(\mathbf{r}^N) = \sum_{\mathrm{i}=1}^{N}\sum_{\mathrm{j}=\mathrm{i}+1}^{N} \mathscr{v}^{(2)}(r_{\mathrm{ij}}) + \sum_{\mathrm{i}=1}^{N}\sum_{\mathrm{j}=\mathrm{i}+1}^{N}\sum_{\mathrm{k}=\mathrm{j}+1}^{N} \mathscr{v}^{(3)}(r_{\mathrm{ij}}, r_{\mathrm{ik}}, r_{\mathrm{jk}}) \tag{3.60}$$

Axilrod and Teller investigated the three-body dispersion contribution, and showed that the leading term is:

$$\mathscr{v}^{(3)}(r_{\mathrm{AB}}\, r_{\mathrm{AC}}, r_{\mathrm{BC}}) = \nu_{\mathrm{A,B,C}} \frac{3\cos\theta_{\mathrm{A}}\cos\theta_{\mathrm{B}}\cos\theta_{\mathrm{C}}}{(r_{\mathrm{AB}} r_{\mathrm{AC}} r_{\mathrm{BC}})^3} \tag{3.61}$$

θ_{A}, θ_{B} and θ_{C} are the internal angles of the triangle with sides of length r_{AB}, r_{AC} and r_{BC} (Figure 3.36). $\nu_{\mathrm{A,B,C}}$ is a constant characteristic of the three species A, B and C. If A, B and C are identical then $\nu_{\mathrm{A,B,C}}$ is approximately related to the Lennard–Jones coefficient C_6 and the polarisability by

$$\nu_{\mathrm{A,B,C}} = -\frac{3\alpha C_6}{4(4\pi\varepsilon_0)} \tag{3.62}$$

The effect of the Axilrod–Teller term (also known as the triple-dipole correction) is to make the interaction energy more negative when three molecules are linear but to weaken it when the molecules form an equilateral triangle. This is because the linear arrangement enhances the correlations of the motions of the electrons whereas the equilateral arrangement reduces it.

The three-body contribution may also be modelled using a term of the form $\mathscr{v}^{(3)}(r_{\mathrm{AB}}, r_{\mathrm{AC}}, r_{\mathrm{BC}}) = K_{\mathrm{A,B,C}}\{\exp(-\alpha r_{\mathrm{AB}})\exp(-\beta r_{\mathrm{AC}})\exp(-\gamma r_{\mathrm{BC}})\}$ where K, α, β and γ are constants describing the interaction between the atoms A, B and C. Such a functional form has been used in simulations of ion–water systems, where

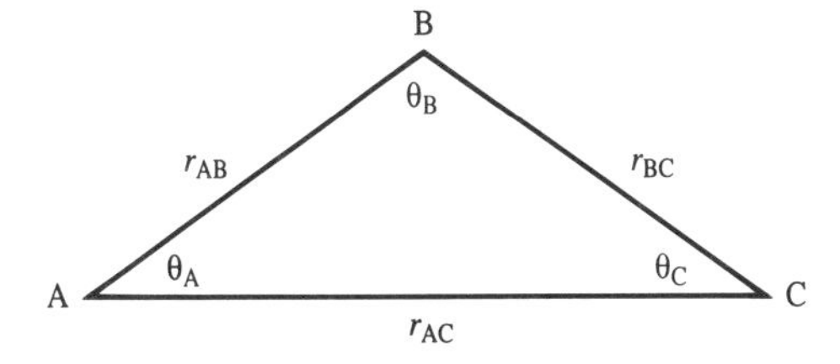

Figure 3.36 Calculating the three-body Axilrod–Teller contribution.

polarisation alone does not exactly model configurations when there are two water molecules close to an ion [Lybrand and Kollman 1985]. The three-body exchange repulsion term is thus only calculated for ion–water–water trimers when the species are close together.

The computational effort is significantly increased if three-body terms are included in the model. Even with a simple pairwise model, the non-bonded interactions usually require by far the greatest amount of computational effort. The number of bond, angle and torsional terms increases approximately with the number of atoms (N) in the system, but the number of non-bonded interactions increases with N^2. There are $N(N-1)/2$ distinct pairs of interactions to evaluate for a pairwise potential. If three-body effects are included then there are $N(N-1)(N-2)/6$ unique three-body interactions. A system with 1000 atoms has 499 500 pairwise interactions and 166 167 000 three-body interactions. In general, there are approximately $N/3$ times more three-body terms than two-body terms and so it is clear why it is often considered preferable to avoid calculating the three-body interactions.

3.11 Effective pair potentials

Fortunately, it is found that a significant proportion of the many-body effects can be incorporated into a pairwise model, if properly parametrised. The pair potentials most commonly used in molecular modelling are thus 'effective' pairwise potentials; they do not represent the true interaction energy between two isolated particles, but are parametrised to include many-body effects in the pairwise energy. Similarly, polarisation effects can be implicitly included in a force field by the simple expedient of enhancing the electrostatic interaction. This can be done by using larger partial charges than those for an isolated molecule. This is manifested in larger multipole moments; the dipole moment of a single water molecule is 1.85 D whereas the dipole moment of many simple water models designed to simulate liquid water are significantly larger (closer to the experimental value for liquid water of 2.6 D).

A notable example of a potential that does include many-body terms is the Barker–Fisher–Watts potential for argon which combines a pairwise potential with an Axilrod–Teller triple potential [Barker *et al.* 1971]. The pair potential is a linear combination of two potentials that each take the following form:

$$\begin{aligned}\mathcal{V}^*(r^*) = {} & e^{\alpha(1-r^*)}[A_0 + A_1(r^*-1) + A_2(r^*-1)^2 + A_3(r^*-1)^3 \\ & + A_4(r^*-1)^4 + A_5(r^*-1)^5] \\ & + \frac{C_6}{\delta + r^{*6}} + \frac{C_8}{\delta + r^{*8}} + \frac{C_{10}}{\delta + r^{*10}}\end{aligned} \tag{3.63}$$

This potential function contains eleven constants: α, $A_0 \ldots A_5$, C_6, C_8, C_{10} and δ. The function is expressed in terms of r^*, which is given by $r^* = r/r_m$ where r_m is the separation at the minimum in the potential. The 'true' interaction energy as

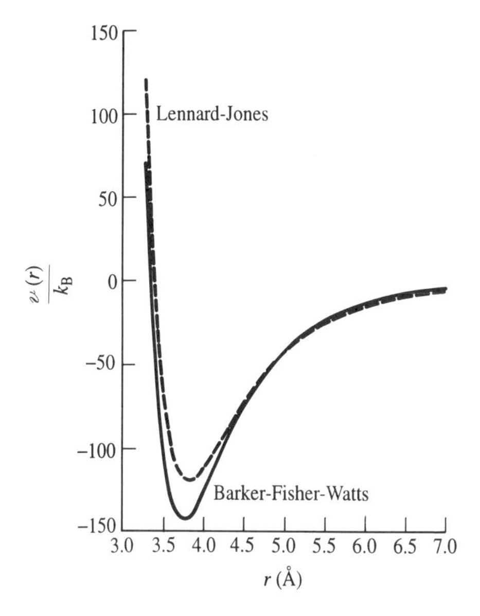

Figure 3.37 Comparison of the Lennard–Jones potential for argon with the Barker–Fisher–Watts pair potential; k_B is Boltzmann's constant.

a function of the separation, r, is then obtained by multiplying $\mathcal{V}^*(r^*)$ by the depth of the potential well, ε:

$$\mathcal{V}(r) = \varepsilon \mathcal{V}^*(r^*) \tag{3.64}$$

A comparison of the pairwise contribution to the Barker–Fisher–Watts potential with the Lennard–Jones potential for argon is shown in Figure 3.37.

3.12 Hydrogen bonding in molecular mechanics

Some force fields replace the Lennard–Jones 6–12 term between hydrogen-bonding atoms by an explicit hydrogen-bonding term, which is often described using a 10–12 Lennard–Jones potential:

$$\mathcal{V}(r) = \frac{A}{r^{12}} - \frac{C}{r^{10}} \tag{3.65}$$

This function is used to model the interaction between the donor hydrogen atom and the heteroatom acceptor atom. Its use is intended to improve the accuracy with which the geometry of hydrogen-bonding systems is predicted. Other force fields incorporate a more complicated hydrogen-bonding function that takes into account deviations from the geometry of the hydrogen bond and is thus dependent upon the coordinates of the donor and acceptor atoms as well as the hydrogen atom. For example, the YETI force field [Vedani 1988] uses the

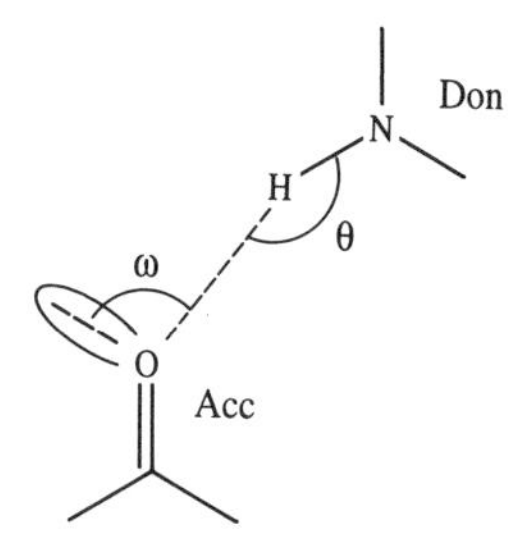

Figure 3.38 Definition of hydrogen-bond geometry used in YETI force field.

following form for its hydrogen bonding term:

$$\mathscr{V}_{\mathrm{HB}} = \left(\frac{A}{r_{\mathrm{H\cdots Acc}}^{12}} - \frac{C}{r_{\mathrm{H\cdots Acc}}^{10}} \right) \cos^2 \theta_{\mathrm{Don\cdots H\cdots Acc}} \cos^4 \omega_{\mathrm{H\cdots Acc-LP}} \tag{3.66}$$

The energy in equation (3.66) depends upon the distance from the hydrogen to the acceptor, the angle subtended at the hydrogen by the bonds to the donor and the acceptor, and the deviation of the hydrogen bond from the closest lone-pair direction at the acceptor atom ($\omega_{\mathrm{H\cdots Acc-LP}}$ in equation (3.66), Figure 3.38).

The GRID program [Goodford 1985] that is used for finding energetically favourable regions in protein binding sites uses a direction-dependent 6–4 function:

$$\mathscr{V}_{\mathrm{HB}} = \left(\frac{C}{d^6} - \frac{D}{d^4} \right) \cos^m \theta \tag{3.67}$$

θ is the angle subtended at the hydrogen and m is usually set to 4.

By no means do all force fields contain explicit hydrogen-bonding terms; most rely upon electrostatic and van der Waals interactions to reproduce hydrogen bonding.

3.13 Force field models for the simulation of liquid water

Many of the concepts that we have considered so far can be illustrated by examining some of the empirical models that have been developed to study water. Despite its small size, water acts as a paradigm for the different force field models that we have discussed. Moreover, many of its properties can be easily determined using computer simulation methods and so readily compared with experiment. It is also one of the most challenging systems to model accurately. A wide range of water models have been proposed. The computational efficiency with which the energy can be calculated using a given model is often an important factor as there may be a very large number of water molecules present, together with a solute; most of the force fields used to simulate liquid water thus use effective pairwise potentials with no explicit three-body terms or polarisation effects.

Water models can be conveniently divided into three types. In the simple interaction site models each water molecule is maintained in a rigid geometry and the interaction between molecules is described using pairwise Coulombic and Lennard–Jones expressions. Flexible models permit internal changes in conformation of the molecule. Finally, models have been developed that explicitly include the effects of polarisation and many body effects.

3.13.1 Simple water models

The 'simple' water models use between three and five interaction sites and a rigid water geometry. The TIP3P [Jorgensen *et al.* 1983] and SPC [Berendsen *et al.* 1981] models use a total of three sites for the electrostatic interactions; the partial positive charges on the hydrogen atoms are exactly balanced by an appropriate negative charge located on the oxygen atom. The van der Waals interaction between two water molecules is computed using a Lennard–Jones function with just a single interaction point per molecule centred on the oxygen atom; no van der Waals interactions involving the hydrogen atoms are calculated. The TIP3P and SPC models differ slightly in the geometry of each water molecule, in the hydrogen charges and in the Lennard–Jones parameters. These differences are indicated in Table 3.3, which also includes data for the SPC/E model [Berendsen *et al.* 1987] that is an updated version of the SPC model. The four-site models such as that of Bernal and Fowler [Bernal and Fowler 1933] (which is now relatively little used but is important for historical reasons as it dates from 1933) and Jorgensen's TIP4P model [Jorgensen *et al.* 1983] shift the negative charge from the oxygen atom to a point along the bisector of the HOH angle towards the hydrogens (Figure 3.39). The parameters for these two models are also given in the table. The most commonly used five-site model is the ST2 potential of Stillinger and Rahman [Stillinger and Rahman 1974]. Here, charges are placed on the hydrogen atoms and on two lone pair sites on the oxygen. The electrostatic contribution is modulated so that for oxygen–oxygen distances below 2.016 Å it is zero and for distances greater than 3.1287 Å it takes its full value. Between these two distances the electrostatic contribution is modulated using a function that smoothly varies from 0.0 at the shorter distance to 1.0 at the longer distance (see section 5.7).

The experimentally determined dipole moment of a water molecule in the gas phase is 1.85 D. The dipole moment of an individual water molecule calculated with any of these simple models is significantly higher; for example, the SPC dipole moment is 2.27 D and that for the TIP4P is 2.18 D. These values are much closer to the effective dipole moment of liquid water, which is approximately 2.6 D. These models are thus all effective pairwise models. The simple water models are usually parametrised by calculating various properties using molecular dynamics or Monte Carlo simulations and then modifying the parameters until the desired level of agreement between experiment and theory is achieved. Thermodynamic and structural properties are usually used in the parametrisation, such as the density, radial distribution function, enthalpy of vaporisation, heat capacity, diffusion coefficient and dielectric constant.* It is found that some properties such as the density and the enthalpy of vaporisation are predicted rather well by all of the models, but there is significant variation in the values for other properties such as the dielectric constant [Jorgensen *et al.* 1983]. When comparing the different models, it is also important to take account of the computational effort each requires. Thus, 9 site–site distances must be calculated for each water dimer using a three-site model; 10 are required for a four-site model, and 17 for the ST2 model.

The use of a rigid model for water is obviously an approximation, and it means that some properties cannot be determined at all. For example, only when internal flexibility is included can the vibrational spectrum be calculated and compared with experiment. Flexibility is most easily incorporated by 'grafting' bond stretching and angle bending terms onto the potential function for a rigid model. Such an approach needs to be done with care. For example, Ferguson has developed a flexible model for water that is based upon the SPC model [Ferguson 1995]. The partial charges and van der Waals parameters in this model were slightly different from those in the rigid model, and flexibility was achieved using cubic and harmonic bond stretching terms and a harmonic angle bending term. The calculated values compared well with experimental results for a wide range of thermodynamic and structural properties, including the dielectric constant and self-diffusion coefficient.

Table 3.3 A comparison of various water models [Jorgensen *et al.* 1983]. For the ST2 potential, q(M) is the charge on the 'lone pairs' which are a distance 0.8 Å from the oxygen atom (see Figure 3.39).

	SPC	SPC/E	TIP3P	BF	TIP4P	ST2
r(OH), Å	1.0	1.0	0.9572	0.96	0.9572	1.0
HOH, deg	109.47	109.47	104.52	105.7	104.52	109.47
$A \times 10^{-3}$, kcal Å^{12}/mol	629.4	629.4	582.0	560.4	600.0	238.7
C, kcal Å^6/mol	625.5	625.5	595.0	837.0	610.0	268.9
q(O)	−0.82	−0.8472	−0.834	0.0	0.0	0.0
q(H)	0.41	0.4238	0.417	0.49	0.52	0.2375
q(M)	0.0	0.0	0.0	−0.98	−1.04	−0.2375
r(OM), Å	0.0	0.0	0.0	0.15	0.15	0.8

* A discussion of the calculation of these properties from computer simulations is given in section 5.2.

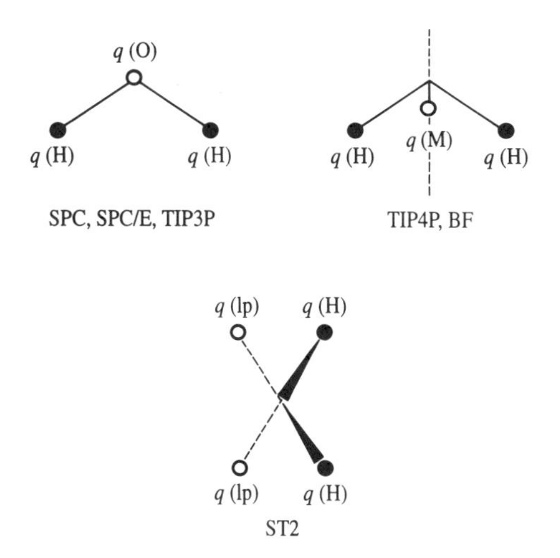

Figure 3.39 Some 'simple' water models (Table 3.3) [Jorgensen *et al.* 1983].

3.13.2 *Polarisable water models*

The simple models give very good results for a wide range of properties of pure liquid water. However, there is some concern that they are not appropriate models to use for the most accurate work. This is especially the case for inhomogeneous systems where there are strong electric field gradients due to the presence of ions, and at the solute–solvent interface. Under such circumstances models that explicitly include polarisation effects and three-body terms are considered to be more appropriate. The inclusion of an explicit polarisation term should also enhance the ability of the model to reproduce the behaviour of water in other phases (e.g. solid and vapour) and at the interface between different phases. The dipole moment of an isolated water molecule in such a model should thus be closer to the gas-phase value rather than to the 'effective' value in liquid water. The simplest way to include polarisation is to use an isotropic molecular polarisability contribution; an alternative is to use atom-centred polarisabilities or the variable charge method. The incorporation of polarisable can significantly increase the computational effort required for a liquid simulation, and even then only the best polarisable models currently compete with the well-established models that use effective pairwise potentials. We have already considered some of the polarisable water models in our discussion of polarisation effects. One early attempt to incorporate such effects into a water model was made by Barnes, Finney, Nicholas and Quinn [Barnes *et al.* 1979]. Their polarisable electropole water model represented the charge distribution by a multipole expansion comprising a dipole of 1.855 D and a quadrupole moment that was determined from quantum mechanical calculations on an isolated molecule. Polarisation effects were calculated using an isotropic molecular polarisability from the electric fields being produced by the dipoles and quadrupoles of surrounding molecules. The model also used a spherically symmetric Lennard–Jones function.

3.13.3 **Ab initio** *potentials for water*

The final category of water model that we shall consider are the '*ab initio*' potentials. These are based upon *ab initio* quantum mechanical calculations on small clusters of water molecules. One example of this type is the NCC model of Nieser, Corongiu and Clementi which combines a two-molecule potential with a polarisation term [Niesar *et al.* 1990]. They had previously tried to explicitly include both three- and four-body effects but found this model computationally too expensive. The two-body model uses partial charges on the hydrogen atoms and a compensating negative charge on a site located along the bisector of the HOH angle, as in the TIP4P model. The equation used is:

$$\begin{aligned}
\mathcal{V}_{\text{two-body}} = {} & q^2\left(\frac{1}{R_{13}} + \frac{1}{R_{14}} + \frac{1}{R_{23}} + \frac{1}{R_{24}}\right) \\
& + \frac{4q^2}{R_{78}} - 2q^2\left(\frac{1}{R_{81}} + \frac{1}{R_{82}} + \frac{1}{R_{73}} + \frac{1}{R_{74}}\right) \\
& + A_{\text{OO}}e^{-B_{\text{OO}}R_{56}} + A_{\text{HH}}(e^{-B_{\text{HH}}R_{13}} + e^{-B_{\text{HH}}R_{14}} \\
& \qquad + e^{-B_{\text{HH}}R_{23}} + e^{-B_{\text{HH}}R_{24}}) \\
& + A_{\text{OH}}(e^{-B_{\text{OH}}R_{53}} + e^{-B_{\text{OH}}R_{54}} + e^{-B_{\text{OH}}R_{61}} + e^{-B_{\text{OH}}R_{62}}) \\
& - A'_{\text{OH}}(e^{-B'_{\text{OH}}R_{53}} + e^{-B'_{\text{OH}}R_{54}} + e^{-B'_{\text{OH}}R_{61}} + e^{-B'_{\text{OH}}R_{62}}) \\
& + A_{\text{PH}}(e^{-B_{\text{PH}}R_{73}} + e^{-B_{\text{PH}}R_{74}} + e^{-B_{\text{PH}}R_{81}} + e^{-B_{\text{PH}}R_{82}}) \\
& + A_{\text{PO}}(e^{-B_{\text{PO}}R_{76}} + e^{-B_{\text{PO}}R_{85}})
\end{aligned} \tag{3.68}$$

The points P are the locations where the negative charge is placed (numbered 7 and 8 in Figure 3.40) and the terms A_{PH} and A_{PO} are used to enhance the performance of the model at short distances. q is the charge on each hydrogen. The polarisation term is calculated in an iterative manner using induced dipoles along each O–H bond. The NCC model was parametrised by fitting to the energies and other properties of 250 trimer and 350 dimer configurations determined with high-level *ab initio* methods and large basis sets. The water trimer data was used to fit the many-body parameters (i.e. the locations of the induced dipole moments and the point charges, together with the polarisability and the value of the hydrogen charge). The dimer data were then used to fit the remaining terms in the potential.

The original NCC potential was designed as a rigid water model and performed well in tests of its ability to reproduce experimental data for both water dimers and for liquid water. A flexible version has also been developed [Corongiu 1992],

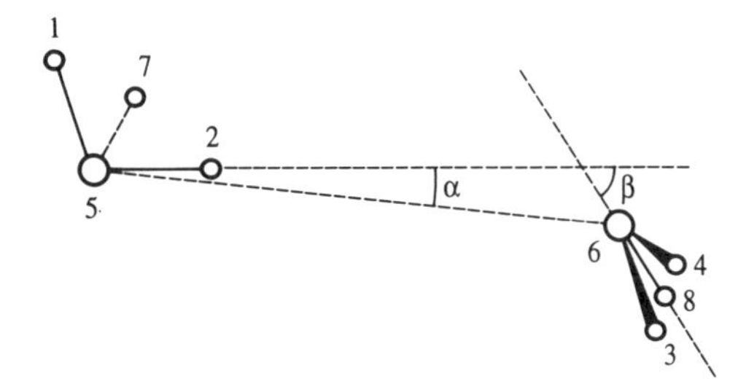

Figure 3.40 The NCC water model. After Corongiu G 1992. Molecular Dynamics Simulation for Liquid Water Using a Polarisable and Flexible Potential. *International Journal of Quantum Chemistry* **42**:1209–1235.

with the energy being expressed as a function of the three internal coordinates (two bond lengths and one angle) with terms up to quartics:

$$\begin{aligned}
\mathscr{V}_{\text{intra}} = {} & \tfrac{1}{2} f_{RR}(\delta_1^2 + \delta_2^2) + \tfrac{1}{2} f_{\theta\theta}(\delta_3^2) + f_{RR'}\delta_1\delta_2 + f_{R\theta}(\delta_1 + \delta_2)\delta_3 \\
& + \frac{1}{R_e}[f_{RRR}(\delta_1^3 + \delta_2^3) + f_{\theta\theta\theta}\delta_3^3 + f_{RRR'}(\delta_1 + \delta_2)\delta_1\delta_2 \\
& \qquad + f_{RR\theta}(\delta_1^2 + \delta_2^2)\delta_3 + f_{RR'\theta}\delta_1\delta_2\delta_3 + f_{R\theta\theta}(\delta_1 + \delta_2)\delta_3^2] \\
& + \frac{1}{R_e^2}[f_{RRRR}(\delta_1^4 + \delta_2^4) + f_{\theta\theta\theta\theta}\delta_3^4 + f_{RRRR'}(\delta_1^2 + \delta_2^2)\delta_1\delta_2 \\
& \qquad + f_{RRR'R'}\delta_1^2\delta_2^2 + f_{RRR\theta}(\delta_1^3 + \delta_2^3)\delta_3] \\
& + \frac{1}{R_e^2}[f_{RRR'\theta}(\delta_1 + \delta_2)\delta_1\delta_2\delta_3 + f_{RR\theta\theta}(\delta_1^2 + \delta_2^2)\delta_3^2 \\
& \qquad + f_{RR'\theta\theta}\delta_1\delta_2\delta_3^2 + f_{R\theta\theta\theta}(\delta_1 + \delta_2)\delta_3^3]
\end{aligned} \tag{3.69}$$

where

$$\begin{aligned}
\delta_1 &= R_1 - R_e \\
\delta_2 &= R_2 - R_e \\
\delta_3 &= R_e(\theta - \theta_e)
\end{aligned}$$

The functional form of the NCC model demonstrates the complexity of some empirical models (and this for a molecule that contains only three atoms!). We should also note that the development of empirical models from *ab initio* quantum mechanical data is an approach that is already well established and looks likely to be a method that is more widely used in the future.

3.14 United atom force fields and reduced representations

In our discussion so far, we have assumed that all of the atoms in the system are explicitly represented in the model. However, as the number of non-bonded interactions scales with the square of the number of interaction sites present, there are clear advantages if the number of interaction sites can be reduced. The simplest way to do this is to subsume some or all of the atoms (usually just the hydrogen atoms) into the atoms to which they are bonded. A methyl group would then be modelled as a single 'pseudo-atom' or 'united atom'. The van der Waals and electrostatic parameters would be modified to take account of the adjoining hydrogen atoms. Considerable computational savings are possible; for example, if butane is modelled as a four-site model rather than one with twelve atoms then the van der Waals interaction between two butane molecules involves the calculation of 16 terms rather than 144. Other hydrocarbons are often represented using united atom models. Many of the earliest calculations on proteins used united atom representations. In this case, not all of the hydrogen atoms in the protein are subsumed into their adjacent atoms, but just those that are bonded to carbon atoms. Hydrogen atoms bonded to polar atoms such as nitrogen and oxygen are able to participate in hydrogen bonding interactions which are modelled much better if these hydrogens are explicitly represented.

One drawback with a united atom force field is that chiral centres may be able to invert during a calculation. This was found to be a problem with the united atom force fields for proteins. The alpha carbon in the peptide unit (C_α in Figure 3.41) is bonded to a hydrogen atom and to the side chain (glycine and proline are slightly different; see section 8.13). A united atom force field model would not explicitly include the alpha hydrogen. Unfortunately, the stereochemistry at the alpha carbon can then invert during a calculation. This should be avoided as the naturally occurring amino acids have a defined stereochemistry (as shown in Figure 3.41). This inversion may be prevented through the use of an improper torsion term (e.g. N–C–C_α–R) to keep the side chain in the correct relative position.

In a united atom force field the van der Waals centre of the united atom is usually associated with the position of the heavy (i.e. non-hydrogen) atom. Thus, for a united CH_3 or CH_2 group the van der Waals centre would be located at the carbon atom. It would be more accurate to associate the van der Waals centre with

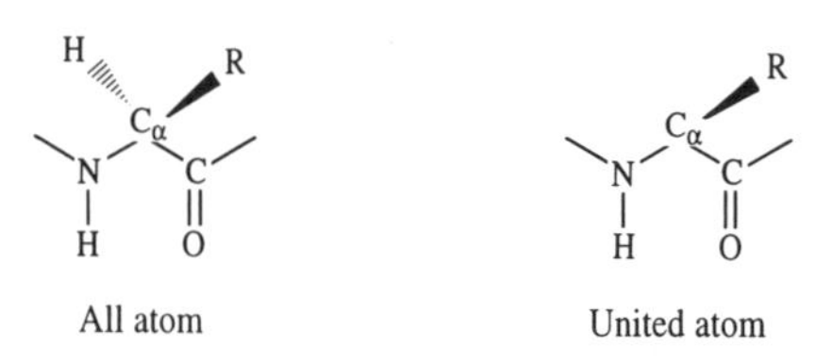

Figure 3.41 Representations of the naturally occurring amino acids.

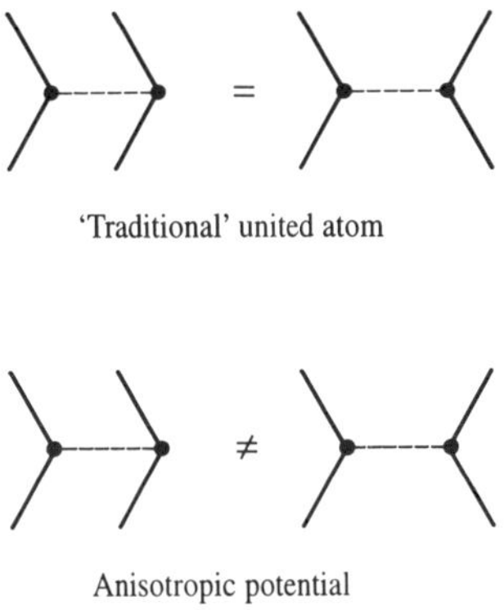

'Traditional' united atom

Anisotropic potential

Figure 3.42 The interaction energy between the two arrangements shown is equal in a 'traditional' united atom force field but different in the Toxvaerd anisotropic model. After Toxvaerd S 1990. Molecular Dynamics Calculations of the Equation of State of Alkanes. *The Journal of Chemical Physics* **93**:4290–4295.

a position that was offset slightly from the carbon position, in order to reflect the presence of the hydrogen atoms. Toxvaerd has developed such a model that gives superior performance for alkanes than do the simple united atom models, particularly for simulations at high pressures [Toxvaerd 1990]. In Toxvaerd's model the interaction sites are located at the geometrical centres of the CH_2 or CH_3 groups. The forces between these sites act on the united atom mass centre, which remains located on the carbon atom (with a mass of 14 for a CH_2 group and 15 for a CH_3 group). As the interaction site is no longer located at an atomic nucleus the forces acting on the masses are more complicated to calculate, but little additional computational expense is required. The effect of using such an anisotropic potential is nicely illustrated by the two arrangements of methylene units shown schematically in Figure 3.42. In the united atom model both arrangements would have the same energies and forces but this is not so with the Toxvaerd anisotropic potential.

3.14.1 Other simplified models

In some force field models, even simpler representations are used than the united atom approach, with entire groups of atoms being modelled as single interaction points. For example, a benzene ring might be modelled as a single site with appropriately chosen parameters.

Yet other models have no obvious relationship to any 'real' molecule but are useful because their simplicity enables larger or more extensive calculations to be performed than would otherwise be possible. The polymer field is full of such models as we shall discuss in section 7.6. Another area where such models have been widely applied is in the study of liquid crystals. Liquid crystals are able to form phases that are characterised by a long-range order of the molecular orientations in at least one dimension. Many of the molecules that exhibit liquid crystalline behaviour are rod-shaped, but disc-like molecules can also form liquid crystalline phases. Some typical examples of molecules that can show such behaviour are shown in Figure 3.43. In the liquid crystalline state the rod-shaped molecules are aligned with their long axes pointing in approximately the same direction. Some very simple computer models have been used to investigate the behaviour of liquid crystals. These simple models enable large simulations to be performed on assemblies of many 'molecules'. One example of such a simplified model is the Gay–Berne potential [Gay and Berne 1981], which models the

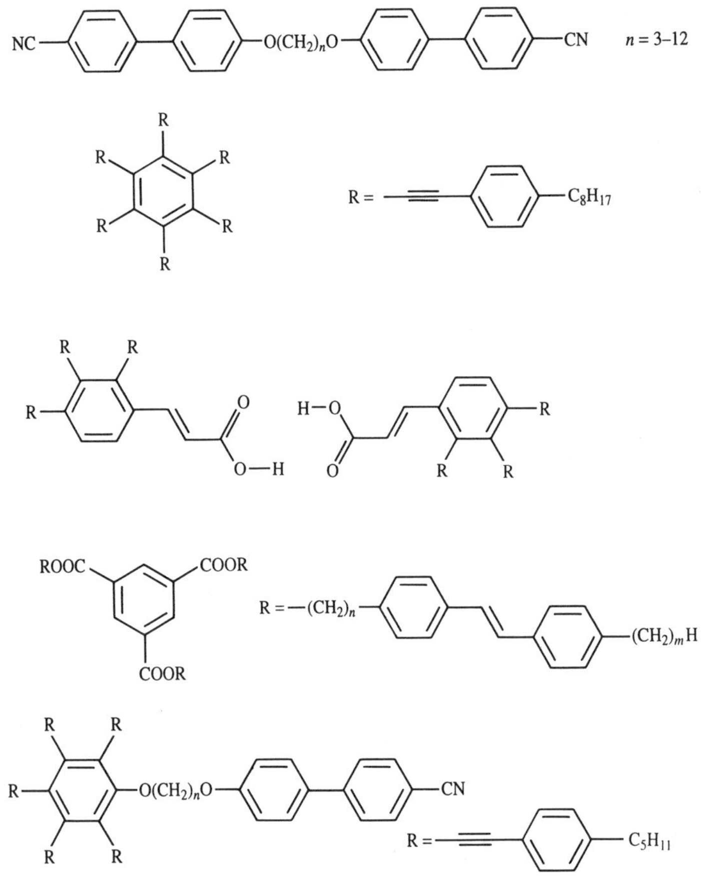

Figure 3.43 Some typical liquid crystal molecules.

anisotropic interaction between two particles as:

$$\mathscr{V}(r_{ij}) = 4\varepsilon(\hat{\mathbf{u}}_i, \hat{\mathbf{u}}_j, \hat{\mathbf{r}})\left\{\left[\frac{\sigma_0}{r_{ij} - \sigma(\hat{\mathbf{u}}_i, \hat{\mathbf{u}}_j, \hat{\mathbf{r}}) + \sigma_0}\right]^{12} - \left[\frac{\sigma_s}{r_{ij} - \sigma(\hat{\mathbf{u}}_i, \hat{\mathbf{u}}_j, \hat{\mathbf{r}}) + \sigma_s}\right]^{6}\right\} \tag{3.70}$$

$\hat{\mathbf{u}}_i$ and $\hat{\mathbf{u}}_j$ are unit vectors that describe the orientations of the two molecules i and j and $\hat{\mathbf{r}}$ is a unit vector along the line connecting their centres, Figure 3.44. The molecules can be considered as ellipsoids which have a shape that is reflected in two size parameters, σ_s and σ_e. σ_e and σ_s are the separations at which the attractive and repulsive terms in the potential cancel for end-to-end and side-by-

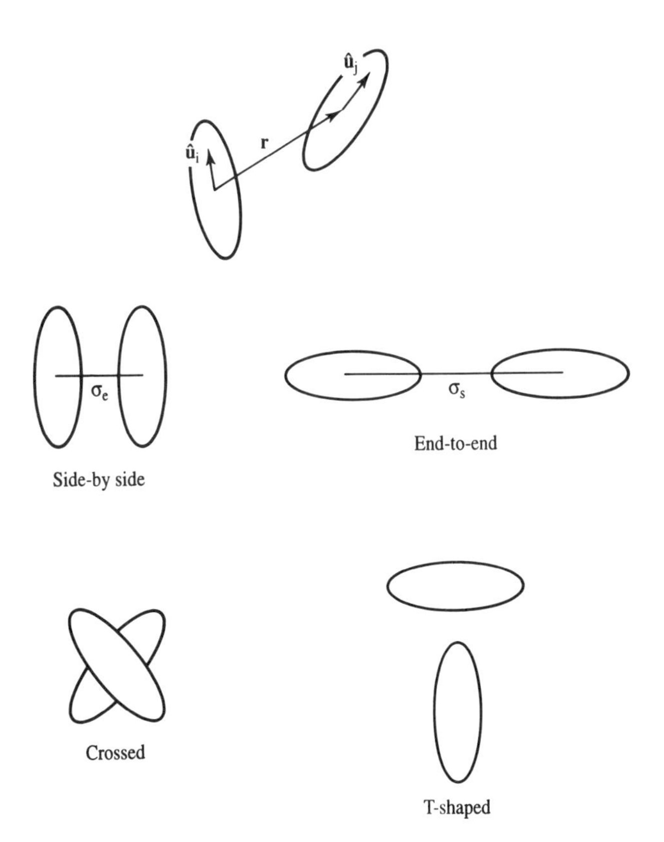

Figure 3.44 The Gay–Berne model for liquid crystal systems and some typical arrangements.

side arrangements respectively. These are incorporated into the potential via the parameter σ:

$$\sigma(\hat{\mathbf{u}}_i, \hat{\mathbf{u}}_j, \hat{\mathbf{r}}) = \sigma_0 \left\{ 1 - \frac{\chi}{2} \left[\frac{(\hat{\mathbf{u}}_i \cdot \hat{\mathbf{r}} + \hat{\mathbf{u}}_j \cdot \hat{\mathbf{r}})^2}{1 + \chi(\hat{\mathbf{u}}_i \cdot \hat{\mathbf{u}}_j)} + \frac{(\hat{\mathbf{u}}_i \cdot \hat{\mathbf{r}} - \hat{\mathbf{u}}_j \cdot \hat{\mathbf{r}})^2}{1 - \chi(\hat{\mathbf{u}}_i \cdot \hat{\mathbf{u}}_j)} \right] \right\}^{-1/2} \tag{3.71}$$

where

$$\chi = \frac{(\sigma_e/\sigma_s)^2 - 1}{(\sigma_e/\sigma_s)^2 + 1} \tag{3.72}$$

χ is the *shape anisotropy parameter*; it is zero for spherical particles and is equal to 1 for infinitely long rods and -1 for infinitely thin discs; σ_0 is typically set equal to σ_s.

The energy term is also orientation dependent and is written as follows:

$$\varepsilon(\hat{\mathbf{u}}_i, \hat{\mathbf{u}}_j, \hat{\mathbf{r}}) = \varepsilon_0 \varepsilon'^{\mu}(\hat{\mathbf{u}}_i, \hat{\mathbf{u}}_j, \hat{\mathbf{r}}) \varepsilon^{\nu}(\hat{\mathbf{u}}_i, \hat{\mathbf{u}}_j) \tag{3.73}$$

where

$$\varepsilon(\hat{\mathbf{u}}_i, \hat{\mathbf{u}}_j) = [1 - \chi^2(\hat{\mathbf{u}}_i \cdot \hat{\mathbf{u}}_j)^2]^{-1/2}$$

$$\varepsilon'(\hat{\mathbf{u}}_i, \hat{\mathbf{u}}_j, \hat{\mathbf{r}}) = \left\{ 1 - \frac{\chi'}{2} \left[\frac{(\hat{\mathbf{u}}_i \cdot \hat{\mathbf{r}} + \hat{\mathbf{u}}_j \cdot \hat{\mathbf{r}})^2}{1 + \chi'(\hat{\mathbf{u}}_i \cdot \hat{\mathbf{u}}_j)} + \frac{(\hat{\mathbf{u}}_i \cdot \hat{\mathbf{r}} - \hat{\mathbf{u}}_j \cdot \hat{\mathbf{r}})^2}{1 - \chi'(\hat{\mathbf{u}}_i \cdot \hat{\mathbf{u}}_j)} \right] \right\} \tag{3.74}$$

χ' measures the anisotropy of the attractive forces:

$$\chi' = \frac{1 - (\varepsilon_e/\varepsilon_s)^{1/\mu}}{(\varepsilon_e/\varepsilon_s)^{1/\mu} + 1}$$

ε_e is the well depth for an end-to-end arrangement of the ellipsoids when the attractive and repulsive contributions cancel, and ε_s is the corresponding well depth for the side-by-side arrangement (Figure 3.44).

The Gay–Berne potential is rather complex but is governed by a relatively small number of parameters, some of which have readily interpretable meanings. The effect of changing the parameters can be most clearly understood by considering certain orientations, such as the side-by-side, end-to-end, crossed and T-shaped structures (Figure 3.44). In the crossed structure the well depth $\varepsilon(\hat{\mathbf{u}}_i, \hat{\mathbf{u}}_j, \hat{\mathbf{r}})$ and the separation $\sigma(\hat{\mathbf{u}}_i, \hat{\mathbf{u}}_j, \hat{\mathbf{r}})$ are independent of χ and χ'. The ratio of the well depths for the end-to-end and side-by-side arrangements is $\varepsilon_e/\varepsilon_s$. The exponents μ and ν are considered adjustable parameters. One way to obtain values for these is to fit the Gay–Berne function to arrangements of Lennard–Jones particles. For example, Luckhurst, Stevens and Phippen determined a value of 1 for ν and a value of 2 for μ by fitting to a linear array of four Lennard–Jones centres [Luckhurst *et al.* 1990].

Depending upon the parameters chosen, simulations performed using the Gay–Berne potential show behaviour typical of liquid crystalline materials. Moreover, by modifying the potential one can determine what contributions affect the liquid crystalline properties, and so help to suggest what types of molecules should be made in order to attain certain properties.

3.15 Derivatives of the molecular mechanics energy function

Many molecular modelling techniques that use force field models require the derivatives of the energy (i.e. the force) to be calculated with respect to the coordinates. It is preferable that analytical expressions for these derivatives are available because they are more accurate and faster than numerical derivatives. A molecular mechanics energy is usually expressed in terms of a combination of internal coordinates of the system (bonds, angles, torsions, etc.) and interatomic distances (for the non-bonded interactions). The atomic positions in molecular mechanics are invariably expressed in terms of Cartesian coordinates (unlike quantum mechanics, where internal coordinates are often used). The calculation of derivatives with respect to the atomic coordinates usually requires the chain rule to be applied. For example, for an energy function that depends upon the separation between two atoms (such as the Lennard–Jones potential, Coulomb electrostatic interaction or bond stretching term) we can write:

$$r_{ij} = \sqrt{(x_i - x_j)^2 + (y_i - y_j)^2 + (z_i - z_j)^2} \tag{3.75}$$

$$\frac{\partial \mathcal{V}}{\partial x_i} = \frac{\partial \mathcal{V}}{\partial r_{ij}} \frac{\partial r_{ij}}{\partial x_i} \tag{3.76}$$

$$\frac{\partial r_{ij}}{\partial x_i} = \frac{(x_i - x_j)}{r_{ij}} \tag{3.77}$$

Thus, for the Lennard–Jones potential:

$$\frac{\partial \mathcal{V}}{\partial r_{ij}} = \frac{24\varepsilon}{r_{ij}} \left[-2\left(\frac{\sigma}{r_{ij}}\right)^{12} + \left(\frac{\sigma}{r_{ij}}\right)^{6} \right] \tag{3.78}$$

The force in the x direction acting on atom i due to its interaction with atom j is given by:

$$\mathbf{f}_{x_i} = (\mathbf{x}_i - \mathbf{x}_j) \frac{24\varepsilon}{r_{ij}^2} \left[2\left(\frac{\sigma}{r_{ij}}\right)^{12} - \left(\frac{\sigma}{r_{ij}}\right)^{6} \right] \tag{3.79}$$

Analytical expressions for the derivatives of the other terms that are commonly found in force fields are also available [Niketic and Rasmussen 1977]. Similar expressions must be derived from scratch when new functional forms are developed.

3.16 Calculating thermodynamic properties using a force field

A molecular mechanics program will return an 'energy value' for any configuration or conformation of the system. This value is properly described as a 'steric energy', and is the energy of the system relative to a zero point that corresponds to a hypothetical molecule in which all of the bond lengths, valence angles, torsions and non-bonded separations are set to their strainless values. It is not necessary to know the actual value of the zero point to calculate the *relative* energies of different configurations or different conformations of the system.

Molecular mechanics can be used to calculate heats of formation. To do so requires the energy to form the bonds in the molecule to be added to the steric energy. These bond energies are typically obtained by fitting to experimentally determined heats of formation and are stored as empirical parameters within the force field. The accuracy with which heats of formation can be predicted with molecular mechanics is, in appropriate cases, comparable with experiment. Thus, the steric energy of a given structure may vary considerably from one force field to another, but its heat of formation should be much closer (if the force fields have been properly parametrised).

A third type of 'energy' that can be obtained from a molecular mechanics calculation is the 'strain energy'. Differences in steric energy are only valid for different conformations or configurations of the same system. Strain energies enable different molecules to be compared. To determine the strain energy it is usual to define some 'strainless' reference point. The reference points can be chosen in many ways and so many different definitions of strain energy have been proposed in the literature. For example, Allinger and co-workers defined the reference point using a set of 'strainless' compounds such as the all-*trans* conformations of the straight-chain alkanes from methane to hexane. From this set of compounds it was possible to derive a set of strainless energy parameters for constituent parts of the molecules. The inherent strain energy of a hydrocarbon is then obtained by subtracting the reference 'strainless' energy from the actual steric energy calculated using the force field. One interesting conclusion of this study was that chair cyclohexane has an inherent strain energy due to the presence of 1,4 van der Waals interactions between the carbon atoms within the ring.

The sources of strain are often quantified by examining the different components (bonds, angles, etc.) of the force field. Such analyses can provide useful information, especially for cases such as highly strained rings. However, in many molecules the strain is distributed among a variety of internal parameters (and in any case is force-field dependent). For intermolecular interactions the interpretation can be easier, for the 'interaction energy' is simply equal to the difference between the energies of the two isolated species and the energy of the intermolecular complex. A good example of this type of calculation and the conclusions that can be drawn from it is the study by Jorgensen and Pratana [Jorgensen and Pratana 1990] of the interaction between analogues of the DNA base pairs. In the double helical structure of DNA the bases pair up adenine (A) with thymine (T) and guanine (G) with cytosine (C), Figure 3.45.

The association constant of the G–C base pair in chloroform is between $10^4\ M^{-1}$ and $10^5\ M^{-1}$ whereas the association between the A–T base pair is significantly weaker, at 40–130 M^{-1}. One obvious reason for this difference is that there are three hydrogen bonds in the G–C base pair and only two in the A–T base pair. However, a simple hydrogen-bond count does not explain all of the data, for synthetic analogues show a significant variation in their association constants, despite having three hydrogen bonds. The weak binding of the uracil-2,6-diaminopyridine (DAP) system (Figure 3.45) could be considered especially anomalous as it contains the same types of hydrogen bond as in G–C ($NH_2 \cdots O$, $NH \cdots N$, $NH_2 \cdots O$). A qualitative explanation for this phenomenon was proposed by Jorgensen and Pratana who examined the secondary interactions in these complexes. As shown in Figure 3.46, the G–C system contains two

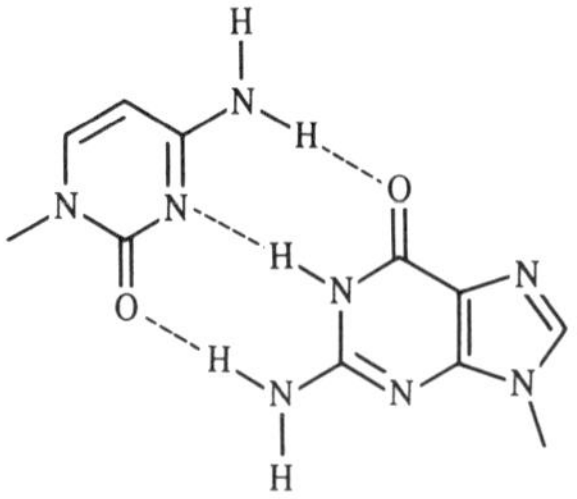

Guanine-cytosine
$K_a = 10^4–10^5\ M^{-1}$

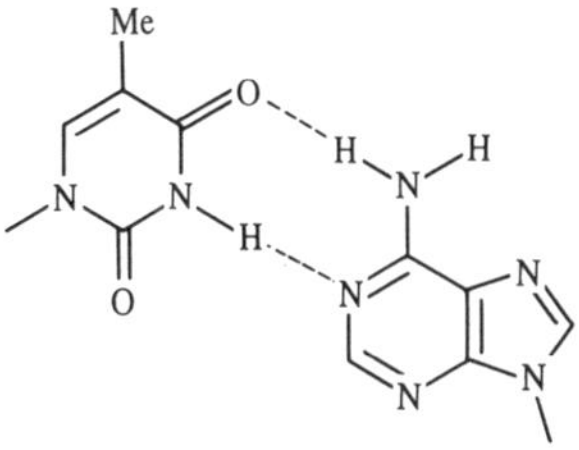

Adenine-thymine
$K_a = 40–130\ M^{-1}$

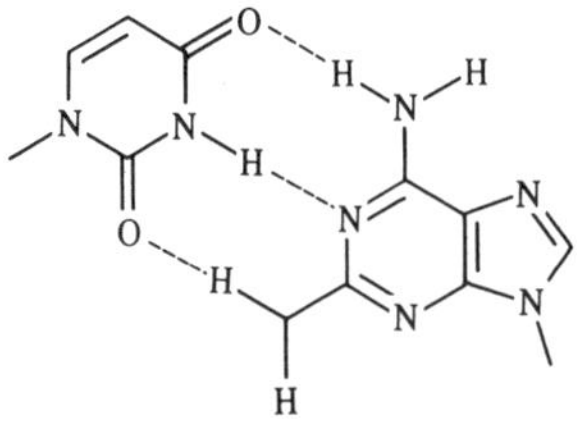

Uracil-DAP
$K_a = 170\ M^{-1}$

Figure 3.45 The DNA base pairs guanine (G), cytosine (C), adenine (A) and thymine (T). The uracil-2,6-diaminopyridine pair can also form three hydrogen bonds but has a much lower association constant than G–C.

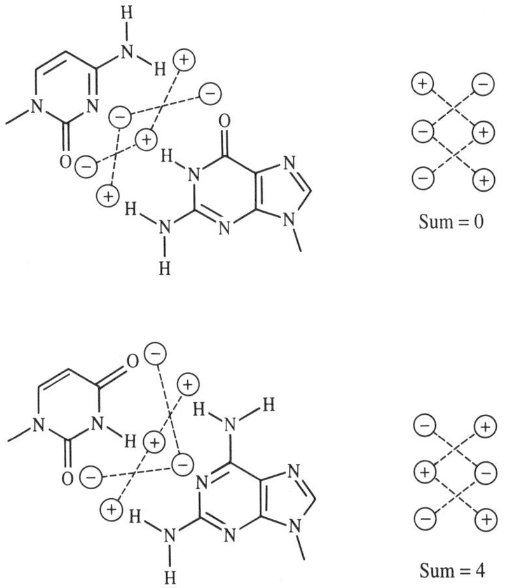

Sum = 0

Sum = 4

Figure 3.46 Secondary interactions in guanine–cytosine and uracil–DAP.

unfavourable secondary interactions and two favourable ones, an overall sum of zero. In the uracil–DAP system, all four secondary interactions are unfavourable.

3.17 Force field parametrisation

A force field can contain a large number of parameters, even if it is intended for calculations on only a small set of molecules. Parametrisation of a force field is not a trivial task. A significant amount of effort is required to create a new force field entirely from scratch, and even the addition of a few parameters to an existing force field in order to model a new class of molecules can be a complicated and time-consuming procedure. The performance of a force field is often particularly sensitive to just a few of the parameters (usually the non-bonded and torsional terms), so it is often sensible to spend more time optimising these parameters rather than others (such as the bond stretching and angle bending terms), the values of which do not greatly affect the results.

The first step is to select the data that are going to be used to guide the parametrisation process. Molecular mechanics force fields may be used to determine a variety of structurally related properties and the parametrisation data should be chosen accordingly. The geometries and relative conformational energies of certain key molecules are usually included in the data set. It is increasingly common to include vibrational frequencies in the parametrisation; these are usually more difficult to reproduce but the incorporation of appropriate cross terms can often help. Some force fields are parametrised to reproduce thermodynamic properties using computer simulation techniques. The OPLS (optimised parameters for liquid simulations [Jorgensen and Tirado-Reeves 1988]) parameters have been obtained in this way.

Unfortunately, experimental data may be non-existent or difficult to obtain for particular classes of molecules. Quantum mechanics calculations are thus increasingly used to provide the data for the parametrisation of molecular mechanics force fields. This is an important development because it greatly extends the range of chemical systems that can be treated using the force field approach. *Ab initio* calculations are able to reproduce experimental results for small representative systems. Clearly, one should be careful to properly validate a force field derived in such a way by testing against experimental data if at all possible.

Once a functional form for the force field has been chosen and the data to be used in the parametrisation identified, there are then two basic methods that can be used to actually obtain the parameters. The first approach is 'parametrisation by trial and error', in which the parameters are gradually refined to give better and better fits to the data. It is difficult to simultaneously modify a large number of parameters in such a strategy and so it is usual to perform the parametrisation in stages. It is important to remember that there is some coupling between all of the degrees of freedom and so for the most sensitive work none of the parameters can truly be taken in isolation. Parameters for the hard degrees of freedom (bond stretching and angle bending) can however often be treated separately from the others (indeed the bond and angle parameters are often transferred from one force field to another without modification). By contrast, the soft degrees of freedom (non-bonded and torsional contributions) are closely coupled and can significantly influence each other. One protocol that can be quite successful is to first establish a series of van der Waals parameters. The electrostatic model is then determined (e.g. by electrostatic potential fitting). Finally, the torsional potentials are determined by ensuring that the torsional barriers are reproduced together with the relative energies of the different conformations. Of course, it may be necessary to modify any of the parameters at any stage should the results be inadequate and so parametrisation is invariably an iterative procedure.

As experimental information on torsional barriers is often sparse or non-existent, quantum mechanical calculations are widely used to determine torsional potentials. The general strategy is as follows. First, a molecular fragment that adequately represents the rotatable bond of interest and its immediate environment is chosen. A series of structures are then generated by rotating about the bond and their energies determined using quantum mechanics. The torsional potential is then fitted to reproduce the energy curve, in conjunction with the van der Waals potential and partial charges. This procedure can be illustrated using the study of Pranata and Jorgensen who wanted to perform some calculations on FK506, a potent immunosuppressant (Figure 3.47) [Pranata and Jorgensen 1991]. FK506 contains a ketoamide functionality that has a *trans* conformation when the molecule is bound to its receptor but which is *cis* in the crystal structure of isolated FK506. NMR experiments suggested that the molecule adopts both *cis* and *trans* conformations in solution. This part of the molecule is clearly implicated in its function and so it was considered important to correctly model the torsional potential about this bond. Pratana and Jorgensen intended to use the AMBER force field for their calculations but the force field contained no parameters for this link.

Molecular orbital calculations were performed on *N,N*-dimethyl-α-ketopropanamide (Figure 3.48, top), which was chosen as an appropriate model system. Semi-empirical calculations using AM1 and *ab initio* calculations using a 6-31G(d) basis set suggested that the minimum energy conformation corresponded to a torsion angle of 124° and 135° respectively with the *anti* conformation being slightly higher in energy (~0.7 kcal/mol). However, an analogous calculation using the 3-21G basis set did predict that the *anti* conformation was at a minimum (Figure 3.48). Crystal structures of compounds containing this fragment revealed that an orthogonal structure was commonly encountered. Torsional parameters were then fitted to the 6-31G(d) potential and evaluated by calculating an energetic profile for rotation in a larger fragment of the FK506 molecule using the force field and comparing it with that obtained using AM1 (Figure 3.48, bottom).

An alternative approach to parametrisation, pioneered by Lifson and coworkers in the development of their 'consistent' force fields, is to use least-squares fitting to determine the set of parameters that gives the optimal fit to the data [Lifson and Warshel 1968]. Again, the first step is to choose a set of experimental data that one wishes the force field to reproduce (or calculate using quantum mechanics, if appropriate). Lifson and colleagues used thermodynamic data, equilibrium conformations and vibrational frequencies. The 'error' for a given set of parameters equals the sum of squares of the differences between the observed and calculated values for the set of properties. The objective is to change the force field parameters to minimise the error. This is done by assuming that the properties can be related to the force field by a Taylor series expansion:

$$\Delta\mathbf{y}(\mathbf{x}+\delta\mathbf{x}) = \Delta\mathbf{y}(\mathbf{x}) + \mathbf{Z}\delta\mathbf{x} + \cdots \tag{3.80}$$

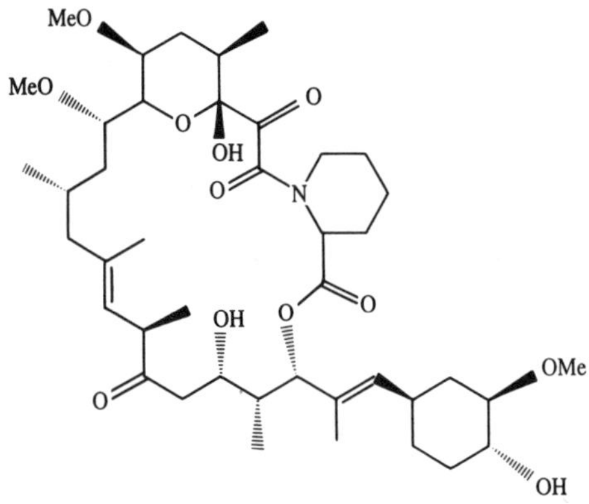

Figure 3.47 The immunosuppressant FK506.

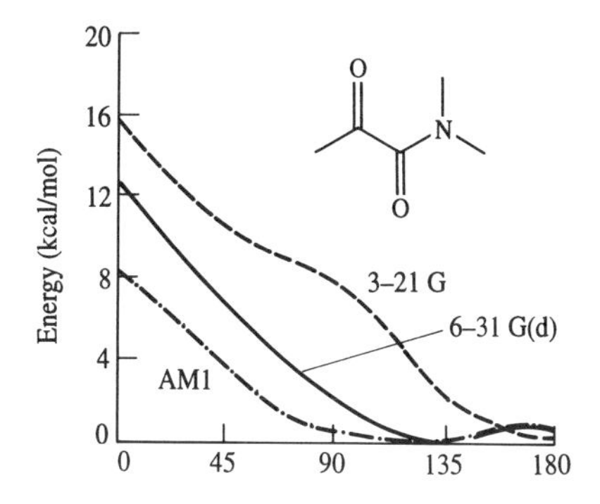

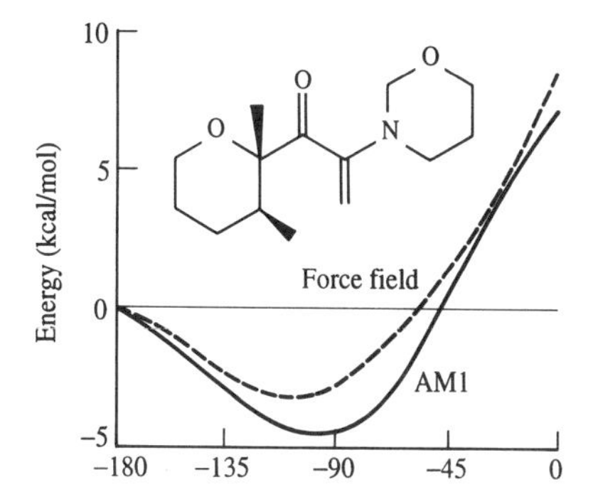

Figure 3.48 Fragments used to derive and evaluate parameters for the ketoamide functionality in FK506. Figure redrawn from J Pranata and W L Jorgensen 1991. Computational Studies on FK506: Conformational Search and Molecular Dynamics Simulations in Water. *The Journal of the American Chemical Society* **113**:9483–9493.

$\Delta\mathbf{y}$ is a vector of the differences between the calculated and experimental data and $\mathbf{x}$ is a vector whose components are the force field parameters. $\mathbf{Z}$ is a matrix whose elements are the derivatives of each property with respect to each of the parameters, $\partial y/\partial x$. An iterative procedure is used to minimise the sum of squares of the differences, $\Delta\mathbf{y}^2$. The method is easily modified to enable various weighting factors to be assigned to the different pieces of experimental data, so that (for example) the thermodynamic data could be given greater importance than the vibrational frequencies.

A well-known application of the least-squares approach to the optimisation of a force field was performed by Hagler, Huler and Lifson, who derived a force field for peptides by fitting to crystal data of a variety of appropriate compounds [Hagler *et al.* 1977; Hagler and Lifson 1974]. A key result of their work was that no explicit hydrogen bond term was required to model the hydrogen bonding interactions, but that a combination of appropriate electrostatic and van der Waals models was sufficient. A group led by Hagler has more recently developed a force field based upon the results of *ab initio* quantum mechanics calculations on small molecules, again using least-squares fitting [Maple *et al.* 1988]. The quantum mechanics calculations were performed not only on small molecules at equilibrium geometries but also on structures that were distorted from equilibrium. For each geometry the energy was calculated together with the first and second derivatives of the energy. This provided a wealth of data for the subsequent fitting procedure. This research has resulted in many new algorithms for the derivation of force field parameters and has also challenged some of the assumptions about the development and functional form of force fields.

3.18 Transferability of force field parameters

The range of systems that have been studied by force field methods is extremely varied. Some force fields have been developed to study just one atomic or molecular species under a wider range of conditions. For example, the chlorine model of Rodger, Stone and Tildesley [Rodger *et al.* 1988] can be used to study the solid, liquid and gaseous phases. This is an anisotropic site model, in which the interaction between a pair of sites on two molecules depends not only upon the separation between the sites (as in an isotropic model such as the Lennard–Jones model) but also upon the orientation of the site–site vector with respect to the bond vectors of the two molecules. The model includes an electrostatic component which contains dipole–dipole, dipole–quadrupole and quadrupole–quadrupole terms and the van der Waals contribution is modelled using a Buckingham-like function.

Other force fields are designed for use with specific classes of molecules; we have already encountered the AMBER force field which is designed for calculations on proteins and nucleic acids. Yet other force fields are intended to be applied to a wide range of molecules and indeed some force fields are designed to model the entire periodic table. Intuitively, one might expect a 'specialised' force field to perform better than a 'general' force field, and while this is certainly true for the best of the specialised force fields, a good general force field can often outperform a poor specific force field.

The ability to transfer parameters from one molecule to another is crucial for any force field. Without it, the task of parametrisation would be impossible, because so many parameters would be required, and the force field would have no predictive ability. Transferability has a number of important consequences for the development and application of force fields. The problem of transferability is often first encountered when a molecular mechanics program fails to run because parameters are missing for the molecule being studied. One must somehow find values for the missing parameters. Some programs automatically 'guess' force field parameters; it is wise to check these assignments as they may be suspect. For the developer of a force field, a compromise must often be found between a complex functional form and a large number of atom types. It is also important to try and ensure that the errors in the force field are balanced, in the sense that it would be silly to spend a lot of time getting (say) the bond stretching terms just right, if the van der Waals parameters give rise to large errors.

Transferability can be helped by using the same parameters for as wide a range of situations as possible. The non-bonded terms are particularly problematic in this regard; it would, in principle, be necessary to have parameters for the non-bonded interactions between all possible pairs of atom types. This would give rise to a very large number of parameters. It is therefore commonly assumed that the same set of van der Waals parameters can be used for most, if not all atoms of the same element. For example, all carbon atoms (sp^3, sp^2, sp, etc.) would be treated with the same set of van der Waals parameters, all nitrogens by a common set, and so on. The torsional terms may also be generalised, so that the torsional parameters depend solely upon the atom types of the two atoms that form the central bond, rather than on all four atoms that comprise the torsion angle, as described in section 3.5 for the AMBER force field.

3.19 The treatment of delocalised π-systems

The bonds in conjugated π-systems are often of different lengths. For example, the central bond in butadiene is approximately 1.47 Å long but the two terminal $CH{=}CH_2$ bonds are approximately 1.34 Å. If butadiene is modelled using a force field in which all four carbon atoms are assigned the same atom type (e.g. 'carbon sp^2') then each bond will be assigned the same bonding parameters and in the equilibrium structure all carbon–carbon bonds will be almost identical in length. A similar situation arises for aromatic systems. For example, not all the bonds in naphthalene are of equal length (unlike benzene). The bond lengths in a delocalised π-system depend upon the bond orders; the higher the bond order, the shorter the bond.

In some cases it may be possible to circumvent this problem by creating a model specific to the conjugated system. For butadiene the central carbon–carbon bond of the π system could be treated in a different manner to the two terminal bonds, for example by using one atom type for the $-CH{=}$ carbon atoms and one for the $=CH_2$ carbon atoms in butadiene. This approach might be acceptable if we wanted to perform an extensive series of calculations on substituted butadienes, but it does compromise the transferability of the force field parameters. An alternative is to incorporate a molecular orbital calculation into the force field. Two variants on this theme have been developed. In one approach, the π and σ systems are treated separately. For a given geometry, a self-consistent field quantum mechanical calculation is performed on the π system, typically with an appropriate semi-empirical theory. Molecular mechanics is simultaneously applied to the σ system. The energies of the quantum mechanical and molecular mechanical calculations are added together, and the geometry is modified to minimise this combined energy. A obvious assumption inherent in this approach is that the π and σ systems can be separated, which may be difficult to justify when deviations from planarity are present.

An alternative approach is exemplified by the MMP2 program (which is an extension of MM2). First, a molecular orbital calculation is performed on the π

system. If the initial conformation of the system is non-planar the calculation is performed on the equivalent planar system. The force field parameters are then modified according to the quantum mechanical bond orders. In MMP2 these parameters are the force constant for the bonds in the π system, the reference bond lengths and the torsional barriers. The system is then subjected to the usual molecular mechanics treatment using the new force field parameters. A linear relationship between the stretching constants and the bond orders, and between the reference bond lengths and the bond orders was found to give good results. Initially, the torsional barriers were assumed to be proportional to the square of the bond orders, but this relationship was modified slightly in subsequent versions of the program.

3.20 Force fields for metals and inorganic systems

It may come as a surprise to many readers to learn that the earliest calculations on inorganic systems were reported at much the same time as the first calculations on organic systems. For example, Corey and Bailar described the use of empirical force field calculations on octahedral complexes of cobalt in 1959 [Corey and Bailar 1959]. The range of metal-containing systems that can be considered by force field methods has steadily expanded since then. Moreover, many systems of commercial interest contain metals or other elements not usually found in 'organic' or 'biochemical' systems.

Some inorganic systems (such as the coordination complexes) are little different to organic systems from a force field point of view; the bonding can be represented in a similar way and many of the force field parameters originally developed for organic systems can be transferred without modification. The major challenge is to properly model the metal; this is complicated by the wide range of geometries that metals adopt. Octahedral and square planar complexes are the simplest to model because of their symmetry, but even here two types of equilibrium angles are present (180° and 90°). The situation can be much more complicated for the other geometries or for structures where the geometry about the metal is a distortion of a regular arrangement. A Urey–Bradley treatment of the bonding about the metal can often be quite successful in achieving the correct geometries. Here, there are no angle bending terms at the metal but van der Waals interactions are computed between pairs of atoms bonded to the metal.

It is much more difficult to use such a force field to model metal π systems, where the bonding between the metal and the ligand is not easily represented by a conventional bonding picture. For example, how should the bonding in ferrocene be represented for a force field calculation? Is there a bond between the iron and each of the carbon atoms in the two cyclopentadienyl rings? Is there a 'bond' from the iron to the centre of each of the rings? Moreover, metal atoms can adopt a wide range of geometries in π complexes which are often significantly distorted from regular structures. Nevertheless, force fields have been developed which can cope with such systems, as well as being able to model more traditional systems such as organic compounds. These force fields often use a rather different functional form from equation (3.1) and the parameters are obtained in a different way. At one extreme, the UFF force field of Rappé, Casewit, Colwell and Goddard [Rappé *et al.* 1992] is designed to model the entire periodic table. The parameters for this force field are calculated from atomic parameters that depend upon the hybridisation state of the atom together with some simple rules. One distinctive feature of this force field, and the SHAPES force field developed by Landis and co-workers [Allured *et al.* 1991], is the way in which angle bending is treated. The harmonic potential that is commonly employed in standard force fields is inappropriate to model the distortion of systems as the angle approaches 180°. UFF uses a cosine Fourier series for each angle ABC:

$$\mathcal{V}(\theta) = K_{\mathrm{ABC}} \sum_{n=0}^{m} C_n \cos n\theta \tag{3.81}$$

The coefficients C_n are chosen to ensure that the function has a minimum at the appropriate reference bond angle. For linear, trigonal, square-planar and octahedral coordination, Fourier series with just two terms are used with a C_0 term and a term for $n = 1$, 2, 3 or 4 term respectively.

The SHAPES angle bending term is very similar:

$$\mathcal{V}(\theta) = K_{\mathrm{ABC}} \sum_{n=0}^{m} [1 + \cos(n\theta - \delta)] \tag{3.82}$$

Landis has more recently developed a formulation for the angle bending term that is based on valence bond theory and which can produce results that compare well with *ab initio* calculations [Landis *et al.* 1995]. For example, using just one set of C–H parameters the H–C–H bond angles in ethene, formaldehyde and both singlet and triplet carbene match closely those found experimentally.

3.20.1 *Force fields for zeolites*

Zeolites are materials generally composed of silicon, aluminium and oxygen which contain channels of molecular dimensions. Zeolites have a multitude of commercial uses including catalysis and separation (e.g. they are used in oil refining to separate linear and branched alkanes). It is therefore natural that molecular modelling techniques should be used to investigate the intrinsic properties of such materials and the way in which they interact with adsorbates.

A simple way to investigate such adsorption processes is to keep the zeolite rigid and to concentrate on the intermolecular interactions. This is often done using a combination of van der Waals and electrostatic terms; a Lennard–Jones potential may be used for the van der Waals component, but a Buckingham-like potential is often preferred. The electrostatic interactions can be very important for zeolites. However, the partial charges used in the various published force fields can vary enormously (from 0.4*e* to as much as 1.9*e* for the silicon atoms in silicates). A correct description of the electrostatic interactions is crucial in such systems. As we shall see in section 5.7, it is common to truncate the non-bonded interactions in order to speed up the calculation of non-bonded interactions. Special techniques to capture the total non-bonded energy (particularly the electrostatic contribution) are therefore widely used in such systems. Such techniques, especially the Ewald summation method, are described in section 5.8.1.

It is an approximation to keep the zeolite rigid, and in more complex models the structure can vary. Many of the force fields that have been developed to model zeolites are very similar to the valence force fields used for organic and biological molecules, typically containing bond stretching, angle bending and torsional terms in addition to the non-bonded interactions. One important consideration when modelling zeolites is that very little energy is required to deform the Si–O–Si bond over an extremely wide range (at least 120° to 180°). This is shown in Figure 3.49, which shows the results of *ab initio* calculations using a 3-21G* basis set for $H_3SiOSiH_3$. The Fourier series expansions used by the UFF and SHAPES force fields for the angle bending terms are designed to cope with such angular variation; Nicholas, Hopfinger, Trouw and Iton suggested the following quartic potential as an alternative specifically for the Si–O–Si angle [Nicholas *et al.* 1991]:

$$\mathscr{v}(\theta) = \frac{k_1}{2}(\theta - \theta_0)^2 + \frac{k_2}{2}(\theta - \theta_0)^3 + \frac{k_3}{2}(\theta - \theta_0)^4 \tag{3.83}$$

With the correct choice of the parameters k_i and θ_0 the *ab initio* data in Figure 3.49 could be reproduced very well. In this force field a Urey–Bradley term was also included between the silicon atoms in such angles to model the lengthening of the Si–O bond as the angle decreased.

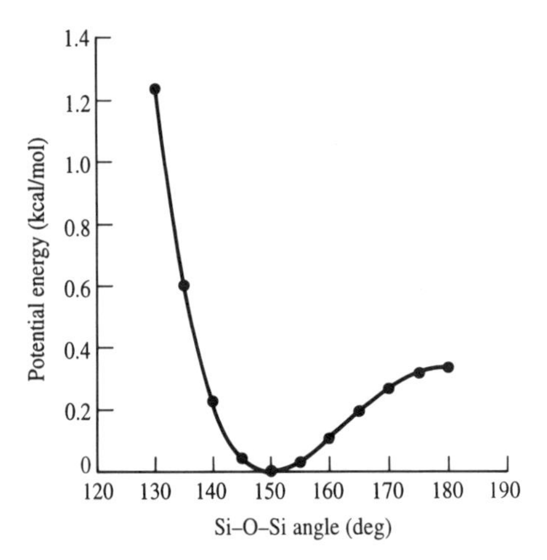

Figure 3.49 Variation in energy with the Si–O–Si angle. Figure redrawn from Grigoras S and T H Lane 1988. Molecular Parameters for Organosilicon Compounds Calculated from *Ab Initio* Computations. *Journal of Computational Chemistry* **9**:25–39.

Appendix 3.1 The interaction between two Drude molecules

In the system comprising two Drude molecules (see section 3.9.1), an additional term must be included in the Hamiltonian [Rigby *et al.* 1986]. This additional term arises from the interactions between the two dipoles. The instantaneous dipole of each molecule is $qz(t)$, where $z(t)$ is the separation of the charges. Thus, if we label the molecules 1 and 2, we can write the dipole–dipole interaction energy as:

$$\mathcal{V}(\mu_1, \mu_2) = -\frac{2\mu_1\mu_2}{4\pi\varepsilon_0 r^3} = -\frac{2z_1z_2q^2}{4\pi\varepsilon_0 r^3} \tag{3.84}$$

r is the separation of the two molecules. The Schrödinger equation for this system is thus:

$$-\frac{\hbar^2}{2m}\frac{\partial^2\psi}{\partial z_1^2} - \frac{\hbar^2}{2m}\frac{\partial^2\psi}{\partial z_2^2} + \left[\frac{1}{2}kz_1^2 + \frac{1}{2}kz_2^2 - \frac{2z_1z_2q^2}{4\pi\varepsilon_0 r^3}\right]\psi = E\psi \tag{3.85}$$

This equation can be solved by making the following substitutions:

$$a_1 = \frac{z_1 + z_2}{\sqrt{2}}; \quad a_2 = \frac{z_1 - z_2}{\sqrt{2}}; \quad k_1 = k - \frac{2q^2}{4\pi\varepsilon_0 r^3}; \quad k_2 = k + \frac{2q^2}{4\pi\varepsilon_0 r^3} \tag{3.86}$$

These reduce equation (3.85) to

$$-\frac{\hbar^2}{2m}\frac{\partial^2\psi}{\partial a_1^2} - \frac{\hbar^2}{2m}\frac{\partial^2\psi}{\partial a_2^2} + [\tfrac{1}{2}k_1a_1^2 + \tfrac{1}{2}k_2a_2^2]\psi = E\psi \tag{3.87}$$

This is the Schrödinger equation for two independent (i.e. non-interacting) oscillators with frequencies given as follows:

$$\omega_1 = \omega\sqrt{1 - \frac{2q^2}{4\pi\varepsilon_0 r^3 k}}, \quad \omega_2 = \omega\sqrt{1 + \frac{2q^2}{4\pi\varepsilon_0 r^3 k}} \tag{3.88}$$

$\omega/2\pi$ is the frequency of an isolated Drude molecule. The ground state energy of the system is therefore just the sum of the zero-point energies of the two oscillators: $E_0 = \frac{1}{2}\hbar(\omega_1 + \omega_2)$.

If we now substitute for ω_1 and ω_2 and expand the square roots using the binomial theorem, then we obtain the following:

$$E_0(r) = \hbar\omega - \frac{q^4\hbar\omega}{2(4\pi\varepsilon_0)^2 r^6 k^2} - \cdots \tag{3.89}$$

The interaction energy of the two oscillators is the difference between this zero-point energy and the energy of the system when the oscillators are infinitely separated and so:

$$\mathcal{V}(r) = -\frac{q^4\hbar\omega}{2(4\pi\varepsilon_0)^2 r^6 k^2} \tag{3.90}$$

The force constant, k, is related to the polarisability of the molecule, α as follows. Suppose a single Drude molecule is exposed to an external electric field

E. In the electric field, a force $q\mathbf{E}$ acts on each charge (in opposite directions as the charges are of opposite sign). This force causes the charges to separate and equilibrium is reached when the restoring force due to the stretching of the bond (kz) is equal to the electrostatic force: $qE = kz$. This separation of the charges is equivalent to a static dipole given by $\mu_{\text{ind}} = qz = q^2E/k$. However, the induced dipole is also related to the polarisability by $\boldsymbol{\mu}_{\text{ind}} = \alpha\mathbf{E}$. Thus the polarisability can be written in terms of the force constant k: $\alpha = q^2/k$. With this substitution the result for the Drude model in two dimensions is:

$$\mathcal{V}(r) = -\frac{\alpha^4 \hbar\omega}{2(4\pi\varepsilon_0)^2 r^6} \tag{3.91}$$

In three dimensions the equivalent result is:

$$\mathcal{V}(r) = -\frac{3\alpha^4 \hbar\omega}{4(4\pi\varepsilon_0)^2 r^6} \tag{3.92}$$

Further reading

Bowen J P and N L Allinger 1991. Molecular Mechanics: The Art and Science of Parameterisation in Lipkowitz K B and D B Boyd (Editors). *Reviews in Computational Chemistry.* Volume 2. New York, VCH Publishers.

Burkert U and N L Allinger 1982. *Molecular Mechanics*. ACS Monograph 177. Washington D.C., American Chemical Society.

Dykstra C E 1993. Electrostatic Interaction Potentials in Molecular Force Fields. *Chemical Reviews* **93**:2339–2353.

Niketic S R and K Rasmussen 1977. *The Consistent Force Field: A Documentation*. Berlin, Springer Verlag.

Rigby M, E B Smith, W A Wakeham and G C Maitland 1981. *Intermolecular Forces: Their Origin and Determination*. Oxford, Clarendon Press.

Rigby M, E B Smith, W A Wakeham and G C Maitland 1986. *The Forces Between Molecules*. Oxford, Clarendon Press.

References

Allinger N L 1977. Conformational Analysis 130. MM2. A Hydrocarbon Force Field Utilizing V_1 and V_2 Torsional Terms. *The Journal of the American Chemical Society* **99**:8127–8134.

Allinger N L, Y H Yuh and J-J Lii 1989. Molecular Mechanics. The MM3 Force Field for Hydrocarbons I. *The Journal of the American Chemical Society* **111**:8551–9556.

Allured V S, C M Kelly and C R Landis 1991. SHAPES Empirical Force-Field —New Treatment of Angular Potentials and Its Application to Square-Planar

Transition-Metal Complexes. *Journal of the American Chemical Society* **113**:1–12.

Barker J A, R A Fisher and R O Watts 1971. Liquid Argon: Monte Carlo and Molecular Dynamics Calculations. *Molecular Physics* **21**:657–673.

Barnes P, J L Finney, J D Nicholas and J E Quinn 1979. Cooperative Effects in Simulated Water. *Nature* **282**:459–464.

Bayly C I, P Cieplak, W D Cornell and P A Kollman 1993. A Well-Behaved Electrostatic Potential Based Method for Deriving Atomic Charges—The RESP Model. *Journal of Physical Chemistry* **97**:10269–10280.

Berendsen H C, J P M Postma, W F van Gunsteren and J Hermans 1981. Interaction Models for Water in Relation to Protein Hydration. In Pullman B (Editor). *Intermolecular Forces*. Dordrecht, Reidel: 331–342.

Berendsen H J C, J R Grigera and T P Straatsma 1987. The Missing Term in Effective Pair Potentials. *The Journal of Physical Chemistry* **91**:6269–6271.

Bernal J D and R H Fowler 1933. A Theory of Water and Ionic Solution, with Particular Reference to Hydrogen and Hydroxyl Ions. *The Journal of Chemical Physics* **1**:515–548.

Bezler B H, K M Merz Jr and P A Kollman 1990. Atomic Charges Derived from Semi-Empirical Methods. *Journal of Computational Chemistry* **11**:431–439.

Breneman C M and K B Wiberg 1990. Determining Atom-Centred Monopoles from Molecular Electrostatic Potentials. The Need for High Sampling Density in Formamide Conformational Analysis. *Journal of Computational Chemistry* **11**:361–373.

Buckingham A D 1959. Molecular Quadrupole Moments. *Quarterly Reviews of the Chemical Society* **13**:183–214.

Chirlian L E and M M Francl 1987. Atomic Charges Derived from Electrostatic Potentials: A Detailed Study. *Journal of Computational Chemistry* **8**:894–905.

Claessens M, M Ferrario and J-P Ryckaert 1983. The Structure of Liquid Benzene. *Molecular Physics* **50**:217–227.

Corey E J and J C Bailar Jr 1959. The Stereochemistry of Complex Inorganic Compounds. XXII. Stereospecific Effects in Complex Ions. *The Journal of the American Chemical Society* **81**:2620–2629.

Cornell W D, P Cieplak, C I Bayly, I R Gould, K M Merz Jr, D M Ferguson, D C Spellmeyer, T Fox, J W Caldwell and P A Kollman 1995. A Second Generation Force Field for the Simulation of Proteins, Nucleic Acids and Organic Molecules. *The Journal of the American Chemical Society* **117**:5179–5197.

Corongiu G 1992. Molecular Dynamics Simulation for Liquid Water Using a Polarisable and Flexible Potential. *International Journal of Quantum Chemistry* **42**:1209–1235.

Cox S R and D E Williams 1981. Representation of the Molecular Electrostatic Potential by a New Atomic Charge Model. *Journal of Computational Chemistry* **2**:304–323.

Dang L X, J E Rice, J Caldwell and P A Kollman 1991. Ion Solvation in Polarisable Water: Molecular Dynamics Simulations. *The Journal of the American Chemical Society* **113**:2481–2486.

Dinur U and A T Hagler 1991. New Approaches to Empirical Force Fields in K B

Lipkowitz and D B Boyd (Editors). *Reviews in Computational Chemistry.* Volume 2. New York, VCH Publishers.

Dinur U and A T Hagler 1995. Geometry-Dependent Atomic Charges: Methodology and Application to Alkanes, Aldehydes, Ketones and Amides. *Journal of Computational Chemistry* **16**:154–170.

Ferenczy G G, C A Reynolds and W G Richards 1990. Semi-Empirical AM1 Electrostatic Potentials and AM1 Electrostatic Potential Derived Charges—A Comparison with *Ab Initio* Values. *Journal of Computational Chemistry* **11**:159–169.

Ferguson D M 1995. Parameterisation and Evaluation of a Flexible Water Model. *Journal of Computational Chemistry* **16**:501–511.

Fowler P W and A D Buckingham 1991. Central or Distributed Multipole Moments? Electrostatic Models of Aromatic Dimers. *Chemical Physics Letters* **176**:11–18.

Gasteiger J and M Marsili 1980. Iterative Partial Equalization of Orbital Electronegativity—Rapid Access to Atomic Charges. *Tetrahedron* **36**:3219–3288.

Gay J G and B J Berne 1981. Modification of the Overlap Potential to Mimic a Linear Site–Site Potential. *The Journal of Chemical Physics* **74**:3316–3319.

Goodford P J 1985. A Computational Procedure for Determining Energetically Favorable Binding Sites on Biologically Important Macromolecules. *Journal of Medicinal Chemistry* **28**:849–857.

Hagler A T and S Lifson 1974. Energy Functions for Peptides and Proteins. II. The Amide Hydrogen Bond and Calculation of Amide Crystal Properties. *The Journal of the American Chemical Society* **96**:5327–5335.

Hagler A T, E Huler and S Lifson 1977. Energy Functions for Peptides and Proteins. I. Derivation of a Consistent Force Field Including the Hydrogen Bond from Amide Crystals. *Journal of the American Chemical Society* **96**:5319–5327.

Hill T L 1948. Steric Effects. I. Van der Waals Potential Energy Curves. *The Journal of Chemical Physics* **16**:399–404.

Hunter C A and J K M Saunders 1990. The Nature of π–π Interactions. *The Journal of the American Chemical Society* **112**:5525–5534.

Jorgensen W L and J Pranata 1990. Importance of Secondary Interactions in Triply Hydrogen Bonded Complexes: Guanine–Cytosine vs Uracil-2,6-Diaminopyridine. *The Journal of the American Chemical Society* **112**:2008–2010.

Jorgensen W L and J Tirado-Rives 1988. The OPLS Potential Functions for Proteins—Energy Minimizations for Crystals of Cyclic-Peptides and Crambin. *Journal of the American Chemical Society* **110**:1666–1671.

Jorgensen W L, J Chandrasekhar, J D Madura, R W Impey and M L Klein 1983. Comparison of Simple Potential Functions for Simulating Liquid Water. *The Journal of Chemical Physics* **79**:926–935.

Landis C R, T Cleveland and T K Firman 1995. Making Sense of the Shapes of Simple Metal Hydrides. *The Journal of the American Chemical Society* **117**:1859–1860.

Lifson S and A Warshel 1968. Consistent Force Field for Calculations of Conformations, Vibrational Spectra and Enthalpies of Cycloalkane and *n*-Alkane Molecules. *The Journal of Chemical Physics* **49**:5116–5129.

London F 1930. Zur Theori und Systematik der Molekularkräfte. *Zeitschrift für Physik* **63**:245–279.

Luckhurst G R, R A Stephens and R W Phippen 1990. Computer Simulation Studies of Anisotropic Systems XIX. Mesophases Formed by the Gay–Berne Model Mesogen. *Liquid Crystals* **8**:451–464.

Luque F J, F Ilas and M Orozco 1990. Comparative Study of the Molecular Electrostatic Potential Obtained from Different Wavefunctions—Reliability of the Semi-Empirical MNDO Wavefunction. *Journal of Computational Chemistry* **11**:416–430.

Lybrand T P and P A Kollman 1985. Water–Water and Water–Ion Potential Functions Including Terms for Many Body Effects. *The Journal of Chemical Physics* **83**:2923–2933.

Maple J R, U Dinur and A T Hagler 1988. Derivation of Force Fields for Molecular Mechanics and Molecular Dynamics from *Ab Initio* Energy Surfaces. *Proceedings of the National Academy of Sciences USA* **85**:5350–5354.

Nicholas J B, A J Hopfinger, F R Trouw and L E Iton 1991. Molecular Modelling of Zeolite Structure. 2. Structure and Dynamics of Silica Sodalite and Silicate Force Field. *The Journal of the American Chemical Society* **113**:4792–4800.

Niesar U, G Corongiu, E Clementi, G R Keller and D K Bhattacharya 1990. Molecular Dynamics Simulations of Liquid Water Using the NCC *Ab Initio* Potential. *The Journal of Physical Chemistry* **94**:7949–7956.

Niketic S R and K Rasmussen 1977. *The Consistent Force Field: A Documentation*. Berlin, Springer Verlag.

Pranata J and W L Jorgensen 1991. Computational Studies on FK506: Computational Search and Molecular Dynamics Simulations in Water. *The Journal of the American Chemical Society* **113**:9483–9493.

Price S L, R J Harrison and M F Guest 1989. An *Ab Initio* Distributed Multipole Study of the Electrostatic Potential Around an Undecapeptide Cyclosporin Derivative and a Comparison with Point Charge Electrostatic Models. *Journal of Computational Chemistry* **10**:552–567.

Rappé A K and W A Goddard III 1991. Charge Equilibration for Molecular Dynamics Simulations. *Journal of Physical Chemistry* **95**:3358–3363.

Rappé A K, C J Casewit, K S Colwell, W A Goddard III and W M Skiff 1992. UFF, a Full Periodic Table Force Field for Molecular Mechanics and Molecular Dynamics Simulations. *The Journal of the American Chemical Society* **114**:10024–10035.

Reynolds C A, J W Essex and W G Richards 1992. Atomic Charges for Variable Molecular Conformations. *The Journal of the American Chemical Society* **114**:9075–9079.

Rigby M, E B Smith, W A Wakeham and G C Maitland 1986. *The Forces Between Molecules*. Oxford, Clarendon Press.

Rodger P M, A J Stone and D J Tildesley 1988. The Intermolecular Potential of Chlorine. A Three Phase Study. *Molecular Physics* **63**:173–188.

Singh U C and P A Kollman 1984. An Approach to Computing Electrostatic Charges for Molecules. *Journal of Computational Chemistry* **5**:129–145.

Smith P E and B M Pettitt 1994. Modelling Solvent in Biomolecular Systems. *Journal of Physical Chemistry* **98**:9700–9711.

Sprik M and M L Klein 1988. A Polarisable Model for Water Using Distributed Charge Sites. *The Journal of Chemical Physics* **89**:7556–7560.

Stillinger F H and A Rahman 1974. Improved Simulation of Liquid Water by Molecular Dynamics. *The Journal of Chemical Physics* **60**:1545–1557.

Stone A J 1981. Distributed Multipole Analysis, or How to Describe a Molecular Charge Distribution. *Chemical Physics Letters* **83**:233–239.

Stone A J and M Alderton 1985. Distributed Multipole Analysis Methods and Applications. *Molecular Physics* **56**:1047–1064.

Toxvaerd S 1990. Molecular Dynamics Calculation of the Equation of State of Alkanes. *The Journal of Chemical Physics* **93**:4290–4295.

Vedani A 1988. YETI: An Interactive Molecular Mechanics Program for Small-Molecular Protein Complexes. *Journal of Computational Chemistry* **9**:269–280.

Vinter J G 1994. Extended Electron Distributions Applied to the Molecular Mechanics of Some Intermolecular Interactions. *Journal of Computer-Aided Molecular Design* **8**:653–668.

Weiner S J, P A Kollman, D A Case, U C Singh, C Ghio, G Alagona, S Profeta and P Weiner 1984. A New Force Field for Molecular Mechanical Simulation of Nucleic Acids and Proteins. *The Journal of the American Chemical Society* **106**:765–784.

Williams D E 1990. Alanyl Dipeptide Potential-Derived Net Atomic Charges and Bond Dipoles, and Their Variation with Molecular Conformation. *Biopolymers* **29**:1367–1386.

4 Energy minimisation and related methods for exploring the energy surface

4.1 Introduction

For all except the very simplest systems the potential energy is a complicated, multidimensional function of the coordinates. For example, the energy of a conformation of ethane is a function of the 18 internal coordinates or 24 Cartesian coordinates that are required to completely specify the structure. As we discussed in section 1.3, the way in which the energy varies with the coordinates is usually referred to as the *potential energy surface* (sometimes called the *hypersurface*). In the interests of brevity all references to 'energy' should be taken to mean 'potential energy' for the rest of this chapter, except where explicitly stated otherwise. For a system with N atoms the energy is thus a function of $3N - 6$ internal or $3N$ Cartesian coordinates. It is therefore impossible to visualise the entire energy surface except for some simple cases where the energy is a function of just one or two coordinates. For example, the van der Waals energy of two argon atoms (as might be modelled using the Lennard–Jones potential function) depends upon just one coordinate: the interatomic distance. Sometimes we may wish to visualize just a part of the energy surface. For example, suppose we take an extended conformation of pentane and rotate the two central carbon–carbon bonds so that the torsion angles vary from 0° to 360°, calculating the energy of each structure generated. The energy in this case is a function of just two variables and can be plotted as a contour diagram or as an isometric plot, as shown in Figure 4.1.

We will use the term energy surface to refer not only to systems in which the bonding remains unchanged, as in these two examples, but also where bonds are

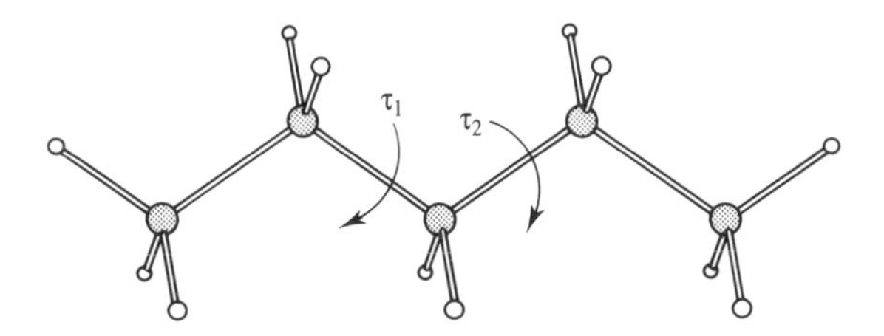

Fig. 4.1 Variation in the energy of pentane with the two torsion angles indicated and represented as a contour diagram and isometric plot. (Continued overleaf.)

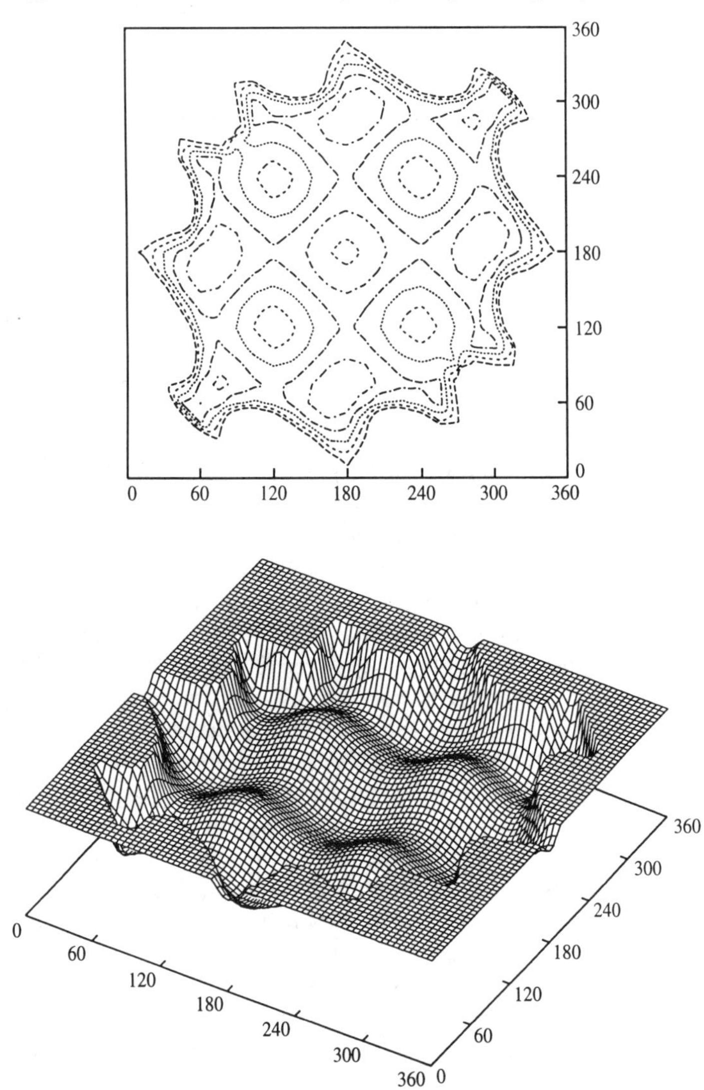

Figure 4.1 Continued. Only the lowest energy regions are shown.

broken and/or formed. Our discussion will be appropriate to both quantum mechanics and molecular mechanics, except where otherwise stated.

In molecular modelling we are especially interested in minimum points on the energy surface. Minimum energy arrangements of the atoms correspond to stable states of the system; any movement away from a minimum gives a configuration with a higher energy. There may be a very large number of minima on the energy surface. The minimum with the very lowest energy is known as the *global energy minimum*. To identify those geometries of the system that correspond to minimum points on the energy surface we use a *minimisation algorithm*. There is a vast literature on such methods and so we will concentrate on those approaches that are most commonly used in molecular modelling. We may also be interested to know how the system changes from one minimum-energy structure to another. For example, how do the relative positions of the atoms vary during a reaction? What structural changes occur as a molecule changes its conformation? The highest point on the pathway between two minima is of especial interest and is known as the *saddle point* with the arrangement of the atoms being the *transition structure*. Both minima and saddle points are stationary points on the energy surface, where the first derivative of the energy function is zero with respect to all the coordinates.

A geographical analogy can be a helpful way to illustrate many of the concepts we shall encounter in this chapter. In this analogy minimum points correspond to the bottom of valleys. A minimum may be described as being in a 'long and narrow valley' or 'a flat and featureless plain'. Saddle points correspond to mountain passes. We refer to algorithms or algorithms taking steps 'uphill' or 'downhill'.

4.1.1 Energy minimisation: statement of the problem

The minimisation problem can be formally stated as follows: given a function f which depends on one or more independent variables $x_1, x_2, \ldots, x_i$, find the values of those variables where f has a minimum value. At a minimum point the first derivative of the function with respect to each of the variables is zero and the second derivatives are all positive:

$$\frac{\partial f}{\partial x_i} = 0; \quad \frac{\partial^2 f}{\partial x_i^2} > 0 \tag{4.1}$$

The functions of most interest to us will be the quantum mechanics or molecular mechanics energy with the variables x_i being the Cartesian or the internal coordinates of the atoms. Molecular mechanics minimisations are nearly always performed in Cartesian coordinates, where the energy is a function of $3N$ variables; it is more common to use internal coordinates (as defined in the Z-matrix) with quantum mechanics. For analytical functions, the minimum of a function can be found using standard calculus methods. However, this is not generally possible for molecular systems due to the complicated way in which the energy varies with the coordinates. Rather, minima are located using numerical methods which gradually change the coordinates to produce configurations with lower and lower energies until the minimum is reached. To illustrate how the various minimisation algorithms operate, we shall consider a simple function of two variables: $f(x, y) = x^2 + 2y^2$. This function is represented as a contour diagram in Figure 4.2. The function has one minimum point, located at the origin. In our examples we will attempt to locate the minimum from the point (9.0, 9.0). Although this is a function of just two variables for the purposes of illustration, all of the methods that we shall consider can be applied to functions of many more variables.

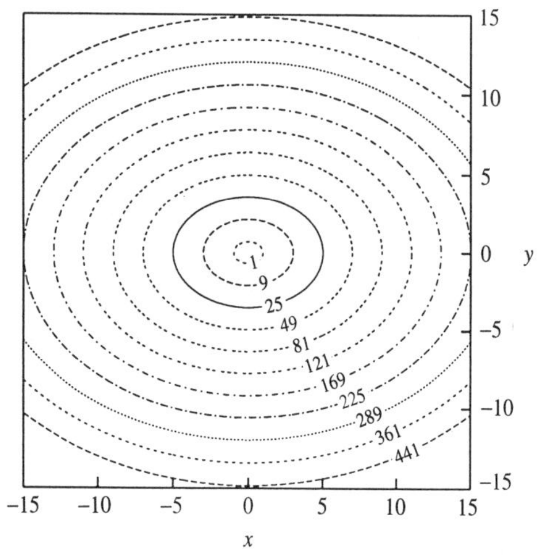

Fig. 4.2 The function $x^2 + 2y^2$.

We can classify minimisation algorithms into two groups: those which use derivatives of the energy with respect to the coordinates and those which do not. Derivatives can be useful because they provide information about the shape of the energy surface, and if used properly they can significantly enhance the efficiency with which the minimum is located. There are many factors that must be taken into account when choosing the most appropriate algorithm (or combination of algorithms) for a given problem; the ideal minimisation algorithm is the one that provides the answer as quickly as possible, using the least amount of memory. No single minimisation method has yet proved to be the best for all molecular modelling problems and so most software packages offer a choice of methods. In particular, a method that works well with quantum mechanics may not be the most suitable for use with molecular mechanics. This is partly because quantum mechanics is usually used to model systems with smaller numbers of atoms than molecular mechanics; some operations that are integral to certain minimisation procedures (such as matrix inversion) are trivial for small systems but formidable for systems containing thousands of atoms. Quantum mechanics and molecular mechanics also require different amounts of computational effort to calculate the energies and the derivatives of the various configurations. Thus an algorithm that takes many steps may be appropriate for molecular mechanics but inappropriate for quantum mechanics.

Most minimisation algorithms can only go downhill on the energy surface and so they can only locate the minimum that is nearest (in a downhill sense) to the starting point. Thus, Figure 4.3 shows a schematic energy surface and the minima that would be obtained by starting from three points A, B and C. The minima can be considered to correspond to the locations where a ball rolling on the energy surface under the influence of gravity would come to rest. To locate more than one minimum or to locate the global energy minimum we therefore usually require a means of generating different starting points, each of which is then minimised. Some specialised minimisation methods can make uphill moves to

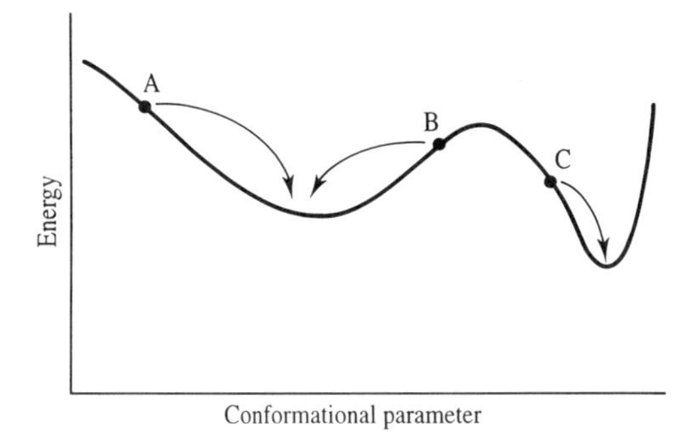

Fig. 4.3 A schematic one-dimensional energy surface. Minimisation methods move downhill to the nearest minimum. The statistical weight of the narrow, deep minimum may be less than a broad minimum which is higher in energy.

seek out minima lower in energy than the nearest one, but no algorithm has yet proved capable of locating the global energy minimum from an arbitrary starting position. The shape of the energy surface may be important if one wishes to calculate the relative populations of the various minimum energy structures. For example, a deep and narrow minimum may be less highly populated than a broad minimum that is higher in energy as the vibration energy levels will be more widely spaced in the deeper minimum and so less accessible. For this reason the global energy minimum may not be the most highly populated minimum. In any case, the 'active' structure (e.g. the biologically active conformation of a drug molecule) may not correspond to the global minimum, nor to the most highly populated conformation, nor even to a minimum energy structure at all.

The input to a minimisation program consists of a set of initial coordinates for the system. The initial coordinates may come from a variety of sources. They may be obtained from an experimental technique, such as X-ray crystallography or NMR. In other cases a theoretical method is employed, such as a conformational search algorithm. A combination of experimental and theoretical approaches may also be used. For example, to study the behaviour of a protein in water one may take an X-ray structure of the protein and immerse it in a solvent 'bath', where the coordinates of the solvent molecules have been obtained from a Monte Carlo or molecular dynamics simulation.

4.1.2 Derivatives

In order to use a derivative minimisation method it is obviously necessary to be able to calculate the derivatives of the energy with respect to the variables (i.e. the Cartesian or internal coordinates, as appropriate). Derivatives may be obtained either analytically or numerically. The use of analytical derivatives is preferable as they are exact, and because they can be calculated more quickly; if only numerical derivatives are available then it may be more effective to use a non-derivative minimisation algorithm. The problems of calculating analytical derivatives with quantum mechanics and molecular mechanics were discussed in sections 2.9.4 and 3.15 respectively.

Nevertheless, under some circumstances it is necessary to use numerical derivatives. These can be calculated as follows. If one of the coordinates x_i is changed by a small change (δx_i) and the energy for the new arrangement is computed then the derivative $\partial E/\partial x_i$ is obtained by dividing the change in energy (δE) by the change in coordinate ($\delta E/\delta x_i$). This strictly gives the derivative at the mid-point between the two points x_i and $x_i + \delta x_i$. A more accurate value of the derivative at the point x_i may be obtained (at the cost of an additional energy calculation) by evaluating the energy at two points, $x_i + \delta x_i$ and $x_i - \delta x_i$. The derivative is then obtained by dividing the difference in the energies by $2\delta x_i$.

4.2 Non-derivative minimisation methods

4.2.1 The simplex method

A *simplex* is a geometrical figure with $M+1$ interconnected vertices, where M is the dimensionality of the energy function. For a function of two variables the simplex is thus triangular in shape. A tetrahedral simplex is used for a function of three variables and so for an energy function of $3N$ Cartesian coordinates the simplex will have $3N+1$ vertices; if internal coordinates are used then the simplex will have $3N-5$ vertices. Each vertex corresponds to a specific set of coordinates for which an energy can be calculated. For our function $f(x, y) = x^2 + 2y^2$ the simplex method would use a triangular simplex.

The simplex algorithm locates a minimum by moving around on the potential energy surface in a fashion that has been likened to the motion of an amoeba. Three basic kinds of moves are possible. The most common type of move is a reflection of the vertex with the highest value through the opposite face of the simplex, in an attempt to generate a new point that has a lower value. If this new point is lower in energy than any of the other points in the simplex then a 'reflection and expansion' move may be applied. When a 'valley floor' is reached then a reflection move will fail to produce a better point. Under such circumstances the simplex contracts along one dimension from the highest point. If this fails to reduce the energy then a third type of move is possible, in which the simplex contracts in all directions, pulling around the lowest point. These three moves are illustrated in Figure 4.4.

To implement the simplex algorithm it is first necessary to generate the vertices of the initial simplex. The initial configuration of the system corresponds to just one of these vertices. The remaining points can be obtained in a variety of ways, but one simple method is to add a constant increment to each coordinate in turn. The energy of the system is calculated at the new point, giving the function value for the relevant vertex.

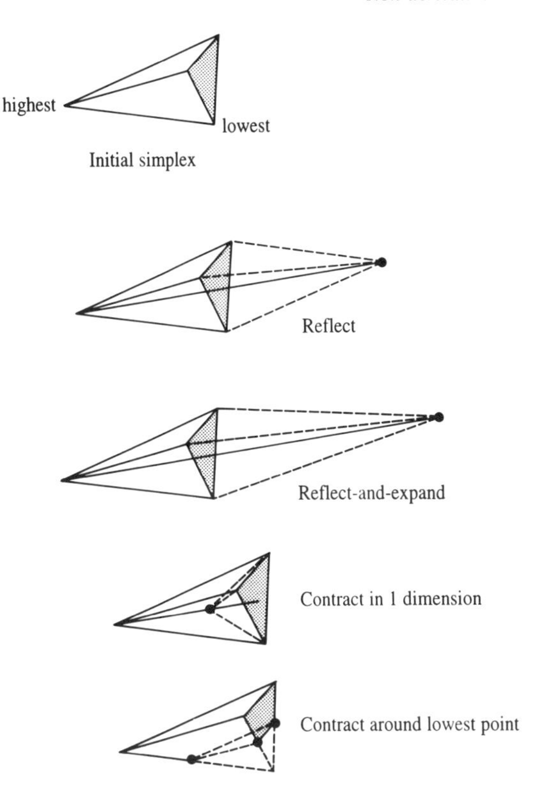

Fig. 4.4 The three basic moves permitted to simplex algorithm (reflection, and its close relation reflect-and-expand; contract in one dimension and contract around the lowest point). Figure adapted from Press W H, B P Flannery, S A Teukolsky and W T Vetterling 1992. *Numerical Recipes in Fortran*. Cambridge, Cambridge University Press.

The simplex method is most useful where the initial configuration of the system is very high in energy, because it rarely fails to find a better solution. However, it can be rather expensive in terms of computer time due to the large number of energy evaluations which are required (merely to generate the initial simplex requires $3N+1$ energy evaluations). For this reason the simplex method is often used in combination with a different minimisation algorithm: a few steps of the simplex method are used to refine the initial structure and then a more efficient method can take over.

Let us consider the application of the simplex method to our quadratic function, $f=x^2+2y^2$ (Figure 4.5). Suppose our initial simplex contains vertices located at the points (9, 9), (11, 9) and (9, 11) which have been generated by adding a constant factor 2 to each of the variables in turn. The values of the function at these points are 243, 283 and 323 respectively. The vertex with the highest function value is at (9, 11) and so in the first iteration this point is reflected through the opposite face of the triangle, to generate a point with

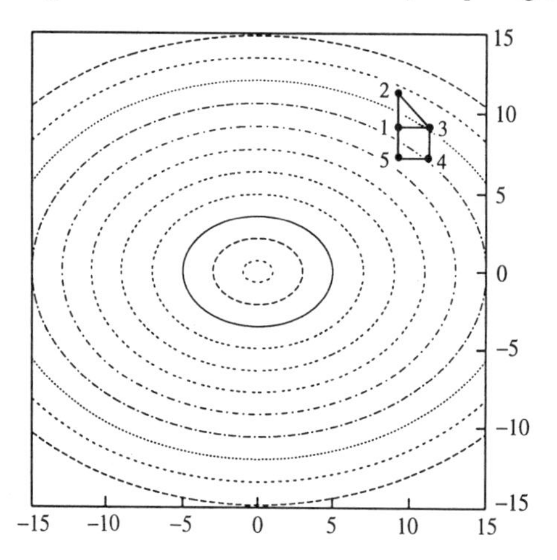

Fig. 4.5 The first few steps of the simplex algorithm with the function $x^2 + 2y^2$. The initial simplex corresponds to the triangle 123. Point 2 has the largest value of the function and the next simplex is the triangle 134. The simplex for the third step is 145.

coordinates (11, 7) and a function value of 219 (we do not use the reflect-and-expand move in our illustration). The highest vertex is now at (11, 9) which is reflected through the opposite face of the simplex to give the point (9, 7) where the function has a value of 179. In fact, for this admittedly artificial problem the simplex algorithm takes more than 30 steps to find a point where the function has a value less than 0.1.

Why does the simplex contain one more vertex than the number of degrees of freedom? The reason is that with fewer than $M+1$ vertices the algorithm cannot explore the whole energy surface. Suppose we use only a two vertex simplex to explore our quadratic energy surface. A simplex with just two vertices is a straight line. The only moves that would be possible in this case would be to other points that lie on this line; none of the energy surface away from the line would be explored. Similarly, if we have a function of three variables and restrict the simplex to a triangle then we will only be able to explore the region of space that lies in the same plane as the triangle whereas the minimum may not lie in this plane.

4.2.2 *The sequential univariate method*

The simplex method is rarely considered suitable for quantum mechanical calculations, due to the number of energy evaluations that must be performed. The sequential univariate method is a non-derivative method that is considered more appropriate in this case. This method systematically cycles through the coordinates in turn. For each coordinate, two new structures are generated by changing the current coordinate (i.e. $x_i + \delta x_i$ and $x_i + 2\delta x_i$). The energies of these two

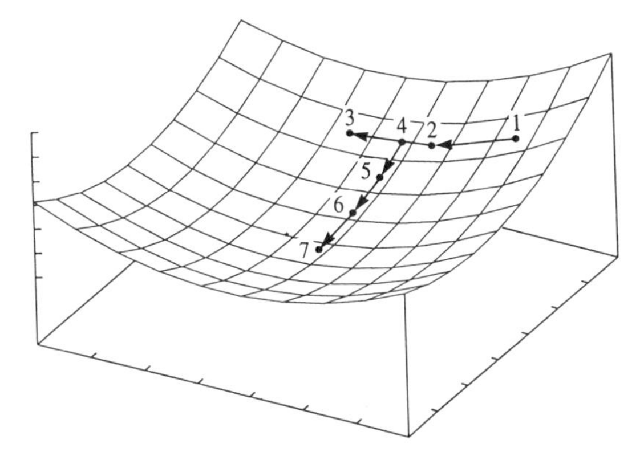

Fig. 4.6 The sequential univariate method. Starting at the point labelled 1 two steps are made along one of the coordinates to give points 2 and 3. A parabola is fitted to these three points and the minimum located (point 4). The same procedure is then repeated along the next coordinate (points 5, 6 and 7). Figure adapted from Schlegel H B 1987. Optimization of Equilibrium Geometries and Transition Structures in Lawley K P (Editor). *Ab Initio Methods in Quantum Chemistry—I.* New York, John Wiley: 249–286.

structures are calculated. A parabola is then fitted through the three points corresponding to the two distorted structures and the original structure. The minimum point in this quadratic function is determined and the coordinate is then changed to the position of the minimum. The procedure is illustrated in Figure 4.6. When the changes in all the coordinates are sufficiently small then the minimum is deemed to have been reached, otherwise a new iteration is performed. The sequential invariate method usually requires fewer function evaluations than the simplex method but it can be slow to converge especially if there is strong coupling between two or more of the coordinates or when the energy surface is analogous to a long narrow valley.

4.3 Introduction to derivative minimisation methods

Derivatives provide information that can be very useful in energy minimisation, and derivatives are used by most popular minimisation methods. The direction of the first derivative of the energy (the gradient) indicates where the minimum lies and the magnitude of the gradient indicates the steepness of the local slope. The energy of the system can be lowered by moving each atom in response to the force acting on it; the force is equal to minus the gradient. Second derivatives indicate the curvature of the function, information that can be used to predict where the function will change direction (i.e. pass through a minimum or some other stationary point).

When discussing derivative methods it is useful to write the function as a Taylor series expansion about the point x_k:

$$\mathscr{V}(x) = \mathscr{V}(x_k) + (x - x_k)\mathscr{V}'(x_k) + (x - x_k)^2\mathscr{V}''(x_k)/2 + \cdots \tag{4.2}$$

For a multidimensional function, the variable x is replaced by the vector $\mathbf{x}$ and matrices are used for the various derivatives. Thus if the potential energy $\mathscr{V}(\mathbf{x})$ is a function of $3N$ Cartesian coordinates, the vector $\mathbf{x}$ will have $3N$ components and $\mathbf{x}_k$ corresponds to the current configuration of the system. $\mathscr{V}'(\mathbf{x}_k)$ is a $3N \times 1$ matrix (i.e. a vector), each element of which is the partial derivative of $\mathscr{V}$ with respect to the appropriate coordinate, $\partial\mathscr{V}/\partial x_i$. We will also write the gradient at the point k as $\mathbf{g}_k$. Each element (i, j) of the matrix $\mathscr{V}''(\mathbf{x}_k)$ is the partial second derivative of the energy function with respect to the two coordinates x_i and x_j, $\partial^2\mathscr{V}/\partial x_i \partial x_j$. $\mathscr{V}''(\mathbf{x}_k)$ is thus of dimension $3N \times 3N$ and is known as the *Hessian* matrix or the *force constant* matrix. The Taylor series expansion can be written in the following form for the multidimensional case:

$$\mathscr{V}(\mathbf{x}) = \mathscr{V}(\mathbf{x}_k) + (\mathbf{x} - \mathbf{x}_k)\mathscr{V}'(\mathbf{x}_k) + (\mathbf{x} - \mathbf{x}_k)^T.\mathscr{V}''(\mathbf{x}_k).(\mathbf{x} - \mathbf{x}_k)/2 + \cdots \tag{4.3}$$

The energy functions used in molecular modelling are rarely quadratic and so the Taylor series expansion, equation (4.3) can only be considered an approximation. There are two important consequences of this. The first consequence is that the performance of a given minimisation method will not be as good for a molecular mechanics or quantum mechanics energy surface as it is for a pure quadratic function. As we shall see, a second derivative method such as the Newton–Raphson algorithm can locate the minimum in a single step for a purely quadratic function, but several iterations are usually required for a typical molecular modelling energy function. The second consequence is that far from the minimum the harmonic approximation is a poor one and some of the less robust methods will fail, even though they may work very well close to a minimum where the harmonic approximation is more valid. For this reason it is important to choose the minimisation protocol with care, possibly using a robust (but perhaps inefficient) method at first, and then a less robust but more efficient method.

The derivative methods can be classified according to the highest order derivative used. First-order methods use the first derivatives (i.e. the gradients) whereas second-order methods use both first and second derivatives. The simplex method can thus be considered a zeroth-order method as it does not use any derivatives.

4.4 First-order minimisation methods

Two first-order minimisation algorithms that are frequently used in molecular modelling are the method of *steepest descents* and the *conjugate gradient* method. These gradually change the coordinates of the atoms as they move the system closer and closer to the minimum point. The starting point for each iteration (k) is the molecular configuration obtained from the previous step which is represented by the multidimensional vector $\mathbf{x}_{k-1}$. For the first iteration the starting point is the initial configuration of the system provided by the user, the vector $\mathbf{x}_1$.

4.4.1 *The steepest descents method*

The steepest descents method moves in the direction parallel to the net force, which in our geographical analogy corresponds to walking straight downhill. For $3N$ Cartesian coordinates this direction is most conveniently represented by a $3N$-dimensional unit vector, $\mathbf{s}_k$. Thus:

$$\mathbf{s}_k = -\mathbf{g}_k/|\mathbf{g}_k| \tag{4.4}$$

Having defined the direction along which to move it is then necessary to decide how far to move along the gradient. Consider the two-dimensional energy surface of Figure 4.7. The gradient direction from the starting point is along the line indicated. If we imagine a cross-section through the surface along the line, the function will pass through a minimum and then increase, as shown in the figure. We can choose to locate the minimum point by performing a *line search* or we can take a step of arbitrary size along the direction of the force.

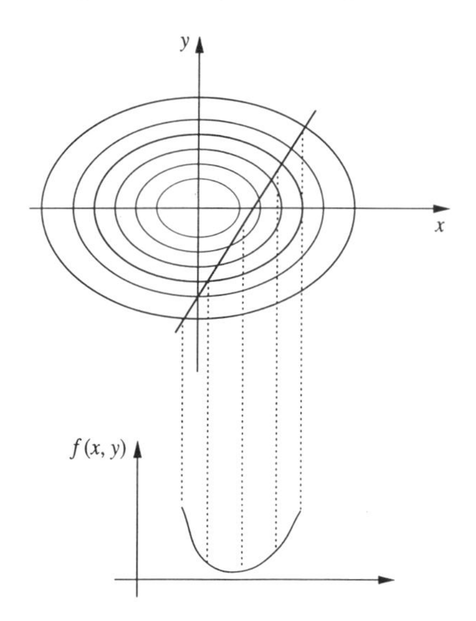

Fig. 4.7 A line search is used to locate the minimum in the function in the direction of the gradient.

4.4.2 Line search in one dimension

The purpose of a line search is to locate the minimum along a specified direction (i.e. along a line through the multidimensional space). The first stage of the line search is to *bracket* the minimum. This entails finding three points along the line such that the energy of the middle point is lower than the energy of the two outer points. If three such points can be found, then at least one minimum must lie between the two outer points. An iterative procedure can then be used to decrease the distance between the three points, gradually restricting the minimum to an even smaller region. This is conceptually an easy process but it may require a considerable number of function evaluations, making it computationally expensive. An alternative is to fit a function such as a quadratic to the three points. Differentiation of the fitted function enables an approximation to the minimum along the line to be identified analytically. A new function can then be fitted to give a better estimate, as shown in Figure 4.8. Higher-order polynomials may give a better fit to the bracketing points but these can give incorrect interpolations when used with functions that change sharply in the bracketed region.

The gradient at the minimum point obtained from the line search will be perpendicular to the previous direction. Thus, when the line search method is used to locate the minimum along the gradient then the next direction in the steepest descents algorithm will be orthogonal to the previous direction (i.e. $\mathbf{g}_k \cdot \mathbf{g}_{k-1} = 0$).

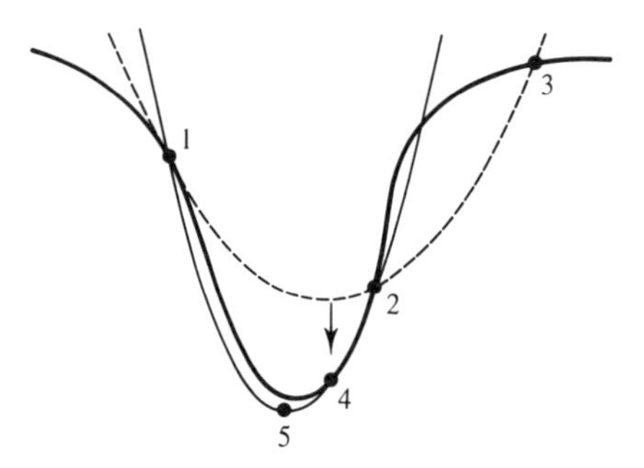

Fig. 4.8 The minimum in a line search may be found more effectively by fitting an analytical function such as a quadratic to the initial set of three points (1, 2 and 3). A better estimate of the minimum can then be found by fitting a new function to the points 1, 2 and 4 and finding its minimum. Figure adapted from Press W H, B P Flannery, S A Teukolsky and W T Vetterling 1992. *Numerical Recipes in Fortran*. Cambridge, Cambridge University Press.

4.4.3 Arbitrary step approach

As the line search may itself be computationally demanding we could obtain the new coordinates by taking a step of arbitrary length along the gradient unit vector $\mathbf{s}_k$. The new set of coordinates after step k would then be given by the equation:

$$\mathbf{x}_{k+1} = \mathbf{x}_k + \lambda_k \mathbf{s}_k \tag{4.5}$$

λ_k is the *step size*. In most applications of the steepest descents algorithm in molecular modelling the step size initially has a predetermined default value. If the first iteration leads to a reduction in energy, the step size is increased by a multiplicative factor (e.g. 1.2) for the second iteration. This process is repeated so long as each iteration reduces the energy. When a step produces an increase in energy, it is assumed that the algorithm has leapt across the valley which contains the minimum and up the slope on the opposite face. The step size is then reduced by a multiplicative factor (e.g. 0.5). The step size depends upon the nature of the energy surface; for a flat surface large step sizes would be appropriate but for a narrow, twisting gully a much smaller step would be more suitable. The arbitrary step method may require more steps to reach the minimum but it can often require fewer function evaluations (and thus less computer time) than the more rigorous line search approach.

The steepest descents method works as follows for our trial function, $f(x, y) = x^2 + 2y^2$. Differentiating the function gives $df = 2x dx + 4y dy$ and so the gradient at any point (x, y) equals $4y/2x$. The direction of the first move from the point (9.0, 9.0) is $(-18.0, -36.0)$ and the equation of the line along which the search is performed is $y = 2x - 9$. The minimum of the function along this line can be obtained using Lagrange multipliers (see section 1.10.5) and is at (4.0, -1.0). The direction of the next move is the vector $(-8, 4)$ and the next line search is performed along the line $y = -0.5x + 1$. The minimum point along this line is (2/3, 2/3) where the function has the value 4/3. The third point found by the steepest descents method is at (0.296,–0.074) where the function has the value 0.099. These moves are illustrated in Figure 4.9.

The direction of the gradient is determined by the largest interatomic forces and so steepest descents is a good method for relieving the highest energy features in an initial configuration. The method is generally robust even when the starting point is far from a minimum where the harmonic approximation to the energy surface is often a poor assumption. However, it suffers from the problem that many small steps will be performed when proceeding down a long narrow valley. The steepest descents method is forced to make a right-angled turn at each point, even though that might not be the best route to the minimum. The path oscillates and continually overcorrects itself as illustrated in Figure 4.10; later steps reintroduce errors that were corrected by earlier moves.

4.4.4 Conjugate gradients minimisation

The conjugate gradients method produces a set of directions which does not show the oscillatory behaviour of the steepest descents method in narrow valleys. In the

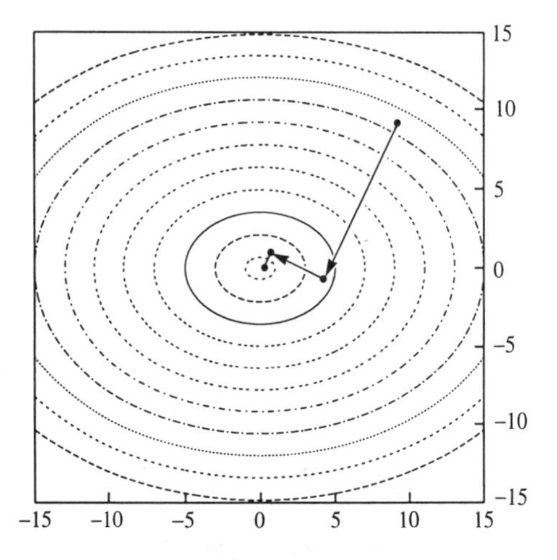

Fig. 4.9 Application of steepest descents to the function $x^2 + 2y^2$.

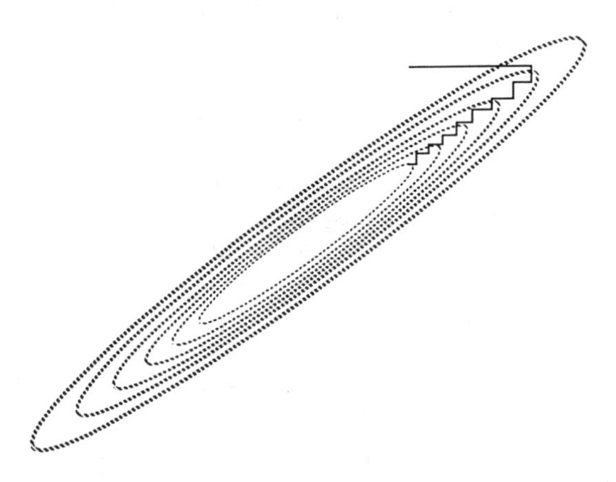

Fig. 4.10 The steepest descents methods can give undesirable behaviour in a long narrow valley.

steepest descents method both the gradients and the direction of successive steps are orthogonal. In conjugate gradients, the gradients at each point are orthogonal but the directions are *conjugate* (indeed, the method is more properly called the conjugate directions method). A set of conjugate directions has the property that for a quadratic function of M variables, the minimum will be reached in M steps. The conjugate gradients method moves in a direction $\mathbf{v}_k$ from point $\mathbf{x}_k$ where $\mathbf{v}_k$ is computed from the gradient at the point and the previous direction vector $\mathbf{v}_{k-1}$:

$$\mathbf{v}_k = -\mathbf{g}_k + \gamma_k \mathbf{v}_{k-1} \tag{4.6}$$

γ_k is a scalar constant given by

$$\gamma_k = \frac{\mathbf{g}_k \cdot \mathbf{g}_k}{\mathbf{g}_{k-1} \cdot \mathbf{g}_{k-1}} \tag{4.7}$$

In the conjugate gradients method all of the directions and gradients satisfy the following relationships:

$$\mathbf{g}_i \cdot \mathbf{g}_j = 0 \tag{4.8}$$

$$\mathbf{v}_i \cdot \mathcal{V}''_{ij} \cdot \mathbf{v}_j = 0 \tag{4.9}$$

$$\mathbf{g}_i \cdot \mathbf{v}_j = 0 \tag{4.10}$$

Clearly equation (4.6) can only be used from the second step onwards and so the first step in the conjugate gradients method is the same as the steepest descents (i.e. in the direction of the gradient). The line search method should ideally be used to locate the one-dimensional minimum in each direction to ensure that each gradient is orthogonal to all previous gradients and that each direction is conjugate to all previous directions. However, an arbitrary step method is also possible.

The conjugate gradients methods deals with our simple quadratic function $f(x, y) = x^2 + 2y^2$ as follows. From the initial point (9, 9) we move to the same point as in steepest descents, (4, −1). To find the direction of the next move, we first determine the negative gradient at the current point. This is the vector (−8, 4). This is then combined with the vector corresponding to minus the gradient at the initial point, (−18, −36) multiplied by γ:

$$\mathbf{v}_k = \begin{pmatrix} -8 \\ 4 \end{pmatrix} + \frac{(-8)^2 + (4)^2}{(-18)^2 + (-36)^2} \begin{pmatrix} -18 \\ -36 \end{pmatrix} = \begin{pmatrix} -80/9 \\ +20/9 \end{pmatrix} \tag{4.11}$$

To locate the second point we therefore need to perform a line search along the line with gradient $-1/4$ that passes through the point $(4, -1)$. The minimum along this line is at the origin, at the true minimum of the function. The conjugate gradients method thus locates the exact minimum of the function exactly in just two moves, as illustrated in Figure 4.11.

Several variants of the conjugate gradients method have been proposed. The formulation given in equation (4.7) is the original Fletcher–Reeves algorithm. Polak and Ribiere proposed an alternative form for the scalar constant γ_k:

$$\gamma_k = \frac{(\mathbf{g}_k - \mathbf{g}_{k-1}) \cdot \mathbf{g}_k}{\mathbf{g}_{k-1} \cdot \mathbf{g}_{k-1}} \tag{4.12}$$

For a purely quadratic function the Polak–Ribiere method is identical to the Fletcher–Reeves algorithm as all gradients will be orthogonal. However, most functions of interest, including those used in molecular modelling, are at best only approximately quadratic. Polak and Riviere claimed that their method

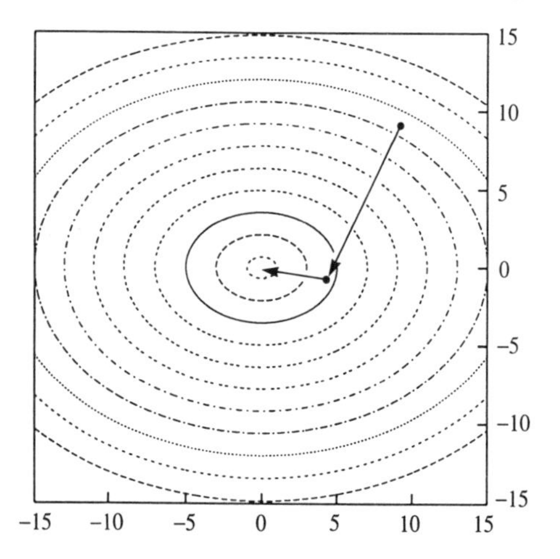

Fig. 4.11 Application of conjugate gradients method to the function $x^2 + 2y^2$.

performed better than the original Fletcher–Reeves algorithm, at least for the functions that they examined.

4.5 Second derivative methods: the Newton–Raphson method

Second-order methods use not only the first derivatives (i.e. the gradients) but also the second derivatives to locate a minimum. Second derivatives provide information about the curvature of the function. The *Newton–Raphson* method is the simplest second-order method. Recall our Taylor series expansion about the point x_k, equation (4.2):

$$\mathcal{V}(x) = \mathcal{V}(x_k) + (x - x_k)\mathcal{V}'(x_k) + (x - x_k)^2\mathcal{V}''(x_k)/2 + \cdots$$

The first derivative of $\mathcal{V}(x)$ is:

$$\mathcal{V}'(x) = x\mathcal{V}'(x_k) + (x - x_k)\mathcal{V}''(x_k) \quad (4.13)$$

If the function is purely quadratic, the second derivative is the same everywhere, and so $\mathcal{V}''(x) = \mathcal{V}''(x_k)$.

At the minimum $(x = x^*)\mathcal{V}'(x^*) = 0$ and so

$$x^* = x_k - \mathcal{V}'(x_k)/\mathcal{V}''(x_k) \quad (4.14)$$

For a multidimensional function: $\mathbf{x}^* = \mathbf{x}_k - \mathcal{V}'(\mathbf{x}_k)\mathcal{V}''^{-1}(\mathbf{x}_k)$.

$\mathcal{V}''^{-1}(\mathbf{x}_k)$ is the inverse Hessian matrix of second derivatives which in the Newton–Raphson method must therefore be inverted. This can be computationally demanding for systems with many atoms and can also require a significant amount of storage. The Newton–Raphson method is thus more suited to small molecules (usually less than 100 atoms or so). For a purely quadratic function the

Newton–Raphson method finds the minimum in one step from any point on the surface, as we will now show for our function $f(x, y) = x^2 + 2y^2$.

The Hessian matrix for this function is:

$$\mathbf{f}'' = \begin{pmatrix} 2 & 0 \\ 0 & 4 \end{pmatrix} \tag{4.15}$$

The inverse of this matrix is:

$$\mathbf{f}''^{-1} = \begin{pmatrix} 1/2 & 0 \\ 0 & 1/4 \end{pmatrix} \tag{4.16}$$

The minimum is obtained using equation (4.14):

$$\mathbf{x}^* = \begin{pmatrix} 9 \\ 9 \end{pmatrix} - \begin{pmatrix} 1/2 & 0 \\ 0 & 1/4 \end{pmatrix} \begin{pmatrix} 18 \\ 36 \end{pmatrix} = \begin{pmatrix} 0 \\ 0 \end{pmatrix} \tag{4.17}$$

In practice, of course, the surface is only quadratic to a first approximation and so a number of steps will be required, at each of which the Hessian matrix must be calculated and inverted. The Hessian matrix of second derivatives must be *positive definite* in a Newton–Raphson minimisation. A positive definite matrix is one for which all the eigenvalues are positive. When the Hessian matrix is not positive definite then the Newton–Raphson method moves to points (e.g. saddle points) where the energy increases. In addition, far from a minimum the harmonic approximation is not appropriate and the minimisation can become unstable. One solution to this problem is to use a more robust method to get near to the minimum (i.e. where the Hessian is positive definite) before applying the Newton–Raphson method.

4.5.1 Variants on the Newton–Raphson method

There are a number of variations on the Newton–Raphson method, many of which aim to eliminate the need to calculate the full matrix of second derivatives. In addition, a family of methods called the quasi-Newton methods require only first derivatives and gradually construct the inverse Hessian matrix as the calculation proceeds. One simple way in which it may be possible to speed up the Newton–Raphson method is to use the same Hessian matrix for several successive steps of the Newton–Raphson algorithm with only the gradients being recalculated at each iteration.

A widely used algorithm is the *block-diagonal Newton–Raphson* method in which just one atom is moved at each iteration. Consequently all terms of the form $\partial^2\mathscr{V}/\partial x_i \partial x_j$, where i and j refer to the Cartesian coordinates of atoms other than the atom being moved, will be zero. This only leaves those terms which involve the coordinates of the atom being moved and so reduces the problem to the trivial one of inverting a 3×3 matrix. However, the block-diagonal approach can be less efficient when the motions of some atoms are closely coupled, such as the concerted movements of connected atoms in a phenyl ring.

4.6 Quasi-Newton methods

The calculation of the inverse Hessian matrix is a potentially time-consuming operation that is a significant drawback to the 'pure' second derivative methods such as Newton–Raphson. Moreover, one may not be able to calculate analytical second derivatives which are preferable. The quasi-Newton methods (also known as variable metric methods) gradually build up the inverse Hessian matrix in successive iterations. That is, a sequence of matrices $\mathbf{H}_k$ is constructed that has the property

$$\lim_{k \to \infty} \mathbf{H}_k = \mathcal{V}''^{-1} \tag{4.18}$$

At each iteration k, the new positions $\mathbf{x}_{k+1}$ are obtained from the current positions $\mathbf{x}_k$, the gradient $\mathbf{g}_k$ and the current approximation to the inverse Hessian matrix $\mathbf{H}_k$:

$$\mathbf{x}_{k+1} = \mathbf{x}_k - \mathbf{H}_k \mathbf{g}_k \tag{4.19}$$

This formula is exact for a quadratic function, but for 'real' problems a line search may be desirable. This line search is performed along the vector $\mathbf{x}_{k+1} - \mathbf{x}_k$. It may not be necessary to locate the minimum in the direction of the line search very accurately, at the expense of a few more steps of the quasi-Newton algorithm. For quantum mechanics calculations the additional energy evaluations required by the line search may prove more expensive than using the more approximate approach. An effective compromise is to fit a function to the energy and gradient at the current point $\mathbf{x}_k$ and at the point $\mathbf{x}_{k+1}$ and determine the minimum in the fitted function.

Having moved to the new positions $\mathbf{x}_{k+1}$, $\mathbf{H}$ is updated from its value at the previous step according to a formula depending upon the specific method being used. The methods of Davidon–Fletcher–Powell (DFP), Broyden–Fletcher–Goldfarb–Shanno (BFGS) and Murtaugh–Sargent (MS) are commonly encountered but there are many others. These methods converge to the minimum, for a quadratic function of M variables, in M steps. The Davidon–Fletcher–Powell (DFP) formula is:

$$\mathbf{H}_{k+1} = \mathbf{H}_k + \frac{(\mathbf{x}_{k+1} - \mathbf{x}_k) \otimes (\mathbf{x}_{k-1} - \mathbf{x}_k)}{(\mathbf{x}_{k+1} - \mathbf{x}_k) \cdot (\mathbf{g}_{k+1} - \mathbf{g}_k)}$$

$$- \frac{[\mathbf{H}_k \cdot (\mathbf{g}_{k+1} - \mathbf{g}_k)] \otimes [\mathbf{H}_k \cdot (\mathbf{g}_{k+1} - \mathbf{g}_k)]}{(\mathbf{g}_{k+1} - \mathbf{g}_k) \cdot \mathbf{H}_k \cdot (\mathbf{g}_{k+1} - \mathbf{g}_k)} \tag{4.20}$$

The symbol $\otimes$ when interposed between two vectors means that a matrix is to be formed. The ij-th element of the matrix $\mathbf{u} \otimes \mathbf{v}$ is obtained by multiplying $\mathbf{u}_i$ by $\mathbf{v}_j$.

The Broyden–Fletcher–Goldfarb–Shanno (BFGS) formula differs from the DFP equation by an additional term:

$$\mathbf{H}_{k+1} = \mathbf{H}_k + \frac{(\mathbf{x}_{k+1} - \mathbf{x}_k) \otimes (\mathbf{x}_{k+1} - \mathbf{x}_k)}{(\mathbf{x}_{k+1} - \mathbf{x}_k) \cdot (\mathbf{g}_{k+1} - \mathbf{g}_k)} - \frac{[\mathbf{H}_k \cdot (\mathbf{g}_{k+1} - \mathbf{g}_k)] \otimes [\mathbf{H}_k \cdot (\mathbf{g}_{k+1} - \mathbf{g}_k)]}{(\mathbf{g}_{k+1} - \mathbf{g}_k) \cdot \mathbf{H}_k \cdot (\mathbf{g}_{k+1} - \mathbf{g}_k)} + [(\mathbf{g}_{k+1} - \mathbf{g}_k) \cdot \mathbf{H}_k \cdot (\mathbf{g}_{k+1} - \mathbf{g}_k)]\mathbf{u} \otimes \mathbf{u} \tag{4.21}$$

where

$$\mathbf{u} = \frac{(\mathbf{x}_{k+1} - \mathbf{x}_k)}{(\mathbf{x}_{k+1} - \mathbf{x}_k) \cdot (\mathbf{g}_{k+1} - \mathbf{g}_k)} - \frac{[\mathbf{H}_k \cdot (\mathbf{g}_{k+1} - \mathbf{g}_k)]}{(\mathbf{g}_{k+1} - \mathbf{g}_k) \cdot \mathbf{H}_k \cdot (\mathbf{g}_{k+1} - \mathbf{g}_k)} \tag{4.22}$$

The Murtaugh–Sargent formula is:

$$\mathbf{H}_{k+1} = \mathbf{H}_k + \frac{[(\mathbf{x}_{k+1} - \mathbf{x}_k) - \mathbf{H}_k(\mathbf{g}_{k+1} - \mathbf{g}_k)] \otimes [(\mathbf{x}_{k+1} - \mathbf{x}_k) - \mathbf{H}_k(\mathbf{g}_{k+1} - \mathbf{g}_k)]}{[(\mathbf{x}_{k+1} - \mathbf{x}_k) - \mathbf{H}_k(\mathbf{g}_{k+1} - \mathbf{g}_k)] \cdot (\mathbf{g}_{k+1} - \mathbf{g}_k)} \tag{4.23}$$

All of these methods use just the new and current points to update the inverse Hessian. The default algorithm used in the Gaussian series of molecular orbital programs [Schlegel 1982] makes use of more of the previous points to construct the Hessian (and thence the inverse Hessian), giving better convergence properties. Another feature of this method is its use of a quartic polynomial that is guaranteed to have just one local minimum in the line search. The DFP, BFGS and MS methods can also be used with numerical derivatives but alternative approaches may be more effective under such circumstances.

The matrix **H** is often initialised to the unit matrix **I**. The performance of the quasi-Newton algorithms can be improved by using a better estimate of the inverse Hessian than just the unit matrix. The unit matrix gives no information about the bonding in the system, nor does it identify any coupling between the various degrees of freedom. For example, a molecular mechanics calculation can be used to provide an initial guess to **H** prior to a quantum mechanical calculation. Alternatively the matrix can be obtained from a quantum mechanical calculation at a lower level of theory (e.g. semi-empirical or with a smaller basis set).

4.7 Which minimisation method should I use?

The choice of minimisation algorithm is dictated by a number of factors, including the storage and computational requirements, the relative speeds with which the various parts of the calculation can be performed, the availability of analytical derivatives and the robustness of the method. Thus, any method that requires the Hessian matrix to be stored (let alone its inverse calculated), may present memory problems when applied to systems containing thousands of atoms. Calculations on systems of this size are invariably performed using molecular mechanics, and so the steepest descents and the conjugate gradients methods are very popular here. For molecular mechanics calculations on small molecules, the Newton–Raphson method may be used, although this algorithm can have problems with structures that are far from a minimum. For this reason it is usual to perform a few steps of minimisation using a more robust method such as the simplex or steepest descents before applying the Newton–Raphson algorithm. Analytical expressions for both first and second derivatives are available for most of the terms found in common force fields.

The performance of the steepest descents and conjugate gradients methods is contrasted in the following example. A model of the antibiotic netropsin (Figure 4.12) bound to DNA was constructed using an automated docking program. This initial model was then subjected to two stages of minimisation. In the first stage, the aim was to produce a structure that did not have any significant high-energy interactions. The structure was then further minimised to give a structure much closer to the minimum. The results are shown in Table 4.1.

This study shows that the steepest descent method can actually be superior to conjugate gradients when the starting structure is some way from the minimum. However, conjugate gradients is much better once the initial strain has been removed.

Quantum mechanical calculations are restricted to systems with relatively small numbers of atoms, and so storing the Hessian matrix is not a problem. As the energy calculation is often the most time-consuming part of the calculation, it is desirable that the minimisation method chosen takes as few steps as possible to reach the minimum. For many levels of quantum mechanics theory analytical first derivatives are available. However, analytical second derivatives are only available for a few levels of theory and can be expensive to compute. The quasi-Newton methods are thus particularly popular for quantum mechanical calculations.

When using internal coordinates in a quantum mechanical minimisation it can be important to use an appropriate Z-matrix as input. For many systems the

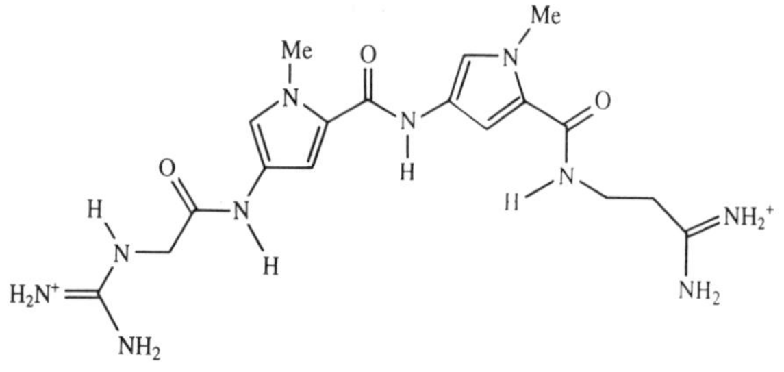

Fig. 4.12 The DNA inhibitor netropsin.

Table 4.1 A comparison of the steepest descents and conjugate gradients methods for an initial refinement and a stringent minimisation.

	Initial refinement (Ave. gradient < 1 kcal Å^{-2})		Stringent minimisation (Ave. gradient < 0.1 kcal Å^{-2})	
Method	Cpu time (s)	Number of iterations	Cpu time (s)	Number of iterations
Steepest descents	67	98	1405	1893
Conjugate gradients	149	213	257	367

Z-matrix can often be written in many different ways as there are many combinations of internal coordinates. If the system is not required to maintain a particular symmetry during the minimisation, there should be $3N-6$ independent coordinates which should all be able to vary during the minimisation. It is also useful if there is no strong coupling between the coordinates. *Dummy atoms* can often help in the construction of an appropriate Z-matrix. A dummy atom is used solely to define the geometry and has no nuclear charge and no basis functions. A simple example of the use of dummy atoms is for a linear molecule such as HN_3 where the angle of 180° would cause problems. The geometry of this molecule can be defined using a dummy atom as illustrated in Figure 4.13; the associated Z-matrix for this system would be:

```
1   N
2   N   1   RN1N2
3   X   1   1.0      2   90.0
4   N   1   RN1N4    3   AN4N1X   2   180.0
5   H   4   RN4H     1   AHN4N1   3   180.0
```

Strong coupling between coordinates can give long 'valleys' in the energy surface, which may also present problems. Care must be taken when defining the Z-matrix for cyclic systems in particular. The natural way to define a cyclic compound would be to number the atoms sequentially round the ring. However, this would then mean that the ring closure bond will be very strongly coupled to all of the other bonds, angles and torsion angles (Figure 4.14). A better definition uses a dummy atom placed at the centre of the ring, Figure 4.14. Some quantum

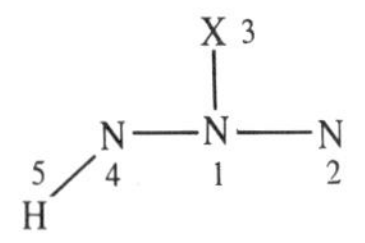

Fig. 4.13 Internal coordinates of HN_3 molecule defined using dummy atom X.

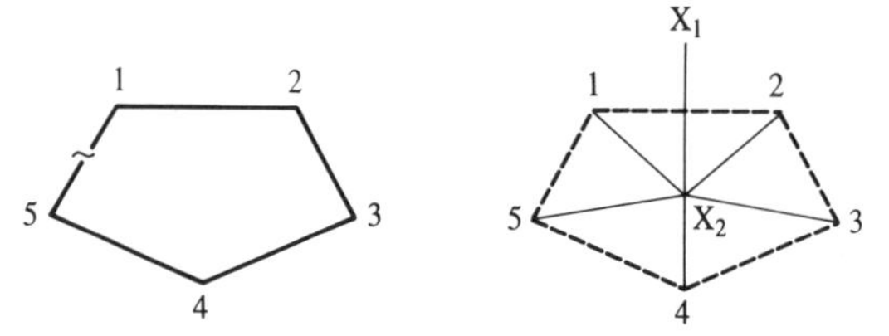

Fig. 4.14 The ring closure bond between atoms 1 and 5 would be strongly coupled to the other internal coordinates (left) unless dummy atoms are used to define the Z-matrix (right).

mechanics programs are now able to convert the input coordinates (be they Cartesian or internal) into the most efficient set for minimisation so removing the problems associated with an incorrect choice of internal coordinates.

4.7.1 Distinguishing between minima, maxima and saddle points

A configuration at which all the first derivatives are zero need not necessarily be a minimum point; this condition holds at both maxima and saddle points as well. From simple calculus we know that the second derivative of a function of one variable, $f'(x)$ is positive at a minimum and negative at a maximum. It is necessary to calculate the eigenvalues of the Hessian matrix to distinguish between minima, maxima and saddle points. At a minimum point there will be 6 zero and $3N - 6$ positive eigenvalues if $3N$ Cartesian coordinates are used. The 6 zero eigenvalues correspond to the translational and rotational degrees of freedom of the molecule (thus these 6 zero eigenvalues are not obtained when internal coordinates are used). At a maximum point all eigenvalues are negative and at a saddle point one or more eigenvalues are negative. We will consider the uses of the eigenvalue and eigenvector information in sections 4.8 and 4.9.

4.7.2 Convergence criteria

In contrast to the simple analytical functions that are often used to illustrate the operation of the various minimisation methods, in 'real' molecular modelling applications it is rarely possible to identify the 'exact' location of minima and saddle points. We can only ever hope to find an approximation to the true minimum or saddle point. Unless instructed otherwise, most minimisation methods would keep going forever, moving ever closer to the minimum. It is therefore necessary to have some means to decide when the minimisation calculation is sufficiently close to the minimum and so can be terminated. Any calculation is of course limited by the precision with which numbers can be stored on the computer, but in most instances it is usual to stop well before this limit is reached. A simple strategy is to monitor the energy from one iteration to the next and to stop when the difference in energy between successive steps falls below a specified threshold. An alternative is to monitor the change in coordinates and to stop when the difference between successive configurations is sufficiently small. A third method is to calculate the root-mean-square gradient. This is obtained by adding the squares of the gradients of the energy with respect to the coordinates, dividing by the number of coordinates and taking the square root:

$$\mathrm{RMS} = \sqrt{\frac{\mathbf{g}^{\mathrm{T}}\mathbf{g}}{3N}} \tag{4.24}$$

It is also useful to monitor the maximum value of the gradient to ensure that the minimisation has properly relaxed all the degrees of freedom and has not left a large amount of strain in one or two coordinates.

4.8 Applications of energy minimisation

Energy minimisation is very widely used in molecular modelling and is an integral part of techniques such as conformational search procedures (Chapter 8). Energy minimisation is also used to prepare a system for other types of calculations. For example, energy minimisation may be used prior to a molecular dynamics or Monte Carlo simulation in order to relieve any unfavourable interactions in the initial configuration of the system. This is especially recommended for simulations of complex systems such as macromolecules or large molecular assemblies. In the following sections we will discuss some techniques that are specifically associated with energy minimisation methods.

4.8.1 Normal mode analysis

The molecular mechanics or quantum mechanics energy at an energy minimum corresponds to a hypothetical, motionless state at 0 K. Experimental measurements are made on molecules at a finite temperature when the molecules undergo translational, rotational and vibration motion. To compare the theoretical and experimental results it is necessary to make appropriate corrections to allow for these motions. These corrections are calculated using standard statistical mechanics formulae. The internal energy $U(T)$ at a temperature T is given by:

$$U(T) = U_{\mathrm{trans}}(T) + U_{\mathrm{rot}}(T) + U_{\mathrm{vib}}(T) + U_{\mathrm{vib}}(0) \tag{4.25}$$

If all translational and rotational modes are fully accessible in accordance with the equipartition theorem, then $U_{\mathrm{trans}}(T)$ and $U_{\mathrm{rot}}(T)$ are both equal to $\frac{3}{2}k_{\mathrm{B}}T$ per molecule (except that $U_{\mathrm{rot}}(T)$ equals $k_{\mathrm{B}}T$ for a linear molecule); k_{B} is Boltzmann's constant. However, the vibrational energy levels are often only partially excited at room temperature. The vibrational contribution to the internal energy at a temperature T thus requires knowledge of the actual vibrational frequencies. The vibrational contribution equals the difference in the vibrational enthalpy at the temperature T and at 0 K and is given by:

$$U_{\text{vib}}(T) = \sum_{\text{i}=1}^{N_{nm}} \left(\frac{h\nu_{\text{i}}}{2} + \frac{h\nu_{\text{i}}}{\exp[h\nu_{\text{i}}/k_{\text{B}}T] + 1} \right) \tag{4.26}$$

N_{nm} is the number of *normal vibrational modes* for the system. Even the zero-point energy ($U_{\text{vib}}(0)$, obtained by summing $\frac{1}{2}h\nu_{\text{i}}$ for each normal mode) can be quite substantial, amounting to about 100 kcal/mol for a six-carbon alkane. Other thermodynamic quantities such as entropies and free energies may also be calculated from the vibrational frequencies using the relevant statistical mechanics expressions.

Normal modes are useful because they correspond to collective motions of the atoms in a coupled system that can be individually excited. The three normal modes of water are schematically illustrated in Figure 4.15; a non-linear molecule with N atoms has $3N - 6$ normal modes. The frequencies of the normal modes together with the displacements of the individual atoms may be calculated from a molecular mechanics force field or from the wavefunction using the Hessian matrix of second derivatives ($\mathcal{V}''$). Of course, if we have used an appropriate

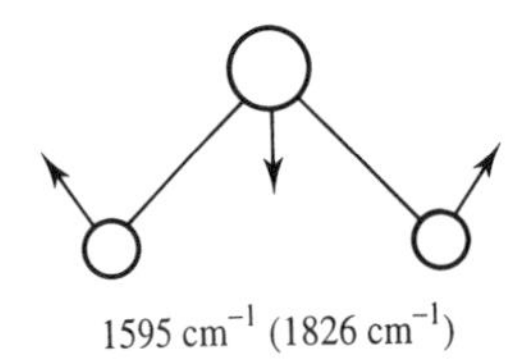

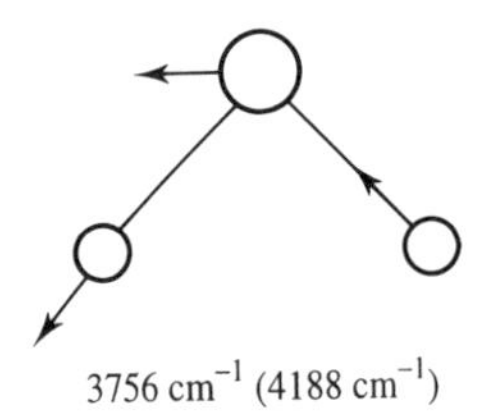

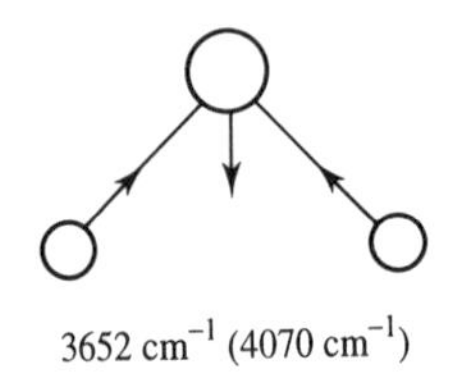

Fig. 4.15 Normal modes of water. Experimental and (calculated) frequencies are shown. Theoretical frequencies calculated using a 6-31G* basis set.

minimisation algorithm then we already know the Hessian. The Hessian must first be converted to the equivalent force constant matrix in *mass-weighted coordinates* (**F**), as follows:

$$\mathbf{F} = \mathbf{M}^{-1/2}\mathcal{V}''\mathbf{M}^{-1/2} \tag{4.27}$$

M is a diagonal matrix of dimension $3N$ by $3N$, containing the atomic masses. All elements of **M** are zero except those on the diagonal; $M_{1,1} = m_1$, $M_{2,2} = m_1$, $M_{3,3} = m_1$, $M_{4,4} = m_2$, $\ldots M_{3N-2,3N-2} = m_N$, $M_{3N-1,3N-1} = m_N$, $M_{3N,3N} = m_N$. Each non-zero element of $\mathbf{M}^{-1/2}$ is thus the inverse square root of the mass of the appropriate atom. The masses of the atoms must be taken into account because a force of a given magnitude will have a different effect upon a larger mass than a smaller one. For example, the force constant for a bond to a deuterium atom is, to a good approximation, the same as to a proton, yet the different mass of the deuteron gives a different motion and a different zero-point energy. The use of mass-weighted coordinates takes care of these problems.

We next solve the secular equation $|\mathbf{F} - \lambda\mathbf{I}| = 0$ to obtain the eigenvalues and eigenvectors of the matrix **F**. This step is usually performed using matrix diagonalisation, as outlined in section 1.10.3. If the Hessian is defined in terms of Cartesian coordinates then six of these eigenvalues will be zero as they correspond to translational and rotational motion of the entire system. The frequency of each normal mode is then calculated from the eigenvalues using the relationship:

$$\nu_i = \frac{\sqrt{\lambda_i}}{2\pi} \tag{4.28}$$

As a simple example of a normal mode calculation consider the linear triatomic system in Figure 4.16. We shall just consider motion along the long axis of the molecule. The displacements of the atoms from their equilibrium positions along this axis are denoted by ξ_i. It is assumed that the displacements are small com-

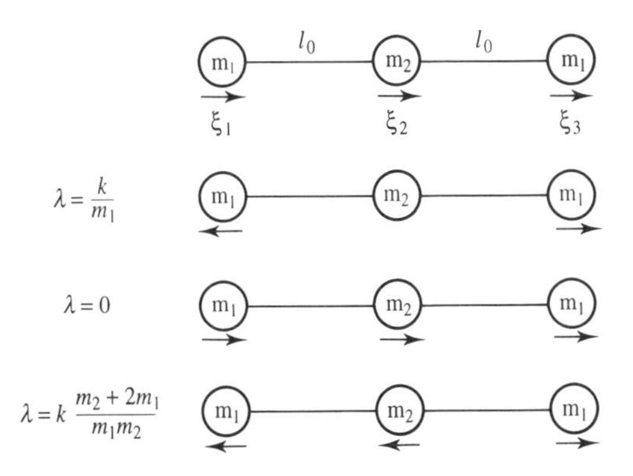

Fig. 4.16 Linear three-atom system with results of normal mode calculation.

pared to the equilibrium values l_0 and the system obeys Hooke's law with bond force constants k. The potential energy is given by:

$$\mathcal{V} = \frac{1}{2}k(\xi_1 - \xi_2)^2 + \frac{1}{2}k(\xi_2 - \xi_3)^2 \tag{4.29}$$

We next calculate the first and then the second derivatives of the potential energy with respect to the three coordinates ξ_1, ξ_2 and ξ_3:

$$\frac{\partial \mathcal{V}}{\partial \xi_1} = k(\xi_1 - \xi_2); \quad \frac{\partial \mathcal{V}}{\partial \xi_2} = -k(\xi_1 - \xi_2) + k(\xi_2 - \xi_3);$$

$$\frac{\partial \mathcal{V}}{\partial \xi_3} = -k(\xi_2 - \xi_3) \tag{4.30}$$

The second derivatives are conveniently represented as a 3×3 matrix:

$$\begin{vmatrix} k & -k & 0 \\ -k & 2k & -k \\ 0 & -k & k \end{vmatrix} \tag{4.31}$$

The mass-weighted matrix is

$$\begin{vmatrix} m_1 & 0 & 0 \\ 0 & m_2 & 0 \\ 0 & 0 & m_3 \end{vmatrix} \tag{4.32}$$

The secular equation to be solved is thus:

$$\begin{vmatrix} \frac{k}{m_1} - \lambda & -\frac{k}{\sqrt{m_1}\sqrt{m_2}} & 0 \\ -\frac{k}{\sqrt{m_1}\sqrt{m_2}} & \frac{2k}{m_2} - \lambda & -\frac{k}{\sqrt{m_1}\sqrt{m_2}} \\ 0 & -\frac{k}{\sqrt{m_1}\sqrt{m_2}} & \frac{k}{m_1} - \lambda \end{vmatrix} = 0 \tag{4.33}$$

This determinant leads to a cubic in λ which has three roots (λ_k), each corresponding to a different mode of motion:

$$\lambda = \frac{k}{m_1}, \quad \lambda = 0, \quad \lambda = k\frac{m_2 + 2m_1}{m_1 m_2} \tag{4.34}$$

The corresponding frequencies can be obtained from equation (4.28). The amplitudes (A) of each normal mode are given by the eigenvector solutions of the secular equation $\mathbf{FA} = \lambda\mathbf{A}$. If A_1, A_2 and A_3 are the amplitudes of each atom then the amplitudes obtained for each eigenvalue are:

$$\lambda = \frac{k}{m_1}: \qquad A_1 = -A_3; \quad A_2 = 0 \tag{4.35}$$

$$\lambda = 0: \qquad A_1 = A_3; \quad A_2 = \sqrt{\frac{m_1}{m_2}}A_1 \tag{4.36}$$

$$\lambda = k\frac{m_2 + 2m_1}{m_1 m_2}: \qquad A_1 = A_3; \quad A_2 = -2\sqrt{\frac{m_2}{m_1}}A_1 \tag{4.37}$$

These normal modes are schematically illustrated in Figure 4.16. They correspond to a symmetric stretch, a translation and an asymmetric stretch respectively.

We have already seen how the results of normal mode calculations can be used to calculate thermodynamic quantities. The frequencies themselves can also be compared with the results of spectroscopic experiments, information which can be used in the parametrisation of a force field. For example, the experimental frequencies for the normal modes of water are shown in Figure 4.15, together with the frequencies determined using a 6-31G* *ab initio* calculation. The calculated values clearly deviate from those obtained experimentally, but the ratio of the experimental and theoretical frequencies is remarkably consistent (at about 1.1). Such empirical scaling factors have been derived which enable frequencies obtained using a given level of theory to be converted to values for experiment or a higher level of theory [Pople *et al.* 1993]. The normal modes of much larger molecules can be calculated using molecular mechanics. For example, the vibrations of a helical polypeptide constructed from a sequence of ten alanine residues (112 atoms) are shown in Figure 4.17. In such cases it is usually the low-frequency vibrations that are of most interest as these correspond to the large-scale conformational motions of the molecule. The results of such analyses can be compared with molecular dynamics simulations from which vibrational contributions can also be extracted [Brooks and Karplus 1983].

A normal mode calculation is based upon the assumption that the energy surface is quadratic in the vicinity of the energy minimum (the harmonic approximation). Deviations from the harmonic model can require corrections to calculated thermodynamic properties. One way to estimate anharmonic corrections is to calculate a force constant matrix using the atomic motions obtained from a molecular dynamics simulation; such simulations are not restricted to movements on a harmonic energy surface. The eigenvalues and eigenvectors are then calculated for this quasi-harmonic force-constant matrix in the normal way, giving a model which implicitly incorporates the anharmonic effects.

The harmonic approximation to the energy surface is found to be appropriate for well-defined energy minima such as the intramolecular degrees of freedom of small molecules and for some small intermolecular complexes. For larger systems such as liquids and large, 'floppy' molecules, the harmonic approximation breaks down. Such systems also have an extraordinarily large number of 'minima' on the energy surface. In such cases it is not possible to calculate accurately

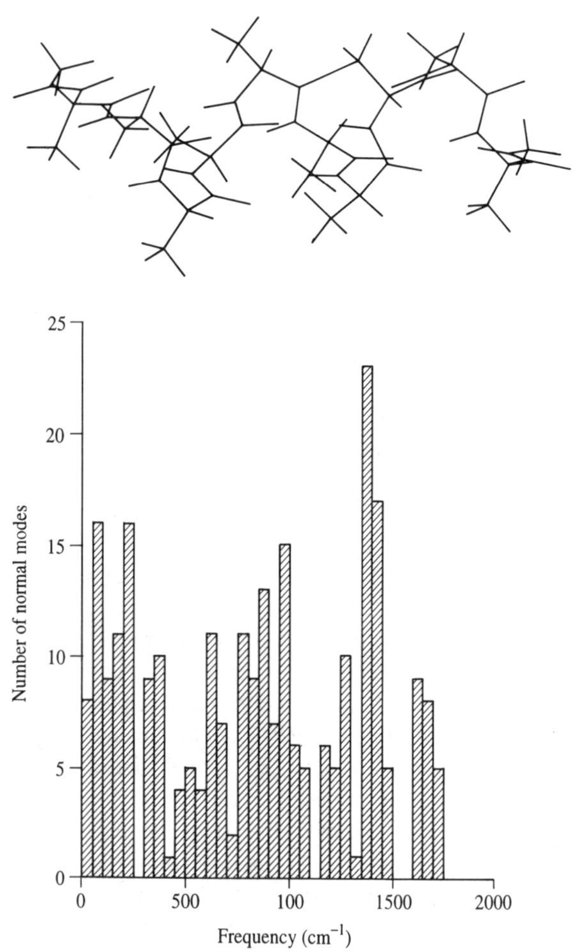

Fig. 4.17 Histogram of the normal modes calculated for a polyalanine polypeptide in an α-helical conformation. The height of each bar indicates the number of normal modes in each 50 cm^{-1} section.

thermodynamic properties using energy minimisation and normal mode calculations. Rather, molecular dynamics or Monte Carlo simulations must be used to sample the energy surface from which properties can be derived as we will discuss in Chapters 5–7.

4.8.2 The study of intermolecular processes

One example of the use of minimisation methods and normal mode analysis is the study by Hagler and co-workers of the binding of the antibacterial drug trimethoprim (Figure 4.18) to the enzyme dihydrofolate reductase (DHFR) [Dauber-

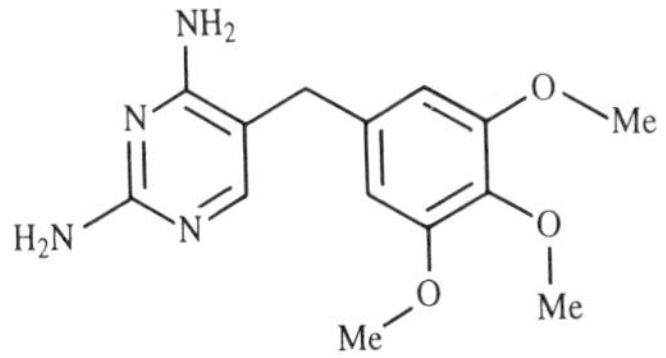

Fig. 4.18 Trimethoprim.

Osguthorpe *et al.* 1988; Fisher *et al.* 1991]. DHFR catalyses the reduction of folic acid and dihydrofolic acid to tetrahydrofolic acid (Figure 4.19) and plays a vital metabolic role in the biosynthesis of nucleic acids in bacteria, protozoa, plants and animals. Trimethoprim exploits the structural differences between bacterial and vertebrate DHFR, binding much more strongly to the former, and is clinically used as an antibacterial agent. Inhibitors of human DHFR are used in cancer therapy. Hagler and colleagues applied energy minimisation to an isolated trimethoprim molecule, to the crystal structure of trimethoprim, to trimethoprim in the presence of water molecules, and to trimethoprim in intermolecular complexes with DHFR from both bacterial and vertebrate sources. An important observation was that the conformation of the trimethoprim when bound to the enzyme was

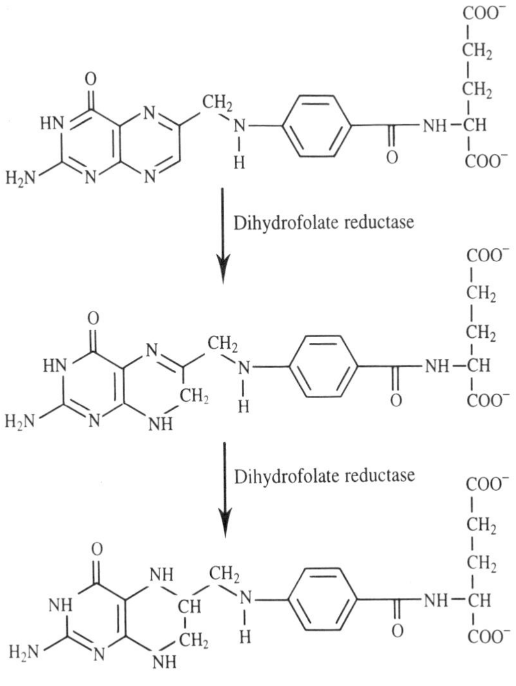

Fig. 4.19 DHFR catalyses the reduction of folic acid to tetrahydrofolic acid.

significantly different from that obtained for the isolated molecule. This reinforces the view that the use of structures obtained from energy minimisation calculations on isolated molecules can lead to misleading conclusions. Intermolecular interactions with the receptor enable the ligand to adopt a conformation whose intramolecular energy is significantly higher than any of its minimum energy structures.

A normal mode analysis on the isolated and bound trimethoprim molecules enabled an estimate to be made of the entropic contribution to binding. Low-frequency modes for the isolated ligand were found to be shifted to higher frequencies for the ligand in the enzyme complex, reflecting a restriction of the motion of the ligand by the protein. This entropic contribution to the free energy of binding was predicted to be quite significant, indicating that conclusions based solely upon energies may be misleading.

4.9 Determination of transition structures and reaction pathways

Chemists are interested not only in the thermodynamics of a process (the relative stability of the various species) but also in its kinetics (the rate of conversion from one structure to another). Knowledge of the minimum points on an energy surface enables thermodynamic data to be interpreted, but for the kinetics it is necessary to investigate the nature of the energy surface away from the minimum points. In particular, we would like to know how the system changes from one minimum to another. What changes in geometry are involved, and how does the energy vary during the transition? The minimum points on the energy surface may be the reactants and products of a chemical reaction, two conformations of a molecule, or two molecules that associate to form a non-covalently bound bimolecular complex. We shall use the term ‘reaction pathway’ to describe the path between two minima, but our use of the word ‘reaction’ does not necessarily mean that bond making and/or breaking is involved. Many methods have been proposed for finding transition structures and elucidating reaction pathways. We do not have space to cover all of the methods, and so we shall restrict our discussion to some of the more common approaches.

As a system moves from one minimum to another, the energy increases to a maximum at the transition structure and then falls. At a saddle point the first derivatives of the potential function with respect to the coordinates are all zero (just as they are at a minimum point). The number of negative eigenvalues in the Hessian matrix is used to distinguish different types of saddle points; an nth order transition or saddle point has n negative eigenvalues. We are usually most interested in first-order saddle points, where the energy passes through a maximum for movement along the pathway that connects the two minima, but is a minimum for displacements in all other directions perpendicular to the path. This is shown schematically for a two-dimensional energy surface in Figure 4.20.

These negative eigenvalues of the Hessian matrix are often referred to as the ‘imaginary’ frequencies for motion of the system over the saddle point. We can

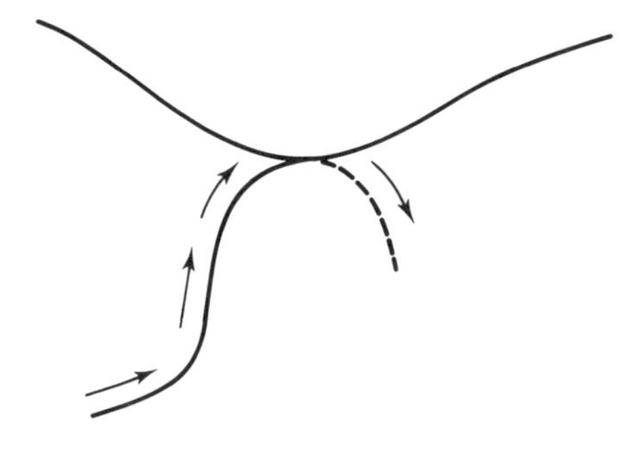

Fig. 4.20 The lowest-energy path from one minimum to another passes through a saddle point.

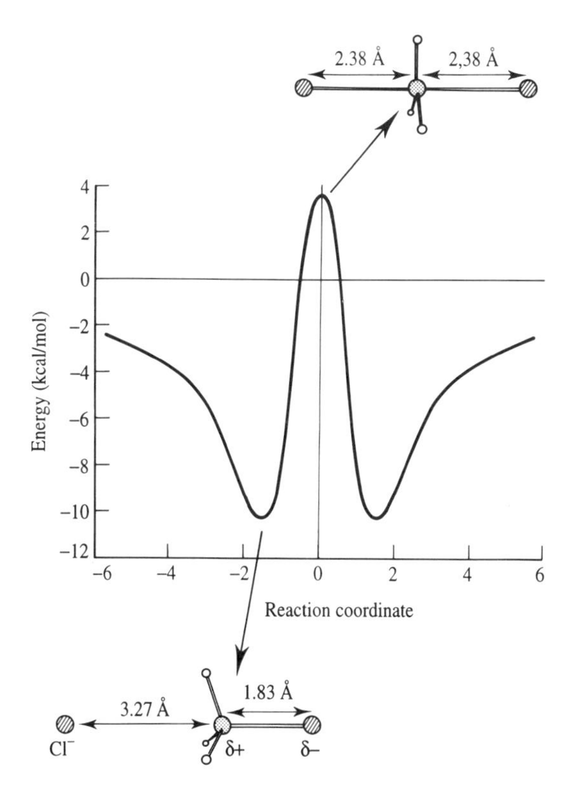

Fig. 4.21 The energy profile for the gas-phase $Cl^- + MeCl$ reaction. Adapted in part from Chandrasekhar J, S F Smith and W L Jorgensen 1985. Theoretical Examination of the S_N2 Reaction Involving Chloride Ion and Methyl Chloride in the Gas Phase and Aqueous Solution. *The Journal of the American Chemical Society* **107**:154–163.

illustrate this concept using the gas-phase S_N2 reaction between Cl^- and CH_3Cl. As the chloride ion approaches the methyl chloride along the line of the C–Cl bond the energy passes through an ion–dipole complex which is at an energy minimum. The energy then rises to a maximum at the pentagonal transition state. The energy profile is drawn in Figure 4.21. The geometries of the minimum and the pentagonal transition state, as determined by an *ab initio* HF/SCF calculation with the 6-31G* basis set are shown in Figure 4.22. The lowest frequency eigenvalues and a representation of the corresponding eigenvectors for the two geometries are also given in Figure 4.22. There are three frequencies in the ion–dipole minimum that are of particularly low energy; two of these correspond to degenerate 'wagging' motions of the system (at 71.3 cm^{-1}). The vibration at 101.0 cm^{-1} is the normal mode that corresponds to motion towards the transition state. At the saddle point there is a single negative eigenvalue (with an imaginary 'frequency' of -415.0 cm^{-1}) that corresponds to vibration along the Cl–C–C axis (i.e. motion along the reaction pathway). The other normal modes at the

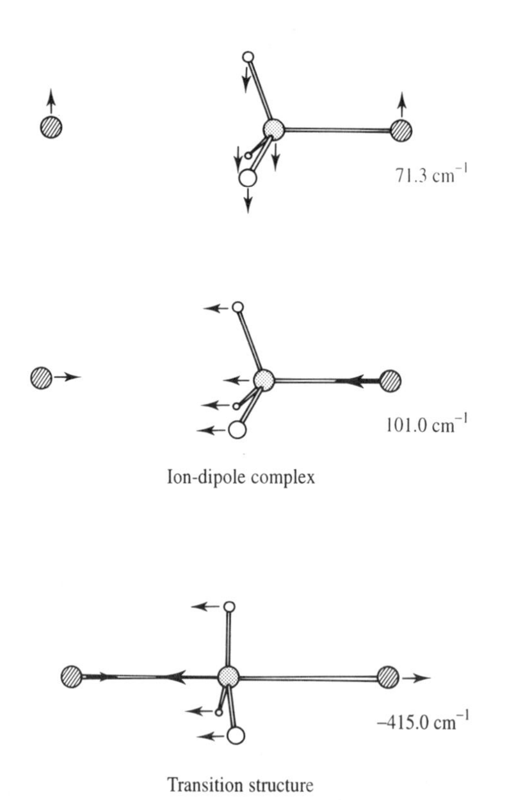

Fig. 4.22 Schematic representation of some of the lower frequencies in the ion–dipole complex for the Cl^- + MeCl reaction and the imaginary frequency of the transition structure, calculated using a 6-31G* basis set.

saddle point all have positive frequencies; the two lowest (at 204.2 cm^{-1}) correspond to wagging motions perpendicular to the Cl–C–Cl axis and the third is a symmetric stretch of the two chlorine atoms along the symmetry axis.

It is important to distinguish the transition *structure* from the transition *state*. The transition structure is the point of highest potential energy along the pathway. By contrast, the transition state is the geometry at the peak in the free energy profile. In many cases the geometry at the transition state is very similar to that of the transition structure. However, the transition state may be different as the free energy of activation includes contributions from sources other than just the potential energy. If the transition state geometry is temperature dependent then entropic factors may be important. An example is the following radical reaction:

$$H^{\cdot} + CH_3CH_2^{\cdot} \rightarrow H_2 + CH_2 = CH_2$$

The calculated geometry of the transition structure resembles the ethyl radical (Figure 4.23) [Doubleday *et al.* 1985]. The entropy change for this reaction is negative and so, as the temperature is increased, the maximum in the free energy profile shifts more towards the products, in the direction of lower entropy.

Methods for finding transition structures and reaction pathways are often closely related. Thus, some methods for finding the reaction pathway start from the transition structure and move down towards a minimum. Such methods must be supplied with the transition structure geometry as the starting point. Conversely, some methods for locating transition structures do so by searching along the reaction pathway, or an approximation to it. Yet other methods require neither the transition structure nor the pathway, but can determine both simultaneously from the two minima. In general, it is more difficult to locate transition structures and determine reaction pathways than to find minimum points. It is therefore crucial to check that the Hessian matrix at any proposed saddle point has the required single negative eigenvalue. Methods for locating saddle points are usually most effective when given as input a geometry that is as close as possible to the transition structure. It can also be helpful to examine the atomic displacements that correspond to the negative eigenvector, to ensure that it corresponds to the correct motion over the saddle point as for the $Cl^- + CH_3Cl$ reaction.

As one approaches the saddle point from a minimum, the Hessian matrix will change from having all positive eigenvalues to including one negative value. The *quadratic region* of a saddle point is that portion of the energy surface surrounding the point where the Hessian contains one negative eigenvalue. Similarly the quadratic region of a minimum is the region where all eigenvalues are positive and the Hessian is positive definite. Some algorithms for finding saddle points require a starting geometry within the quadratic region. We can illustrate the concept of a quadratic region by considering the function $f(x, y) = x^4 + 4x^2y^2 - 2x^2 + 2y^2$, which is drawn in Figure 4.24. This function has two minima at $(1, 0)$ and $(-1, 0)$ and one saddle point at $(0, 0)$. In this case it

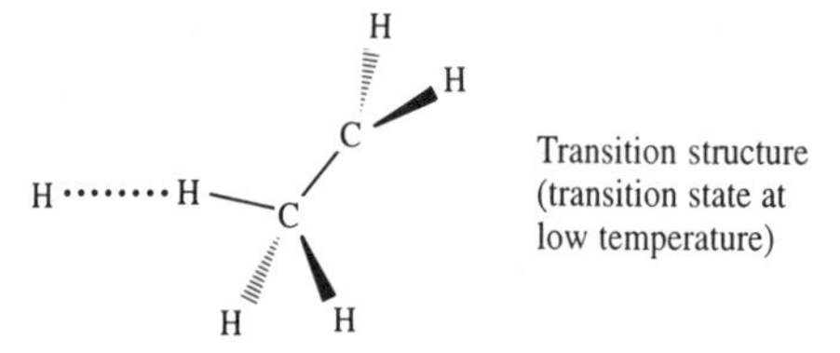

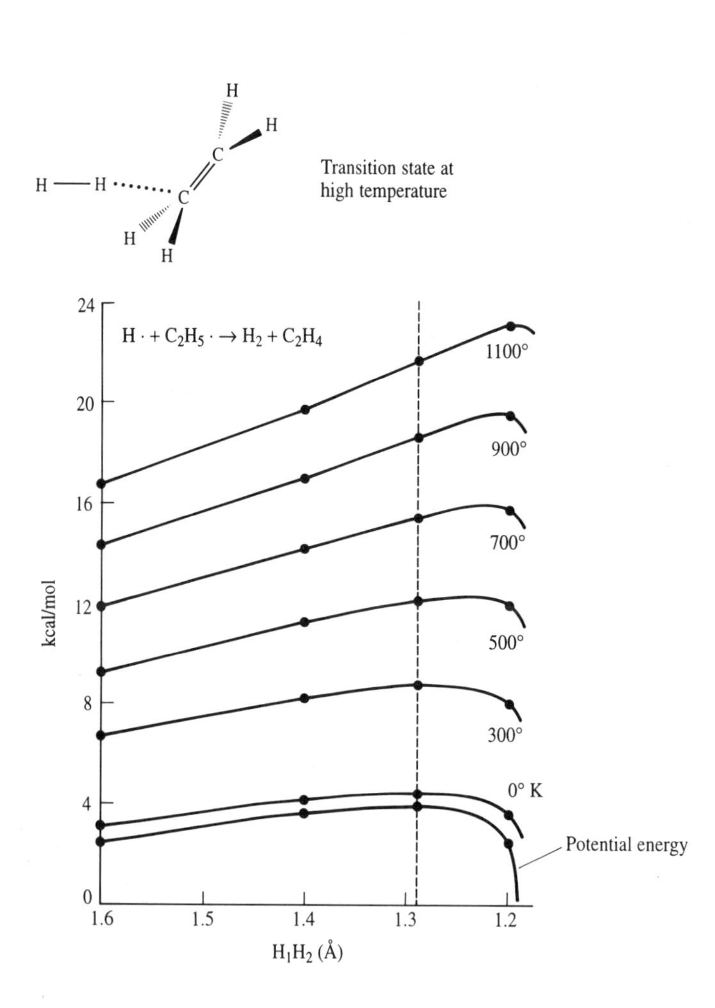

Fig. 4.23 The transition structure for the $H^{\cdot} + CH_3CH_2^{\cdot} \rightarrow H_2 + CH_2{=}CH_2$ reaction. At low temperature the transition structure corresponds to the transition state (maximum of free energy). At high temperature the transition state moves closer to the products, as can be seen from the graph. Redrawn from Doubleday C, J McIver, M Page and T Zielinski 1985. Temperature Dependence of the Transition-State Structure for the Disproportionation of Hydrogen Atom with Ethyl Radical. *The Journal of the American Chemical Society* **107**:5800–5801.

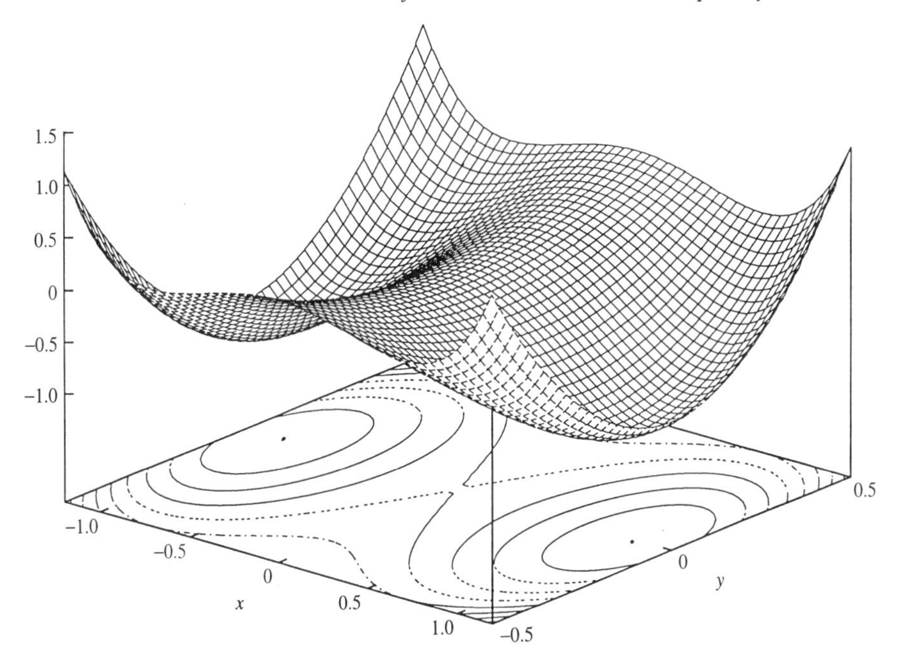

Fig. 4.24 The function $f(x, y) = x^4 + 4x^2y^2 - 2x^2 + 2y^2$ has a saddle point at $(0, 0)$ and minima at $(1, 0)$ and $(-1, 0)$.

is possible to derive and characterise the stationary points analytically. The Hessian matrix of second derivatives for this function is:

$$\begin{pmatrix} 12x^2 + 8y^2 - 4 & 16xy \\ 16xy & 8x^2 + 4 \end{pmatrix} \tag{4.38}$$

At the point $(1, 0)$ the Hessian matrix is thus

$$\begin{pmatrix} 8 & 0 \\ 0 & 4 \end{pmatrix}.$$

The eigenvalues of this matrix are obtained by setting the secular determinant to zero:

$$\begin{vmatrix} 8 - \lambda & 0 \\ 0 & 4 - \lambda \end{vmatrix} = 0 \tag{4.39}$$

The eigenvalues are $\lambda = 4$ and $\lambda = 8$. Thus both eigenvalues are positive and the point is a minimum. At the point $(0, 0)$ the Hessian matrix is

$$\begin{pmatrix} -4 & 0 \\ 0 & 4 \end{pmatrix}$$

with one negative and one positive eigenvalue (-4 and $+4$). The normalised eigenvectors corresponding to these eigenvalues are $(0, 1)$ for the eigenvalue

$\lambda = 4$ and $(1, 0)$ for the eigenvalue $\lambda = -4$. These eigenvectors indicate the directions in which the gradient of the function changes sign. Thus along the line $x = 0$ the function passes through a minimum, as can be seen from Figure 4.24. By contrast, if one progresses from $(-1, 0)$ to $(1, 0)$ through the origin then the function passes through a maximum. As one progresses through a transition structure the eigenvector of the negative eigenvalue corresponds to the concerted motions of the atoms that give rise to motion through the saddle point. If we move along the x axis from the minimum at $(1, 0)$ to the saddle point at the origin, both eigenvalues will be positive so long as $12x^2 + 8y^2 - 4 > 0$. Thus, so long as x is larger than $1/\sqrt{3}$ the eigenvalues of the Hessian matrix will be positive. When x becomes smaller than $1/\sqrt{3}$ there will be one negative and one positive eigenvalue. In this case the quadratic region would correspond to all points where the absolute value of x was less than $1/\sqrt{3}$.

4.9.1 Methods to locate saddle points

In some simple cases such as the chloride/methyl chloride reaction the geometry of the transition structure can be predicted by inspection. In other cases a *grid search* can be used to scan the energy surface in order to locate the approximate position of the transition state. In a grid search, the coordinates are systematically varied to generate a set of structures, for each of which the energy is calculated. It may then be possible to fit an analytical expression to these points, from which the saddle point can be predicted by standard calculus methods. The grid search method is widely used for constructing potential energy surfaces but is restricted to systems with a very small number of atoms or where only a limited number of degrees of freedom are being explored such as the $H + H_2 \rightarrow H_2 + H$ reaction. An advantage of the grid search is that it does provide information about the energy surface away from the pathway, which can be important if one wishes to investigate the dynamics of a reaction and the interconversion of energy between different modes. The grid search method is not the method of choice for all but the smallest systems due to the number of energy evaluations that are required. In any case it does not directly provide the transition structure.

The conversion of one minimum energy structure into another may sometimes occur primarily along just one or two coordinates. In such cases, an approximation to the reaction pathway can be obtained by gradually changing the coordinate(s), allowing the system to relax at each stage using minimisation while keeping the chosen coordinate(s) fixed. The point of highest energy on the path is an approximation to the saddle point and the structures generated during the course of the calculation can be considered to represent a sequence of points on the interconversion pathway. When coordinate driving methods are applied to conformational changes that occur primarily via rotation about bonds, the procedure is often referred to as *adiabatic mapping* or *torsion angle driving*. Adiabatic mapping can be used to study quite large systems such as the energetics of rotation of tyrosine and phenalanine side chains in the interior of proteins [Gelin and Karplus 1975]. The twofold symmetry of these rings (see Figure 4.25) means that there are two energetically equivalent mininum energy orientations, but we would like to know the magnitude of the energy barrier for converting one conformation into the other. Simply rotating the ring while keeping the rest of the protein fixed will give a wildly inaccurate value of the barrier due to high-energy interactions with the rest of the molecule. The protein would be expected to modify its conformation to accommodate different orientations of the aromatic ring. A more accurate estimation of the energetics of the change can be obtained if the rest of the system is permitted to relax (using energy minimisation) for each orientation of the ring. In adiabatic mapping this calculation is repeated for many positions of the ring, each obtained from its predecessor by rotating the bond by a few degrees. For example, the torsion angle labelled χ_2 in Figure 4.25 would be changed through 180° in (say) 10° steps. To fix the torsion angle at the desired value during the minimisation step an extra torsional term is included in the potential function of the form:

$$\mathscr{V}(\omega) = V/2\{1 - \cos(\omega - \theta)\} \tag{4.40}$$

If the 'barrier' V is large enough then ω will be forced to adopt the value θ.

Torsion angle driving has also been widely used in conjunction with molecular mechanics to deduce the pathways between the minimum energy conformations of molecules such as chair and twist-boat cyclohexane rings. It is usually assumed that the minimum energy pathway can be calculated by changing just one or two torsion angles. The calculation is performed by forcing the chosen torsion angles to adopt the desired values through the use of an additional torsional term such as equation (4.40). In many cases, the true reaction coordinate has contributions from other internal coordinates including angle bending. A frequently observed phenomenon in torsion angle driving is a 'lagging behind' of other coordinates. This can be observed even in simple cases such as ethane. If one of the HCCH torsion angles in ethane is chosen as the reaction coordinate and is driven from a value of 60° (staggered) towards 0° (eclipsed), then as the transition structure is approached, the other HCCH torsion angles remain closer to 60° than the driven dihedral, Figure 4.26. At the same time, the angles to the hydrogens open to compensate. More serious flaws have been reported for conformational transitions in larger systems.

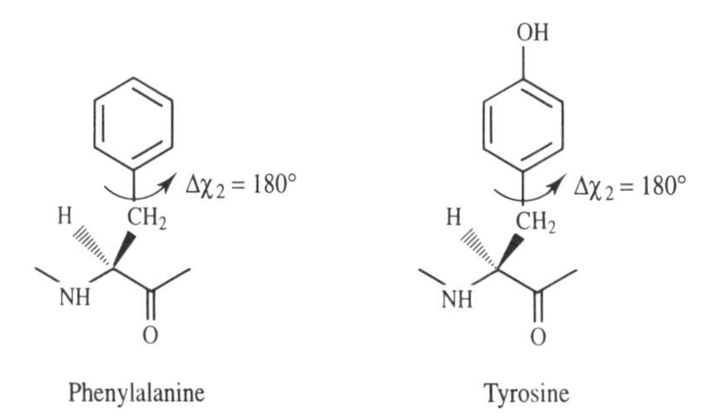

Fig. 4.25 Rotation about χ_2 in phenylalanine and tyrosine.

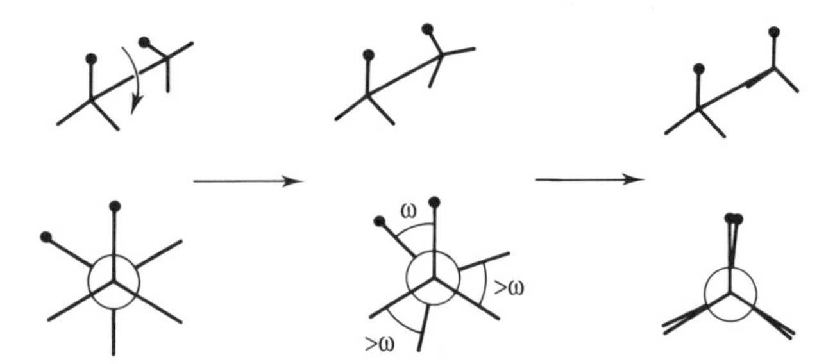

Fig. 4.26 Torsion angle driving in ethane: the H–C–C–H torsion angles lag behind the driven torsion.

The coordinate driving method can be used with quantum mechanics in appropriate cases. An alternative to moving along a predefined set of coordinates is to gradually move up the valley to the saddle point along the 'least-steep path', or 'path of shallowest ascent'. The difficulty lies in determining the correct direction to proceed at each step and the size of the step. As might be expected, there is frequently a conflict between the need for a small step size and the computational demands; too large a step and the transition structure may be missed, as shown in Figure 4.27. Too small and the calculation takes too long. If the Hessian matrix of second derivatives is available then the appropriate direction to take is uphill along the eigenvector of the smallest eigenvalue when all eigenvalues are positive and downhill along the eigenvector corresponding to the negative eigenvalue when within the quadratic region of the saddle point [Baker 1986].

As we have stated frequently, at a saddle point the gradient is zero (as it is for a minimum). It might therefore be imagined that a minimisation algorithm (or some variant) could be used to locate saddle points. Some minimisation algorithms can occasionally incorrectly converge to a saddle point, especially if the starting structure is close to the transition structure. A simple example is the Newton–Raphson method which will converge to a transition structure when giving a starting position that is within the quadratic region. Other minimisation algorithms can also be modified so that they consistently locate saddle points when provided with an initial structure within the quadratic region [Schlegel 1982].

Dewar, Healy and Stewart have described a simple method for locating transition structures that starts from the reactant and product species [Dewar *et al.* 1984]. Suppose A and B are the two species in question, and that a_i are the coordinates of A and b_i are the coordinates of B where i is one of the atoms. The 'reaction coordinate distance' between A and B is then:

$$R = \sqrt{\sum_\mathrm{i}[a_\mathrm{i} - b_\mathrm{i}]^2} \tag{4.41}$$

As A is transformed into B and B into A the distance R between them is gradually reduced. At each iteration the higher-energy structure is chosen as the 'fixed' structure. The lower-energy structure is then minimised, subject to the

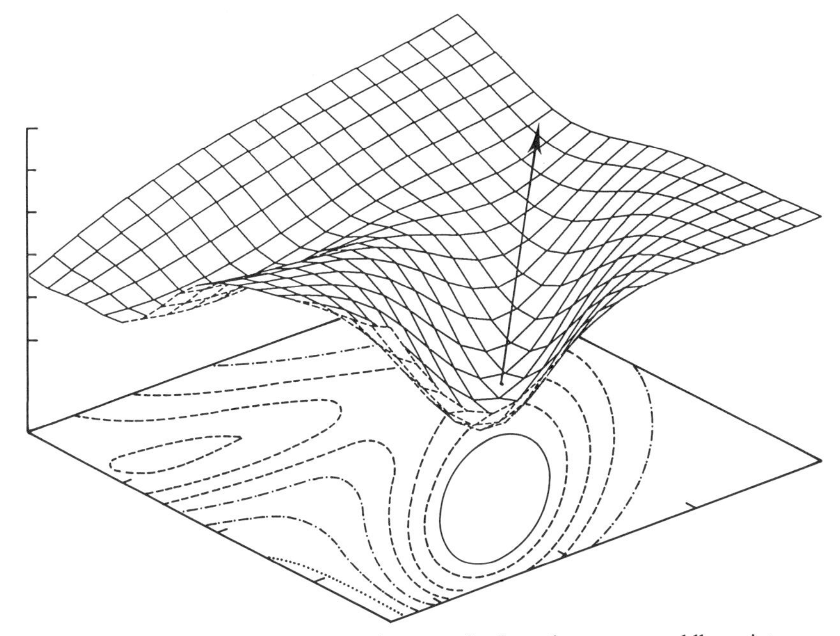

Fig. 4.27 Too large a step size may lead to the wrong saddle point or an inefficient algorithm.

constraint that the distance *R* from the fixed structure is some fraction of the original distance (say 95%). The 'fixed' structure for the next iteration is chosen as the structure that is higher in energy. This method gradually reduces the distance *R* until it is sufficiently small, in which case the two structures should be reasonable approximations to the transition structure.

4.9.2 Reaction path following

The traditional way to elucidate the reaction path is to move downhill from a saddle point to the two associated minima. There may be many different paths that could be followed from the saddle point to the associated minima. The *intrinsic reaction coordinate* (IRC) is the path that would be followed by a particle moving along the steepest descents paths with an infinitely small step from the transition structure down to each minimum when the system is described using mass-weighted coordinates (as in a normal mode calculation) [Fukui 1981]. The initial directions towards each minimum can be obtained directly from the eigenvector that corresponds to the imaginary frequency at the transition structure. A simple steepest descents algorithm with a reasonable step size will usually give a path that oscillates about the true minimum energy path, as illustrated in Figure 4.28. This is perfectly acceptable in a minimisation, where the objective is to locate the minimum as efficiently as possible and where we are not interested in

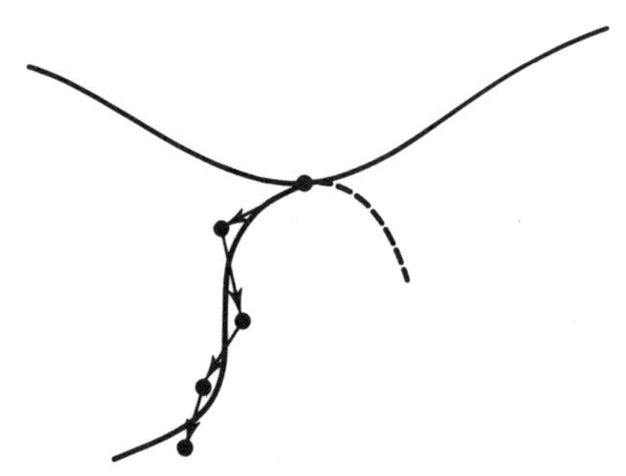

Fig. 4.28 A steepest descents minimisation algorithm produces a path that oscillates about the true reaction pathway from the transition structure to a minimum.

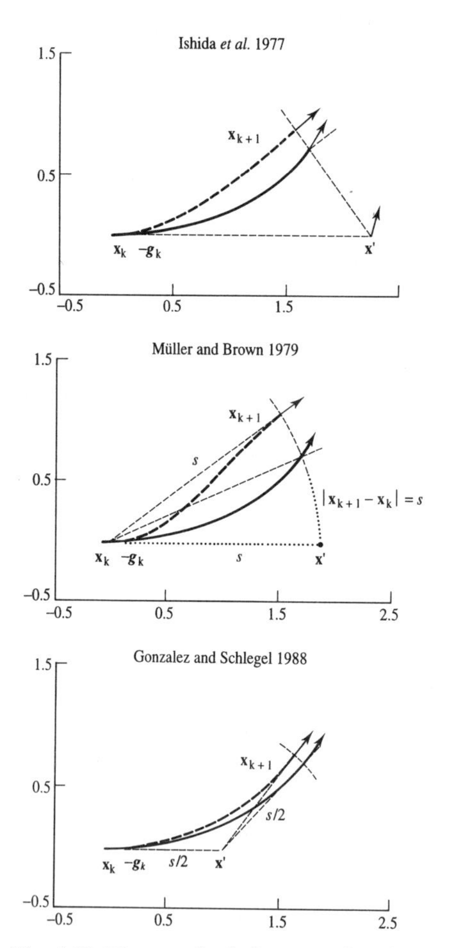

Fig. 4.29 Three methods for correcting the path followed by a steepest descents algorithm to generate the intrinsic reaction coordinate. Figure redrawn from Gonzalez C and H B Schlegel 1988. An Improved Algorithm for Reaction Path Following. *The Journal of Chemical Physics* **90**:2154–2161.

the intermediate structures. To determine the true reaction pathway (or a better approximation to it) it is necessary to 'correct' the path taken by the steepest descents algorithm. These corrective methods are especially useful when the path is curved.

Three closely related algorithms for finding the IRC are illustrated and compared in Figure 4.29. Each of these methods first calculates the gradient at the current point, $\mathbf{x}_k$. The method of Ishida, Morokumz and Komornicki [Ishida *et al.* 1977] takes a step of a predefined size (s) to give a new point $\mathbf{x}'$. The gradient at the point $\mathbf{x}'$ is calculated and then a minimisation is performed in the direction along the bisector of the initial gradient (at $\mathbf{x}_k$) and the gradient at the new point ($\mathbf{x}'$) to give the next point ($\mathbf{x}_{k+1}$) on the approximation to the reaction path. Müller and Brown's method [Müller and Brown 1979] also takes a step (of size s) along the gradient. The next point on the reaction path is obtained by minimising from this point while keeping a constant distance (s) from the original point. Gonzalez and Schlegel [Gonzalez and Schlegel 1988] take a step of $s/2$ along the gradient to give a new point ($\mathbf{x}'$). The next point on the reaction path is obtained by minimising the energy subject to the constraint that the distance between $\mathbf{x}'$ and the new point on the reaction path ($\mathbf{x}_{k+1}$) is $s/2$. The reaction path is then approximated by a circle that passes through both $\mathbf{x}_k$ and $\mathbf{x}_{k+1}$ and whose tangents at those two points are in the directions of the gradients. This third algorithm has been widely used due to its incorporation into the Gaussian program.

4.9.3 *Transition structures and reaction pathways for large systems*

Most of the algorithms we have discussed so far with the possible exception of adiabatic mapping were originally designed to be used with quantum mechanics where relatively small numbers of atoms are involved. It is often difficult to apply these methods to the study of conformational transitions. There are several reasons for this, but one important feature is that it is assumed that there is only one saddle point between the initial and final states. There may be a number of transition structures along the pathway between two conformations of a complex molecule. Here we will discuss two related methods that were originally designed to tackle this problem using molecular mechanics.

In the self penalty walk (SPW) method of Czerminski and Elber [Czerminski and Elber 1990; Nowak *et al.* 1991] a 'polymer' is constructed that consists of a series of $M+2$ 'monomers'. Each monomer is a complete copy of the actual system and so there are $(M+2)N$ atoms present in the calculation. The two ends of the polymer correspond to the two minima between which we are trying to elucidate the pathway (the 'reactant' and the 'product'). The M intermediate points in the polymer each comprise a set of monomer coordinates, which are approximations to the points on the reaction pathway. The simplest way to obtain these is via straight-line interpolation in Cartesian coordinate space. The 'potential energy', $\mathscr{S}$, of the polymer is a sum of intra- and intermonomer energies:

$$\mathscr{S} = \sum_{\mathrm{i}=1}^{M} \mathscr{V}(\mathbf{r}_{\mathrm{i}}^{N}) + \sum_{\mathrm{i}=0}^{M} \mathscr{B}_{\mathrm{i}} + \sum_{\mathrm{i}=0}^{M} \sum_{\mathrm{j}=\mathrm{i}+1}^{M} \mathscr{R}_{\mathrm{ij}} \tag{4.42}$$

The intramonomer energies $\mathscr{V}(\mathbf{r}_{\mathrm{i}}^{N})$ are the individual energies of the copies of the physical system. $\mathscr{B}_{\mathrm{i}}$ is the 'bond energy' between successive pairs of monomers:

$$\mathscr{B}_{\mathrm{i}} = \gamma(d_{\mathrm{i,i+1}} - \langle d \rangle)^2 \tag{4.43}$$

$$d_{\mathrm{i,j}} = [(\mathbf{r}_{\mathrm{i}}^{N} - \mathbf{r}_{\mathrm{j}}^{N})^2]^{1/2} \quad \text{and} \quad \langle d \rangle = \left(\frac{1}{M+1} \sum_{\mathrm{i}=0}^{M} d_{\mathrm{i,i+1}}^2 \right)^{1/2} \tag{4.44}$$

The index 0 refers to the fixed 'reactant' and the index $M+1$ to the fixed 'product'. γ is equivalent to a force constant for the 'bond' between monomers. The second term in equation (4.42) is thus used to ensure that the distances between adjacent monomers are approximately equal and so we end up with a set of copies of the physical system that are approximately adjacent on the reaction pathway. The final term in the equation (4.42) comprise 'non-bonded' monomer–monomer energies ($\mathscr{R}_{\mathrm{ij}}$) that ensure adequate sampling in the region of the transition structure(s). These depend upon two parameters, ρ and λ:

$$\mathscr{R}_{\mathrm{ij}} = \rho \exp\left(-\frac{d_{\mathrm{ij}}^2}{(\lambda\langle d \rangle)^2} \right) \tag{4.45}$$

In the SPW method the 'energy function' $\mathscr{S}$ is minimised to give a series of monomers that represent configurations of the system along the pathway.

The SPW method and variants on it have been applied to a wide variety of problems, many of which contain large numbers of atoms. For example, when the pathway between two conformations of the protein myoglobin was elucidated the change was found to be distributed over many, but not all, of the atoms in the molecule [Elber and Karplus 1987]. Even though the difference between the two conformations was 0.26 Å some residues changed by as much as 1 Å. Subsequent calculations investigated the diffusion of carbon monoxide through leghaemoglobin, suggesting a two-stage mechanism in which the ligand first 'hops' from the haeme pocket to another cavity, and then 'hops' outside the protein.

The SPW method does not directly provide the transition structures, which must therefore be located using an alternative method. The conjugate peak refinement algorithm of Fischer and Karplus has been proposed as a robust method for locating transition structures for systems with many atoms [Fischer and Karplus 1992]. The steps taken by this algorithm are illustrated in Figure 4.30. First, a line is drawn between the two minima and the location of a maximum along this line is determined. A line minimisation is performed along a conjugate vector from this point. Lines are then drawn from this point to the two

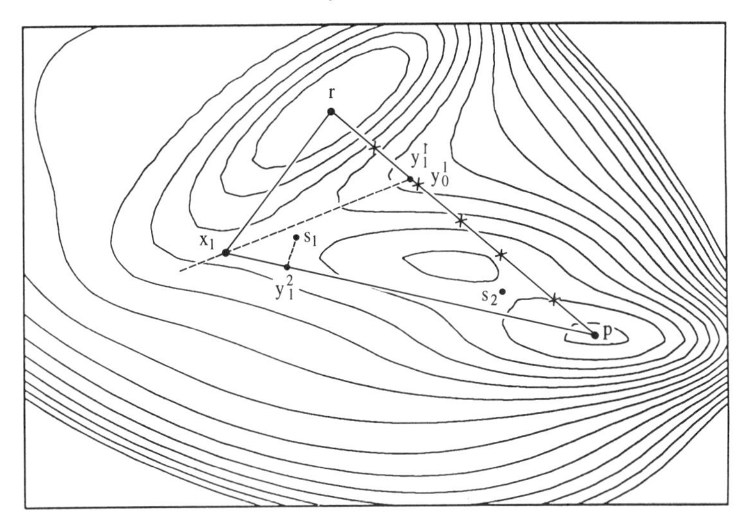

Fig. 4.30 The conjugate peak refinement method. r and p are the initial minima (the reactants and the products). A coarse step search along the line connecting r and p suggests that there is a maximum near the point y_0^1. Minimisation along the line connecting r to p gives the point y_1^1. A line minimisation is then performed along the conjugate vector to give the point x_1. In the second iteration the procedure is repeated for the lines $r - x_1$ and $x_1 - p$. A maximum is found at y_1^2 which after minimisation along the conjugate vector gives the saddle point s_1. Subsequent iterations of the algorithm enable the second saddle point s_2 to be identified. Figure adapted in part from Fischer S and M Karplus 1992. Conjugate Peak Refinement: An Algorithm for Finding Reaction Paths and Accurate Transition States in Systems with Many Degrees of Freedom. *Chemical Physical Letters* **194**:252– 261.

initial minima, and the process repeated. In this way the saddle points between the two minima can be identified, as shown in Figure 4.30.

The conjugate peak refinement technique has been applied to a variety of problems including the interconversion of conformations of calix[4]arenes [Fischer *et al.* 1995]. These molecules contain four phenyl rings connected by methylene bridges and have four characteristic conformations, as shown in Figure 4.31. A variety of substitution patterns are possible for these molecules; initially we will concentrate on the results for the parent compound which has X = H and R = H. The transition structures connecting the partial cone conformation to the other three conformations were elucidated using conjugate peak refinement and the results illustrated schematically in Figure 4.32. The results suggested that the lowest energy pathway from the cone to its energetically equivalent inverse was via the 1,2-alternate structure with the rate-limiting step being the cone → partial cone transition. This result was confirmed by performing a conjugate peak refinement calculation for the cone → inverted cone transition with the same result being obtained.

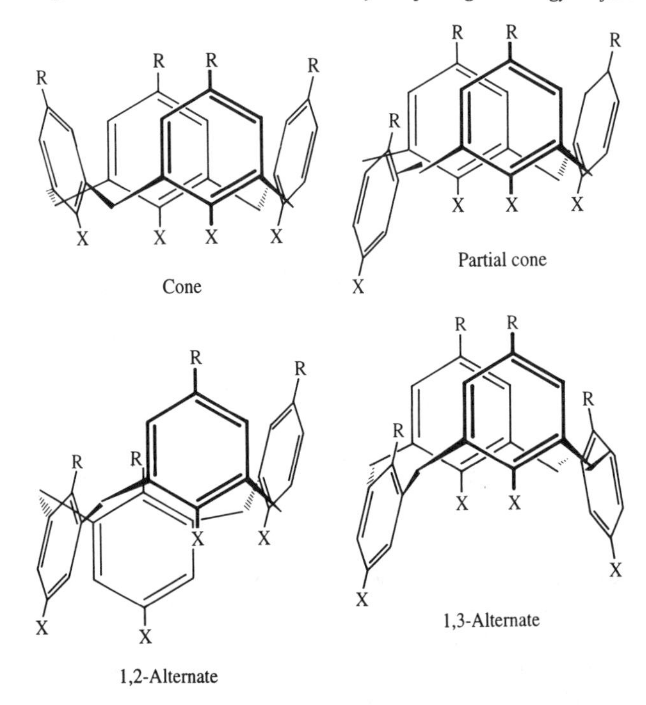

Fig. 4.31 Possible conformations of the calix[4]arene systems. After Fischer S, P D J Groothenuis, L C Groenen, W P van Hoorn, F C J M van Geggel, D N Reinhoudt and M Karplus 1995. Pathways for Conformational Interconversion of Calix[4]arenes. *The Journal of the American Chemical Society* **117**:1611–1620.

For the calix[4]arene with X = OH, the cone conformation is very much more stable, due to hydrogen bonding between the hydroxyl groups. A variety of mechanisms had previously been proposed for the interconversion of the cone to the inverted cone structure in such compounds; some of these suggestions involved flips of individual phenyl rings whereas others proposed a concerted process in which more than one phenyl ring rotated simultaneously. The pathway calculations suggested that in fact only single ring flips were involved, giving the energy diagram in Figure 4.32. The predicted activation barrier of 14.5 kcal/mol for the cone → inverted cone transition was in very good agreement with the experimentally determined value of 14.2 kcal/mol. Much of the barrier (9.1 kcal/mol) was due to the need to break two hydrogen bonds; the remainder was due to the need to deform some bond angles such as those of the bridging methylene carbons.

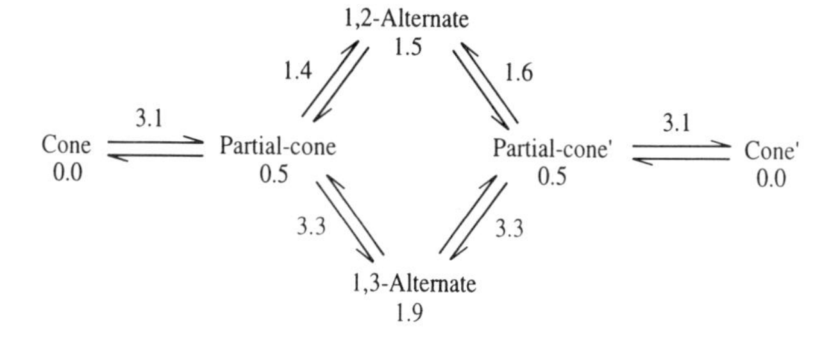

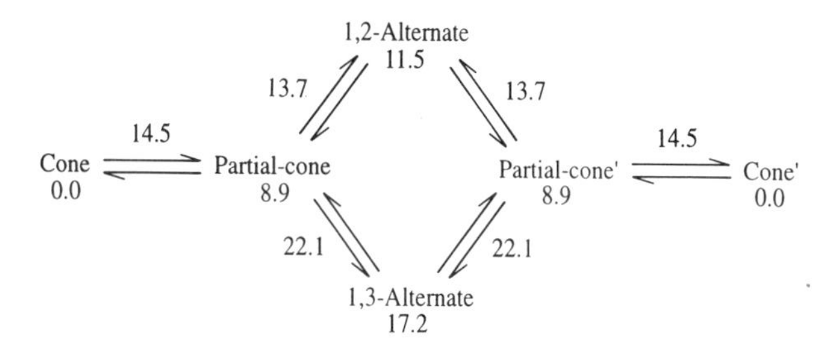

Fig. 4.32 Interconversion between various conformations of calix[4]arenes. X = H, R = H (top); X = OH, R = H (bottom). Energies in kcal/mol.

4.9.4 The transition structures of pericyclic reactions

One of the most celebrated examples of the use of quantum mechanical methods in understanding chemical reactivity is the work of Woodward and Hoffmann [Woodward and Hoffmann 1969] who were able to explain the experimentally observed nature of certain types of concerted reaction. The reactions which they studied include cycloadditions, sigmatropic rearrangements, cheletropic reactions, electrocyclic reactions and the ene reaction (Figure 4.33) and are collectively known as pericyclic reactions. The products obtained from such reactions can be understood in terms of simple mechanistic arguments, but such arguments cannot explain some aspects. In particular, the reactions are often highly stereospecific with the reaction rates and the stereoselectivity changes dramatically with the reaction conditions. Woodward and Hoffmann successfully employed molecular orbital theory to rationalise the existing data and their theory has also been very successful in predicting the outcome of similar reactions. The basic principle applied by Woodward and Hoffmann was that of the conservation of orbital symmetry and as a consequence of their work a series of rules (often called the Woodward–Hoffmann rules) were developed. The Woodward–Hoffmann rules only apply to concerted reactions and are based upon the principle that maximum bonding is maintained throughout the course of a reaction. Fukui also discovered the importance of orbital symmetry and suggested that the majority of chemical reactions should take place at the position of, and in the direction of maximum overlap between the highest occupied molecular orbital (HOMO) of one species

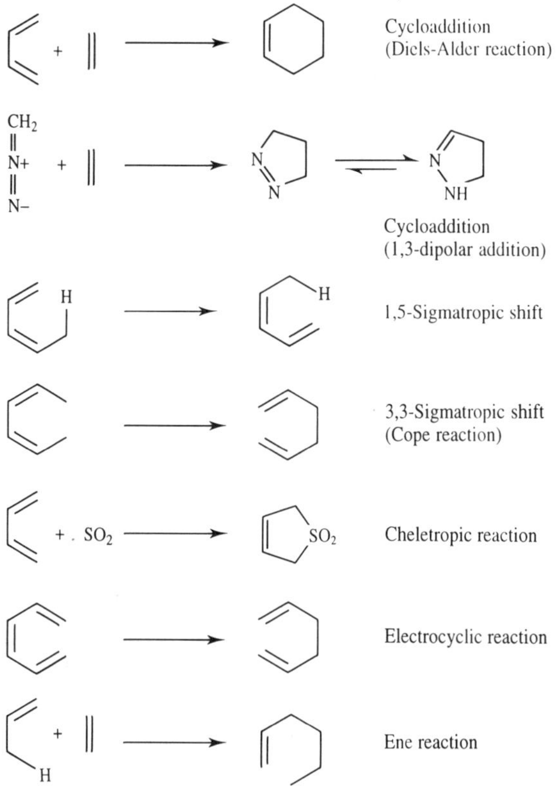

Cycloaddition (Diels-Alder reaction)

CH_2 = N+ = N− + ‖ → N N; N NH

Cycloaddition (1,3-dipolar addition)

1,5-Sigmatropic shift

3,3-Sigmatropic shift (Cope reaction)

+ . SO_2 → SO_2 Cheletropic reaction

Electrocyclic reaction

Ene reaction

Fig. 4.33 Typical pericyclic reactions.

and the lowest unoccupied molecular orbital (LUMO) of the other component [Fukui 1971]. These orbitals are collectively known as the *frontier orbitals*.

The HOMO–LUMO interaction depends on various factors, including the geometry of approach (which affects the amount of overlap), the phase relationship of the orbitals and their energy separation. For example, the HOMO and LUMO of ethene are illustrated pictorially in Figure 4.34. The most obvious mode of interaction between the two molecules involves suprafacial attack shown in Figure 4.33 to give cyclobutane. However, the symmetries of the overlapping orbitals must have the same phase for a favourable interaction to occur and this is not possible for ethene unless an energetically unfavourable antarafacial approach is adopted. By contrast, the interaction between ethene and the butadiene does occur in a suprafacial sense with both HOMO/LUMO pairs of orbitals having the appropriate phase relationship, Figure 4.34.

The Woodward–Hoffmann rules state what the outcome of a pericyclic reaction will be, but they do not state the mechanism by which the reaction occurs. Many theoretical techniques have been applied to the study of these

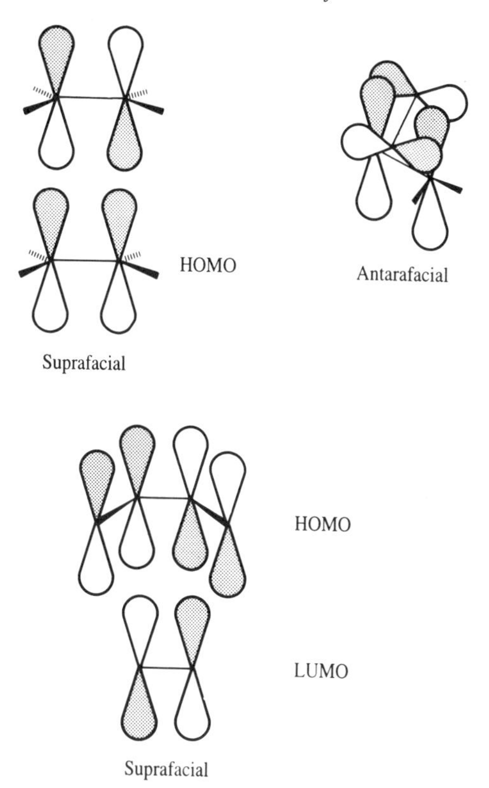

Fig. 4.34 Suprafacial attack of one ethene molecule on another (top) is not permitted by the Woodward–Hoffmann rules and the alternative antarafacial mode of attack is sterically unfavourable. Suprafacial attack is however permitted for the Diels–Alder reaction between butadiene and ethene (bottom).

problems over the years [Houk *et al.* 1992] and a passionate debate has ensued on the nature of the transition structures involved in these reactions. The debate has been fuelled by the fact that different theoretical treatments (especially semi-empirical methods) give different results. For example, at one extreme the Diels–Alder reaction between butadiene and ethene would proceed via a two-step mechanism involving a biradical transition structure. At the other extreme the reaction would involve a symmetrical transition state formed in a concerted, synchronous reaction. *Ab initio* calculations at various levels of theory suggest the concerted transition structure. The geometry obtained for the prototypical Diels–Alder reaction between butadiene/ethene using a CASSSCF calculation and a 6-31G* basis set is shown in Figure 4.35 [Houk *et al.* 1995]. The alternative biradial structure is also shown in Figure 4.35; this is predicted to be 6 kcal/mol higher in energy than the symmetrical transition structure.

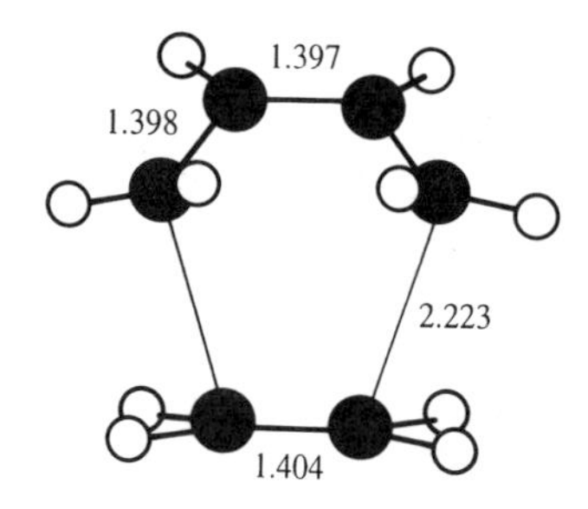

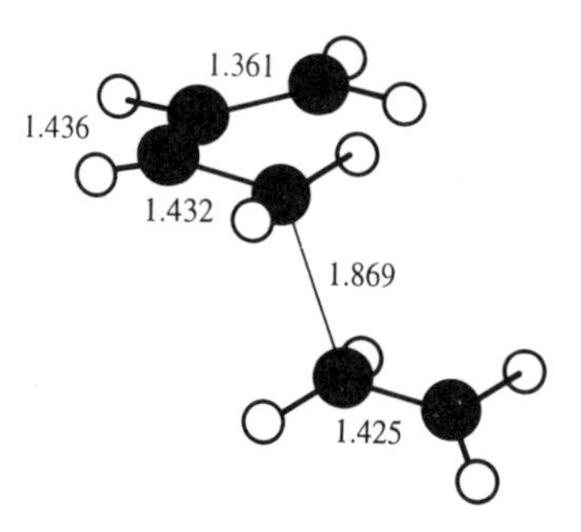

Fig. 4.35 Geometry predicted by CASSCF *ab initio* calculations of the two possible transition structure geometries for the Diels–Alder reaction between ethene and butadiene. Figure adapted from Houk K N, J González and Y Li 1995. Pericyclic Reaction Transition States: Passions and Punctilios 1935–1995. *Accounts of Chemical Research* **28**:81–90.

Further reading

Gill P E and W Murray 1981. *Practical Optimization*. London, Academic.

Press W H, B P Flannery, S A Teukolsky, W T Vetterling 1992. *Numerical Recipes in Fortran*. Cambridge, Cambridge University Press.

Schlegel H B 1987. Optimization of Equilibrium Geometries and Transition Structures in Lawley K P (Editor). *Ab Initio Methods in Quantum Chemistry- I*. New York, John Wiley: 249–286.

Schlegel H B 1989. Some Practical Suggestions for Optimizing Geometries and Locating Transition States in Bertrán J and I G Csizmadia (Editors). *New Theoretical Concepts for Understanding Organic Reactions*. Kluwer Academic Publishers: 33–53.

Williams I H 1993. Interplay of Theory and Experiment in the Determination of Transition-State Structures. *Chemical Society Reviews* Volume 1, 277–283.

References

Baker J 1986. An Algorithm for the Location of Transition States. *Journal of Computational Chemistry* **7**:385–395.

Brooks B and M Karplus 1983. Harmonic Dynamics of Proteins: Normal Modes and Fluctuations in Bovine Pancreatic Trypsin Inhibitor. *Proceedings of the National Academy of Sciences USA* **80**:6571–6575.

Czerminski R and R Elber 1990. Self-Avoiding Walk Between 2 Fixed-Points as a Tool to Calculate Reaction Paths in Large Molecular Systems. *International Journal of Quantum Chemistry* **S24**:167–186.

Dauber-Osguthorpe P, V A Roberts, D J Osguthorpe, J Wolff, M Genest and A T Hagler 1988. Structure and Energetics of Ligand Binding to Proteins: *Escherichia coli* Dihydrofolate Reductase-Trimethoprim, A Drug-Receptor System. *Proteins: Structure, Function and Genetics* **4**:31–47.

Dewar M J S, E F Healy and J J P Stewart 1984. Location of Transition States in Reaction Mechanisms. *Journal of the Chemical Society Faraday Transactions 2* **80**:227–233.

Doubleday C, J McIver, M Page and T Zielinski 1985. Temperature Dependence of the Transition-State Structure for the Disproportionation of Hydrogen Atom with Ethyl Radical. *The Journal of the American Chemical Society* **107**:5800–5801.

Elber R and M Karplus 1987. A Method for Determining Reaction Paths in Large Molecules: Application to Myoglobin. *Chemical Physics Letters* **139**:375–380.

Fischer S and M Karplus 1992. Conjugate Peak Refinement: An Algorithm for Finding Reaction Paths and Accurate Transition States in Systems with Many Degrees of Freedom. *Chemical Physics Letters* **194**:252–261.

Fischer S, P D J Groothenuis, L C Groenen, W P van Hoorn, F C J M van Geggel, D N Reinhoudt and M Karplus 1995. Pathways for Conformational Interconversion of Calix[4]arenes. *The Journal of the American Chemical Society* **117**:1611–1620.

Fisher C L, V A Roberts and A T Hagler 1991. Influence of Environment on the Antifolate Drug Trimethoprim: Energy Minimization Studies. *Biochemistry* **30**:3518–3526.

Fukui K 1971. Recognition of Stereochemical Paths by Orbital Interaction. *Accounts of Chemical Research* **4**:57–64.

Fukui K 1981. The Path of Chemical Reactions—The IRC Approach. *Accounts of Chemical Research* **14**:368–375.

Gelin B R and M Karplus 1975. Sidechain Torsional Potential and Motion of Amino Acids in Proteins: Bovine Pancreatic Trypsin Inhibitor. *Proceedings of the National Academy of Science USA* **72**:2002–2006.

Gonzalez C and H B Schlegel 1988. An Improved Algorithm for Reaction Path Following. *The Journal of Chemical Physics* **90**:2154–2161.

Houk K N, Y Li and J D Evanseck 1992. Transition Structures of Hydrocarbon Pericyclic Reactions. *Angewandte Chemie International Edition in English* **31**:682–708.

Houk K N, J González and Y Li 1995. Pericyclic Reaction Transition States: Passions and Punctilios 1935–1995. *Accounts of Chemical Research* **28**:81–90.

Ishida K, K Morokuma and A Komornicki 1977. The Intrinsic Reaction Coordinate. An *Ab Initio* Calculation for HNC $\rightarrow$ HCN and $H^- + CH_4 \rightarrow CH_4 + H^{-*}$. *The Journal of Chemical Physics* **66**:2153–2156.

Müller K and L D Brown 1979. Location of Saddle Points and Minimum Energy Paths by a Constrained Simplex Optimisation Procedure. *Theoretica Chemica Acta* **53**:75–93.

Nowak W, R Czerminski and R Elber 1991. Reaction Path Study of Ligand Diffusion in Proteins: Application of the Self Penalty Walk (SPW) Method to Calculate Reaction Coordinates for the Motion of CO through Leghemoglobin. *The Journal of the American Chemical Society* **113**:5627–5737.

Pople J A, A P Scott, M W Wong and L Radom 1993. Scaling Factors for Obtaining Fundamental Vibrational Frequencies and Zero-Point Energies from HF/6-31G* and MP2/6-31G* Harmonic Frequencies. *Israel Journal of Chemistry* **33**:345–350.

Schlegel H B 1982. Optimisation of Equilibrium Geometries and Transition Structures. *Journal of Computational Chemistry* **3**:214–218.

Woodward R B and R Hoffmann 1969. The Conservation of Orbital Symmetry. *Angewandte Chemie International Edition in English* **8**:781–853.

5 Computer simulation methods

5.1 Introduction

Energy minimisation generates individual minimum energy configurations of a system. In some cases the information provided by energy minimisation can be sufficient to predict accurately the properties of a system. If all minimum configurations on an energy surface can be identified then statistical mechanical formulae can be used to derive a partition function from which thermodynamic properties can be calculated. However, this is possible only for relatively small molecules or small molecular assemblies in the gas phase. The molecular modeller more often wants to understand and to predict the properties of liquids, solutions and solids, to study complex processes such as the adsorption of molecules onto surfaces and into solids and to investigate the behaviour of macromolecules which have many closely separated minima. In such systems the experimental measurements are made on macroscopic samples that contain extremely large numbers of atoms or molecules, with an enormous number of minima on their energy surfaces. A full quantification of the energy surfaces of such systems is not possible, nor is it ever likely to be. Computer simulation methods enable us to study such systems and predict their properties through the use of techniques that consider small replications of the macroscopic system with manageable numbers of atoms or molecules. A simulation generates representative configurations of these small replications in such a way that accurate values of structural and thermodynamic properties can be obtained with a feasible amount of computation. Simulation techniques also enable the time-dependent behaviour of atomic and molecular systems to be determined, providing a detailed picture of the way in which a system changes from one conformation or configuration to another. Simulation techniques are also widely used in some experimental procedures, such as the determination of protein structures from X-ray crystallography.

In this chapter we shall discuss some of the general principles involved in the two most common simulation techniques used in molecular modelling: the molecular dynamics and the Monte Carlo methods. We shall also discuss several concepts that are common to both of these methods. A more detailed discussion of the two simulation methods can be found in Chapters 6 and 7.

5.1.1 Time averages, ensemble averages and some historical background

Suppose we wish to determine experimentally the value of a property of a system such as the pressure or the heat capacity. In general, such properties will depend upon the positions and momenta of the N particles that comprise the system. The instantaneous value of the property A can thus be written as $A(\mathbf{p}^N(t), \mathbf{r}^N(t))$, where $\mathbf{p}^N(t)$ and $\mathbf{r}^N(t)$ represent the N momenta and positions respectively at time t (i.e. $A(\mathbf{p}^N(t), \mathbf{r}^N(t)) \equiv A(p_{1x}, p_{1y}, p_{1z}, p_{2x}, \ldots, x_1, y_1, z_1, x_2, \ldots, t)$ where p_{1x} is the momentum of particle 1 in the x direction and x_1 is its x coordinate). Over time, the instantaneous value of the property A fluctuates as a result of interactions between the particles. The value that we measure experimentally is an average of A over the time of the measurement and is therefore known as a *time average*. As the time over which the measurement is made increases to infinity, so the value of the following integral approaches the 'true' average value of the property:

$$A_{\text{ave}} = \lim_{\tau \to \infty} \frac{1}{\tau} \int_{t=0}^{\tau} A(\mathbf{p}^N(t), \mathbf{r}^N(t))\, \mathrm{d}t \tag{5.1}$$

To calculate average values of the properties of the system, it would therefore appear to be necessary to simulate the dynamical behaviour of the system (i.e. to determine values of $A(\mathbf{p}^N(t), \mathbf{r}^N(t))$, based upon a model of the intra- and intermolecular interactions present). In principle, this is relatively straightforward to do. For any arrangement of the atoms in the system, the force acting on each atom due to interactions with other atoms can be calculated by differentiating the energy function. From the force on each atom it is possible to determine its acceleration via Newton's second law. Integration of the equations of motion should then yield a trajectory that describes how the positions, velocities and accelerations of the particles vary with time, and from which the average values of properties can be determined using the numerical equivalent of equation (5.1). The difficulty is that for 'macroscopic' numbers of atoms or molecules (of the order 10^{23}) it is not even feasible to determine an initial configuration of the system, let alone integrate the equations of motion and calculate a trajectory. Recognising this problem, Boltzmann and Gibbs developed statistical mechanics, in which a single system evolving in time is replaced by a large number of replications of the system that are considered simultaneously. The time average is then replaced by an *ensemble average*:

$$\langle A \rangle = \iint \mathrm{d}\mathbf{p}^N \mathrm{d}\mathbf{r}^N A(\mathbf{p}^N, \mathbf{r}^N)\rho(\mathbf{p}^N, \mathbf{r}^N) \tag{5.2}$$

The angle brackets $\langle\ \rangle$ indicate an ensemble average, or *expectation value*; that is the average value of the property A over all replications of the ensemble generated by the simulation. Equation (5.2) is written as a double integral for convenience but in fact there should be $6N$ integral signs on the integral for the $6N$ positions and momenta of all the particles. $\rho(\mathbf{p}^N \mathbf{r}^N)$ is the *probability density* of the ensemble; that is the probability of finding a configuration with momenta

$\mathbf{p}^N$ and positions $\mathbf{r}^N$. The ensemble average of the property A is then determined by integrating over all possible configurations of the system. In accordance with the *ergodic hypothesis*, which is one of the fundamental axioms of statistical mechanics, the ensemble average is equal to the time average. Under conditions of constant number of particles, volume and temperature, the probability density is the familiar Boltzmann distribution:

$$\rho(\mathbf{p}^N\mathbf{r}^N) = \exp(-\mathrm{E}(\mathbf{p}^N, \mathbf{r}^N)/k_\mathrm{B}T)/Q \tag{5.3}$$

In equation (5.3), $\mathrm{E}(\mathbf{p}^N, \mathbf{r}^N)$ is the energy, Q is the partition function, k_B is Boltzmann's constant and T is the temperature. The partition function is more generally written in terms of the Hamiltonian, $\mathscr{H}$; for a system of N identical particles the partition function for the canonical ensemble is as follows:

$$Q_{NVT} = \frac{1}{N!}\frac{1}{h^{3N}}\iint \mathrm{d}\mathbf{p}^N \mathrm{d}\mathbf{r}^N \exp\left[-\frac{\mathscr{H}(\mathbf{p}^N, \mathbf{r}^N)}{k_\mathrm{B}T}\right] \tag{5.4}$$

The canonical ensemble is the name given to an ensemble for constant temperature, number of particles and volume. For our purposes $\mathscr{H}$ can be considered the same as the total energy, $\mathrm{E}(\mathbf{p}^N, \mathbf{r}^N)$, which equals the sum of the kinetic energy ($\mathscr{K}(\mathbf{p}^N)$) of the system that depends upon the momenta of the particles and the potential energy ($\mathscr{V}(\mathbf{r}^N)$) that depends upon the positions. The factor $N!$ arises from the indistinguishability of the particles and the factor $1/h^{3N}$ is required to ensure that the partition function is equal to the quantum mechanical result for a particle in a box. A short discussion of some of the key results of statistical mechanics is provided in Appendix 5.1 and further details can be found in standard textbooks.

The first computer simulations of fluids were performed in 1952 by Metropolis, Rosenbluth, Rosenbluth, Teller and Teller who developed a scheme for sampling from the Boltzmann distribution to give ensemble averages. This gave rise to the Monte Carlo simulation method. Not long afterwards (in 1957) Alder recognised that it was in fact possible to integrate the equations of motion for a relatively small number of particles, and to mimic the behaviour of a real system using periodic boundary conditions. This led to the first molecular dynamics simulations of molecular systems.

5.1.2 A brief description of the molecular dynamics method

Molecular dynamics calculates the 'real' dynamics of the system, from which time averages of properties can be calculated. Sets of atomic positions are derived in sequence by applying Newton's equations of motion. Molecular dynamics is a *deterministic* method, by which we mean that the state of the system at any future time can be predicted from its current state. The first molecular dynamics simulations were performed using very simple potentials such as the hard-sphere potential. The behaviour of the particles in this potential is similar to that of billiard or snooker balls: the particles move in straight lines at constant velocity between collisions. The collisions are perfectly elastic and occur when the separation between a pair of spheres equals the sum of their radii. After a collision, the new velocities of the colliding spheres are calculated using the principle of conservation of linear momentum. The hard-sphere model has provided many useful results but is obviously not ideal for simulating atomic or molecular systems. In potentials such as the Lennard–Jones potential the force between two atoms or molecules changes continuously with their separation. By contrast, in the hard-sphere model there is no force between particles until they collide. The continuous nature of the more realistic potentials requires the equations of motion to be integrated by breaking the calculation into a series of very short time steps (typically between 1 femtosecond and 10 femtoseconds; 10^{-15} s to 10^{-14} s). At each step, the forces on the atoms are computed and combined with the current positions and velocities to generate new positions and velocities a short time ahead. The force acting on each atom is assumed to be constant during the time interval. The atoms are then moved to the new positions, an updated set of forces is computed and so on. In this way a molecular dynamics simulation generates a *trajectory* that describes how the dynamic variables change with time. Molecular dynamics simulations are typically run for tens or hundreds of picoseconds (a 100 ps simulation using a 1 fs time step requires 100 000 steps). Thermodynamic averages are obtained from molecular dynamics as time averages using numerical integration of equation (5.2):

$$\langle A \rangle = \frac{1}{M} \sum_{\mathrm{i}=1}^{M} A(\mathbf{p}^N, \mathbf{r}^N) \tag{5.5}$$

M is the number of time steps. Molecular dynamics is also extensively used to investigate the conformational properties of flexible molecules as will be discussed in Chapters 6 and 8.

5.1.3 The basic elements of the Monte Carlo method

In a molecular dynamics simulation the successive configurations of the system are connected in time. This is not the case in a Monte Carlo simulation, where each configuration depends only upon its predecessor and not upon any other of the configurations previously visited. The Monte Carlo method generates configurations randomly, and uses a special set of criteria to decide whether or not to accept each new configuration. These criteria ensure that the probability of obtaining a given configuration is equal to its Boltzmann factor, $\exp\{-\mathscr{V}(\mathbf{r}^N)/k_{\mathrm{B}}T\}$, where $\mathscr{V}(\mathbf{r}^N)$ is calculated using the potential energy function. States with a low energy are thus generated with a higher probability than configurations with a higher energy. For each configuration that is accepted the values of the desired properties are calculated and at the end of the calculation the average of these properties is obtained by simply averaging over the number of values calculated, M:

$$\langle A \rangle = \frac{1}{M} \sum_{i=1}^{M} A(\mathbf{r}^N) \tag{5.6}$$

Most Monte Carlo simulations of molecular systems are more properly referred to as Metropolis Monte Carlo calculations after Metropolis and his colleagues who reported the first such calculation. The distinction can be important because there are other ways in which an ensemble of configurations can be generated. As we shall see in Chapter 7, the Metropolis scheme is only one of a number of possibilities, though it is by far the most popular.

In a Monte Carlo simulation each new configuration of the system may be generated by randomly moving a single atom or molecule. In some cases new configurations may also be obtained by moving several atoms or molecules, or by rotating about one or more bonds. The energy of the new configuration is then calculated using the potential energy function. If the energy of the new configuration is lower than the energy of its predecessor then the new configuration is accepted. If the energy of the new configuration is higher than the energy of its predecessor then the *Boltzmann factor* of the energy difference is calculated: $\exp[(\mathscr{V}_{\text{new}}(\mathbf{r}^N) - \mathscr{V}_{\text{old}}(\mathbf{r}^N))/k_B T]$. A random number is then generated between 0 and 1 and compared with this Boltzmann factor. If the random number is higher than the Boltzmann factor then the move is rejected and the original configuration is retained for the next iteration; if the random number is lower then the move is accepted and the new configuration becomes the next state. This procedure has the effect of permitting moves to states of higher energy. The smaller the uphill move (i.e. the smaller the value of $\mathscr{V}_{\text{new}}(\mathbf{r}^N) - \mathscr{V}_{\text{old}}(\mathbf{r}^N)$) the greater is the probability that the move will be accepted.

5.1.4 Differences between the molecular dynamics and Monte Carlo methods

The molecular dynamics and Monte Carlo simulation methods differ in a variety of ways. The most obvious difference is that molecular dynamics provides information about the time dependence of the properties of the system whereas there is no temporal relationship between successive Monte Carlo configurations. In a Monte Carlo simulation the outcome of each trial move only depends upon its immediate predecessor, whereas in molecular dynamics it is possible to predict the configuration of the system at any time in the future – or indeed at any time in the past. Molecular dynamics has a kinetic energy contribution to the total energy whereas in a Monte Carlo simulation the total energy is determined directly from the potential energy function. The two simulation methods also sample from different ensembles. Molecular dynamics is traditionally performed under conditions of constant number of particles (N), volume (V) and energy (E) (the microcanonical or constant NVE ensemble) whereas a traditional Monte Carlo simulation samples from the canonical ensemble (constant N, V and temperature, T). Both the molecular dynamics and Monte Carlo techniques can be modified to sample from other ensembles; for example, molecular dynamics can be adapted to simulate from the canonical ensemble. Two other ensembles are common:

isothermal–isobaric: fixed N, T, P (pressure)
grand canonical: fixed μ (chemical potential), V, T

In the canonical, microcanonical and isothermal–isobaric ensembles the number of particles is constant but in a grand canonical simulation the composition can change (i.e. the number of particles can increase or decrease). The equilibrium states of each of these ensembles are characterised as follows:

canonical ensemble: minimum Helmholtz free energy (A)
microcanonical ensemble: maximum entropy (S)
isothermal–isobaric ensemble: minimum Gibbs function (G)
grand canonical ensemble: maximum pressure $\times$ volume (PV)

5.2 Calculation of simple thermodynamic properties

A wide variety of thermodynamic properties can be calculated from computer simulations; a comparison of experimental and calculated values for such properties is an important way in which the accuracy of the simulation and the underlying energy model can be quantified. Simulation methods also enable predictions to be made of the thermodynamic properties of systems for which there is no experimental data, or for which experimental data is difficult or impossible to obtain. Simulations can also provide structural information about the conformational changes in molecules and the distributions of molecules in a system. The emphasis in our discussion will be on those properties that are routinely calculated in computer simulations and on the way in which they are obtained. It is important to recognise that the results we derive are for the canonical ensemble. Sometimes the equivalent expressions in other ensembles are provided. The result obtained from one ensemble may also be transformed to another ensemble, though this is strictly only possible in the limit of an infinitely large system. The expressions follow from standard statistical mechanical relationships which are given in standard texts and summarised in Appendix 5.1.

5.2.1 Energy

The internal energy is easily obtained from a simulation as the ensemble average of the energies of the states that are examined during the course of the simulation:

$$U = \langle E \rangle = \frac{1}{M} \sum_{\mathrm{i}=1}^{M} E_{\mathrm{i}} \tag{5.7}$$

5.2.2 Heat capacity

At a phase transition the heat capacity will often show a characteristic dependence upon the temperature (a first-order phase transition is characterised by an infinite heat capacity at the transition but in a second-order phase transition the heat capacity changes discontinuously). Monitoring the heat capacity as a function of temperature may therefore enable phase transitions to be detected. Calculations of the heat capacity can also be compared with experimental results and so be used to check the energy model or the simulation protocol.

The heat capacity is formally defined as the partial derivative of the internal energy with respect to temperature:

$$C_V = \left(\frac{\partial U}{\partial T}\right)_V \tag{5.8}$$

The heat capacity can therefore be calculated by performing a series of simulations at different temperatures, and then differentiating the energy with respect to the temperature. The differentiation can be done numerically or by fitting a polynomial to the data and then analytically differentiating the fitted function. The heat capacity may also be calculated from a single simulation by considering the instantaneous fluctuations in the energy as follows:

$$C_V = \{\langle E^2\rangle - \langle E\rangle^2\}/k_B T^2 \tag{5.9}$$

An alternative way to write this expression uses the relationship

$$\langle (E - \langle E\rangle)^2\rangle = \langle E^2\rangle - \langle E\rangle^2 \tag{5.10}$$

giving

$$C_V = \langle (E - \langle E\rangle)^2\rangle / k_B T^2 \tag{5.11}$$

A derivation of this result is provided in Appendix 5.2.

The heat capacity can therefore be obtained by keeping a running count of E^2 and E during the simulation, from which their expectation values $\langle E^2\rangle$ and $\langle E\rangle$ can be calculated at the end of the calculation. Alternatively, if the energies are stored during the simulation then the value of $\langle (E - \langle E\rangle)^2\rangle$ can be calculated once the simulation has finished. This second approach may be more accurate due to round-off errors; $\langle E^2\rangle$ and $\langle E\rangle^2$ are usually both large numbers and so there may be a large uncertainty in their difference.

5.2.3 Pressure

The pressure is usually calculated in a computer simulation via the virial theorem of Clausius. The *virial* is defined as the expectation value of the sum of the products of the coordinates of the particles and the forces acting on them. This is usually written $W = \sum x_i \dot{p}_{x_i}$ where x_i is a coordinate (e.g. the x or y coordinate of an atom) and $\dot{p}_{x_i}$ is the first derivative of the momentum along that coordinate ($\dot{p}_i$

is the force, by Newton's second law). The virial theorem states that the virial is equal to $-3Nk_BT$.

In an ideal gas, the only forces are those due to interactions between the gas and the container and it can be shown that the virial in this case equals $-3PV$. This result can also be obtained directly from $PV = Nk_BT$.

Forces between the particles in a real gas or liquid affect the virial, and thence the pressure. The total virial for a real system equals the sum of an ideal gas part ($-3PV$) and a contribution due to interactions between the particles. The result obtained is:

$$W = -3PV + \sum_{i=1}^{N}\sum_{j=i+1}^{N} r_{ij}\frac{d\mathscr{V}(r_{ij})}{dr_{ij}} = -3Nk_BT \tag{5.12}$$

The real gas part is derived in Appendix 5.3. If $d\mathscr{V}(r_{ij})/dr_{ij}$ is written as f_{ij}, the force acting between atoms i and j, then we have the following expression for the pressure:

$$P = \frac{1}{V}\left[Nk_BT - \frac{1}{3k_BT}\sum_{i=1}^{N}\sum_{j=i+1}^{N} r_{ij}f_{ij}\right] \tag{5.13}$$

The forces are calculated as part of a molecular dynamics simulation, and so little additional effort is required to calculate the virial and thus the pressure. The forces are not routinely calculated during a Monte Carlo simulation, and so some additional effort is required to determine the pressure by this route. When calculating the pressure it is also important to check that the components of the pressure in all three directions are equal.

5.2.4 Temperature

In a canonical ensemble the total temperature is constant. In the microcanonical ensemble, however, the temperature will fluctuate. The temperature is directly related to the kinetic energy of the system as follows:

$$\mathscr{K} = \sum_{i=1}^{N}\frac{|\mathbf{p}_i|^2}{2m_i} = \frac{k_BT}{2}(3N - N_c) \tag{5.14}$$

In this equation, $\mathbf{p}_i$ is the total momentum of particle i and m_i is its mass. According to the theorem of the equipartition of energy each degree of freedom contributes $k_BT/2$. If there are N particles, each with three degrees of freedom, then the kinetic energy should equal $3Nk_BT/2$. N_c in equation (5.14) is the number of constraints on the system. In a molecular dynamics simulation the total linear momentum of the system is often constrained to a value of zero, which has the effect of removing three degrees of freedom from the system and so N_c would be equal to 3. Other types of constraint are also possible as we shall discuss in section 6.5.

5.2.5 Radial distribution functions

Radial distribution functions are a useful way to describe the structure of a system, particularly of liquids. Consider a spherical shell of thickness δr at a distance r from a chosen atom (Figure 5.1). The volume of the shell is given by:

$$\begin{aligned} V &= \tfrac{4}{3}\pi(r+\delta r)^3 - \tfrac{4}{3}\pi r^3 \\ &= 4\pi r^2\delta r + 4\pi r\delta r^2 + \tfrac{4}{3}\pi\delta r^3 \approx 4\pi r^2\delta r \end{aligned} \tag{5.15}$$

If the number of particles per unit volume is ρ, then the total number in the shell is $4\pi\rho r^2\delta r$ and so the number of atoms in the volume element varies as r^2.

The pair distribution function, $g(r)$, gives the probability of finding an atom (or molecule, if simulating a molecular fluid) a distance r from another atom (or molecule) compared to the ideal gas distribution. $g(r)$ is thus dimensionless. Higher radial distribution functions (e.g. the triplet radial distribution function) can also be defined, but are rarely calculated and so references to the 'radial distribution function' are usually taken to mean the pairwise version. In a crystal, the radial distribution function has an infinite number of sharp peaks whose separations and heights are characteristic of the lattice structure.

The radial distribution function of a liquid is intermediate between the solid and the gas, with a small number of peaks as short distances, superimposed on a steady decay to a constant value at longer distances. The radial distribution function calculated from a molecular dynamics simulation of liquid argon (shown in Figure 5.2) is typical. For short distances (less than the atomic diameter) $g(r)$ is zero. This is due to the strong repulsive forces. The first (and largest) peak occurs at $r \approx 3.7$ Å with $g(r)$ having a value of about 3. This means that it is three times more likely that two molecules would have this separation than in the ideal gas. The radial distribution function then falls and passes through a minimum value around $r \approx 5.4$ Å. The chances of finding two atoms with this separation are less than for the ideal gas. At long distances, $g(r)$ tends to the ideal gas value, indicating that there is no long-range order.

To calculate the pair distribution function from a simulation, the neighbours around each atom or molecule are sorted into distance 'bins', or histograms. The number of neighbours in each bin is then averaged over the entire simulation. For

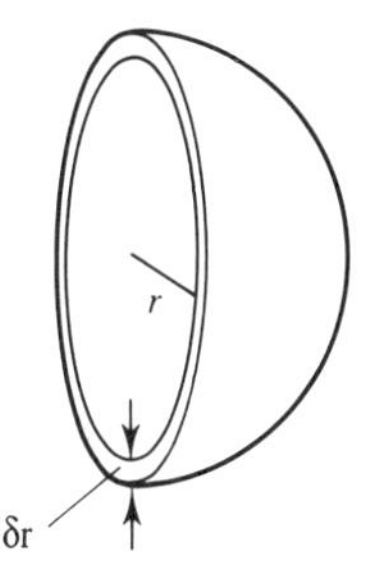

Fig. 5.1 Radial distribution functions use a spherical shell of thickness δr.

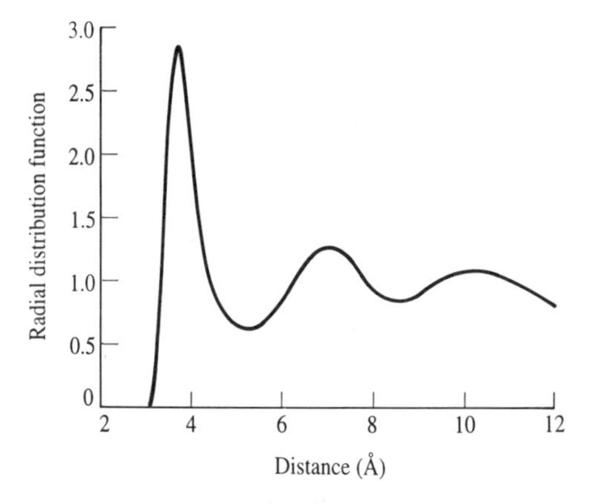

Fig. 5.2 Radial distribution function determined from a 100 ps molecular dynamics simulation of liquid argon at a temperature of 100 K and a density of 1.396 g cm^{-3}.

example, a count is made of the number of neighbours between (say) 2.5 Å and 2.75 Å, 2.75 Å and 3.0 Å and so on for every atom or molecule in the simulation. This count can be performed during the simulation itself or by analysing the configurations that are generated.

Radial distribution functions can be measured experimentally using X-ray diffraction. The regular arrangement of the atoms in a crystal gives the characteristic X-ray diffraction pattern with bright, sharp spots. For liquids, the diffraction pattern has regions of high and low intensity but no sharp spots. The X-ray diffraction pattern can be analysed to calculate an experimental distribution function, which can then be compared with that obtained from the simulation.

Thermodynamic properties can be calculated using the radial distribution function, if pairwise additivity of the forces is assumed. These properties are usually given as an ideal gas part plus a real gas part. For example, to calculate the energy of a real gas, we consider the spherical shell of volume $4\pi r^2 \delta r$ that contains $4\pi r^2 \rho g(r) \delta r$ particles. If the pair potential at a distance r has a value $\mathscr{v}(r)$ then the energy of interaction between the particles in the shell and the central particle is $4\pi r^2 \rho g(r) \mathscr{v}(r) \delta r$. The total potential energy of the real gas is obtained by integrating this between 0 and ∞ and multiplying the result by $N/2$ (the factor $1/2$ ensures that we only count each interaction once). The total energy is then given by:

$$E = \frac{3}{2} N k_{\mathrm{B}} T + 2\pi N \rho \int_0^\infty r^2 \mathscr{v}(r) g(r) \mathrm{d}r \tag{5.16}$$

In a similar way the following expression for the pressure can be derived:

$$PV = Nk_{\mathrm{B}}T - \frac{2\pi N\rho}{3k_{\mathrm{B}}T}\int_{0}^{\infty} r^2 r \frac{\mathrm{d}\mathscr{V}(r)}{\mathrm{d}r} g(r)\mathrm{d}r \tag{5.17}$$

It is usually more accurate to calculate such properties directly, partly because the radial distribution function is not obtained as a continuous function, but is derived by dividing the space into small but discrete bins.

For molecules, the orientation must be taken into account if the true nature of the distribution is to be determined. The radial distribution function for molecules is usually measured between two fixed points, such as between the centres of mass. This may then be supplemented by an orientational distribution function. For linear molecules, the orientational distribution function may be calculated as the angle between the axes of the molecule, with values ranging from $-180°$ to $+180°$. For more complex molecules it is usual to calculate a number of site–site distribution functions. For example, for a three-site model of water, three functions can be defined (g(O–O), g(O–H) and g(H–H)). An advantage of the site–site models is that they can be directly related to information obtained from the X-ray scattering experiments. The O–O, O–H and H–H radial distribution functions have been particularly useful for refining the various potential models for simulating liquid water.

5.3 Phase space

An important concept in computer simulation is that of the *phase space*. For a system containing N atoms, $6N$ values are required to define the state of the system (3 coordinates per atom and 3 components of the momentum). Each combination of $3N$ positions and $3N$ momenta defines a point in the $6N$-dimensional phase space; an ensemble can thus be considered to be a collection of points in phase space. Molecular dynamics generates a sequence of points in phase space that are connected in time. These points correspond to the successive configurations of the system generated by the simulation. A molecular dynamics simulation performed in the microcanonical (constant NVE) ensemble will sample phase space along a contour of constant energy. There is no momentum component in a Monte Carlo simulation and such simulations sample from the $3N$ dimensional space corresponding to the positions of the atoms. It might seem odd that thermodynamic properties can be obtained from Monte Carlo simulations, given that there is no momentum contribution and so $3N$ degrees of freedom are not explored. In fact, all of the deviations from ideal gas behaviour are a consequence of interactions between the atoms and are encapsulated in the potential function, $\mathscr{V}(\mathbf{r}^N)$, which only depends upon the positions of the atoms. A Monte Carlo simulation does sample from the positional degrees of freedom and so can be used to provide the deviations of thermodynamic properties from ideal gas behaviour, which is what we want to calculate. We shall return to this point in Chapter 7.

If it were possible to visit all the points in phase space then the partition function could be calculated by summing the values of $\exp(-E/k_B T)$. The phase-space trajectory in such a case would be termed *ergodic* and the results would be independent of the initial configuration. For the systems that are typical of those studied using simulation methods the phase space is immense, and an ergodic trajectory is not achievable (indeed, even for relatively small systems with only a few tens of atoms the time that would be required to cycle round all of the points in phase space is longer than the age of the universe). A simulation can thus only ever provide an estimate of the 'true' energies and other thermodynamic properties and so a sequence of simulations using different starting conditions would be expected to give similar, but different results.

The thermodynamic properties that we have considered so far, such as the internal energy, the pressure and the heat capacity are collectively known as the mechanical properties and can be routinely obtained from a Monte Carlo or molecular dynamics simulation. Other thermodynamic properties are difficult to determine accurately without resorting to special techniques. These are the so-called entropic or thermal properties: the free energy, the chemical potential and the entropy itself. The difference between the mechanical and thermal properties is that the mechanical properties are related to the derivative of the partition function whereas the thermal properties are directly related to the partition function itself. To illustrate the difference between these two classes of properties, let us consider the internal energy, U, and the Helmholtz free energy, A. These are related to the partition function by:

$$U = \frac{k_B T^2}{Q}\frac{\partial Q}{\partial T} \tag{5.18a}$$

$$A = -k_B T \ln Q \tag{5.18b}$$

Q is given by equation (5.4) for a system of identical particles. We shall ignore any normalisation constants in our treatment here, to enable us to concentrate on the basics and so it does not matter whether the system consists of identical or distinguishable particles. We also replace the Hamiltonian by the energy, E. The internal energy is obtained by differentiating equation (5.18a):

$$\begin{aligned} U &= k_B T^2 \frac{1}{Q}\iint \mathrm{d}\mathbf{p}^N \mathrm{d}\mathbf{r}^N \frac{E(\mathbf{p}^N, \mathbf{r}^N)}{k_B T^2} \exp(-E(\mathbf{p}^N, \mathbf{r}^N)/k_B T) \\ &= \iint \mathrm{d}\mathbf{p}^N \mathrm{d}\mathbf{r}^N E(\mathbf{p}^N, \mathbf{r}^N) \frac{\exp(-E(\mathbf{p}^N, \mathbf{r}^N)/k_B T)}{Q} \end{aligned} \tag{5.19}$$

Now consider the probability of the state with energy $E(\mathbf{p}^N, \mathbf{r}^N)$:

$$\frac{\exp(-E(\mathbf{p}^N, \mathbf{r}^N)/k_B T)}{Q}$$

This probability is written $\rho(\mathbf{p}^N, \mathbf{r}^N)$; the internal energy is thus given by

$$U = \iint \mathrm{d}\mathbf{p}^N \mathrm{d}\mathbf{r}^N E(\mathbf{p}^N, \mathbf{r}^N)\rho(\mathbf{p}^N \ \mathbf{r}^N) \tag{5.20}$$

The crucial point about equation (5.20) is that high values of $E(\mathbf{p}^N, \mathbf{r}^N)$ have a very low probability and make an insignificant contribution to the integral. The Monte Carlo and molecular dynamics methods preferentially generate states of low energy, which are the states that make a significant contribution to the integral in equation (5.20). These methods sample from phase space in a way that is representative of the equilibrium state and are able to generate accurate estimates of properties such as the internal energy, heat capacity and so on.

Let us now consider the problem of calculating the Helmholtz free energy of a molecular liquid. Our aim is to express the free energy in the same functional form as the internal energy, that is as an integral which incorporates the probability of a given state. First, we substitute for the partition function in equation (5.18b):

$$A = -k_{\mathrm{B}}T \ln Q = k_{\mathrm{B}}T \ln\left(\frac{N!h^{3N}}{\iint \mathrm{d}\mathbf{p}^N \mathrm{d}\mathbf{r}^N \exp(-E(\mathbf{p}^N, \mathbf{r}^N)/k_{\mathrm{B}}T)}\right) \tag{5.21}$$

Next we recognise that the following integral is equal to one:

$$1 = \frac{1}{(8\pi^2 V)^N}\iint \mathrm{d}\mathbf{p}^N \mathrm{d}\mathbf{r}^N \exp\left(-\frac{E(\mathbf{p}^N, \mathbf{r}^N)}{k_{\mathrm{B}}T}\right)\exp\left(\frac{E(\mathbf{p}^N, \mathbf{r}^N)}{k_{\mathrm{B}}T}\right) \tag{5.22}$$

Inserting this into the expression for the free energy and ignoring the constants (which act to change the zero point from which the free energy is calculated) gives:

$$A = k_{\mathrm{B}}T \ln\left(\frac{\iint \mathrm{d}\mathbf{p}^N \mathrm{d}\mathbf{r}^N \exp\left(-\frac{E(\mathbf{p}^N, \mathbf{r}^N)}{k_{\mathrm{B}}T}\right)\exp\left(+\frac{E(\mathbf{p}^N, \mathbf{r}^N)}{k_{\mathrm{B}}T}\right)}{\iint \mathrm{d}\mathbf{p}^N \mathrm{d}\mathbf{r}^N \exp(-E(\mathbf{p}^N, \mathbf{r}^N)/k_{\mathrm{B}}T)}\right) \tag{5.23}$$

We can now substitute for the probability density, $\rho(\mathbf{p}^N, \mathbf{r}^N)$ in this equation, leading to the final result (in which we have again ignored the normalisation factors):

$$A = k_{\mathrm{B}}T \ln\left(\iint \mathrm{d}\mathbf{p}^N \mathrm{d}\mathbf{r}^N \exp\left(+\frac{E(\mathbf{p}^N, \mathbf{r}^N)}{k_{\mathrm{B}}T}\right)\rho(\mathbf{p}^N, \mathbf{r}^N)\right) \tag{5.24}$$

The important feature of this result is that the configurations with a high energy make a significant contribution to the integral due to the presence of the exponential term $\exp(+E(\mathbf{p}^N, \mathbf{r}^N)/k_{\mathrm{B}}T)$. A Monte Carlo or molecular dynamics simulation preferentially samples the *lower energy* regions of phase space. An ergodic trajectory would of course visit all of these high-energy regions, but in practice these will never be adequately sampled by a real simulation. The results for the free energy and other entropic properties will as a consequence be poorly converged and inaccurate.

To reiterate a point that we made earlier, these problems of accurately calculating the free energy and entropy do not arise for isolated molecules that have a small number of well characterised minima which can all be enumerated. The partition function for such systems can be obtained by standard statistical mechanical methods involving a summation over the minimum energy states, taking care to include contributions from internal vibrational motion.

5.4 Practical aspects of computer simulation

5.4.1 Setting up and running a simulation

There are significant differences between the molecular dynamics and Monte Carlo simulation methods, but the same general strategies are used to set up and run either type of simulation. The first task is to decide which energy model is to be used to describe the interactions within the system. Simulations are usually performed with relatively large numbers of atoms over many iterations or time steps. The intra- and intermolecular interactions are therefore almost always described using an empirical (i.e. molecular mechanics) energy model. Faster computers and new theoretical techniques do now enable simulations to be performed using models based only on quantum mechanics or mixed models based on molecular mechanics/quantum mechanics are discussed in sections 9.12 and 9.13. Having chosen an energy model, the simulation itself can be broken into four distinct stages. First, an initial configuration for the system must be established. An *equilibration phase* is then performed, during which the system evolves from the initial configuration. Thermodynamic and structural properties are monitored during the equilibration until stability is achieved. Several distinct steps may be required during the equilibration, particularly for inhomogeneous systems. At the end of the equilibration the *production phase* commences. It is during the production phase that simple properties of the system are calculated. At regular intervals the configuration of the system (i.e. the atomic coordinates) is output to a disk file. Finally, the simulation is analysed; properties not calculated during the simulation are determined and the configurations are examined, not only to discover how the structure of the system changed but also to check for any unusual behaviour that might indicate a problem with the simulation.

5.4.2 Choosing the initial configuration

Before a simulation can be performed it is obviously necessary to select an initial configuration of the system. This should be done with some care as the initial arrangement can often determine the success or failure of a simulation. For simulations of systems at equilibrium (the most common sort) it is wise to choose an initial configuration that is close to the state which it is desired to simulate. For example, it would be unwise to initiate a simulation of a face-centred cubic solid from a body-centred cubic starting point. It is also good practice to ensure that the initial configuration does not contain any high-energy interactions as these may cause instabilities in the simulation. Such 'hot spots' can often be eradicated by performing energy minimisation prior to the simulation itself.

To simulate homogeneous liquids which contain large numbers of the same molecule, a standard lattice structure is often chosen as the starting configuration. If an experimentally determined arrangement is available (e.g. an X-ray structure) then this could be used, provided that it was appropriate to the simulation being performed. When no experimental structure is available the initial configuration can be chosen from one of the common crystallographic lattices (simply placing molecules at random can often give rise to high-energy overlaps and instabilities). The most common lattice is the face-centred cubic lattice (fcc), shown in Figure 5.3. This structure contains $4M^3$ points ($M = 2, 3, 4, \ldots$). For this reason, simulations are often performed using 108, 256, 525, 784 ... atoms or molecules. The lattice size is chosen so that the density is appropriate to that of the system under study. For simulations of molecules it is also necessary to assign an orientation to each molecule. For small linear molecules, the solid structure of CO_2 is often chosen as the initial configuration. This is a face-centred cubic lattice with the molecules oriented in a regular fashion along the four diagonals of the unit cell. Alternatively, the orientations may be chosen completely at random or by making small random changes from the orientation in a regular lattice. At high densities non-physical overlaps may result, particularly if the molecules are large; in such cases it is more important to use an initial configuration that is close to the expected equilibrium distribution. For example, simulations of rod-shaped molecules such as liquid crystals are usually initiated from a configuration in which the molecules are all aligned approximately in the same direction.

For simulations of inhomogeneous systems comprising a solute molecule or intermolecular complex immersed in a solvent, the starting conformation of the solute may be obtained from an experimental technique such as X-ray crystallography or NMR, or may be generated by theoretical modelling. The

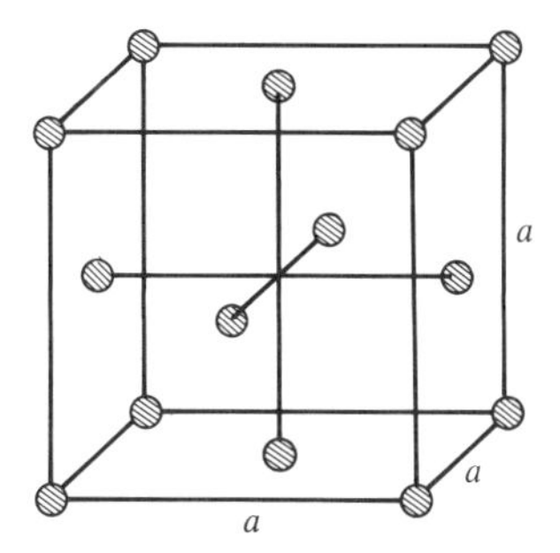

Fig. 5.3 The face-centred cubic cell.

coordinates of some solvent molecules may be known if the structure is obtained from X-ray crystallography but it is usually necessary to add other solvent molecules to give the appropriate solvent density. A typical approach is to use the coordinates obtained from a previous simulation of the pure solvent. The solute is immersed in the solvent 'bath' and any solvent molecules that are too close to the solute are then discarded before the calculation proceeds.

5.5 Boundaries

The correct treatment of boundaries and boundary effects is crucial to simulation methods because it enables 'macroscopic' properties to be calculated from simulations using relatively small numbers of particles. The importance of boundary effects can be illustrated by considering the following simple example. Suppose we have a cube of volume 1 litre which is filled with water at room temperature. The cube contains approximately 3.3×10^{25} molecules. Interactions with the walls can extend up to 10 molecular diameters into the fluid. The diameter of the water molecule is approximately 2.8 Å and so the number of water molecules that are interacting with the boundary is about 2×10^{19}. So only about 1 in 1.5 million water molecules are influenced by interactions with the walls of the container. The number of particles in a Monte Carlo or molecular dynamics simulation is far fewer than 10^{25}–10^{26}, and is frequently less than 1000. In a system of 1000 water molecules most, if not all of them would be within the influence of the walls of the boundary. Clearly, a simulation of 1000 water molecules in a vessel would not be an appropriate way to derive 'bulk' properties. The alternative is to dispense with the container altogether. Now, approximately three quarters of the molecules would be at the surface of the sample rather than being in the bulk. Such a situation would be relevant to studies of liquid drops, but not to studies of bulk phenomena.

5.5.1 Periodic boundary conditions

Periodic boundary conditions enable a simulation to be performed using a relatively small number of particles, in such a way that the particles experience forces as if they were in bulk fluid. Imagine a cubic box of particles which is replicated in all directions to give a periodic array. A two-dimensional box is shown in Figure 5.4. In the two-dimensional example each box is surrounded by 8 neighbours; in three dimensions each box would have 26 nearest neighbours. The coordinates of the particles in the image boxes can be computed simply by adding or subtracting integral multiples of the box sides. Should a particle leave the box during the simulation then it is replaced by an image particle that enters from the opposite side, as illustrated in Figure 5.4. The number of particles within the central box thus remains constant.

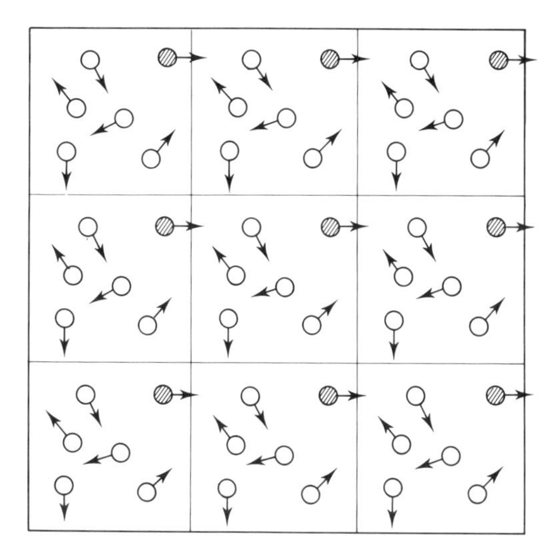

Fig. 5.4 Periodic boundary conditions in two dimensions.

The cubic cell is the simplest periodic system to visualise and to program. However, a cell of a different shape might be more appropriate for a given simulation. This may be particularly important for simulations of systems which comprise a single molecule or intermolecular complex surrounded by solvent molecules. In such systems it is usually the behaviour of the central solute molecule that is of most interest and so it is desirable that as little of the computer time is spent simulating the solvent far from the solute. In principle, any cell shape can be used provided it fills all of space by translation operations of the central box in three dimensions. Five shapes satisfy this condition: the cube (and its close relation, the parallelepiped), the hexagonal prism, the truncated octahedron, the rhombic dodecahedron and the 'elongated' dodecahedron (Figure 5.5) [Adams 1983]. It is often sensible to choose a periodic cell that reflects the underlying geometry of the system. For example, a rectangular cell is not the ideal choice to simulate an approximately spherical molecule. The truncated octahedron and the rhombic dodecahedron provide periodic cells that are approximately spherical and so may be more appropriate for simulations of spherical molecules. The distance between adjacent cells in the truncated octahedron or the rhombic dodecahedron is larger than the conventional cube for a system with a given number of particles and so a simulation using one of the spherical cells will require fewer particles than a comparable simulation using a cubic cell. Of the two approximately spherical cells, the truncated octahedron is often preferred as it is somewhat easier to program. The hexagonal prism can be used to simulate molecules with a cylindrical shape such as DNA.

Of the five possible shapes, the cube/parallelepiped and the truncated octahedron have been most widely used with some simulations in the hexagonal prism. The formulae used to translate a particle back into the central simulation box for these three shapes are given in Appendix 5.4. It may be preferable to use

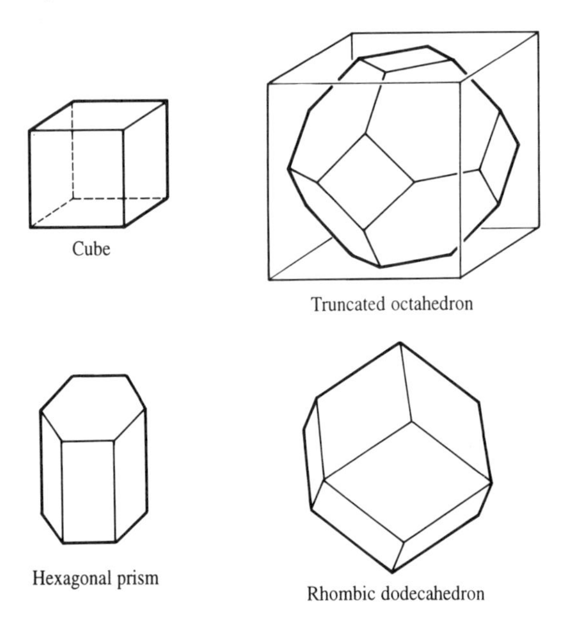

Fig. 5.5 Periodic cells used in computer simulations: the cube, parallelepiped, truncated octahedron, hexagonal prism and rhombic dodecahedron.

one of the more common periodic cells even if there are aesthetic reasons for using an alternative. This is because the expressions for calculating the images may be difficult and inefficient to implement, even though the simulation would use fewer atoms.

For some simulations it is inappropriate to use standard periodic boundary conditions in all directions. For example, when studying the adsorption of molecules onto a surface, it is clearly inappropriate to use the usual periodic boundary conditions for motion perpendicular to the surface. Rather, the surface is modelled as a true boundary, for example, by explicitly including the atoms in the surface. The opposite side of the box must still be treated; when a molecule strays out of the top side of the box it is reflected back into the simulation cell as indicated in Figure 5.6. Usual periodic boundary conditions apply to motion parallel to the surface.

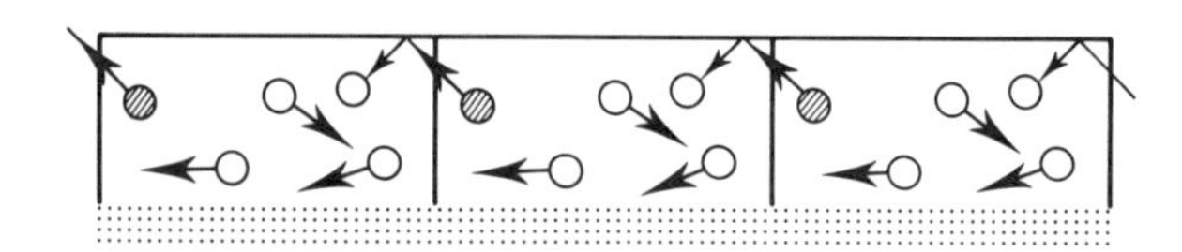

Fig. 5.6 Periodic boundary conditions for surface simulations. After Allen M P and D J Tildesley 1987. *Computer Simulation of Liquids*. Oxford, Oxford University Press.

Periodic boundaries are widely used in computer simulations, but they do have some drawbacks. A clear limitation of the periodic cell is that it is not possible to achieve fluctuations that have a wavelength greater than the length of the cell. This can cause problems in certain situations, such as near the liquid–gas critical point. The range of the interactions present in the system is also important; if the cell size is large compared to the range over which the interactions act then there should be no problems. For example, for the relatively short-range Lennard–Jones potential the cell should have a side greater than approximately 6σ which corresponds to about 20 Å for argon. For longer-range electrostatic interactions the situation is more difficult and it is often necessary to accept that some long-range order will be imposed upon the system. The effects of imposing a periodic boundary can be evaluated empirically by comparing the results of simulations performed using a variety of cell shapes and sizes.

5.5.2 Non-periodic boundary methods

Periodic boundary conditions are not always used in computer simulations. Some systems, such as liquid droplets or van der Waals clusters, inherently contain a boundary. Periodic boundary conditions may also cause difficulties when simulating inhomogeneous systems or systems that are not at equilibrium. In other cases the use of periodic boundary conditions would require a prohibitive number of atoms to be included in the simulation. This particularly arises in the study of the structural and conformational behaviour of macromolecules such as proteins and protein-ligand complexes. The first simulations of such systems ignored all solvent molecules due to the limited computational resources then available. This corresponds to the unrealistic situation of simulating an isolated protein *in vacuo* and then comparing the results with experimental data obtained in solution. Vacuum calculations can lead to significant problems. A vacuum boundary tends to minimise the surface area and so may distort the shape of the system if it is non-spherical. Small molecules may adopt more compact conformations when simulated *in vacuo* due to favourable intramolecular electrostatic and van der Waals interactions which would be dampened in the presence of a solvent.

As computer power has increased it has become possible to incorporate explicitly some solvent molecules and thereby simulate a more realistic system. The simplest way to do this is to surround the molecule with a 'skin' of solvent molecules. If the skin is sufficiently deep then the system is equivalent to a solute molecule inside a 'drop' of solvent. The number of solvent molecules in such cases is usually significantly fewer than would be required in the analogous periodic boundary simulation where the solute molecule is positioned at the centre of the cell and the empty space is filled with solvent. Boundary effects should be transferred from the molecule–vacuum interface to the solvent–vacuum interface, and so might be expected to result in a more realistic treatment of the solute. To illustrate these three situations, we can consider dihydrofolate reductase, which is a small enzyme that contains approximately 2500 atoms. If this enzyme is surrounded by water molecules in a cubic periodic system such that the surface of the protein is at least 10 Å from any side of the box, then the number of atoms rises to almost 20 000. If a shell 10 Å thick is used then the number of atoms falls to 14 700, and with a 5 Å shell the system contains 8900 atoms.

Sometimes we are only interested in a specific part of the solute, such as the active site of an enzyme. It has been common practice in such cases to divide the system into two regions (Figure 5.7). One region, often called the *reaction zone*, contains all atoms or groups within a given radius R_1 of the site of interest. The atoms in the reaction zone are subjected to the full simulation method. The second region (the *reservoir region*) contains all atoms outside the reaction zone but within a distance R_2 of the active site. The atoms in the reservoir region may be kept fixed in their initial positions, or may be restrained so that they stay within the shell defined by R_1 and R_2. Alternatively, they may be restrained to their initial positions using a harmonic potential. Any atoms further away from the active site than R_2 are discarded or may be kept fixed in their initial positions. It is important to be aware that restraining or fixing atoms in this way may prevent natural changes from occurring and so lead to artificial behaviour. A variety of schemes for performing simulations using such *stochastic boundary conditions* have been proposed. However, such methods can be rather complicated to implement and if not used properly can give anomalous results. If at all possible, a periodic boundary is the 'safest' way to ensure that boundary effects are minimised, but sometimes an alternative may be the only practical course.

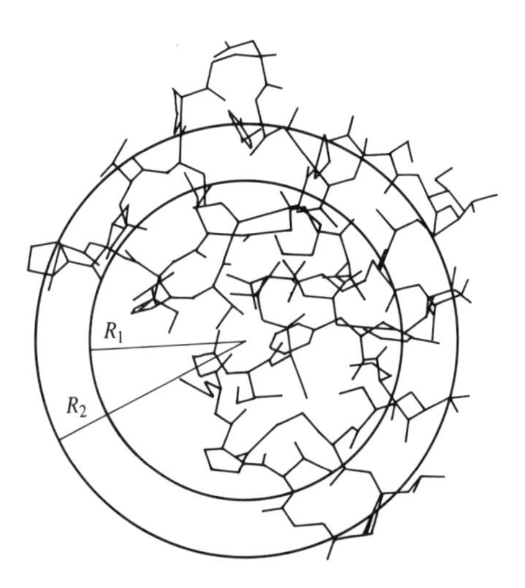

Fig. 5.7 Division into reaction zone and reservoir regions in a simulation using stochastic boundary conditions.

5.6 Monitoring the equilibration

The purpose of the equilibration phase is to enable the system to evolve from the starting configuration to reach equilibrium. Equilibration should continue until the values of a set of monitored properties become stable. The properties to be monitored usually include thermodynamic quantities such as the energy, temperature and pressure and also structural properties. Many simulations of the liquid state involve a starting configuration that corresponds to a solid lattice. It is therefore important to establish that the lattice has 'melted' before the production phase begins. *Order parameters* can be used to determine that the liquid state has been reached. An order parameter is a measure of the degree of order (or, equivalently, disorder) in the system. During a simulation of a crystal lattice the atoms would be expected to remain in approximately the same positions throughout and thereby maintain a high degree of order. In a liquid, however, we would expect considerable mobility of the species present, giving rise to translational disorder. One way to measure translational order in a system initially in a face-centred cubic lattice was suggested by Verlet, whose order parameter λ is:

$$\lambda = \tfrac{1}{3}[\lambda_x + \lambda_y + \lambda_z] \tag{5.25}$$

$$\lambda_x = \frac{1}{N}\sum_{\mathrm{i}=1}^{N} \cos\left(\frac{4\pi x_{\mathrm{i}}}{a}\right) \tag{5.26}$$

a is the length of one edge of the unit cell. Initially, all of the coordinates x_{i}, y_{i} and z_{i} are multiples of $a/2$ and so the order parameter has a value of one. As the simulation proceeds the order parameter should gradually decrease to a value of zero, indicating that the atoms are distributed randomly. When equilibrium has been reached the fluctuations in the order parameter should be proportional to $1/\sqrt{N}$, where N is the size of the system. A typical result is shown in Figure 5.8 for an argon simulation.

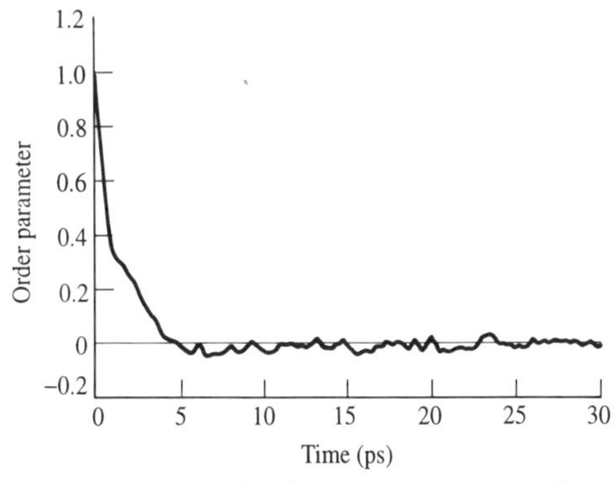

Fig. 5.8 Variation in Verlet order parameter during the equilibration phase of a molecular dynamics simulation of argon.

For molecules, it is also necessary to consider their orientations, which can be monitored using a rotational order parameter. For some systems, such as carbon monoxide or water, complete disorder would be expected in the liquid state at equilibrium. However, if we were simulating a dense fluid of rod-shaped molecules which form a liquid crystalline phase then we might expect that, on average, the molecules would tend to line up in a common direction. The Viellard–Baron rotational order parameter for linear molecules is calculated using the following formula:

$$P_1 = \frac{1}{N}\sum_{\mathrm{i}=1}^{N} \cos \gamma_{\mathrm{i}} \tag{5.27}$$

γ_{i} is the angle between the current and original direction of the molecular axis of molecule i. A value of 1 indicates that the molecules are perfectly aligned. Rotational disorder is indicated by a value of zero. The fluctuations about the average value should again be proportional to $1/\sqrt{N}$. For non-linear molecules, a number of rotational order parameters can be defined and each monitored.

The *mean squared displacement* also provides a means to establish whether a solid lattice has melted. The mean squared displacement is given by:

$$\Delta r^2(t) = \frac{1}{N}\sum_{\mathrm{i}=1}^{N} [\mathbf{r}_{\mathrm{i}}(t) - \mathbf{r}_{\mathrm{i}}(0)]^2 \tag{5.28}$$

For a fluid, with no underlying regular structure, the mean squared displacement gradually increases with time, Figure 5.9. For a solid, however, the mean squared displacement oscillates about a mean value.

The radial distribution function can also be used to monitor the progress of the equilibration. This function is particularly useful for detecting the presence of two phases. Such a situation is characterised by a larger than expected first peak and by the fact that $g(r)$ does not decay towards a value of 1 at long distances. If two-phase behaviour is inappropriate then the simulation should probably be

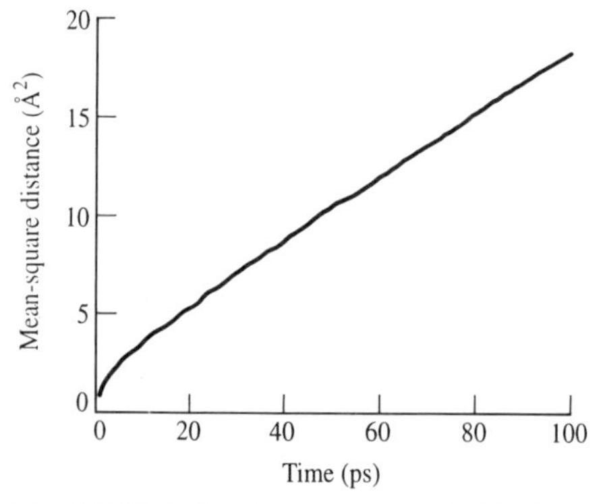

Fig. 5.9 Variation in mean squared displacement during the initial steps of a molecular dynamics simulation of argon.

terminated and examined. If, however, a two-phase system is desired, then a long equilibration phase is usually required.

5.7 Truncating the potential and the minimum image convention

The most time-consuming part of a Monte Carlo or molecular dynamics simulation (or, indeed, of an energy minimisation) is the calculation of the non-bonded energies and/or forces. The numbers of bond stretching, angle bending and torsional terms in a force field model are all proportional to the number of atoms but the number of non-bonded terms that need to be evaluated increases as the square of the number of atoms (for a pairwise model) and is thus of order N^2. In principle, the non-bonded interactions are calculated between every pair of atoms in the system. However, for many interaction models this is not justified. The Lennard–Jones potential falls off very rapidly with distance: at 2.5σ the Lennard–Jones potential has just 1% of its value at σ. This reflects the r^{-6} distance dependence of the dispersion interaction. The most popular way to deal with the non-bonded interactions is to use a *non-bonded cutoff* and to apply the *minimum image convention*. In the minimum image convention, each atom 'sees' at most just one image of every other atom in the system (which is repeated infinitely via the periodic boundary method). The energy and/or force is calculated with the closest atom or image, as illustrated in Figure 5.10. When a cutoff is employed, the interactions between all pairs of atoms that are further apart than the cutoff value are set to zero taking into account the closest image. When periodic boundary conditions are being used, the cutoff should not be so large that a particle sees its own image, or indeed the same molecule twice. This has the effect of limiting the cutoff to no more than half the length of the cell

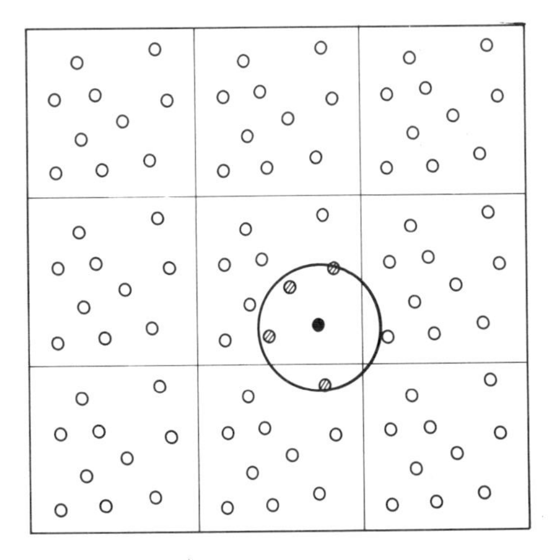

Fig. 5.10 The spherical cutoff and the minimum image convention.

when simulating atomic fluids in a cubic cell. For rectangular cells the cutoff should be no greater than half the length of the shortest side. For simulations of molecules the upper limit on the cutoff may also be affected by the size of the molecules, as we shall see below in section 5.7.2. In simulations where the Lennard–Jones potential is the only non-bonded interaction, a cutoff of 2.5σ gives rise to a relatively small error. However, when long-range electrostatic interactions are involved, the cutoff should be much greater and indeed there is evidence to suggest that using any cutoff leads to errors. A value of at least 10 Å is generally recommended but even this may be insufficient. More comprehensive methods have been devised for dealing with the electrostatic interactions which are considered in section 5.8.

5.7.1 Non-bonded neighbour lists

By itself, the use of a cutoff may not dramatically reduce the time taken to compute the number of non-bonded interactions. This is because we would still have to calculate the distance between every pair of atoms in the system, simply to decide whether they are close enough to calculate their interaction energy. Calculating all the $N(N-1)$ distances takes almost as much time as calculating the energy itself.

In simulations of fluids, an atom's neighbours (i.e. those atoms that are within the cutoff distance) do not change significantly over 10 or 20 molecular dynamics time steps or Monte Carlo iterations. If we 'knew' which atoms to include in the non-bonded calculation (for example, by storing them in an array), then it would be possible to identify directly each atom's neighbours without having to calculate the distances to all the other atoms in the system. The *non-bonded neighbour* list (first proposed by Verlet) is just such a device. The Verlet neighbour list [Verlet 1967] stores all atoms within the cutoff distance, together with all atoms that are slightly further away than the cutoff distance. This is most efficiently done using a large neighbour list array L, and a pointer array, P. The pointer array indicates where in the neighbour list the first neighbour for that atom is located. The last neighbour of atom i is stored in element $P[i+1]-1$ of the neighbour list as shown in Figure 5.11. Thus the neighbours of atom i are stored in elements $L[P[i]]$ through $L[P[i+1]-1]$ of the array L. The neighbour list is updated at regular intervals throughout the simulation. Between updates, the neighbour and pointer lists are used to directly identify the nearest neighbours of each atom i. The distance used to calculate each atom's neighbours should be larger than the actual non-bonded cutoff distance so that no atom, initially outside the neighbour cutoff, approaches closer than the non-bonded cutoff distance before the neighbour list is updated again.

It is important to update the neighbour list at the correct frequency. If the update frequency is too high the procedure is inefficient; too low and the energies and forces may be calculated incorrectly due to atoms moving within the non-bonded cutoff region. An update frequency between 10 and 20 steps is common. An algorithm that can automatically update the neighbour list and so circumvent

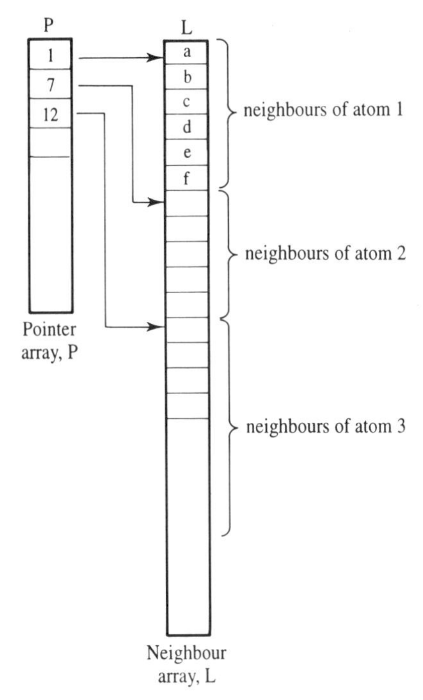

Fig. 5.11 Pointer and neighbour arrays can be used to implement the Verlet neighbour list.

these problems is as follows [Thompson 1983]. An array element is set to zero for each atom whenever the neighbour list is updated. This array is used to store the displacement of each atom or molecule in subsequent steps. When the sum of the maximum displacements of any two atoms exceeds the difference between the non-bonded cutoff distance and the neighbour list distance, then it is time to update the neighbour list again.

There are no fixed rules that determine how much larger the neighbour list cutoff should be than the non-bonded cutoff. Clearly there will be a trade-off between the size of the cutoff and the frequency at which the neighbour list must be updated: the larger the difference, the less frequently will the neighbour list have to be updated. There may also be storage implications if the list is too large.

When the number of molecules in the simulation is very large, it can require a significant computational effort just to update the neighbour list. This is because the standard way to update the neighbour list requires the distance between all pairs of atoms in the system to be calculated. When the size of the system is much larger than the cutoff distance, a *cell index method* can be used to make the updating procedure more efficient. In the cell index method, the simulation box is divided into a number of cells. The length of each cell is longer than the non-bonded cutoff distance. All of the neighbours of an atom will then be found either in the cell containing the atom, or in one of the surrounding cells. If the entire system is divided into M^3 cells, there will be an average of N/M^3 molecules in each cell. To determine the neighbours of a given atom or molecule, it is then necessary to consider just $27N/M^3$ atoms rather than N. The cell index method requires a mechanism for identifying the atoms or molecules in each cell. Two arrays can be used to do this: a linked list array L and a pointer array P. The pointer array indicates the location of one of the atoms or molecules in a given cell. Thus, P[1] would indicate the number of the 'first' atom or molecule in cell 1 and P[2] is the number of the 'first' atom in cell 2. Each element of the linked list array then gives the number of the 'next' atom or molecule in the cell. Thus the value stored in L[1] is the number of the second atom in the first cell. Suppose P[1] is atom 10. Then the value stored in L[10] is the second atom in the cell. If this second atom is number 15, then L[15] contains the third atom in the cell. The last molecule in the sequence is identified by the fact that its array element is zero. The cell index method clearly requires a mechanism for updating the pointer and linked list arrays when atoms or molecules move from one cell to another, which can add to the complexity.

When simulating species with a significant electrostatic contribution, it may be desirable to use different cutoffs for the electrostatic and van der Waals interactions. This is because the electrostatic interaction has a much longer range. Using a longer cutoff for the electrostatic interactions will of course significantly increase the number of pairs that must be calculated. A compromise is to use a *twin-range method*, in which two cutoffs are specified. All interactions below the lower cutoff are calculated as normal at each step. Interactions due to atoms between the lower and upper cutoffs are evaluated only when the neighbour list is updated and are kept constant between these updates. The rationale here is that the contribution of the atoms that are further away will not vary much between updates.

The use of a cutoff is amply justified in many cases, if only on the grounds of expediency, but its use will always lead to some fraction of the potential energy being ignored. This lost energy can be easily captured at the end of the simulation, if it is assumed that the radial distribution function takes the value of 1 at distances greater than the cutoff. The calculation is analogous to that used to determine the total energy from the radial distribution function, equation (5.16) but the integration is now performed between the cutoff distance r_c and infinity and $g(r)$ is now taken to be 1 in this range. For N particles, the correction is:

$$E_{\text{correction}} = 2\pi\rho N \int_{r_c}^{\infty} r^2 \mathscr{V}(r)\mathrm{d}r \qquad (5.29)$$

For the Lennard–Jones potential the long-range contribution can be determined analytically:

$$E_{\text{correction}} = 8\pi\rho N\varepsilon\left[\frac{\sigma^{12}}{9r^9} - \frac{\sigma^6}{3r^3}\right] \qquad (5.30)$$

5.7.2 Group-based cutoffs

When simulating large molecular systems, it is often advantageous to use a group-based cutoff (sometimes called a residue-based cutoff). Here, the large molecules are divided into 'groups', each of which contains a relatively small number of connected atoms. If the calculation involves small solvent molecules then each solvent molecule is also conveniently regarded as a single unconnected group. Why are groups useful? Let us consider the electrostatic interaction between two water molecules. In the popular TIP3P model there is a charge of $-0.834e$ on the oxygen, and $0.417e$ on each hydrogen. The electrostatic interaction between two water molecules is calculated as the sum of nine distinct site–site interactions. If we start from the minimum energy arrangement for the water dimer shown in Figure 5.12 and gradually move one water molecule relative to the other as indicated then the electrostatic energy varies as shown in the graph in Figure 5.13.

Although the overall interaction energy is relatively small beyond 6 Å or so, each of these energies is the sum of several rather large terms; for example, at an O–O separation of 8 Å, the overall interaction energy is about -0.27 kcal/mol but this comprises an oxygen–oxygen interaction of approximately 29 kcal/mol, oxygen–hydrogen interactions of -59.4 kcal/mol and hydrogen–hydrogen interactions of 29.2 kcal/mol. Suppose that a simple atom-based non-bonded cutoff is applied to the water dimer. The interaction energy then fluctuates violently near the cutoff distance, as shown in Figure 5.14 for a cutoff of 8 Å. This is because only some of the pairwise interactions are included in this case. Clearly such a model would almost certainly lead to serious problems for any simulation. This problem can be avoided by collecting all of the atoms from each water molecule into a single group, and by calculating the interactions on a group–group basis, even though some of the atom pairs may have a separation larger than the cutoff.

How should a molecule be divided into groups? In some cases there may appear to be a chemically obvious way to define the groups, especially when the molecule is a polymer that is constructed from distinct chemical residues. A particularly desirable feature is that each group should, if possible, be of zero charge. The reason for this can be understood if we recall how the different electrostatic interactions vary with distance from section 3.8.1:

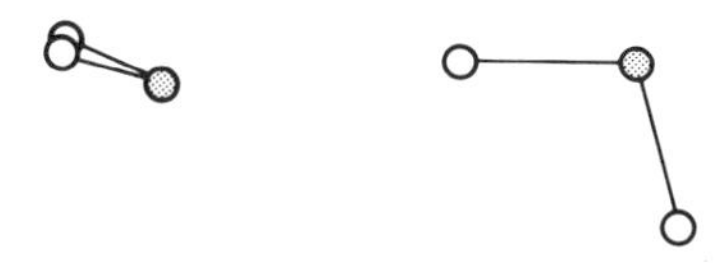

Fig. 5.12 Minimum energy structure for water dimer with TIP3P model.

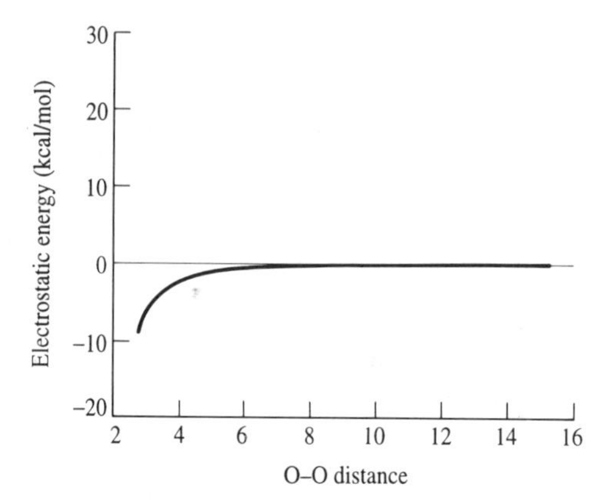

Fig. 5.13 The variation in the electrostatic interaction energy of the water dimer as a function of the O–O distance without a cutoff.

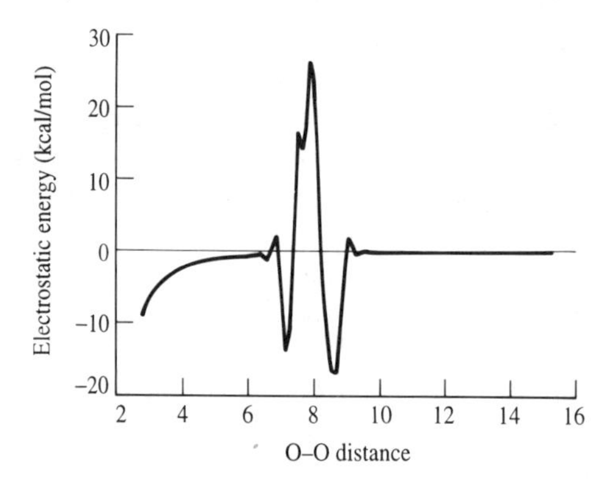

Fig. 5.14 The variation in interaction of the water dimer as a function of the O–O distance with an 8 Å atom-based cutoff.

charge–charge $\sim 1/r$
charge–dipole $\sim 1/r^2$
dipole–dipole $\sim 1/r^3$
dipole–quadrupole $\sim 1/r^4$
charge–induced dipole $\sim 1/r^4$
dipole–induced dipole $\sim 1/r^6$

If the groups are electrically neutral, then the leading term in the electrostatic interaction between a pair of groups is the dipole–dipole interaction, which is dependent upon $1/r^3$. By comparison, the charge–charge terms vary as $1/r$. Of course, it is not always possible to arrange atoms in neutral groups as occurs when some of the species are charged.

A further question with the group-based scheme is: How do we decide whether a particular group–group interaction needs to be considered? In other words, how are cutoffs included into the group scheme? One strategy is to include a particular group–group interaction if any pair of atoms in the two groups are closer than the cutoff distance. Alternatively, a 'marker' atom may be nominated within the group; when the marker atoms come closer than the cutoff then the appropriate group–group interaction is included. When using marker atoms, it is important that the groups are not too large; thus the groups used by some simulation programs are much smaller than the 'chemically obvious' groupings. For example the most obvious choice for proteins and peptides is to define each entire amino acid residue as a single group. However, this is not necessarily the most appropriate strategy. Consider the situation in which two arginine residues are spatially close together, Figure 5.15. Arginine has a long side chain that is comparable in length to the non-bonded cutoff distances often employed. Suppose the alpha-carbon atom (marked C_α in Figure 5.15) in each arginine residue is chosen as the marker atom. If the distance between the alpha-carbons of two arginine residues is greater than the cutoff, no interactions between any atoms in the arginine residues would be calculated, despite the fact that the positively charged ends of the residues could be very close, as shown in Figure 5.15. Were the alpha-carbons to approach closer than the cutoff, there would then be a dramatic increase in the energy due to the unfavourable interaction between the two side chains, inevitably leading to an unstable simulation. It may therefore be appropriate to define 'charge groups' that contain smaller numbers of atoms than are in the chemically obvious scheme. For example, the groups that are used by the GROMOS simulation program for the amino acids arginine and asparagine are shown in Figure 5.16.

5.7.3 *Problems with cutoffs and how to avoid them*

A cutoff introduces a discontinuity in both the potential energy and the force near the cutoff value. This creates problems, especially in molecular dynamics simulations where energy conservation is required. There are several ways that the effects of this discontinuity can be counteracted. One approach is to use a shifted potential, in which a constant term is subtracted from the potential at all values, Figure 5.17:

$$\mathscr{v}'(r) = \mathscr{v}(r) - \mathscr{v}_c \qquad r \leq r_c \tag{5.31a}$$

$$\mathscr{v}'(r) = 0 \qquad r > r_c \tag{5.31b}$$

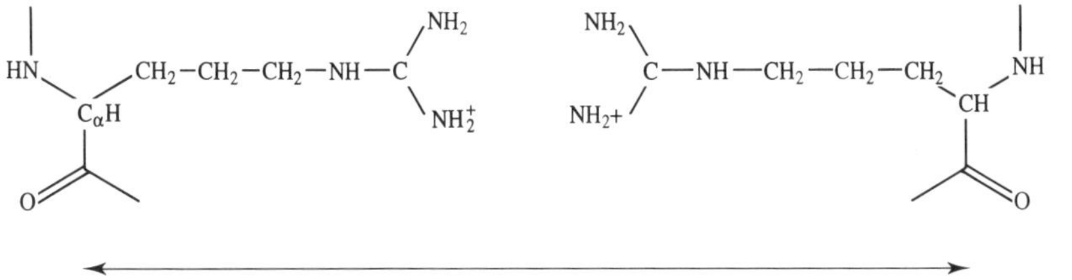

HN–C_αH–CH_2–CH_2–CH_2–NH–C(NH_2)(NH_2^+) NH_2(NH_2+)C–NH–CH_2–CH_2–CH_2–CH–NH

Fig. 5.15 The use of a marker atom on the alpha-carbon in an arginine residue may lead to a significant electrostatic interaction being neglected because the distance between the marker atoms exceeds the cutoff.

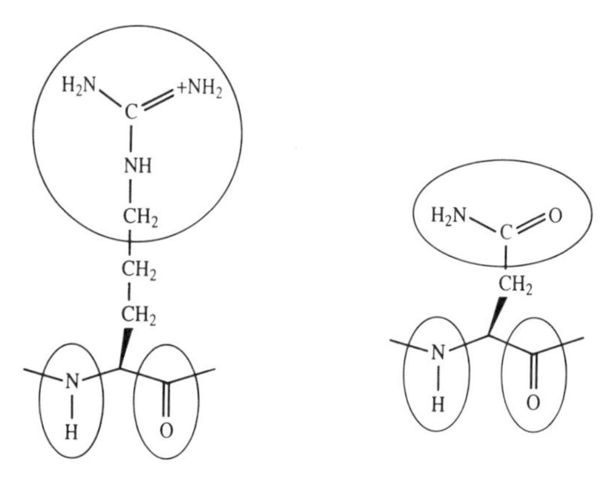

Fig. 5.16 The charge groups used in the GROMOS simulation program for simulating proteins [van Gunsteren and Berendsen 1986], illustrated using the amino acid acids arginine and asparaginine. The CH_2 groups have zero charge.

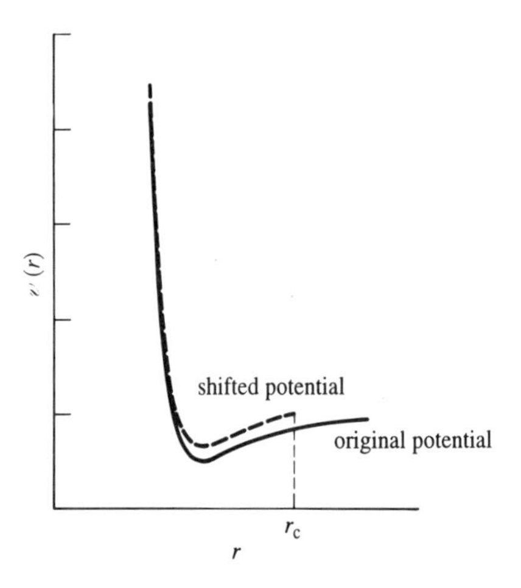

Fig. 5.17 A shifted Lennard–Jones potential.

r_c is the cutoff distance and $\mathscr{V}_c$ is equal to the value of the potential at the cutoff distance. As the additional term is constant, it disappears when the potential is differentiated and so does not affect the force calculation in molecular dynamics. Use of the shifted potential does improve energy conservation, though as the number of atom pairs separated by a distance smaller than the cutoff varies, so the contribution of the shifted potential to the total energy will change. An additional problem is that there is a discontinuity in the force with the shifted potential; at the cutoff distance, the force will have a finite value which drops suddenly to zero just beyond the cutoff. This can also give instabilities in a simulation. To avoid this, a linear term can be added to the potential, making the derivative zero at the cutoff:

$$\upsilon'(r) = \upsilon(r) - \upsilon_c - \left(\frac{d\upsilon(r)}{dr}\right)_{r=r_c}(r - r_c) \qquad r \le r_c \qquad (5.32a)$$

$$\upsilon'(r) = 0 \qquad r > r_c \qquad (5.32b)$$

The shift makes the potential deviate from the 'true' potential, and so any calculated thermodynamic properties will be changed. The 'true' values can be retrieved but it is difficult to do so, and the shifted potential is thus rarely used in 'real' simulations. Moreover, while it is relatively straightforward to implement for a homogeneous system under the influence of a simple potential such as the Lennard–Jones potential, it is not easy for inhomogeneous systems containing many different types of atoms.

An alternative way to eliminate discontinuities in the energy and force equations is to use a *switching function*. A switching function is a polynomial function of the distance by which the potential energy function is multiplied. Thus the switched potential $\upsilon'(r)$ is related to the true potential $\upsilon(r)$ by $\upsilon'(r) = \upsilon(r)S(r)$. Some switching functions are applied to the entire range of the potential up to the cutoff point. One such function is:

$$\upsilon'(r) = \upsilon(r)\left[1 - 2\left(\frac{r}{r_c}\right)^2 + \left(\frac{r}{r_c}\right)^4\right] \qquad (5.33)$$

The switching function has a value of 1 at $r = 0$ and a value of 0 at $r = r_c$, the cutoff distance. Between these two values it varies as shown in Figure 5.18, which also shows how the potential function is affected.

A switching function applied to the potential function over the entire range does have drawbacks; for example, equilibrium structures are affected (the minimum energy separation for the argon dimer decreases slightly). A more acceptable alternative is to gradually taper the potential between two cutoff values. The potential takes its usual value until the lower cutoff distance. Between the lower (r_l) and upper cutoff distance (r_u) the potential is multiplied by the switching function, which takes the value 1 at the lower cutoff distance and 0 at the upper cutoff distance. The lower cutoff distance is typically relative close to the upper cutoff distance (for example, r_l might be 9 Å and r_u 10 Å). A simple switching function has the following linear form:

$$S = 1.0 \qquad r_{ij} < r_l \qquad (5.34a)$$

$$S = (r_u - r_{ij})/(r_u - r_l) \qquad r_l \le r_{ij} \le r_u \qquad (5.34b)$$

$$S = 0.0 \qquad r_u < r_{ij} \qquad (5.34c)$$

This suffers from discontinuities in both the energy and the force at the two cutoff values. A more acceptable switching function smoothly changes from a

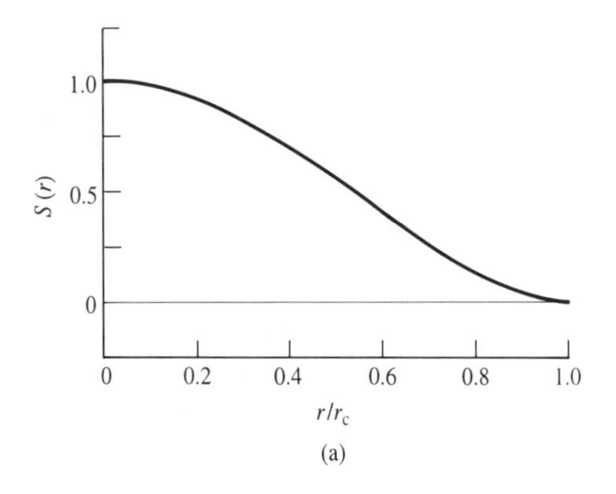

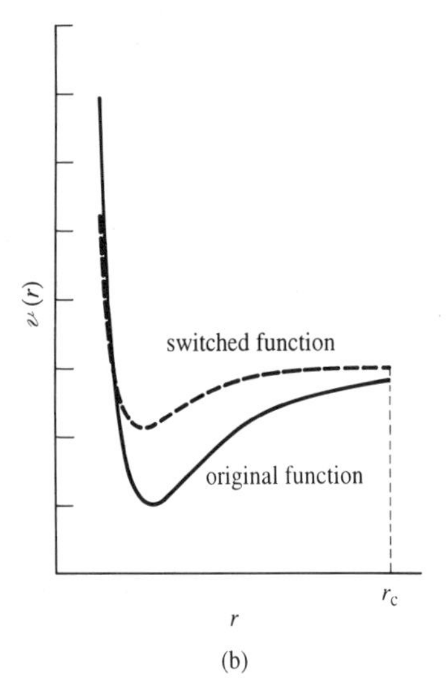

Fig. 5.18 (a) The effect of a switching function that applies over the entire range and (b) its effect on the Lennard–Jones potential.

value of 1 to a value of 0 (Figure 5.19) between r_l and r_u and satisfies the following requirements:

$$S_{r=r_l} = 1; \quad \left(\frac{\mathrm{d}S}{\mathrm{d}r}\right)_{r=r_l} = 0; \quad \left(\frac{\mathrm{d}^2S}{\mathrm{d}r^2}\right)_{r=r_l} = 0 \tag{5.35a}$$

$$S_{r=r_u} = 0; \quad \left(\frac{\mathrm{d}S}{\mathrm{d}r}\right)_{r=r_u} = 0; \quad \left(\frac{\mathrm{d}^2S}{\mathrm{d}r^2}\right)_{r=r_u} = 0 \tag{5.35b}$$

By ensuring that the first derivative is zero at the endpoints the force also approaches zero smoothly. A continuous second derivative is required to ensure

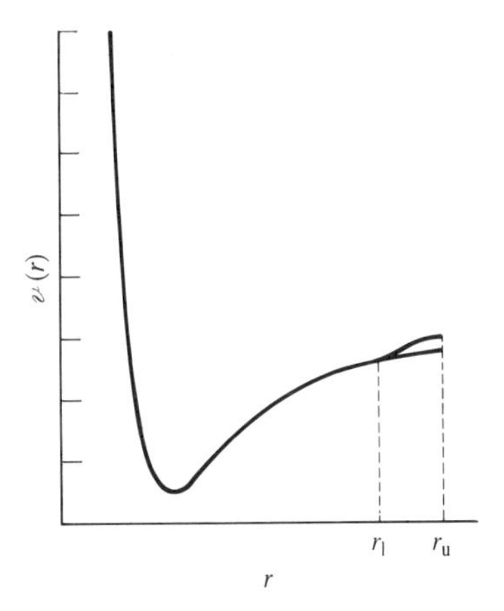

Fig. 5.19 A switching function that applies over a narrow range near the cutoff and its effect on the Lennard–Jones potential.

that the integration algorithm works properly. If the switch function is assumed to take the following form:

$$S(r) = c_0 + c_1\left[\frac{r-r_l}{r_u-r_l}\right] + c_2\left[\frac{r-r_l}{r_u-r_l}\right]^2 + c_3\left[\frac{r-r_l}{r_u-r_l}\right]^3 + c_4\left[\frac{r-r_l}{r_u-r_l}\right]^4 + c_5\left[\frac{r-r_l}{r_u-r_l}\right]^5 \quad (5.36)$$

then the following values of the coefficients $c_0 \ldots c_5$ satisfy the six requirements in equations (5.35a) and (5.35b):

$$c_0 = 1; \quad c_1 = 0; \quad c_2 = 0; \quad c_3 = -10; \quad c_4 = 15; \quad c_5 = -6 \quad (5.37)$$

When using a switching function in a molecular simulation with a residue-based cutoff it is important that the function has the same value for all pairs of atoms in the two interacting groups. Otherwise, severe fluctuations in the energy can arise when the separation is within the cutoff region. These two contrasting situations can be formally expressed as follows:

$$\textit{atom based}: \mathcal{V}_{\mathrm{AB}} = \sum_{\mathrm{i}=1}^{N_{\mathrm{A}}}\sum_{\mathrm{j}=1}^{N_{\mathrm{B}}} S_{\mathrm{ij}}(r_{\mathrm{ij}})\upsilon_{\mathrm{ij}}(r_{\mathrm{ij}}) \quad (5.38a)$$

$$\textit{residue or molecule based}: \mathcal{V}_{\mathrm{AB}} = S_{\mathrm{AB}}(|\mathbf{r}_{\mathrm{A}} - \mathbf{r}_{\mathrm{B}}|)\sum_{\mathrm{i}=1}^{N_{\mathrm{A}}}\sum_{\mathrm{j}=1}^{N_{\mathrm{B}}} \upsilon_{\mathrm{ij}}(r_{\mathrm{ij}}) \quad (5.38b)$$

N_{A} and N_{B} are the numbers of atoms in the two groups A and B and S is the switching function. With the group-based switching function, it is necessary to define the 'distance' between the two groups (i.e. the two points $\mathbf{r}_{\mathrm{A}}$ and $\mathbf{r}_{\mathrm{B}}$). There is no definitive way to do this. As with cutoffs, a special marker atom can be nominated within each residue, or the centre of mass, centre of geometry or centre of charge may be used.

Group-based switching functions have several advantages. Better energy conservation can be achieved, and there are advantages when performing energy minimisation since the potential is defined analytically at all points. However, it is important to beware of possible problems with group-based switching functions when the groups are large. We have already seen how this can arise when an ordinary group-based cutoff is used. Let us re-examine our arginine problem (Figure 5.15) when a switching function is employed. When the two marker atoms have a separation only slightly less than the upper switch cutoff, the switching function will be close to zero, and so there will not be the dramatic increase in energy that is observed with the simple cutoff. Nevertheless, although the switching function does help to prevent the simulation from 'blowing up', the representation of the energy and the forces in the system is still unsatisfactory. The only real alternative being to make the groups smaller or dispense with cutoffs altogether.

5.8 Long-range forces

Those interactions that decay no faster than r^{-n}, where n is the dimensionality of the system, can be a problem as their range is often greater than half the box length. The charge–charge interaction, which decays as r^{-1} is particularly problematic in molecular simulations. There is increasing evidence that it is important to properly model these long-range forces which are particularly acute when simulating charged species such a molten salts (when it is not possible to construct neutral groups). A proper treatment of long-range forces can also be important when calculating certain properties, such as the dielectric constant. One way to tackle the errors introduced by an inadequate treatment of long-range forces would be to use a much larger simulation cell, but this is usually impractical. Nevertheless, increasing computer power does mean that more rigorous ways of dealing with long-range forces can be considered, even in simulations of large systems. A variety of methods have been developed to handle long-range forces. The methods that we will discuss in detail are the Ewald summation, the reaction field method and the cell multiple method.

5.8.1 The Ewald summation method

The Ewald sum was first devised by Ewald [Ewald 1921] to study the energetics of ionic crystals. In this method, a particle interacts with all the other particles in the simulation box and with all of their images in an infinite array of periodic cells. Figure 5.20 illustrates how the array of simulation cells is constructed; in the limit the cell array is considered to have a spherical shape. The position of each image box (assumed for simplicity to be cube of side L containing N

charges) can be related to the central box by specifying a vector, each of whose components is an integral multiple of the length of the box, $(\pm \mathrm{i}L, \pm \mathrm{j}L, \pm \mathrm{k}L)$; i, j, k = 0, 1, 2, 3, etc. The charge–charge contribution to the potential energy due to all pairs of charges in the central simulation box can be written:

$$\mathscr{V} = \frac{1}{2}\sum_{\mathrm{i}=1}^{N}\sum_{\mathrm{j}=1}^{N}\frac{q_\mathrm{i}q_\mathrm{j}}{4\pi\varepsilon_0 r_\mathrm{ij}} \tag{5.39}$$

r_ij is the minimum distance between the charges i and j. There are six boxes at a distance L from the central box with coordinates ($\mathbf{r}_\mathrm{box}$) given by $(0, 0, L)$, $(0, 0, -L)$, $(0, L, 0)$, $(0, -L, 0)$, $(L, 0, 0)$ and $(-L, 0, 0)$ (only four of these are shown in the two-dimensional picture in Figure 5.20). The contribution of the charge–charge interaction between the charges in the central box and all images of all particles in these six surrounding boxes is given by:

$$\mathscr{V} = \frac{1}{2}\sum_{\mathrm{nbox}=1}^{6}\sum_{\mathrm{i}=1}^{N}\sum_{\mathrm{j}=1}^{N}\frac{q_\mathrm{i}q_\mathrm{j}}{4\pi\varepsilon_0|\mathbf{r}_\mathrm{ij} + \mathbf{r}_\mathrm{box}|} \tag{5.40}$$

In general, for a box which is positioned at a cubic lattice point $\mathbf{n}$ $(=(n_xL, n_yL, n_zL)$ with n_x, n_y, n_z being integers):

$$\mathscr{V} = \frac{1}{2}\sum_{\mathbf{n}}\sum_{\mathrm{i}=1}^{N}\sum_{\mathrm{j}=1}^{N}\frac{q_\mathrm{i}q_\mathrm{j}}{4\pi\varepsilon_0|\mathbf{r}_\mathrm{ij} + \mathbf{n}|} \tag{5.41}$$

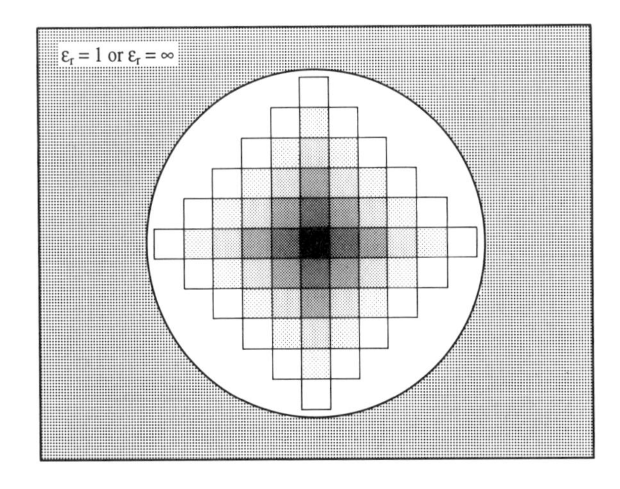

Fig. 5.20 The construction of a system of periodic cells in the Ewald method. Figure adapted from Allen M P and D J Tildesley 1987. *Computer Simulation of Liquids*. Oxford, Oxford University Press.

$|\mathbf{n}|$ thus takes the values 1, $\sqrt{2}, \ldots$ This expression is often written in such a way to incorporate the interactions between pairs of charges in the central box (for which $|\mathbf{n}| = 0$):

$$\mathscr{V} = \frac{1}{2} \sum_{|\mathbf{n}|=0}^{\infty}{}' \sum_{i=1}^{N} \sum_{j=1}^{N} \frac{q_i q_j}{4\pi\varepsilon_0 |\mathbf{r}_{ij} + \mathbf{n}|} \tag{5.42}$$

The prime on the first summation indicates that the series does not include the interaction i = j for $\mathbf{n} = 0$.

The total energy thus comprises a contribution from the interactions in the central box together with the interactions between the central box and all image boxes. There is also a contribution from the interaction between the spherical array of boxes and the surrounding medium. The problem is that the summation in equation (5.42) converges extremely slowly and in fact it is *conditionally convergent*. A conditionally convergent series contains a mixture of positive and negative terms such that the positive terms alone form a divergent series (i.e. a series which does not have a finite sum) as do the negative terms when taken alone. The sum of a conditionally convergent series depends on the order in which its terms are considered.

The trick when calculating the Ewald sum is to convert the summation into two series, each of which converges much more rapidly. This is done by considering each charge to be surrounded by a neutralising charge distribution of equal

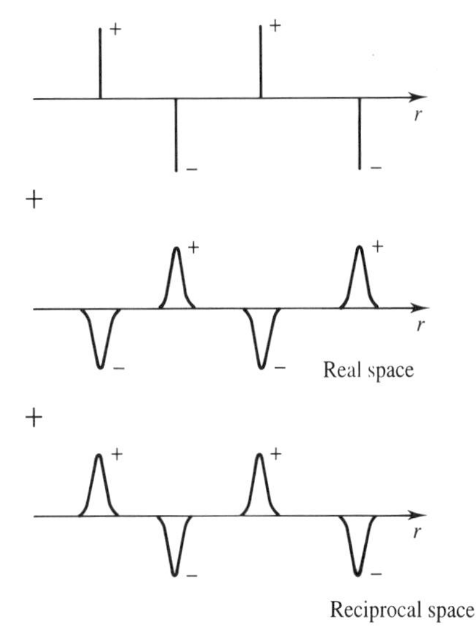

Fig. 5.21 In the Ewald summation method the initial set of charges are surrounded by a Gaussian distribution (calculated in real space) to which a cancelling charge distribution must be added (calculated in reciprocal space).

magnitude but of opposite sign as shown in Figure 5.21. A Gaussian charge distribution of the following functional form is commonly used:

$$\rho_i(\mathbf{r}) = \frac{q_i\alpha^3}{\pi^{\frac{3}{2}}}\exp(-\alpha^2 r^2) \tag{5.43}$$

The sum over point charges is now converted to a sum of the interactions between the charges *plus* the neutralising distributions. This dual summation (the 'real space' summation) is given by:

$$\mathscr{V} = \frac{1}{2}\sum_{i=1}^{N}\sum_{j=1}^{N}\sum_{|\mathbf{n}|=0}^{\infty}{}' \frac{q_i q_j}{4\pi\varepsilon_0}\frac{\text{erfc}(\alpha|\mathbf{r}_{ij}+\mathbf{n}|)}{|\mathbf{r}_{ij}+\mathbf{n}|} \tag{5.44}$$

erfc is the complementary error function, which is:

$$\text{erfc}(x) = \frac{2}{\sqrt{\pi}}\int_x^{\infty}\exp(-t^2)dt \tag{5.45}$$

The crucial point is that this new summation involving the error function converges very rapidly. The rate of convergence depends upon the width of the cancelling Gaussian distributions; the wider the Gaussian, the faster the series converges. Specifically, α should be chosen so that the only terms in the series (5.44) are those for which $|\mathbf{n}| = 0$ (i.e. only pairwise interactions involving charges in the central box, or if a cutoff is used α is chosen so that only interactions with other charges within the cutoff are included). A second charge distribution is now added to the system which exactly counteracts the first neutralising distribution (Figure 5.21). The contribution from this second charge distribution is:

$$\mathscr{V} = \frac{1}{2}\sum_{k\neq 0}\sum_{i=1}^{N}\sum_{j=1}^{N}\frac{1}{\pi L^3}\frac{q_i q_j}{4\pi\varepsilon_0}\frac{4\pi^2}{k^2}\exp\left(-\frac{k^2}{4\alpha^2}\right)\cos(\mathbf{k}\cdot\mathbf{r}_{ij}) \tag{5.46}$$

This summation is performed in *reciprocal space*, the details of which need not concern us here. The vectors $\mathbf{k}$ are reciprocal vectors and are given by $\mathbf{k} = 2\pi\mathbf{n}/L^2$. This reciprocal sum also converges much more rapidly than the original point-charge sum. However, the number of terms that must be included increases with the width of the Gaussians. There is thus a clear need to balance the real-space and reciprocal-space summations; the former converges more rapidly for large α whereas the latter converges more rapidly for small α. A value for α of $5/L$ and 100–200 reciprocal vectors $\mathbf{k}$ have been suggested as providing acceptable results.

The sum of Gaussian functions in real space includes the interaction of each Gaussian with itself. A third self-term must be subtracted:

$$\mathscr{V} = -\frac{\alpha}{\sqrt{\pi}}\sum_{k=1}^{N}\frac{q_k^2}{4\pi\varepsilon_0} \tag{5.47}$$

A fourth correction term may also be required, depending upon the medium that surrounds the sphere of simulation boxes. If the surrounding medium has an infinite relative permittivity (e.g. if it is a conductor) then no correction term is required. However, if the surrounding medium is a vacuum (with a relative permittivity of 1) then the following energy must be added:

$$\mathscr{V}_{\text{correction}} = \frac{2\pi}{3L^3}\left|\sum_{i=1}^{N}\frac{q_i}{4\pi\varepsilon_0}\mathbf{r}_i\right|^2 \tag{5.48}$$

The final expression is thus:

$$\mathscr{V} = \frac{1}{2}\sum_{i=1}^{N}\sum_{j=1}^{N}\left\{\begin{array}{l} \sum\limits_{|\mathbf{n}|=0}^{\infty}{}' \dfrac{q_i q_j}{4\pi\varepsilon_0}\dfrac{\operatorname{erfc}(\alpha|\mathbf{r}_{ij}+\mathbf{n}|)}{|\mathbf{r}_{ij}+\mathbf{n}|} \\ + \sum\limits_{k\neq 0}\dfrac{1}{\pi L^3}\dfrac{q_i q_j}{4\pi\varepsilon_0}\dfrac{4\pi^2}{k^2}\exp\left(-\dfrac{k^2}{4\alpha^2}\right)\cos(\mathbf{k}.\mathbf{r}_{ij}) \\ -\dfrac{\alpha}{\sqrt{\pi}}\sum\limits_{k=1}^{N}\dfrac{q_k^2}{4\pi\varepsilon_0} \\ +\dfrac{2\pi}{3L^3}\left|\sum\limits_{k=1}^{N}\dfrac{q_k}{4\pi\varepsilon_0}r_k\right|^2 \end{array}\right. \tag{5.49}$$

The Ewald sum is the most ‘correct’ way to accurately include all the effects of long-range forces into a computer simulation. It has been extensively used in simulations involving highly charged systems (such as ionic melts and in studies of processes in and on solids) and is increasingly being applied to other systems where electrostatic effects are important, such as lipid bilayers, proteins and DNA. Nevertheless, the Ewald method is not without problems and it tends to reinforce artifacts that arise from imposing periodic boundary conditions. For example, the method artificially results in each charge–charge interaction being minimised at a separation of half the box length. Instantaneous fluctuations in the simulation cell tend to be replicated throughout the infinite system rather than being damped out. The Ewald summation is computationally quite expensive to implement, though it is possible to speed up the calculation considerably. For example, effective ways have been devised for evaluating the summation in reciprocal space, so enabling a more appropriate Gaussian width α to be chosen.

5.8.2 *The reaction field and image charge methods*

In the reaction field method, a sphere is constructed around the molecule with a radius equal to the cutoff distance. The interaction with molecules that are within the sphere is calculated explicitly. To this is added the energy of interaction with the medium beyond the sphere, which is modelled as a homogeneous medium of

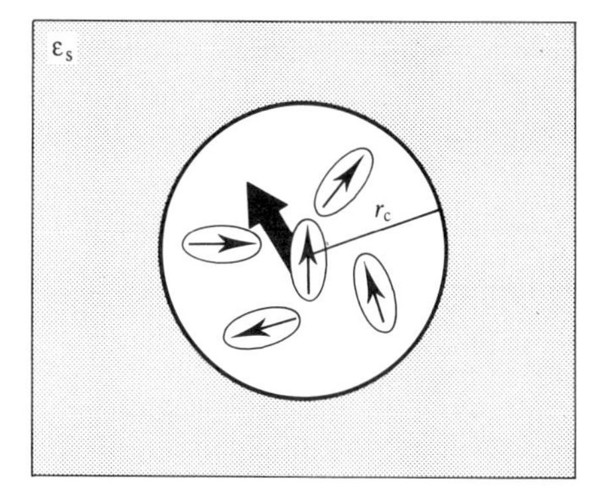

Fig. 5.22 The reaction field method. The shaded arrow represents the sum of the dipoles of the other molecules within the cutoff sphere.

dielectric constant ε_s (Figure 5.22). The electrostatic field due to the surrounding dielectric is given by:

$$\mathbf{E}_i = \frac{2(\varepsilon_s - 1)}{\varepsilon_s + 1}\left(\frac{1}{r_c^3}\right)\sum_{j;\, r_{ij} \leq r_c} \boldsymbol{\mu}_j \tag{5.50}$$

$\boldsymbol{\mu}_j$ are the dipoles of the neighbouring molecules that are within the cutoff distance (r_c) of the molecule i. The interaction between the molecule i and the reaction field equals $\mathbf{E}_i \cdot \boldsymbol{\mu}_i$, which is added to the short-range molecule–molecule interaction. Problems with the reaction field method may arise from discontinuities in the energy and/or force when the number of molecules j within the cavity of the molecule i changes. These problems can be avoided by employing a switching function for molecules that are near the reaction field boundary.

Similar approaches employ a single boundary for the entire system. This boundary may be spherical or may have a more complicated shape that better approximates the true molecular surface of the molecule. In the *image charge method*, a spherical boundary is employed and the reaction field due to a charge inside the boundary is considered to arise from a so-called image charge situated in the continuous dielectric beyond the sphere (Figure 5.23) [Friedman 1975]. If the position of the charge is $\mathbf{r}_i$ then the image charge is located at $(R/r_i)^2\mathbf{r}_i$ (where R is the radius of the bounding sphere) and has magnitude:

$$q_{im} = -\frac{(\varepsilon_s - \varepsilon_r)}{(\varepsilon_s + \varepsilon_r)}\frac{q_i R}{r_i} \tag{5.51}$$

ε_r and ε_s are the dielectric constants inside and outside the boundary respectively. This expression holds if the dielectric constant beyond the boundary is much greater than that inside ($\varepsilon_s \gg \varepsilon_r$). A drawback with this method is that as a charge approaches the boundary, so too does its image and the method breaks down.

The reaction field and image charge methods have the advantage of being conceptually simple, are relatively easy to implement and are computationally

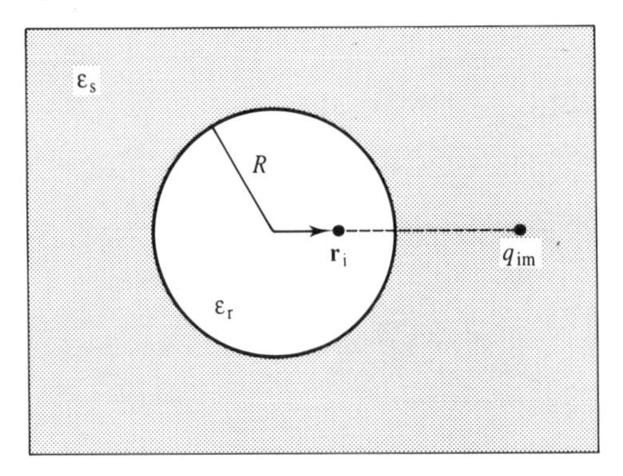

Fig. 5.23 The image charge method.

efficient. However, they do rely upon the assumption that molecules beyond the cutoff can be modelled as a continuous dielectric. This is not necessarily the case, but is often a reasonable assumption for homogeneous fluids. A value for the dielectric of the surrounding continuum must also be specified. This can be taken from experimental data, but the dielectric constant may be the property that one is trying to calculate! In practice, it is often necessary only to ensure that $\varepsilon_r \leq \varepsilon_s \leq \infty$. There are several ways in which the dielectric constant can be calculated from a computer simulation. A common approach is via the average of the square of the total dipole moment of the system, $\langle \mathbf{M}^2 \rangle$. With a reaction field boundary the dielectric constant ε_r is given by:

$$\frac{4\pi}{9} \frac{\langle \mathbf{M}^2 \rangle}{V k_B T} = \frac{(\varepsilon_r - 1)}{3} \frac{(2\varepsilon_s + 1)}{(2\varepsilon_s + \varepsilon_r)} \tag{5.52}$$

V is the volume of the simulation system. Even though the value of $\langle \mathbf{M}^2 \rangle$ can vary quite considerably with the reaction field dielectric ε_s, almost identical values of ε_r are obtained. An alternative approach is to determine the polarisation response of the liquid to an electric field $\mathbf{E}_0$. If the average dipole moment per unit volume along the direction of the applied field is $\langle \mathbf{P} \rangle$ then the dielectric constant is given by:

$$\frac{4\pi}{3} \frac{\langle \mathbf{P} \rangle}{\mathbf{E}_0} = \frac{(\varepsilon_r - 1)}{3} \frac{(2\varepsilon_s + 1)}{(2\varepsilon_s + \varepsilon_r)} \tag{5.53}$$

This perturbation method is claimed to be more efficient than the fluctuating dipole method, at least for certain water models [Alper and Levy 1989], but it is important to ensure that the polarisation $\langle \mathbf{P} \rangle$ is linear in the electric field strength to avoid problems with dielectric saturation.

5.9 The cell multipole method for non-bonded interactions

The cell multipole method (also called the fast multipole method) is an algorithm that enables *all* $N(N-1)$ pairwise non-bonded interactions to be enumerated in a time that scales linearly with N, rather than N^2 [Greengard and Roklin 1987; Ding *et al.* 1992a; Ding *et al.* 1992b]. The cell multipole method can be used to evaluate interactions that can be expressed in the following general form:

$$\sum_{\mathrm{i}} \sum_{j>i} \frac{q_{\mathrm{i}} q_{\mathrm{j}}}{|\mathbf{r}_{\mathrm{i}} - \mathbf{r}_{\mathrm{j}}|^{p}} \tag{5.54}$$

Both the Coulomb and Lennard–Jones potentials can be considered examples of this type. In the cell multipole method the simulation space is divided into uniform cubic cells. The multipole moments (charge, dipole, quadrupole) of each cell are then calculated by summing over the atoms contained within the cell. The interaction between all of the atoms in the cell and another atom outside the cell (or indeed another cell) can then be calculated using an appropriate multipole expansion (see section 3.8.1).

This multipole expansion is only valid if the separation between the interacting particles (be they atoms, molecules or cells) is larger than the sum of the radii of convergence of the multipoles. In the cell multipole method, the multipole expansion is used for interactions that are more than one cell distance away. For interactions that are within one cell distance the usual atomic pairwise interaction method is employed.

Consider an atom in a cell, C_0. The interactions with atoms in nearby cells are calculated using the usual pairwise formulae. There are 27 such cells (i.e. the cell in which the atom is positioned and with the surrounding 26 cells). The interaction between the atom and all of the atoms in each of the far-away cells is then calculated using the multipole expansion. The potential due to a far-away cell will be approximately constant for all atoms in the cell of current interest, C_0 (the cells are usually small, containing on average 4 atoms). Thus the potential due to each far-away cell can be represented as a Taylor series expansion about the centre of C_0. If there are M cells in total then there are $M-27$ far-away cells then the calculation of these cell–cell interactions for the entire system will be of order $M(M-27)$. As the number of cells is approximately equal to the number of atoms, this still leaves us with a quadratic dependency upon the number of atoms present (though it does now vary as about $N^2/16$, if there is an average of four atoms per cell).

The algorithm can be converted to one which shows linear dependency by recognising that in the method we have just described, the interactions due to very far-away cells are calculated with the same accuracy as interactions with cells that are much closer. This level of accuracy can be considered unnecessary as any error is largely due to the closer cells. The small cells are thus grouped into larger cells, with the cell size increasing with the distance from the cell of interest, C_0. The accuracy of the calculation remains approximately constant, if the ratio of the cell size to the distance remains constant. This grouping scheme is illustrated in

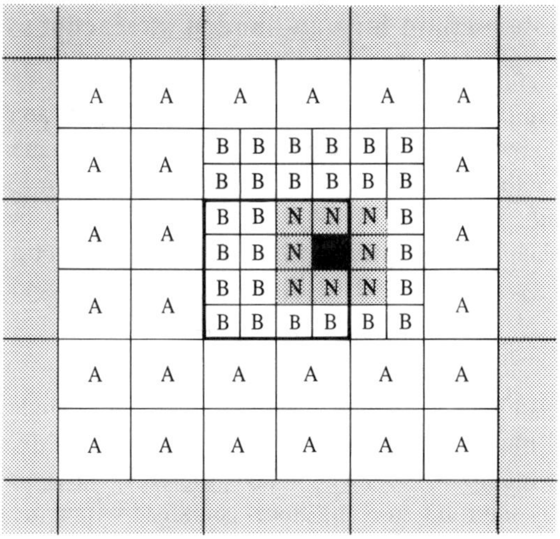

Fig. 5.24 The hierarchy of cells used in the cell multipole method. For an atom in the black cell, the interactions with atoms in the 26 nearby cells (N) are calculated explicitly. Interactions with the atoms in cells labelled A and B are calculated using a Taylor series multipole expansion. Figure adapted from Ding H-Q, N Karasawa and W A Goddard III, 1992b. The Reduced Cell Multipole Method for Coulomb Interactions in Periodic Systems with Million-Atom Unit Cells. *Chemical Physics Letters* **196**:6–10.

Figure 5.24. The multipoles for each of the larger cells are calculated by translating and adding the moments of its constituent smaller cells. The use of multipole expansions and Taylor series approximations does mean that there will be a degree of truncation error, though this can be reduced by simply including more terms in the multipole expansion. The cell multipole algorithm requires an amount of bookkeeping to keep track of the hierarchy of the cells, which means that up to a certain size of problem the exact N^2 algorithm is faster. The algorithm then suddenly switches to a linear dependence. There is some debate over the point at which this occurs, with estimates ranging from 300 particles to 3000. Nevertheless, for calculations on systems with thousands, if not millions of atoms, the approach appears very promising.

5.10 Analysing the results of a simulation and estimating errors

A simulation can generate an enormous amount of data that should be properly analysed to extract relevant properties and to check that the calculation has behaved properly. The primary reason for undertaking a particular simulation may be to calculate just a single physical or thermodynamic property or to investigate the conformational properties of a molecule. However, it is also advisable to check that other aspects of the simulation have performed as expected. Some properties can be calculated during the simulation itself (such as the energy and the virial), but it is often sensible not to impose too severe a burden on the simulation program itself. In part this is because many properties do not vary significantly from one step to another, but can be calculated at less frequent intervals. The configurations (i.e. positions of each atom or molecule in the system) do not change much from one step to another, in either a molecular dynamics or Monte Carlo simulation and so it is usual to store configurations every 5 to 25 steps, depending upon the nature of the system and the disk space available. It is good practice to visually examine configurations selected from throughout the simulation, to ensure that no strange or unexpected behaviour is present. In many simulations of molecular systems the major objective is to investigate the structural behaviour of the system, rather than to calculate thermodynamic properties, and so the focus of the analysis will change accordingly.

A computer simulation is subject to error, and this error should be properly calculated and assessed. Of course, computers only do what they are told to by the programmer, and so a program will always give the same results for the same set of initial conditions (if not, some serious fault should be suspected!). The results of a computer simulation may be subject to two kinds of error, just as any other scientific experiment. These are systematic errors and statistical errors. Systematic error results in a constant bias from the 'proper' behaviour. The most obvious effect of a systematic error is to displace the average property from its proper value. Systematic errors are sometimes due to a fault in the simulation algorithm or in the energy model, and may be relatively easy to spot, especially if they have an obvious or even catastrophic effect on the simulation. Systematic errors may also arise from approximations inherent in the algorithm such as truncation (all finite difference methods used in molecular dynamics generate only an approximation to the true integral of the equations of motion), and round-off errors (due to the limited precision with which numbers can be stored in a computer). Such errors can be more difficult to detect. One way to detect systematic error is to compare the distribution of the values of simple thermodynamic properties about their average values. The distribution of such properties about their average values should be normal (i.e. Gaussian), such that the probability of finding a particular value for the property A is given by:

$$p(A) = \frac{1}{\sigma\sqrt{2\pi}} \exp[-(A - \langle A \rangle)^2 / 2\sigma^2] \qquad (5.55)$$

σ^2 is the variance, given by $\sigma^2 = \langle (A - \langle A \rangle)^2 \rangle$. The standard deviation is the square root of the variance. More information on these statistical terms can be found in section 1.10.7.

The chi-squared test can be used to provide a quantitative estimate of the deviation of a calculated distribution from that expected. Suppose that the value of some property (A) has been calculated from the simulation at regular intervals to give a total of M values. The average value of the property A is determined together with the standard deviation. The data, comprising all of the A values from the simulation are then divided into bins such that the number of values in each bin (M_i) is approximately the same. The number of values that would be expected in the ith bin is:

$$n_i = \frac{M}{\sigma\sqrt{2\pi}} \int_{A_i - \Delta A/2}^{A_i + \Delta A/2} \exp\left[\frac{-(A_i - \langle A \rangle)^2}{2\sigma^2}\right] \tag{5.56}$$

A_i is the value of the property in the ith bin and ΔA is the width of each bin. The number of values that would be expected in each bin, n_i, does not have to be integral, though the actual number as determined from the simulation (M_i) will of course be an integer. The chi-squared function is given by:

$$\chi^2 = \sum_i \frac{(M_i - n_i)}{n_i} \tag{5.57}$$

If χ^2 is large (bigger than unity) then it is unlikely that the two distributions are the same. Any significant deviations from the expected behaviour should be investigated further to try and eliminate as much of the systematic error as possible. It is good practice to vary as many of the parameters as possible: using different computers, different compilers, different algorithms and different ways of implementing a given algorithm, and different simulation methods (Monte Carlo and molecular dynamics) not only to test the component parts of the simulation but also the software used to perform the calculation.

If all sources of systematic error can be eliminated, there will still remain statistical errors. These errors are often reported as standard deviations. What we would particularly like to calculate is the error in the average value, $\langle A \rangle$. Clearly, the average value obtained from a simulation of a finite number of steps will not necessarily be the same as the 'true' average that would be obtained from a simulation with an infinite number of steps. The error in the calculated value is related to the true energy as follows:

$$\sigma_{\langle A \rangle} \approx \frac{\sigma}{\sqrt{M}} \tag{5.58}$$

$\sigma_{\langle A \rangle}$ is the standard deviation of the average value $\langle A \rangle$ obtained from M data values with respect to the 'true' average, σ, that would be obtained from an infinitely long simulation. $\sigma_{\langle A \rangle}$ is the standard deviation of the M values from the average as calculated in the simulation. Thus the standard deviation of the calculated average is inversely proportional to the square root of the number of values. An important feature of equation (5.58) is that it applies to independent (i.e. random) samples. Thus the number M in the denominator is not simply equal to the number of steps in the simulation. This is because there is a high degree of correlation between successive configurations in either a Monte Carlo or molecular dynamics simulation. What we need to know is the correlation or relaxation 'time' of the simulation; this is the number of steps required for the system to lose its 'memory' of previous configurations. In molecular dynamics,

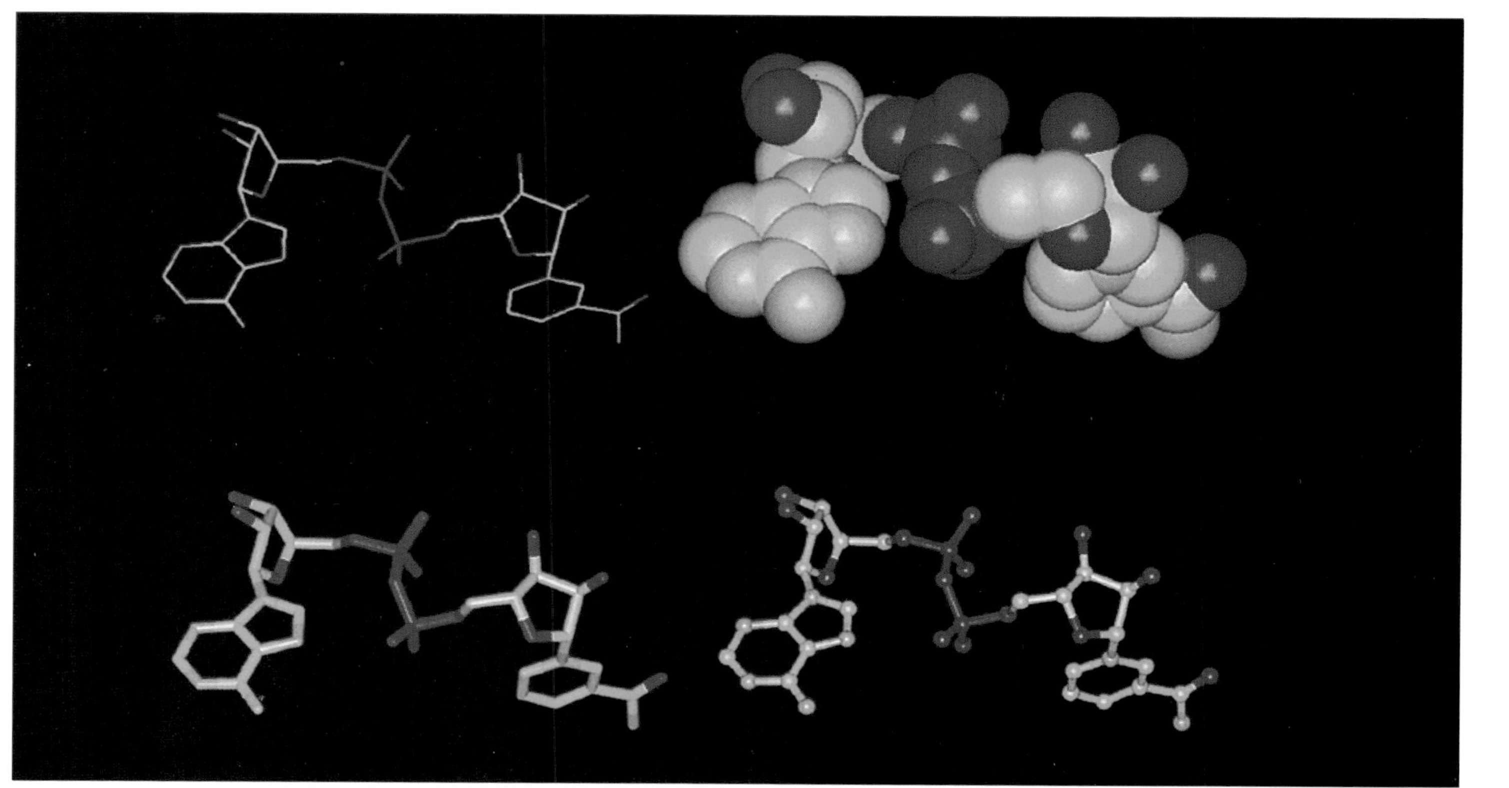

Fig 1.4 Some of the common molecular graphics representations of molecules, illustrated using the crystal structure of nicotinamide adenine dinucleotide phosphate (NADPH)[Reddy *et al.* 1981]. Clockwise, from top left: stick, CPK/space filling, 'balls and stick' and 'tube'.

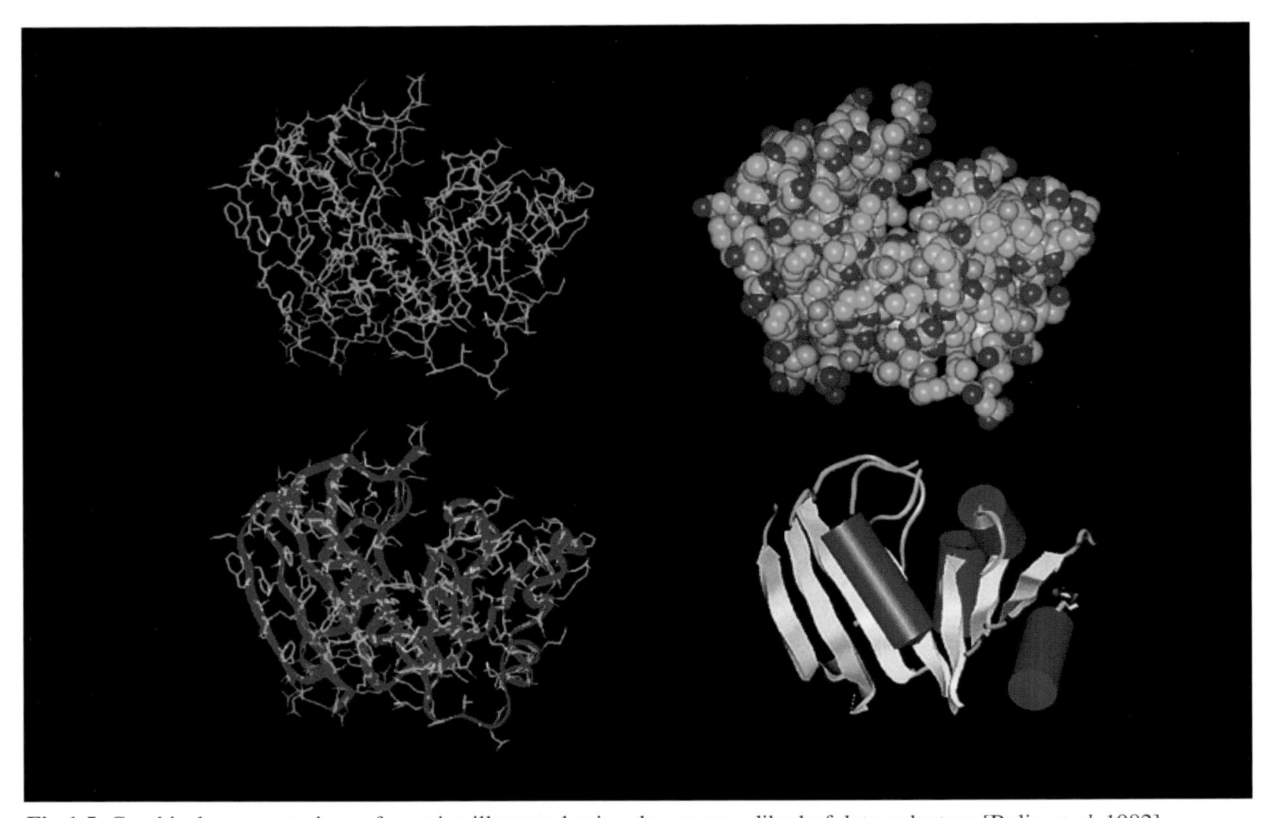

Fig 1.5 Graphical representations of proteins illustrated using the enzyme dihydrofolate reductase [Bolin *et al.* 1982]. Clockwise from top left: stick, CPK, 'cartoon' and 'ribbon'.

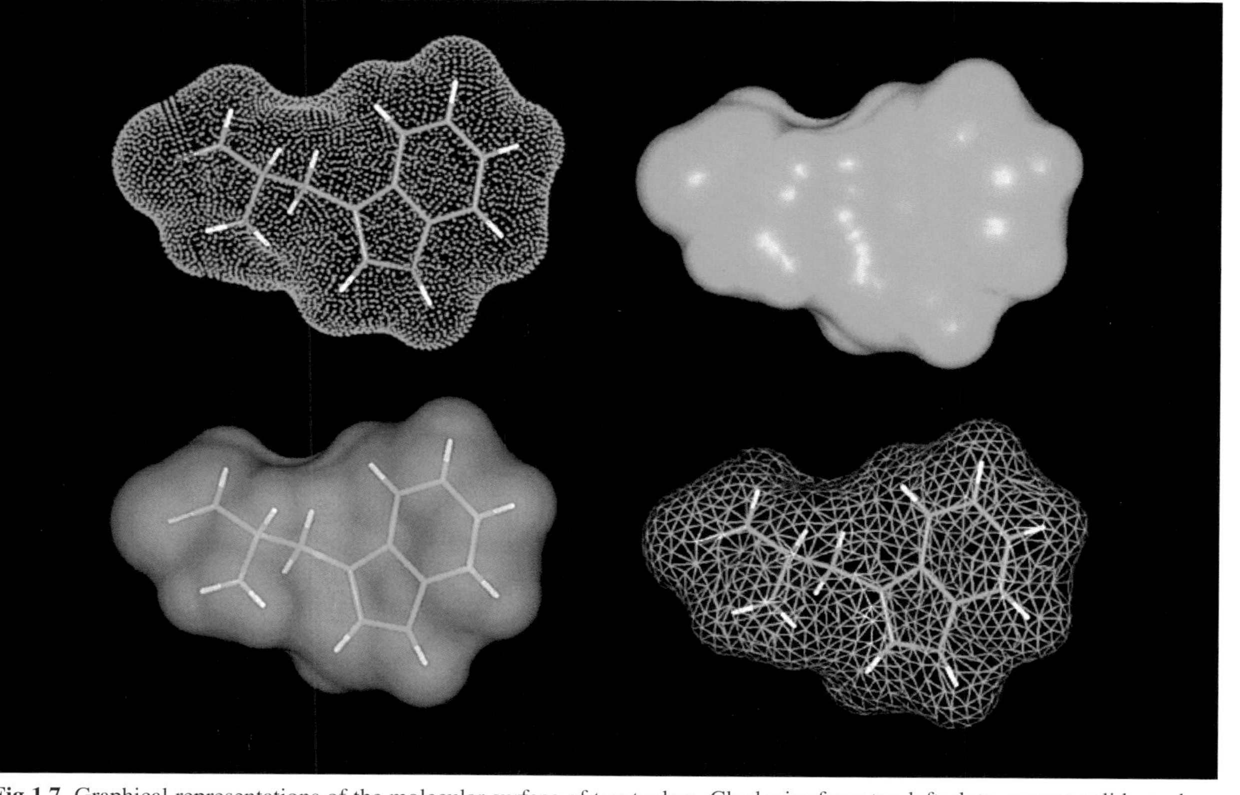

Fig 1.7 Graphical representations of the molecular surface of tryptophan. Clockwise from top left: dots, opaque solid, mesh, transluscent solid.

The UCSF Computer Graphics Laboratory

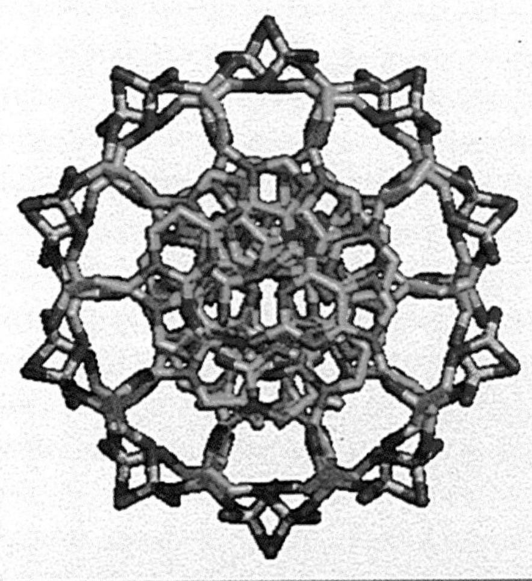

Axial view of DNA from MidasPlus

Overview:

The UCSF Computer Graphics Laboratory, funded primarily by the National Center for Research Resources of the National Institutes of Health, provides access to state-of-the-art computer graphics hardware and software for research on biomolecular structures and interactions. The laboratory is used extensively not only within the Department of Pharmaceutical Chemistry and University of California, San Francisco campus, but by visitors from around the world. Equipment includes computers for numerical and symbolic computing as well as high performance interactive three-dimensional color graphics systems for visualizations of complex molecular structures. All computers employee the UNIX operating system. Software developed in the laboratory, particularly the molecular display program MidasPlus, is widely distributed to other sites and is in use in applications including molecular modeling, drug design and protein engineering. The laboratory director and principal investigator is Dr. Thomas Ferrin.

This year, the laboratory is helping to organize the first Pacific Symposium on Biocompting. In addition, the laboratory runs a Sequence Analysis and Consulting Service, providing subscribers with access to, and consulting on, DNA and protein sequence analysis software and databases.

Fig 1.8 A sample WWW page. Items highlighted in blue are hyperlinks to other documents, data, graphics, etc.

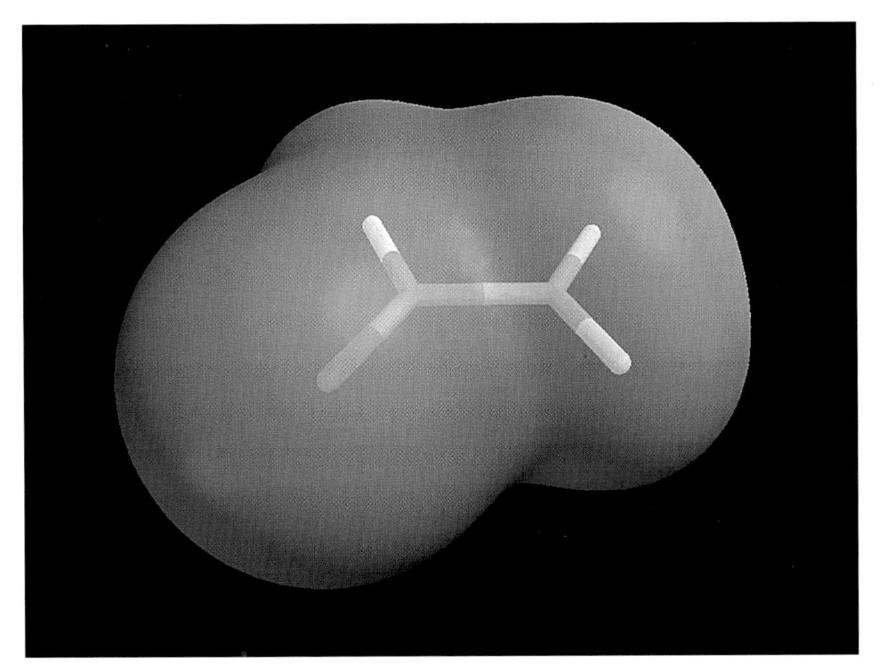

Fig 2.17 Surface representation of electron density around formamide at a contour of 0.0001 au (electrons/bohr^3).

Fig 2.18 HOMO of formamide. The red contour indicates the negative part of the wavefunction and blue the positive part of the wavefunction. The formamide molecule is oriented with the oxygen atom on the left pointing towards the viewer, as in Fig. 2.17.

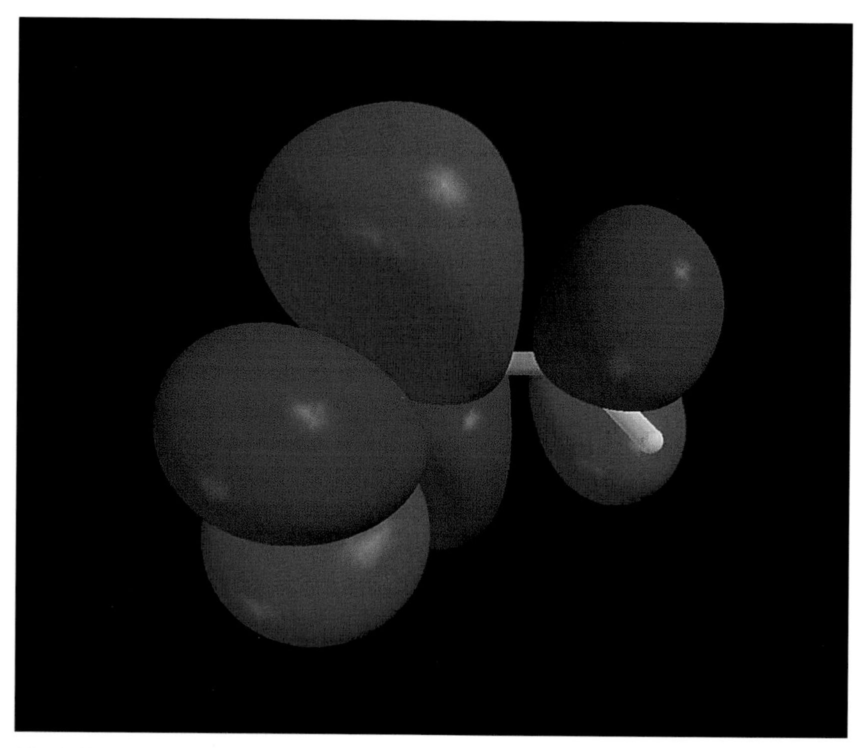

Fig 2.19 LUMO of formamide.

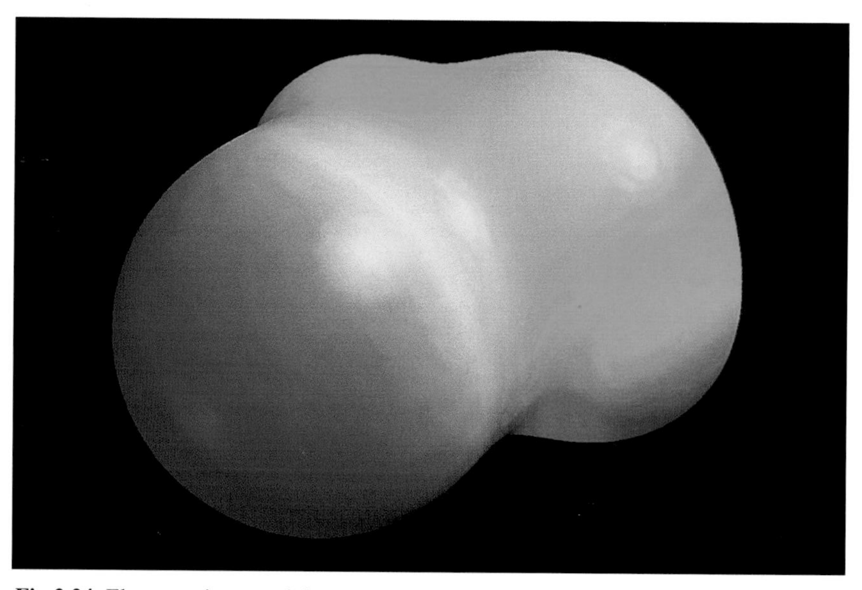

Fig 2.24 Electrostatic potential mapped onto the electron density surface for formamide. The orientation of the molecule is as in Fig 2.17. Red indicates negative electrostatic potential and blue is positive potential.

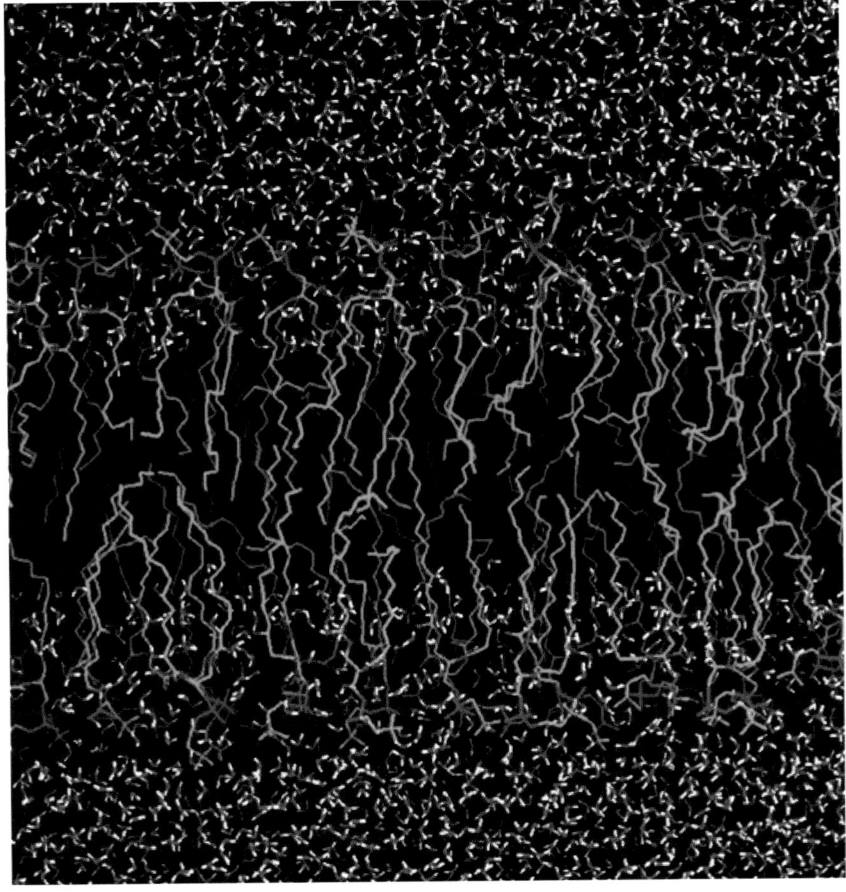

Fig 6.21 Snapshot from a molecular dynamics simulation of a solvated lipid bilayer [Robinson *et al.* 1994]. The disorder of the alkyl chains can be clearly seen.

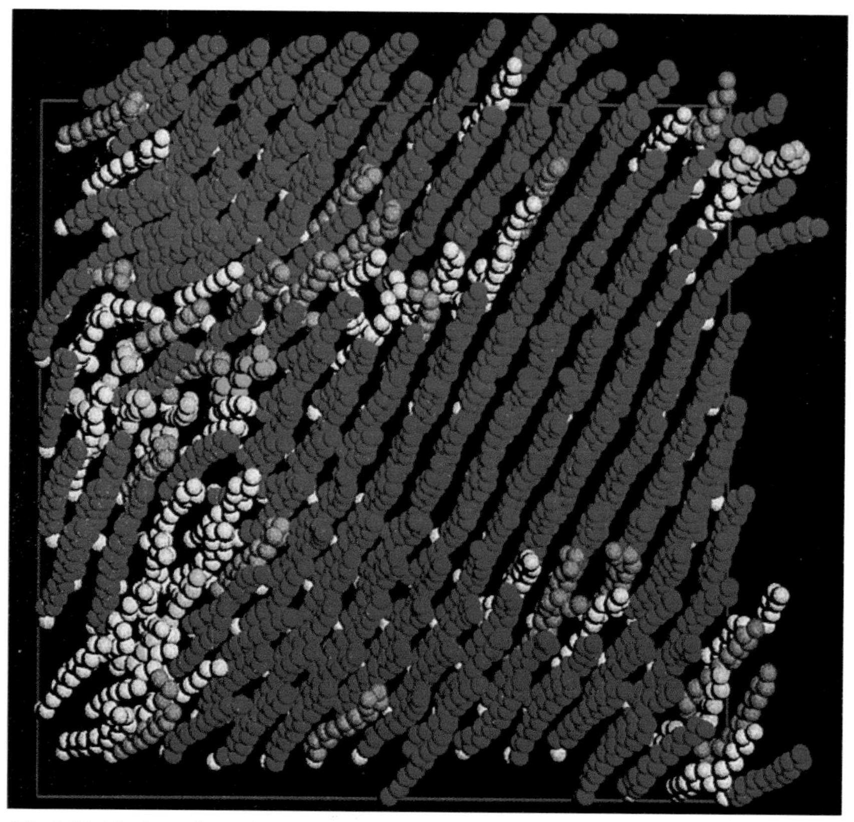

Fig 7.19 Final configuration obtained from a Configurational Bias Monte Carlo simulation of thioalkanes absorbed on a gold surface [Siepmann and MacDonald 1993a]. The system contains 224 molecules which are colour coded according to the number of gauche defects, with red chains being all trans, yellow chains containing three gauche bonds and green chains containing five gauche bonds.

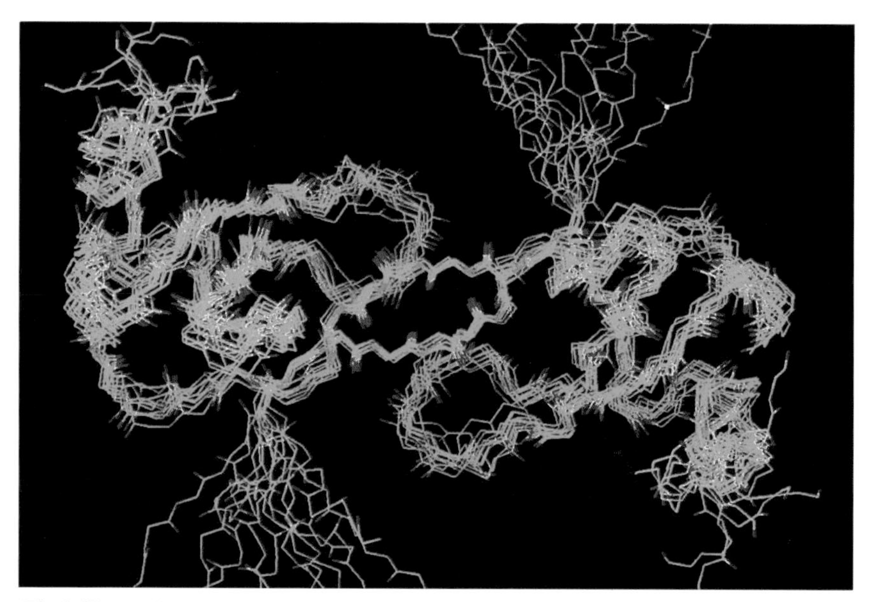

Fig 8.19 Twelve conformations of the chemokine RANTES generated from NMR data using distance geometry [Chung *et al.* 1995].

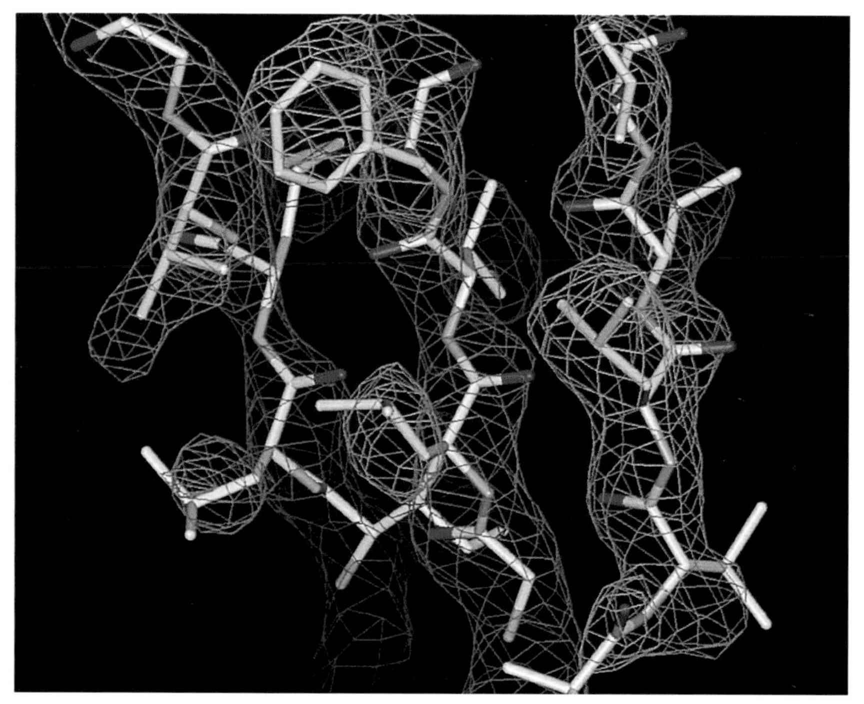

Fig 8.22 Fitting a polypeptide chain to the electron density when determining the structure of a protein using X-ray crystallography. The figure shows part of the density for rat ADP-ribosylation factor-1 (ARF-1) [Greasley *et al.* 1995].

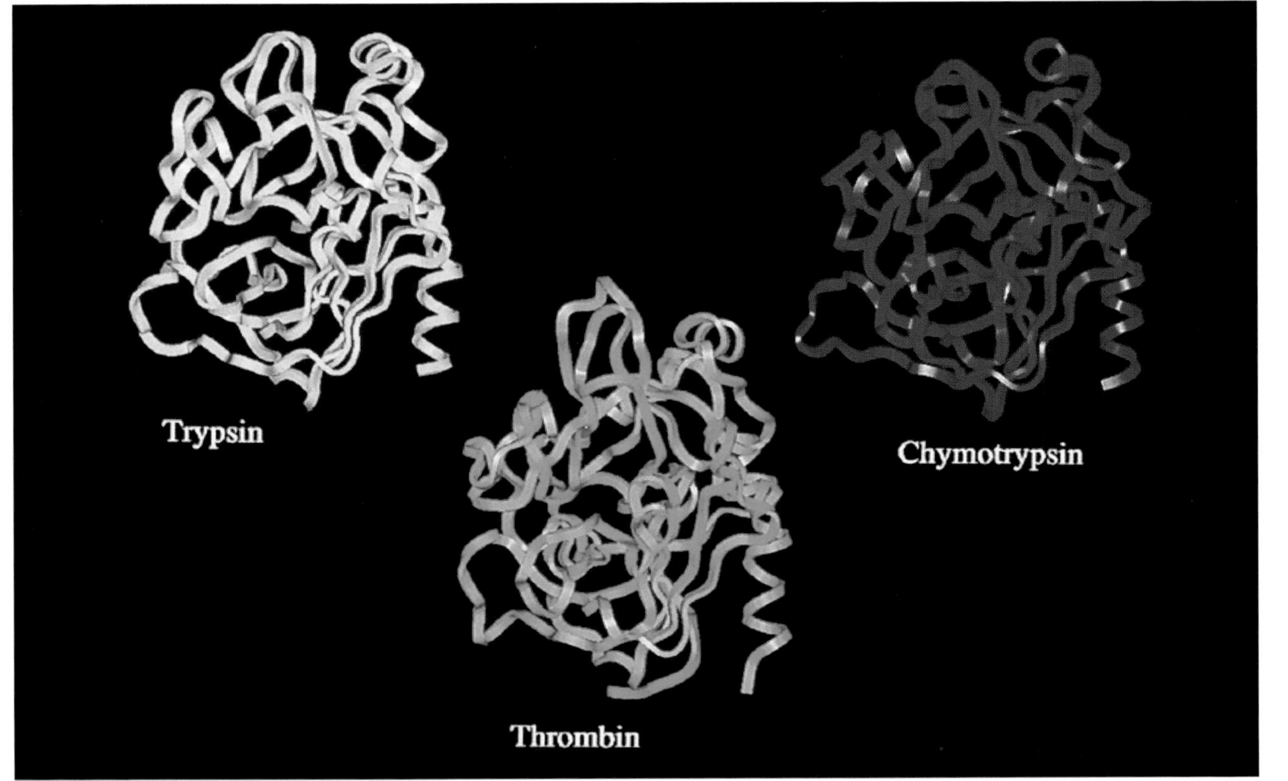

Fig 8.39 Trypsin (top left) [Turk *et al.* 1991], chymotrypsin (top right) [Birktoft and Blow 1972] and thrombin (bottom) [Turk *et al.* 1992] have similar three-dimensional structures.

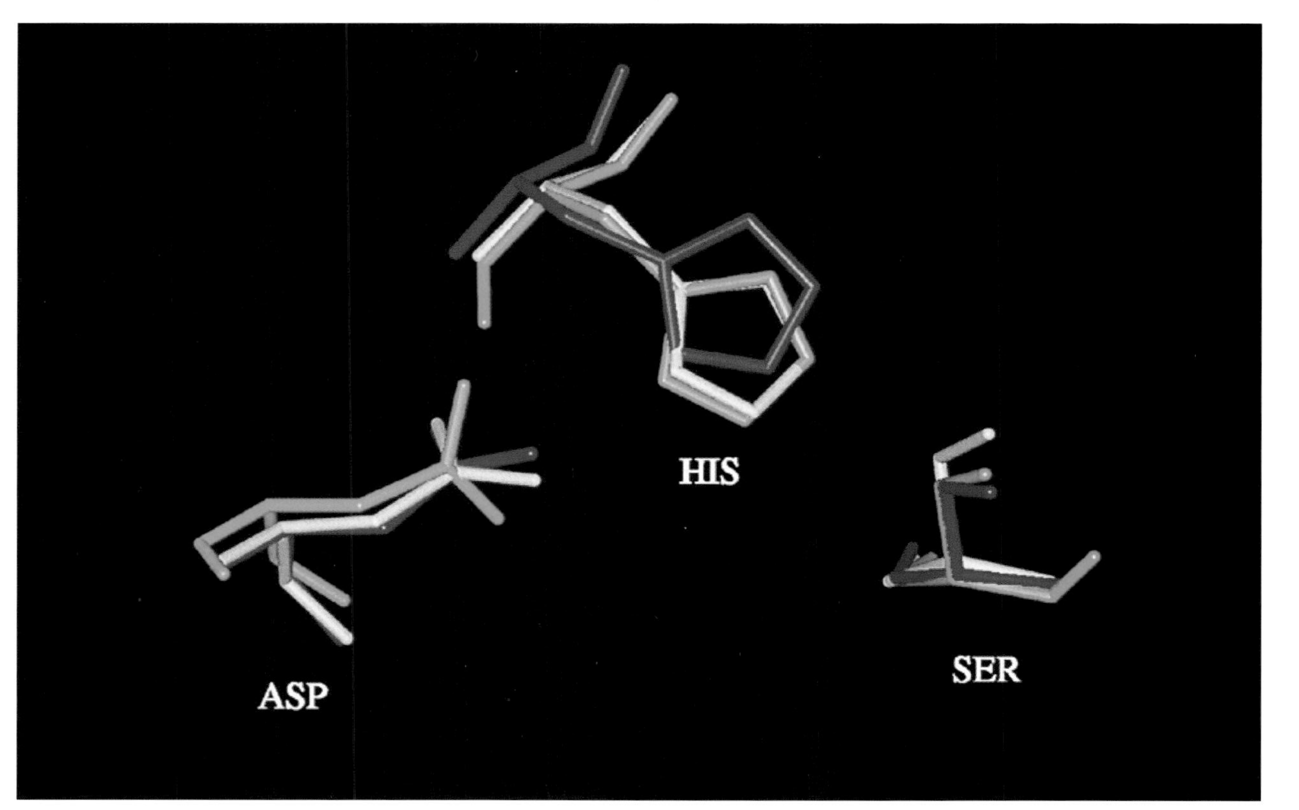

Fig 8.41 A superposition of the aspartic acid, histidine and serine amino acids in the active sites of trypsin (yellow), chymotrypsin (red) and thrombin (green).

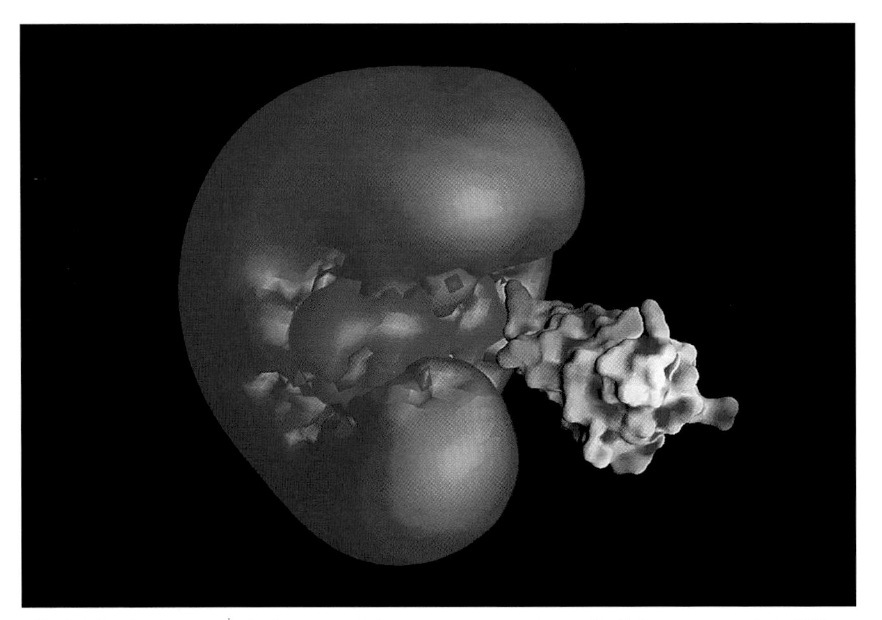

Fig 9.22 3D Electrostatic isopotential contours around trypsin [Marquart *et al.* 1983]. Contours are drawn at $-1k_BT$ (red) and $+1k_BT$ (blue). The trypsin inhibitor is also represented with its electrostatic potential mapped onto its molecular surface.

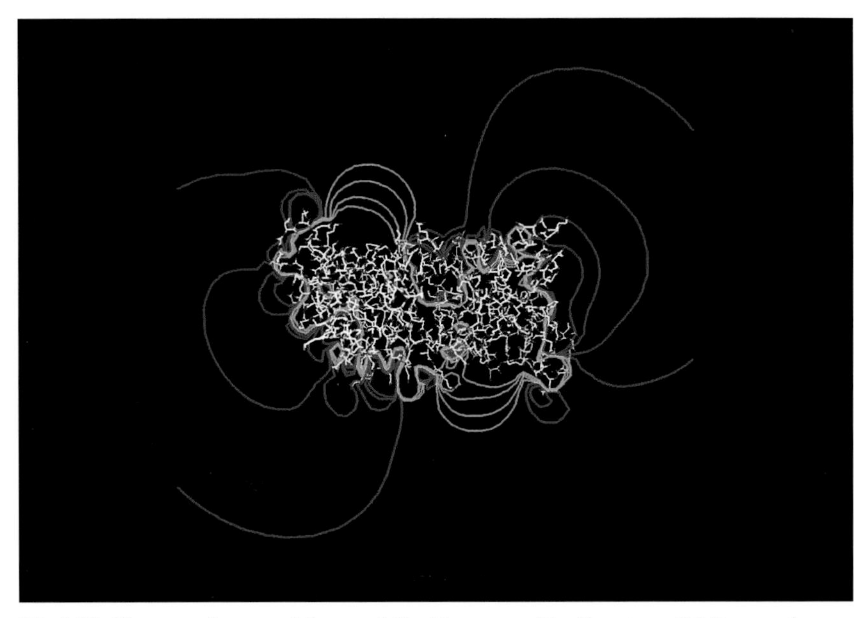

Fig 9.23 Electrostatic potential around Cu–Zn superoxide dismutase [McRee *et al.* 1990]. Red contours indicate negative electrostatic potential and blue contours indicate positive electrostatic potential. Two active sites are present in each dimer, at the top left and bottom right of the figure where there is a significant concentration of positive electrostatic potential.

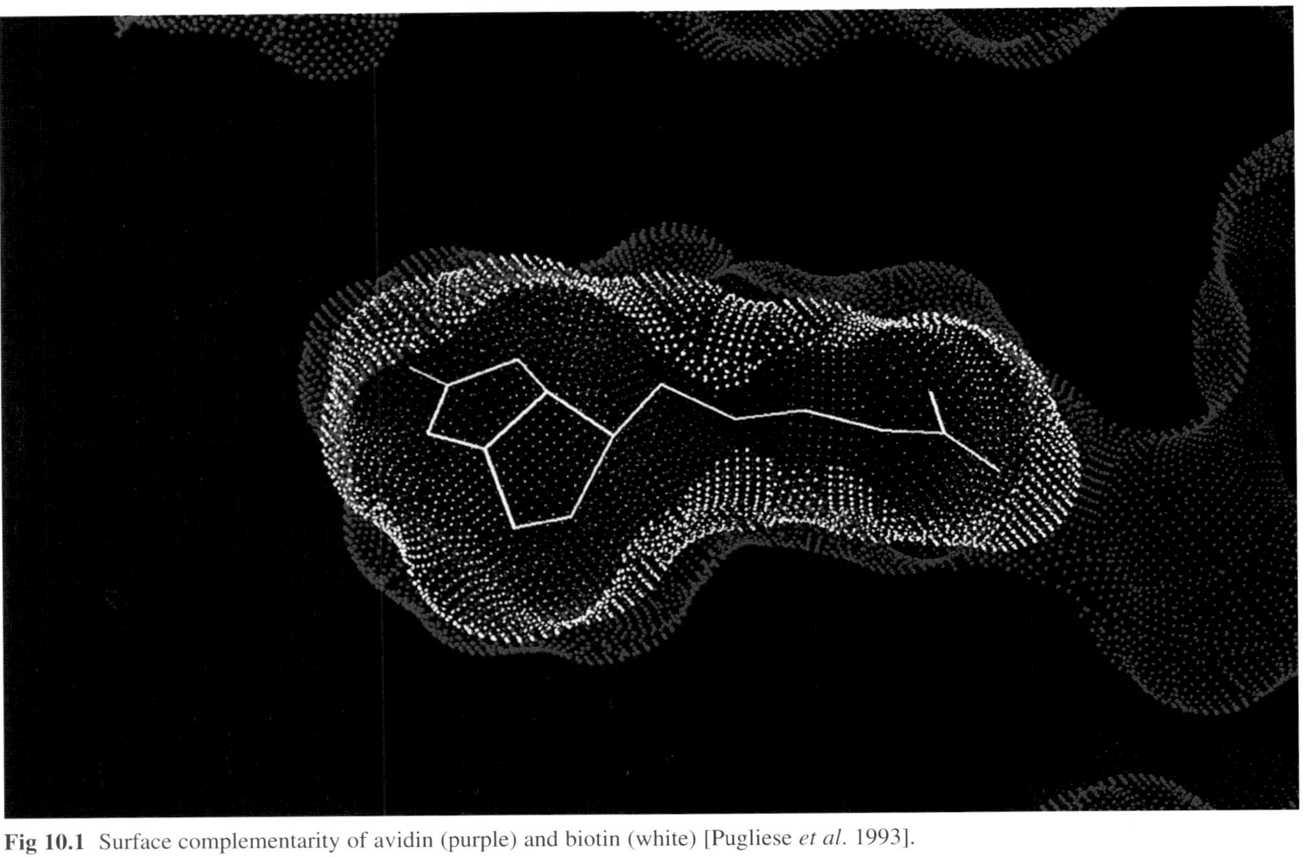

Fig 10.1 Surface complementarity of avidin (purple) and biotin (white) [Pugliese *et al.* 1993].

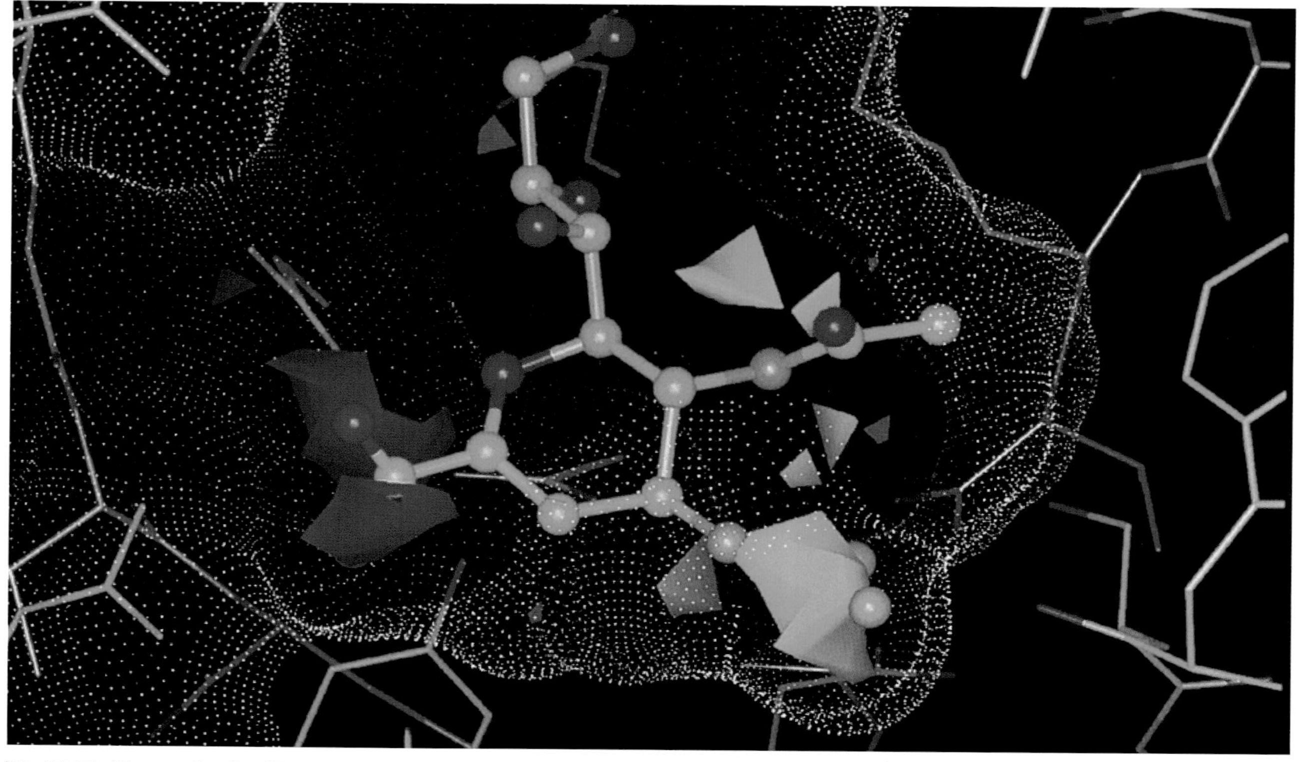

Fig 10.21 The result of a GRID calculation using carboxylate and amidine probes in the binding site of neuraminidase. The regions of minimum energy are contoured (carboxylate red; amidine blue). Also shown is the inhibitor 4-guanidino-Neu5Ac2en which contains two such functional groups [von Itzstein *et al.* 1993].

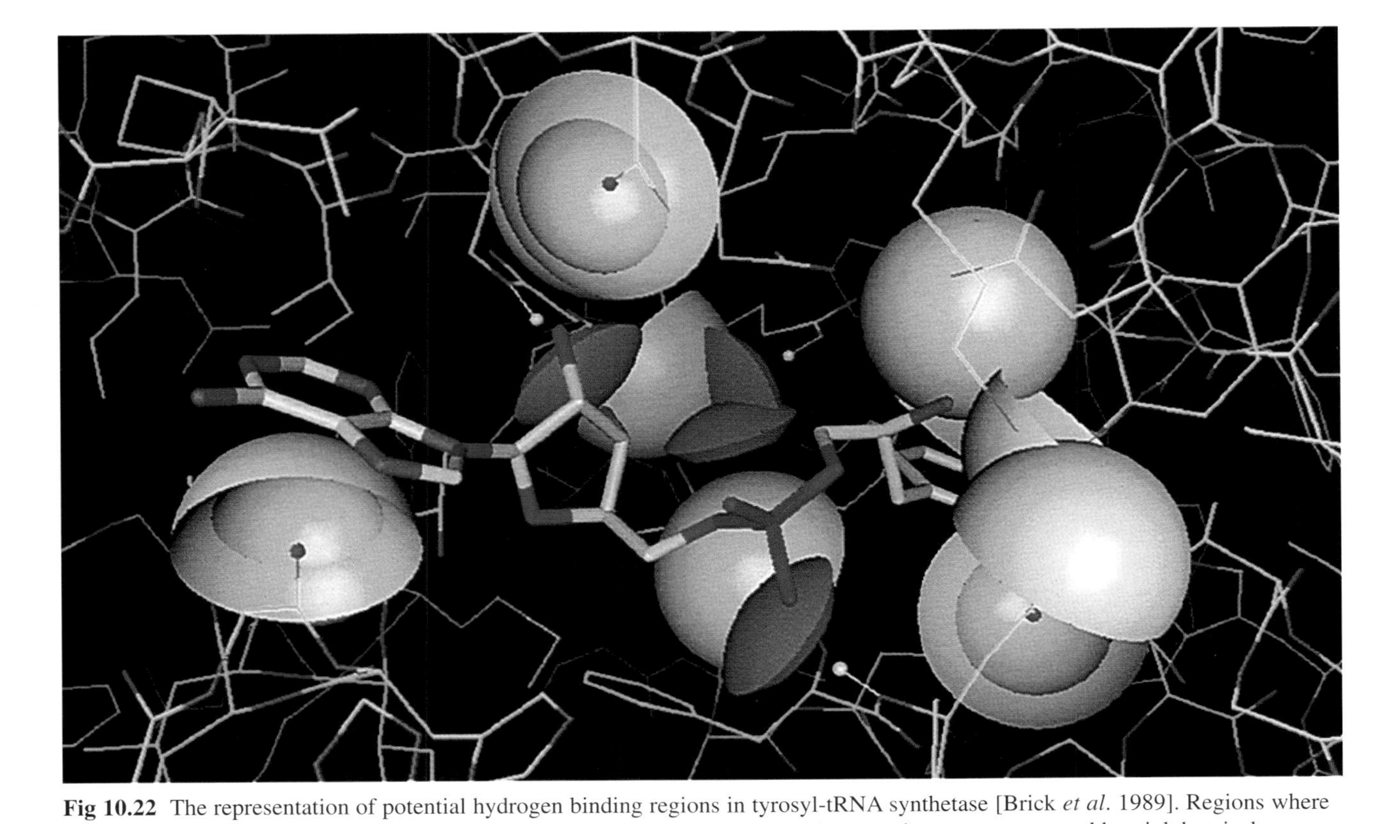

Fig 10.22 The representation of potential hydrogen binding regions in tyrosyl-tRNA synthetase [Brick *et al.* 1989]. Regions where a hydrogen bond acceptor could be positioned are indicated by red disks and donor regions are represented by pink hemispheres. The ligand (tyrosyl adenylate) is also shown.

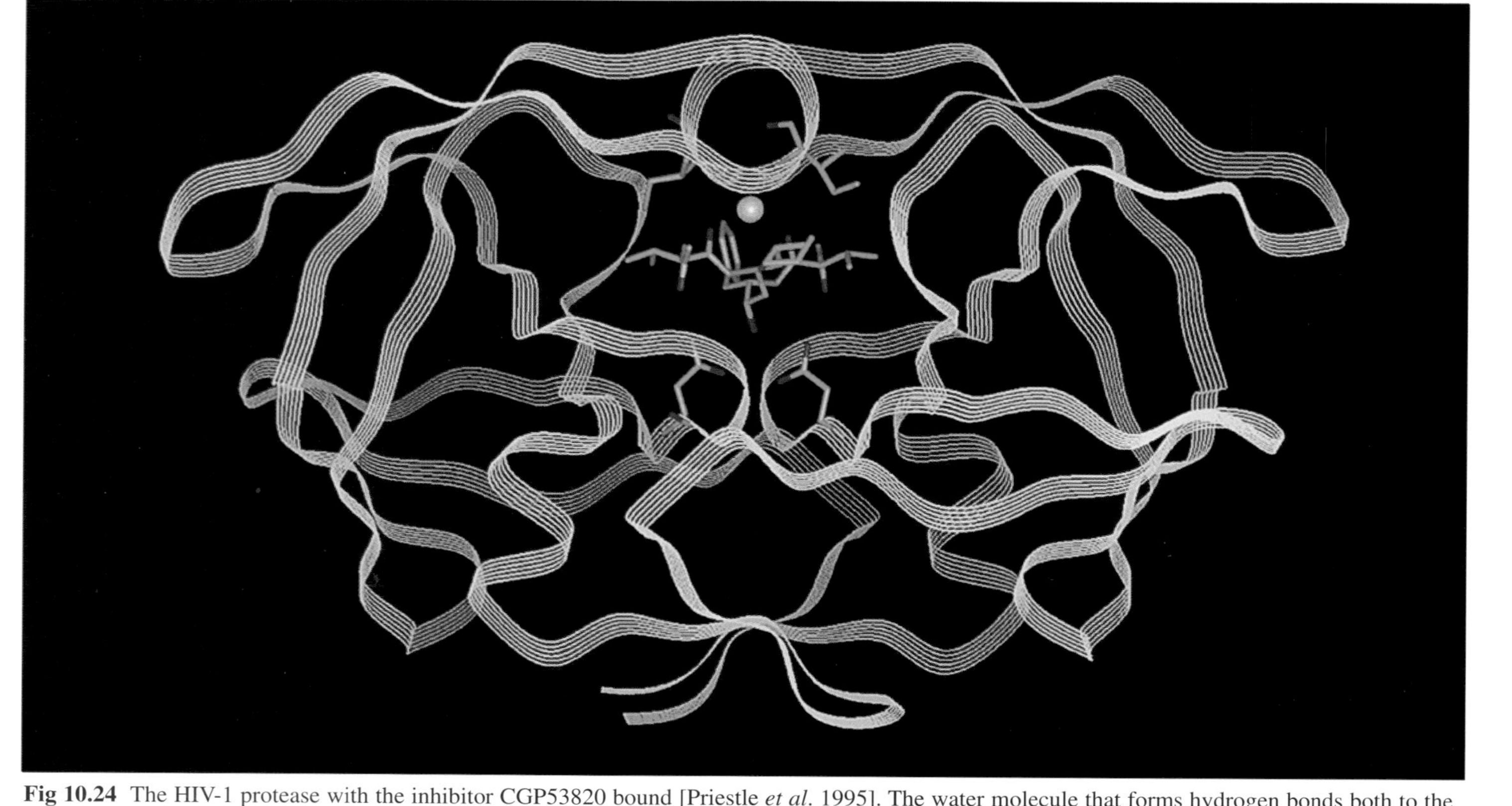

Fig 10.24 The HIV-1 protease with the inhibitor CGP53820 bound [Priestle *et al.* 1995]. The water molecule that forms hydrogen bonds both to the inhibitor and to the 'flaps' of the protein is drawn as a white sphere and the catalytic aspartate groups are also represented.

where successive steps are related in a temporal fashion, the correlation 'time' is a true time and will be discussed in more detail in section 6.6. Usually, the correlation time will be unknown prior to the simulation but it can be estimated as follows. First, the configurations are broken down into a series of blocks. Suppose each block contains b successive steps and that there are n_b blocks (so the total simulation contains bn_b steps, as shown in Figure 5.25). The average value of the property is calculated for each block:

$$\langle A \rangle_b = \frac{1}{b}\sum_{\mathrm{i}=1}^{b} A_{\mathrm{i}} \tag{5.59}$$

As the number of steps n_b in each block increases, so it would be expected that the block averages become uncorrelated. When this is the case, then the variance of the block averages, $\sigma^2(\langle A \rangle_b)$ will become inversely proportional to n_b. $\sigma^2(\langle A \rangle_b)$ is calculated as follows:

$$\sigma^2(\langle A \rangle_b) = \frac{1}{n_b}\sum_{b=1}^{n_{\mathrm{b}}}(\langle A \rangle_b - \langle A \rangle_{\mathrm{total}})^2 \tag{5.60}$$

$\langle A \rangle_{\mathrm{total}}$ is the average over the entire simulation. The limiting number of steps to obtain uncorrelated configurations (the statistical inefficiency, s) can be calculated using:

$$s = \lim_{n_b \to \infty} \frac{n_b \sigma^2(\langle A \rangle_b)}{\sigma^2(A)} \tag{5.61}$$

To determine s, $n_b\sigma^2(\langle A \rangle_b)/\sigma^2(A)$ is plotted against n_b or $\sqrt{n_b}$. The graph should show a steep rise for low n_b and then level off to give a plateau, as shown in Figure 5.26. The plateau value is the limiting value that gives the correlation time ($s \approx 23$ in this case).

Having determined the value of s, the 'true' standard deviation of the average value is related to the 'true' error for an infinite simulation by:

$$\sigma_{\langle A \rangle} \approx \sigma\sqrt{\frac{s}{M}} \tag{5.62}$$

M here is the actual number of steps or iterations in the simulation. If the value of s can be reduced, then a more accurate average value can be calculated for a given length of simulation. This should be an important consideration when deciding what simulation protocol to use. For example, it may be more appropriate to use a complex simulation algorithm than a simpler one if the statistical inefficiency is significantly reduced.

If the relaxation time is known, then sample averages are often best calculated using the block method, Figure 5.25. Each block should contain more steps than

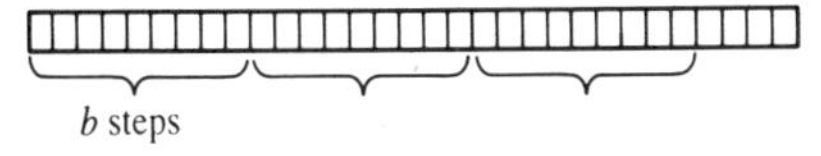

Fig. 5.25 Blocking a simulation to calculate the statistical error.

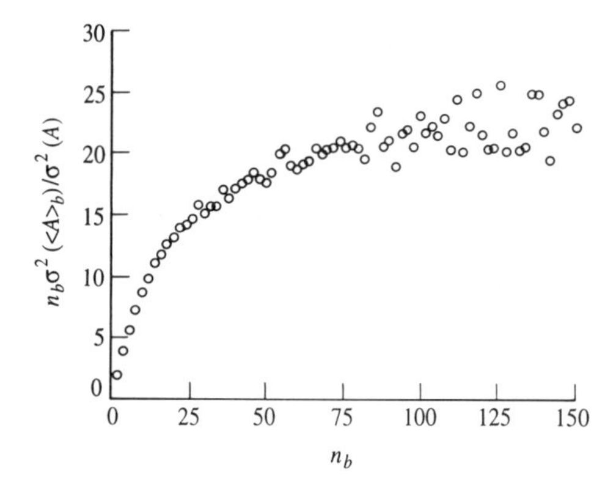

Fig. 5.26 Calculating the statistical efficiency, s. A plot of $n_b\sigma^2(\langle A\rangle_b)/\sigma^2(A)$ against n_b shows a steep rise before levelling off. Here the property A corresponds to the pressure calculated from the molecular dynamics simulation of argon.

the relaxation time. The sample average for the whole run can be obtained in a variety of ways:

1. Stratified systematic sampling, in which a single value of the property is taken from each block;
2. Stratified random sampling, in which a single value is taken at random from each block;
3. Coarse-graining, in which the average value for each block is determined and then the average for the run is calculated by averaging the coarse-grain averages.

The coarse-graining approach is commonly used for thermodynamic properties whereas the systematic or random sampling methods are appropriate for static structural properties such as the radial distribution function.

Another way to improve the error in a simulation, at least for properties such as the energy and the heat capacity that depend on the size of the system (the extensive properties), is to increase the number of atoms or molecules in the calculation. The standard deviation of the average of such a property is proportional to $1/\sqrt{N}$. Thus, more accurate values can be obtained by running longer simulations on larger systems. In computer simulation it is unfortunately the case that the more effort that is expended the better the results that are obtained. Such is life!

Appendix 5.1 Basic statistical mechanics

The Boltzmann distribution is fundamental to statistical mechanics. The Boltzmann distribution is derived by maximising the entropy of the system (in accordance with the second law of thermodynamics) subject to the constraints on the system. Let us consider a system containing N particles (atoms or molecules) such that the energy levels of the particles are ε_1, ε_2, ... If there are n_1 particles in the energy level ε_1, n_2 particles in ε_2 and so on, then there are W ways in which this distribution can be achieved:

$$W(n_1, n_2, \ldots) = N!/n_1!n_2!\ldots \tag{5.63}$$

The most favourable distribution is the one with the highest weight and this corresponds to the configuration with just one particle in each energy level ($W = N!$). However, there are two important constraints on the system. First, the total energy is fixed:

$$\sum_{\rm i} n_{\rm i}\varepsilon_{\rm i} = E \tag{5.64}$$

The second constraint arises from the fact that the total number of particles is fixed:

$$\sum_{\rm i} n_{\rm i} = N$$

The Boltzmann distribution gives the number of particles $n_{\rm i}$ in each energy level $\varepsilon_{\rm i}$ as:

$$\frac{n_{\rm i}}{N} = \frac{\exp(-\varepsilon_{\rm i}/k_{\rm B}T)}{\sum_{\rm i}\exp(-\varepsilon_{\rm i}/k_{\rm B}T)} \tag{5.65}$$

The denominator in this expression is the molecular partition function:

$$q = \sum_{\rm i}\exp(-\varepsilon_{\rm i}/k_{\rm B}T) \tag{5.66}$$

For translational, rotational and vibrational motion the partition function can be calculated using standard results obtained by solving the Schrödinger equation:

$$\textit{translation}: \quad q^{\rm t} = \left(\frac{2\pi m k_{\rm B}T}{h^2}\right)^{3/2} V \tag{5.67}$$

V is the volume

$$\textit{rotation}: \quad q^{\rm r} \approx \left(\frac{\pi^{1/2}}{\sigma}\right)\left(\frac{2I_{\rm A}k_{\rm B}T}{\hbar^2}\right)\left(\frac{2I_{\rm B}k_{\rm B}T}{\hbar^2}\right)\left(\frac{2I_{\rm C}k_{\rm B}T}{\hbar^2}\right) \tag{5.68}$$

$I_{\rm A}$, $I_{\rm B}$, $I_{\rm C}$ are the moments of inertia and σ is the symmetry number (2 for H_2O, 3 for NH_3, 12 for benzene)

$$\textit{vibration}: \quad r^{\rm v} = \frac{1}{1 - \exp(-\hbar\omega/k_{\rm B}T)} \tag{5.69}$$

ω is the angular frequency: $\omega = \sqrt{(k/\mu)}$, where μ is the reduced mass. This form of the vibrational partition function is measured relative to the zero-point energy.

In computer simulations we are particularly interested in the properties of a system comprising a number of particles. An ensemble is a collection of such systems, as might be generated using a molecular dynamics or a Monte Carlo simulation. Each member of the ensemble has an energy and the distribution of the system within the ensemble follows the Boltzmann distribution. This leads to the concept of the ensemble partition function, Q.

Various thermodynamic properties can be calculated from the partition function. Here we simply state some of the most common:

$$\textit{internal energy}: \quad U = \frac{k_B T^2}{Q}\left(\frac{\partial Q}{\partial T}\right)_V = k_B T^2\left(\frac{\partial \ln Q}{\partial T}\right)_V \tag{5.70}$$

$$\textit{enthalpy}: \quad H = k_B T^2\left(\frac{\partial \ln Q}{\partial T}\right)_V + k_B T V\left(\frac{\partial \ln Q}{\partial V}\right)_T \tag{5.71}$$

$$\textit{Helmholtz free energy}: \quad A = -k_B T \ln Q \tag{5.72}$$

$$\textit{Gibbs free energy}: \quad G = -k_B T \ln Q + k_B T V\left(\frac{\partial \ln Q}{\partial V}\right)_T \tag{5.73}$$

Appendix 5.2 Heat capacity and energy fluctuations

The heat capacity is related to the internal energy U by

$$C_V = \left(\frac{\partial U}{\partial T}\right)_V$$

If we differentiate the expression for the internal energy, equation (5.18a), we can obtain the heat capacity in terms of the partition function:

$$C_V = \frac{\partial}{\partial T}\left(\frac{k_B T^2}{Q}\frac{\partial Q}{\partial T}\right)_V = \frac{k_B T^2}{Q}\frac{\partial^2 Q^2}{\partial T^2} + \frac{2k_B T}{Q}\frac{\partial Q}{\partial T} - \frac{k_B T^2}{Q^2}\left(\frac{\partial Q}{\partial T}\right)^2 \tag{5.74}$$

The desired expression is obtained by writing each of these three terms as a function of the average energy, $\langle E \rangle$. The internal energy is just the expectation value of the energy, $\langle E \rangle$, and so:

$$\langle E \rangle = \frac{k_B T^2}{Q}\frac{\partial Q}{\partial T} \tag{5.75}$$

Thus for the second term in equation (5.74) we have

$$\frac{2k_B T}{Q}\frac{\partial Q}{\partial T} = \frac{2\langle E \rangle}{T} \tag{5.76}$$

We can also rewrite the third term in equation (5.74):

$$k_B T\left(\frac{1}{Q}\frac{\partial Q}{\partial T}\right)^2 = \frac{\langle E \rangle^2}{k_B T} \tag{5.77}$$

For the first term, we need to do a little more work. The starting point is:

$$\frac{\partial}{\partial T}\left(\frac{\langle E\rangle}{k_B T^2}\right) = \frac{\partial}{\partial T}\left\{\frac{1}{Q}\left(\frac{\partial Q}{\partial T}\right)\right\} \tag{5.78}$$

or

$$-2\frac{\langle E\rangle}{k_B T^3} = \frac{1}{Q}\frac{\partial^2 Q}{\partial T^2} + \frac{\partial Q}{\partial T}\frac{\partial}{\partial T}\left(\frac{1}{Q}\right) \tag{5.79}$$

We can use the chain rule as follows:

$$\frac{\partial Q}{\partial T}\frac{\partial}{\partial T}\left(\frac{1}{Q}\right) = \frac{\partial Q}{\partial T}\frac{\partial Q}{\partial T}\frac{\partial}{\partial Q}\left(\frac{1}{Q}\right) = -\left(\frac{\partial Q}{\partial T}\right)^2\left(\frac{1}{Q}\right)^2 \tag{5.80}$$

Thus

$$\frac{k_B T^2}{Q}\frac{\partial^2 Q}{\partial T^2} = -2\frac{\langle E\rangle}{k_B T^3} + \frac{\langle E^2\rangle}{k_B^2 T^4} \tag{5.81}$$

So

$$C_V = k_B T^2\left\{-2\frac{\langle E\rangle}{k_B T^3} + \frac{\langle E^2\rangle}{k_B^2 T^4}\right\} + 2\frac{\langle E\rangle}{T} - \frac{\langle E\rangle^2}{k_B T^2} \tag{5.82}$$

or

$$C_V = \frac{(\langle E^2\rangle - \langle E\rangle^2)}{k_B T^2} \tag{5.83}$$

Appendix 5.3 The real gas contribution to the virial

If the gas particles interact through a pairwise potential, then the contribution to the virial from the intermolecular forces can be derived as follows. Consider two atoms i and j separated by a distance r_{ij}.

$$r_{ij} = \sqrt{(x_i - x_j)^2 + (y_i - y_j)^2 + (z_i - z_j)^2} \tag{5.84}$$

The contribution to the virial from the interaction $\mathscr{v}(r_{ij})$ between atoms i and j is given by:

$$W_{real} = \left[x_i\frac{\partial}{\partial x_i} + x_j\frac{\partial}{\partial x_j} + y_i\frac{\partial}{\partial y_i} + y_j\frac{\partial}{\partial y_j} + z_i\frac{\partial}{\partial z_i} + z_j\frac{\partial}{\partial z_j}\right]\mathscr{v}(r_{ij}) \tag{5.85}$$

Since

$$x_i\frac{\partial r_{ij}}{\partial x_i} = x_i\frac{(x_i - x_j)}{r_{ij}} \quad \text{and} \quad x_j\frac{\partial r_{ij}}{\partial x_i} = -x_j\frac{(x_i - x_j)}{r_{ij}},$$

and similarly for the y and z coordinates, we can apply the chain rule, $\partial/\partial x_i = (\partial/\partial r_{ij})(\partial r_{ij}/\partial x_i)$, as follows:

$$W_{\text{real}} = \left[\frac{(x_i - x_j)^2}{r_{ij}} + \frac{(y_i - y_j)^2}{r_{ij}} + \frac{(z_i - z_j)^2}{r_{ij}}\right]\frac{d\mathcal{V}(r_{ij})}{dr_{ij}} = r_{ij}\frac{d\mathcal{V}(r_{ij})}{dr_{ij}} \tag{5.86}$$

When we include the contributions from all pairs of atoms, we obtain:

$$W_{\text{real}} = \sum_{i=1}^{N}\sum_{j=i+1}^{N} r_{ij}\frac{d\mathcal{V}(r_{ij})}{dr_{ij}} \tag{5.87}$$

Appendix 5.4 Translating particle back into central box

Table 5.1 Formulae to translate particles back into the central box for three box shapes. From Smith W 1983. The Periodic Boundary Condition in Non-Cubic MD Cells: Wigner–Seitz Cells with Reflection Symmetry. *CCP5 Quarterly* **10**:37–42. Table 5.1 contains references to several built-in FORTRAN function. The AINT function returns the integral part of its argument, e.g. AINT(3.4) = 3.0; AINT(4.7) = 4.0; AINT(−0.5) = 0.0) and AINT(−1.7) = −1.0. ANINT returns the nearest integer, so ANINT(0.49) = 0.0 and ANINT(0.51) = 1.0. SIGN(x, y) returns $|x|$ if $y \geq 0$ and $-|x|$ if $y < 0$. ABS(x) returns the absolute value of x, $|x|$.

Rectangular box, side 2a (x) by 2b (y) by 2c (z)	$x = x - 2 \times a \times$ AINT(x/a) $y = y - 2 \times b \times$ AINT(y/b) $z = z - 2 \times c \times$ AINT(z/c) A common alternative is: $x = x - a \times$ ANINT(x/a) $y = y - b \times$ ANINT(y/b) $z = z - c \times$ ANINT(z/c)

Truncated octahedron derived from cube of side 2a	$x = x - 2 \times a \times$ AINT(x/a)
	$y = y - 2 \times b \times$ AINT(y/a)
	$z = z - 2 \times c \times$ AINT(z/a)
	if (ABS(x) + ABS(y) + ABS(z)) $\geq 1.5 \times$ A then
	  $x = x -$ SIGN(a, x)
	  $y = y -$ SIGN(a, y)
	  $z = z -$ SIGN(a, z)
	endif

Hexagonal prism of length 2a (in z direction) and distance between opposite faces of the hexagon 2b	$z = z - 2 \times a \times$ AINT(z/a)
	$x = x - 2 \times b \times$ AINT(x/b)
	if (ABS(x) + $\sqrt{3} \times$ ABS(y)) $\geq 2 \times$ B then
	  $x = x -$ SIGN(b, x)
	  $y = y -$ SIGN($\sqrt{3} \times b$, y)
	endif

Further reading

Allen M P and D J Tildesley 1987. *Computer Simulation of Liquids*. Oxford, Oxford University Press.

Bradbury T C 1968. Theoretical Mechanics. Krieger.

Chandler D 1987. *Introduction to Modern Statistical Mechanics*. New York, Oxford University Press.

Hansen J P and I R McDonald 1976. *Theory of Simple Liquids*. London, Academic Press.

Smith P E and van Gunsteren W F 1993. Methods for the Evaluation of Long Range Electrostatic Forces. In van Gunsteren W F, P K Weiner and A J Wilkinson (Editors). *Computer Simulation of Biomolecular Systems*. Leiden, ESCOM.

van Gunsteren W F and H J C Berendsen 1990. Computer Simulation of Molecular Dynamics: Methodology, Applications and Perspectives in Chemistry. *Angewandte Chemie International Edition in English* **29**:992–1023.

References

Adams D J 1983. Alternatives to the Periodic Cube in Computer Simulation. *CCP5 Quarterly* **10**:30–36.

Alper H E and R M Levy 1989. Computer-Simulations of the Dielectric-Properties of Water—Studies of the Simple Point-Charge and Transferable Intermolecular Potential Models. *Journal of Chemical Physics* **91**:1242–1251.

Ding H-Q, N Karasawa and W A Goddard III, 1992a. Atomic Level Simulations on a Million Particles: The Cell Multipole Method for Coulomb and London Nonbonding Interactions. *The Journal of Chemical Physics* **97**:4309–4315.

Ding H-Q, N Karasawa and W A Goddard III, 1992b. The Reduced Cell Multipole Method for Coulomb Interactions in Periodic Systems with Million-Atom Unit Cells. *Chemical Physics Letters* **196**:6–10.

Ewald P 1921. Due Berechnung optischer und elektrostatischer Gitterpotentiale. *Ann. Phys.* **64**:253–287.

Friedman H L 1975. Image Approximation to the Reaction Field. *Molecular Physics* **29**:1533–1543.

Greengard L and V I Roklin 1987. A Fast Algorithm for Particle Simulations. *Journal of Computational Physics* **73**:325–348.

Thompson S M 1983. Use of Neighbour Lists in Molecular Dynamics. *CCP5 Quarterly* **8**:20–28.

van Gunsteren W F and H J C Berendsen 1986. *GROMOS User Guide*.

Verlet L 1967. Computer 'Experiments' on Classical Fluids. II. Equilibrium Correlation Functions. *Physical Review* **165**:201–204.

6 Molecular dynamics simulation methods

6.1 Introduction

In molecular dynamics, successive configurations of the system are generated by integrating Newton's laws of motion. The result is a trajectory that specifies how the positions and velocities of the particles in the system vary with time. Newton's laws of motion can be stated as follows:

1. A body continues to move in a straight line at constant velocity unless a force acts upon it.
2. Force equals the rate of change of momentum.
3. To every action there is an equal and opposite reaction.

The trajectory is obtained by solving the differential equations embodied in Newton's second law ($F = ma$):

$$\frac{d^2 x_i}{dt^2} = \frac{F_{x_i}}{m_i} \tag{6.1}$$

This equation describes the motion of a particle of mass m_i along one coordinate (x_i) with F_{x_i} being the force on the particle in that direction.

It is helpful to distinguish three different types of problem to which Newton's laws of motion may be applied. In the simplest case, no force acts on each particle between collisions. From one collision to the next, the position of the particle thus changes by $\mathbf{v}_i \delta t$, where $\mathbf{v}_i$ is the (constant) velocity and δt is the time between collisions. In the second situation, the particle experiences a constant force between collisions. An example of this type of motion would be that of a charged particle moving in a uniform electric field. In the third case, the force on the particle depends on its position relative to the other particles. Here the motion is often very difficult, if not impossible to describe analytically, due to the coupled nature of the particles' motions.

6.2 Molecular dynamics using simple models

The first molecular dynamics simulation of a condensed phase system was performed by Alder and Wainwright in 1957 using a hard-sphere model [Alder and

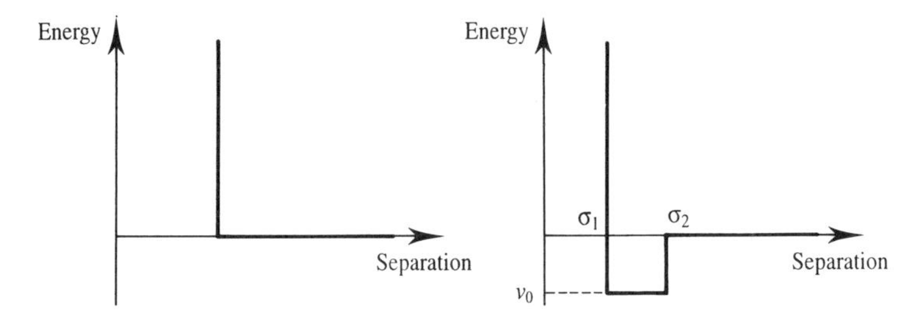

Fig. 6.1 The hard-sphere and square-well potentials.

Wainwright 1957]. In this model, the spheres move at constant velocity in straight lines between collisions. All collisions are perfectly elastic and occur when the separation between the centres of the spheres equals the sphere diameter. The pair potential thus has the form shown in Figure 6.1. Some early simulations also used the square-well potential, where the interaction energy between two particles is zero beyond a cutoff distance σ_2; infinite below a smaller cutoff distance σ_1; and equal to $\mathscr{v}_0$ between the two cutoff values (Figure 6.1). The steps involved in the hard-sphere calculation are as follows:

1. Identify the next pair of spheres to collide and calculate when the collision will occur.
2. Calculate the positions of all the spheres at the collision time.
3. Determine the new velocities of the two colliding spheres after the collision.
4. Repeat from 1 until finished.

The new velocities of the colliding spheres are calculated by applying the principle of conservation of linear momentum.

Simple interaction models such as the hard-sphere potential obviously suffer from many deficiencies but have nevertheless provided many useful insights into the microscopic nature of fluids. The early workers were particularly keen to quantify the differences between the solid and fluid phases; it is interesting to note that such investigations were facilitated by early molecular graphics systems which enabled the trajectories of the particles to be represented simultaneously (Figure 6.2).

6.3 Molecular dynamics with continuous potentials

In more realistic models of intermolecular interactions, the force on each particle will change whenever the particle changes its position, or whenever any of the other particles with which it interacts changes position. The first simulation using

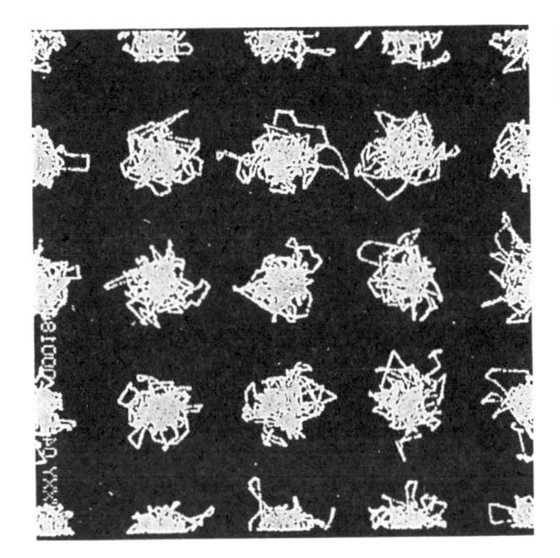
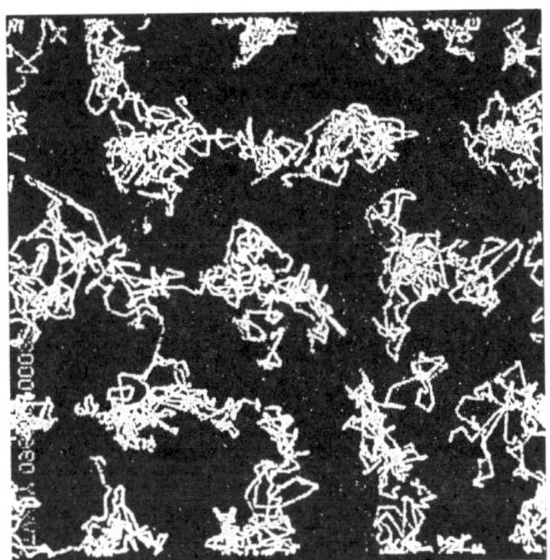

Fig. 6.2 Molecular graphics representation of the paths generated by 32 hard spherical particles in the solid (left) and fluid (right) phase. Reproduced from Alder B J and TE Wainwright 1959. Studies in Molecular Dynamics. I. General Method. *The Journal of Chemical Physics* **31**: 459–466.

continuous potentials was of argon by A Rahman [Rahman 1964], who also performed the first simulation of a molecular liquid (water) [Rahman and Stillinger 1971]) and made many other important methodological contributions in molecular dynamics. Under the influence of a continuous potential the motions of all the particles are coupled together, giving rise to a many-body problem that cannot be solved analytically. Under such circumstances the equations of motion are integrated using a *finite difference method.*

6.3.1 Finite difference methods

Finite difference techniques are used to generate molecular dynamics trajectories with continuous potential models, which we will assume to be pairwise additive. The essential idea is that the integration is broken down into many small stages, each separated in time by a fixed time δt. The total force on each particle in the configuration at a time t is calculated as the vector sum of its interactions with other particles. From the force we can determine the accelerations of the particles which are then combined with the positions and velocities at a time t to calculate the positions and velocities at a time $t+\delta t$. The force is assumed to be constant during the time step. The forces on the particles in their new positions are then determined, leading to new positions and velocities at time $t+2\delta t$, and so on.

There are many algorithms for integrating the equations of motion using finite difference methods, several of which are commonly used in molecular dynamics calculations. All algorithms assume that the positions and dynamical properties (velocities, accelerations, etc.) can be approximated as Taylor series expansions:

$$\mathbf{r}(t+\delta t) = \mathbf{r}(t) + \delta t\mathbf{v}(t) + \tfrac{1}{2}\delta t^2\mathbf{a}(t) + \tfrac{1}{6}\delta t^3\mathbf{b}(t) + \tfrac{1}{24}\delta t^4\mathbf{c}(t) + \cdots \tag{6.2}$$

$$\mathbf{v}(t+\delta t) = \mathbf{v}(t) + \delta t\mathbf{a}(t) + \tfrac{1}{2}\delta t^2\mathbf{b}(t) + \tfrac{1}{6}\delta t^3\mathbf{c}(t) + \cdots \tag{6.3}$$

$$\mathbf{a}(t+\delta t) = \mathbf{a}(t) + \delta t\mathbf{b}(t) + \tfrac{1}{2}\delta t^2\mathbf{c}(t) \cdots \tag{6.4}$$

$$\mathbf{b}(t+\delta t) = \mathbf{b}(t) + \delta t\mathbf{c}(t) + \cdots \tag{6.5}$$

$\mathbf{v}$ is the velocity (the first derivative of the positions with respect to time); $\mathbf{a}$ is the acceleration (the second derivative), $\mathbf{b}$ is the third derivative, and so on. The *Verlet algorithm* [Verlet 1967] is probably the most widely used method for integrating the equations of motion in a molecular dynamics simulation. The Verlet algorithm uses the positions and accelerations at time t, and the positions from the previous step, $\mathbf{r}(t-\delta t)$ to calculate the new positions at $t+\delta t$, $\mathbf{r}(t+\delta t)$. We can write down the following relationships between these quantities and the velocities at time t:

$$\mathbf{r}(t+\delta t) = \mathbf{r}(t) + \delta t\mathbf{v}(t) + \tfrac{1}{2}\delta t^2\mathbf{a}(t) + \cdots \tag{6.6}$$

$$\mathbf{r}(t-\delta t) = \mathbf{r}(t) - \delta t\mathbf{v}(t) + \tfrac{1}{2}\delta t^2\mathbf{a}(t) - \cdots \tag{6.7}$$

Adding these two equations gives

$$\mathbf{r}(t+\delta t) = 2\mathbf{r}(t) - \mathbf{r}(t-\delta t) + \delta t^2\mathbf{a}(t) \tag{6.8}$$

The velocities do not explicitly appear in the Verlet integration algorithm. The velocities can be calculated in a variety of ways; a simple approach is to divide the difference in positions at times $t+\delta t$ and $t-\delta t$ by $2\delta t$:

$$\mathbf{v}(t) = [\mathbf{r}(t+\delta t) - \mathbf{r}(t-\delta t)]/2\delta t \tag{6.9}$$

Alternatively, the velocities can be estimated at the half-step, $t+\frac{1}{2}\delta t$:

$$\mathbf{v}(t+\tfrac{1}{2}\delta t) = [\mathbf{r}(t+\delta t) - \mathbf{r}(t)]/\delta t \tag{6.10}$$

Implementation of the Verlet algorithm is straightforward and the storage requirements are modest, comprising two sets of positions ($\mathbf{r}(t)$ and $\mathbf{r}(t-\delta t)$) and the accelerations $\mathbf{a}(t)$. One of its drawbacks is that the positions $\mathbf{r}(t+\delta t)$ are obtained by adding a small term ($\delta t^2\mathbf{a}(t)$) to the difference of two much larger terms, $2\mathbf{r}(t)$ and $\mathbf{r}(t-\delta t)$. This may lead to a loss of precision. The Verlet algorithm has some other disadvantages. The lack of an explicit velocity term in the equations makes it difficult to obtain the velocities, and indeed the velocities are not available until the positions have been computed at the next step. In addition, it is not a self-starting algorithm; the new positions are obtained from the current positions $\mathbf{r}(t)$ and the positions from the previous time step, $\mathbf{r}(t-\delta t)$. At $t=0$ there is obviously only one set of positions and so it is necessary to employ some other means to obtain positions at $t-\delta t$. One way to obtain $\mathbf{r}(t-\delta t)$ is to use the

Taylor series, equation (6.2), truncated after the first term. Thus, $\mathbf{r}(-\delta t) = \mathbf{r}(0) - \delta t\mathbf{v}(0) - \delta t\mathbf{v}(0)$.

Several variations on the Verlet algorithm have been developed. The *leap-frog* algorithm [Hockney 1970] uses the following relationships:

$$\mathbf{r}(t + \delta t) = \mathbf{r}(t) + \delta t\mathbf{v}(t + \tfrac{1}{2}\delta t) \tag{6.11}$$

$$\mathbf{v}(t + \tfrac{1}{2}\delta t) = \mathbf{v}(t - \tfrac{1}{2}\delta t) + \delta t\mathbf{a}(t) \tag{6.12}$$

To implement the leap-frog algorithm, the velocities $\mathbf{v}(t + \frac{1}{2}t)$ are first calculated, from the velocities at time $t - \frac{1}{2}\delta t$ and the accelerations at time t. The positions $\mathbf{r}(t + \delta t)$ are then deduced from the velocities just calculated together with the positions at time $\mathbf{r}(t)$ using equation (6.11). The velocities at time t can be calculated from

$$\mathbf{v}(t) = \tfrac{1}{2}[\mathbf{v}(t + \tfrac{1}{2}\delta t) + \mathbf{v}(t - \tfrac{1}{2}\delta t)] \tag{6.13}$$

The velocities thus 'leap-frog' over the positions to give their values at $t + \frac{1}{2}\delta t$ (hence the name). The positions then leap over the velocities to give their new values at $t + \delta t$, ready for the velocities at $t + \frac{3}{2}\delta t$, and so on. The leap-frog method has two advantages over the standard Verlet algorithm: it explicitly includes the velocity and also does not require the calculation of the differences of large numbers. However, it has the obvious disadvantage that the positions and velocities are not synchronised. This means that it is not possible to calculate the kinetic energy contribution to the total energy at the same time as the positions are defined (from which the potential energy is determined).

The *velocity Verlet* method [Swope *et al.* 1982] gives positions, velocities and accelerations at the same time and does not compromise precision:

$$\mathbf{r}(t + \delta t) = \mathbf{r}(t) + \delta t\mathbf{v}(t) + \tfrac{1}{2}\delta t^2\mathbf{a}(t) \tag{6.14}$$

$$\mathbf{v}(t + \delta t) = \mathbf{v}(t) + \tfrac{1}{2}\delta t[\mathbf{a}(t) + \mathbf{a}(t + \delta t)] \tag{6.15}$$

Beeman's algorithm [Beeman 1976] is also related to the Verlet method:

$$\mathbf{r}(t + \delta t) = \mathbf{r}(t) + \delta t\mathbf{v}(t) + \tfrac{2}{3}\delta t^2\mathbf{a}(t) - \tfrac{1}{6}\delta t^2\mathbf{a}(t - \delta t) \tag{6.16}$$

$$\mathbf{v}(t + \delta t) = \mathbf{v}(t) + \tfrac{1}{3}\delta t\mathbf{a}(t) + \tfrac{5}{6}\delta t\mathbf{a}(t) - \tfrac{1}{6}\delta t\mathbf{a}(t - \delta t) \tag{6.17}$$

The Beeman integration scheme uses a more accurate expression for the velocity. As a consequence it often gives better energy conservation, because the kinetic energy is calculated directly from the velocities. However, the expressions used are more complex than those of the Verlet algorithm and so it is computationally more expensive.

We have already encountered four different integration methods, with more to come! Why should we use one method in preference to another? What features characterise a 'good' integration method? As with any other computer algorithm, an ideal integration scheme should be fast, require minimal memory and be easy to program. However, for most molecular dynamics simulations these issues are of secondary importance; most calculations do not make significant memory demands of even a modest workstation, and the time required for the integration is usually trivial compared to the other parts of the calculation. The most demanding part of a molecular dynamics simulation is invariably the calculation of the force on each particle in the system. More important considerations are that the integration algorithm should conserve energy and momentum, be time-reversible, and should permit a long time step, δt, to be used. The size of the time step is particularly relevant to the computational demands as a simulation using a long time step will require fewer iterations to cover a given amount of phase space. A less important requirement is that the integration algorithm should give the same results as an exact, analytical trajectory (this can be tested using simple problems for which an analytical solution can be derived). We would in any case expect the calculated trajectory to deviate from the exact trajectory because the computer can only store numbers to a given precision.

The *order* of an integration method is the degree to which the Taylor series expansion, equation (6.2) is truncated: it is the lowest term that is not present in the expansion. The order may not always be apparent from the formulae used. For example, the highest order derivative that appears in the Verlet formulae is the second, $\mathbf{a}(t)$, yet the Verlet algorithm is in fact a fourth order method. This is because the third-order terms, which cancel when equation (6.6) is added to equation (6.7), are still implied in the expansion:

$$\mathbf{r}(t+\delta t) = \mathbf{r}(t) + \delta t\mathbf{v}(t) + \tfrac{1}{2}\delta t^2\mathbf{a}(t) + \tfrac{1}{6}\delta t^3\mathbf{b}(t) + \tfrac{1}{24}\delta t^4\mathbf{c}(t) \tag{6.18}$$

$$\mathbf{r}(t-\delta t) = \mathbf{r}(t) - \delta t\mathbf{v}(t) + \tfrac{1}{2}\delta t^2\mathbf{a}(t) - \tfrac{1}{6}\delta t^3\mathbf{b}(t) + \tfrac{1}{24}\delta t^4\mathbf{c}(t) \tag{6.19}$$

6.3.2 *Predictor–corrector integration methods*

The predictor–corrector methods [Gear 1971] form a general family of integration algorithms from which one can select a scheme that is correct to a given order. These methods have three basic steps. First, new positions, velocities, accelerations and higher-order terms are predicted according to the Taylor expansion, equations (6.2)–(6.4). In the second stage, the forces are evaluated at the new positions to give accelerations $\mathbf{a}(t+\delta t)$. These accelerations are then compared with the accelerations that are predicted from the Taylor series expansion, $\mathbf{a}^c(t+\delta t)$. The difference between the predicted and calculated accelerations are then used to 'correct' the positions, velocities etc. in the correction step:

$$\Delta\mathbf{a}(t+\delta t) = \mathbf{a}^c(t+\delta t) - \mathbf{a}(t+\delta t) \tag{6.20}$$

Then

$$\mathbf{r}^c(t+\delta t) = \mathbf{r}(t+\delta t) + c_0\Delta\mathbf{a}(t+\delta t) \tag{6.21}$$

$$\mathbf{v}^c(t+\delta t) = \mathbf{v}(t+\delta t) + c_1\Delta\mathbf{a}(t+\delta t) \tag{6.22}$$

$$\mathbf{a}^c(t+\delta t)/2 = \mathbf{a}(t+\delta t)/2 + c_2\Delta\mathbf{a}(t+\delta t) \tag{6.23}$$

$$\mathbf{b}^c(t+\delta t)/6 = \mathbf{b}(t+\delta t)/6 + c_3\Delta\mathbf{a}(t+\delta t) \tag{6.24}$$

Gear has suggested 'best' values of the coefficients c_0, $c_1 \ldots$ The set of coefficients to use depends upon the order of the Taylor series expansion. In equations (6.21)–(6.24) the expansion has been truncated after the third derivative of the positions (i.e. $\mathbf{b}(t)$). The appropriate set of coefficients to use in this case is $c_0 = 1/6$, $c_1 = 5/6$, $c_2 = 1$ and $c_3 = 1/3$.

The storage required for the Gear predictor–corrector algorithm is $3 \times (O+1)N$, where O is the highest order differential used in the Taylor series expansion and N is the number of atoms. Thus the storage required for our example is $15N$, which is rather more than for the Verlet algorithm which uses $9N$. More importantly, the Gear algorithm requires two time-consuming force evaluations per time step, though this is not necessarily a disadvantage as it may permit a time step more than twice as long as an alternative algorithm.

There are many variants of the 'predictor–corrector' theme; of these we will only mention the algorithm used by Rahman in the first molecular dynamics simulations with continuous potentials [Rahman 1964]. In this method, the first step is to predict new positions as follows:

$$\mathbf{r}(t + \delta t) = \mathbf{r}(t - \delta t) + 2\delta t \mathbf{v}(t) \tag{6.25}$$

New accelerations are calculated at these new positions, in the usual way. These accelerations are then used to generate a set of new velocities, and then corrected positions:

$$\mathbf{v}(t + \delta t) = \mathbf{v}(t) + \tfrac{1}{2}\delta t(\mathbf{a}(t + \delta t) + \mathbf{a}(t)) \tag{6.26}$$

$$\mathbf{r}^{\mathrm{c}}(t + \delta t) = \mathbf{r}(t) + \tfrac{1}{2}\delta t(\mathbf{v}(t) + \mathbf{v}(t + \delta t)) \tag{6.27}$$

The acceleration can then be recalculated at the new corrected positions, to given new velocities. The method then iterates over the two equations (6.26) and (6.27). Two or three passes are usually required to achieve consistency, with a force evaluation at each step. The computational demands of this scheme mean that it is now rarely used, though it does give accurate solutions of the equations of motion.

6.3.3 Which integration algorithm is most appropriate?

The wide variety of integration schemes available can make it difficult to decide which is the most appropriate one to use. Various factors may need to be taken into account when deciding which is most appropriate. Clearly, the computational effort required is a major consideration. As we have already indicated, an algorithm that is nominally more expensive (for example, because it requires more than one force evaluation per iteration) may permit a significantly longer time step to be used and so in fact be more cost-effective. One of the most important considerations is energy conservation; this can be calculated as the root-mean-square fluctuation and is often plotted against the time step as shown in Figure 6.3. In Appendix 6.1 we show why energy conservation would be expected in a molecular dynamics simulation. The kinetic and potential energy components would be expected to fluctuate in equal and opposite directions; this is also shown in Figure 6.3.

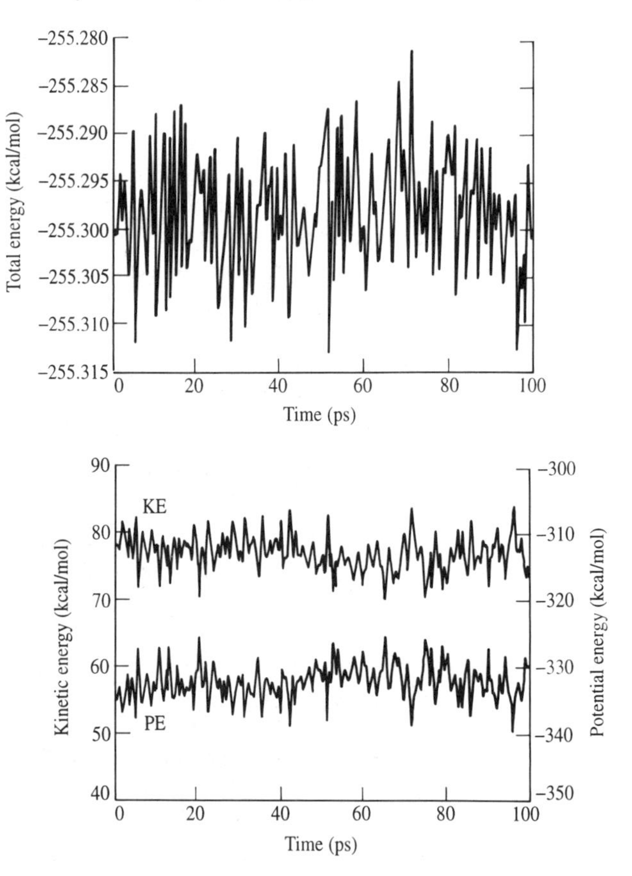

Fig. 6.3 Variation in total energy versus time for the production phase of a molecular dynamics simulation of 256 argon atoms at a temperature of 100K and a density of 1.396 g cm^{-3}(top). The time step was 10 fs and the equations of motion were integrated using the velocity Verlet algorithm. The variations in the kinetic and potential energies are also shown. The graphs have different scales.

As the time step increases, so the RMS energy fluctuation also increases. For the argon simulation reported in Figure 6.3, the RMS fluctuation in the total energy is approximately 0.006 kcal/mol and the RMS fluctuations in the kinetic and potential energies are approximately 2.5 kcal/mol. With a time step of 25 fs the RMS fluctuation rises to 0.04 kcal/mol and with a time step of 5 fs the value is 0.002 kcal/mol. A variation of 1 part in 10^4 is generally considered acceptable. The different algorithms may vary in the rate at which the error varies with the time step. For example, it has been shown that for short time steps the predictor–corrector methods may be more accurate, but for longer time steps the Verlet algorithm may be better [Fincham and Heyes 1982]. Other factors that may be important when choosing an integration algorithm include the memory required; the synchronisation of positions and velocities; whether they are self-starting (some methods require properties at $t - \delta t$, which obviously do not exist); and whether it is possible to perform simulations in other ensembles such as the isothermal–isobaric ensemble.

6.3.4 Choosing the time step

There are no hard and fast rules for calculating the most appropriate time step to use in a molecular dynamics simulation; too small and the trajectory will cover only a limited proportion of the phase space; too large and instabilities may arise in the integration algorithm due to high energy overlaps between atoms. These two extremes are illustrated in Figure 6.4. Such instabilities would certainly lead to a violation of energy and linear momentum conservation and could result in a program failure due to numerical overflow. The effect of changing the time step on the dynamics can be illustrated using a very simple system consisting of two argon atoms interacting under the Lennard–Jones potential. The behaviour of this system can be determined analytically and so compared with the numerical integration. Suppose the argon atoms are moving towards each other along the x axis with initial velocities of 353 ms^{-1} (this corresponds to the most probable speed of argon at 300 K). We can then plot how the interatomic distance varies with time and compare it to the analytical potential. The results obtained using two time steps (10 fs and 50 fs) are shown in Figure 6.5. In both cases the numerical trajectory initially lags behind the analytical one, but then as the atoms pass

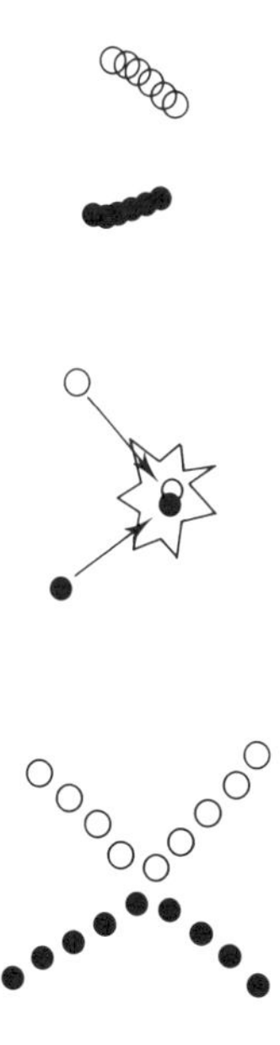

Fig. 6.4 With a very small time step (top) phase space is covered very slowly; a large time step (middle) gives instabilities. With an appropriate time step (bottom) phase space is covered efficiently and collisions occur smoothly.

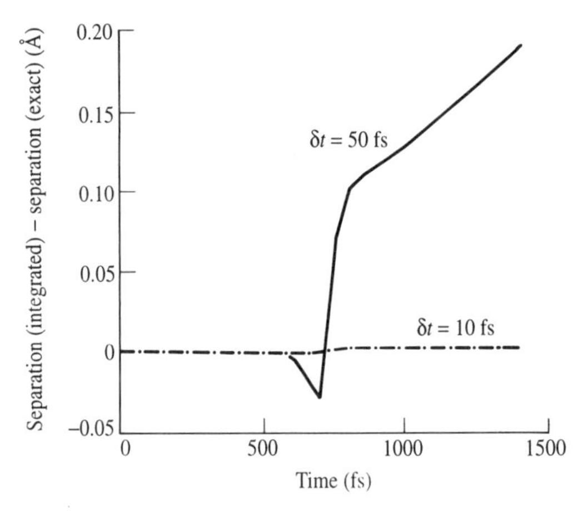

Fig. 6.5 Difference between the exact and numerical trajectories for the approach of two argon atoms with time steps of 10 fs and 50 fs.

through their minimum energy separation and move up the repulsive barrier the atoms 'jump through' the energy barrier. This leads to a gain in energy and the atoms then move apart with velocities that are slightly too high. In both numerical trajectories the total energy rises after the collision. Unfortunately, the atoms move most quickly and take the largest steps in the very region (i.e. near the energy minimum) where it would be best to take the smallest steps. The total error is correlated with the time step, with the largest errors arising for the largest time steps. Of course, with a small time step much more computer time will be required for a given length of calculation; the aim is to find the correct balance between simulating the 'correct' trajectory and covering the phase space. If the time step is too large, then the trajectory will 'blow up', as can be seen for the argon dimer system with a time step of 100 fs, Figure 6.6.

When simulating an atomic fluid the time step should be small compared to the mean time between collisions. When simulating flexible molecules a useful guide is that the time step should be approximately one tenth the time of the shortest period of motion. In flexible molecules, the highest frequency vibrations are due to bond stretches, especially those of bonds to hydrogen atoms. A C–H bond vibrates with a repeat period of approximately 10 fs. The timescales of some typical motions together with appropriate time steps are shown in Table 6.1 which can be used to choose an appropriate time step.

The requirement that the time step is approximately one order of magnitude smaller than the shortest motion is clearly a severe restriction, particularly as these high-frequency motions are usually of relatively little interest and have a minimal effect on the overall behaviour of the system. One solution to this problem is to 'freeze out' such vibrations by constraining the appropriate bonds to their equilibrium values while still permitting the rest of the degrees of freedom to vary under the intramolecular and intermolecular forces present. This enables a longer time step to be used. We will consider such constraint dynamics methods in section 6.5.

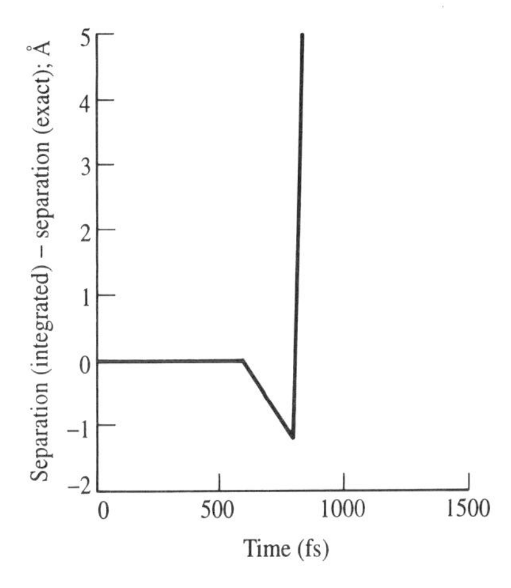

Fig. 6.6 Difference between exact and numerical trajectory for the approach of two argon atoms for a time step of 100 fs. The simulation 'blows up'.

6.4 Setting up and running a molecular dynamics simulation

In this section we will examine some of the steps involved in performing a molecular dynamics simulation in the microcanonical ensemble. First it is necessary to establish an initial configuration of the system. As discussed in section 5.4.2, the initial configuration may be obtained from experimental data, from a theoretical model, or from a combination of the two. It is also necessary to assign initial velocities to the atoms. This can be done by randomly selecting from a Maxwell–Boltzmann distribution at the temperature of interest:

$$p(v_{ix}) = \left(\frac{m_i}{2\pi k_B T}\right)^{1/2} \exp\left[-\frac{1}{2}\frac{m_i v_{ix}^2}{k_B T}\right] \tag{6.28}$$

The Maxwell–Boltzmann equation provides the probability that an atom i of mass m_i has a velocity v_{ix} in the x direction at a temperature T. A Maxwell–Boltzmann distribution is a Gaussian distribution, which can be obtained using a

Table 6.1 The different types of motion present in various systems together with suggested time steps.

System	Types of motion present	Suggested time step(s)
Atoms	translation	10^{-14}
Rigid molecules	translation and rotation	5×10^{-15}
Flexible molecules, rigid bonds	translation, rotation, torsion	2×10^{-15}
Flexible molecules, flexible bonds	translation, rotation, torsion, vibration	10^{-15} or 5×10^{-16}

random number generator. Most random number generators are designed to produce random numbers that are uniform in the range 0 to 1. However, it is relatively straightforward to convert such a random number generator to sample from a Gaussian distribution (or indeed from one of several other distributions [Rubinstein 1981]). The probability of generating a value from a Gaussian (normal) distribution with mean $\langle x\rangle$ and variance σ^2 ($\sigma^2 = \langle(x - \langle x\rangle)^2\rangle$) is:

$$p(x) = \frac{1}{\sqrt{2\pi\sigma^2}}\exp\left[-\frac{(x - \langle x\rangle)^2}{2\sigma^2}\right] \tag{6.29}$$

One option is to first generate two random numbers ξ_1 and ξ_2 between 0 and 1. The corresponding two numbers from the normal distribution are then calculated using

$$x_1 = \sqrt{(-2\ln\xi_1)}\cos(2\pi\xi_2) \text{ and } x_2 = \sqrt{(-2\ln\xi_1)}\sin(\pi\xi_2) \tag{6.30}$$

An alternative approach is to generate 12 random numbers $\xi_1 \ldots \xi_{12}$ and then calculate

$$x = \sum_{i=1}^{12} \xi_i - 6 \tag{6.31}$$

These two methods generate random numbers in the normal distribution with zero mean and unit variance. A number (x) generated from this distribution can be related to its counterpart (x') from another Gaussian distribution with mean $\langle x'\rangle$, and variance σ using

$$x' = \langle x'\rangle + \sigma x \tag{6.32}$$

The initial velocities may also be chosen from a uniform distribution or from a simple Gaussian distribution. In either case the Maxwell–Boltzmann distribution of velocities is usually rapidly achieved.

The initial velocities are often adjusted so that the total momentum of the system is zero. Such a system then samples from the constant *NVE***P** ensemble. To set the total linear momentum of the system to zero, the sum of the components of the atomic momenta along the *x*, *y* and *z* axes are calculated. This gives the total momentum of the system in each direction which, when divided by the total mass, is subtracted from the atomic velocities to give an overall momentum of zero.

Having set up the system and assigned the initial velocities, the simulation proper can commence. At each step the force on each atom must be calculated by differentiating the potential function. The force on an atom may include contributions from the various terms in the force field such as bonds, angles, torsional terms and non-bonded interactions. The force is straightforward to calculate for two atoms interacting under the Lennard–Jones potential:

$$\mathbf{f}_{ij} = \frac{\mathbf{r}_{ij}}{|\mathbf{r}_{ij}|}\frac{24\varepsilon}{\sigma}\left[2\left(\frac{\sigma}{r_{ij}}\right)^{13} - \left(\frac{\sigma}{r_{ij}}\right)^{7}\right] \tag{6.33}$$

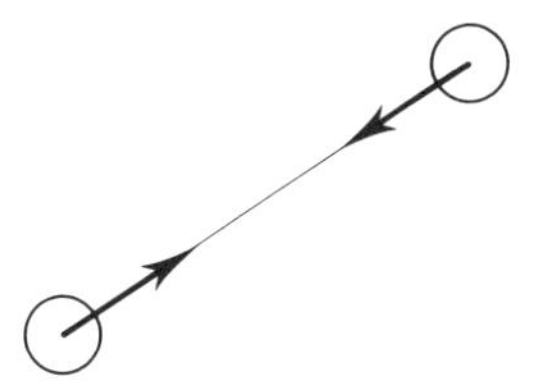

Fig. 6.7 The force between two particles acts along the line joining their centres of mass, in accordance with Newton's third law.

The force between the two atoms is equal in magnitude and opposite in direction and applies along the line connecting the two nuclear centres, in accordance with Newton's third law, Figure 6.7. It is necessary to calculate the force between each atom pair just once. This is most easily achieved by arranging to compute the force between an atom and those atoms with a higher index (i.e. for an atom i the forces are calculated with atoms i + 1, i + 2, ... *N*). Having calculated the force between an atom i and an atom with a higher index j, minus the force is added to the accumulating sum of the forces on j. The force calculation is most easily implemented using two loops, as outlined in the following pseudocode:

```
set elements on force array to zero
while atom1 = 1 to N − 1
        while atom2 = atom1 + 1 to N
                calculate force on atom1 due to interaction with atom2
                add the force to the array element, atom1
                subtract the force from the appropriate force array element atom2
        enddo
enddo
```

At the end of these two loops, the total force on each atom is known. A consequence of the fact that the force between the two atoms is equal and opposite is that the neighbour list for each atom need only contain those atoms with a higher number as the force on an atom due to interactions with lower numbered atoms will be calculated earlier in the loop. This organisation of the neighbour list contrasts with the structure used for Monte Carlo simulations, where all the neighbours of each atom (with both lower and higher indices) must be stored.

Analytical expressions for the forces due to other terms in the molecular mechanics potential function have been published for most of the functional forms encountered in common force fields. These expressions can seem rather complicated because the intramolecular terms (e.g. bonds, angles, torsions) are calculated in terms of the internal coordinates whereas molecular dynamics is typically performed using Cartesian coordinates. The chain rule must therefore be used to obtain the desired functional forms. However, the resulting expressions are relatively easy to implement in a computer program.

The first stage of a molecular dynamics simulation is the equilibration phase, the purpose of which is to bring the system to equilibrium from the starting configuration. During equilibration, various parameters are monitored together with the actual configurations. When these parameters achieve stable values then the production phase can commence. It is during the production phase that thermodynamic properties and other data are calculated. The parameters that are used to characterise whether equilibrium has been reached depend to some extent on the system being simulated, but invariably include the kinetic, potential and total energies, the velocities, the temperature and the pressure. As we have indicated, the kinetic and potential energies would be expected to fluctuate in a simulation in the microcanonical ensemble but the total energy should remain constant. The components of the velocities should describe a Maxwell–Boltzmann distribution (in all three directions x, y and z) and the kinetic energy should be equally distributed among the three directions x, y and z. It is usually desired to perform a simulation at a specified temperature and so it is common practice to adjust the temperature of the system by scaling the velocities (see section 6.7.1) during the equilibration phase. During the production phase the temperature is a variable of the system. Order parameters may be calculated to monitor changes in structure which can supplement visual examination of the evolving trajectory.

When simulating an inhomogeneous system a more detailed equilibration procedure is usually desirable. A typical procedure suitable for a molecular dynamics simulation of a macromolecular solute such as a protein in solution would be as follows. First, the solvent alone together with any mobile counterions is subject to energy minimisation with the solute kept fixed in its initial conformation. The solvent and any counterions are then allowed to evolve using either a molecular dynamics (or indeed Monte Carlo) simulation, again keeping the structure of the solute molecule fixed. This solvent equilibration phase should be sufficiently extensive to allow the solvent to completely readjust to the potential field of the solute. For molecular dynamics this implies that the length of this solvent equilibration phase should be longer than the relaxation time of the solvent (the time taken for a molecule to lose any 'memory' of its original orientation, which for water is about 10 ps). Next, the entire system (solute and solvent) is minimised. Only then does the molecular dynamics simulation of the whole system commence.

At the start of the production phase all counters are set to zero and the system is permitted to evolve. In a microcanonical ensemble no velocity scaling is performed during the production phase and so the temperature becomes a calculated property of the system. Various properties are routinely calculated and stored during the production phase for subsequent analysis and processing. Careful monitoring of these properties during the simulation can show whether the simulation is 'well behaved' or not; it may be necessary to restart a simulation if problems are encountered. It is also usual to store the positions, energies and velocities of configurations at regular intervals (e.g. every 5–20 time steps), from which other properties can be determined once the simulation has finished.

6.4.1 Calculating the temperature

Many thermodynamic properties can be calculated from a molecular dynamics simulation. Most of these were dealt with in section 5.2; here we just discuss the calculation of temperature. The instantaneous value of the temperature is related to the kinetic energy via the particles' momenta as follows:

$$\mathcal{K} = \sum_{i=1}^{N} \frac{|\mathbf{p}_i|^2}{2m_i} = \frac{k_B T}{2}(3N - N_c) \tag{6.34}$$

N_c is the number of constraints and so $3N - N_c$ is the total number of degrees of freedom. For an isolated system (i.e. for a simulation of a system *in vacuo*) the total translational momentum of the system and the total angular momentum are conserved and can be made equal to zero by an appropriate choice of initial velocities. For a simulation performed using periodic boundary conditions, the total linear momentum is conserved but the total angular momentum is not. It is common practice to choose a set of initial velocities that ensures that the total linear momentum and the total angular momentum are zero. As the system evolves, the linear momentum should remain zero but the angular momentum will not. Molecular dynamics with periodic boundary conditions thus strictly samples from the constant $NVE\mathbf{P}$ ensemble where $\mathbf{P}$ is the total linear momentum. This differs trivially from the standard microcanonical ensemble but it should be remembered that the appropriate number of degrees of freedom must be subtracted from the total when calculating the kinetic energy per degree of freedom. Specifically, for a system *in vacuo* where the total linear and angular momenta have been set to zero, six degrees of freedom need to be subtracted. For a simulation using periodic boundary conditions three degrees of freedom need to be subtracted if the centre-of-mass motion of the system is removed. In constraint dynamics, discussed in the next section, rather more degrees of freedom may be fixed and N_c must be calculated accordingly.

6.5 Constraint dynamics

The earliest molecular dynamics simulations using 'realistic' potentials were of atoms interacting under the Lennard–Jones potential. In such calculations the only forces on the atoms are those due to non-bonded interactions. It is rather more difficult to simulate molecules because the interaction between two non-spherical molecules depends upon their relative orientation as well as the distance between them. If the molecules are flexible then there will also be intramolecular interactions that give rise to changes in conformation. Clearly, the simplest model is to treat the species present as rigid bodies with no intramolecular conformational freedom. In such cases the dynamics of each molecule can often be considered in terms of translations of its centre of mass and rotations about its centre of mass. The force on the molecule equals the vector sum of all the forces acting at the centre of mass, and the rotational motion is determined by the torque about the centre of mass. To deal with these rotational motions is considerably more complicated than for the translational motions, but in favourable cases they can be programmed quite efficiently.

When the simulation involves conformationally flexible molecules then the motion is inevitably described in terms of the atomic Cartesian coordinates. The conformational behaviour of a flexible molecule is usually a complex superposition of different motions. The high frequency motions (e.g. bond vibrations) are usually of less interest than the lower frequency modes which often correspond to major conformational changes. Unfortunately, the time step of a molecular dynamics simulation is dictated by the highest frequency motion present in the system. It would therefore be of considerable benefit to be able to increase the time step without prejudicing the accuracy of the simulation. Constraint dynamics enables individual internal coordinates or combinations of specified coordinates to be constrained, or 'fixed' during the simulation without affecting the other internal degrees of freedom.

Before we consider in detail the use of constraint dynamics, it is helpful to establish the difference between *constraints* and *restraints*; we shall discuss the method of restrained molecular dynamics in a later chapter (see section 8.7). A constraint is a requirement that the system is forced to satisfy. As we shall see, in constraint dynamics bonds or angles are forced to adopt specific values throughout a simulation. When a bond or angle is restrained then it *is* able to deviate from the desired value; the restraint only acts to 'encourage' a particular value. Restraints are most easily incorporated using additional terms in the force field which impose a penalty for deviations from the reference value. An additional difference is that restrained degrees of freedom still have an energy $k_BT/2$ associated with them whereas constrained degrees of freedom do not.

The most commonly used method for applying constraints, particularly in molecular dynamics, is the SHAKE procedure of Ryckaert, Ciccotti and Berendsen [Ryckaert *et al.* 1977]. In constraint dynamics the equations of motion are solved while simultaneously satisfying the imposed constraints. Constrained systems have been much studied in classical mechanics; we shall illustrate the general principles using a simple system comprising a box sliding down a frictionless slope in two dimensions (Figure 6.8). The box is constrained to remain on the slope and so the box's x and y coordinates must always satisfy the equation of the slope (which we shall write as $y = mx + c$). If the slope were not present then the box would fall vertically downwards. Constraints are often categorised as *holonomic* or *non-holonomic*. Holonomic constraints can be expressed in the form

$$f(q_1, q_2, q_3, \ldots, t) = 0 \tag{6.35}$$

q_1, q_2, etc. are the coordinates of the particles. Non-holonomic constraints cannot be expressed in this way. For example, the motion of a particle constrained to lie on the surface of a sphere is subject to a holonomic constraint, but if the particle is able to fall off the sphere under the influence of gravity then the constraint becomes non-holonomic. A holonomic constraint that keeps a particle on the surface of a sphere can be written:

$$r^2 - a^2 = 0 \tag{6.36}$$

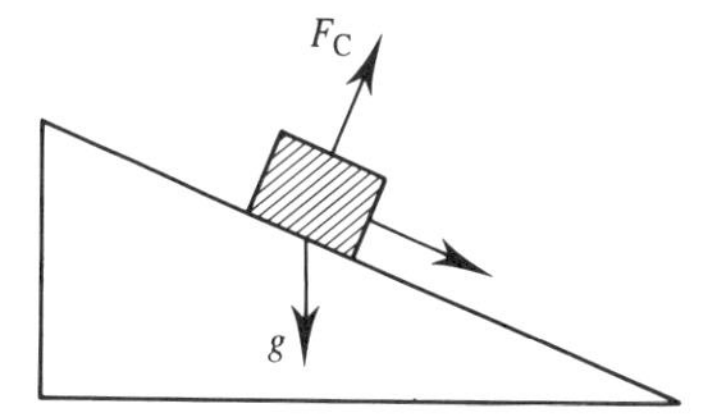

Fig. 6.8 A box sliding down a slope under the influence of gravityis subject to the constraint that it must remain on the slope. The constraint force F_c acts perpendicular to the direction of motion.

r is the distance of the particle from the origin where the sphere of radius a is centred. The equivalent non-holonomic constraint is written as an inequality:

$$r^2 - a^2 \geq 0 \tag{6.37}$$

SHAKE uses holonomic constraints. In a constrained system the coordinates of the particles are not independent and the equations of motion in each of the coordinate directions are connected. A second difficulty is that the magnitude of the constraint forces are unknown. Thus in the case of the box on the slope, the gravitational force acting on the box is in the y direction whereas the motion is down the slope. The motion is thus not in the same direction as the gravitational force. As such, the total force on the box can be considered to arise from two sources: one due to gravity and the other a constraint force that is perpendicular to the motion of the box (Figure 6.8). As there is no motion perpendicular to the surface of the slope, the constraint force does no work.

As we know, the motion of a system of N particles can be described in terms of $3N$ independent coordinates or degrees of freedom. If there are k holonomic constraints then the number of degrees of freedom is reduced to $3N - k$. It is possible, in principle at least, to find $3N - k$ independent coordinates (the *generalised coordinates*) that can then be used to solve the problem directly. For example, the motion of the box can be described using a single coordinate, q, along the direction of the slope. The component of the gravitational force that acts along the slope is $Mg \sin\theta$ and so the acceleration down the slope is $g \sin\theta$. The position at any time t can thus be obtained by integrating the following equation of motion:

$$\frac{d^2q}{dt^2} = g \sin\theta \tag{6.38}$$

The solution to this equation is:

$$q(t) = q(0) + t\dot{q}(0) + \frac{t^2}{2} g \sin\theta \tag{6.39}$$

$q(0)$ is the value of ξ at time $t = 0$ and $\dot{q}(0)$ is the initial velocity of the box along the slope. In this simple example it is quite easy to identify the single generalised coordinate that can be used to describe the motion in the constrained system. When there are many constraints, it can be difficult to determine the generalised coordinates. In any case, it is usually desirable to work with the atomic Cartesian coordinates. The motion of the box can be more generally described in terms of the Cartesian (x, y) coordinates of the box as follows.

Newton's equations in the x and y directions are:

$$M\frac{d^2x}{dt^2} = F_{cx} \tag{6.40}$$

$$M\frac{d^2y}{dt^2} = -Mg + F_{cy} \tag{6.41}$$

F_{cx} and F_{cy} are the components of the as yet unknown constraint force in the x and the y directions respectively. We know that the constraint force acts perpendicular to the slope, and so the ratio of its x and y components must be:

$$\frac{F_{cx}}{F_{cy}} = -m \tag{6.42}$$

The constraint force can be introduced into Newton's equations as a Lagrange multiplier (see section 1.10.5). To achieve consistency with the usual Lagrangian notation, we write F_{cy} as $-\lambda$ and so F_{cx} equals λm. Thus:

$$M\frac{d^2x}{dt^2} = \lambda m \tag{6.43}$$

$$M\frac{d^2y}{dt^2} = -Mg - \lambda \tag{6.44}$$

The two equations (6.43) and (6.44) contain three unknowns (d^2x/dt^2, d^2y/dt^2 and λ). A third equation that links x and y is the equation of the slope, which can be written in the following form:

$$\sigma = mx - y + c = 0 \tag{6.45}$$

This constraint equation is expressed in terms of x and y rather than their second derivatives. However, as $\sigma(x, y) = 0$ holds for all x, y it follows that $d\sigma = 0$ and $d^2\sigma = 0$ also. Consequently, the constraint equation can be written

$$m\frac{d^2x}{dt^2} - \frac{d^2y}{dt^2} = 0 \tag{6.46}$$

Solving the three equations gives

$$\frac{d^2x}{dt^2} = -g\frac{m}{1 + m^2} \tag{6.47}$$

$$\frac{d^2y}{dt^2} = -g\frac{m^2}{1 + m^2} \tag{6.48}$$

The x and y coordinates at time t are thus given by:

$$x(t) = x(0) + t\frac{dx(0)}{dt} - g\frac{t^2}{2}\frac{m}{(1+m^2)} \tag{6.49}$$

$$y(t) = y(0) + t\frac{dy(0)}{dt} - g\frac{t^2}{2}\frac{m^2}{(1+m^2)} \tag{6.50}$$

In the general case, the equations of motion for a constrained system involve two types of force: the 'normal' forces arising from the intra- and intermolecular interactions, and the forces due to the constraints. We are particularly interested in the case where the constraint σ_k requires the bond between atoms i and j to remain fixed. The constraint influences the Cartesian coordinates of atoms i and j. The force due to this constraint can be written as follows:

$$F_{ckx} = \lambda_k \frac{\partial \sigma_k}{\partial x} \tag{6.51}$$

λ_k is the Lagrange multiplier and x represents one of the Cartesian coordinates of the two atoms. Applying equation (6.51) to the above example, we would write $F_{cx} = \lambda \partial\sigma/\partial x = \lambda m$ and $F_{cy} = \lambda\partial\sigma/\partial y = -\lambda$. If an atom is involved in a number of constraints (because it is involved in more than one constrained bond) then the total constraint force equals the sum of all such terms. The nature of the constraint for a bond between atoms i and j is:

$$\sigma_{ij} = (\mathbf{r}_i - \mathbf{r}_j)^2 - d_{ij}^2 = 0 \tag{6.52}$$

The constraint force lies along the bond at all times. For each constrained bond, there is an equal and opposite force on the two atoms that comprise the bond. The overall effect is that the constraint forces do no work. Suppose the constraint k corresponds to the bond length between atoms i and j. The constraint forces are obtained by differentiating the constraint with respect to the coordinates of atoms i and j and multiplying by an (as yet) undetermined multiplier:

$$\partial\sigma_k/\partial\mathbf{r}_i = 2(\mathbf{r}_i - \mathbf{r}_j) \text{ so } F_{ci} = \lambda(\mathbf{r}_i - \mathbf{r}_j) \tag{6.53}$$

$$\partial\sigma_k/\partial\mathbf{r}_j = -2(\mathbf{r}_i - \mathbf{r}_j) \text{ and } F_{cj} = -\lambda(\mathbf{r}_i - \mathbf{r}_j) \tag{6.54}$$

The factor 2 that arises when we differentiate the square term has been incorporated into the Lagrange multiplier λ. The above expression for the forces can be incorporated into the Verlet algorithm as follows:

$$\mathbf{r}_i(t+\delta t) = 2\mathbf{r}_i(t) - \mathbf{r}_i(t-\delta t) + \frac{\delta t^2}{m_i}\mathbf{F}_i(t) + \sum_k \frac{\lambda_k \delta t^2}{m_i}\mathbf{r}_{ij}(t) \tag{6.55}$$

Recall that the positions that would be obtained from the Verlet algorithm without constraints are: $\mathbf{r}_i'(t+\delta t) = 2\mathbf{r}_i(t) - \mathbf{r}_i(t-\delta t) + \delta t^2\mathbf{F}_i(t)/m_i$. The summation in equation (6.55) is over all constraints k that affect atom i. These constraints perturb the positions that would otherwise have been obtained from the integration algorithm and so the above expression can be written:

$$\mathbf{r}_i(t+\delta t) = \mathbf{r}'_i(t+\delta t) + \sum_k \frac{\lambda_k \delta t^2}{m_i} \mathbf{r}_{ij}(t) \tag{6.56}$$

The next problem is to determine the multipliers λ_k that enable all the constraints to be satisfied simultaneously. This can be done algebraically in simple cases. Suppose we wish to fix the bond in a diatomic molecule. There is just one constraint in this case and if the atoms are labelled 1 and 2 we can write:

$$\mathbf{r}_1(t+\delta t) = \mathbf{r}'_1(t+\delta t) + \lambda_{12}(\delta t^2/m_1)(\mathbf{r}_1(t) - \mathbf{r}_2(t)) \tag{6.57}$$

$$\mathbf{r}_2(t+\delta t) = \mathbf{r}'_2(t+\delta t) - \lambda_{12}(\delta t^2/m_2)(\mathbf{r}_1(t) - \mathbf{r}_2(t)) \tag{6.58}$$

A third equation is derived from the requirement that the new positions keep the bond at the required distance:

$$|\mathbf{r}_1(t+\delta t) - \mathbf{r}_2(t+\delta t)|^2 = |\mathbf{r}_1(t) - \mathbf{r}_2(t)|^2 = d_{12}^2 \tag{6.59}$$

We now have three equations and three unknowns ($\mathbf{r}_1(t+\delta t)$, $\mathbf{r}_2(t+\delta t)$ and λ_{12}). Subtracting, and putting $\mathbf{r}_{12}(t) = (\mathbf{r}_1(t) - r_2(t))$ and $\mathbf{r}'_{12}(t+\delta t) = \mathbf{r}'_1(t+\delta t) - \mathbf{r}'_2(t+\delta t)$ gives:

$$\mathbf{r}_1(t+\delta t) - \mathbf{r}_2(t+\delta t) = \mathbf{r}'_{12}(t+\delta t) + \lambda_{12}\delta t^2(1/m_1 + 1m_2)\mathbf{r}_{12}(t) \tag{6.60}$$

Squaring both sides and imposing the constraint gives:

$$\mathbf{r}'_{12}(t+\delta t)^2 + 2\lambda_{12}\delta t^2(1/m_1 + 1/m_2)\mathbf{r}_{12}(t) + \lambda_{12}^2\delta t^4(1/m_1 + 1/m_2)^2\mathbf{r}_{12}(t)^2 = d_{12}^2 \tag{6.61}$$

Solving this quadratic equation for λ_{12} enables the new positions $\mathbf{r}_1(t+\delta t)$ and $\mathbf{r}_2(t+\delta t)$ to be determined.

In the case of a triatomic molecule with two bonds (between atoms 1, 2 and 2, 3) two constraint equations are obtained:

$$\mathbf{r}_{12}(t+\delta t) = \mathbf{r}'_{12}(t+\delta t) + \delta t^2(1/m_1 + 1/m_2)\lambda_{12}\mathbf{r}_{12}(t) - (\delta t^2/m_2)\lambda_{23}\mathbf{r}_{23}(t) \tag{6.62}$$

$$\mathbf{r}_{23}(t+\delta t) = \mathbf{r}'_{23}(t+\delta t) + \delta t^2(1/m_2 + 1/m_3)\lambda_{23}\mathbf{r}_{23}(t) - (\delta t^2/m_2)\lambda_{12}\mathbf{r}_{12}(t) \tag{6.63}$$

These expressions could be solved algebraically but even in this simple case the algebra becomes rather complicated. A solution can be obtained by ignoring the terms that are quadratic in λ as this produces equations which are linear in the Lagrange multipliers λ. When there are many constraints, the problem is equivalent to inverting a $k \times k$ matrix, even when the quadratic terms are ignored. The SHAKE method uses an alternative approach, in which each constraint is considered in turn and solved. Satisfying one constraint may cause another constraint to be violated, and so it is necessary to iterate round the constraints until they are all satisfied to within some tolerance. The tolerance should be tight enough to ensure that the fluctuations in the simulation due to the SHAKE algorithm are much smaller than the fluctuations due to other sources, such as the use of cutoffs. Another important requirement is that the constrained degrees of freedom should be only weakly coupled to the remaining degrees of freedom, so that the motion of the molecule is not affected by the application of the constraints. The sampling of unconstrained degrees of freedom should also be unaffected. For example, if the bond lengths and angles are constrained in butane then the only degree of freedom remaining is the torsion angle. It is important that this torsion is able to explore its entire range of values, in a way that is not biased because of the SHAKE procedure.

Our discussion so far has considered the use of SHAKE with the Verlet algorithm. Versions have also been derived for other integration schemes, such as the leap-frog algorithm, the predictor–corrector methods and the velocity Verlet algorithm. In the case of the velocity Verlet algorithm, the method has been named RATTLE [Anderson 1983]. When velocities appear in the integration algorithm these must be corrected as well as the positions.

Angle constraints can be easily accommodated in the SHAKE scheme by recognising that an angle constraint simply corresponds to an additional distance constraint. The angle in a triatomic molecule could thus be maintained at the desired value by requiring the distance between the two end atoms to adopt the appropriate value. This is how some small molecules (e.g. water) are maintained in a rigid geometry. For example, the simple point charge (SPC) model of water uses three distance constraints. However, it is generally accepted that to constrain the bond angles in simulations of conformationally flexible molecules can have a deleterious effect on the efficiency with which the system explores configurational space. This is because many conformational transitions involve some opening or closing of angles as well as rotations about bonds. The most common use of SHAKE is for constraining bonds involving hydrogen atoms due to their much higher vibrational frequencies. This can enable the time step in a molecular dynamics simulation to be increased (e.g. from 1 fs to 2 fs).

The SHAKE method has been extended by Tobias and Brooks [Tobias and Brooks 1988] to enable constraints to be applied to arbitrary internal coordinates. This enables the torsion angle of a rotable bond to be constrained to a particular value during a molecular dynamics simulation which is particularly useful when used in conjunction with methods for calculating free energies (see section 9.7).

6.6 Time-dependent properties

Molecular dynamics generates configurations of the system that are connected in time and so an MD simulation can be used to calculate time dependent properties. This is a major advantage of molecular dynamics over the Monte Carlo method. Time dependent properties are often calculated as *time correlation coefficients*.

6.6.1 Correlation functions

Suppose we have two sets of data values, x and y and we wish to determine what correlation (if any) exists between them. For example, imagine that we are performing a simulation of a fluid in a capillary, and that we wish to determine the correlation between the absolute velocity of an atom and its distance from the wall of the tube. One way to do this would be to plot the sets of values as a graph. A correlation function (also known as a correlation coefficient) provides a numerical value that encapsulates the data and quantitates the strength of the correlation. A series of simulations with different capillary diameters could then be compared by examining the correlation coefficients. A variety of correlation functions can be defined, a commonly used one being:

$$C_{xy} = \frac{1}{M}\sum_{\mathrm{i}=1}^{M} x_{\mathrm{i}}y_{\mathrm{i}} \equiv \langle x_{\mathrm{i}}y_{\mathrm{i}}\rangle \tag{6.64}$$

We have assumed that there are M values of x_{i} and y_{i} in the data sets. This correlation function can be normalised to a value between -1 and $+1$ by dividing by the root-mean-square values of x and y:

$$c_{xy} = \frac{\frac{1}{M}\sum_{\mathrm{i}=1}^{M} x_{\mathrm{i}}y_{\mathrm{i}}}{\sqrt{\left(\frac{1}{M}\sum_{\mathrm{i}=1}^{M} x_{\mathrm{i}}^2\right)\left(\frac{1}{M}\sum_{\mathrm{i}=1}^{M} y_{\mathrm{i}}^2\right)}} = \frac{\langle x_{\mathrm{i}}y_{\mathrm{i}}\rangle}{\langle x_{\mathrm{i}}^2\rangle\langle y_{\mathrm{i}}^2\rangle} \tag{6.65}$$

A value of 0 indicates no correlation and an absolute value of 1 indicates a high degree of correlation (which may be positive or negative). We will use a lower-case c to indicate a normalised correlation coefficient.

Sometimes the quantities x and y will fluctuate about non-zero mean values $\langle x\rangle$ and $\langle y\rangle$. Under such circumstances it is typical to consider just the fluctuating part and to define the correlation function as:

$$c_{xy} = \frac{\frac{1}{M}\sum_{\mathrm{i}=1}^{M}(x_{\mathrm{i}} - \langle x\rangle)(y_{\mathrm{i}} - \langle y\rangle)}{\sqrt{\left(\frac{1}{M}\sum_{\mathrm{i}=1}^{M}(x_{\mathrm{i}} - \langle x\rangle)^2\right)\left(\frac{1}{M}\sum_{\mathrm{i}=1}^{M}(y_{\mathrm{i}} - \langle y\rangle)^2\right)}}$$

$$= \frac{\langle(x_{\mathrm{i}} - \langle x\rangle)(y_{\mathrm{i}} - \langle y\rangle)\rangle}{\sqrt{\langle(x_{\mathrm{i}} - \langle x\rangle)^2\rangle\langle(y_{\mathrm{i}} - \langle y\rangle)^2\rangle}} \tag{6.66}$$

c_{xy} can also be written in the following useful way:

$$c_{xy} = \frac{\sum_{i=1}^{M} x_i y_i - \frac{1}{M}\left(\sum_{i=1}^{M} x_i\right)\left(\sum_{i=1}^{M} y_i\right)}{\sqrt{\left[\sum_{i=1}^{M} x_i^2 - \frac{1}{M}\left(\sum_{i=1}^{M} x_i\right)^2\right]\left[\sum_{i=1}^{M} y_i^2 - \frac{1}{M}\left(\sum_{i=1}^{M} y_i\right)^2\right]}} \tag{6.67}$$

Equation (6.67) does not require the mean values $\langle x \rangle$ and $\langle y \rangle$ to be determined before the correlation coefficient can be calculated and so values can be accumulated as the simulation proceeds.

A molecular dynamics simulation provides data values at specific times. This enables the value of some property at some instant to be correlated with the value of the same or another property at a later time t. The resulting values are known as *time correlation coefficients*. The correlation function is then written:

$$C_{xy}(t) = \langle x(t)y(0) \rangle \tag{6.68}$$

The following two results are useful:

$$\lim t \to 0 \qquad C_{xy}(0) = \langle xy \rangle \tag{6.69}$$

$$\lim t \to \infty \qquad C_{xy}(t) = \langle x \rangle \langle y \rangle \tag{6.70}$$

If the quantities x and y are different, then the correlation function is sometimes referred to as a *cross-correlation function*. When x and y are the same then the function is usually called an *autocorrelation function*. An autocorrelation function indicates the extent to which the system retains a 'memory' of its previous values (or, conversely, how long it takes the system to 'lose' its memory). A simple example is the velocity autocorrelation coefficient whose value indicates how closely the velocity at a time t is correlated with the velocity at time 0. Some correlation functions can be averaged over all the particles in the system (as can the velocity autocorrelation function) whereas other functions are a property of the entire system (e.g. the dipole moment of the sample). The value of the velocity autocorrelation coefficient can be calculated by averaging over the N atoms in the simulation:

$$C_{vv}(t) = \frac{1}{N}\sum_{i=1}^{N} \mathbf{v}_i(t) \cdot \mathbf{v}_i(0) \tag{6.71}$$

To normalise the function, we divide by $\langle \mathbf{v}_i(0) \cdot \mathbf{v}_i(0) \rangle$:

$$c_{vv}(t) = \frac{1}{N}\sum_{i=1}^{N} \frac{\langle \mathbf{v}_i(t) \cdot \mathbf{v}_i(0) \rangle}{\langle \mathbf{v}_i(0) \cdot \mathbf{v}_i(0) \rangle} \tag{6.72}$$

In general, an autocorrelation function such as the velocity autocorrelation coefficient has an initial value of 1, and at long times has the value 0. The time taken to lose the correlation is often called the *correlation time*, or the *relaxation time*. If the length of the simulation is significantly longer than the relaxation time

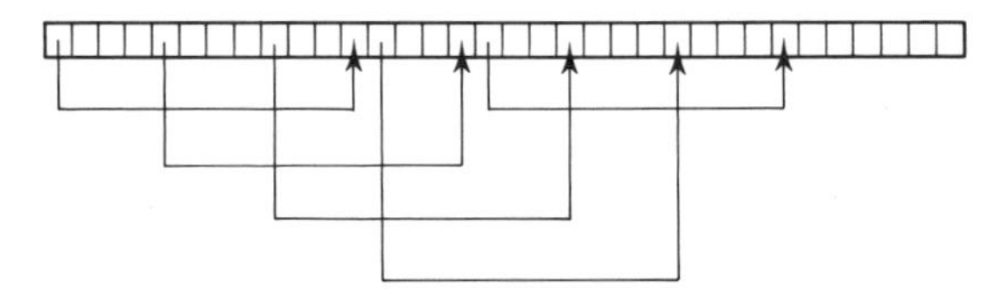

Fig. 6.9 The use of different time origins improves the accuracy with which time correlation functions can be calculated.

(as it should be), then many sets of data can be extracted from the simulation to calculate the correlation function and to reduce the uncertainty in the calculation. If P steps of molecular dynamics are required for complete relaxation, and the simulation has been run for a total of Q steps, then $(Q - P)$ different sets of values could be used to calculate a value for the correlation function. The first set would run from step 1 to step N; the second from step 2 to step $N+1$, and so on (Figure 6.9). In fact, as we saw in section 5.10 the high degree of correlation between successive time steps means that it is common to use time origins that are separated by several time steps as shown in Figure 5.26. If we use M time origins (t_j) then the velocity autocorrelation function is given by:

$$C_{vv}(t) = \frac{1}{MN} \sum_{j=1}^{M} \sum_{i=1}^{N} \mathbf{v}_i(t_j) \cdot \mathbf{v}_i(t_j + t) \tag{6.73}$$

Quantities with small relaxation times can thus be determined with greater statistical precision, as it will be possible to include a greater number of data sets from a given simulation. Moreover, no quantity with a relaxation time greater than the length of the simulation can be determined accurately.

The velocity autocorrelation functions obtained from molecular dynamics simulations of argon at two different densities are shown in Figure 6.10. The correlation coefficient has an initial value of 1 and is then quadratic at short times, a result that is predicted theoretically. The subsequent behaviour depends upon the density of the fluid. For a low density fluid, the velocity autocorrelation coefficient gradually decreases to zero. At high densities $c_{vv}(t)$ crosses the axis, and then becomes negative. A negative correlation coefficient simply means that the particle is now moving in the direction opposite to that at $t=0$. This result has been interpreted in terms of a 'cage' structure of the liquid; the atom hits the side of the cage formed by its nearest neighbours and rebounds, reversing the direction of its motion. At both low density and high density the decay towards zero is significantly slower than the exponential decay predicted by kinetic theory. In fact, the decay varies as $t^{-3/2}$. This was one of the most interesting results obtained from the early molecular dynamics simulations, and can be observed even with a hard-sphere model [Alder and Wainwright 1970]. The phenomenon is ascribed to the formation of a 'hydrodynamic vortex'. As the atom moves through the fluid it pushes other atoms out of the way. These atoms circle round and eventually give it a final 'push' so resulting in a less rapid decrease to zero (Figure 6.11).

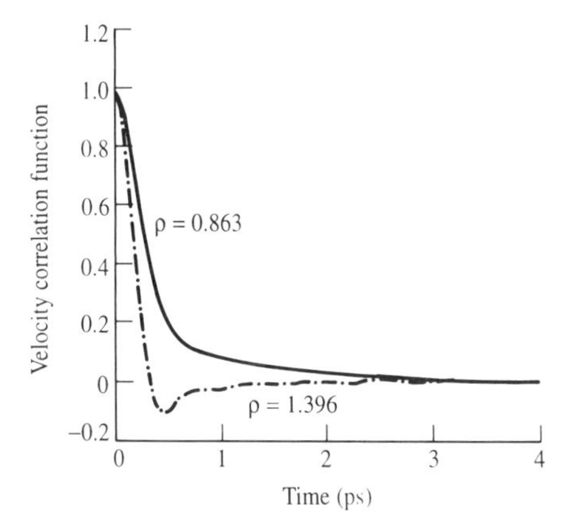

Fig. 6.10 Velocity autocorrelation functions for liquid argon at densities of 1.396 g cm^{-3} and 0.863 g cm^{-3}.

The slow decay of the velocity autocorrelation function can present practical problems when deriving other properties such as the transport coefficients that require the correlation function to be integrated between $t = 0$ and $t = \infty$. The so-called 'long time-tail' of the autocorrelation function makes a significant contribution to the integral, but unfortunately the statistical uncertainty with which this part of the function can be calculated is greater as fewer segments of the appropriate length can be extracted from the simulation.

The velocity autocorrelation function is an example of a single-particle correlation function, in which the average is calculated not only over time origins but

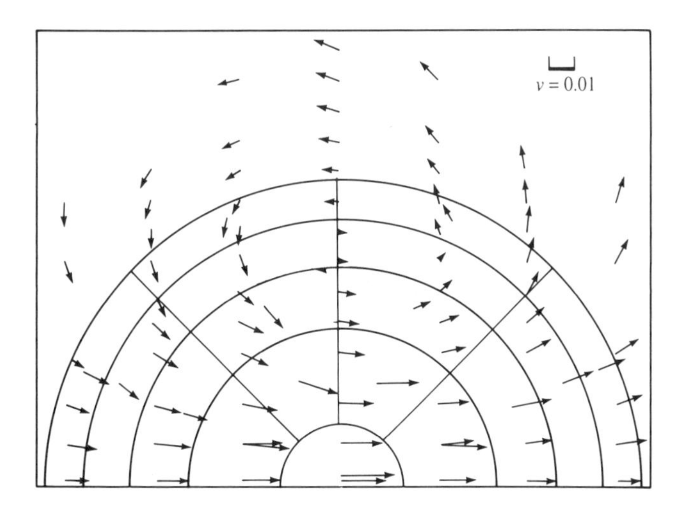

Fig. 6.11 The slow decay of the velocity autocorrelation function towards zero can be explained in terms of the formation of a hydrodynamic vortex. Figure adapted from Alder B J and T E Wainwright 1970. Decay of the Velocity Autocorrelation Function. *Physical Review A* **1**:18–21.

also over all the atoms. Some properties are calculated for the entire system. One such property is the net dipole moment of the system, which is the vector sum of all the individual dipoles of the molecules in the system (clearly the dipole moment of the system can be non-zero only if each individual molecule has a dipole). The magnitude and orientation of the net dipole will change with time and is given by:

$$\boldsymbol{\mu}_{\text{tot}}(t) = \sum_{i=1}^{N} \boldsymbol{\mu}_i(t) \tag{6.74}$$

$\boldsymbol{\mu}_i(t)$ is the dipole moment of molecule i at time t. The total dipolar correlation function is given by:

$$c_{\text{dipole}}(t) = \frac{\langle \boldsymbol{\mu}_{\text{tot}}(t) \cdot \boldsymbol{\mu}_{\text{tot}}(0) \rangle}{\langle \boldsymbol{\mu}_{\text{tot}}(0) \cdot \boldsymbol{\mu}_{\text{tot}}(0) \rangle} \tag{6.75}$$

The dipole correlation time of the system is a valuable quantity to calculate as it is related to the sample's absorption spectrum. Liquids usually absorb in the infrared region of the electromagnetic spectrum and a typical spectrum is shown in Figure 6.12. As can be seen, the spectrum is very broad with none of the sharp peaks characteristic of a well-resolved spectrum for a species in the gas phase. This is because the overall dipole of a liquid does not change at a constant rate, but rather there is a distribution of frequencies. The intensity of absorption at any frequency depends upon the relative contribution of that frequency to the overall distribution. If, on average, the overall dipole changes very rapidly (i.e. the relaxation time is short) then the maximum in the absorption spectrum will occur at a higher frequency than if the relaxation time is long. To predict the spectrum from the correlation function it is therefore necessary to extract the relative contribution of each dipole fluctuation. This is done using Fourier analysis techniques, in which the correlation function is transformed from the time domain into the frequency domain (an introduction to Fourier analysis is provided in Appendix 6.2). The Fourier analysis picks out the intensity of dipole fluctuation at each frequency using the following relationship:

$$\hat{c}_{\text{dipole}}(\nu) = \int_{-\infty}^{\infty} c_{\text{dipole}}(t)\exp(-2\pi\nu t)\mathrm{d}t \tag{6.76}$$

Having calculated the Fourier transform it is then possible to plot the simulated spectrum and compare it to that obtained by experiment, as in Figure 6.12.

6.6.2 Orientational correlation functions

Other orientational correlation coefficients can be calculated in the same way as the correlation coefficients that we have discussed already. Thus, the reorientational correlation coefficient of a single rigid molecule indicates the degree to which the orientation of a molecule at a time t is related to its orientation at time

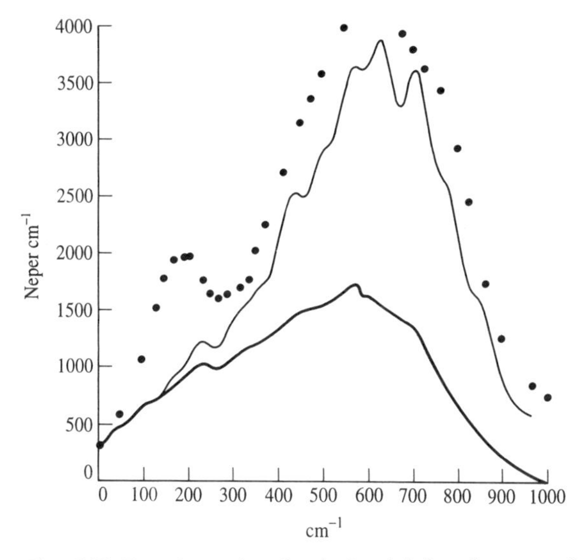

Fig. 6.12 Experimental and calculated infrared spectra for liquid water. The black dots are the experimental values. The heavy curve is the classical profile produced by the molecular dynamics simulation. The thin curve is obtained by applying quantum corrections. Figure taken from Guillot B 1991. A Molecular Dynamics Study of the Infrared Spectrum of Water. *The Journal of Chemical Physics* **95**:1543–1551.

0. The angular velocity autocorrelation function is the rotational equivalent of the velocity correlation function:

$$c_{\omega\omega}(t) = \frac{\langle \omega_i(t) \cdot \omega_i(0) \rangle}{\langle \omega_i(0) \cdot \omega_i(0) \rangle} \tag{6.77}$$

In a liquid the rotation of a molecule is influenced by neighbouring molecules and over time the correlation will decay to zero. The information embodied in the orientational correlation functions can be compared to a variety of spectroscopic experiments including infrared, Raman and NMR spectra. For non-spherical molecules it can be useful to derive separate autocorrelation functions for the angular velocity along each of the principal axes of rotation. For example, for a spherical molecule such as CBr_4 neighbouring molecules have a relatively small influence on the loss in correlation in the angular velocity. By contrast, a linear molecule such as CS_2 experiences significant torques as it rotates. This has the effect of damping the rotational motion more severely than for the spherical case, and indeed the correlation function can change sign, indicating that the molecule is now rotating in the opposite direction. For some molecules such as water the presence of specific interactions between molecules (for example, due to hydrogen bonding) can give rise to very rapid damping and several minima in $c_{\omega\omega}(t)$.

6.6.3 Transport properties

Transport refers to a phenomenon that gives rise to a flow of material from one region to another. For example, if a solution is prepared with a non-equilibrium solute distribution, then the solute diffuses until the concentration is equal throughout. If a thermal gradient is created energy flows until the temperature is equalised. A momentum gradient gives rise to viscosity. The very presence of transport implies that the system is not an equilibrium. Techniques have been developed to perform non-equilibrium molecular dynamics simulations from which transport properties can be calculated, but we will not consider these here. We are thus faced with the problem of calculating non-equilibrium properties from equilibrium simulations. This is possible by virtue of the microscopic local fluctuations that occur even in systems at equilibrium. We should be aware, however, that non-equilibrium molecular dynamics simulations can be a more efficient way to calculate transport properties and other quantities [Allen and Tildesley 1987].

To a first approximation the rate at which transport of the relevant quantity occurs (called the flux) is proportional to the gradient of the property with the constant of proportionality being the relevant transport property coefficient. For example, the flux of matter J_z (i.e. the amount passing through unit area in unit time) equals the diffusion coefficient (D) multiplied by the concentration gradient; this is Fick's first law of diffusion:

$$J_z = -D(\mathrm{d}\mathcal{N}/\mathrm{d}z) \tag{6.78}$$

$\mathcal{N}$ is the number density (the number of atoms per unit volume). Equation (6.78) refers to diffusion in the z direction. The minus sign indicates that flux increases in the direction of negative concentration gradient. The time dependence of diffusive behaviour (which applies if a distribution is established at some time and is then allowed to evolve) is governed by Fick's second law, which gives the rate of change of concentration with time:

$$\frac{\partial \mathcal{N}(z, t)}{\partial t} = D\frac{\partial^2 \mathcal{N}(z, t)}{\partial z^2} \tag{6.79}$$

To solve Fick's second law equation it is necessary to impose two boundary conditions for the spatial dependence and one boundary condition for the temporal dependence (the equation is second order in space and first order in time). For example, we might require that at time zero all N_0 particles have $z = 0$. The solution to the equation is then:

$$\mathcal{N}(z, t) = \frac{N_0}{A\sqrt{\pi D t}}\exp\left[-\frac{z^2}{4Dt}\right] \tag{6.80}$$

A is the cross-sectional area of the sample. Equation (6.8) is a Gaussian function which initially has a sharp peak at $z = 0$ but which gradually becomes more spread out as time progresses. When the material being simulated is a pure liquid then the coefficient D is often referred to as a *self-diffusion coefficient*. The diffusion coefficient is related to the mean square distance, $\langle | [\mathbf{r}(t) - \mathbf{r}(0)]^2 | \rangle$, which

Einstein showed was equal to $2Dt$. In three dimensions the mean square displacement is given by:

$$3D = \lim_{t\to\infty} \frac{\langle |\mathbf{r}(t) - \mathbf{r}(0)|^2 \rangle}{2t} \tag{6.81}$$

As indicated, the relationship strictly holds only in the limit as $t \to \infty$.

The Einstein relationship can thus be used to calculate the diffusion coefficient from an equilibrium simulation, by plotting the mean square displacement as a function of time and then attempting to obtain the limiting behaviour as $t \to \infty$ (Fick's law is inapplicable at short times). The quantity $|\mathbf{r}(t) - \mathbf{r}(0)|$ can be averaged over the particles in the system, to reduce the statistical error. It is also usual practice to average over time origins where possible. When using this method for calculating the diffusion coefficient the mean-squared distances should not be limited by the edges of the periodic box. In other words, we require a set of positions that have not been translated back into the central simulation cell. This can be achieved either by storing a set of 'uncorrected' positions, or indeed by not correcting any of the positions during the simulation, and simply generating the appropriate minimum image positions as required for the calculation of the energy or forces.

Einstein relationships hold for other transport properties, e.g. the shear viscosity, the bulk viscosity and the thermal conductivity. For example, shear viscosity η is given by:

$$\eta_{xy} = \frac{1}{Vk_{\mathrm{B}}T} \lim_{t\to\infty} \frac{\left\langle \left(\sum_{i=1}^{N} m\dot{x}_i(t)y_i(t) - \sum_{i=1}^{N} m\dot{y}_i(t)x_i(t) \right)^2 \right\rangle}{2t} \tag{6.82}$$

The shear viscosity is a tensor quantity, with components η_{xy}, η_{xz}, η_{yx}, η_{yz}, η_{zx}, η_{zy}. It is a property of the whole sample rather than of individual atoms and so cannot be calculated with the same accuracy as the self-diffusion coefficient. For a homogeneous fluid the components of the shear viscosity should be all equal and so the statistical error can be reduced by averaging over the six components. An estimate of the precision of the calculation can then be determined by evaluating the standard deviation of these components from the average. Unfortunately, equation (6.82) cannot be directly used in periodic systems, even if the positions have been unfolded, because the 'unfolded' distance between two particles may not correspond to the distance of the minimum image that is used to calculate the force. For this reason alternative approaches are required.

One alternative approach to the calculation of the diffusion and other transport coefficients is via an appropriate autocorrelation function. For example, the diffusion coefficient depends upon the way in which the atomic position $\mathbf{r}(t)$ changes with time. At a time t the difference between $\mathbf{r}(t)$ and $\mathbf{r}(0)$ is given by:

$$|\mathbf{r}(t) - \mathbf{r}(0)| = \int_0^{t'} \mathbf{v}(t')\mathrm{d}t' \tag{6.83}$$

If both sides of equation (6.83) are now squared then we obtain the following for the mean-square value:

$$\langle|\mathbf{r}(t)-\mathbf{r}(0)|\rangle^2 = \int_0^t dt' \int_0^t dt'' \langle\mathbf{v}(t')\cdot\mathbf{v}(t'')\rangle \tag{6.84}$$

The crucial feature to recognise is that the relevant correlation functions are unaffected by changing the origin, which means that the following holds:

$$\langle\mathbf{v}(t')\cdot\mathbf{v}(t'')\rangle = \langle\mathbf{v}(t''-t')\cdot\mathbf{v}(0)\rangle \tag{6.85}$$

Integration of the double integral, equation (6.84) leads to the *Green–Kubo* formula:

$$\frac{\langle|\mathbf{r}(t)-\mathbf{r}(0)|^2\rangle}{2t} = \int_0^t \langle\mathbf{v}(\tau)\cdot\mathbf{v}(0)\rangle\left(1-\frac{\tau}{t}\right)d\tau \tag{6.86}$$

In the limit:

$$\int_0^\infty \langle\mathbf{v}(\tau)\cdot\mathbf{v}(0)\rangle d\tau = \lim_{t\to\infty}\frac{\langle|\mathbf{r}(t)-\mathbf{r}(0)|^2\rangle}{2t} = 3D \tag{6.87}$$

We can now see why long time-tails in the autocorrelation functions are so important. The area under the curve during the slow decay towards zero may be a significant part of the integral in the Green–Kubo formula. In practice these integrals are determined numerically. The long time-tail may be dealt with by fitting a function to the curve and then attempting to integrate to infinity.

6.7 Molecular dynamics at constant temperature and pressure

Molecular dynamics is traditionally performed in the constant *NVE* (or *NVE***P**) ensemble. Although thermodynamic results can be transformed between ensembles, this is strictly only possible in the limit of infinite system size ('the thermodynamic limit'). It may therefore be desired to perform the simulation in a different ensemble. The two most common alternative ensembles are the constant *NVT* and the constant *NPT* ensembles. In this section we will therefore consider how molecular dynamics simulations can be performed under conditions of constant temperature and/or constant pressure.

6.7.1 *Constant temperature dynamics*

There are several reasons why we might want to maintain or otherwise control the temperature during a molecular dynamics simulation. Even in a constant *NVE* simulation it is common practice to adjust the temperature to the desired value during the equilibration phase. A constant temperature simulation may be required if we wish to determine how the behaviour of the system changes with temperature, such as the unfolding of a protein or glass formation. Simulated annealing algorithms require the temperature of the system to be reduced gradually while the system explores its degrees of freedom. Simulated annealing is used in searching conformational space and in the elucidation of macromolecular structure from NMR and X-ray data and is discussed in section 8.7.2.

The temperature of the system is related to the time average of the kinetic energy which for an unconstrained system is given by:

$$\langle \mathcal{K} \rangle_{NVT} = \tfrac{3}{2} N k_B T \tag{6.88}$$

An obvious way to alter the temperature of the system is thus to scale the velocities [Woodcock 1971]. If the temperature at time t is $T(t)$ and the velocities are multiplied by a factor λ, then the associated temperature change can be calculated as follows:

$$\Delta T = \frac{1}{2}\sum_{i=1}^{N} \frac{2}{3} \frac{m_i(\lambda v_i)^2}{N k_B} - \frac{1}{2}\sum_{i=1}^{N} \frac{2}{3} \frac{m_i v_i^2}{N k_B} \tag{6.89}$$

$$\Delta T = (\lambda^2 - 1)T(t) \tag{6.90}$$

$$\lambda = \sqrt{(T_{new}/T(t))} \tag{6.91}$$

The simplest way to control the temperature is thus to multiply the velocities at each time step by the factor $\lambda(=\sqrt{(T_{req}/T_{curr})}$, where T_{curr} is the current temperature as calculated from the kinetic energy and T_{req} is the desired temperature).

An alternative way to maintain the temperature is to couple the system to an external heat bath that is fixed at the desired temperature [Berendsen *et al.* 1984]. The bath acts as a source of thermal energy, supplying or removing heat from the system as appropriate. The velocities are scaled at each step, such that the rate of change of temperature is proportional to the difference in temperature between the bath and the system:

$$\frac{dT(t)}{dt} = \frac{1}{\tau}(T_{bath} - T(t)) \tag{6.92}$$

τ is a coupling parameter whose magnitude determines how tightly the bath and the system are coupled together. This method gives an exponential decay of the system towards the desired temperature. The change in temperature between successive time steps is:

$$\Delta T = \frac{\delta t}{\tau}(T_{bath} - T(t)) \tag{6.93}$$

The scaling factor for the velocities is thus:

$$\lambda^2 = 1 + \frac{\delta t}{\tau}\left(\frac{T_{bath}}{T(t)} - 1\right) \tag{6.94}$$

If τ is large, then the coupling will be weak. If τ is small, the coupling will be strong and when the coupling parameter equals the time step ($\tau = \delta t$) then the algorithm is equivalent to the simple velocity scaling method. A coupling constant of approximately 0.4 ps has been suggested as an appropriate value to use when the time step is 1 fs, giving $\delta t/\tau \approx 0.0025$. The advantage of this approach is that it does permit the system to fluctuate about the desired temperature.

These two relatively simple temperature scaling methods do not generate rigorous canonical averages. Velocity scaling artificially prolongs any temperature differences among the components of the system, which can lead to the phenomenon of 'hot solvent, cold solute', in which the 'temperature' of the solute is lower than that of the solvent, even though the overall temperature of the system is at the desired value. One 'solution' to this problem is to apply temperature coupling separately to the solute and to the solvent but the problem of unequal distribution of energy between the various components (and between the various modes of motion) may still remain. Two methods that do generate rigorous canonical ensembles if properly implemented are the *stochastic collisions* method and the *extended system* method.

In the stochastic collisions method a particle is randomly chosen at intervals and its velocity is reassigned by random selection from the Maxwell–Boltzmann distribution [Anderson 1980]. This is equivalent to the system being in contact with a heat bath that randomly emits 'thermal particles' which collide with the atoms in the system. Between each collision the system is simulated at constant energy and so the overall effect is equivalent to a series of microcanonical simulations, each performed at a slightly different energy. The distribution of the energies of these 'mini microcanonical' simulations will be a Gaussian function. The stochastic collisions method does not of course generate a smooth trajectory which may be a drawback. By calculating the energy change due to a collision, Anderson showed that the mean rate (ν) at which each particle should suffer a stochastic collision is given by:

$$\nu = \frac{2a\kappa}{3k_B \mathcal{N}^{1/3} N^{2/3}} \tag{6.95}$$

a is a dimensionless constant, κ is the thermal conductivity and $\mathcal{N}$ is the number density of the particles. If the thermal conductivity is not known then a suitable value of ν can be obtained from the intermolecular collision frequency ν_c:

$$\nu = \nu_c / N^{2/3} \tag{6.96}$$

If the collision rate is too low then the system does not sample from a canonical distribution of energies. If it is too high then the temperature control algorithm dominates and the system does not show the expected fluctuations in kinetic energy. The velocity of more than one particle can be changed in the stochastic collision method; in the limit the velocities of all the particles are changed simultaneously, though it is preferable to do this at quite long intervals. A distinction can thus be made between 'minor' collisions, in which only one (or a few) particles are affected, and 'major' (or 'massive') collisions, where the velocities of all particles are changed. It is also possible to use a combined approach, with minor collisions occurring relatively frequently and major collisions at longer intervals.

Extended system methods, originally introduced for performing constant temperature molecular dynamics by Nosé [Nosé 1984] and subsequently developed by Hoover [Hoover 1985], consider the thermal reservoir to be an integral part of the system. The reservoir is represented by an additional degree of freedom, labelled s. The reservoir has potential energy $(f+1)k_BT \ln s$, where f is the number of degrees of freedom in the physical system and T is the desired temperature. The reservoir also has kinetic energy $(Q/2)(\mathrm{d}s/\mathrm{d}t)^2$. Q is a parameter with the dimensions of energy × (time)^2 and can be considered the (fictitious) mass of the extra degree of freedom. The magnitude of Q determines the coupling between the reservoir and the real system and so influences the temperature fluctuations.

Each state of the extended system that is generated by the molecular dynamics simulation corresponds to a unique state of the real system. There is not, however, a direct correspondence between the velocities and the time in the real and the extended systems. The velocities of the atoms in the real system are given by:

$$\mathbf{v}_i = s\frac{\mathrm{d}\mathbf{r}_i}{\mathrm{d}t} \tag{6.97}$$

$\mathbf{r}_i$ is the position of particle i in the simulation and $\mathbf{v}_i$ is considered to be the real velocity of the particle. The time step $\delta t'$ is related to the time step in 'real time' δt by

$$\delta t = s\delta t' \tag{6.98}$$

The value of the additional degree of freedom s can change and so the time step in real time can fluctuate. Thus regular time intervals in the extended system correspond to a trajectory of the real system which is unevenly spaced in time.

The parameter Q controls the energy flow between the system and the reservoir. If Q is large then the energy flow is slow; in the limit of infinite Q conventional molecular dynamics is regained and there is no energy exchange between the reservoir and the real system. However, if Q is too small then the energy oscillates, resulting in equilibration problems. Nosé has suggested that Q should be proportional to fk_BT; the constant of proportionality can then be obtained by performing a series of trial simulations for a test system and observing how well the system maintains the desired temperature.

6.7.2 Constant pressure dynamics

Just as one may wish to specify the temperature in a molecular dynamics simulation, so it may be desired to maintain the system at a constant pressure. This enables the behaviour of the system to be explored as a function of the pressure, enabling one to study phenomena such as the onset of pressure-induced phase transitions. Many experimental measurements are made under conditions of constant temperature and pressure, and so simulations in the isothermal–isobaric ensemble are most directly relevant to experimental data. Certain structural rearrangements may be achieved more easily in an isobaric simulation than in a simulation at constant volume. Constant pressure conditions may also be important when the number of particles in the system changes (as in some of the 'test particle' methods for calculating free energies and chemical potentials; see section 7.9).

The pressure often fluctuates much more than quantities such as the total energy in a constant *NVE* molecular dynamics simulation. This is as expected because the pressure is related to the virial, which is obtained as the product of the positions and the derivative of the potential energy function. This product, $r_{ij}d\mathscr{V}(r_{ij})/dr_{ij}$, changes more quickly with r than does the internal energy, hence the greater fluctuation in the pressure.

A macroscopic system maintains constant pressure by changing its volume. A simulation in the isothermal–isobaric ensemble also maintains constant pressure by changing the volume of the simulation cell. The amount of volume fluctuation is related to the isothermal compressibility, κ:

$$\kappa = -\frac{1}{V}\left(\frac{\partial V}{\partial P}\right)_T \tag{6.99}$$

An easily compressible substance has a larger value of κ so larger volume fluctuations occur at a given pressure than in a more incompressible substance. Conversely, in a constant volume simulation a less compressible substance shows larger fluctuations in the pressure. The isothermal compressibility is the pressure analogue of the heat capacity, which is related to the energy fluctuations.

A volume change in an isobaric simulation can be achieved by changing the volume in all directions, or in just one direction. It is instructive to consider the range of volume changes that one might expect to observe in a constant pressure simulation of a 'typical' system. The isothermal compressibility is related to the mean square volume displacement by:

$$\kappa = \frac{1}{k_B T}\frac{\langle V^2\rangle - \langle V\rangle^2}{\langle V^2\rangle} \tag{6.100}$$

The isothermal compressibility of an ideal gas is approximately 1 atm^{-1}. So for a simulation in a box of side 20 Å (volume 8000 Å^3) at 300 K, the root mean square change in the volume is approximately 18 100 Å^3. This is larger than the initial size of the box! For a relatively incompressible substance such as water ($\kappa = 44.75 \times 10^{-6}\ \text{atm}^{-1}$) the fluctuation is 121 Å^3, which corresponds to the box only changing by about 0.1 Å in each direction. These values have clear implications for the appropriate size of the simulation system.

Many of the methods used for pressure control are analogous to those used for temperature control. Thus, the pressure can be maintained at a constant value by simply scaling the volume. An alternative is to couple the system to a 'pressure bath', analogous to a temperature bath [Berendsen *et al.* 1984]. The rate of change of pressure is given by:

$$\frac{dP(t)}{dt} = \frac{1}{\tau_p}(P_{bath} - P(t)) \tag{6.101}$$

τ_p is the coupling constant, P_{bath} is the pressure of the 'bath', and $P(t)$ is the actual pressure at time t. The volume of the simulation box is scaled by a factor λ, which is equivalent to scaling the atomic coordinates by a factor $\lambda^{1/3}$. Thus:

$$\lambda = 1 - \kappa \frac{\delta t}{\tau_p}(P - P_{bath}) \tag{6.102}$$

The new positions are given by:

$$\mathbf{r}_i' = \lambda^{1/3}\mathbf{r}_i \tag{6.103}$$

The constant κ can be combined with the relaxation constant τ_p as a single constant. This expression can be applied isotropically (i.e. such that the scaling factor is equal for all three directions) or anisotropically (where the scaling factor is calculated independently for each of the three axes). In general, it is best to use the anisotropic approach as this enables the box dimensions to change independently. Unfortunately, it has not been possible to determine from which ensemble this method samples.

In the extended pressure-coupling system methods, first introduced by Anderson [Anderson 1980], an extra degree of freedom, corresponding to the volume of the box, is added to the system. The kinetic energy associated with this degree of freedom (which can be considered to be equivalent to a piston acting on the system) is $\frac{1}{2}Q(\mathrm{d}V/\mathrm{d}t)^2$, where Q is the 'mass' of the piston. The piston also has potential energy PV, where P is the desired pressure and V is the volume of the system. A piston of small mass gives rise to rapid oscillations in the box whereas a large mass has the opposite effect. An infinite mass returns normal molecular dynamics behaviour. The volume can vary during the simulation with the average volume being determined by the balance between the internal pressure of the system and the desired external pressure. The extended-system temperature scaling method of Nosé uses a scaled time; in the extended pressure method the coordinates of the extended system are related to the 'real' coordinates by:

$$\mathbf{r}_i' = V^{-1/3}\mathbf{r}_i \tag{6.104}$$

6.8 Incorporating solvent effects into molecular dynamics: potentials of mean force and stochastic dynamics

In many simulations of solute–solvent systems the primary focus is the behaviour of the solute; the solvent is of relatively little interest, particularly in regions far from the solute molecule. The use of non-rectangular periodic boundary conditions, stochastic boundaries and 'solvent shells' can all help to reduce the number of solvent molecules required and enable a larger proportion of the computing time to be spent simulating the solute. In this section we consider a group of techniques that incorporate the effects of solvent without requiring any explicit specific solvent molecules to be present.

One approach to this problem is to use a *potential of mean force* (PMF), which describes how the free energy changes as a particular coordinate (such as the

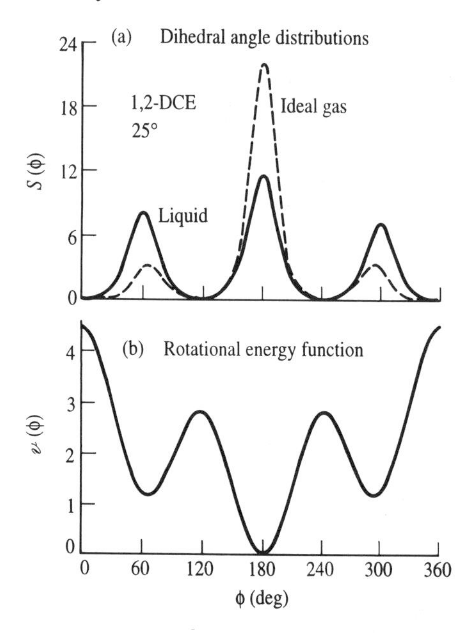

Fig. 6.13 Population distribution for 1,2-dichloroethane in the gas and liquid phases. Figure redrawn from Jorgensen W L, R C Binning Jr and B Bigot 1981. Structures and Properties of Organic Liquids: *n*-Butane and 1,2-Dichloroethane and Their Conformational Equilibria. *The Journal of the American Chemical Society* **103**:4393–4399.

separation of two atoms or the torsion angle of a bond) is varied. The free energy change described by the potential of mean force includes the averaged effects of the solvent.

Potentials of mean force may be determined using a molecular dynamics or Monte Carlo simulation using the techniques of umbrella sampling or free energy perturbation, which will be discussed in section 9.7. Here we illustrate the concept using an example. The energy difference between the *trans* and *gauche* conformations for an isolated molecule of 1,2-dichloroethane (i.e. in the gas phase) is approximately 1.14 kcal mol^{-1} with a population containing 77% *trans* and 23% *gauche* conformers. In liquid 1,2-dichloroethane, however, the relative population of the *gauche* conformer is significantly increased relative to the *trans* conformer by comparison with the isolated molecule, with 44% *trans* and 56% *gauche*. These experimental results were reproduced by Jorgensen (see Figure 6.13) using Monte Carlo simulations [Jorgensen *et al.* 1981]. The potential of mean force would be designed to reproduce this new population and so enable a single 1,2-dichloroethane molecule to be simulated as if it were present in the liquid.

A simulation performed using a potential of mean force enables the modulating effects of the solvent to be taken into account. The solvent also influences the dynamical behaviour of the solute via random collisions, and by imposing a frictional drag on the motion of the solute through the solvent. The Langevin equation of motion is the starting point for the *stochastic dynamics* models, in which these two effects are also incorporated. In stochastic dynamics the force on a particle is considered to arise from three sources. The first component is due to interactions between the particle and other particles. This force ($\mathbf{F}_i$) depends upon the position of the particle relative to the other particles and is modelled using a potential of mean force. The second force arises from the motion of the particle through the solvent and is equivalent to the frictional drag on the particle due to the solvent. This frictional force is proportional to the speed of the particle with the constant of proportionality being the friction coefficient:

$$\mathbf{F}_{\text{frictional}} = -\xi \mathbf{v} \tag{6.105}$$

$\mathbf{v}$ is the velocity and ξ is the friction coefficient. The friction coefficient is related to the collision frequency (γ) by $\gamma = \xi/m$ (m is the mass of the particle). γ^{-1} can be considered as the time taken for the particle to lose memory of its initial velocity (the velocity relaxation time). For a spherical particle the friction coefficient is related to the diffusion constant D by:

$$\xi = k_B T/D \tag{6.106}$$

If the radius of the spherical particle is a then the frictional force is given by Stokes law:

$$\mathbf{F}_{\text{frictional}} = 6\pi a \eta \mathbf{v} \tag{6.107}$$

η is the viscosity of the fluid.

The third contribution to the force on the particle is due to random fluctuations caused by interactions with solvent molecules. We will write this force $\mathbf{R}(t)$. The Langevin equation of motion for a particle i can therefore be written:

$$m_i \frac{d^2 x_i(t)}{dt^2} = \mathbf{F}_i\{x_i(t)\} - \gamma_i \frac{dx_i(t)}{dt} m_i + \mathbf{R}_i(t) \tag{6.108}$$

A number of simulation methods based on equation (6.108) have been described. These differ in the assumptions that are made about the nature of frictional and random forces. A common simplifying assumption is that the collision frequency γ is independent of time and position. The random force $\mathbf{R}(t)$ is often assumed to be uncorrelated with the particle velocities, positions and the forces acting on them, and to obey a Gaussian distribution with zero mean. The force $\mathbf{F}_i$ is assumed to be constant over the time step of the integration.

Three different situations can be considered, depending upon the relative magnitudes of the integration time step and the velocity relaxation time. The first category corresponds to timescales that are short relative to the velocity relaxation time ($\gamma \delta t \ll 1$). Under such circumstances the solvent does not activate or deactivate the particle to any significant extent. In the limit of zero γ (when there are no effects due to solvent) then the Langevin equation (6.108) reduces to that obtained from Newton's laws of motion. At the other extreme the velocity relaxation time is much smaller than the time step. This corresponds to the diffusive regime, where the motion is rapidly damped by the solvent. The third situation is intermediate between these two extremens. Various methods have been proposed for integrating the Langevin equation of motion in these three regions.

In the region where $\gamma\delta t \ll 1$ the following is a simple integration algorithm [van Gunsteren *et al.* 1981]:

$$x_{i+1} = x_i + v_i\delta t + \tfrac{1}{2}(\delta t)^2\{-\gamma v_i + m^{-1}(F_i + R_i)\} \tag{6.109}$$

$$v_{i+1} = v_i + (\delta t)\{-\gamma v_i + m^{-1}(F_i + R_i)\} \tag{6.110}$$

The average random force over the time step is taken from a Gaussian with a variance $2mk_BT\gamma(\delta t)^{-1}$. x_i is one of the $3N$ coordinates at time step i_i; F_i and R_i are the relevant components of the frictional and random forces at that time; v_i is the velocity component.

An alternative expression is based on the following finite difference approximations [Brunger *et al.* 1984]:

$$d^2x/dt^2 \approx (x_{i+1} - 2x_i + x_{i-1})/\delta t^2 \tag{6.111}$$

$$dx/dt \approx (x_{i+1} - x_{i-1})/2\delta t \tag{6.112}$$

This leads to the following expressions for the coordinates x_{i+1}:

$$x_{i+1} = x_i + (x_i - x_{i-1})\frac{1 - \frac{1}{2}\gamma\delta t}{1 + \frac{1}{2}\gamma\delta t} + \left(\frac{\delta t^2}{m}\right)\frac{F_i + R_i}{1 + \frac{1}{2}\gamma\delta t} \tag{6.113}$$

In the region where $\gamma\delta t \gg 1$ then if the interparticle force is assumed to be constant over the integration time step the following result is obtained [van Gunsteren *et al.* 1981]:

$$x_{i+1} = x_i + F_i(m\gamma)^{-1}\delta t + X_i(\delta t) \tag{6.114}$$

X_i is a Gaussian distribution with zero mean and a variance of $2k_BT(m\gamma)^{-1} = 2D\delta t$. An extension of this treatment is to permit the force F_i to vary linearly over the time step, giving:

$$x_{i+1} = x_i + \frac{\delta t}{m\gamma}\left(F_i + \tfrac{1}{2}\dot{F}_i\delta t\right) + X_i \tag{6.115}$$

$\dot{F}_i$ is the derivative of the force at the time step i and is obtained numerically:

$$\dot{F}_i = (F_i - F_{i-1})/\delta t \tag{6.116}$$

In the intermediate region, where there are no restrictions on $\gamma\delta t$ then integration of the equations of motion gives the following rather complicated result [van Gunsteren and Berendsen 1982]:

$$x_{i+1} = x_i + v_i\gamma^{-1}(1 - \exp(-\gamma\delta t)) + F_i(m\gamma)^{-1}[\delta t - \gamma^{-1}(1 - \exp(-\gamma\delta t))] + (mg)^{-1}\int_{t_i}^{t_{i+1}} [1 - \exp(-\gamma(t_{i+1} - t'))]R(t')dt' \tag{6.117a}$$

$$v_{i+1} = v_i \exp(-\gamma\delta t) + F_i(m\gamma)^{-1}(1 - \exp(-\gamma\delta t)) + (m)^{-1}\int_{t_i}^{t_{i+1}} \exp(-\gamma(t_{i+1} - t'))R(t')dt' \tag{6.117b}$$

The important feature of these two equations is that the new positions and the new velocities both depend upon an integral over the random force, $R(t)$ (the final terms in equations (6.117a) and (6.117b). As both of these integrals depend upon $R_i(t)$ they are correlated. Specifically, they obey a *bivariate* Gaussian distribution. Such a distribution provides the probability that a particle located at x_i at time t with velocity v_i and experiencing a force F_i will be at x_{i+1} at time $t + \delta t$ with velocity v_{i+1}. In practice, this means that the distribution for the second variable depends upon the value selected for the first variable. It can be difficult to properly sample from such distributions but van Gunsteren and Berendsen showed that the equations can be reformulated in terms of sampling from two independent Gaussian functions.

More complex stochastic dynamics treatments are possible; our treatment has only provided a rather simple treatment of solvent effects. For example, we have assumed that the frictional force at a given instant is proportional only to its velocity at the same time. A more realistic model assumes that the frictional forces are correlated; they have a 'memory' of previous values. The friction coefficient can also be made to depend on the coordinates of the other particles.

6.8.1 Practical aspects of stochastic dynamics simulations

A stochastic dynamics simulation requires a value to be assigned to the collision frequency friction coefficient γ. For simple particles such as spheres this can be related to the diffusion constant in the fluid. For a simulation of a rigid molecule it may be possible to derive γ via the diffusion coefficient from a standard molecular dynamics situation. In the more general case we require the friction coefficient of each atom. For simple molecules such as butane the friction coefficient can be considered to be the same for all atoms. The optimal value for γ can be determined by trial and error, performing a stochastic dynamics simulation for different values of γ and comparing the results with those from experiment (where available) or from molecular dynamics simulations. For large molecules the atomic friction coefficient is reconsidered to depend upon the degree to which each atom is in contact with the solvent and is usually taken to be proportional to the accessible surface area of the atom (defined in section 1.5).

One of the main advantages of the stochastic dynamics methods is that dramatic time savings can be achieved, which enables much longer stimulations to be performed. For example, Widmalm and Pastor performed 1 ns molecular dynamics and stochastic dynamics simulations of an ethylene glycol molecule in aqueous solution of the solute and 259 water molecules [Widmalm and Pastor 1992]. The molecular dynamics simulation required 300 hours whereas the stochastic dynamics simulation of the solute alone required just 24 minutes. The dramatic reduction in time for the stochastic dynamics calculation is due not only to the very much smaller number of molecules present, but also the fact that longer time steps can often be used in stochastic dynamics simulations.

Stochastic dynamics has been widely used to study the behaviour of long-chain molecules and polymers. The advantages of stochastic dynamics are especially

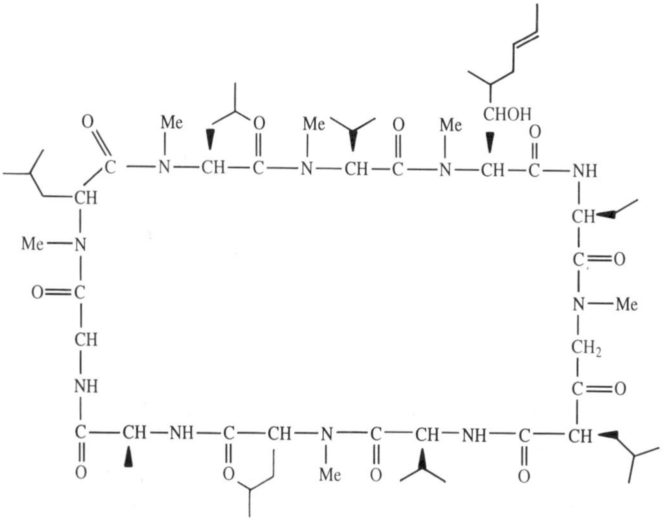

Fig. 6.14 Cyclosporin.

important for polymers [Helfand 1984], where many interesting phenomena occur over relatively long time periods, so putting them beyond the scope of conventional molecular dynamics. However, one must take care with the Langevin method when simulating systems in which specific solute–solvent interactions are present. For example, Yun-Yu, Lu and van Gunsteren used both stochastic dynamics and molecular dynamics to study the immunosuppressant drug cyclosporin (Figure 6.14) in two solvents: carbon tetrachloride and water [Yun-Yu *et al.* 1988]. The time-averaged structures obtained from each method were compared, to determine the similarity between the average structure obtained for each simulation. Fluctuations in torsion angles were also compared. The analysis showed that the structures obtained from the molecular dynamics and stochastic dynamics simulations of cyclosporin in carbon tetrachloride were very similar, but that the results were very different for the Langevin and molecular dynamics simulations performed in water. This was due to an excessive degree of internal hydrogen bonding in the stochastic dynamics simulation; the equivalent molecular dynamics simulation contained much more hydrogen bonding between cyclosporin and the solvent.

6.9 Conformational changes from molecular dynamics simulations

Molecular dynamics can provide information about the conformational properties of molecular systems, and the way in which the conformation changes with time. Molecular graphics programs can facilitate the analysis of such simulations by displaying the structural parameters of interest in a manner that enables the time dimension to be taken into account. Perhaps the most direct way to demonstrate the conformational behaviour of the system is as a movie, where coordinate sets saved at regular intervals are displayed in sequence. For publication purposes, time-dependent data can be displayed graphically, with one of the axes corresponding to the time, such as the plots of energy or autocorrelation function versus time (Figures 6.3 and 6.10). The representation of bond rotations is difficult using x/y plots due to the 2π periodicity of a torsion angle. Lavery and Sklenar have developed a method to represent torsion data as a polar plot [Lavery and Sklenar 1988], where the distance from the origin corresponds to the time, Figure 6.15. Such 'dials' are very useful for detecting the presence of correlated conformational changes.

When viewing a movie of a molecular dynamics simulation of a complex molecule one is often struck by the chaotic nature of the motion. This should be expected; the motion of complex molecules *is* chaotic, but there are often underlying low-frequency motions which correspond to more significant and more interesting conformational changes. Fourier analysis techniques can be used to filter out the unwanted high-frequency motions, enabling the important low-frequency changes to be observed unhindered. Here we describe the filtering method of Dauber-Osguthorpe and Osguthorpe [Dauber-Osguthorpe and Osguthorpe 1990; Dauber-Osguthorpe and Osguthorpe 1993].

A Fourier transform enables one to convert the variation of some quantity as a function of time into a function of frequency, and vice versa. Thus, if we represent the quantity that varies in time as $x(t)$, then Fourier analysis enables us to also represent that quantity as a function $X(\nu)$, where ν is the frequency ($-\infty < \nu < \infty$). Fourier analysis is usually introduced by considering functions that vary in a periodic manner with time which can be written as a superposition of sine and cosine functions (a Fourier series; see Appendix 6.2). If the period of the function $x(t)$ is τ then the cosine and sine terms in the Fourier series are functions of frequencies $2\pi n/\tau$, where n can take integer values 1, 2, 3 ...

A Fourier series is rarely relevant to the interpretation of a molecular dynamics simulation as the movement of the atoms is not periodic but chaotic. The *Fourier transform* enables a non-periodic function to be converted into the equivalent frequency function (and vice versa). The Fourier transform can be developed from the Fourier series simply by considering the effect of increasing the period of a periodic function to infinity. The frequency function obtained from a Fourier transform is a continuous function rather than one written as a series of discrete frequencies. Further details are provided in Appendix 6.2.

At each step of the Fourier analysis of a molecular dynamics simulation the variation with time of one of the Cartesian coordinates of one of the atoms in the system is converted into the correponding frequency function. Special methods called fast Fourier techniques are usually employed for this step. The frequency spectrum can then be filtered to remove high frequencies. This is achieved simply by setting the coefficients of the unwanted frequencies in the frequency function to zero. The resulting spectrum is then converted back to the time domain to give a new set of coordinate values at each of the time steps in the trajectory. This new

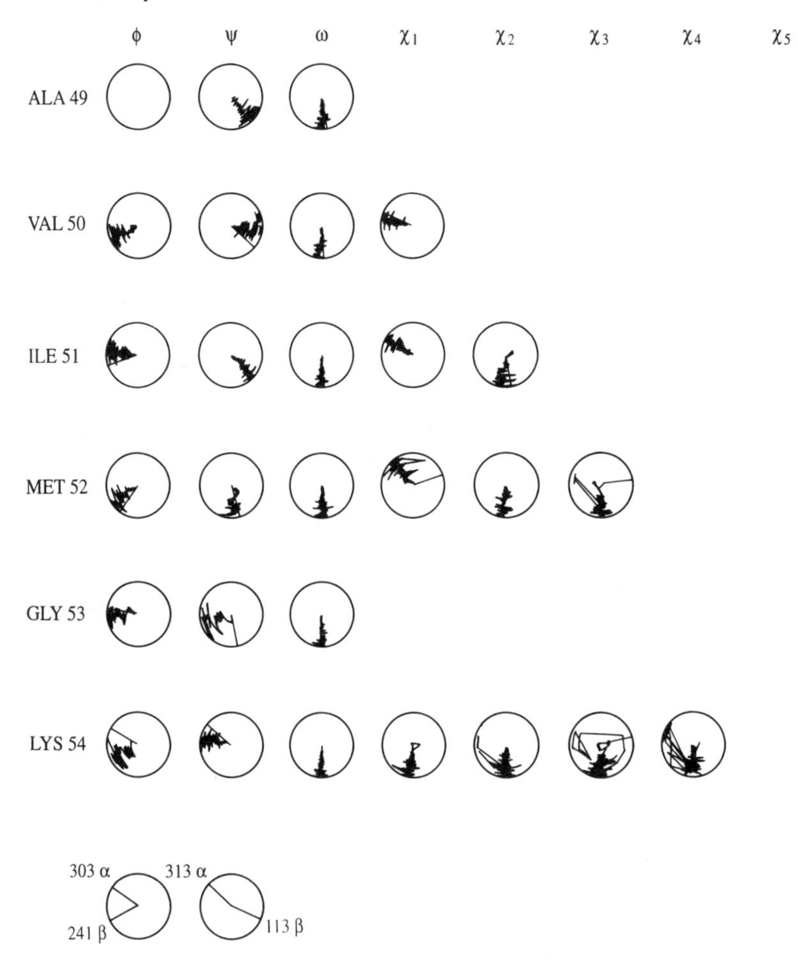

Fig. 6.15 The variation in torsion angles can be effectively represented as a series of 'dials' where the time corresponds to the distance from the centre of the dial. Data from a molecular dynamics simulation of an intermolecular complex between the enzyme dihydrofolate reductase and a triazine inhibitor [Leach and Klein, 1995].

coordinate set only includes the selected frequencies. This process can be repeated for the three coordinates of each atom to give a filtered trajectory for the entire system. It is also possible to select just a single frequency (i.e. a single normal mode) from the frequency spectrum and view this in isolation.

6.10 Molecular dynamics simulations of chain amphiphiles

The molecular dynamics technique is widely used for simulating large molecular systems, some of which have many degrees of conformational freedom. In this section we will examine the application of molecular dynamics to chain amphiphiles, a class of molecules of interest to both the 'biological' and 'materials science' communities. These molecules have a polar head group attached to one or more hydrocarbon chains. Some examples are shown in Figure 6.16. The head group has a high affinity for water whereas the hydrocarbon tail prefers to exist in a hydrophobic environment. The molecules therefore exist in both phases at a water/oil interface. A characteristic feature of these molecules is their ability to form extended layer structures. Monolayers, bilayers and multiple layers are all possible. A monolayer at the water/air interface is known as a Langmuir film; when this is transferred to a solid substrate it is known as a Langmuir–Blodgett film. Langmuir–Blodgett films with many layers can be constructed in the laboratory but most simulation studies of these systems have been restricted to monolayers or bilayers. The ability to control the thickness of a Langmuir–Blodgett

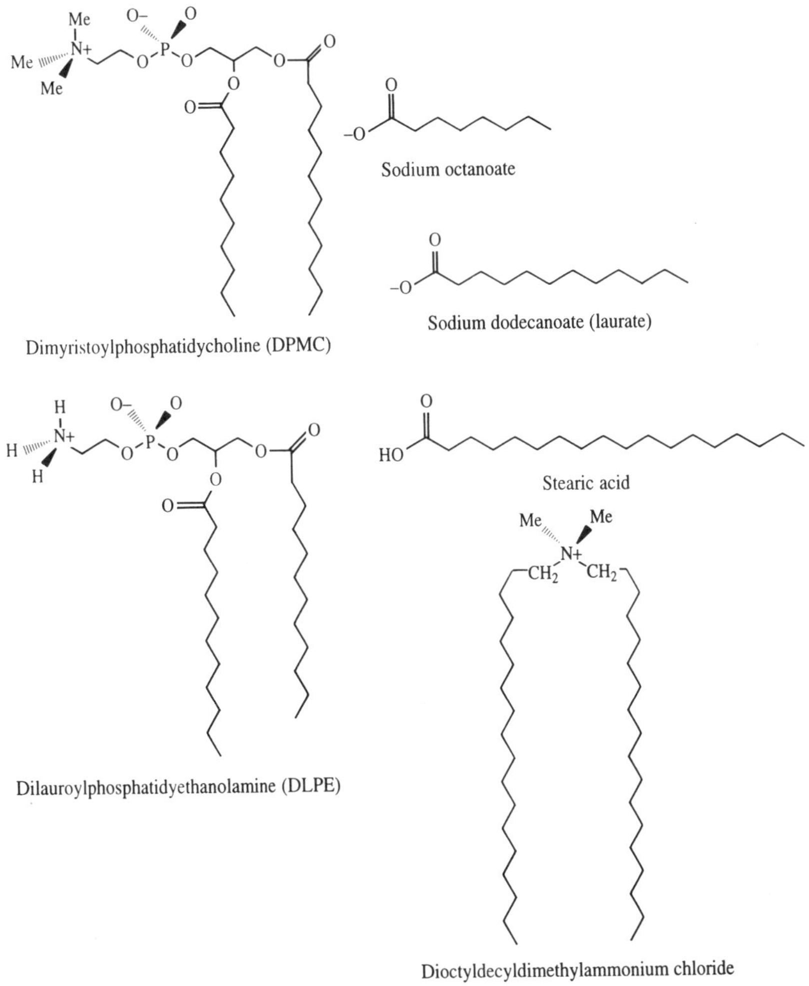

Dimyristoylphosphatidycholine (DPMC)

Sodium octanoate

Sodium dodecanoate (laurate)

Dilauroylphosphatidyethanolamine (DLPE)

Stearic acid

Dioctyldecyldimethylammonium chloride

Fig. 6.16 Some typical amphiphiles.

film and their high degree of order means that they are intensively investigated as insulators in semiconductors, filtration devices and as anti-reflective coatings. Amphiphiles are important in biology as cell membranes are formed from lipid bilayers. At a high enough concentration some amphiphiles can form micelles, which are globular structures that have the head groups all pointing into solution and the tails inside (Figure 6.17).

Amphiphiles often have a complex phase behaviour with several liquid crystalline phases. These liquid crystalline phases are often characterised by long-range order in one direction together with the formation of a layer structure. The molecules may nevertheless be able to move laterally within the layer and perpendicular to the surface of the layer. Structural information can be obtained using spectroscopic techniques including X-ray and neutron diffraction and NMR. The quadrupolar splitting in the deuterium NMR spectrum can be used to determine an *order parameter* for the carbon atoms on the hydrocarbon tail. The order parameter is defined as:

$$S = 0.5\langle 3 \cos\theta_{\mathrm{i}} \cos\theta_{\mathrm{j}} - \delta_{\mathrm{ij}} \rangle \tag{6.118}$$

θ_{i} is the angle between the *i*th molecular axis and the *director*, which is the average of the molecular axes over the sample. For a bilayer in the $L\alpha$ phase (the one present in cell membranes) the director is the same as the bilayer normal and is conventionally taken to be the z axis; see Figure 6.18. δ_{ij} is the Kronecker delta function ($\delta_{\mathrm{ij}} = 1$ if $\mathrm{i} = \mathrm{j}$; $\delta_{\mathrm{ij}} = 0$ if $\mathrm{i} \neq \mathrm{j}$). The expression for S is averaged over time and over molecules. The deuterium NMR experiment provides the order parameter S_{CD}, which indicates the average orientation of the C–D bond vector with respect to the bilayer normal. The experimental order parameters S_{CD} can range from 1.0 (indicating full order along the bilayer normal) to −0.5 (full order perpendicular to the bilayer normal) [Seelig and Seelig 1974]. A value of zero is considered to indicate full isotropic motion of the group. Experimental values are determined using molecules with deuterium substituted methylene groups at positions along the hydrocarbon chain. Many simulations of amphiphiles are performed using united atom models for the hydrocarbon chains and it is therefore necessary to be able to relate the experimental order parameters to values that can be calculated from a simulation. This is done as follows [Essex *et al.* 1994]. Molecular axes are defined for each CH_2 unit in the chain as shown in Figure 6.19. These molecular axes are defined for the *n*th CH_2 unit as follows:

z: vector from C_{n-1} to C_{n+1}
y: vector perpendicular to z and in the plane through C_{n-1}, C_n and C_{n+1}
x: perpendicular to y and z

Using these definitions, components of the molecular order parameter tensor can be determined (for example, S_{zz} is determined by measuring the angle between the molecular z axis and the bilayer normal). The experimental order parameter can be related to the molecular order parameter using the equation:

$$S_{\mathrm{CD}} = 2S_{xx}/3 + S_{yy}/3 \tag{6.119}$$

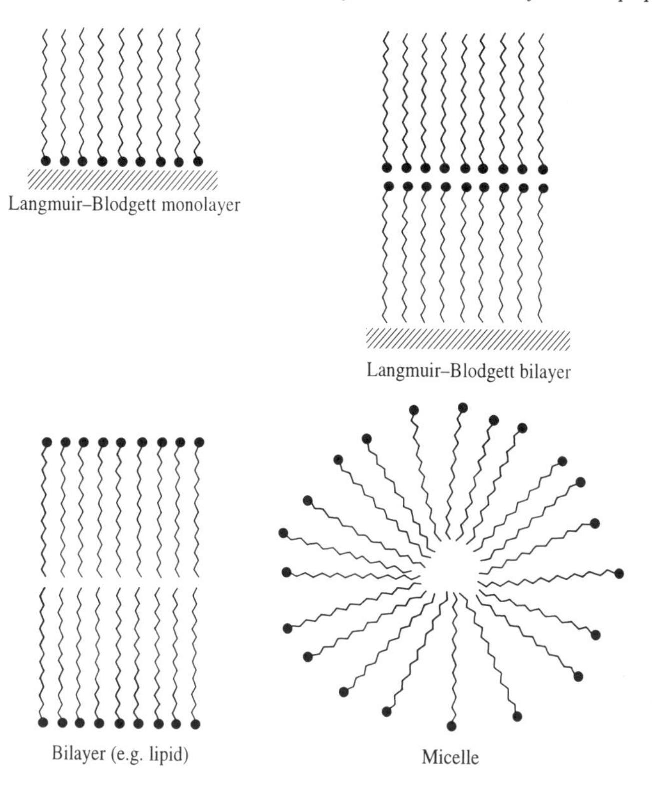

Fig. 6.17 Some of the various phases that amphiphiles may form.

With all-atom simulations the locations of the hydrogen atoms are known and so the order parameters can be calculated directly. Another structural property of interest is the ratio of *trans* conformations to *gauche* conformations for the CH_2–CH_2 bonds in the hydrocarbon tail. The *trans*:*gauche* ratio can be estimated using a variety of experimental techniques such as Raman, infrared and NMR spectroscopy.

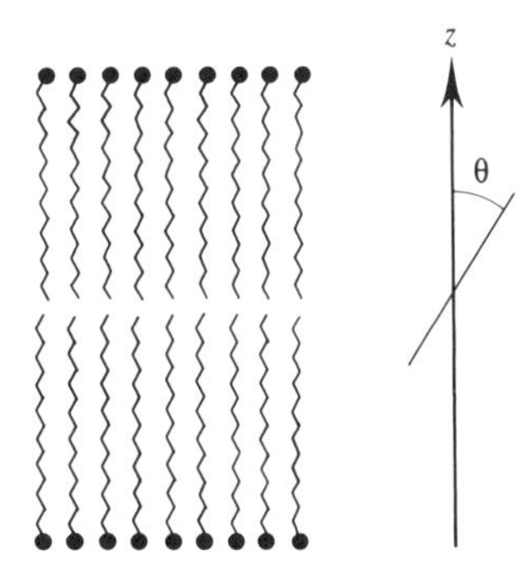

Fig. 6.18 Definition of the order parameter.

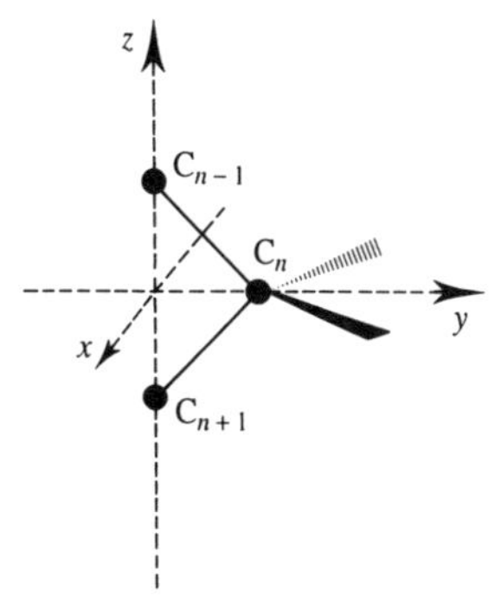

Fig. 6.19 Calculation of the order parameter for united atom simulations.

6.10.1 Simulation of lipids

There has been considerable interest in the simulation of lipid bilayers due to their biological importance. Early calculations on amphiphilic assemblies were limited by the computing power available, and so relatively simple models were employed. One of most important of these is the mean field approach of Marcelja [Marcelja 1973; Marcelja 1974], in which the interaction of a single hydrocarbon chain with its neighbours is represented by two additional contributions to the energy function. The energy of a chain in the mean field is given by:

$$\mathscr{V}_{\text{tot}} = \mathscr{V}_{\text{int}} + \mathscr{V}_{\text{disp}} + \mathscr{V}_{\text{rep}} \tag{6.120}$$

$\mathscr{V}_{\text{int}}$ is the internal energy of a chain, which can be calculated using standard force field methods. $\mathscr{V}_{\text{disp}}$ simulates the van der Waals interactions with the neighbouring molecules. It is often modelled using a Maier–Saupe potential:

$$\mathscr{V}_{\text{disp}} = -\Phi \sum_{\text{i}=1}^{\text{carbons}} \frac{1}{2}(3\cos^2\theta_{\text{i}} - 1) \tag{6.121}$$

The summation runs over all carbon atoms in the chain. θ_{i} is the angle between the bilayer normal and the molecular axis, as discussed above. Φ is the field strength; this may be parametrised to reproduce appropriate experimental data such as the deuterium NMR order parameters or it may be obtained by a self-consistent protocol, as described below. In his work on lipid bilayers Marcelja used a slightly different expression for $\mathscr{V}_{\text{disp}}$ which involved the fraction of *trans* bonds in the system:

$$\mathscr{V}_{\text{disp}} = -\Phi \frac{n_{\text{trans}}}{n} \sum_{\text{i}=1}^{\text{carbons}} \frac{1}{2}(3\cos^2\theta_{\text{i}} - 1) \tag{6.122}$$

This additional factor was introduced to ensure the proper behaviour over both liquid crystalline and solid phases. In simulations of the liquid crystalline phase alone this term may be omitted for computational efficiency.

The repulsive contribution, $\mathcal{V}_{\text{rep}}$, is due to lateral pressure on each chain. In Marcelja's original treatment, this was set equal to the product of the lateral pressure, γ, and the cross-sectional area of the chain. The cross-sectional area was approximated by:

$$A = A_0 l_0 / l \tag{6.123}$$

l_0 and A_0 are the length and cross-sectional area respectively of the hydrocarbon chain in a fully extended conformation. l is the length of the chain in the current conformation, projected onto the bilayer normal. If the bilayer normal is along the z axis then l is taken to be the z coordinate of the last carbon atom in the hydrocarbon chain. In other mean field models [Pastor *et al.* 1988] the product $\gamma A_0 / l_0$ is replaced with a single adjustable parameter Γ and so $\mathcal{V}_{\text{rep}}$ is given by:

$$\mathcal{V}_{\text{rep}} = \sum_{\text{chains}} \frac{\Gamma}{(z_n - z_0)} \tag{6.124}$$

z_n is the z coordinate of the last carbon in the chain and z_0 is the coordinate of the surface of the monolayer or bilayer. This force acts to keep the last carbon away from the surface; the closer it gets the larger the force pulling it away.

In his calculations, Marcelja generated all possible conformations of the hydrocarbon chain, restricting each carbon–carbon bond to the *trans* and *gauche* conformations. The energy of each conformation was evaluated. From the ensemble of conformations a partition function can be computed:

$$Z = \sum_{\text{all conformations}} \exp[-\mathcal{V}_{\text{tot}} / k_{\text{B}} T] \tag{6.125}$$

The molecular field is related to the partition function:

$$\Phi = \sum_{\text{all conformations}} \left\{ \frac{\dfrac{n_{\text{trans}}}{n} \displaystyle\sum_{\text{i}=1}^{\text{carbons}} \frac{1}{2}(3\cos^2\theta_{\text{i}} - 1) \exp[-\mathcal{V}_{\text{tot}} / k_{\text{B}} T]}{Z} \right\} \tag{6.126}$$

The molecular field is thus related to the partition function and so it is possible to generate a self-consistent value of the molecular field, Φ. Thermodynamic properties can then be calculated from the partition function. For example, Marcelja calculated the pressure as a function of the area per polar head group for surface monolayers at a variety of temperatures. His results showed good qualitative agreement with experimental results for such systems.

The mean field approach can be incorporated into a molecular dynamics simulation. It is particularly useful when used in conjunctin with Langevin dynamics, as very long simulations can be performed. For example, Pearce and Harvey were able to perform simulations of three unsaturated phospholipids for 100 ns (i.e.

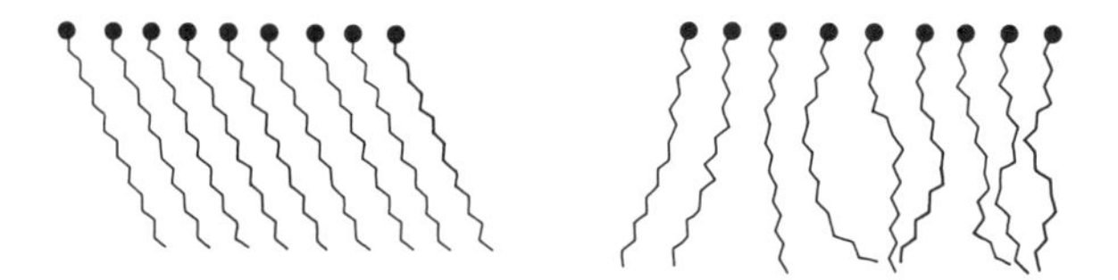

Fig. 6.20 Variation in alignment of chains in lipid simulation with tilt angle [van der Ploeg and Berendsen 1982].

0.1 μs) in single-molecule Langevin dynamics calculations [Pearce and Harvey 1993]. An extension of this strategy is to use a central 'core' containing one or more molecules that are simulated using molecular dynamics. This core is surrounded by a shell of molecules that are simulated using Langevin dynamics with the mean field. In this way one attempts to simulate a more 'realistic' system without incurring the computational penalty of a full molecular dynamics simulation of the entire system [De Loof *et al.* 1991].

The first molecular dynamics simulations of a lipid bilayer which used an explicit representation of all the molecules was performed by van der Ploeg and Berendsen in 1982 [van der Ploeg and Berendsen 1982]. Their simulation contained 32 decanoate molecules arranged in two layers of 16 molecules each. Periodic boundary conditions were employed and a united atom force potential was used to model the interactions. The head groups were restrained using a harmonic potential of the form:

$$\mathcal{V}(z) = \frac{k_{\mathrm{h}}}{2}(z - \langle z \rangle)^2 \tag{6.127}$$

By writing the restraint in terms of the average z coordinates of the head groups ($\langle z \rangle$) van der Ploeg and Berendsen ensured that the bilayer was able to change its thickness to reach its equilibrium value. This restraining potential was designed to reproduce the interactions between the head groups and the water layer, neither of which was explicitly included in the calculation. A key feature of the simulation was the long equilibration time required. By explicitly representing all the molecules in the system it was possible to determine the collective motion of the system as a whole. One distinct feature was a slowly fluctuating collective tilt of the molecules away from the normal to the bilayer surface, Figure 6.20. The degree to which the molecules were aligned with each other was also correlated with the tilt angle. When the average tilt angle reached a maximum the chains were much more likely to be well aligned, but when the average tilt angle was close to zero (i.e. such that the average orientation of the chains was almost normal to the bilayer surface) much less order was observed. In their original simulations this collective tilt phenomenon was observed to extend over the entire simulation cell, suggesting that the cell dimensions were too small and that the use of periodic boundary conditions was enhancing the long-range correlations. Simulations using a larger system subsequently showed that this collective tilt could be observed for subsets of the molecules.

Faster computers have enabled more realistic simulations of lipid bilayers to be performed, of larger systems, with more accurate models and for longer times [Stouch 1993]. The trend is very much towards simulations that use full representations of all species present (i.e. 'all atom' models with explicit solvent and counterions). The charged and highly polar nature of lipid head groups means that a proper representation of the long-range electrostatic forces can be critical, using a method such as the Ewald summation. Equilibration of such systems often requires hundreds of picoseconds and certain phenomena are only observed on a nanosecond timescale. These simulations have revealed many hitherto unknown features of the behaviour of such systems. For example, considerable conformational mobility of the hydrocarbon chains is often observed in the liquid crystalline phases. This is illustrated in Figure 6.21 (colour plate section), which shows a snapshot of a lipid bilayer after a molecular dynamics simulation of several hundred picoseconds. The considerable degree of disorder in the hydrocarbon chains near the middle of the bilayer is clear from this figure and is very different to the idealised, 'text-book' pictures in which the chains are perfectly aligned in completely extended conformations. The distribution of *gauche* conformations tends to be higher towards the end of the chain, though in some systems a *gauche* link is required near the head group to enable the chain to lie perpendicular to the interface. 'Kinks' are often observed in the chains; these are arrangements of three successive bonds with *gauche*(+)–*trans*–*gauche*(−) torsion angle which enable the chain to remain perpendicular to the surface.

6.10.2 Simulations of Langmuir–Blodgett films

The simulations of Langmuir–Blodgett systems can be difficult due to the need to correctly model the solid support. To illustrate the procedure we will describe the calculations of Kim, Moller, Tildesley and Quirke [Kim *et al.* 1994a] who simulated stearic acid ($CH_3(CH_2)_{16}COOH$) adsorbed onto graphite. The surface was modelled using a Lennard–Jones 9–3 potential that depends upon the height of the atom (α) above the surface (z_α):

$$\mathcal{V}_{\alpha s}(z_\alpha) = \frac{2\pi\rho}{3}\varepsilon_{ss}\left[\frac{2}{15}\left(\frac{\sigma_{\alpha s}}{z_\alpha}\right)^9 - \left(\frac{\sigma_{\alpha s}}{z_\alpha}\right)^3\right] \tag{6.128}$$

ρ is the number of density of the solid and ε_{xx} and δ_{ss} are its Lennard–Jones parameters. An image change method was also applied to the acid head group with the interaction between a charge and its image being:

$$\mathcal{V}_{\mathrm{ic}}(z) = \frac{1}{2}\frac{(\varepsilon - \varepsilon')}{(\varepsilon + \varepsilon')}\left[\frac{q_\alpha^2}{8\pi\varepsilon_0(z - z_{\mathrm{ip}})}\right] \tag{6.129}$$

ε' is the relative permittivity of the solid (taken to be 4.0) and ε is the permittivity above the surface ($\varepsilon = 1.0$). The image plane is located at $z_{\mathrm{ip}} = \sigma_{ss}/2$. Each charge interacts with its own image and with the images of other charges, but there are

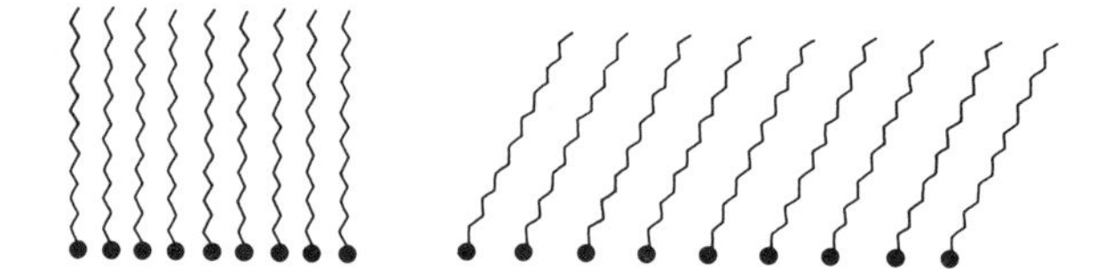

Fig. 6.22 Simulations of a Langmuir–Blodgett film [Kim *et al.* 1994a]: as the area per head group increases the chains tilt away from the normal.

no interactions between the image charges themselves. The hydrocarbon chain of the stearic acid was modelled using an all-atom model, with explicit hydrogen atoms.

A molecular dynamics simulation of 64 molecules with periodic boundary conditions confirmed the presence of a transition in which the collective tilt of the chains changed from being upright (i.e. perpendicular to the surface) to having an angle of around 20°, Figure 6.22. This transition was induced by increasing the area per head group. The proportion of molecules in the all-*trans* conformation decreased significantly as the head group area was increased (97.7% of molecules were fully extended for a head group area of 20.6 Å^2 but only 66.9% for an area of 21.2 Å^2). The bond linking the chain to the acid head growth showed a considerable degree of rotational disorder.

Bilayers of stearic acid were also simulated on a hydrophobic surface [Kim *et al.* 1994b]. In the bilayer the molecules are arranged head to head, with the hydrocarbon tail on the surface. In this arrangement hydrogen bonds form between the head groups (Figure 6.23). The bilayer also showed the tilt angle transition that was observed for the monolayer, though the degree of tilt was considerably less for the bilayer, suggesting that hydrogen bonding between the head groups was important in controlling the orientation of the molecules.

An extension of these calculations to cationic dialkylamide salts required an even more complex model [Adolf *et al.* 1995]. These molecules have the general

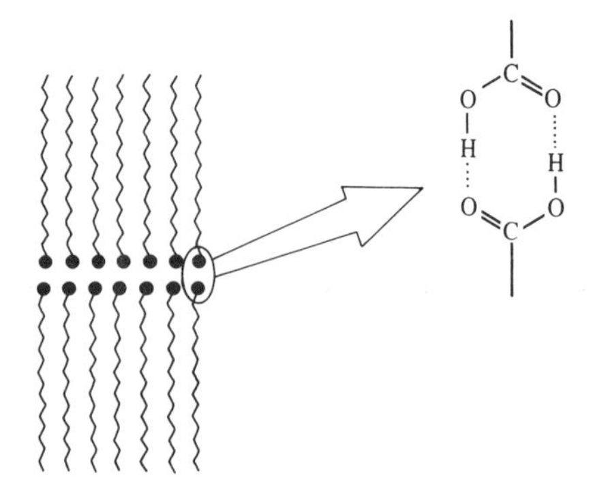

Fig. 6.23 In simulations of stearic acid on a hydrophobic surface hydrogen bonding between the head groups is important in controlling the orientation of the molecules [Kim *et al.* 1994b].

formula $(CH_3)_2N^+[(CH_2)_{n-1}CH_3][(CH_2)_{m-1}CH_3]Cl^-$ and the isomer with $m=n=18$ is one of the main active ingredients in commerical fabric softeners. The presence of two long alkyl chains and an ionic head group means that these molecules are also structurally similar to phospholipids. A modified Ewald method was used to calculate electrostatic interactions in the two dimensions parallel to the surface, and the anisotropic potential model of Toxvaerd (see section 3.14) was employed to retain the computational savings of a united atom model. This system also showed a variation in the tilt with head group area, though the results at the highest head group densities were not as 'solid like' as was suggested by the experimental data. Nevertheless, there were some areas where the model could be improved, including the need to incorporate water molecules and use a more appropriate representation of the chloride anion.

Appendix 6.1 Energy conservation in molecular dynamics

The total energy is the sum of the kinetic $\mathscr{K}(t)$ and potential energies $\mathscr{V}(t)$

$$E(t) = \mathscr{K}(t) + \mathscr{V}(t)$$

We want to derive an expression for the rate of change of the energy with time, $\mathrm{d}E/\mathrm{d}t$. First, we differentiate the kinetic energy term with respect to time:

$$\frac{\mathrm{d}\mathscr{K}}{\mathrm{d}t} = \sum_{\mathrm{i}=1}^{N} \frac{\mathrm{d}}{\mathrm{d}t}\left(\frac{1}{2} m_{\mathrm{i}} \mathrm{v}_{\mathrm{i}}^2\right) = \sum_{\mathrm{i}=1}^{N} m_{\mathrm{i}} v_{\mathrm{i}} \frac{\mathrm{d}\mathrm{v}_{\mathrm{i}}}{\mathrm{d}t} \tag{6.130}$$

As $m_{\mathrm{i}} \dfrac{\mathrm{d}v_{\mathrm{i}}}{\mathrm{d}t}$ is equal to the force on the atom i, the result can be written:

$$\frac{\mathrm{d}\mathscr{K}}{\mathrm{d}t} = \sum_{\mathrm{i}=1}^{N} \mathrm{v}_{\mathrm{i}} f_{\mathrm{i}} \tag{6.131}$$

f_{i} is the force on atom i.

The potential energy is written as a series of pairwise interaction terms:

$$\mathscr{V}(t) = \sum_{\mathrm{i}=1}^{N} \sum_{\mathrm{j}=\mathrm{i}+1}^{N} \upsilon(r_{\mathrm{ij}}(t)) \tag{6.132}$$

The derivative of the potential energy with respect to time can be written:

$$\frac{\mathrm{d}\mathscr{V}}{\mathrm{d}t} = \sum_{\mathrm{i}=1}^{N} \sum_{\mathrm{j}=\mathrm{i}+1}^{N} \frac{\partial \mathscr{V}}{\partial \upsilon(r_{\mathrm{ij}})} \frac{\mathrm{d}\upsilon(r_{\mathrm{ij}})}{\mathrm{d}t} \tag{6.133}$$

$\partial \mathscr{V} / \partial \upsilon(r_{\mathrm{ij}})$ equals 1 for each pairwise combination i and j. Each term $\upsilon(r_{\mathrm{ij}})$ is a function of the positions of atom i and j ($\mathbf{r}_{\mathrm{i}}$ and $\mathbf{r}_{\mathrm{j}}$) and we can then write:

$$\frac{\mathrm{d}\upsilon(r_{\mathrm{ij}})}{\mathrm{d}t} = \frac{\mathrm{d}\upsilon(r_{\mathrm{ij}})}{\mathrm{d}\mathbf{r}_{\mathrm{i}}} \frac{\mathrm{d}\mathbf{r}_{\mathrm{i}}}{\mathrm{d}t} + \frac{\mathrm{d}\upsilon(r_{\mathrm{ij}})}{\mathrm{d}\mathbf{r}_{\mathrm{j}}} \frac{\mathrm{d}\mathbf{r}_{\mathrm{j}}}{\mathrm{d}t} \tag{6.134}$$

For a given atom i, there will be a total of $N - 1$ terms of the form $\mathscr{v}(r_{ij})$ in the expression for the potential energy due to the interactions between i and all other atoms j. Hence we can write $d\mathscr{V}/dt$ as follows:

$$\frac{d\mathscr{V}}{dt} = \sum_{i=1}^{N} \sum_{j=1;j\neq i}^{N} \frac{\partial \mathscr{v}(r_{ij})}{\partial \mathbf{r}_i} \frac{d\mathbf{r}_i}{dt} = \sum_{i=1}^{N} \frac{d\mathbf{r}_i}{dt} \sum_{j=1;j\neq i}^{N} \frac{\partial \mathscr{v}(r_{ij})}{\partial \mathbf{r}_i} \tag{6.135}$$

The force on atom i due to its interaction with atom j equals minus the gradient with respect to $\mathbf{r}_i$, or $-dv(r_{ij})/d\mathbf{r}_i$. Thus the total force on the atom is equal to

$$-\sum_{j=1;j\neq i}^{N} \frac{\partial \mathscr{v}(r_{ij})}{\partial \mathbf{r}_i}$$

and so we have:

$$\frac{d\mathscr{V}}{dt} = -\sum_{i=1}^{N} \frac{d\mathbf{r}_i}{dt} f_i = -\sum_{i-1}^{N} v_i f_i \tag{6.136}$$

Thus $\dfrac{d\mathscr{V}}{dt} + \dfrac{d\mathscr{K}}{dt} = \dfrac{dE}{dt} = 0$ which implies that the energy is constant (though in practice, the total energy fluctuates about a constant value).

Appendix 6.2 Fourier series and Fourier analysis

Consider a periodic function $x(t)$ that repeats between $-\tau/2$ and $+\tau/2$ (i.e. has period τ). Even though $x(t)$ may not correspond to an analytical expression it can

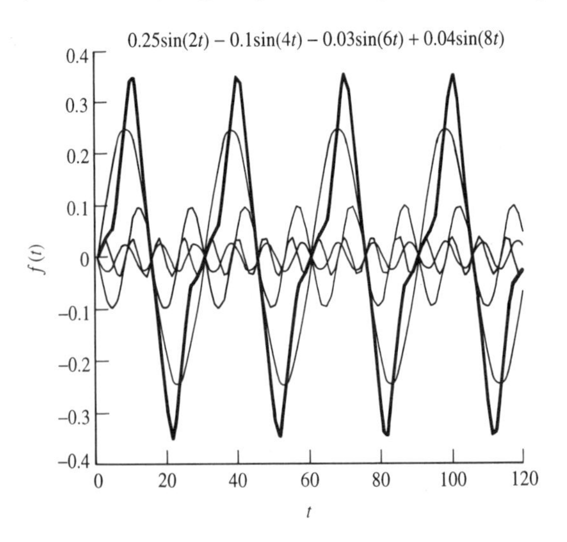

Fig. 6.24 In a Fourier series a periodic function is expressed as a sum of sine and cosine functions.

be written as the superposition of simple sine and cosine functions or *Fourier series*, Figure 6.24.

$$x(t) = a_0 + a_1 \cos \omega_0 t + a_2 \cos 2\omega_0 t + \cdots$$
$$+ b_1 \sin \omega_0 t + b_2 \sin 2\omega_0 t + \cdots \qquad (6.137a)$$

$$x(t) = a_0 + \sum_{n=1} (a_n \cos n\omega_0 t + b_n \sin n\omega_0 t) \qquad (6.137b)$$

ω_0 is related to the period of the function by $\omega_0 = 2\pi/\tau$ and to the frequency of the function by $\omega_0 = 2\pi\nu_0$. The frequencies of the contributing harmonics are thus $n\nu_0$ and are separated by $1/\tau$.

The coefficients a_n and b_n can be obtained as follows:

$$a_0 = \frac{1}{\tau}\int_{-\tau/2}^{\tau/2} x(t)dt \qquad (6.138)$$

$$a_n = \frac{2}{\tau}\int_{-\tau/2}^{\tau/2} x(t)\cos(2n\pi x/\tau)dx \qquad (6.139)$$

$$b_n = \frac{2}{\tau}\int_{-\tau/2}^{\tau/2} x(t)\sin(2n\pi x/\tau)dx \qquad (6.140)$$

An alternative way to express a Fourier series makes use of the following relationships:

$$\sin \omega_0 t = [\exp(i\omega_0 t) - \exp(-i\omega_0 t)]/2i \qquad (6.141a)$$
$$\cos \omega_0 t = [\exp(i\omega_0 t) + \exp(-i\omega_0 t)]/2 \qquad (6.141b)$$

The Fourier series is then written

$$x(t) = \sum_{-\infty}^{+\infty} c_n \exp(in\omega_0 t) \qquad (6.142)$$

with

$$c_n = \frac{1}{\tau}\int_{-\tau/2}^{\tau/2} x(t)\exp(in\omega_0 t)dt \qquad (6.143)$$

A Fourier series represents a function that is periodic with time period τ in terms of frequencies $n\omega_0 = 2\pi n/\tau$. The *Fourier transform* is used when the function has no periodicity. There is a close relationship between the Fourier series and the Fourier transform. One way to demonstrate the gradual change from a Fourier series to a Fourier transform is to consider the effect on the distribution of contributing frequencies as the period increases. This is illustrated in Figure 6.25, where the period of a square wave function is gradually increased. Also shown are the frequency contributions. It can be seen that an increasing number of frequency components are needed to describe the function as the period increases, and that when the period is infinite, the frequency spectrum is continuous.

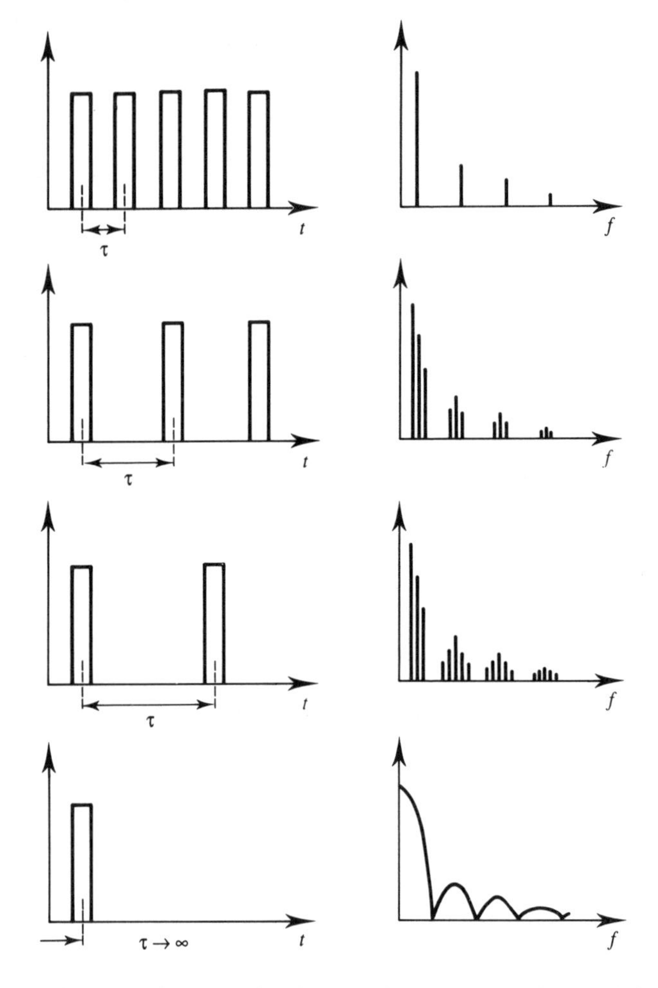

Fig. 6.25 The connection between the Fourier transform and the Fourier series can be established by gradually increasing the period of the function. When the period is infinite a continuous spectrum is obtained. Figure adapted from Ramirez R W 1985 *The FFT Fundamentals and Concepts*. Englewood Cliffs, NJ, Prentice-Hall.

The Fourier transform relationship between a function $c(t)$ and the corresponding frequency function $X(v)$ is:

$$x(t) = \int_{-\infty}^{+\infty} X(v) \exp(2\pi ivt) dv \tag{6.144}$$

The frequency function $X(v)$ is given by

$$X(v) = \int_{-\infty}^{+\infty} x(t) \exp(-2\pi ivt) dt \tag{6.145}$$

In practical applications, $x(t)$ is not a continuous function, and the data to be transformed are usually obtained as discrete values obtained by sampling at intervals. Under such circumstances, the discrete Fourier transform (DFT) is used to obtain the frequency function. Let us suppose that the time-dependent data values are obtained by sampling at regular intervals separated by δt and that a total of M samples are obtained (starting at $t=0$). From M samples, a total of M frequency coefficients can be obtained using the DFT expression [Press *et al.* 1992]:

$$X(k\delta v) = \delta t \sum_{n=0}^{M-1} x(n\delta t) \exp[-2\pi ink/M] \tag{6.146}$$

Here, $x(n\delta t)$ $(n=0, 1, \ldots M-1)$ are the experimental values obtained and $X(k\delta v)$ is the set of Fourier coefficients $(k=0, 1, \ldots M-1)$. The separation between the frequencies, δv, depends on the number of samples and the time between samples: $\delta v = 1/M\delta t$. An expression for converting frequency data into the time domain is also possible:

$$x(n\delta t) = \frac{1}{M} \sum_{k=0}^{M-1} X(k\delta v) \exp[2\pi ink/M] \tag{6.147}$$

To compute each Fourier coefficient $X(k\delta T)$ (of which there are M) it is therefore necessary to evaluate the summation $\sum_{n=0}^{M-1} x(n\delta t) \exp[-2\pi i/nkM]$ for that value of k. There will be M terms in the summation. A simple algorithm to determine the frequency spectrum would scale with the square of the number of measurements, M. This is a severe limitation for many problems involve an extremely large number of pieces of data. It is for this reason that the fast Fourier Transform (FFT) (ascribed to Cooley and Tukey [Cooley and Tukey 1965] but in fact using methods developed much earlier) has made such an impact. The FFT algorithm scales as $M\ln M$. With the FFT algorithm it is possible to derive the Fourier transforms even with a considerable number of data points.

Further reading

Allen M P and D J Tildesley 1987. *Computer Simulation of Liquids*. Oxford, Oxford University Press.

Berendsen H C and W F van Gunsteren 1984. Molecular Dynamics Simulations: Techniques and Approached in Barnes A J, W J Orville-Thomas and J Yarwood (Editors). *Molecular Liquids, Dynamics and Interactions*. NATO ASI Series C135, New York, Reidel: 475–600.

Berendsen H C and W F van Gunsteren 1986. Practical Algorithms for Dynamic Simulations. Molecular Dynamics Simulation of Statistical Mechanical Systems. *Proceedings of the Enrico Fermi Summer School Varenna Soc. Italian di Fiscia. Bologna*: 43–65.

Brooks C L III M Karplus and B M Pettitt 1988. Proteins. A Theoretical Perspective of Dynamics, Structure and Thermodynamics. *Advances in*

Chemical Physics. Volume LXXI. New York, John Wiley.
Goldstein H 1980. *Classical Mechanics (2nd Edition)*. Reading, MA, Addison-Wesley.
Haile J M 1992. *Molecular Dynamics Simulation. Elementary Methods*. New York, John Wiley.
McCammon J A and S C Harvey 1987. *Dynamics of Proteins and Nucleic Acids*. Cambridge, Cambridge University Press.
Ramirez R W 1985. *The FFT Fundamentals and Concepts*. Englewood Cliffs, NJ, Prentice-Hall.
van Gunsteren W F 1994. Molecular Dynamics and Stochastic Dynamics Simulations: A Primer in van Gunsteren W F, P K Weiner and A J Wilkinson (Editors). *Computer Simulations of Biomolecular Systems*. Volume 2. Leiden, ESCOM.
van Gunsteren W F and H J C Berendsen 1990. Computer Simulation of Molecular Dynamics: Methodology, Applications and Perspectives in Chemistry. *Angewandte Chemie International Edition in English* **29**: 992–1023.

References

Adolf D B, D J Tildesley, M R S Pinches, J B Kingdon, T Madden and A Clark 1995. Molecular Dynamics Simulations of Dioctadecyldimethylammonium Chloride Monolayers. *Langmuir* **11**:237–246.
Alder B J and T E Wainwright 1957. Phase Transition for a Hard-Sphere System. *The Journal of Chemical Physics* **27**:1208–1209.
Alder B J and T E Wainwright 1970. Decay of the Velocity Autocorrelation Function. *Physical Review A* **1**:18–21.
Allen M P and D J Tildesley 1987. *Computer Simulation of Liquids*. Oxford, Oxford University Press.
Anderson H C 1980. Molecular Dynamics Simulations at Constant Pressure and/or Temperature. *The Journal of Chemical Physics* **72**:2384–2393.
Anderson H C 1983. Rattle: A 'Velocity' Version of the Shake Algorithm for Molecular Dynamics Calculations. *Journal of Computational Physics* **52**:24–34.
Beeman D 1976. Some Multistep Methods for Use in Molecular Dynamics Calculations. *Journal of Computational Physics* **20**:130–139.
Berendsen H J C, J P M Postma, W F van Gunsteren, A Di Nola and J R Haak 1984. Molecular Dynamics with Coupling to an External Bath. *The Journal of Chemical Physics* **81**:3684–3690.
Brunger A, C B Brooks and M Karplus 1984. Stochastic Boundary Conditions for Molecular Dynamics Simulations of ST2 Water. *Chemical Physics Letters* **105**:495–500.
Cooley J W and J W Tukey 1965. An Algorithm for the Machine Calculation of Complex Fourier Series. *Mathematics of Computation.* **19**:297–301.

Dauber-Osguthorpe P and D J Osguthorpe 1990. Analysis of Intramolecular Motions by Filtering Molecular Dynamics Trajectories. *The Journal of the American Chemical Society* **112**:7921–7935.

Dauber-Osguthorpe P and D J Osguthorpe 1993. Partitioning the Motion in Molecular Dynamics Simulations into Characteristic Modes of Motion. *Journal of Computational Chemistry* **14**:1259–1271.

De Loof H, S C Harvey, J P Segrest and R W Pastor 1991. Mean Field Stochastic Boundary Molecular Dynamics Simulation of a Phospholipid in a Membrane. *Biochemistry* **30**:2099–2113.

Essex J W, M M Hann and W G Richards 1994. Molecular Dynamics of a Hydrated Phospholipid Bilayer. *Philosophical Transactions of the Royal Society of London B* **344**:239–260.

Fincham D and Heyes D M 1982. Integration Algorithms in Molecular Dynamics. *CCP5 Quarterly* **6**:4–10.

Gear C W 1971. *Numerical Initial Value Problems in Ordinary Differential Equations*. Englewood Cliffs, NJ, Prentice-Hall.

Helfand E 1984. Dynamics of Conformational Transitions in Polymers. *Science* **226**:647–650.

Hockney R W 1970. The Potential Calculation and Some Applications. *Methods in Computational Physics* **9**:136–211.

Hoover W G 1985. Canonical Dynamics: Equilibrium Phase-Space Distributions. *Physical Review A* **31**:1695–1697.

Jorgensen W L, R C Binning Jr and B Bigot 1981. Structures and Properties of Organic Liquids: *n*-Butane and 1,2-Dichloroethane and Their Conformational Equilibria. *The Journal of the American Chemical Society* **103**:4393-4399.

Kim K S, M A Moller, D J Tildesley and N Quirke 1994a. Molecular Dynamics Simulations of Langmuir–Blodgett Monolayers with Explicit Head-Group Interactions. *Molecular Simulation* **13**:77–99.

Kim K S, D J Tildesley and N Quirke 1994b. Molecular Dynamics of Langmuir–Blodgett Films: II. Bilayers. *Molecular Simulation* **13**:101–114.

Lavery R and H Sklenar 1988. The Definition of Generalized Helicoidal Parameters and of Axis Curvature for Irregular Nucleic-Acids. *Journal of Biomolecular Structure and Dynamics* **6**:63–91.

Leach A R and T E Klein 1995. A Molecular Dynamics Study of the Inhibitors of Dihydrofolate Reductase by a Phenyl Triazine. *Journal of Computational Chemistry* **16**:1378–1393.

Marcelja S 1973. Molecular Model for Phase Transition in Biological Membranes. *Nature* **241**:451–453.

Marcelja S 1974. Chain Ordering in Liquid Crystals. II. Structure of Bilayer Membranes. *Biochimica et Biophysica Acta* **367**:165–176.

Nosé S 1984. A Molecular Dynamics Method for Simulations in the Canonical Ensemble. *Molecular Physics* **53**:255–268.

Pastor R W, R M Venable and M Karplus 1988. Brownian Dynamics Simulation of a Lipid Chain in a Membrane Bilayer. *The Journal of Chemical Physics* **89**:1112–1127.

Pearce L L and S C Harvey 1993. Langevin Dynamics Studies of Unsaturated

Phospholipids in a Membrane Environment. *Biophysical Journal* **65**:1084–1092.

Press W H, B P Flannery, S A Teukolsky, W T Vetterling 1992. *Numerical Recipes in Fortran*. Cambridge, Cambridge University Press: 490–529.

Rahman A 1964. Correlations in the Motion of Atoms in Liquid Argon. *Physical Review* **136**:A405–A411.

Rahman A and F H Stillinger 1971. Molecular Dynamics Study of Liquid Water. *The Journal of Chemical Physics* **55**:3336–3359.

Robinson A J, W G Richards, P J Thomas and M M Hann 1994. Head Group and Chain Behaviour in Biological Membranes—A Molecular Dynamics Simulation. *Biophysical Journal* **67**:2345–2354.

Rubinstein R Y 1981. *Simulation and Monte Carlo Methods*. New York, Wiley.

Ryckaert J P, G Cicotti and H J C Berensden 1977. Numerical Integration of the Cartesian Equations of Motion of a System with Constraints: Molecular Dynamics of *n*-Alkanes. *Journal of Computational Physics* **23**:327–341.

Seelig A and J Seelig 1974. The Dynamic Structure of Fatty Acyl Chains in a Phospholipid Bilayer Measured by Deuterium Magnetic Resonance. *Biochemistry* **13**:4839–4845.

Stouch T R 1993. Lipid Membrane Structure and Dynamics Studied by All-Atom Molecular Dynamics Simulations of Hydrated Phospholipid Bilayers. *Molecular Simulation* **10**:335–362.

Swope W C, H C Anderson, P H Berens and K R Wilson 1982. A Computer Simulation Method for the Calculation of Equilibrium Constants for the Formation of Physical Clusters of Molecules: Application to Small Water Clusters. *Journal of Chemical Physics* **76**:637–649.

Tobias D J and C L Brooks III 1988. Molecular Dynamics with Internal Coordinate Constraints. *The Journal of Chemical Physics* **89**:5115–5126.

van der Ploeg P and H J C Berendsen 1982. Molecular Dynamics Simulation of a Bilayer Membrane. *The Journal of Chemical Physics* **76**:3271–3276.

van Gunsteren W F and H J C Berendsen 1982. Algorithms for Brownian Dynamics. *Molecular Physics* **45**:637–547.

van Gunsteren W F, H J C Berendsen and J A C Rullmann 1981. Stochastic Dynamics for Molecules with Constraints. Brownian Dynamics of *n*-Alkanes. *Molecular Physics* **44**:69–95.

Verlet L 1967. Computer 'Experiments' on Classical Fluids. I. Thermodynamical Properties of Lennard–Jones Molecules. *Physical Review* **159**:98–103.

Widmalm G and R W Pastor 1992. Comparison of Langevin and Molecular Dynamics Simulations. *Journal of the Chemical Society Faraday Transactions* **88**:1747–1754.

Woodcock L V 1971. Isothermal Molecular Dynamics Calculations for Liquid Salts. *Chemical Physics Letters* **10**:257–261.

Yun-Yu S, W Lu and W F van Gunsteren 1988. On the Approximation of Solvent Effects on the Conformation and Dynamics of Cyclosporin A by Stochastic Dynamics Simulation Techniques. *Molecular Simulation* **1**:369–383.

7 Monte Carlo simulation methods

7.1 Introduction

The Monte Carlo simulation method occupies a special place in the history of molecular modelling, as it was the technique used to perform the first computer simulation of a molecular system. A Monte Carlo simulation generates configurations of a system by making random changes to the positions of the species present, together with their orientations and conformations where appropriate. Many computer algorithms are said to use a 'Monte Carlo' method, meaning that some kind of random sampling is employed. In molecular simulations 'Monte Carlo' is almost always used to refer to methods that use a technique called *importance sampling*. Importance sampling methods are able to generate states of low energy, as this enables properties to be calculated accurately. We can calculate the potential energy of each configuration of the system, together with the values of other properties, from the positions of the atoms. The Monte Carlo method thus samples from a $3N$-dimensional space of the positions of the particles. There is no momentum contribution in a Monte Carlo simulation, in contrast to a molecular dynamics simulation. How then can a Monte Carlo simulation be used to calculate thermodynamic quantities, given that phase space is $6N$-dimensional?

To resolve this difficulty, let us return to the canonical ensemble partition, Q, which for a system of N identical particles of mass m can be written:

$$Q_{NVT} = \frac{1}{N!}\frac{1}{h^{3N}}\iint \mathrm{d}\mathbf{p}^N \mathrm{d}\mathbf{r}^N \exp\left[-\frac{\mathscr{H}(\mathbf{p}^N, \mathbf{r}^N)}{k_\mathrm{B}T}\right] \tag{7.1}$$

The factor $N!$ disappears when the particles are no longer indistinguishable. $\mathscr{H}(\mathbf{p}^N, \mathbf{r}^N)$ is the Hamiltonian that corresponds to the total energy of the system. The value of the Hamiltonian depends upon the $3N$ positions and $3N$ momenta of the particles in the system (one position and one momentum for each of the three coordinates of each particle). The Hamiltonian can be written as the sum of the kinetic and potential energies of the system:

$$\mathscr{H}(\mathbf{p}^N, \mathbf{r}^N) = \sum_{\mathrm{i}=1}^{N} \frac{|\mathbf{p}_\mathrm{i}|^2}{2m} + \mathscr{V}(\mathbf{r}^N) \tag{7.2}$$

The crucial point to recognise is that the double integral in equation 7.1 can be separated into two separate integrals, one over positions and the other over the momenta:

$$Q_{NVT} = \frac{1}{N!}\frac{1}{h^{3N}}\int d\mathbf{p}^N \exp\left[-\frac{|\mathbf{r}|^2}{2mk_BT}\right]\int d\mathbf{r}^N \exp\left[-\frac{\mathcal{V}(\mathbf{r}^N)}{k_BT}\right] \tag{7.3}$$

This separation is possible only if the potential energy function, $\mathcal{V}(\mathbf{r}^N)$, is not dependent upon the velocities (this is a safe assumption for almost all potential functions in common use). The integral over the momenta can now be performed analytically, the result being:

$$\int d\mathbf{r}^N \exp\left[-\frac{|\mathbf{p}|^2}{2mk_BT}\right] = (2\pi mk_BT)^{\frac{3N}{2}} \tag{7.4}$$

The partition function can thus be written:

$$Q_{NVT} = \frac{1}{N!}\left(\frac{2\pi mk_BT}{h^2}\right)^{\frac{3N}{2}}\int d\mathbf{r}^N \exp\left(-\frac{\mathcal{V}(\mathbf{r}^N)}{k_BT}\right) \tag{7.5}$$

The integral over the positions is often referred to as the *configurational integral*, Z_{NVT}:

$$Z_{NVT} = \int d\mathbf{r}^N \exp\left(-\frac{\mathcal{V}(\mathbf{r}^N)}{k_BT}\right) \tag{7.6}$$

In an ideal gas there are no interactions between the particles and so the potential energy function, $\mathcal{V}(\mathbf{r}^N)$, equals zero. $\exp(-\mathcal{V}(\mathbf{r}^N)/k_BT)$ is therefore equal to 1 for every gas particle in the system. The integral of 1 over the coordinates of each atom is equal to the volume, and so for N ideal gas particles the configurational integral is given by V^N ($V \equiv$ volume). This leads to the following result for the canonical partition function of an ideal gas:

$$Q_{NVT} = \frac{V^N}{N!}\left(\frac{2\pi k_BTm}{h^2}\right)^{\frac{3N}{2}} \tag{7.7}$$

This is often written in terms of the *de Broglie thermal wavelength*, Λ:

$$Q_{NVT} = \frac{V^N}{N!\Lambda^{3N}} \tag{7.8}$$

where $\Lambda = \surd(h^2/2\pi k_BTm)$

By combining equations (7.4) and (7.6) we can see that the partition function for a 'real' system has a contribution due to ideal gas behaviour (the momenta) and contribution due to the interactions between the particles. Any deviations from ideal gas behaviour are due to interactions within the system as a consequence of these interactions. This enables us to write the partition function as:

$$Q_{NVT} = Q_{NVT}^{\text{ideal}}Q_{NVT}^{\text{excess}} \tag{7.9}$$

The excess part of the partition function is given by:

$$Q_{NVT}^{\text{excess}} = \frac{1}{V^N} \int d\mathbf{r}^N \exp\left[-\frac{\mathcal{V}(\mathbf{r}^N)}{k_B T}\right] \tag{7.10}$$

A consequence of writing the partition function as a product of a real gas and an ideal gas part is that thermodynamic properties can be written in terms of an ideal gas value and an excess value. The ideal gas contributions can be determined analytically by integrating over the momenta. For example, the Helmholtz free energy is related to the canonical partition function by:

$$A = -k_B T \ln Q_{NVT} \tag{7.11}$$

Writing the partition function as the product, equation (7.9), leads to

$$A = A^{\text{ideal}} + A^{\text{excess}} \tag{7.12}$$

The important conclusion is that all of the deviations from ideal gas behaviour are due to the presence of interactions between the atoms in the system, as calculated using the potential energy function. This energy function is dependent only upon the positions of the atoms and not their momenta, and so a Monte Carlo simulation is able to calculate the excess contributions that give rise to deviations from ideal gas behaviour.

7.2 Calculating properties by integration

Having established that we can indeed explore configurational phase space and derive useful thermodynamic properties, let us consider how we might achieve this in practice. For example, the average potential energy can, in principle at least, be determined by evaluating the integral:

$$\langle \mathcal{V}(\mathbf{r}^N) \rangle = \int d\mathbf{r}^N \mathcal{V}(\mathbf{r}^N) \rho(\mathbf{r}^N) \tag{7.13}$$

This is a multidimensional integral over the $3N$ degrees of freedom of the N particles in the system. $\rho(\mathbf{r}^N)$ is the probability of obtaining the configuration $\mathbf{r}^N$ and is given by

$$\rho(\mathbf{r}^N) = \frac{\exp[-\mathcal{V}(\mathbf{r}^N)/k_B T]}{Z} \tag{7.14}$$

The denominator, Z is the configurational integral (equation (7.6)). For the potential functions commonly used in molecular modelling, it is not possible to evaluate these integrals analytically. However, we could attempt to obtain values for the integrals using numerical methods. One simple numerical integration method is the trapezium rule. This approximates the integral as a series of trapeziums between the two limits, as illustrated for a one-dimensional problem in Figure 7.1. In this case we have divided the integral onto 10 trapeziums which requires 11 function evaluations. Simpson's rule involves a similar procedure and may provide a more accurate value of the integral [Stephenson 1973]. For a func-

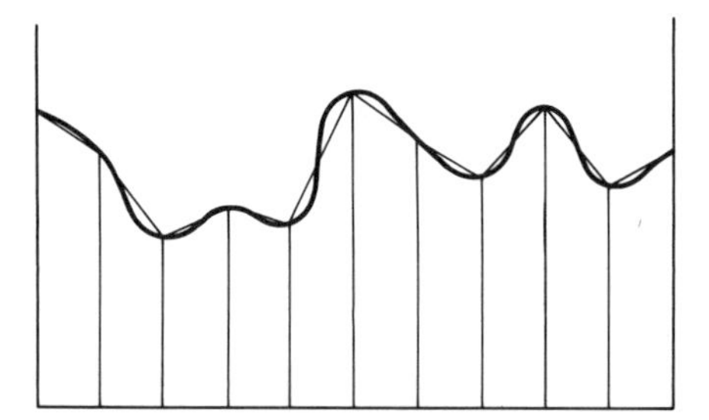

Fig. 7.1 Evaluation of a one-dimensional integral using the trapezium rule. The area under the curve is approximated as the sum of the areas of the trapeziums.

tion of two variables ($f(x, y)$), it is necessary to square the number of function evaluations required. For a $3N$-dimensional integral the total number of function evaluations required to determine the integral would be m^{3N}, where m is the number of points needed to determine the integral in each dimension. This number is enormous even for very small numbers of particles. For example, with just 50 particles and three points per dimension, a total of 3^{150} ($\sim 10^{71}$) evaluations would be required. Integration using the trapezium rule or Simpson's rule is clearly not a feasible approach.

We could consider a random method as a possible alternative. The general principle can be illustrated using the function shown in Figure 7.2. To determine the area under the curve in Figure 7.2 a series of random points would be generated within the bounding area. The area under the curve is then calculated by multiplying the bounding area A by the ratio of the number of trial points that lie under the curve to the total number of points generated. An estimate of π can be determined in this way, as illustrated in Figure 7.2.

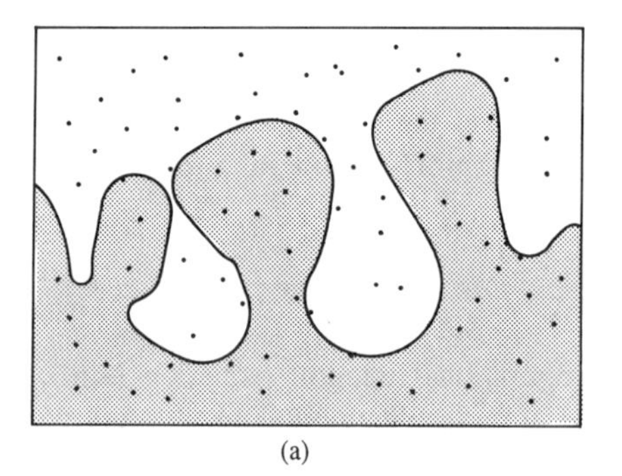

(a)

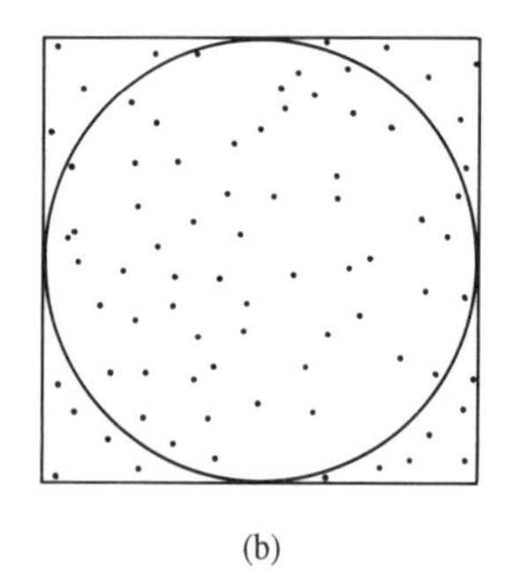

(b)

Fig. 7.2 Simple Monte Carlo integration. (a) The shaded area under the irregular curve equals the ratio of the number of random points under the curve to the total number of points, multiplied by the area of the bounding area. (b) An estimate of π can be obtained by generating random numbers within the square. π then equals the number of points within the circle divided by the total number of points within the square, multiplied by 4.

To calculate the partition function for a system of N atoms using this simple Monte Carlo integration method would involve the following steps:

1. Obtain a configuration of the system by randomly generating $3N$ Cartesian coordinates which are assigned to the particles.
2. Calculate the potential energy of the configuration, $\mathcal{V}(\mathbf{r}^N)$
3. From the potential energy, calculate the Boltzman factor, $\exp(-\mathcal{V}(\mathbf{r}^N)k_B T)$
4. Add the Boltzmann factor to the accumulated sum of Boltzmann factors and the potential energy contribution to its accumulated sum and return to step 1.
5. After a number, N_{trial}, of iterations, the mean value of the potential energy would be calculated using:

$$\langle \mathcal{V}(\mathbf{r}^N) \rangle = \frac{\sum_{i=1}^{N_{\text{trial}}} \mathcal{V}_i(\mathbf{r}^N) \exp[-\mathcal{V}_i(\mathbf{r}^N)/k_B T]}{\sum_{i=1}^{N_{\text{trial}}} \exp[-\mathcal{V}_i(\mathbf{r}^N)/k_B T]} \tag{7.15}$$

Unfortunately, this is not a feasible approach for calculating thermodynamic properties due to the large number of configurations that have extremely small (effectively zero) Boltzmann factors caused by high-energy overlaps between the particles. This reflects the nature of the phase space, most of which corresponds to non-physical configurations with very high energies. Only a very small proportion of the phase space corresponds to low-energy configurations where there are no overlapping particles and where the Boltzmann factor has an appreciable value. These low-energy regions coincide with the physically observed phases such as solid, liquid, etc.

One way around this impasse is to generate configurations that make a large contribution to the integral (7.15), which is the strategy adopted in importance sampling and which is the essence of the method described by Metropolis, Rosenbluth, Rosenbluth, Teller and Teller in 1953 [Metropolis *et al.* 1953]. For many thermodynamic properties of a molecular system, those states with a high probability ρ are also the ones that make a significant contribution to the integral (there are some notable exceptions to this, such as the free energy). The Metropolis method has become so widely adopted that in the simulation and molecular modelling communities it is usually referred to as 'the Monte Carlo method'. Fortunately, there is rarely any confusion with the simple Monte Carlo methods. The crucial feature of the Metropolis approach is that it biases the generation of configurations towards those that make the most significant contribution to the integral. Specifically, it generates states with a probability $\exp(-\mathcal{V}(\mathbf{r}^N)/k_B T)$ and then counts each of them equally. By contrast, the simple Monte Carlo integration method generates states with equal probability (both high and low energy) and then assigns them a weight $\exp(-\mathcal{V}(\mathbf{r}^N)/k_B T)$.

7.3 Some theoretical background to the Metropolis method

The Metropolis algorithm generates a *Markov chain* of states. A Markov chain satisfies the following two conditions:

1. The outcome of each trial depends only upon the preceding trial and not upon any previous trials.
2. Each trial belongs to a finite set of possible outcomes.

Condition (1) provides a clear distinction between the molecular dynamics and Monte Carlo methods, for in a molecular dynamics simulation all of the states are connected in time. Suppose the system is in a state m. We denote the probability of moving to state n as π_{mn}. The various π_{mn} can be considered to constitute an $N \times N$ matrix $\boldsymbol{\pi}$ (the transition matrix) where N is the number of possible states. Each row of the transition matrix sums to 1 (i.e. the sum of the probabilities π_{mn} for a given m equals 1). The probability that the system is in a particular state is represented by a probability vector $\boldsymbol{\rho}$:

$$\boldsymbol{\rho} = (\rho_1, \rho_2, \ldots \rho_m, \rho_n, \ldots \rho_N) \tag{7.16}$$

Thus ρ_1 is the probability that the system is in state 1 and ρ_m the probability that the system is in state m. If $\boldsymbol{\rho}(1)$ represents the initial (randomly chosen) configuration, then the probability of the second state is given by:

$$\boldsymbol{\rho}(2) = \boldsymbol{\rho}(1)\boldsymbol{\pi} \tag{7.17}$$

The probability of the third state is:

$$\boldsymbol{\rho}(3) = \boldsymbol{\rho}(2)\boldsymbol{\pi} = \boldsymbol{\rho}(1)\boldsymbol{\pi}\boldsymbol{\pi} \tag{7.18}$$

The equilibrium distribution of the system can be determined by considering the result of applying the transition matrix an infinite number of times. This limiting distribution of the Markov chain is given by $\boldsymbol{\rho}_{\text{limit}} = \lim_{N\to\infty} \boldsymbol{\rho}(1)\boldsymbol{\pi}^{\text{N}}$.

One feature of the limiting distribution is that it is independent of the initial guess $\boldsymbol{\rho}(1)$. The limiting or equilibrium distribution for a molecular or atomic system is one in which the probabilities of each state are proportional to the Boltzmann factor. We can illustrate the use of the probability distribution and the transition matrix by considering a two-level system in which the energy levels are such that the ratio of the Boltzmann factors is 2:1. The expected limiting distribution thus corresponds to a configuration vector $\left(\frac{2}{3}, \frac{1}{3}\right)$. The following transition matrix enables the limiting distribution to be achieved:

$$\boldsymbol{\pi} = \begin{pmatrix} 0.5 & 0.5 \\ 1 & 0 \end{pmatrix} \tag{7.19}$$

We can illustrate the use of this transition matrix as follows. Suppose the initial probability vector is (1 0) and so the system starts with a 100% probability of being in state 1 and no probability of being in state 2. Then the second state is given by:

$$\boldsymbol{\rho}(2) = (1 \quad 0)\begin{pmatrix} 0.5 & 0.5 \\ 1 & 0 \end{pmatrix} = (0.5 \quad 0.5) \tag{7.20}$$

The third state is $\boldsymbol{\rho}(3) = (0.75, 0.25)$. Successive applications of the transition matrix give the limiting distribution (2/3, 1/3).

When the limiting distribution is reached then application of the transition matrix must return the same distribution back:

$$\boldsymbol{\rho}_{\text{limit}} = \boldsymbol{\rho}_{\text{limit}}\boldsymbol{\pi} \tag{7.21}$$

Thus, if an ensemble can be prepared that is at equilibrium, then one Metropolis Monte Carlo step should return an ensemble that is still at equilibrium. A consequence of this is that the elements of the probability vector for the limiting distribution must satisfy:

$$\sum_m \rho_m \pi_{mn} = \rho_n \tag{7.22}$$

This can be seen to hold for our simple two-level example:

$$(2/3 \quad 1/3)\begin{pmatrix} 1/2 & 1/2 \\ 1 & 0 \end{pmatrix} = (2/3 \quad 1/3) \tag{7.23}$$

We will henceforth use the symbol ρ to refer to the limiting distribution.

Closely related to the transition matrix is the *stochastic matrix*, whose elements are labelled α_{mn}. This matrix gives the probability of choosing the two states m and n between which the move is to be made. It is often known as the *underlying matrix* of the Markov chain. If the probability of accepting a trial move from m to n is p_{mn} then the probability of making a transition from m to n (π_{mn}) is given by multiplying the probability of choosing states m and n (α_{mn}) by the probability of accepting the trial move (p_{mn}):

$$\pi_{mn} = \alpha_{mn} p_{mn} \tag{7.24}$$

It is often assumed that the stochastic matrix $\boldsymbol{\alpha}$ is symmetrical (i.e. the probability of choosing the states m and n is the same whether the move is made from m to n or from n to m). If the probability of state n is greater than that of state m in the limiting distribution (i.e. if the Boltzmann factor of n is greater than that of m because the energy of n is lower than the energy of m) then in the Metropolis recipe, the transition matrix element π_{mn} for progressing from m to n equals the probability of selecting the two states in the first place (i.e. $\pi_{mn} = \alpha_{mn}$ if $\rho_n \geq \rho_m$). If the Boltzmann weight of the state n is less than that of state m, then the probability of permitting the transition is given by multiplying the stochastic matrix element α_{mn} by the ratio of the probabilities of the state n to the previous state m. This can be written:

$$\pi_{mn} = \alpha_{mn} \; (\rho_n \geq \rho_m) \tag{7.25}$$

$$\pi_{mn} = \alpha_{mn} \; (\rho_n/\rho_m) \; (\rho_n < \rho_m) \tag{7.26}$$

These two conditions apply if the initial and final states m and n are different. If m and n are the same state, then the transition matrix element is calculated from the fact that the rows of the stochastic matrix sum to one:

$$\pi_{mm} = 1 - \sum_{m \neq n} \pi_{mn} \tag{7.27}$$

Let us now try to reconcile the Metropolis algorithm as outlined in section 5.1.3 with the more formal approach that we have just developed. We recall that in the Metropolis method a new configuration n is accepted if its energy is lower than the original state m. If the energy is higher, however, then we would like to choose the move with a probability according to equation (7.24). This is achieved by comparing the Boltzmann factor $\exp(-\Delta\mathscr{V}(\mathbf{r}^N)/k_BT)$ $(\Delta\mathscr{V}(\mathbf{r}^N) = [\mathscr{V}(\mathbf{r}^N)_n - \mathscr{V}(\mathbf{r}^N)_m])$ to a random number between 0 and 1. If the Boltzmann factor is greater than the random number then the new state is accepted. If it is smaller then the new state is rejected. Thus if the energy of the new state (n) is very close to that of the old state (m) then the Boltzmann factor of the energy difference will be very close to 1, and so the move is likely to be accepted. If the energy difference is very large, however, then the Boltzmann factor will be close to zero and the move is unlikely to be accepted. The Metropolis method is derived by imposing the condition of microscopic reversibility: at equilibrium the transition between two states occurs at the same rate. The rate of transition from a state m to a state n equals the product of the population (ρ_m) and the appropriate element of the transition matrix (π_{mn}). Thus, at equilibrium we can write:

$$\pi_{mn}\rho_m = \pi_{nm}\rho_n \tag{7.28}$$

The ratio of the transition matrix elements thus equals the ratio of the Boltzmann factors of the two states:

$$\frac{\pi_{mn}}{\pi_{nm}} = \exp[-(\mathscr{V}(\mathbf{r}^N)_n - \mathscr{V}(\mathbf{r}^N)_m)/k_BT] \tag{7.29}$$

7.4 Implementation of the Metropolis Monte Carlo method

A Monte Carlo program to simulate an atomic fluid is quite simple to construct. At each iteration of the simulation a new configuration is generated. This is usually done by making a random change to the Cartesian coordinates of a single randomly chosen particle using a random number generator. If the random number generator produces numbers (ξ) in the range 0 to 1, moves in both positive and negative directions are possible if the coordinates are changed as follows:

$$x_{\text{new}} = x_{\text{old}} + (2\xi - 1)\delta r_{\max} \tag{7.30}$$

$$y_{\text{new}} = y_{\text{old}} + (2\xi - 1)\delta r_{\max} \tag{7.31}$$

$$z_{\text{new}} = z_{\text{old}} + (2\xi - 1)\delta r_{\max} \tag{7.32}$$

A unique random number is generated for each of the three directions x, y and z. δr_{max} is the maximum possible displacement in any direction. The energy of the new configuration is then calculated; this need not require a complete recalculation of the energy of the entire system, but only those contributions involving the particle that has just been moved. As a consequence, the neighbour list used by a Monte Carlo simulation must contain *all* the neighbours of each atom, because it is necessary to identify all the atoms which interact with the moving atom (recall that in molecular dynamics the neighbour list for each atom only contains neighbours with a higher index). Proper account should be taken of periodic boundary conditions and the minimum image convention when generating new configurations and calculating their energies. If the new configuration is lower in energy than its predecessor then the new configuration is retained as the starting point for the next iteration. If the new configuration is higher in energy than its predecessor then the Boltzmann factor, $\exp(-\Delta\mathcal{V}/k_BT)$ is compared to a random number between 0 and 1. If the Boltzmann factor is greater than the random number then the new configuration is accepted; if not then it is rejected and the initial configuration is retained for the next move. This acceptance condition can be written in the following concise fashion:

$$\text{rand}(0, 1) \leq \exp(-\Delta\mathcal{V}(\mathbf{r}^N)/k_BT) \tag{7.33}$$

The size of the move at each iteration is governed by the maximum displacement, δr_{max}. This is an adjustable parameter whose value is usually chosen so that approximately 50% of the trial moves are accepted. If the maximum displacement is too small then many moves will be accepted but the states will be very similar and the phase space will only be explored very slowly. Too large a value of δr_{max} and many trial moves will be rejected because they lead to unfavourable overlaps. The maximum displacement can be adjusted automatically while the program is running to achieve the desired acceptance ratio by keeping a running score of the proportion of moves that are accepted. Every so often the maximum displacement is then scaled by a few percent: if too many moves have been accepted then the maximum displacement is increased; too few and δr_{max} is reduced.

As an alternative to the random selection of particles it is possible to move the atoms sequentially (this requires one fewer call to the random number generator per iteration). Alternatively, several atoms can be moved at once; if an appropriate value for the maximum displacement is chosen then this may enable phase space to be covered more efficiently.

As with a molecular dynamics simulation, a Monte Carlo simulation comprises an equilibration phase followed by a production phase. During equilibration, appropriate thermodynamic and structural quantities such as the total energy (and the partitioning of the energy among the various components), mean square displacement and order parameters (as appropriate) are monitored until they achieve stable values whereupon the production phase can commence. In a Monte Carlo simulation from the canonical ensemble, the temperature and volume are of course fixed. In a constant pressure simulation the volume will change and should therefore also be monitored to ensure that a stable system density is achieved.

7.4.1 Random number generators

The random number generator at the heart of every Monte Carlo simulation program is accessed a very large number of times, not only to generate new configurations but also to decide whether a given move should be accepted or not. Random number generators are also used in other modelling applications; for example, in a molecular dynamics simulation the initial velocities are normally assigned using a random number generator. The numbers produced by a random number generator are not, in fact, truly random; the same sequence of numbers should always be generated when the program is run with the same initial conditions (if not, then a serious error in the hardware or software must be suspected!). The sequences of numbers are thus often referred to as 'pseudo-random' numbers as they possess the statistical properties of 'true' sequences of random numbers. Most random number generators are designed to generate different sequences of numbers if a different 'seed' is provided. In this way several independent runs can be performed using different seeds. One simple strategy is to use the time and/or date as the seed; this is information that can often be obtained automatically by the program from the operating system.

The numbers produced by a random number generator should satisfy certain statistical properties. This requirement usually supercedes the need for a computationally very fast algorithm as other parts of a Monte Carlo simulation take much time (such as calculating the change in energy). One useful and simple test of a random number generator is to break a sequence of random numbers into blocks of k numbers which are taken to be coordinates in a k-dimensional space. A good random number should give a random distribution of points. Many of the common generators do not satisfy this test because the points lie on a plane or because they show clear correlations [Sharp and Bays 1992].

The *linear congruential* method is widely used for generating random numbers. Each number in the sequence is generated by taking the previous number, multiplying by a constant (the multiplier, a), adding a second constant (the increment, b), and taking the remainder when dividing by a third constant (the modulus, m). The first value is the seed, supplied by the user. Thus:

$$\xi[1] = \text{seed} \tag{7.34}$$

$$\xi[\text{i}] = \text{MOD}\{(\xi[\text{i}-1] \times a + b), m\} \tag{7.35}$$

The MOD function returns the remainder when the first argument is divided by the second (for example, MOD(14, 5) equals 4). If the constants are chosen carefully, the linear congruential method generates all possible integers between 0 and $m-1$ and the period (i.e. the number of iterations before the sequence starts to repeat itself) will be equal to the modulus. The period cannot of course be greater than m. The linear congruential method generates integral values, which can be

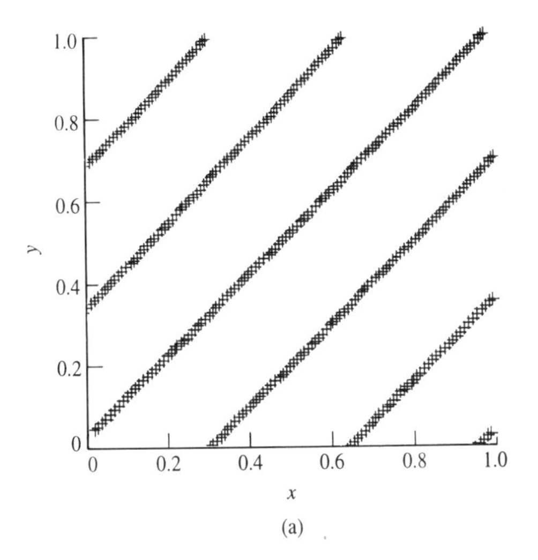

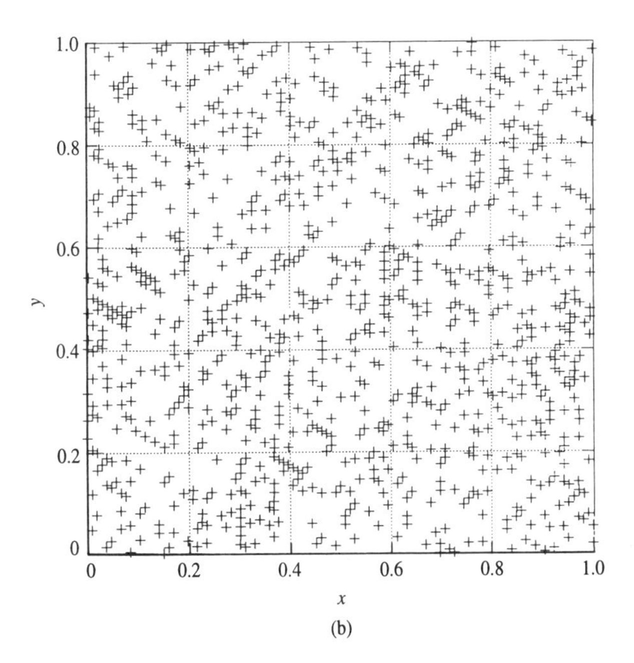

Fig. 7.3 Two 'random' distributions obtained by plotting pairs of values from a linear congruential random generator. The distribution (a) was obtained using $m = 32\,769$, $a = 10\,924$, $b = 11\,830$. The distribution (b) was obtained using $m = 6075$, $a = 106$, $b = 1283$. [Data from Sharp and Bays 1992.]

converted to real numbers between 0 and 1 by dividing by m. The modulus is often chosen to be the largest prime number that can be represented in a given number of bits (usually chosen to be the number of bits per word; $2^{31} - 1$ is thus a common choice on a 32-bit machine).

Although popular, by virtue of the ease with which it can be programmed, the linear congruential method does not satisfy all of the requirements that are now regarded as important in a random number generator. For example, the points obtained from a linear congruential generator lie on $(k - 1)$ dimensional planes rather than uniformly filling up the space. Indeed, if the constants a, b and m are chosen inappropriately then the linear congruential method can give truly terrible results, as shown in Figure 7.3. One random number generator that is claimed to perform well in all of the standard tests is that of G Marsaglia which is described in Appendix 7.1.

7.5 Monte Carlo simulation of molecules

The Monte Carlo method is most easily implemented for atomic systems because it is only necessary to consider the translational degrees of freedom. The algorithm is easy to implement and accurate results can be obtained from relatively short simulations of a few tens of thousands of steps. There can be practical problems in applying the method to molecular systems, and especially to molecules which have a significant degree of conformational flexibility. This is because in such systems it is necessary to permit the internal degrees of freedom to vary. Unfortunately, such changes often lead to high-energy overlaps either within the molecule or between the molecule and its neighbours and thus a high rejection rate.

7.5.1 Rigid molecules

For rigid, non-spherical molecules, the orientations of the molecules must be varied as well as their positions in space. It is usual to translate and rotate one molecule during each Monte Carlo step. Translations are usually described in terms of the position of the centre of mass. There are various ways to generate a new orientation of a molecule. The simplest approach is to choose one of the three Cartesian axes (x, y or z) and to rotate about the chosen axis by a randomly chosen angle $\delta\omega$, chosen to lie within the maximum angle variation, $\delta\omega_{max}$ [Barker and Watts 1969]. The rotation is achieved by applying routine trigonometric relationships. For example, if the vector ($x\mathbf{i}$, $y\mathbf{j}$, $z\mathbf{k}$) describes the orientation of a molecule then the new vector ($x'\mathbf{i}$, $y'\mathbf{j}$, $z'\mathbf{k}$) that corresponds to rotation by $\delta\omega$ about the x axis is calculated as follows:

$$\begin{pmatrix} x' \\ y' \\ z' \end{pmatrix} = \begin{pmatrix} 1 & 0 & 0 \\ 0 & \cos\delta\omega & \sin\delta\omega \\ 0 & -\sin\delta\omega & \cos\delta\omega \end{pmatrix} \begin{pmatrix} x \\ y \\ z \end{pmatrix} \tag{7.36}$$

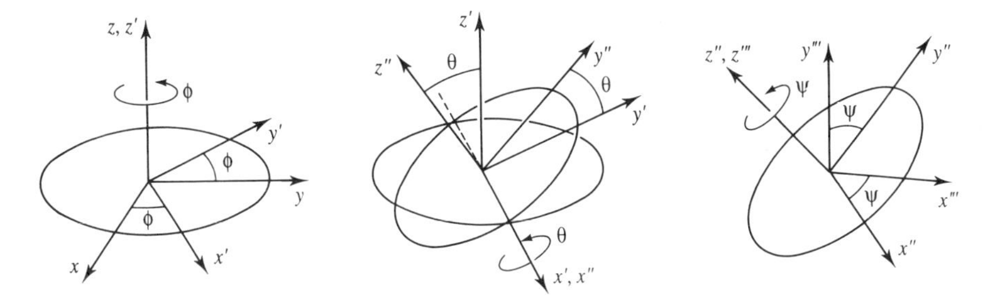

Fig. 7.4 The Euler angles ϕ, θ and ψ.

The *Euler angles* are often used to describe the orientations of a molecule. There are three Euler angles; ϕ, θ and ψ. ϕ is a rotation about the Cartesian z axis; this has the effect of moving the x and y axes. θ is a rotation about the new x axis. Finally, ψ is a rotation about the new z axis (Figure 7.4). If the Euler angles are randomly changed by small amounts $\delta\phi$, $\delta\theta$ and $\delta\psi$ then a vector $\mathbf{v}_{\text{old}}$ is moved according to the following matrix equation:

$$\mathbf{v}_{\text{new}} = \mathbf{A}_{\text{old}}\mathbf{v}_{\text{old}} \tag{7.37}$$

where the matrix $\mathbf{A}$ is

$$\begin{pmatrix} \cos\delta\phi\cos\delta\psi - \sin\delta\phi\cos\delta\theta\sin\delta\psi & \sin\delta\phi\cos\delta\psi + \cos\delta\psi\cos\delta\theta\sin\delta\psi & \sin\delta\theta\sin\delta\psi \\ -\cos\delta\phi\sin\delta\psi - \sin\delta\phi\cos\delta\theta\cos\delta\psi & -\sin\delta\phi\sin\delta\psi + \cos\delta\phi\cos\delta\theta\cos\delta\psi & \sin\delta\theta\cos\delta\psi \\ \sin\delta\phi\sin\delta\theta & -\cos\delta\phi\sin\delta\theta & \cos\delta\theta \end{pmatrix} \tag{7.38}$$

It is important to note that simply sampling displacements of the three Euler angles does not lead to a uniform distribution; it is necessary to sample from $\cos\theta$ rather than θ, Figure 7.5.

The preferred approach is to sample directly in $\cos\theta$ as follows:

$$\phi_{\text{new}} = \phi_{\text{old}} + 2(\xi - 1)\delta\phi_{\text{max}} \tag{7.39}$$

$$\cos\theta_{\text{new}} = \cos\theta_{\text{old}} + (2\xi - 1)\delta(\cos\theta)_{\text{max}} \tag{7.40}$$

$$\psi_{\text{new}} = \psi_{\text{old}} + 2(\xi - 1)\delta\psi_{\text{max}} \tag{7.41}$$

The alternative is to sample in θ and to modify the acceptance or rejection criteria as follows:

$$\theta_{\text{new}} = \theta_{\text{old}} + (2\xi - 1)\delta\theta_{\text{max}} \tag{7.42}$$

$$\frac{\rho_{\text{new}}}{\rho_{\text{old}}} = \exp(-\Delta\mathcal{V}/k_{\text{B}}T)\frac{\sin\theta_{\text{new}}}{\sin\theta_{\text{old}}} \tag{7.43}$$

This second approach may give problems if θ_{old} equals zero.

A disadvantage of the Euler angle approach is that the rotation matrix contains a total of six trigonometric functions (sine and cosine for each of the three Euler

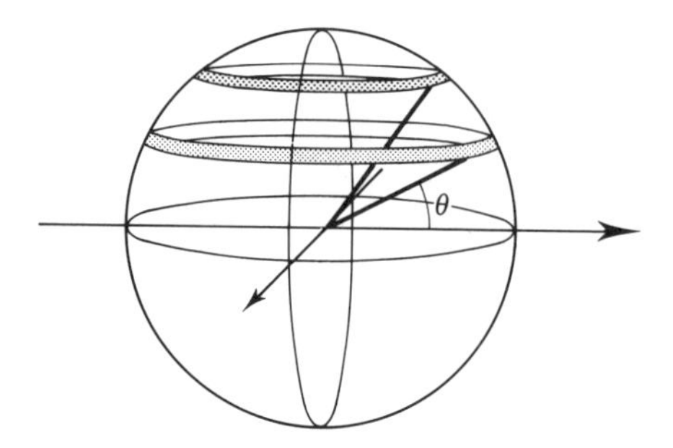

Fig. 7.5 To achieve a uniform distribution of points over the surface of a sphere it is necessary to sample from $\cos\theta$ rather than θ. If the sampling is uniform in θ then the number of points per unit area increases with θ, leading to an uneven distribution over the sphere.

angles). These trigonometric functions are computationally expensive to calculate. An alternative is to use *quaternions*. A quaternion is a four-dimensional vector such that its components sum to one: $q_0^2 + q_1^2 + q_2^2 + q_3^2 = 1$. The quaternion components are related to the Euler angles as follows:

$$q_0 = \cos\tfrac{1}{2}\,\theta\cos\tfrac{1}{2}\,(\phi + \psi) \tag{7.44}$$

$$q_1 = \sin\tfrac{1}{2}\,\theta\cos\tfrac{1}{2}\,(\phi + \psi) \tag{7.45}$$

$$q_2 = \sin\tfrac{1}{2}\,\theta\sin\tfrac{1}{2}\,(\phi + \psi) \tag{7.46}$$

$$q_3 = \cos\tfrac{1}{2}\,\theta\sin\tfrac{1}{2}\,(\phi + \psi) \tag{7.47}$$

The Euler angle rotation matrix can then be written

$$\mathbf{A} = \begin{pmatrix} q_0^2 + q_1^2 - q_2^2 - q_3^2 & 2(q_1q_2 + q_0q_3) & 2(q_1q_3 - q_0q_2) \\ 2(q_1q_2 - q_0q_3) & q_0^2 - q_1^2 + q_2^2 - q_3^2 & 2(q_2q_3 + q_0q_1) \\ 2(q_1q_3 + q_0q_2) & 2(q_2q_3 - q_0q_1) & q_0^2 - q_1^2 - q_2^2 + q_3^2 \end{pmatrix} \tag{7.48}$$

To generate a new orientation, it is necessary to rotate the quaternion vector to a new (random) orientation. As it is a four-dimensional vector, the orientation must be performed in four-dimensional space. This can be achieved as follows [Vesely 1982]:

1. Generate pairs of random numbers (ξ_1, ξ_2) between -1 and 1 until $S_1 = \xi_1^2 + \xi_2^2 < 1$.
2. Do the same for pairs ξ_3 and ξ_4 until $S_2 = \xi_3^2 + \xi_4^2 < 1$.
3. Form the random unit four-dimensional vector $(\xi_1\ \xi_2\ \xi_3\sqrt{(1-S_1)/S_2}\ \xi_4\sqrt{(1-S_1)/S2})$.

To achieve an appropriate acceptance rate the angle between the two vectors that describe the new and old orientations should be less than some value; this corresponds to sampling randomly and uniformly from a region on the surface of a sphere.

The introduction of an orientational component as well as a translational component increases the number of maximum displacement parameters that determine the acceptance ratio. It is important to check that the desired acceptance ratio is achieved, and also that an appropriate proportion of orientational and translational moves are made. Trial and error is often the most effective way to find the best combination of parameters.

7.5.2 Monte Carlo simulations of flexible molecules

Monte Carlo simulations of flexible molecules are often difficult to perform successfully unless the system is small, or some of the internal degrees of freedom are frozen out, or special models or methods are employed. The simplest way to generate a new configuration of a flexible molecule is to perform random changes to the Cartesian coordinates of individual atoms, in addition to translations and rotations of the entire molecule. Unfortunately, it is often found that very small atomic displacements are required to achieve an acceptable acceptance ratio, which means that the phase space is covered very slowly. For example, even small movements away from an equilibrium bond length will cause a large increase in the energy. One obvious tactic is to freeze out some of the internal degrees of freedom, usually the 'hard' degrees of freedom such as the bond lengths and the bond angles. Such algorithms have been extensively used to investigate small molecules such as butane. However, for large molecules, even relatively small bond rotations may cause large movements of atoms down the chain. This invariably leads to high-energy configurations as illustrated in Figure 7.6. The rigid bond and rigid angle approximation must be used with care, for freezing out some of the internal degrees of freedom can affect the distributions of other internal degrees of freedom.

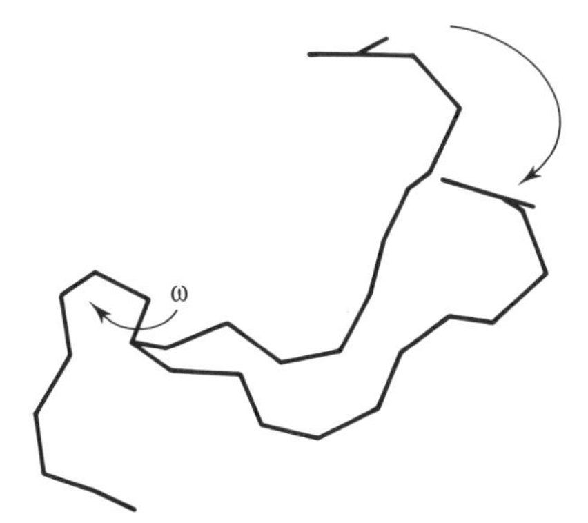

Fig. 7.6 A bond rotation in the middle of a molecule may lead to a large movement at the end.

7.6 Models used in Monte Carlo simulations of polymers

A polymer is a macromolecule that is constructed by chemically linking together a sequence of molecular fragments. In simple synthetic polymers such as polyethylene or polystyrene all of the molecular fragments comprise the same basic unit (or monomer). Other polymers contain mixtures of monomers. Proteins, for example are polypeptide chains in which each unit is one of the twenty amino acids. Cross-linking between different chains gives rise to yet further variations in the constitution and structure of a polymer. All of these features may affect the overall properties of the molecule, sometimes in a dramatic way. Moreover, one may be interested in the properties of the polymer under different conditions, such as in solution, in a polymer melt or in the crystalline state. Molecular modelling can help to develop theories for understanding the properties of polymers, and can also be used to predict their properties.

A wide range of time and length scales are needed to completely describe a polymer's behaviour. The time scale ranges from approximately 10^{-14} s (i.e. the period of a bond vibration) through to seconds, hours or even longer for collective phenomena. The size scale ranges from the 1–2 Å of chemical bonds to the diameter of a coiled polymer that can be several hundreds of angstroms. Many kinds of models have been used to represent and simulate polymeric systems and predict their properties. Some of these models are based upon very simple ideas about the nature of the intra- and intermolecular interactions within the system, but have nevertheless proved to be extremely useful. One famous example is Flory's rotameric state model [Flory 1969]. Increasing computer performance now makes it possible to use techniques such as molecular dynamics and Monte Carlo simulations to study polymer systems.

Most simulations on polymers are performed using empirical energy models (though with faster computers and new methods it is becoming possible to apply quantum mechanics to larger and larger systems). Moreover, there are various ways in which the configurational and conformational degrees of freedom may be restricted so as to produce a computationally more efficient model. The simplest models use a lattice representation in which the polymer is constructed from connected interaction centres that are required to occupy the vertices of a lattice. At the next level of complexity are the bead models, where the polymer is composed of a sequence of connected 'beads'. Each bead represents an 'effective monomer' and interacts with the other beads to which it is bonded and also with other nearby beads. The ultimate level of detail is achieved with the atomistic models, in which each non-hydrogen atom is explicitly represented (and sometimes all of the hydrogens as well). Our aim here is to give a flavour of the way in which Monte Carlo methods can be used to investigate polymeric systems. We divide the discussion into lattice and continuum models but recognise that there is a spectrum of models from the simplest to the most complex.

7.6.1 Lattice models of polymers

Lattice models have provided many insights into the behaviour of polymers despite the obvious approximations involved. The simplicity of a lattice model means that many states can be generated and examined very rapidly. Both two-dimensional and three-dimensional lattices are used. The simplest models use cubic or tetrahedral lattices in which successive monomers occupy adjacent lattice points, Figure 7.7. The energy models are usually very simple, in part to reflect the simplicity of the representation but also to permit the rapid calculation of the energy.

More complex models have been developed in which the lattice representation is closer to the 'true' geometry of the molecule. For example, in Figure 7.8 we show the bond fluctuation model of polyethylene, in which the 'bond' between successive monomers on the lattice represent three bonds in the actual molecule [Baschnagel *et al.* 1991]. In this model each monomer is positioned at the centre of a cube within the lattice and five different distances are possible for the monomer–monomer bond lengths.

Lattices can be used to study a wide variety of polymeric systems, from single polymer chains to dense mixtures. The simplest type of simulation is a 'random walk', in which the chain is randomly grown in the lattice until it contains the desired number of bonds, Figure 7.9. In this model the chain is free to cross itself (i.e. excluded volume effects are ignored). Various properties can be calculated from such simulations, by averaging the results over a large number of trials. For example, a simple measure of the size of a polymer is the mean square end-to-end distance, $\langle R_n^2 \rangle$. For the random walk model $\langle R_n^2 \rangle$ is related to the number of bonds (n) and the length of each bond (l) by:

$$\langle R_n^2 \rangle = nl^2 \tag{7.49}$$

The radius of gyration is another commonly calculated property; this is the root mean square distance of each atom (or monomer) from the centre of mass. For the random walk model the radius of gyration $\langle s^2 \rangle$ is given in the asymptotic limit by:

$$\langle s^2 \rangle = \langle R_n^2 \rangle / 6 \tag{7.50}$$

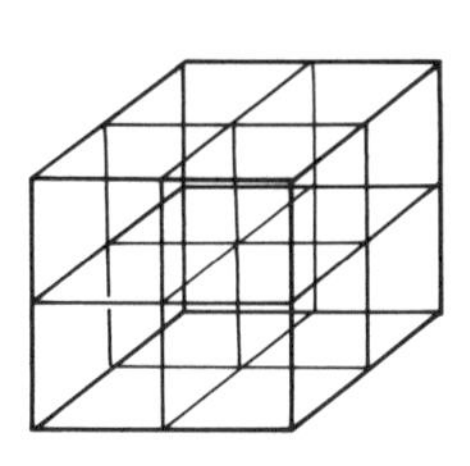

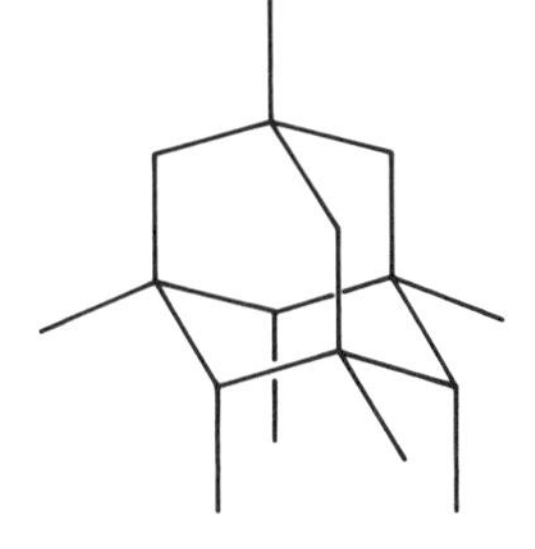

Fig. 7.7 Cubic and tetrahedral (diamond) lattices that are commonly used for lattice simulations of polymers.

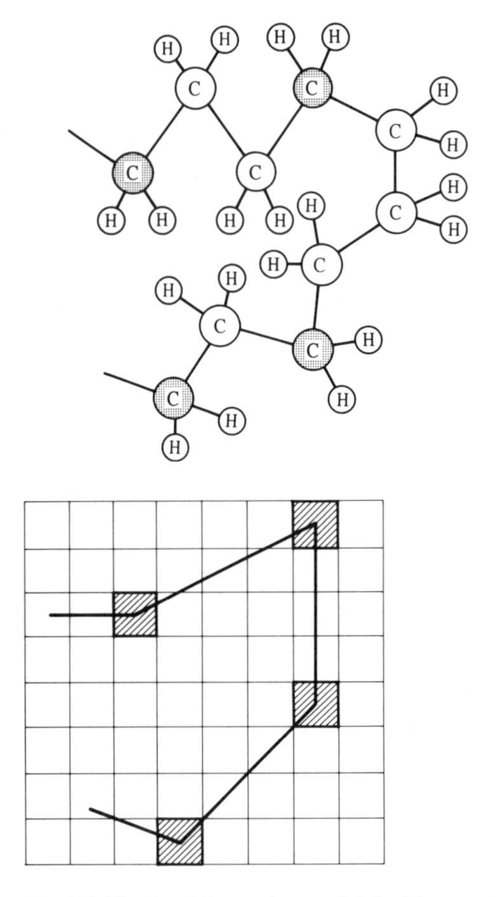

Fig. 7.8 The bond fluctuation model. In this example three bonds in the polymer are incorporated into a single 'effective bond' between 'effective monomers'. Figure adapted from Baschnagel J, K Binder, W Paul, M Laso, U Suter, I Batoulis, W Jilge and T Bürger 1991. On the Construction of Coarse-Grained Models for Linear Flexible Polymer-Chains—Distribution-Functions for Groups of Consecutive Monomers. *The Journal of Chemical Physics* **95**:6014–6025.

The ability of the chain to cross itself in the random walk may seem to be a serious limitation, but it is found to be valid under some circumstances. Excluded volume effects can be taken into account by generating a 'self-avoiding walk' of the chain in the lattice, Figure 7.10. In this model only one monomer can occupy each lattice site. Self-avoiding walks have been used to exhaustively enumerate all possible conformations for a chain of a given length on the lattice. If all states are known then the partition function can be determined and thermodynamic quantities calculated. The 'energy' of each state may be calculated using an appropriate interaction model. For example, the energy may be proportional to the number of adjacent pairs of occupied lattice sites. A variation on this is to use

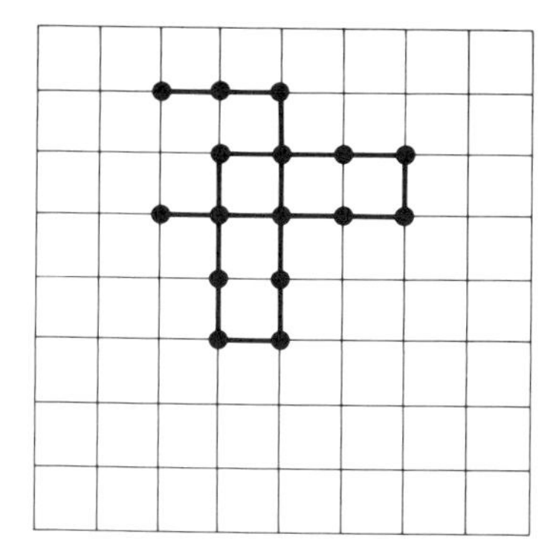

Fig. 7.9 In a random walk on a square lattice the chain can cross itself.

polymers consisting of two types of monomer (A and B) which have up to three different energy values: A-A, B-B and A-B. Again, the energy is determined by counting the number of occupied, adjacent lattice sites. The relationship between the mean square end-to-end distance and the length of the chain (n) has been investigated intensively; with the self-avoiding walk the result obtained is different from the random walk, with $\langle R_n^2 \rangle$ being proportional to $n^{1.18}$ in the asymptotic limit.

Having grown a polymer onto the lattice, we now have to consider the generation of alternative configurations. Motion of the entire polymer chain or large-scale conformational changes are often difficult, especially for densely packed polymers. In variants of the Verdier–Stockmayer algorithm (Verdier and Stockmayer 1962] new configurations are generated using combinations of 'crankshaft', 'kink jump' and 'end rotation' moves, Figure 7.11. Another widely used algorithm in Monte Carlo simulations of polymers (not just in lattice models) is the 'slithering snake' model. Motion of the entire polymer chain is

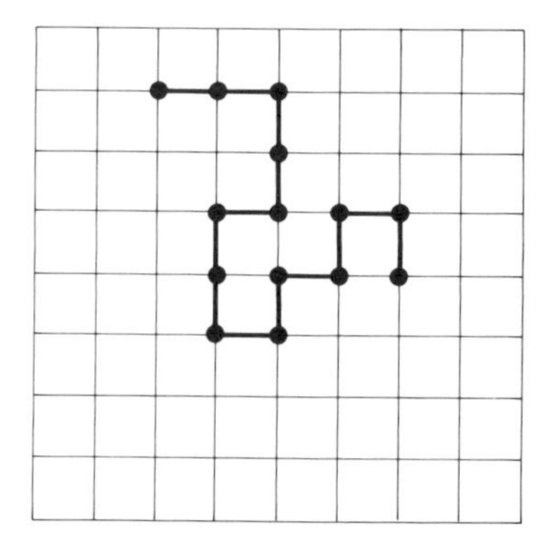

Fig. 7.10 Self-avoiding walk: only one monomer can occupy each lattice site.

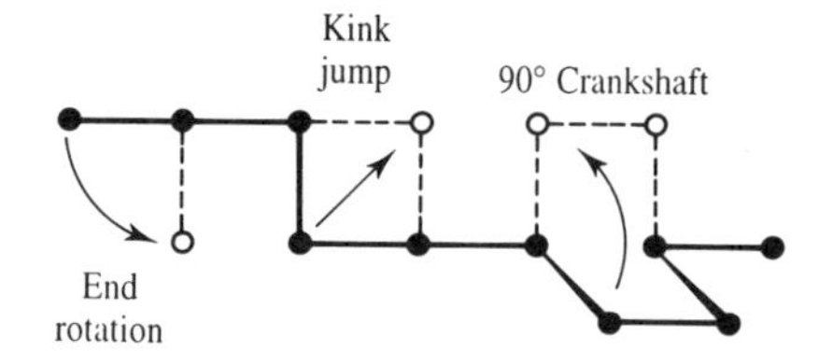

Fig. 7.11 The 'crankshaft', 'kink jump' and 'end rotation' moves used in Monte Carlo simulations of polymers.

very difficult, especially for densely packed polymers, and one way in which the polymer can move is by wriggling around obstacles, a process known as *reptation*. To implement a slithering snake algorithm, one end of the polymer chain is randomly chosen as the 'head' and an attempt is made to grow a new bead at one of the available adjacent lattice positions. Each of the remaining beads is then advanced to that of its predecessor in the chain as illustrated in Figure 7.12. The procedure is then repeated. Even if it is impossible to move the chosen 'head', the configuration must still be included when ensemble averages are calculated.

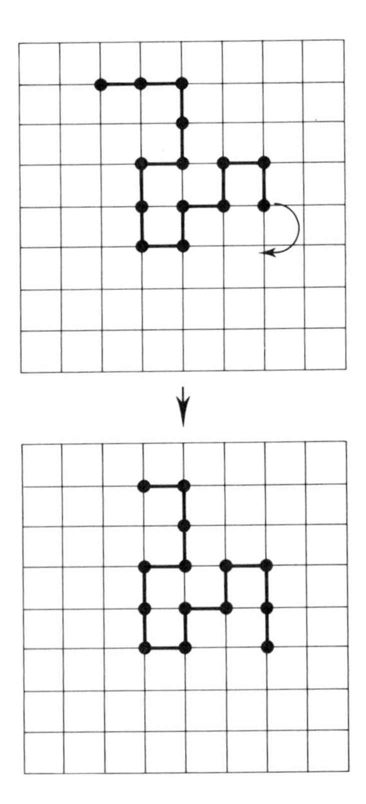

Fig. 7.12 The 'slithering snake' algorithm.

7.6.2 'Continuous' polymer models

The simplest of the continuous models of a polymer consists of a string of connected beads (Figure 7.13). The beads are freely jointed and interact with the other beads via a spherically symmetric potential such as the Lennard–Jones potential. The beads should not be thought of as being identical to the monomers in the polymer, though they are often referred to as such ('effective monomers' is a more appropriate term). Similarly, the links between the beads should not be thought of as bonds. The links may be modelled as rods of a fixed and invariant length, or may be permitted to vary using a harmonic potential function.

In Monte Carlo studies with bead models the beads can sample from a continuum of positions. The *pivot algorithm* is one way that new configurations can be generated. Here, a segment of the polymer is randomly selected and rotated by a random amount, as illustrated in Figure 7.13. For isolated polymer chains the pivot algorithm can give a good sampling of the configurational/conformational space. However, for polymers in solution or in the melt, the proportion of accepted moves is often very small due to high energy steric interactions.

The ultimate level of detail in polymer modelling is achieved with the atomistic models, which as the name implies explicitly represent the atoms in the system. An atomistic model is clearly the closest to 'reality', and is necessary if one wishes to calculate accurately certain properties. One of the major problems with simulations of polymers that particularly arises for the atomistic models is to generate an initial configuration of the system. Amorphous polymers by definition do not adopt a characteristic and reproducible three-dimensional structure. One must therefore attempt to generate a suitable configuration prior to the simulation. It is important that the properties of this initial configuration are similar to the state one wishes to simulate else the computer time needed to move to the required state can be prohibitive. For short chains containing approximately 20–30 backbone bonds it is feasible to start from a regular crystalline structure which can then be melted, but to 'melt' a long chain may require a prohibitive amount of computer time. For longer chains an initial configuration may be generated using a random walk and periodic boundary conditions, Figure 7.14. Such an arrangement will inevitably contain high-energy overlaps. These unfavourable

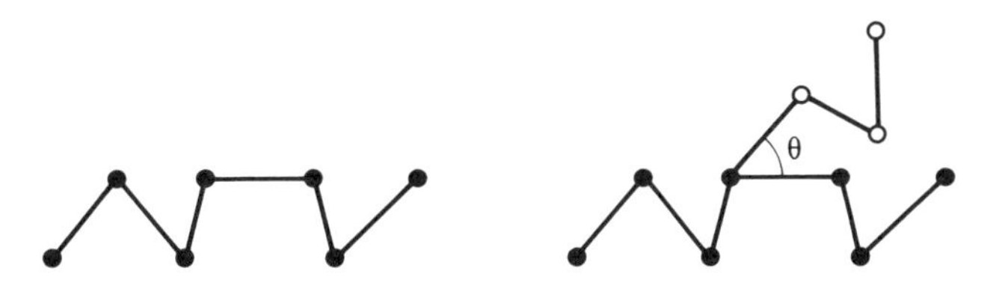

Fig. 7.13 The bead model for polymer simulations. The beads may be connected by stiff rods or by harmonic springs.

Fig. 7.14 Generation of an initial configuration of a polymer using periodic boundary conditions.

interactions may be removed by relaxing the system using minimisation and/or computer simulation, during which the force field potentials are gradually 'turned on'.

7.7 'Biased' Monte Carlo methods

In some situations one is particularly interested in the behaviour of just a part of the system. For example, if we were simulating a solute–solvent system that contained a single solute molecule surrounded by a large number of solvent molecules then the behaviour of the solute and its interactions with the solvent would be of most interest. Solvent molecules far from the solute would be expected to behave almost like bulk solvent. A variety of techniques have been developed which can enhance the ability of the Monte Carlo method to explore the most important regions of phase space in such cases. One relatively simple procedure is *preferential sampling*, where the molecules in the vicinity of the solute are moved more frequently than those further away. This can be implemented by defining a cutoff region around the solute; molecules outside the cutoff region are moved less frequently than those inside the region as determined by a probability parameter p. At each Monte Carlo iteration a molecule is randomly chosen. If the molecule is inside the cutoff region then it is moved; if it is outside the region then a random number is generated between 0 and 1 and compared to the probability p. If p is greater than the random number a trial move is attempted; otherwise no move is made, no averages are accumulated, and a new molecule is randomly selected. The closer p is to zero, the more often 'closer' molecules are moved than 'further' molecules.

An alternative to the use of a fixed cutoff region is to relate the probability of choosing a solvent molecule to its distance from the solute, usually by some inverse power of the distance:

$$p \propto r^{-n} \tag{7.51}$$

In preferential sampling it is necessary to ensure that the correct procedures are followed when deciding whether to accept or reject a move in a manner that is consistent with the principle of microscopic reversibility. Suppose a molecule inside the cutoff region is moved outside the cutoff region. In the preferential sampling scheme the chances of the molecule now being selected for an out $\rightarrow$ in move are less than for the original in $\rightarrow$ out move and this must be taken into account when determining the acceptance criteria.

The *force-bias* Monte Carlo method [Pangali *et al.* 1978; Rao *et al.* 1979] biases the movement according to the direction of the forces on it. Having chosen an atom or a molecule to move, the force on it is calculated. The force corresponds to the direction in which a 'real' atom or molecule would move. In the force-bias Monte Carlo method the random displacement is chosen from a probability distribution function that peaks in the direction of this force. The *smart Monte Carlo method* [Rossky *et al.* 1978] also requires the forces on the moving atom to be calculated. The displacement of an atom or molecule in this method has two components; one component is the force, and the other is a random vector $\delta \mathbf{r}^G$:

$$\delta \mathbf{r}_i = \frac{A\mathbf{f}_i}{k_B T} + \delta \mathbf{r}_i^G \tag{7.52}$$

$\mathbf{f}_i$ is the force on the atom and A is a parameter. The random displacement $\delta \mathbf{r}_i^G$ is chosen from a normal distribution with zero mean and variance equal to $2A$.

The main difference between the force-bias and the smart Monte Carlo methods is that the latter does not impose any limit on the displacement that an atom may undergo. The displacement in the force-bias method is limited to a cube of the appropriate size centred on the atom. However, in practice the two methods are very similar and there is often little to choose betwen them. In suitable cases they can be much more efficient at covering phase space and are better able to avoid bottlenecks in phase space than the conventional Metropolis Monte Carlo algorithm. The methods significantly enhance the acceptance rate of trial moves, thereby enabling larger moves to be made as well as simultaneous moves of more than one particle. However, the need to calculate the forces makes the methods much more elaborate, and comparable in complexity to molecular dynamics.

7.8 Monte Carlo sampling from different ensembles

A Monte Carlo simulation traditionally samples from the constant *NVT* (canonical) ensemble, but the technique can also be used to sample from different ensembles. A common alternative is the isothermal–isobaric, or constant *NPT* ensemble. To simulate from this ensemble, it is necessary to have a scheme for changing the volume of the simulation cell in order to keep the pressure constant. This is done by combining random displacements of the particles with random changes in the volume of the simulation cell. The size of each volume change is governed by the maximum volume change, δV_{max}. Thus a new volume is generated from the old volume as follows:

$$V_{new} = V_{old} + \delta V_{max}(2\xi - 1) \tag{7.53}$$

As usual, ξ is a random number between 0 and 1. When the volume is changed, it is in principle necessary to recalculate the interaction energy of the entire system, not just the interactions involving the one atom or molecule that has been displaced. However, for simple interatomic potentials the change in energy associated with a volume change can be calculated very rapidly by using *scaled coordinates*. For a set of particles that are modelled by a Lennard–Jones potential in a cubic box of length L_{old}, the potential energy can be written:

$$\mathscr{V}_{old}(\mathbf{r}^N) = 4\varepsilon \sum_{i=1}^{N} \sum_{j=i+1}^{N} \left(\frac{\sigma}{L_{old}s_{ij}}\right)^{12} - 4\varepsilon \sum_{i=1}^{N} \sum_{j=i+1}^{N} \left(\frac{\sigma}{L_{old}s_{ij}}\right)^{6} \tag{7.54}$$

s_{ij} is a scaled coordinate which is related to the actual interatomic distance by $s_{ij} = L_{old}^{-1} r_{ij}$. It is necessary to write the energy as the sum of two components, one from the repulsive part of the Lennard–Jones potential and the other from the attractive part:

$$\mathscr{V}_{old}(\mathbf{r}^N) = \mathscr{V}_{old}(12) + \mathscr{V}_{old}(6) \tag{7.55}$$

The advantage of using scaled coordinates is that they are independent of the size of the simulation box. Thus the energy of the configuration in a different sized box (with side L_{new}) is:

$$\mathscr{V}_{new}(\mathbf{r}^N) = 4\varepsilon \sum_{i=1}^{N} \sum_{j=i+1}^{N} \left(\frac{\sigma}{L_{new}s_{ij}}\right)^{12} - 4\varepsilon \sum_{i=1}^{N} \sum_{j=i+1}^{N} \left(\frac{\sigma}{L_{new}s_{ij}}\right)^{6} \tag{7.56}$$

The energy $\mathscr{V}_{new}(\mathbf{r}^N)$ is related to the energy $\mathscr{V}_{old}(\mathbf{r}^N)$ as follows:

$$\mathscr{V}_{new}(\mathbf{r}^N) = \mathscr{V}_{old}(12)\left\{\frac{L_{old}}{L_{new}}\right\}^{12} + \mathscr{V}_{old}(6)\left\{\frac{L_{old}}{L_{new}}\right\}^{6} \tag{7.57}$$

The change in energy from the old to the new system is thus:

$$\Delta\mathscr{V}(\mathbf{r}^N) = \mathscr{V}_{old}(12)\left\{\frac{L_{old}}{L_{new}} - 1\right\}^{12} + \mathscr{V}_{old}(6)\left\{\frac{L_{old}}{L_{new}} - 1\right\}^{6} \tag{7.58}$$

Any long-range corrections to the potential must also be taken into account when the volume changes. One way to deal with these is to assume that the non-bonded cutoff scales with the box length. Under such circumstances the long-range corrections to both the repulsive and attractive parts of the potential scale in exactly the same manner as the short-range interactions. However, the use of this assumption can give rise to serious problems, particularly for techniques such as the Gibbs ensemble Monte Carlo simulation (see section 7.11) where two coupled simulation boxes of different dimensions are involved. The boxes contain identical particles, but this would be compromised by the use of different non-bonded cutoffs and long-range corrections.

This simple scaling method cannot be used when simulating molecules, for a change in the scaled coordinates would have the effect of introducing large and energetically unfavourable changes in the internal coordinates such as the bond lengths. It is therefore necessary to recalculate the total interaction energy of the system each time a volume change is made. This is computationally expensive to do, but it is in any case advisable to change the volume relatively infrequently compared to the rate at which the particles are moved. One way to speed up the energy calculation associated with a volume change is to write the potential energy change as a Taylor series expansion of the box size.

The criterion used to accept or reject a new configuration is slightly different for the isothermal–isobaric simulation than for a simulation in the canonical ensemble. The following quantity is used:

$$\Delta H(\mathbf{r}^N) = \mathscr{V}_{\text{new}}(\mathbf{r}^N) - \mathscr{V}_{\text{old}}(\mathbf{r}^N) + P(V_{\text{new}} - V_{\text{old}}) - Nk_{\text{B}}T\ln\left(\frac{V_{\text{new}}}{V_{\text{old}}}\right) \tag{7.59}$$

If ΔH is negative then the move is accepted; otherwise $\exp(-\Delta H/k_{\text{B}}T)$ is compared to a random number between 0 and 1 and the move accepted according to:

$$\text{rand}(0, 1) \leq \exp(-\Delta H/k_{\text{B}}T) \tag{7.60}$$

To check that an isothermal–isobaric simulation is working properly, the pressure can be calculated from the virial as outlined in section 5.2.3, including the appropriate long-range correction (which will not of course be constant as the volume of the box changes). Its value should be equal to the input pressure that appears in equation (7.59).

7.8.1 Grand canonical Monte Carlo simulations

In the grand canonical ensemble the conserved properties are the chemical potential, the volume and the temperature. It can sometimes be more convenient to perform a grand canonical simulation at constant activity, z, which is related to the chemical potential μ by:

$$\mu = k_{\text{B}}T\ln\Lambda^3 z \tag{7.61}$$

Λ is the de Broglie wavelength given by $\Lambda = \sqrt{(h^2/2\pi m k_{\text{B}}T)}$.

The key feature about the grand canonical Monte Carlo method is that the number of particles may change during the simulation. There are three basic moves in a grand canonical Monte Carlo simulation:

1. A particle is displaced, using the usual Metropolis method.
2. A particle is destroyed.
3. A particle is created at a random position.

The probability of creating a particle should be equal to the probability of destroying a particle. To determine whether to accept a destruction move the following quantity is calculated:

$$\Delta D = \frac{[\mathscr{V}_{\text{new}}(\mathbf{r}^N) - \mathscr{V}_{\text{old}}(\mathbf{r}^N)]}{k_B T} - \ln\left(\frac{N}{zV}\right) \tag{7.62}$$

For a creation step the equivalent quantity is:

$$\Delta C = \frac{[\mathscr{V}_{\text{new}}(\mathbf{r}^N) - \mathscr{V}_{\text{old}}(\mathbf{r}^N)]}{k_B T} - \ln\left(\frac{zV}{N+1}\right) \tag{7.63}$$

If ΔD or ΔC is negative then the move is accepted; if positive then the exponential $\exp(-\Delta D/k_B T)$ or $\exp(-\Delta C/k_B T)$ as appropriate is calculated and compared with a random number between 0 and 1 in the usual way.

It is important that the possibility of creating a new particle is the same as the probability of destroying an old one. The ratio of particle creation/destruction moves to translation/rotation moves can vary but the most rapid convergence is often achieved if all types of move occur with approximately equal frequency.

In grand canonical Monte Carlo simulations of liquids there can be some practical problems in achieving statistically accurate results. This is because the probability of achieving a successful creation or destruction step is often very small. Creation steps fail because the fluid is so dense that it is difficult to insert a new particle without causing significant overlaps with neighbouring particles. Destruction steps fail because the particles in a fluid often experience significant attractive interactions which are lost when the particle is removed. These problems are particularly acute for long-chain molecules. However, some of the newer Monte Carlo techniques such as the configurational bias Monte Carlo method do enable such systems to be simulated and accurate results obtained. These techniques will be discussed in section 7.10.

7.8.2 Grand canonical Monte Carlo simulations of adsorption processes

One application of the grand canonical Monte Carlo simulation method is in the study of the adsorption and transport of fluids through porous solids. Mixtures of gases or liquids can be separated by the selective adsorption of one component in an appropriate porous material. The efficacy of the separation depends to a large extent upon the ability of the material to adsorb one component in the mixture much more strongly than the other component. The separation may be performed over a range of temperatures and so it is useful to be able to predict the adsorption isotherms of the mixtures.

A typical example of such a calculation is the simulation of Cracknell, Nicholson and Quirke [Cracknell *et al.* 1994] who studied the adsorption of a mixture of methane and ethane onto a microporous graphite surface. Four types of move were employed in their simulations: particle moves, particle deletions, particle creations and attempts to exchange particles. Methane was modelled as a single Lennard–Jones particle and ethane as two Lennard–Jones particles separated by a fixed bond length. The graphite surfaces were modelled as Lennard–Jones atoms with a slit-shaped pore being constructed from two layers of graphite separated by an appropriate distance. Triangle-shaped pores can also be used. The simulations were used to calculate the selectivity of the solid for the two components as the ratio of the mole fractions in the pore to the ratio of the mole fractions in the bulk. The selectivity was determined as a function of the pressure for different pore sizes to give some indication of the effect of changing the physical nature of the solid. The pressure can be calculated directly from the input chemical potential using the following standard relationship (for an ideal gas):

$$P = \{\exp(\mu/k_B T)k_B T\}/\Lambda^3 \tag{7.64}$$

Λ^3 is the thermal de Broglie wavelength given by $\Lambda = \sqrt{(h^2/2\pi k_B T m)}$. The selectivity showed a complicated dependence upon the pore size, Figure 7.15.

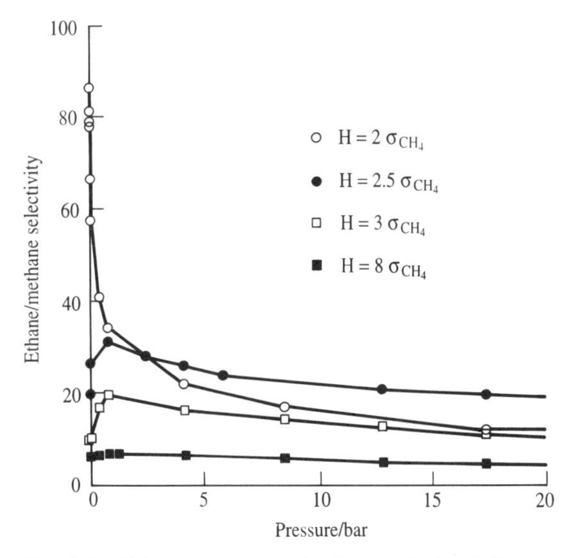

Fig. 7.15 Ethane/methane selectivity calculated from grand canonical Monte Carlo simulations of mixtures in slit pores at a temperature of 296 K. The selectivity is defined as the ratio of the mole fractions in the pore to the ratio of the mole fractions in the bulk. H is the slit width defined in terms of the methane collision diameter σ_{CH_4}. Figure redrawn from Cracknell R F, D Nicholson and N Quirke 1994. A Grand Canonical Monte Carlo Study of Lennard–Jones Mixtures in Slit Pores; 2: Mixtures of Two-Centre Ethane with Methane. *Molecular Simulation* **13**:161–175.

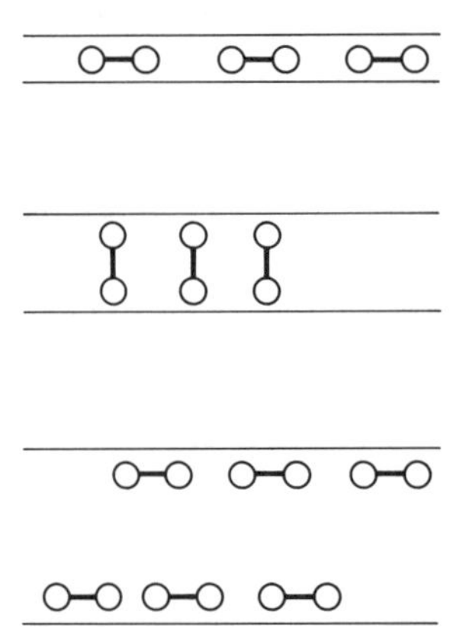

Fig. 7.16 Schematic illustration of the arrangements of ethane molecules in slits of varying sizes. In the slit of width $2.5\sigma_{CH_4}$ each methyl group is able to occupy a potential minimum from the pore (middle).

The selectivity is best interpreted by considering the interactions between ethane molecules and the walls of the pore. For the smallest pore sizes, the molecules are restricted to the centre of the pore and the ethane molecules are forced to lie flat. As the pore size increases, it becomes possible for ethane to adopt a particularly favourable orientation perpendicular to the walls, with each methyl group being in a potential energy minimum for interaction with the pore atoms. This particular pore size ($2.5\sigma_{CH_4}$) thus has the greatest selectivity for ethane over methane. As the pore size increases further the distribution of ethane becomes more complex, with some layers of ethane lying flat on the pore wall and some in the centre of the pore, with ethane molecules spanning the space between. These arrangements are shown in Figure 7.16.

7.9 Calculating the chemical potential

In a grand canonical simulation the chemical potential is constant. One may also wish to determine how the chemical potential varies during a simulation. The chemical potential is usually determined using an approach due to Widom [Widom 1963], in which a 'test' particle is inserted into the system and the resulting change in potential energy is calculated. The Widom approach is applicable to both molecular dynamics and Monte Carlo simulations. Consider a system containing $N-1$ particles, into which we insert another particle at a random position. The inserted particle causes the internal potential energy to change by an amount $\mathscr{V}(\mathbf{r}^{\text{test}})$ i.e. $\mathscr{V}(\mathbf{r}^N) = \mathscr{V}(\mathbf{r}^{N-1}) + \mathscr{V}(\mathbf{r}^{\text{test}})$. Then the configurational integral for the $N\ \mathscr{V}$ particle system is given by:

$$Z_N = \int d\mathbf{r}^N \exp[-\mathscr{V}(\mathbf{r}^N)/k_B T] \tag{7.65}$$

or

$$Z_N = \int d\mathbf{r}^N \exp[-\mathscr{V}(\mathbf{r}^{\text{test}})/k_B T] \exp[-\mathscr{V}(\mathbf{r}^{N-1})/k_B T] \tag{7.66}$$

By substituting unity in the form Z_{N-1}/Z_{N-1} it is possible to show that $Z_N = Z_{N-1} V \langle \exp[-\mathscr{V}(\mathbf{r}^{\text{test}})/k_B T] \rangle$.

The excess chemical potential, that is the difference between the actual value and that of the equivalent ideal gas system, is given by:

$$\mu_{\text{excess}} = -k_B T \ln \langle \exp[-\mathscr{V}(\mathbf{r}^{\text{test}})/k_B T] \rangle \tag{7.67}$$

The excess chemical potential is thus determined from the average of $\exp[-\mathscr{V}(\mathbf{r}^{\text{test}})/k_B T]$. In ensembles other than the canonical ensemble the expressions for the excess chemical potential are slightly different. The ghost particle does not remain in the system and so the system is unaffected by the procedure. To achieve statistically significant results many Widom insertion moves may be required. However, practical difficulties are encountered when applying the Widom insertion method to dense fluids and/or to systems containing molecules, because the proportion of insertions that give rise to low values of $\mathscr{V}(\mathbf{r}^{\text{test}})$ falls dramatically. This is because it is difficult to find a 'hole' of the appropriate size and shape.

7.10 The configurational bias Monte Carlo method

Various techniques have been developed to tackle the problem of calculating the chemical potential in cases where the routine Widom method does not give converged results [Allen and Tildesley 1987]. Of these methods, the configurational bias Monte Carlo method (CBMC) which was originally introduced by Siepmann [Siepmann 1990] is particularly exciting as it can be applied to assemblies of chain molecules. The configurational bias Monte Carlo method also provides a way to overcome the problems associated with Monte Carlo simulations of assemblies of chain molecules, where many proposed moves are rejected because of high-energy overlaps. The problem of calculating the chemical potential in such cases is clear from the following example. The probability of successfully inserting a single monomer into a fluid of typical liquid density is of the order of 0.5%, or 1 in 200. If one wishes to insert a molecule consisting of n such monomers, the probability is thus approximately 1 in 200^n. For an eight-segment molecule, this probability is less than 1 in 10^{18}, making such calculations impractical. The configurational bias Monte Carlo simulation technique can dramatically improve the chances of making a successful insertion.

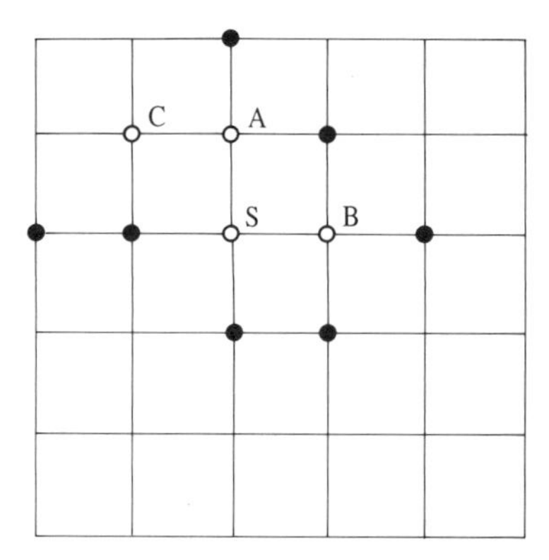

Fig. 7.17 The insertion of a three-unit molecule onto the lattice shown, starting at point S can be achieved in only one way (see text). Figure adapted from Siepmann J I 1990. A Method for the Direct Calculation of Chemical Potentials for Dense Chain Systems, *Molecular Physics* **70**:1145–1158.

The essence of the configurational bias Monte Carlo method is that a growing molecule is preferentially directed (i.e. biased) towards acceptable structures. The effects of these biases can then be removed by modifying the acceptance rules. The configurational bias methods are based upon work published in 1955 by Rosenbluth and Rosenbluth [Rosenbluth and Rosenbluth 1955] and can be applied to both lattice models and to systems with arbitrary molecular potentials and conformations. The method is most easily explained using a two-dimensional lattice model. Suppose we wish to insert a three-unit molecule onto the lattice shown in Figure 7.17. First we consider how the conventional approach would tackle this problem. The initial step is to select a lattice point at random. Suppose we select the lattice point labelled S in Figure 7.17. We then choose one of the four neighbours of S at random. Of the four neighbouring sites, two are occupied and two are free (A and B in Figure 7.17). There is thus a 50% probability that the move will be rejected at this stage. If we select site B then the molecule can be grown no further as all of the adjacent sites are filled. Were we to grow onto A then we would select one of the three neighbouring sites at random. Of these only site C is available. On average, only one trial in twelve will successfully grow a molecule from S using a conventional Monte Carlo algorithm.

Let us now consider how the configurational bias Monte Carlo method would tackle this problem. Again, the first site (S) is chosen at random. We next consider where to place the second unit. The sites adjacent to S are examined to see which is free. In this case, only two of the four sites are free. One of these free sites is chosen at random. Note that the conventional Monte Carlo procedure selected from all four adjoining sites at random, irrespective of whether it is occu-

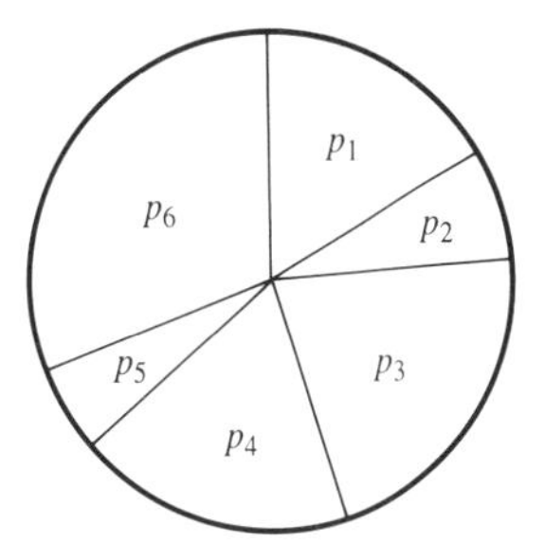

Fig. 7.18 A biased roulette wheel chooses states according to their probabilities.

pied or not. A *Rosenbluth weight* for the move is then calculated. The Rosenbluth weight for each step i is given by:

$$W_\mathrm{i} = \frac{n'}{n} W_{\mathrm{i}-1} \tag{7.68}$$

$W_{\mathrm{i}-1}$ is the weight for the previous step ($W_0 = 1$), n' is the number of available sites and n is the total number of neighbouring sites (not including the one occupied by the previous unit). In the case of our lattice, $W_1 = 2/4 = 1/2$. If site B is chosen then there is no site available for the third unit, and so the attempt has to be abandoned. If site A is chosen, its adjacent sites are examined to see which are free. In this case there is only one free site where the third and final unit can be placed. The Rosenbluth weight for this step is $1/3 \times 1/2 = 1/6$. The overall statistical weight for the move is obtained by multiplying the number of successful trials by the Rosenbluth weight of each trial; as half the trials succeed, the statistical weight is therefore $1/2 \times 1/6 = 1/12$. This is exactly the same result that would be obtained with a conventional sampling scheme, though recall that in a conventional scheme only one trial in twelve results in a successful insertion. By contrast, with the configurational bias method the proportion of successful trials is one in two.

The configurational bias algorithm can be extended to take account of intermolecular interactions between the growing chain and its lattice neighbours. If the energy of segment i when occupying a particular site Γ is $\mathscr{v}_\Gamma(\mathrm{i})$ then that site is chosen with a probability given by:

$$p_\Gamma(\mathrm{i}) = \frac{\exp[-\mathscr{v}_\Gamma(\mathrm{i})/k_\mathrm{B}T]}{Z_\mathrm{i}} \tag{7.69}$$

Z_i is the sum of the Boltzmann factors for all of the b positions considered:

$$Z_\mathrm{i} = \sum_{\Gamma=1}^{b} \exp[-\mathscr{v}_\Gamma(\mathrm{i})/k_\mathrm{B}T] \tag{7.70}$$

The site can be chosen using a biased roulette-wheel algorithm, in which the interval between 0 and 1 is divided into b adjacent segments each with a size proportional to the probabilities $p_1(\mathrm{i}), p_2(\mathrm{i}) \ldots p_b(\mathrm{i})$ (Figure 7.18). The site within whose interval a random number between 0 and 1 lies is the one chosen. The chain is thus biased towards those sites with a higher Boltzmann weighting; the sum of the Boltzmann factors plays the role of n' in equation (7.68). The Rosenbluth weight for the entire chain (of length l) can be calculated as:

$$W_l = \exp[-\mathscr{V}_{\mathrm{tot}}(l)/k_{\mathrm{B}}T] \prod_{\mathrm{i}=2}^{l} \frac{Z_{\mathrm{i}}}{b} \tag{7.71}$$

$\mathscr{V}_{\mathrm{tot}}(l)$ is the total energy of the chain, equal to the sum of the individual segment energies $\upsilon_{\Gamma}(\mathrm{i})$. The average Rosenbluth weight is directly related to the excess chemical potential:

$$\mu_{\mathrm{ex}} = k_{\mathrm{B}}T \ln\langle W_l \rangle \tag{7.72}$$

If a segment has a zero Rosenbluth weight then growth of the chain is terminated. However, such chains must still be included in the averaging used to determine the excess chemical potential.

So far, we have only considered a fixed number of neighbouring sites for each segment. The method can be extended to cover fully flexible chains, where the set of possible neighbouring positions is infinite [De Pablo *et al.* 1992; De Pablo *et al.* 1993]. When growing each segment, a subset containing k random directions is chosen. These trial directions need not be uniformly distributed in space. For each of these orientations the energy $\upsilon_{\Gamma}(\mathrm{i})$ is calculated and so is the Boltzmann factor. An orientation is then chosen with probability:

$$p_{\Gamma}(\mathrm{i}) = \frac{\exp[-\upsilon_{\Gamma}(\mathrm{i})/k_{\mathrm{B}}T]}{\sum_{\Gamma=1}^{k} \exp[-\upsilon_{\Gamma}(\mathrm{i})/k_{\mathrm{B}}T]} \tag{7.73}$$

The Rosenbluth factor is accumulated as follows:

$$W_{\mathrm{i}} = W_{\mathrm{i}-1} \frac{1}{k} \sum_{\Gamma=1}^{k} \exp[-\upsilon_{\Gamma}(\mathrm{i})/k_{\mathrm{B}}T] \tag{7.74}$$

To implement this method it is necessary to determine the appropriate number of trial directions, k. If $k = 1$ then the method is equivalent to the original Widom particle insertion method. If k is too large then too much time is taken calculating the Rosenbluth factors for trial positions that are very close in phase space. Frenkel and colleagues have investigated how the choice of k influences the accuracy of the results and efficiency with which those results were obtained [Frenkel *et al.* 1991]. The system they examined was of a flexible chain containing up to 20 segments in a moderately dense atomic fluid. The conventional particle insertion method failed completely for this system. Not surprisingly, the results showed that as the length of the chain increases so the number of random orientations that need to be considered also increases. At least four trial orientations were used at each step and k was chosen to increase logarithmically with the number of segments to be grown. The limiting value of k was considered to be reached when so many trials were required that the configurational bias method was no more efficient than alternative methods of regrowing chains, such as reptation algorithms. For example, for a 6-segment chain the proportion of accepted configurations (once the initial monomer had been inserted successfully) was 0.000 01% for $k=1$, 3.2% for $k=10$ and 35% for $k=50$. For a 20-segment chain the proportion of accepted configurations was 0.000 1% for $k=20$, 0.66% for $k=50$ and 2.0% for $k=100$.

The Rosenbluth algorithm can also be used as the basis for a more efficient way to perform Monte Carlo sampling for fully flexible chain molecules [Siepmann and Frenkel 1992] which as we have seen is difficult to do as bond rotations often give rise to high energy overlaps with the rest of the system.

The configurational bias Monte Carlo method involves three types of move. Two of these are translational or rotational moves of the entire molecule, which are performed in the conventional way. The third type of move is a conformational change. A chain is selected at random and one of the segments within it is also randomly chosen. That part of the chain that lies above or below the segment (chosen with equal probability) is discarded and an attempt is made to regrow the discarded portion. Let us consider first the case where each segment is restricted to a given number of discrete orientations, either because the chain is restricted to a lattice, or because the model discretely samples the conformational space (e.g. it only permits *gauche* and *trans* conformations to a hydrocarbon chain). At each stage, the Boltzmann weights of the b discrete conformations are determined and one of the sites is chosen with a probability given by equation (7.73). The Rosenbluth weight is determined for the growing chain using equation (7.74).

Having generated a trial conformation it must be decided whether to accept it or not. To do this a random number if generated in the range 0–1 and compared with the ratio of the Rosenbluth weights for the trial conformation ($W_{l,\mathrm{trial}}$) and the old conformation ($W_{l,\mathrm{old}}$). The new chain is then accepted using the following criterion:

$$\mathrm{rand}(0, 1) \leq \frac{W_{l,\mathrm{trial}}}{W_{l,\mathrm{old}}} \tag{7.75}$$

A similar approach can be adopted with continuous chains. Here it is also possible to enhance the sampling by guiding the choice of trial sites towards those with a particularly favourable intramolecular energy. This can be achieved by generating random vectors on the surface of a sphere of unit radius for each segment. The potential energy (angle bending and torsional) for a bond directed along this vector is calculated. The vector is then accepted or rejected using the Metropolis criterion. If it is accepted, the vector is scaled to the appropriate bond length. This procedure continues until the desired number of trial sites have been generated. A trial site is then selecting using Boltzmann factors which only consider the intra- and intermolecular non-bonded interactions of the sites with the chain and with the rest of the system. The Rosenbluth weights are similarly calculated and the move is accepted according to the ratio of the new and old Rosenbluth weights. Again, the correct choice of the number of trial sites is crucial to the efficiency of the method.

7.10.1 Applications of the configurational bias Monte Carlo method

The CBMC method has been used to investigate a number of systems involving long-chain alkanes. Siepmann and McDonald examined a monolayer of 90 $CH_3(CH_2)_{15}SH$ molecules chemisorbed onto a gold surface [Siepmann and McDonald 1993a; Siepmann and McDonald 1993b]. The thiol group forms a bond with the gold surface atoms, thus producing a high degree of surface ordering of the adsorbed molecules. Spectroscopic experiments indicated that the chains were tilted relative to the surface and adopted a predominantly *trans* conformation for the alkyl links. Both discrete and continuous versions of the configurational bias Monte Carlo method were employed; in the discrete model each CH_2CH_2 segment was restricted to the *trans* and two *gauche* conformations. In the continuous simulation six trial sites were used for each segment. The molecules were initially placed on a triangular lattice in an extended conformation perpendicular to the surface.

In the structure obtained at the end of the simulation the chains were ordered in an approximately hexagonal pattern. During equilibration *gauche* conformations were introduced into the alkyl chains, causing the system to tilt. However, once the molecules had all tilted the *gauche* defects were gradually squeezed out to give chains with predominantly *trans* links. The final configuration is shown in Figure 7.19 (colour plate section).

The configurational bias Monte Carlo method has also been used to investigate the adsorption of alkanes in zeolites. Such systems are of especial interest in the petrochemical industry. One interesting experimental result obtained for the zeolite silicalite was that short-chain alkanes (C_1 to C_5) and long chain alkanes (C_{10}) have simple adsorption isotherms but hexane and heptane show kinked isotherms. Such systems are obvious candidates for theoretical investigations because the experimental data is difficult and time-consuming to obtain. Moreover, the simulation can often provide a detailed molecular explanation for the observed behaviour. The simulation of such systems is difficult using conventional methods; the Monte Carlo method suffers from the problems of low acceptance ratios or a very slow exploration of phase space, and long simulation times would be required with molecular dynamics as the diffusion of long chain alkanes is very slow. The configurational bias Monte Carlo method enabled effective and efficient simulations to be performed, providing both thermodynamic properties and the spatial distribution of the molecules within the zeolite [Smit and Siepmann 1994; Smit and Maesen 1995]. The adsorption isotherms were calculated using grand canonical simulations in which the zeolite was coupled to a reservoir at constant temperature and chemical potential.

Silicalite has both straight and zig-zag channels which are connected via intersections (Figure 7.20). An analysis of the configurations showed that the distribution of a short alkane such as butane was approximately equal between the two types of channel. However, as the length of the alkane chain was increased, so there was a greater probability of finding it in a straight channel than in a zig-zag channel. Hexane is an interesting case, for its length is almost equal to the period of the zig-zag channels. At low pressure the hexane molecules move freely in the

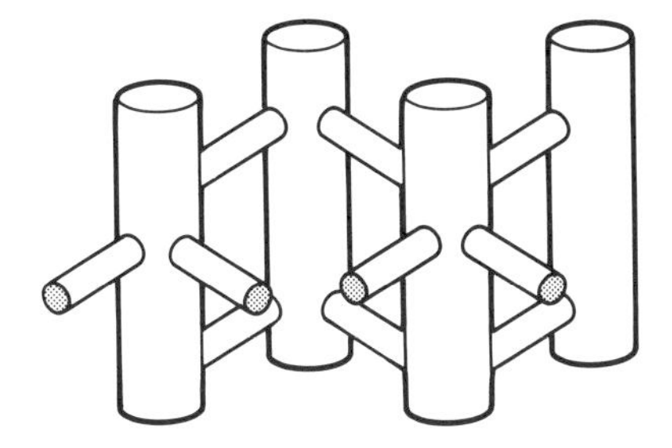

Fig. 7.20 Schematic structure of the zeolite silicalite showing the straight and zigzag channels. Figure adapted from Smit B and J I Siepmann 1994. Simulating the Adsorption of Alkanes in Zeolites. *Science* **264**:1118–1120.

zig-zag channels and occupy the intersections for part of the time. To fill the zeolite with hexane it is first necessay for the alkane molecules to occupy just the zig-zag channels and not the intersections. This is accompanied by a loss of entropy which must be compensated by a higher chemical potential and so gives the kinked isotherm. The straight channels can then be filled with hexane. Different behaviour is observed for smaller alkanes because more than one molecule can occupy the zig-zag channels. Longer alkanes always partially occupy the intersection and so there is no benefit from freezing the molecules in the zig-zag channels.

7.11 Simulating phase equilibria by Gibbs ensemble Monte Carlo method

The most 'obvious' way to investigate phase equilibria is to set up an appropriate system with a conventional simulation technique. Unfortunately, simulations of systems with more than one phase usually require inordinate amounts of computer time. There are several reasons why the use of conventional simulation methods to investigate phase equilibria is difficult. First, it would take a very long time to equilibrate such a system, which would need to separate into its two phases (e.g. liquid and vapour). The properties of the fluid in the interfacial region differ substantially from the properties in the bulk and so to obtain a 'bulk' measurement all of the interfacial atoms must be ignored. Smit has calculated the percentage of the number of particles in the interfacial region for systems of varying sizes; these percentages range from 10% in the interfacial region for a system of 50 000 particles to 95% for a system of 100 particles [Smit 1993]. To simulate phase equilibria directly would thus require long simulations to be performed on systems containing many particles.

The Gibbs ensemble Monte Carlo simulation method, invented by Panagiotopoulos [Panagiotopoulos 1987], enables phase equilibria to be directly studied using small numbers of particles. Rather than trying to form an interface within a single simulation, two simulation boxes are used, each representing one of the two phases. There is no physical interface between the two boxes, which are subject to the usual periodic boundary conditions (Figure 7.21). Three types

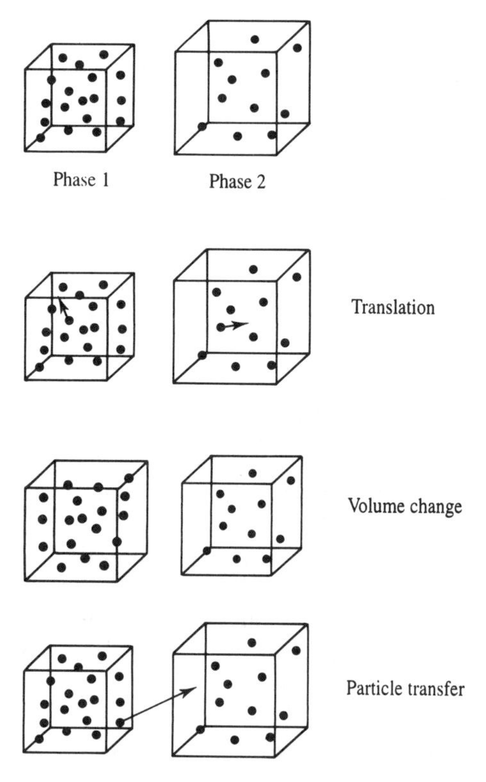

Fig. 7.21 The Gibbs Ensemble Monte Carlo simulation method uses one box for each of the two phases. Three types of move are permitted: translations within either box; volume changes (keeping the total volume constant) and transfer of a particle from one box to the other.

of move are possible. The first type of move comprises particle displacements within each box, as in a conventional Monte Carlo simulation. The second type of move involves volume changes of the two boxes by equal and opposite amounts so that the total volume of the system remains constant. The third type of move involves the removal of a particle from one box and its attempted placement in the other box. This is identical to the Widom insertion method for calculating the chemical potential. Indeed, as the energy of the inserted particle must be calculated it is possible to determine the chemical potential in the Gibbs ensemble without any additional computational cost. These three types of move are often performed in strict order, but it may be better to choose each type of move at random, ensuring that on average the appropriate numbers of each type of move are made.

The properties of the Gibbs ensemble Monte Carlo simulation method have been examined in great detail using simple systems such as the Lennard–Jones fluid and simple gases. A particularly exciting development is the use of the configurational bias Monte Carlo method in conjunction with the Gibbs ensemble method to construct the phase diagrams of complex, long-chain molecules. For example, the vapour–liquid phase equilibria of *n*-pentane and *n*-octane have been investigated by Siepmann, Karaborni and Smit using this combined approach on systems containing 200 pentane or 160 octane molecules [Siepmann *et al.* 1993a]. The calculated properties of these two systems agreed very well with the available experimental data, particularly for the shorter alkane. Their studies were subsequently extended to much longer alkanes (up to C_{48}) [Siepmann *et al.* 1993b]. One particularly noteworthy result was that the density at the critical point increased with the length of the carbon chain up to *n*-octane but then *decreased* as the chain increased in length. Until shortly before the simulations were performed it had been assumed that the critical density for longer chains could be extrapolated from the experimental data obtained with short chains under the assumption that the critical density increased with the length of the chain. Later experiments were able to examine longer chains and did indeed demonstrate that the critical point density passed through a maximum at octane and then decreased for shorter chain lengths.

7.12 Monte Carlo or molecular dynamics?

In principle, the modeller has the choice of using either the Monte Carlo or molecular dynamics technique for a given simulation. In practice one technique must be chosen over the other. Sometimes the decision is a trivial one, for example because a suitable program is readily available. In other cases there are clear reasons for choosing one method instead of the other. For example, molecular dynamics is required if one wishes to calculate time-dependent quantities such as transport coefficients. Conversely, Monte Carlo is often the most appropriate method to investigate systems in certain ensembles; for example, it is much easier to perform simulations at exact temperatures and pressures with the Monte Carlo method than using the sometimes awkward and ill-defined constant temperature and constant pressure molecular dynamics simulation methods. The Monte Carlo method is also well suited to certain types of models such as the lattice models.

The two methods can differ in their ability to explore phase space. A Monte Carlo simulation often gives much more rapid convergence of the calculated thermodynamic properties of a simple molecular liquid (modelled as a rigid molecule), but it may explore the phase space of large molecules very slowly due to the need for small steps unless special techniques such as the configurational bias Monte Carlo method are employed. However, the ability of the Monte Carlo method to make non-physical moves can significantly enhance its capacity to explore phase space in appropriate cases. This may arise for simulations of isolated molecules, where there are a number of minimum energy states separated by high barriers. Molecular dynamics may not be able to cross the barriers between the conformations sufficiently often to ensure that each conformation is sampled according to the correct statistical weight. Molecular dynamics advances the positions and velocities of all the particles simultaneously and it can be very useful for exploration of the local phase space whereas the Monte Carlo method may be more effective for conformational changes which jump to a completely different area of phase space.

Given that the two techniques in some ways complement each other in their ability to explore phase space, it is not surprising that there has been some effort to combine the two methods. Some of the techniques that we have considered in this chapter and in Chapter 6 incorporate elements of the Monte Carlo and molecular dynamics techniques. Two examples are the stochastic collisions method for performing constant temperature molecular dynamics, and the force bias Monte Carlo method. More radical combinations of the two techniques are also possible.

An obvious way to combine Monte Carlo with molecular dynamics is to use each technique for the most appropriate part of a simulation. For example, when simulating a solvated macromolecule, the equilibration phase is usually performed in a series of stages. In the first stage the solute is kept fixed while the solvent molecules (and any ions, if present) are allowed to move under the influence of the solute's electrostatic field. This solvent equilibration may often be performed more effectively using a Monte Carlo simulation as the solvent and ions do not have any appreciable conformational flexibility. To simulate the whole system, molecular dynamics is then the most appropriate method. Such a protocol has been used to perform long simulations of DNA molecules [Swaminathan *et al.* 1991].

A variety of hybrid molecular dynamics/Monte Carlo methods have been devised, in which the simulation algorithm alternates between molecular dynamics and Monte Carlo. The aim of such methods is to achieve better sampling, and thereby more rapid convergence of thermodynamic properties. *In extremis*, each molecular dynamics (or stochastic dynamics) step is followed by a Monte Carlo step, the velocities being unaffected by acceptance or rejection of the Monte Carlo step. Such a method has been devised by Guarnieri and Still [Guarnieri and Still 1994]. An alternative is to perform a block of molecular dynamics steps to generate a new state, which is then accepted or rejected on the basis of the total energy (potential plus kinetic) using the usual Metropolis criterion. If the new coordinates are rejected then the original coordinates from the start of the block are restored and molecular dynamics is run again, but with an entirely new set of velocities that is chosen from a Gaussian distribution. This approach is very similar to the stochastic collisions method for temperature control discussed in section 6.7.1 but with the addition of a Monte Carlo acceptance or rejection step [Duane *et al.* 1987]. A simulation using this hybrid algorithm samples from the canonical ensemble (constant temperature) and was shown by Clamp and colleagues to be more effective than conventional molecular dynamics or Monte Carlo methods for exploring phase space of both simple model systems and proteins [Clamp *et al.* 1994].

Appendix 7.1 The Marsaglia random number generator

The Marsaglia random number generator [Marsaglia *et al.* 1990] is known as a *combination generator* because it is constructed from two different generators. It has a period of about 2^{144}. The first generator is a lagged Fibonacci generator that performs the following binary operation on two real numbers x and y:

$$x \bullet y = x - y \text{ if } x \geq y; x \bullet y = x - y - 1 \text{ if } x < y \tag{7.76}$$

The values x and y are chosen from numbers earlier in the sequence, so that the nth value in the sequence is calculated by:

$$x_n = x_{n-r} \bullet x_{n-s} \tag{7.77}$$

r and s are the *lags*, which are chosen to give numbers that are satisfactorily random and have a long period. Marsaglia chose $r = 97$ and $s = 33$. The algorithm does therefore require the last 97 numbers to be stored at all stages.

The second generator is an arithmetic sequence method that generates random numbers using the following mathematical operation:

$$c \circ d = c - d \text{ if } c \geq d; c \circ d = c - d + 16\,777\,213/16\,777\,216 \text{ if } c < d \tag{7.78}$$

The nth value in this sequence is given by:

$$c_n = c_{n-1} \circ (7654321/16777216) \tag{7.79}$$

The nth number, U_n, in the combined sequence is then obtained as

$$U_n = x_n \circ c_n \tag{7.80}$$

The c sequence requires one initial seed value and the x sequence requires 97 initial seeds (which should themselves be reasonably random). These can be supplied by the user but in the published algorithm these 97 values were obtained from another combination generator comprising a lagged Fibonacci generator and a congruential algorithm.

Further reading

Adams D J 1983. Introduction to Monte Carlo Simulation Techniques. In *Physics of Superionic Conductors and Electrode Materials*, Perran J W (Editor). New York, Plenum: 177–195.

Allen M P and D J Tildesley 1987. *Computer Simulation of Liquids*. Oxford, Oxford University Press.

Colbourn E A (Editor) 1994. *Computer Simulation of Polymers*. Harlow, Longman.

Frenkel D. Monte Carlo Simulations: A Primer in van Gunsteren W F, P K Weiner and A J Wilkinson (Editors). *Computer Simulation of Biomolecular Systems*. Volume 2. Leiden, ESCOM: 37–66.

Kaols M H and P A Whitlock 1986. *Monte Carlo Methods, Volume 1: Basics*. New York, Wiley.

Kermer K 1993. Computer Simulation of Polymers in Allen M P and D J Tildesley (Editors). *Computer Simulation in Chemical Physics*. Dordrecht, Kluwer Academic Publishers. NATO ASI Series 397: 397–459.

Rubinstein R Y 1981. *Simulation and Monte Carlo Methods*. New York, Wiley.

References

Allen M P and D J Tildesley 1987. *Computer Simulations of Liquids*. Oxford, Oxford University Press.

Barker J A and R O Watts 1969. Structure of Water; A Monte Carlo Calculation. *Chemical Physics Letters* **3**:144–145.

Baschnagel J, K Binder, W Paul, M Laso, U Suter, I Batoulis, W Jilge and T Bürger 1991. On the Construction of Coarse-Grained Models for Linear Flexible Polymer-Chains—Distribution-Functions for Groups of Consecutive Monomers. *The Journal of Chemical Physics* **95**:6014–6025.

Clamp M E, P G Baker, C J Stirling and A Brass 1994. Hybrid Monte Carlo: An Efficient Algorithm for Condensed Matter Simulation. *Journal of Computational Chemistry* **15**:838–846.

Cracknell R F, D Nicholson and N Quirke 1994. A Grand Canonical Monte-Carlo Study of Lennard–Jones Mixtures in Slit Pores; 2: Mixtures of Two-Centre Ethane with Methane. *Molecular Simulation* **13**:161–175.

De Pablo J J, M Laso, and U W Suter 1992. Estimation of the Chemical Potential of Chain Molecules by Simulation *The Journal of Chemical Physics* **96**:6157–6162.

De Pablo J J, M Laso, J I Siepmann and U W Suter 1993. Continuum-Configurational Bias Monte Carlo Simulations of Long-Chain Alkanes. *Molecular Physics* **80**:55–63.

Duane S, A D Kennedy and B J Pendleton 1987. Hybrid Monte Carlo. *Physics Letters B* **195**:216–222.

Flory P J 1969. *Statistical Mechanics of Chain Molecules*. New York, Interscience.

Frenkel D, D C A M Mooij and B Smit 1991. Novel Scheme to Study Structural and Thermal Properties of Continuously Deformable Materials. *Journal of Physics Condensed Matter* **3**:3053–3076.

Guarnieri F and W C Still 1994. A Rapidly Convergent Simulation Method: Mixed Monte Carlo/Stochastic Dynamics. *Journal of Computational Chemistry* **15**:1302–1310.

Marsaglia G, A Zaman and W W Tsang 1990. Towards a Universal Random Number Generator. *Statistics and Probability Letters* **8**:35–39.

Metropolis N, A W Rosenbluth, M N Rosenbluth, A H Teller and E Teller 1953. Equation of State Calculations by Fast Computing Machines. *The Journal of Chemical Physics* **21**:1087–1092.

Panagiotopoulos A Z 1987. Direct Determination of Phase Coexistence Properties of Fluids by Monte-Carlo Simulation in a New Ensemble. *Molecular Physics* **61**:813–826.

Pangali C, M Rao and B J Berne 1978. On a Novel Monte Carlo Scheme for Simulating Water and Aqueous Solutions. *Chemical Physics Letters* **55**:413–417.

Rao M and B J Berne 1979. On the Force Bias Monte Carlo Simulation of Simple Liquids. *The Journal of Chemical Physics* **71**:129–132.

Rosenbluth M N and A W Rosenbluth 1955. Monte Carlo Calculation of the Average Extension of Molecular Chains. *The Journal of Chemical Physics* **23**:356–359.

Rossky P J, J D Doll and H L Friedman 1978. Brownian Dynamics as Smart Monte Carlo Simulation. *The Journal of Chemical Physics* **69**:4628–4633.

Sharp W E and C Bays 1992. A Review of Portable Random Number Generators. *Computers and Geosciences* **18**:79–87.

Siepmann J I 1990. A Method for the Direct Calculation of Chemical Potentials for Dense Chain Systems. *Molecular Physics* **70**:1145–1158.

Siepmann J I and D Frenkel 1992. Configurational Bias Monte Carlo: A New Sampling Scheme for Flexible Chains *Molecular Physics* **75**:59–70.

Siepmann J I and I R McDonald 1993a. Domain Formation and System-Size Dependence in Simulations of Self-Assembled Monolayers. *Langmuir* **9**:2351–2355.

Siepmann J I and I R McDonald 1993b. Monte Carlo Study of the Properties of Self-Assembled Monolayers Formed by Adsorption of $CH_3(CH_2)_{15}SH$ on the (111) Surface of Gold. *Molecular Physics* **79**:457–473.

Siepmann J I, S Karaborni and B Smit 1993a. Vapor–Liquid Equilibria of Model Alkanes. *The Journal of the American Chemical Society* **115**:6454–6455.

Siepmann J I, S. Karaborni and B Smit 1993b. Simulating the Cricial Behaviour of Complex Fluids: *Nature* **365**:330–332.

Smit B 1993. Computer Simulation in the Gibbs Ensemble in Allen M P and D J Tildesley (Editors). *Computer Simulation in Chemical Physics*. Dordrecht, Kluwer Academic Publishers. NATO ASI Series 397: 173–210.

Smit B and T L M Maesen 1995. Commensurate 'Freezing' of Alkanes in the Channels of a Zeolite. *Nature* **374**:42–44.

Smit B and J I Siepmann 1994. Simulating the Adsorption of Alkanes in Zeolites. *Science* **264**:1118–1120.

Stephenson G 1973. *Mathematical Methods for Science Students*. London, Longman.

Swaminathan S, G. Ravishanker and D L Beveridge 1991. Molecular-Dynamics of B-DNA Including Water and Counterions—A 140-ps Trajectory for d(CGCGAATTCGCG) based on the Gromos Force-Field. *Journal of the American Chemical Society* **113**:5027–5040.

Verdier P H and W H Stockmayer 1962. Monte Carlo Calculations on the Dynamics of Polymers in Dilute Solution. *The Journal of Chemical Physics* **36**:227–235.

Vesely F J 1982. Angular Monte Carlo Integration Using Quaternion Parameters: A Spherical Reference Potential for CCl_4. *Journal of Computational Physics* **47**:291–296.

Widom B 1963. Topics in the Theory of Fluids. *The Journal of Chemical Physics* **39**:2808–2812.

8 Conformational analysis

8.1 Introduction

The physical, chemical and biological properties of a molecule often depend critically upon the three-dimensional structures, or *conformations*, that it can adopt. Conformational analysis is the study of the conformations of a molecule and their influence on its properties. The development of modern conformational analysis is often attributed to D H R Barton, who showed in 1950 that the reactivity of substituted cyclohexanes was influenced by the equatorial or axial nature of the substituents [Barton 1950]. An equally important reason for the development of conformational analysis at that time was the introduction of analytical techniques such as infrared spectroscopy, NMR and X-ray crystallography which actually enabled the conformation to be determined.

The conformations of a molecule are traditionally defined as those arrangements of its atoms in space that can be interconverted purely by rotation about single bonds. This definition is usually relaxed in recognition of the fact that small distortions in bond angles and bond lengths often accompany conformational changes, and that rotations can occur about bonds in conjugated systems that have an order between one and two.

A key component of a conformational analysis is the *conformational search*, the objective of which is to identify the 'preferred' conformations of a molecule; those conformations that determine its behaviour. This usually requires us to locate conformations that are at minimum points on the energy surface. Energy minimisation methods therefore play a crucial role in conformational analysis. An important feature of methods for performing energy minimisation is that they move to the minimum point that is closest to the starting structure. For this reason, it is necessary to have a separate algorithm which generates the initial starting structures for subsequent minimisation. It is these algorithms for generating initial structures that will be a major focus of this chapter. It is important to recognise the difference between a conformational search and a molecular dynamics or Monte Carlo simulation; the conformational search is concerned solely with locating minimum energy structures whereas the simulation generates an ensemble of states that includes structures not at energy minima. However, as we shall see, both the Monte Carlo and molecular dynamics methods can be used as part of a conformational search strategy.

If possible, it is desirable to identify all minimum energy conformations on the energy surface. However, the number of minima may be so large that it is impractical to contemplate finding them all. Under such circumstances it is usual to try and find all the accessible minima. The relative populations of a molecule's conformations can be calculated using statistical mechanics via the Boltzmann distribution, though it is important to remember that the statistical weights involve contributions from all the degrees of freedom, including the vibrations as well as the energies. Solvation effects may also be important, and various schemes are now available for calculating the solvation free energy of a conformation which can be added to the intramolecular energy. These solvation schemes (which will be discussed in more detail in section 9.8) provide computationally efficient ways to include the effects of the solvent on conformational equilibria. For some molecules such as proteins there are so many minima on the energy surface that it is impractical to try and find them all. Under such circumstances, it is often assumed that the native (i.e. naturally occurring) conformation is the one with the very lowest value of the energy function. This conformation is usually referred to as the *global minimum energy conformation*. One should usually be wary of algorithms which only find a single conformation. For example, even though the global minimum energy conformation has the lowest energy, it may not be the most highly populated because of the contribution of the vibrational energy levels to the statistical weight of each structure. Moreover, the global minimum energy conformation may not be the active (i.e. functional) structure. Indeed, in some cases it is possible that the active conformation does not correspond to any minimum on the energy surface of the isolated molecule. It may even be necessary for a molecule to adopt more than one conformation. For example, a substrate might bind in one conformation to an enzyme and then adopt a different conformation prior to reaction.

Conformational search methods can be conveniently divided into the following categories: systematic search algorithms, model building methods, random approaches, distance geometry and molecular dynamics. Before discussing these methods, we should note that conformational analysis can sometimes be performed quite effectively using Dreiding or CPK mechanical models. The invention of these models should be regarded as an important development in conformational analysis and molecular modelling. Mechanical models do, however, have some shortcomings. For example, they provide no quantitative information about the relative energies of the various conformations. It is often quite difficult to make accurate measurements of a molecule's internal coordinates such as the distance between two atoms that are on opposite sides of the structure. They are subject to the forces of gravity (which makes them unwieldy for large molecules) and the hands of marauding colleagues! It is also difficult to construct models that have significant deviations from standard bond lengths and angles. Nevertheless, manual models can be very useful, particularly as they are portable and because they can be easily manipulated in a way that is often not possible with computer images (although this may change with the development of 'virtual reality' molecular modelling systems).

8.2 Systematic methods for exploring conformational space

As the name suggests, a systematic search explores the conformational space by making regular and predictable changes to the conformation. The simplest type of systematic search (often called a *grid search*) is as follows. First, all rotatable bonds in the molecule are identified. The bond lengths and angles remain fixed throughout the calculation. Each of these bonds is then systematically rotated through 360° using a fixed increment. Every conformation so generated is subjected to energy minimisation to derive the associated minimum energy conformation. The search stops when all possible combinations of torsion angles have been generated and minimised. To illustrate the grid search algorithm, let us consider the conformational energy surface for the 'alanine dipeptide' $CH_3CONHCHMeCONHCH_3$ (Figure 8.1) which is used as a model for the conformational behaviour of amino acids in proteins. If we assume that the bond lengths and bond angles are fixed and that the amide bonds adopt *trans* conformations then only the two torsion angles labelled ϕ and ψ in Figure 8.1 can vary. The energy is then a function of just these two variables and as such, it can be represented as a contour diagram as shown in Figure 8.2. This contour plot is known as a *Ramachandran map*, after G N Ramachandran who showed that the amino acids were restricted to a limited range of conformations [Ramachandran *et al.* 1963]. The accessible areas on the contour maps calculated by Ramachandran do indeed correspond to those conformations that are observed in X-ray structures of proteins, Figure 8.3. Two regions are particularly important; these correspond to the α-helix and β-strand structures, which will be discussed in more detail in section 8.13.1. The amino acid glycine, which has no side chain (Figure 8.1), has a wider range of accessible conformations than the other amino acids as can be seen from the Ramachandran map in Figure 8.2.

To perform a grid search of the conformational space of the alanine dipeptide, a series of conformations would be generated by systematically varying ϕ and ψ between 0° and 360°. This is equivalent to drawing a two-dimensional grid over the Ramachandran contour diagram in Figure 8.2; each grid point corresponds to a conformation generated by the grid search with some combination of ϕ and ψ. It can readily be seen that even for a relatively large torsional increment the number of conformations generated by the grid search is much larger than the number of minima on the surface; many of the initial conformations minimise to

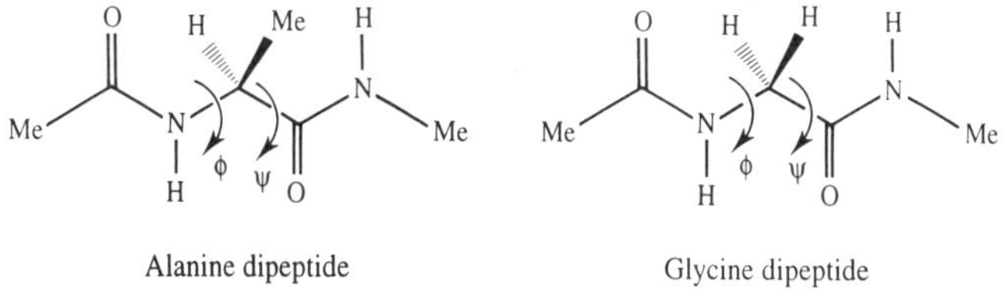

Alanine dipeptide

Glycine dipeptide

Fig. 8.1 The alanine dipeptide and the glycine dipeptide.

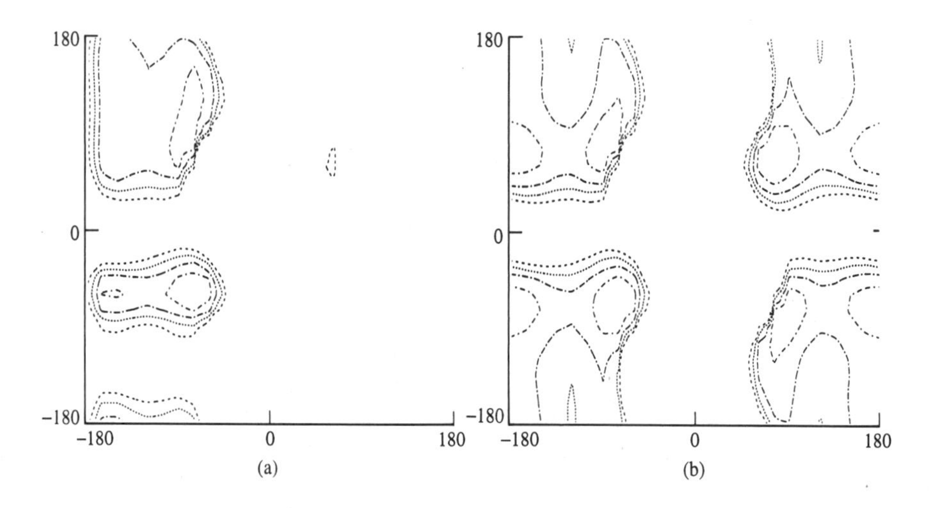

Fig. 8.2 Ramachandran map for the alanine dipeptide (a) and glycine dipeptide (b), calculated using the AMBER force field [Weiner *et al.* 1984]. In both cases contours are drawn at 0.5, 1.0, 1.5 and 2.0 kcal/mol above the lowest-energy conformation found.

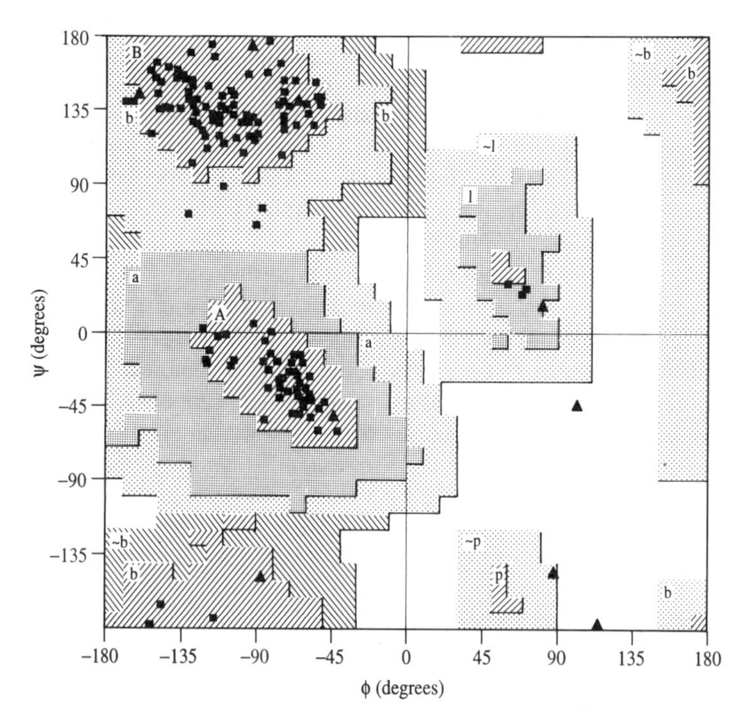

Fig. 8.3 Experimentally observed distribution of (ϕ, ψ) angles in dihydrofolate reductase. The symbols are the actual values; the shaded areas correspond to the (ϕ, ψ) distribution averaged over many protein structures.

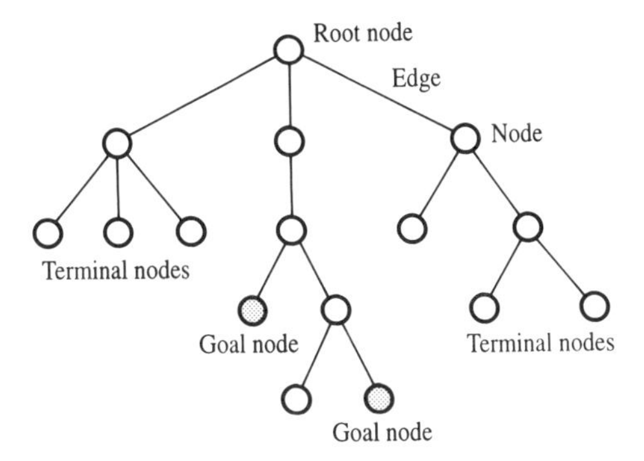

Fig. 8.4 Schematic illustration of a search tree.

the same minimum energy structure. Moreover many of the initial conformations are very high in energy.

A major drawback of the grid search is that the number of structures to be generated and minimised increases in an exceptional fashion with the number of rotatable bonds, a phenomenon known as a *combinatorial explosion*. The number of structures generated is given by:

$$\text{Number of conformations} = \prod_{i=1}^{N} \frac{360}{\theta_i} \tag{8.1}$$

θ_i is the dihedral increment chosen for bond i. For example, if there are five bonds and an increment of 30° is used for each bond, then 248 832 structures will be generated. If the number of bonds is increased to seven, then the number of structures increases to almost 36 million. To put these figures into context, suppose each structure takes just one second to minimise. The five-bond problem will then require 69 hours to complete and the seven-bond problem will require 415 days. Despite this apparent limitation, systematic search algorithms are routinely employed to consider problems involving 10–15 bonds. This is achieved by eliminating from the time-consuming energy minimisation stage structures that have a very high energy or some other problem. The best way to describe these enhanced systematic search methods is to use a *tree representation* of the problem.

Search trees are widely used to represent the different states that a problem can adopt. An example is shown in Figure 8.4 from which it should be clear where the name derives, especially if the page is turned upside down. A tree contains *nodes* that are connected by *edges*. The presence of an edge indicates that the two nodes it connects are related in some way. Each node represents a state that the system may adopt. The *root node* represents the initial state of the system. *Terminal nodes* have no child nodes. A *goal node* is a special kind of terminal node that corresponds to an acceptable solution to the problem.

Suppose we wish to use a grid search to explore the conformational space of a simple alkane, *n*-hexane. We will assume that rotation of the terminal methyl

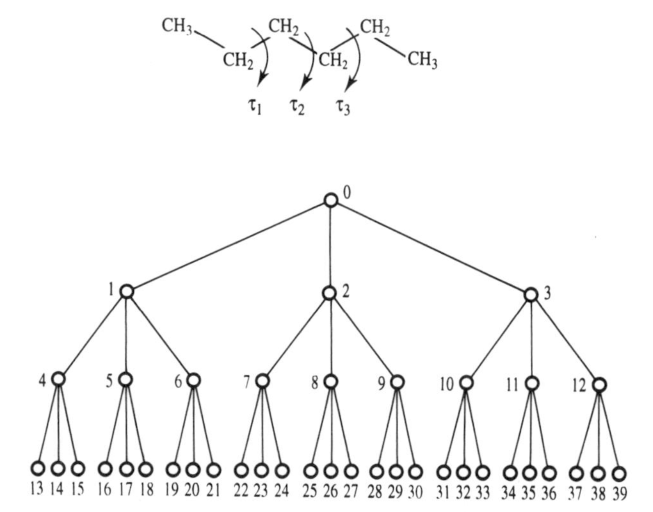

Fig. 8.5 Tree representation of the conformation search problem for hexane.

groups can be ignored and so just three bonds can vary. If we permit each of the variable bonds to adopt just three values, corresponding to the *trans*, *gauche*(+) and *gauche*(−) conformations* then the search tree for this problem contains 27 terminal nodes ($\equiv 3 \times 3 \times 3$) and is shown in Figure 8.5. The root node represents the starting point, where none of the rotatable bonds have been assigned a torsion angle. When the first rotatable bond is set to its first value (i.e. the *trans* conformation with a torsion angle of 180°), this corresponds to moving from the root node to the node numbered 1 in the tree. The second bond is now set to a *trans* conformation; this corresponds to a move to node 4. When the third bond is assigned *trans* we reach node 13, which is a terminal node and corresponds to a conformation ready for minimisation. To generate further conformations, it is necessary to change one of the three torsion angles. The most convenient way to do this is to assign a new value to the last bond to be set (i.e. bond 3). Setting bond 3 to a *gauche*(+) conformation is equivalent to moving back up the tree from node 13 to node 4 and then down to the terminal node 14. This gives a second completed conformation. By proceeding in this fashion through the search tree (a process called *backtracking*) all conformations of the molecule can be generated. The search algorithm we have described is known as the *depth-first search*.

The efficiency of a depth-first search can be enhanced by discarding structures that violate some energetic or geometric criterion. Structures with high-energy steric interactions are then rejected before the energy minimisation stage. We can further enhance the efficiency of the systematic search by checking partially constructed conformations, before all the torsion angles have been assigned. Suppose

* The *trans* conformation corresponds to a torsion angle of 180°, the *gauche*(+) conformation to one of +60° and the *gauche*(−) conformation to −60°. These approximately correspond to the torsion angles of the three minimum energy conformations of butane.

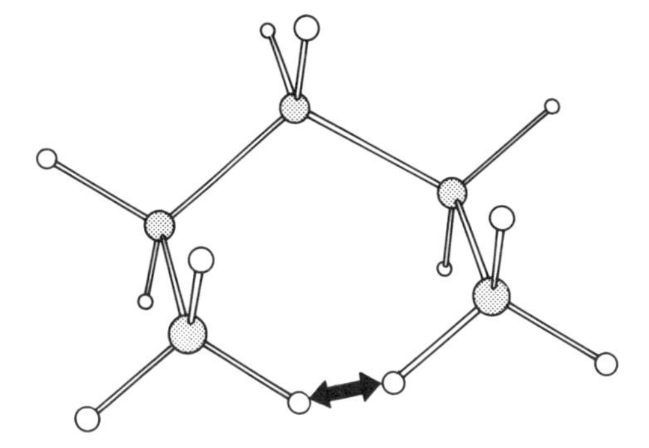

Fig. 8.6 A pentane violation arises from when there are successive *gauche*(+) and *gauche*(−) torsion angles in an alkane chain.

we generate a partial structure containing two non-bonded atoms that are very close in space. In hexane such a high-energy structure is generated if the first rotatable bond is set to a *gauche*(+) conformation and the second rotatable bond is assigned *gauche*(−) (a pentane violation, Figure 8.6). Whatever value is assigned to the third torsion angle, this high-energy steric problem will remain. All structures that lie below that node in the search tree (number 9 in Figure 8.5) can thus be eliminated, or *pruned*. It is important to stress that this is only possible if those parts of the molecule that are in violation will not be moved relative to each other by a subsequent torsional assignment.

Cyclic molecules are often quite difficult to analyse using a systematic search. The usual strategy is to break the ring, giving a 'pseudo-acyclic' molecule that can then be treated as a normal acyclic molecule. This process is illustrated for cyclohexane in Figure 8.7. When searching the conformational space of cyclic molecules additional checks must be included to ensure that the rings are properly formed. For example, an all-*trans* structure is a perfectly acceptable conformation of *n*-hexane, but is not an acceptable conformation of cyclohexane due to the unreasonable bond length between the ring closure atoms. It is therefore common practice to check several intramolecular parameters when using a systematic search to explore the conformational space of a cyclic system; these parameters usually include the bond length between the ring closure bonds together with the bond angles at these atoms (Figure 8.8). In some programs other internal parameters (e.g. the torsion angles adjacent to the ring closure bond) are also checked. The main reason why rings are problematic for the systematic search is that these checks can often be applied only very late in the analysis; it is often necessary to almost complete the ring before the structure is rejected or accepted.

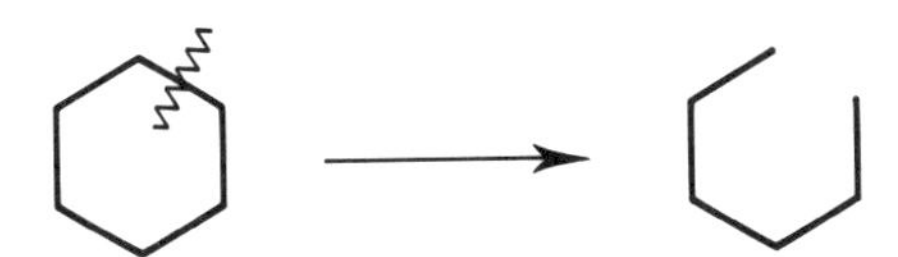

Fig. 8.7 A 'pseudo-acyclic' molecule is generated by breaking the ring.

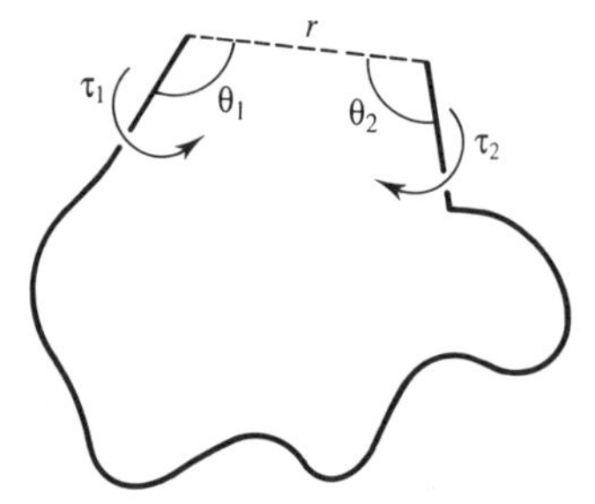

Fig. 8.8 The intramolecular parameters that may be checked when exploring the conformational space of a ring system.

One simple check that can be used when constructing cyclic molecules is to ensure that at all stages the distance of the growing chain from the start atom is short enough to enable the remaining bonds to close the ring.

The systematic search is most efficient when the rotatable bonds are processed in a unidirectional fashion. This ensures that once an atom has been fixed in space relative to the atoms already considered then it will not be moved again. For an acyclic molecule this means that the search starts at one end of the molecule and moves down the chain. Molecules containing rings are treated by opening the cycles to give the pseudo-acyclic molecule as described above and then processed (Figure 8.9).

Any systematic search ultimately requires a balance to be made between the resolution of the grid and the available computer resources. Too fine a grid and the search may take too long; too coarse and important minima may be missed. The non-bonded criteria, which determine whether a structure is to be rejected, must also be assigned. The non-bonded criterion is often referred to as a 'bump check' and is usually set to a modest value (say 2.0 Å) as the energy minimisation

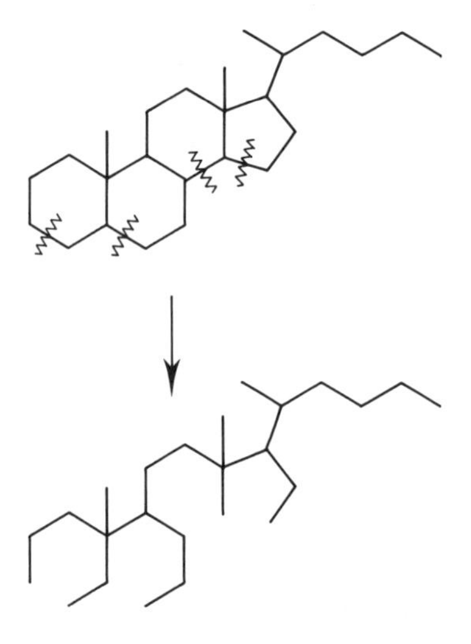

Fig. 8.9 Creation of pseudo-acyclic molecule for a system with many rings.

step will be able to remove minor problems in the structure. For cyclic molecules the ring closure criteria may also affect the results. It must also be remembered that the various cutoffs are interdependent, so that changing one may require others to be reassigned.

8.3 Model-building approaches

One way to at least partially alleviate the combinatorial explosion that inevitably accompanies a systematic search is to use larger 'building blocks', or *molecular fragments* to construct the conformations [Gibson and Scheraga 1987; Leach *et al.* 1988]. Fragment- or model-building approaches to conformational analysis construct conformations of a molecule by joining together three-dimensional structures of molecular fragments. This approach would be expected to be more efficient than the normal systematic search because there are usually many fewer combinations of fragment conformations than combinations of torsion angle values. This is particularly so for cyclic fragments, which are in any case problematic for the systematic search method. For example, the molecule in Figure 8.10 could be constructed by joining together the fragments indicated. Many molecular modelling systems offer a facility for constructing structures from molecular fragments, though the user usually has to manually specify which fragments are to be joined and how this is to be achieved. Clearly, if each fragment can adopt a number of conformations, then it is impractical to tackle the problem manually and some means of automating the method is required.

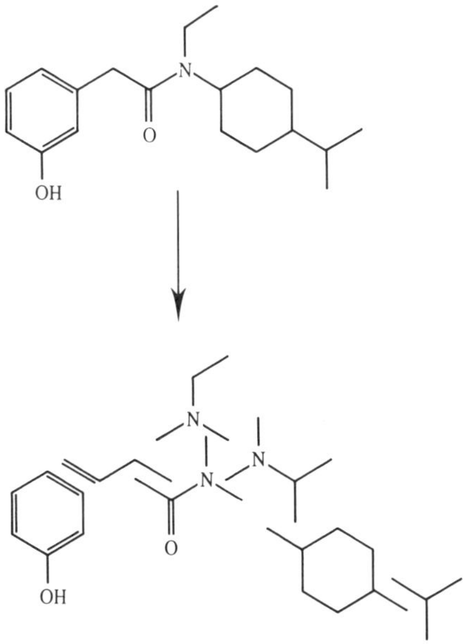

Fig. 8.10 A conformation may be obtained by joining appropriate fragments.

A program to automatically explore conformational space using the fragment-building approach must first decide which fragments are needed to construct the molecule [Leach *et al.* 1990]. This is done using a *substructure search algorithm* which determines whether each of the fragments that the program 'knows about' is present in the molecule and how the atoms in the fragment match onto the atoms in the molecule. Having identified the fragments that are required, conformations can be generated. The conformations available to each fragment (often called templates) should span the range of conformations the fragment can adopt. For example, cyclohexane rings adopt the chair, twist-boat and boat conformations in molecules and so templates corresponding to these structures should be available. A conformation of the molecule is constructed by assigning a template to each fragment and then attempting to join the templates together. The search problem can be represented as a tree, just for a systematic search, and so all of the usual tree-searching algorithms are applicable. The search can be significantly enhanced by tree pruning.

The fragment-based approach to conformational analysis relies upon two assumptions. The first assumption is that each fragment must be conformationally independent of the other fragments in the molecule. The second assumption is that the conformations stored for each fragment must cover the range of structures that are observed in fully constructed molecules. The fragment conformations can be obtained from a variety of sources; two common approaches are by analysing a structural database (see section 8.9) or from some other conformational search method. A third limitation is that one can obviously only analyse molecules for which there are fragments available.

8.4 Random search methods

A random search is in many ways the antithesis of a systematic search. A systematic search explores the energy surface of the molecule in a predictable fashion, whereas it is not possible to predict the order in which conformations will be generated by a random method. A random search can move from one region of the energy surface to a completely unconnected region in a single step. A random search can explore conformational space by changing either the atomic Cartesian coordinates or the torsion angles of rotatable bonds. Both types of algorithm use a similar approach that is outlined in flow chart form in Figure 8.11. At each iteration, a random change is made to the 'current' conformation. The new structure is then refined using energy minimisation. If the minimised conformation has not been found previously, it is stored. The conformation to be used as the starting point for the next iteration is then chosen and the cycle starts again. The procedure continues until a given number of iterations have been performed or until it is decided that no new conformations can be found.

The Cartesian and dihedral versions of the random search differ in the way in which each new structure is generated. The Cartesian method adds a random amount to the x, y and z coordinates of all the atoms in the molecule [Saunders

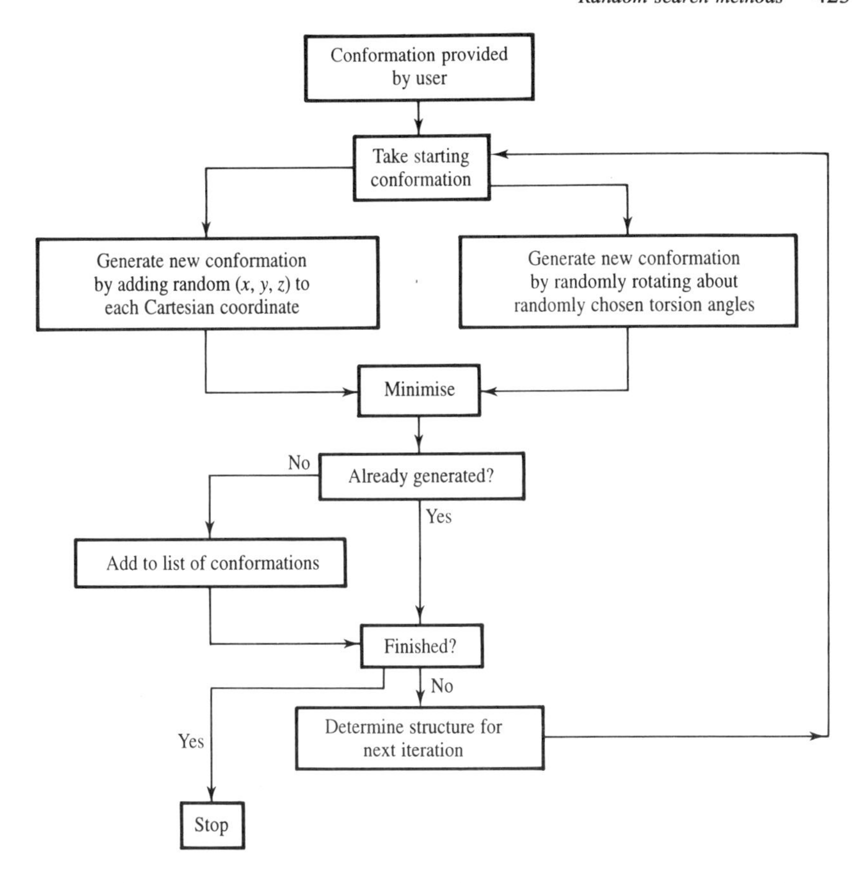

Fig. 8.11 Flow chart steps followed by a random conformational search.

1987; Ferguson and Raber 1989] whereas the dihedral method generates new conformations by making random changes to the rotatable bonds, with the bond lengths and bond angles being kept fixed [Li and Scheraga 1987; Chang *et al.* 1989]. The Cartesian method is extremely simple to implement, but does rely heavily upon the minimisation step as the initial structures that are generated by the randomisation procedure can be extremely high in energy; in some implementations the coordinates can change by up to 3 Å or more and so the resulting structures are extremely distorted. The advantage of the dihedral search method is that many fewer degrees of freedom need to be considered. However, special procedures are required when applying the dihedral method to molecules that contain rings; typically these are broken in a manner similar to the systematic search described above to give a pseudo-acyclic molecule (Figure 8.9). After randomisation each ring is checked to ensure that any ring closure constraints are satisfied. In the random dihedral method it is possible to change all dihedrals or just a randomly chosen subset of them.

There are many ways in which the structure for input to the next iteration of the search can be selected. A simple approach is to take the structure obtained from the previous step. An alternative is to select randomly a structure from those generated previously, weighting the choice towards those structures that have been selected the least. A third method is to use the lowest energy structure found so far, or to bias the selection towards the lowest energy structures. The Metropolis Monte Carlo scheme is often used to make the choice. Each newly generated structure (after energy minimisation) is accepted as the starting point for the next iteration if it is lower in energy than the previous structure or if the Boltzmann factor of the energy difference, $\exp[-(\mathscr{V}_{\text{new}}(\mathbf{r}^N) - \mathscr{V}_{\text{old}}(\mathbf{r}^N))/k_BT]$ is larger than a random number between 0 and 1. If not, the previous structure is retained for the next iteration. There is no fundamental reason why any of these methods should be preferred over another but some are reported to be more efficient at exploring the conformational space or finding the global minimum energy conformation.

In a systematic search there is a defined endpoint to the procedure, which is reached when all possible combinations of bond rotations have been considered. In a random search, there is no natural endpoint; one can never be absolutely sure that all of the minimum energy conformations have been found. The usual strategy is to generate conformations until no new structures can be obtained. This usually requires each structure to be generated many times and so the random methods inevitably explore each region of the conformational space a large number of times.

8.5 Genetic algorithms

Generic algorithms are a class of methods based on biological evolution that are designed to find optimal solutions to problems [Goldberg 1989]. The first step is to create a 'population' of possible solutions. In conformational analysis this initial population would correspond to a set of randomly generated conformations of the molecule. The 'fitness' of each member of the population is then calculated. Finally, a new population is then generated from the old one, with a bias towards the fitter members. Each member of the population is coded by a 'chromosome'. This is usually stored as a linear string of bits (i.e. 0 s and 1 s). The chromosome codes for the values of the torsion angles of the rotatable bonds in the molecule, Figure 8.12. The initial population is most easily obtained by randomly setting bits to 0 or 1 in the chromosomes. After decoding each chromosome and assigning the torsion angles to the appropriate values in the molecule, the fitness of each member of the population can be calculated. The value of the fitness function indicates the 'quality' of each conformation. In conformational analysis an appropriate fitness function would be the internal energy, as might be calculated using molecular mechanics. A new population is then generated using *operators*. Three commonly used operators are *reproduction, crossover* and *mutation*. The reproduction operator simply copies individual chromosomes according to their fitness, with the highest-scoring individuals having a greater probability of being reproduced than their weaker counterparts. A simple way to implement this uses a biased roulette wheel in which each individual has a slot

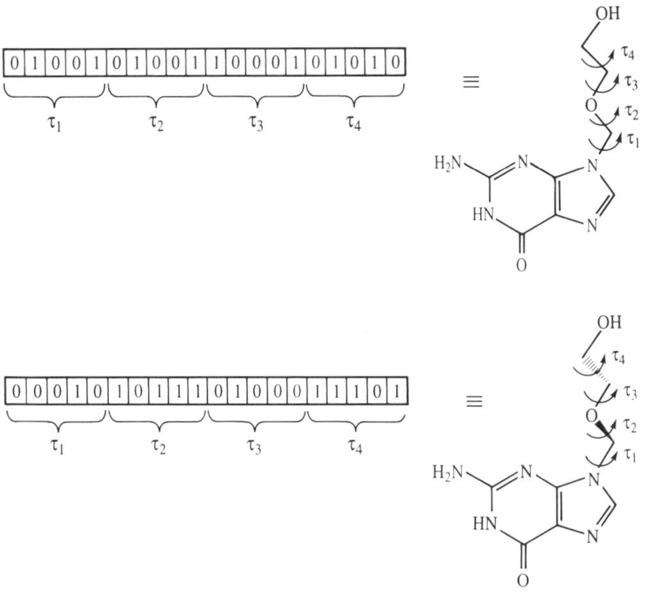

Fig. 8.12 The chromosome in a genetic algorithm codes for the torsion angles of the rotatable bonds.

size proportional to its value of the fitness function. The crossover operator first randomly selects pairs of individuals for 'mating'. A crossover position is then randomly selected at a position i ($1 \leq i \leq l - 1$), where l is the length of the chromosome). Two new strings are then created by swapping the bits between positions $i + 1$ and l. For example, suppose we have the following two chromosomes:

```
00100011110001

11000011001100
```

and the crossover point is chosen to be 6. Then the two new strings are:

```
00100011001100

11000011110001
```

The third operator is mutation. Here, bits are randomly selected and changed (0 to 1 and vice versa). The mutation operator is usually assigned a low probability.

Many variants on these basic operations are possible. For example, it is common practice to carry forward the highest ranking individuals unchanged: this is often referred to as an 'elitist' strategy. This ensures that the best individuals are not lost, as would occur if crossover or mutation were applied to every member of the new population.

Genetic algorithms were originally introduced as a technique for performing global optimisation. If this were so then only the global minimum energy conformation would be obtained from each run. In fact, in the conformational analysis of complex molecules one cannot guarantee to reach the global minimum energy conformation and the genetic algorithm approach is primarily used as a very effective method for generating a large number of 'good' solutions [Judson *et al.* 1993; McGarrah and Judson 1993].

8.6 Distance geometry

One way to describe the conformation of a molecule other than by Cartesian or internal coordinates is in terms of the distances between all pairs of atoms. There are $N(N-1)/2$ interatomic distances in a molecule, which are most conveniently represented using an $N \times N$ symmetric matrix. In such a matrix the elements (i, j) and (j, i) contain the distance between atoms i and j and the diagonal elements are all zero. Distance geometry explores conformational space by randomly generating many distance matrices which are then converted into conformations in Cartesian space. The crucial feature about distance geometry (and the reason why it works) is that it is not possible to arbitrarily assign values to the interatomic distances in a molecule and always obtain a low-energy conformation. Rather, the interatomic distances are closely interrelated and indeed many combinations of distances are geometrically impossible. This can be illustrated using a simple three-atom molecule (ABC). Simple trigonometry requires that the sum of the distances AB and AC must be greater than or equal to the distance BC. Thus, a conformation in which the distances are AB = 1.5 Å, AC = 1.4 Å and BC = 3.5 Å is not geometrically possible.

Distance geometry uses a four stage process to derive a conformation of a molecule [Crippen 1981, Crippen and Havel 1988]. First, a matrix of upper and lower interatomic *distance bounds* is calculated. This matrix contains the maximum and minimum values permitted to each interatomic distance in the molecule. Values are then randomly assigned to each interatomic distance between its upper and lower bounds. In the third step, the distance matrix is converted into a trial set of Cartesian coordinates, which in the fourth step are then refined.

Some of the interatomic distance bounds can be determined from simple chemical principles. For example, X-ray crystallographic studies have shown that bond lengths are restricted to a small range of values that are determined primarily by the atomic number and hybridisation of the two atoms. The distance between two atoms which are both bonded to a third atom (i.e. are in a 1,3 relationship) is also severely restricted and can be calculated from the angle at the central atom and the lengths of the two bonds. The distance between two atoms that are separated by three bonds (i.e. are in a 1,4 relationship) can vary with the torsion angle of the central bond, the minimum distance corresponding to a torsion angle of 0° and the maximum distance to a torsion angle of 180°. These three cases are shown in Figure 8.13. It is not so easy to determine limits on the other interatomic distances (i.e. between atoms in a $1, n$ relationship where $n > 4$) but it is usual to require that such atom pairs do not approach closer than the sum

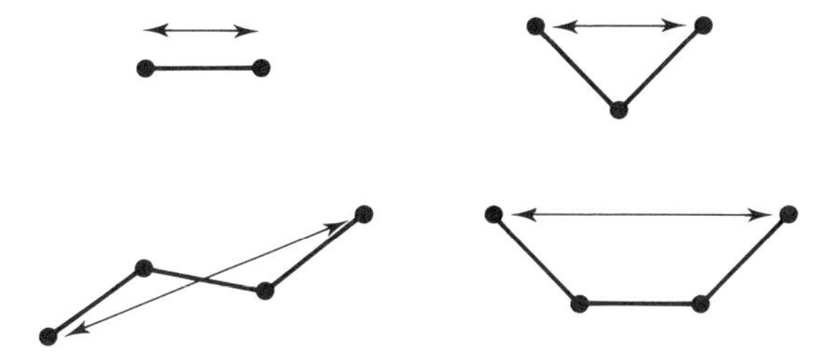

Fig. 8.13 The upper and lower distance bounds for atoms in 1,2, in 1,3 and 1,4 relationships can be derived from simple chemical principles.

of the van der Waals radii of the two atoms. The upper bound is then usually assigned an arbitrarily large value.

A procedure called *triangle smoothing* is then used to refine the initial set of distance bounds. Triangle smoothing uses two simple trigonometrical restrictions on groups of three atoms that are illustrated in Figure 8.14. The first restriction is that the distance between two atoms A and C can be no greater than the sum of the maximum values of the distances AB and BC. This can be written:

$$u_{AC} \leq u_{AB} + u_{BC} \tag{8.2}$$

u_{AB} indicates the upper bound on the AB distance. The second restriction is that the minimum value of the AC distance can be no less than the difference between the lower bound on AB and the upper bound on BC:

$$l_{AC} \geq l_{AB} - u_{BC} \tag{8.3}$$

l_{AB} is used to indicate the lower bound distance. These two inequalities are repeatedly applied to the set of distances bounds until the entire set of distance bounds is self-consistent and all possible interatomic distance triplets satisfy both inequalities. Triangle smoothing need only be performed once for each molecule.

We can now proceed to the generation of conformations. First, random values are assigned to all the interatomic distances between the upper and lower bounds to give a trial distance matrix. This distance matrix is now subjected to a process called *embedding*, in which the 'distance space' representation of the conforma-

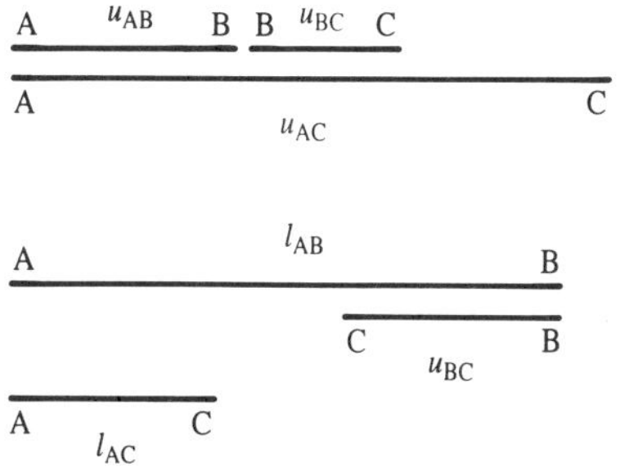

A u_{AB} B B u_{BC} C

A C

u_{AC}

l_{AB}

A B

C B

u_{BC}

A C

l_{AC}

Fig. 8.14 The two triangle inequalities used in distance geometry.

tion is converted to a set of atomic Cartesian coordinates by performing a series of matrix operations. We calculate the *metric matrix*, **G**, each of whose elements (i, j) is equal to the scalar product of the vectors from the origin to atoms i and j:

$$G_{\mathrm{ij}} = \mathbf{i}.\mathbf{j} \tag{8.4}$$

The elements G_{ij} can be calculated from the distance matrix using the cosine rule:

$$G_{\mathrm{ij}} = (d_{\mathrm{io}}^2 + d_{\mathrm{jo}}^2 - d_{\mathrm{ij}}^2)/2 \tag{8.5}$$

d_{io} is the distance from the origin to atom i and d_{ij} is the distance between atoms i and j.

It is usual to take the centre of the molecule as the origin of the coordinate system. The distance of each atom from the centre can be calculated directly from the interatomic distances using the following expression:

$$d_{\mathrm{io}}^2 = \frac{1}{N}\sum_{\mathrm{j=1}}^{N} d_{\mathrm{ij}}^2 - \frac{1}{N^2}\sum_{\mathrm{j=2}}^{N}\sum_{\mathrm{k=1}}^{\mathrm{j-1}} d_{\mathrm{jk}}^2 \tag{8.6}$$

The metric matrix **G** is a square symmetric matrix. A general property of such matrices is that they can be decomposed as follows:

$$\mathbf{G} = \mathbf{V}\mathbf{L}^2\mathbf{V}^{\mathrm{T}} \tag{8.7}$$

The diagonal elements of $\mathbf{L}^2$ are the eigenvalues of **G** and the columns of **V** are its eigenvectors. The atomic coordinates can be derived from the metric matrix by rewriting equation (8.4) as

$$\mathbf{G} = \mathbf{X}\mathbf{X}^{\mathrm{T}} \tag{8.8}$$

X is a matrix containing the atomic coordinates. Equating equations (8.7) and (8.8) gives $\mathbf{X} = \mathbf{VL}$

$$\mathbf{X} = \mathbf{VL} \tag{8.9}$$

As **L** has only diagonal entries, the matrix **L** is identical to its transpose: $\mathbf{L} = \mathbf{L}^{\mathrm{T}}$. The atomic coordinates are thus obtained by multiplying the square roots of the eigenvalues by the eigenvectors.

The triangle smoothing and embedding steps of distance geometry are best understood using a specific example. Let us consider a five-atom, all-carbon fragment (Figure 8.15), in which all of the carbon–carbon bonds are assumed to have an optimal length of 1.3 Å and all internal angles are 120°. If we further assume

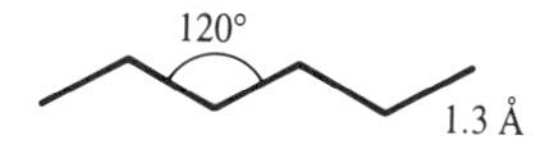

Fig. 8.15 Five-carbon fragment to illustrate distance geometry algorithm.

that the carbon van der Waals radius is 1.4 Å, then the initial bounds matrix is as follows:

$$\begin{pmatrix} 0.0 & 1.3 & 2.2517 & 3.4395 & 99.0 \\ 1.3 & 0.0 & 1.3 & 2.2517 & 3.4395 \\ 2.2517 & 1.3 & 0.0 & 1.3 & 2.2517 \\ 2.6 & 2.2517 & 1.3 & 0.0 & 1.3 \\ 2.8 & 2.6 & 2.2517 & 1.3 & 0.0 \end{pmatrix} \tag{8.10}$$

Note that the lower bound for the distance between atoms 1 and 5 equals the sum of their van der Waals radii and that the upper bound has been set to an arbitrarily large value of 99Å. All other distances have been allocated on the basis of geometric arguments. In a 'real' example the upper and lower bounds for bonded atoms would usually be slightly different (by approximately 0.1 Å) to reflect the fact that bond lengths in real molecules do vary a little. The 1,3 bounds would also be made slightly different. After triangle smoothing only one distance bound is changed, the upper bound of the distance between atoms 1 and 5. This distance is changed to a value that is equal to the sum of the upper bounds of the distances between atoms 1 and 3 and 3 and 5. The smoothed bounds matrix that results is:

$$\begin{pmatrix} 0.0 & 1.3 & 2.2517 & 3.4395 & 4.5033 \\ 1.3 & 0.0 & 1.3 & 2.2517 & 3.4395 \\ 2.2517 & 1.3 & 0.0 & 1.3 & 2.2517 \\ 2.6 & 2.2517 & 1.3 & 0.0 & 1.3 \\ 2.8 & 2.6 & 2.2517 & 1.3 & 0.0 \end{pmatrix} \tag{8.11}$$

Suppose interatomic distances are now randomly assigned between the lower and upper bounds to give the following distance matrix:

$$\begin{pmatrix} 0.0 & 1.3 & 2.25 & 3.11 & 3.42 \\ & 0.0 & 1.3 & 2.25 & 2.85 \\ & & 0.0 & 1.3 & 2.25 \\ & & & 0.0 & 1.3 \\ & & & & 0.0 \end{pmatrix} \tag{8.12}$$

The corresponding metric matrix is:

$$\begin{pmatrix} 3.571 & 1.569 & -0.427 & -2.276 & -2.436 \\ 1.569 & 1.256 & 0.105 & -1.122 & -1.808 \\ -0.427 & 0.105 & 0.644 & 0.261 & -0.583 \\ -2.276 & -1.122 & 0.261 & 1.569 & 1.569 \\ -2.436 & -1.808 & -0.583 & 1.569 & 3.259 \end{pmatrix} \quad (8.13)$$

The eigenvalues of this matrix are 8.18, 1.74, 0.26, 0.10 and 0.0 with the matrix of eigenvectors being:

$$\mathbf{W} = \begin{pmatrix} 0.621 & 0.455 & -0.425 & 0.164 \\ 0.355 & -0.184 & 0.800 & 0.020 \\ 0.0 & -0.573 & -0.368 & -0.580 \\ -0.408 & -0.287 & -0.153 & 0.727 \\ -0.567 & 0.590 & 0.145 & -0.330 \end{pmatrix} \quad (8.14)$$

The 'best' three-dimensional structure is obtained by taking the eigenvectors that correspond to the three largest eigenvalues, providing they are all positive. If these eigenvalues are λ_1, λ_2 and λ_3 and $\mathbf{W}$ is the matrix containing the associated eigenvectors, then the Cartesian coordinates (x_i, y_i, z_i) of each atom i are calculated as follows:

$$x_i = \sqrt{\lambda_1} W_{i1} \quad (8.15)$$

$$y_i = \sqrt{\lambda_2} W_{i2} \quad (8.16)$$

$$z_i = \sqrt{\lambda_3} W_{i3} \quad (8.17)$$

For our five-carbon example, the coordinates obtained using the three highest eigenvalues are:

Atom	*x coordinate*	*y coordinate*	*z coordinate*
1	1.777	0.601	− 0.218
2	1.014	− 0.244	0.410
3	− 0.001	− 0.757	− 0.188
4	− 1.166	− 0.379	− 0.079
5	− 1.623	0.799	0.075

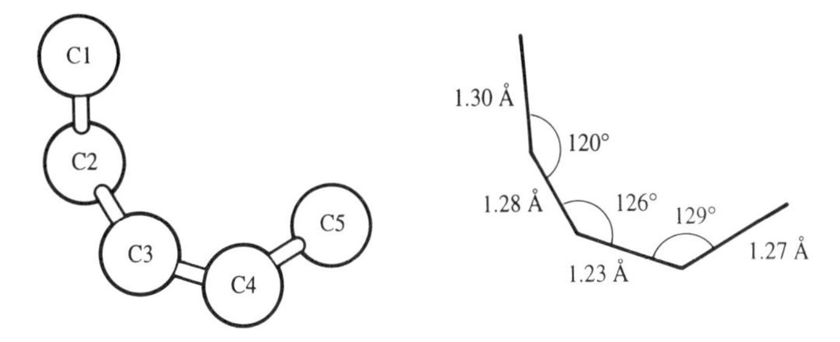

Fig. 8.16 Conformation of the five-carbon fragment generated by distance geometry.

This conformation is schematically illustrated in Figure 8.16. The interatomic distance matrix for this conformation is:

$$\begin{pmatrix} 0.0 & 1.299 & 2.24 & 3.10 & 3.42 \\ & 0.0 & 1.29 & 2.24 & 2.85 \\ & & 0.0 & 1.23 & 2.25 \\ & & & 0.0 & 1.25 \\ & & & & 0.0 \end{pmatrix} \tag{8.18}$$

Note that the distances in this conformation do not equal the original randomly chosen distances, nor do they all lie between the upper and lower bound values in the bounds matrix. For example, the distance between the atoms 4 and 5 is 1.25 Å rather than 1.3 Å. This is because it may be necessary to use more than just three dimensions to find a conformation that satisfies the distances in the initial distance matrix. The number of non-zero eigenvectors of the metric matrix equals the dimensionality of the space in which a solution can be found. In general, if there are N interatomic distances then a solution can be found in $N - 1$ dimensions. This is in part a consequence of the fact that three-dimensional objects must not only satisfy triangle inequalities, but also tetrangle, pentangle and hexangle relationships. Moreover, triangle smoothing is often only applied to the bounds matrix; the distance matrix that is used as input to the embedding stage may in fact contain combinations of distances that violate the triangle inequalities. Improved sampling of conformational space can be achieved if the trial distances are selected so that they do satisfy the triangle inequalities (a process known as *metrisation*), but for reasons of compuational cost it is often not used in its full form.

If we add the coordinates corresponding to the fourth eigenvalue, then the original distance matrix is reproduced exactly. These fourth-dimensional coordinates are as follows:

Atom	*4th coordinate*
1	0.053
2	0.006
3	– 0.188
4	0.235
5	– 0.107

The distance between atoms 4 and 5 in this four-dimensional space is exactly 1.3 Å.

In the final step of the distance geometry algorithm the coordinates are refined so that the conformation better satisfies the initial distance bounds. A conjugate gradients minimisation algorithm is often employed for this step. The function to be minimised has a positive value for distances that are outside the permitted range but is zero otherwise. The penalty functions most commonly used are:

$$E = \sum_{i} \sum_{j>i} \begin{cases} (d_{ij}^2 - u_{ij}^2)^2 & d_{ij} > u_{ij} \\ 0 & l_{ij} \le d_{ij} \le u_{ij} \\ (l_{ij}^2 - u_{ij}^2)^2 & d_{ij} < l_{ij} \end{cases} \quad (8.19)$$

$$E = \sum_{i} \sum_{j>i} \begin{cases} \left[(d_{ij}^2 - u_{ij}^2)/u_{ij}^2\right]^2 & d_{ij} > u_{ij} \\ 0 & l_{ij} \le d_{ij} \le u_{ij} \\ \left[(l_{ij}^2 - u_{ij}^2)/d_{ij}^2\right]^2 & d_{ij} < l_{ij} \end{cases} \quad (8.20)$$

u_{ij} is the upper bound distance between atoms i and j and l_{ij} is the lower bound distance. The first function weights long distances more than short distances whereas the second error function weights all distances equally. Both functions are zero when all the distances are between the upper and lower bounds. A conformation in which all distance bounds are satisfied is not necessarily at an energy minimum, and so the final structure may subsequently be subjected to force field energy minimisation to derive the associated minimum energy structure.

During the optimisation of the structure against the distance constraints it is usual to incorporate *chiral constraints*. These are used to ensure that the final conformation is the desired stereoisomer. Chiral constraints are necessary because the interatomic distances in two enantiomeric conformations are idential and as a consequence the 'wrong' isomer may quite legitimately be generated. Chiral constraints are usually incorporated into the error function as a chiral volume, calculated as a scalar triple product. For example, to maintain the correct stereochemistry about the tetrahedral atom number 4 in Figure 8.17, the following scalar triple product must be positive:

$$(\mathbf{v}_1 - \mathbf{v}_4).[(\mathbf{v}_2 - \mathbf{v}_4) \times (\mathbf{v}_3 - \mathbf{v}_4)] \quad (8.21)$$

The other stereoisomer corresponds to a negative chiral volume. Chiral constraints are included in the penalty function by adding terms of the following form:

$$(V_{ch} - V_{ch}^*)^2 \quad (8.22)$$

V_{ch}^* is the desired value of the chiral constraint. Chiral constraints can also be used to force groups of atoms to lie in the same plane by requiring the chiral volume to have a value of zero. More than one such constraint may be required

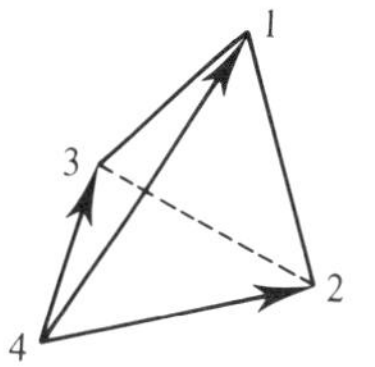

Fig. 8.17 The stereochemistry about tetrahedral atoms can be maintained with an appropriate chiral constraint.

for each planar group. For example, to force all six atoms about the double bond in Figure 8.18 to lie in the same plane, the three sets of chiral volumes defined by the atoms 1, 2, 3, 4; 2, 4, 5, 6; 1, 2, 4, 5 must be zero. A commonly used strategy in many distance geometry programs is to perform the first few steps of refinement using a conformation defined in four dimensions, as this can help to invert any incorrect chiral centres. The minimisation then swiches to three-dimensional conformations for the final stages of the distance geometry refinement. A force field energy minimisation may also be used. When the conformation has been refined, the next structure is generated, starting with the assignment of random distances.

Many enhancements have been made to this basic distance geometry method. Some of the most useful enhancements result from the incorporation of chemical information. For example, if the lower bound for the 1,4 distances is set to a value equivalent to a torsion angle of 60° rather than one of 0° then eclipsed conformations can be avoided. Similarly, amide bonds can be forced to adopt a nearly planar structure by an appropriate choice of distance bounds and chiral constraints.

8.6.1 The use of distance geometry in NMR

One of the most important uses of distance geometry is for deriving conformations that are consistent with experimental distance information, especially distances obtained from NMR experiments. The NMR spectroscopist has at his or her disposal a range of experiments that can provide a wealth of information about the conformation of a molecule. Two of the most commonly used NMR

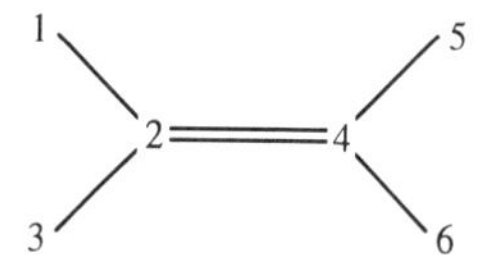

Fig. 8.18 A double bond can be forced to adopt a planar conformation through the use of appropriate chiral constraints.

experiments that provide such conformationally dependent information are the 2D-NOESY (nuclear Overhauser enhancement spectroscopy) and the 2D-COSY (correlated spectroscopy) experiments [Derome 1987]. NOESY provides information about the distances between atoms which are close together in space but may be separated by many bonds. The strength of the NOESY signal is inversely proportional to the sixth power of the distance and so by analysing the nuclear Overhauser spectrum it is possible to calculate approximate values for the distance between relevant pairs of atoms. COSY experiments are often used to provide information about atoms which are covalently separated by three bonds (i.e. torsion angles). Both types of experiments provide information about interatomic distances, and so distance geometry is a natural technique to use for generating conformations that are consistent with the experimental data. Distance geometry has been particularly useful for solving the structures of proteins and nucleic acids, where the amount of data is so large that it is impossible to perform the task manually. The distance information provided by NMR experiments does of course supplement the geometrical constraints on the interatomic distances that are derived from the internal geometry (i.e. bond lengths and angles).

Distance geometry is at heart a random technique. It is therefore usual to generate more than one conformation, in order to try and explore the conformational space that is consistent with the experimentally derived distances. The resulting set of structures is often displayed as a superimposed set; this enables the similarities and differences between the structures to be easily identified. For example, in Figure 8.19 [Chung *et al.* 1995] we show an ensemble of conformations of RANTES, a small protein called a chemokine that is implicated in inflammation (colour plate section). It is often found that some parts of the molecule adopt very similar conformations in all the structures whereas other regions show considerable variation. This is often interpreted as an indication of conformational flexibility, but it is important to remember that it may also indicate a lack of experimental data for those atoms.

8.7 Exploring conformational space using simulation methods

The Monte Carlo and molecular dynamics simulation methods can be used to explore the conformational space of molecules. During such a simulation the system is able to overcome energy barriers and so explore different regions of the conformational space. A distinction must be made between a 'true' Monte Carlo simulation (Chapter 7) and the minimisation-based random search methods discussed in section 8.4. A true Monte Carlo simulation does not include any energy minimisation, and each randomly generated conformation would be accepted or rejected using the Metropolis criterion. There are difficulties in applying the Monte Carlo technique in simulations of flexible molecules, as we have discussed in Chapter 7.

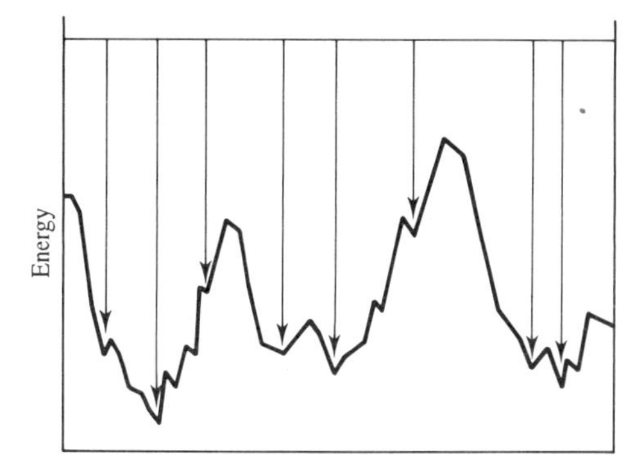

Fig. 8.20 Schematic illustration of an energy surface. A high-temperature molecular dynamics simulation may be able to overcome very high energy barriers and so explore conformational space. On minimisation, the appropriate minimum energy conformation is obtained (arrows).

Molecular dynamics is widely used for exploring conformational space. A common strategy is to perform the simulation at a very high, physically unrealistic temperature. The additional kinetic energy enhances the ability of the system to explore the energy surface and can prevent the molecule from getting stuck in a localised region of conformational space. This is schematically illustrated in Figure 8.20. Structures are then selected at regular intervals from the trajectory for subsequent energy minimisation.

8.7.1 Simulated annealing

Annealing is the process in which the temperature of a molten substance is slowly reduced until the material crystallises to give a large single crystal. It is a technique that is widely used in many areas of manufacture, such as the production of silicon crystals for computer chips. A key feature of annealing is the use of very careful temperature control at the liquid–solid phase transition. The perfect crystal that is eventually obtained corresponds to the global minimum of the free energy. Simulated annealing is a computational method that mimics this process in order to find the 'optimal' or 'best' solutions to problems which have a large number of possible solutions [Kirkpatrick *et al.* 1983].

In simulated annealing, a cost function takes the role of the free energy in physical annealing and a control parameter corresponds to the temperature. To use simulated annealing in conformational analysis the cost function would be the internal energy. At a given temperature the system is allowed to reach 'thermal equilibrium' using a molecular dynamics or Monte Carlo simulation. At high temperatures, the system is able to occupy high energy regions of conformational space and to pass over high energy barriers. As the temperature falls, the lower energy states becomes more probable in accordance with the Boltzmann distribution. At absolute zero the system should occupy the lowest energy state (i.e. the global minimum energy conformation). To guarantee that the globally optimal solution is actually reached would require an infinite number of temperature steps, at each of which the system would have to come to thermal equilibrium. Careful temperature control is required when the energy of the system is comparable to the height of the barriers that separate one region of conformational space from another. This is often difficult to achieve in practice and so simulated annealing cannot *guarantee* to find the global minimum, much as a genetic algorithm cannot guarantee to identify the globally optimal solution. However, if the same answer is obtained from several different runs then there is a high probability that it corresponds to the true global minimum. Several simulated annealing runs may enable a series of low energy conformations of a molecule to be obtained.

8.7.2 Solving protein structures by restrained molecular dynamics refinement

A particularly important application of the molecular dynamics/simulated annealing method is in the refinement of X-ray and NMR data, to determine the three-dimensional structures of large biological molecules such as proteins. The aim of such refinement is to determine the conformation (or conformations) that best explain the experimental data. In both X-ray crystallography and NMR a modified form of molecular dynamics called *restrained molecular dynamics* is often used. In a restrained molecular dynamics simulation additional terms called *penalty functions* are added to the potential energy function. These extra terms have the effect of penalising conformations that do not agree with the experimental data. Molecular dynamics is used to explore the conformational space in order to find a conformation (or conformations) that not only have a low intrinsic energy but are also consistent with the experimental data. Simulated annealing is often a convenient way to ensure that the conformational space is explored effectively.

8.7.3 X-ray crystallographic refinement

X-ray crystallography is a powerful technique for elucidating the structures of molecules. An X-ray diffraction pattern arises because of constructive and destructive interference between X-rays scattered from different parts of the crystal. An X-ray beam scattered by an electron at a point $\mathbf{r}$ travels a different distance to the detector than a beam scattered by an electron at the origin, Figure 8.21. As a consequence, the two scattered X-ray beams will have different phases and will interfere. As the detector is moved through different scattering angles, θ (Figure 8.21), the intensity of the scattered radiation will fluctuate between zero (destructive interference) and twice the amplitude of the original beam (constructive interference). In a real sample the amplitude of the scattered radiation from a point is proportional to the electron density at that point. The total

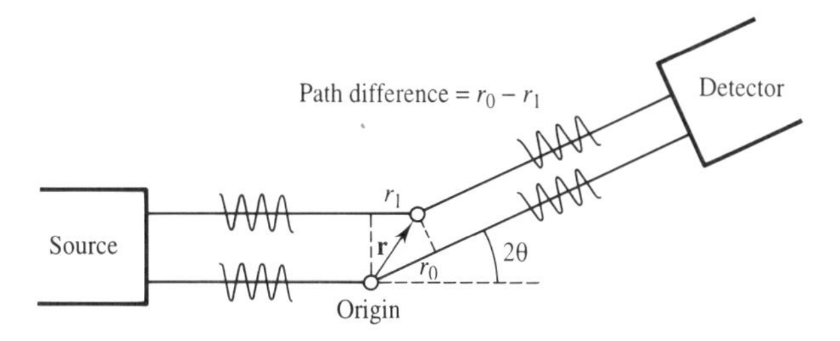

Fig. 8.21 Schematic illustration of an X-ray scattering experiment. The X-ray beam travels a different distance when scattered by an electron at the origin compared to an electron situated at **r**.

signal reaching the detector is obtained by integrating the electron density over the whole crystal and is expressed as the structure factor, F. The structure factor is a complex number that can be written $F = |F|e^{i\phi}$ where $|F|$ is the amplitude and $e^{i\phi}$ is the phase. If the electron distribution is known (i.e. if we know the three-dimensional structure) it is possible to determine the structure factor for all scattering angles, and so we can calculate the X-ray diffraction pattern. The X-ray crystallographer is faced with the reverse problem: to determine the electron distribution (and thereby the three-dimensional structure) from the diffraction pattern. The difficulty is that it is only possible to measure the intensities of the spots (which are equal to the amplitudes $|F|^2$), but not the phases; this is the famous *phase problem*, which is one of the major obstacles in solving an X-ray structure.

To obtain the electron density distribution it is necessary to guess, calculate or indirectly estimate the phases. Various methods have been developed to tackle the phase problem. For proteins the most common strategy is multiple isomorphous replacement in which the protein crystals are soaked in solutions containing salts of heavy metals such as mercury, platinum or silver. These heavy atoms may bind to specific parts of the protein (e.g. mercury ions may react with exposed SH groups). By comparing the diffraction patterns of the native crystals and the crystals of the heavy-atom derivatives it can be possible to estimate some of the phases (under the assumption that no structural change occurs). Once some of the phases are known, others can be determined, and eventually an initial electron density map can be obtained. The electron density map is often represented as a three-dimensional surface by contouring at a constant value (Figure 8.22, colour plate section). An initial model of the molecule is then fitted to the electron density. When the diffraction experiment is performed at high resolution then the locations of individual atoms are often easy to identify. However, at lower resolution it can be difficult to find the optimal fit of the atoms in the model to the electron density as individual features are not so well defined. This is often the case in proteins.

The objective of the refinement is to obtain a structure that gives the best possible agreement with the experimental data. This is done by gradually changing the structure to give better and better agreement between the calculated and observed structure factor amplitudes. This degree of agreement is quantified by the value of the crystallographic R factor, which is defined as the difference between the observed ($|F_{obs}|$) and calculated ($|F_{calc}|$) structure factor amplitudes:

$$R = \frac{\sum ||F_{obs}| - |F_{calc}||}{\sum |F_{obs}|} \tag{8.23}$$

Traditionally, least-squares methods have been used to refine protein crystal structures. In this method a set of simultaneous equations is set up whose solutions correspond to a minimum of the R factor with respect to each of the atomic coordinates. Least-squares refinement requires an $N \times N$ matrix to be inverted, where N is the number of parameters. It is usually necessary to visually examine an evolving model every few cycles of the refinement to check that the structure looks reasonable. During visual examination it may be necessary to alter a model to give a better fit to the electron density, and prevent the refinement from falling into an incorrect local minimum. X-ray refinement is time consuming, requires substantial human involvement and is a skill which usually takes several years to acquire.

Jack and Levitt introduced molecular modelling techniques into the refinement in the form of an energy minimisation step (using a force field function) that was performed alternately with the least-squares refinement [Jack and Levitt 1978]. This approach was shown to give convergence to better structures. More recently, restrained molecular dynamics methods were introduced by Brunger, Kuriyan and Karplus [Brunger *et al.* 1987]. These methods have had a dramatic impact on the refinement of X-ray and NMR structure of proteins.

In the restrained molecular dynamics approach the total 'potential energy' is written as the sum of the usual potential energy and the penalty term, as usual:

$$E_{tot} = \mathcal{V}(\mathbf{r}^N) + E_{sf} \tag{8.24}$$

The additional penalty function that is added to the empirical potential energy function in restrained dynamics X-ray refinement has the form:

$$E_{sf} = S \sum [|F_{obs}| - |F_{calc}|]^2 \tag{8.25}$$

E_{sf} describes the differences between the observed structure factor amplitudes and those calculated from the atomic model. S is a scale factor which is chosen so that the gradient of E_{sf} is comparable to the gradient of the potential energy part of the function. The conformational space is explored using molecular dynamics with simulated annealing; very high temperatures are used in the initial stages to permit the system to range widely over the energy surface. The temperature is then gradually reduced as the structure settles into a conformation which not only has a low energy but also a low R factor.

8.7.4 Molecular dynamics refinement of NMR data

We have already discussed in section 8.6.1 the type of information that NMR experiments can provide about the conformation of a molecule and the use of distance geometry for determining structures that are consistent with the experimental data. Restrained molecular dynamics can also be used to tackle this problem. In the simplest approach, we could incorporate harmonic restraint terms of the form $k(d - d_0)^2$ where d is the distance between the atoms in the current conformation and d_0 is the desired distance derived from the NMR spectrum. k is a force constant, the value of which determines how tightly the restraint should be applied. The information provided by the COSY experiment can also be expressed as a torsion angle via the Karplus equation; torsional restraints may be incorporated into the molecular dynamics energy function as an alternative to the use of distances. There are many other ways in which the restraints can be incorporated; for example, some practitioners prefer to penalise a structure only if the distance exceeds the target:

$$\mathcal{V}(d) = k(d - d_0)^2 \qquad d > d_0 \tag{8.26}$$

$$\mathcal{V}(d) = 0 \qquad d \leq d_0 \tag{8.27}$$

The atoms are prevented from coming too close by the van der Waals terms in the force field. More sophisticated functional forms have also been used which try to take into account the imprecise nature of the experimental values. A simple Hooke's law relationship implies that an exact value is known for the distance, whereas there can be significant uncertainty about its value. A more appropriate functional form has the following form:

$$\mathcal{V}(d) = k_l(d - d_l)^2 \qquad d < d_l \tag{8.28}$$

$$\mathcal{V}(d) = 0 \qquad d_l \leq d \leq d_u \tag{8.29}$$

$$\mathcal{V}(d) = k_u(d - d_u)^2 \qquad d_u < d \tag{8.30}$$

This potential is shown schematically in Figure 8.23. d_l and d_u are the lower and upper distances that are considered to be consistent with the experimental data. $(d_l + d_u)/2$ is thus the assigned target distance obtained from a measurement

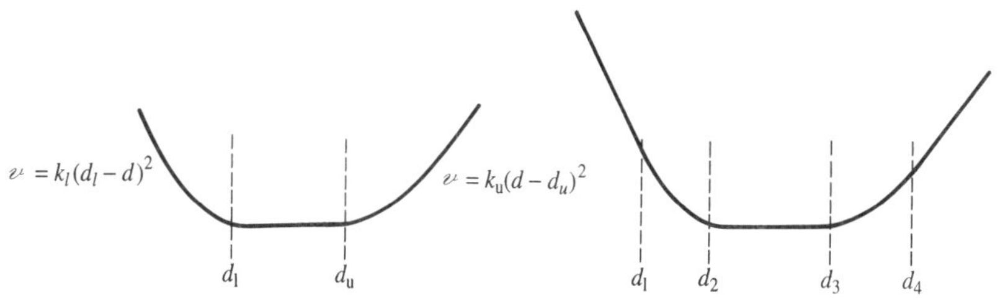

Fig. 8.23 A restraining potential that does not penalise structures in which the distance lies between the lower and upper distances d_l and d_u and uses harmonic functions outside this range (left). The harmonic potentials may also be replaced by linear restraints further from this region (right).

of the NOESY intensity and the error associated with that measurement is $\pm(d_u - d_l)/2$. A distance between d_l and d_u incurs no penalty. Outside this region the restraint is applied using two harmonic potentials. These restraining potentials may have different force constants and so be of different steepness. In some functional forms, the harmonic potential is eventually replaced by a linear function, as illustrated in Figure 8.23.

8.7.5 Time-averaged NMR refinement

If the molecule interconverts between two or more conformations on a timescale that is rapid compared to the chemical shift timescale then the NMR spectrum only shows the average of the signals from the individual conformations. This behaviour is schematically illustrated in Figure 8.24, where an atom or group (such as a leucine side chain) interconverts between two energy minima within the protein. The NMR spectrum comprises a single peak that is a weighted

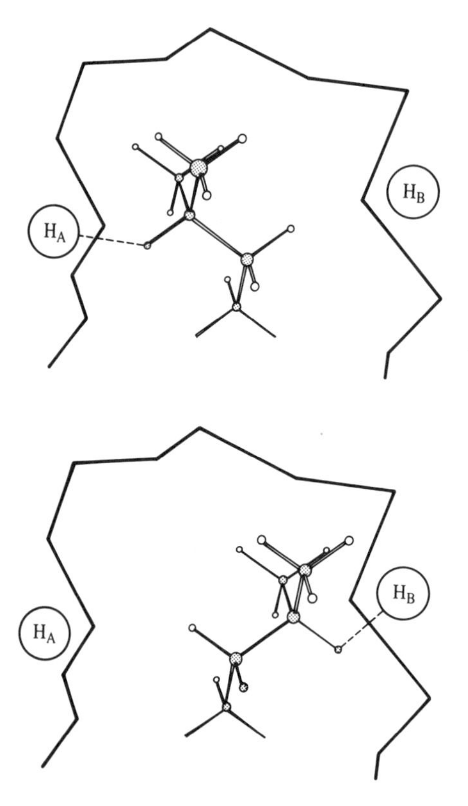

Fig. 8.24 If the leucine side chain interconverts rapidly between two conformations then the NMR spectrum will be an average of them. With a traditional refinement this leads to a structure that simultaneously tries to satisfy all restraints and is at the top of the energy barrier between the two minima.

average of the resonances from the two individual conformations. If the two conformations make distinct interactions then two sets of distance restraints can be derived and a standard refinement procedure would attempt to satisfy both sets of restraints simultaneously. This would lead to a conformation in which the group is positioned at the top of the barrier between the two minima. This incorrect result is a consequence of assuming that one single structure is consistent with all of the experimental data, rather than recognising that the experimental data may arise from more than one conformation.

Time-averaged restraints may be able to overcome this problem [Torda *et al.* 1990]. Rather than using the instantaneous value of a distance in the restraint function, the time-averaged restraint method uses a value that is averaged over time. The simple harmonic error function then becomes:

$$\mathcal{V}(d) = k(\langle d(t)\rangle - d_0)^2 \tag{8.31}$$

$\langle d(t)\rangle$ is the time-averaged value of the distance, as obtained from the molecular dynamics simulation. At a time t', $\langle d(t')\rangle$ is given by:

$$\langle d(t')\rangle = \frac{1}{t'}\int_0^{t'} d(t)\mathrm{d}t \tag{8.32}$$

As the intensity of the NOESY signal is proportional to the inverse sixth power of the distance, the 'distance' to use in this case is actually given by:

$$d_{\mathrm{NOESY}} = \langle d(t')^{-6}\rangle^{-1/6} \tag{8.33}$$

The time-averaged value of the distance that should be used in the error function is thus

$$\langle d(t')\rangle = \left[\frac{1}{t'}\int_0^{t'} d(t)^{-6}\mathrm{d}t\right]^{-1/6} \tag{8.34}$$

One way to implement time-averaged restraints is to evaluate $\langle d(t')\rangle$ using equation (8.34) as the simulation proceeds, and incorporate the value in the error function equation (8.31). If the simulation is run for long enough, all the accessible conformational states should be visited and be included in the calculation of the average distances. However, it is rarely possible to achieve the length of simulation needed to ensure that the conformation space has been adequately covered. We require a method that can supply an accurate picture of the dynamics of the molecule with minimal computational effort. The normalisation factor $1/t'$ in equation (8.34) becomes progressively larger as the simulation proceeds, thus making $\langle d(t')\rangle$ increasingly less sensitive to the current value of the distance. What we require is a means to bias the instantaneous value of $\langle d(t')\rangle$ towards the values from the most recent part of the simulation. In this way, if the 'current' value of $\langle d(t')\rangle$ is incompatible with the restraint, then the penalty function should be proportionately increased. This can be achieved using an exponential 'memory function' which has the effect of weighting the recent history more heavily. Various memory functions are possible; one functional form is:

$$\langle d(t')\rangle = \left(\frac{\int_0^{t'} e^{(t-t')/\tau} d(t)^{-6} dt}{\int_0^{t'} e^{(t-t')} dt}\right)^{-1/6} \tag{8.35}$$

τ is the time constant for the exponential damping factor. Small values of τ give a higher weighting to recent values of the distance. If τ is infinite, all the past history of the simulation is given equal weight.

The time-averaged restraint method is quite complicated to implement, and some skill is required when choosing the most appropriate functional form and the damping constant. The data produced by the simulation must also be interpreted with care. The technique is only truly applicable where the conformations are relatively close together, so that interconversion between the different conformations can be achieved relatively easily. Nevertheless, the technique does provide a more accurate representation of the dynamics of the real system, and it does enable the conformation to fluctuate more. One drawback of any restraint method is that the additional penalty terms represent an unnatural perturbation of the forces within the molecule. When using 'static' restraints the size of the force constants for the restraint terms can be quite large, which can often cause the conformations to have rather high energies. Smaller force constants can often be used with time-averaged restraints, which means that the conformations generally have lower energies.

8.8 Which conformational search method should I use?

With such an array of methods for exploring conformational space, it can be difficult to decide which to choose. Each method has its own strengths and weaknesses. Systematic searches are subject to the effects of combinatorial explosion, and they are not naturally suited to molecules with rings. However, they do have a definite endpoint; when the search has finished one can be guaranteed to have found all conformations for a given dihedral increment. Random search methods can require long runs to ensure that the conformational space has been covered, and they can generate the same structure many times. Distance geometry is particularly useful when experimental information can be incorporated, as is restrained molecular dynamics.

A comparison of various methods for searching conformational space has been performed for cycloheptadecane ($C_{17}H_{34}$) [Saunders *et al.* 1990]. The methods compared were the systematic search, random search (both Cartesian and torsional), distance geometry and molecular dynamics. The number of unique minimum energy conformations found with each method within 3 kcal/mol of the global minimum after 30 days of computer processing were determined (the study was performed in 1990 on what would now be considered a very slow computer). The results are shown in Table 8.1.

Table 8.1 A comparison of five different conformational searching algorithms. [Data from Saunders *et al.* 1990.]

Method	Total unique conformers found after 30 days, processing
Systematic search	211
Random Cartesian search	222
Random dihedral search	249
Distance geometry	176
Molecular dynamics	169

Combining the results from all the different methods revealed a grand total of 262 conformations within 3 kcal/mol of the global energy minimum. No one method found all of them but the random dihedral search did give the best performance in this case. The global energy minimum would be expected to constitute only about 8% of the total population of conformational states for this molecule if it is assumed that the entropies of all the conformations are the same.

8.9 Structural databases

Experimental information about the structures of molecules can often be extremely useful for forming theories of conformational analysis and helping to predict the structures of molecules for which no experimental information is available. The most important technique currently available for determining the three-dimensional structure of molecules is X-ray crystallography. The international crystallographic community has established centres where crystallographic data is collected and then distributed in electronic form. Two particularly important databases for the molecular modeller are the Cambridge Structural Database (CSD) [Allen *et al.* 1979] which contains crystal structures of organic and organometallic molecules; and the Protein Databank (PDB) [Bernstein *et al.* 1977] which contains structures of proteins and some DNA fragments. Other databases are also available, such as the Inorganic Structural Database of inorganic compounds and complexes [Bergerhoff *et al.* 1983].

A database is of little use without software tools to search, extract and manipulate the data. A simple use of a database is for extracting information about a particular molecule or group of molecules. For example, one may wish to retrieve the crystal structure of ranitidine (Figure 8.25). The molecule(s) may be specified in a variety of ways, such as by name, molecular formula or by literature citation. The data may also be identified by creating a two-dimensional representation of the molecule (as in Figure 8.25) and using a *substructure search program* to search the database (in fact, the CSD contains two entries for ranitidine, one is

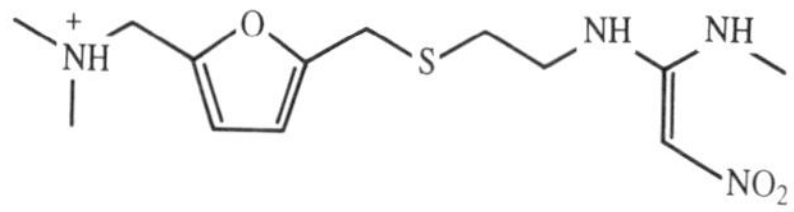

Fig. 8.25 Ranitidine.

the crystal structure of the hydrochloride salt and the other is the structure of the oxalate salt). Crystallographic databases have also been used to develop an understanding of the factors that influence the conformations of molecules, and of the ways in which molecules interact with each other. For example, the CSD has been comprehensively analysed to characterise how the lengths of chemical bonds depend upon the atomic numbers, hybridisations and environment of the atoms involved [Allen *et al.* 1987]. Analyses of intermolecular hydrogen bonding have revealed distinct distance and angular preferences [Murray-Rust and Glusker 1984; Glusker 1995]. A major use of the CSD is substructure searching for molecules which contain a particular fragment, in order to investigate the conformation(s) that the fragment adopts.

The protein database has provided much useful information about the structures that proteins adopt and the PDB has been extensively analysed to try and understand the principles that determine why a given amino acid sequence folds into one specific conformation. We shall discuss the use of molecular modelling methods to predict protein structure in more detail in section 8.13. Here we shall just mention one interesting way in which the information contained in the protein database has been put to a practical purpose. One of the steps in determining the structure of a protein by X-ray crystallography involves fitting the polypeptide chain to the electron density. This can be a complex and time-consuming task, even with today's sophisticated molecular graphics. Computer programs have been developed which extract the conformations of short polypeptide fragments (up to four amino acids long) from known X-ray structures [Jones and Thirup 1986]. These fragments are then used to generate a chain that fits the electron density. This method is feasible because a given segment of polypeptide chain often adopts a limited selection of conformations in protein structures, and a significant proportion of an unknown protein structure can often be constructed using such 'spare parts' taken from other proteins.

It should be remembered that a crystallographic database can only provide information about the crystal state of matter, and that the possible influence of crystal packing forces should always be taken into account. This is less of a concern for proteins than for 'small' molecules as protein crystals contain a large amount of water and indeed NMR studies have established that proteins have approximately the same structure in solution as in the crystal. A second, more subtle, bias is that crystallographic databases only contain molecules that can be crystallised and indeed only those molecules whose X-ray structures were considered important enough to be published. The structures in a crystallographic database may therefore not necessarily be a wholly representative set.

8.10 Molecular fitting

Fitting is the procedure whereby two or more conformations of the same or different molecules are oriented in space so that particular atoms or functional groups are optimally superimposed upon each other. Fitting methods are widely used in molecular modelling. For example, fitting is an integral part of many conformational search algorithms, particularly those that require each conformation to be compared with those generated previously in order to check for duplicates.

A molecular fitting algorithm requires a numerical measure of the 'difference' between two structures when they are positioned in space. The objective of the fitting procedure is to find the relative orientations of the molecules in which this function is minimised. The most common measure of the fit between two structures is the root-mean-square-distance between pairs of atoms, or RMSD:

$$\mathrm{RMSD} = \sqrt{\frac{\sum_{\mathrm{i}=1}^{N_{\mathrm{atoms}}} d_{\mathrm{i}}^2}{N_{\mathrm{atoms}}}} \tag{8.36}$$

N_{atoms} is the number of atoms over which the RMSD is measured and d_{i} is the distance between the coordinates of atom i in the two structures, when they are overlaid.

When fitting two structures, the aim is to find the relative orientations of the two molecules in which the RMSD is a minimum. Many methods have been devised to perform this seemingly innocuous calculation. Some algorithms, such as that described by Ferro and Hermans [Ferro and Hermans 1977] use an iterative procedure in which the one molecule is moved relative to the other, gradually reducing the RMSD. Other methods locate the best fit directly, such as Kabsch's algorithm [Kabsch 1978].

If the molecules are flexible then a better fit might be achieved if one or both of the molecules can change their conformation (for example, by rotating about single bonds). This is often referred to as *flexible fitting* or *template forcing*. In its simplest form flexible fitting is achieved by minimising the RMSD using a special minimisation algorithm that permits rotation about single bonds as well as translation and rotation in space. An alternative approach is to use restrained molecular dynamics which may enable a more thorough exploration of the conformational space in order to find the best fit. Here, restraints are placed on the distance between pairs of matched atoms which are incorporated into the energy function as additional penalty terms.

8.11 Clustering algorithms and pattern recognition techniques

Molecular modelling programs can generate a large amount of data which must often be processed and analysed. Many of the conformational search algorithms that we have considered can generate conformations that are very similar, if not

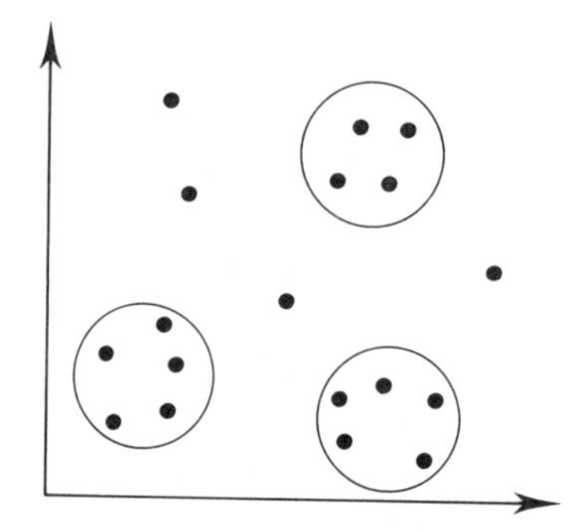

Fig. 8.26 The aim of cluster analysis is to group together 'similar' objects.

identical. Under such circumstances it is desirable to be able to select from the data set a smaller, 'representative' set of conformations for subsequent analysis. This can be done using cluster analysis, which groups together 'similar' objects, from which the representatives can be extracted, Figure 8.26.

There is no 'correct' method of performing cluster analysis and a large number of algorithms have been devised from which one must choose the most appropriate approach. There can also be a wide variation in the efficiency of the various cluster algorithms which may be an important consideration if the data set is large.

A cluster analysis requires a measure of the 'similarity' (or dissimilarity) between pairs of objects. When comparing conformations, the RMSD would be an obvious measure to use. Alternatively, the 'distance' between two conformations can be measured in terms of their torsion angles. Here, there may be more than one way in which the 'distance' can be calculated. The Euclidean distance between two conformations would be calculated using:

$$d_{\mathrm{ij}} = \sqrt{\sum_{m=1}^{N_{\mathrm{tor}}} (\omega_{m,\mathrm{i}} - \omega_{m,\mathrm{j}})} \tag{8.37}$$

$\omega_{m,\mathrm{i}}$ is the value of torsion angle m in conformation i. N_{tor} is the total number of torsion angles. An alternative is the Manhattan distance or city-block distance (Figure 8.27):

$$d_{\mathrm{ij}} = \sum_{m=1}^{n_{\mathrm{tor}}} |\omega_{m,\mathrm{i}} - \omega_{m,\mathrm{j}}| \tag{8.38}$$

When using torsion angles to calculate 'distances' between conformations it is important to remember that a torsion angle is a cyclic measure and that the difference should be measured along the shortest path, in either a clockwise or an anticlockwise direction. The clusters produced using the RMSD and the torsion angle measures may be very different. This is due to a 'leverage' effect when using torsion angles, which arises because small changes in the torsion angles in the middle of a molecule can give rise to large movements near the ends. The RMSD produces clusters in which the molecules have a similar shape.

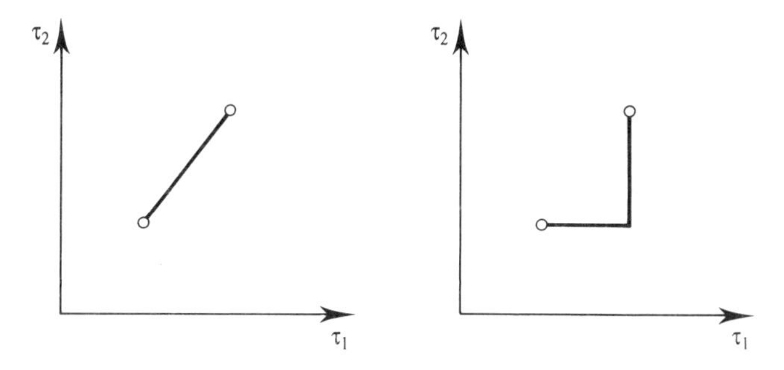

Fig. 8.27 Euclidean and 'city block' distance measures of torsional similarity.

One family of relatively straightforward clustering algorithms are the *linkage methods*. These algorithms first require the distance between each pair of conformations to be calculated. At the start of the cluster analysis the data set contains as many clusters as there are conformations; each cluster contains just a single conformation. At each step the total number of clusters is reduced by one by merging the 'closest' or 'most similar' pair of clusters into a single cluster. In the first step the closest two conformations are merged into a single cluster. In the next step, the closest two clusters are merged, and so on. Clustering continues until the distance between the closest pair of clusters exceeds a predetermined value, until the number of clusters falls below a specified maximum number, or until all the conformations have been merged into a single cluster. Such algorithms are referred to as *agglomerative* methods, in contrast to *divisive* clustering algorithms which start with a single cluster containing all of the data that is then partitioned into clusters. A representative conformation is then chosen from each cluster; for example, the conformation that is closest to the average structure of the cluster. The linkage methods differ in the way in which they calculate the distance between two clusters.

In the *single linkage* method the 'distance' between a pair of clusters is equal to the shortest distance between any two members, one from each cluster. The *complete linkage* method is the logical opposite of the single linkage method, in that it considers the furthest pair of objects in a pair of clusters. The *average linkage* method computes the average of the similarities between all pairs of objects in the two clusters. We can contrast these methods using the data shown in Figure 8.28, which were obtained by searching the Cambridge Structural Database for the ribose phosphate fragment also shown in Figure 8.28. A total of 44 molecules were found to contain this fragment. The values of the two torsion angles τ_1 and τ_2 indicated in Figure 8.28 were determined for each occurrence of the fragment (some molecules contained more than one representative of the fragment). For simplicity, the results for just eight fragments are plotted in Figure 8.28. The similarity matrix for these eight sets of two torsion angles, calculated using a Euclidean measure is given in Table 8.2.

All of the clustering methods first join the two structures that are 'closest' (conformations 3 and 4), to which is then added conformation 7. In the third step

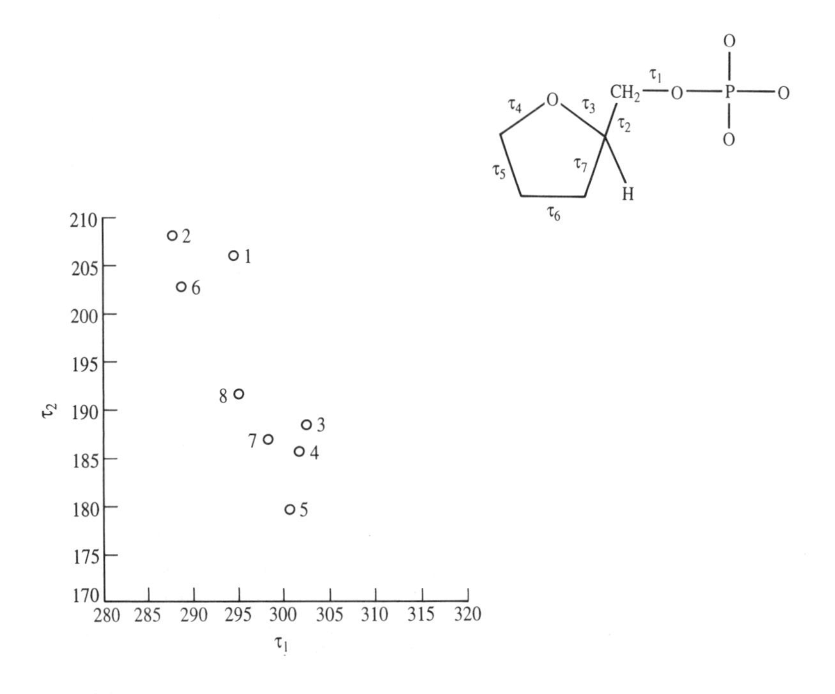

Fig. 8.28 Ribose phosphate fragment used to extract data from Cambridge Structural Database and eight sets of torsion angle values for τ_1 and τ_2.

conformations 2 and 6 are connected. In the fourth step the single linkage and complete linkage methods differ; the single linkage algorithm joins conformation 8 to the cluster 3–4–7 whereas the complete linkage method joins conformation 1 to the cluster 2–6. The order in which the points are clustered together for the three linkage methods is given in Table 8.3. As can be seen, the three linkage methods form the clusters in a similar but not identical order.

Table 8.2 Similarity matrix for eight ribose phosphate fragments.

	1	2	3	4	5	6	7	8
1	0.0	7.2	19.3	21.4	26.9	6.6	19.4	14.2
2	7.2	0.0	24.6	26.3	31.1	5.3	23.6	17.9
3	19.3	24.6	0.0	2.7	8.8	19.9	4.5	8.1
4	21.4	26.3	2.7	0.0	6.1	21.4	3.7	8.9
5	26.9	31.1	8.8	6.1	0.0	26.0	7.6	13.2
6	6.6	5.3	19.9	21.4	26.0	0.0	18.5	12.8
7	19.4	23.6	4.5	3.7	7.6	18.5	0.0	5.7
8	14.2	17.9	8.1	8.9	13.2	12.8	5.7	0.0

Table 8.3 A comparison of the single linkage, complete linkage and average linkage cluster methods using the data in Table 8.2. The figures in parentheses indicate the 'distance' between the clusters as they are formed.

Step number	Single linkage	Complete linkage	Average linkage
1	3–4 (2.7)	3–4 (2.7)	3–4 (2.7)
2	3–4–7 (3.7)	3–4–7 (4.5)	3–4–7 (4.1)
3	2–6 (5.3)	2–6 (5.3)	2–6 (5.3)
4	3–4–7–8 (5.7)	2–6–1 (7.2)	2–6–1 (6.9)
5	3–4–7–8–5 (6.1)	3–4–7–5 (8.8)	3–4–7–5 (7.5)
6	2–6–1 (6.6)	3–4–7–5–8 (13.2)	3–4–7–5–8 (9.0)
7	2–6–1–3–4–7–8–5 (12.8)	2–6–1–3–4–7–5–8 (31.1)	2–6–1–3–4–7–5–8 (21.3)

These three linkage methods are all *hierarchical* clustering methods, because there is a specific order in which the clusters are formed and amalgamated. These methods have the advantage of being simple to program, and they also produce a clustering that is independent of the order in which the objects are stored. However, they do suffer from some drawbacks. The need to calculate the $M \times M$ similarity matrix can severely limit their applicability when clustering large data sets. In addition, the commonly used single linkage method tends to produce long, elongated clusters. For these reasons, alternative clustering methods may be required.

A hierarchical clustering can be visually represented by constructing a *dendrogram* which indicates the relationship between the items in the data set. A sample dendrogram is shown in Figure 8.29 for the single linkage clustering described above. Along the x axis are represented the M individual objects. The y axis indi-

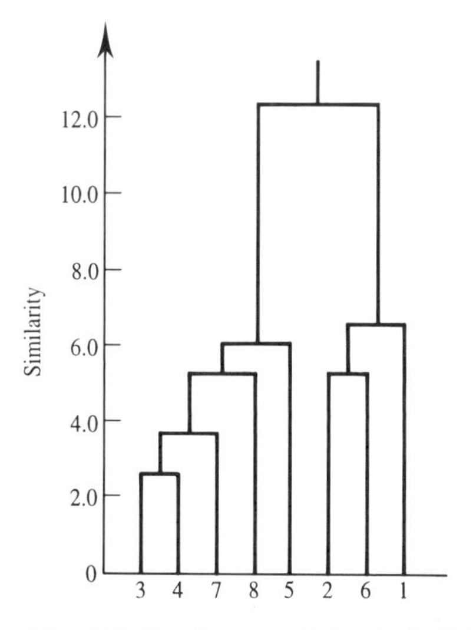

Fig. 8.29 Dendrogram of the single linkage data in Table 8.3

cates the intercluster distance. The dendrogram enables us to identify how many clusters there are at any stage, and what the members of those clusters are. A dendrogram can thus be a very useful way to show the underlying structure of the data and for suggesting the appropriate number of clusters to choose. A line drawn across the dendrogram enables one to read off how many clusters there are at that particular distance measure. For example, there are four clusters at a value of 6.0. For the data in Figure 8.28 it would probably be decided that the data fall into two clusters, one containing conformations 1, 2 and 6 and the other containing 3, 4, 5, 7, 8. As this example illustrates, deciding how many clusters there are can be somewhat subjective; a small threshold can lead to a large number of 'tight' clusters (and frequently many clusters with just one member) whereas a larger threshold can produce clusters that are spread out.

An example of a non-hierarchical clustering method is the Jarvis–Patrick algorithm [Jarvis and Patrick 1973]. The Jarvis–Patrick method uses a 'nearest neighbours' approach. The nearest neighbours of each conformation are the other conformations that are the shortest 'distance' away. Two conformations are considered to be in the same cluster in the Jarvis–Patrick method if they satisfy the following criteria:

1. They are in each other's list of m nearest neighbours.
2. They have p (where $p < m$) nearest neighbours in common.

Conformations can thus be placed in clusters and clusters fused together (because any two individual elements satisfy the two criteria) without any hierarchical relationships. The Jarvis–Patrick method can also be extended to take account not only of the number of nearest neighbours but also the position of each conformation within the neighbour list.

To illustrate the use of the Jarvis–Patrick method, let us consider the data in Figure 8.28 once more. The three nearest neighbours of each fragment are given in Table 8.4. Suppose we require that two out of the three nearest neighbours should be common. If we examine the pair 1, 2 we find that each neighbour list contains the other fragment and the remaining two nearest neighbours are the same (i.e. 6, 8). These objects would therefore be placed in the same cluster. However, fragments 2 and 6 would not be considered in the same cluster

Table 8.4 The three nearest neighbours of each fragment in Figure 8.28.

Fragment	Nearest neighbours
1	2, 6, 8
2	1, 6, 8
3	4, 7, 8
4	3, 5, 7
5	3, 4, 7
6	1, 2, 8
7	3, 4, 8
8	3, 4, 7

according to these criteria; although they are in each other's list they do not have two out of three nearest neighbours in common. One advantage of the Jarvis–Patrick algorithm is that it can be used to cluster very large data sets which cannot be considered using any of the hierarchical methods. A common use is in selecting a set of representative molecules from a large chemical database [Downs *et al.* 1994].

8.12 Reducing the dimensionality of a data set

The *dimensionality* of a data set is the number of variables that are used to describe each object. For example, a conformation of a cyclohexane ring might be described in terms of the six torsion angles in the ring. However, it is often found that there are significant correlations between these variables. Under such circumstances a cluster analysis is often facilitated by reducing the dimensionality of a data set to eliminate these correlations. *Principal components analysis* (PCA) is a commonly used method for reducing the dimensionality of a data set.

8.12.1 Principal components analysis

Consider the data shown in Figure 8.30. It is easy to see that there is a high degree of correlation between the x and the y values. If we were to define a new variable, $z = x + y$, then we could express most of the variation in the data as the values of this new variable z. The new variable is called a *principal component.* In general, a principal component is a linear combination of the variables:

$$p_i = \sum_{j=1}^{v} c_{i,j} x_j \tag{8.39}$$

p_i is the ith principal component and $c_{i,j}$ is the coefficient of the variable x_j. There are v such variables. The first principal component of a data set corresponds to

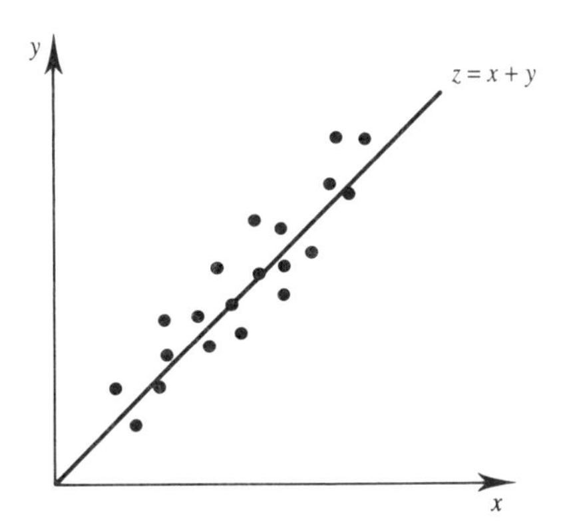

Fig. 8.30 Most of the variance in this set of highly correlated data values can be explained in terms of a new variable $z = x + y$.

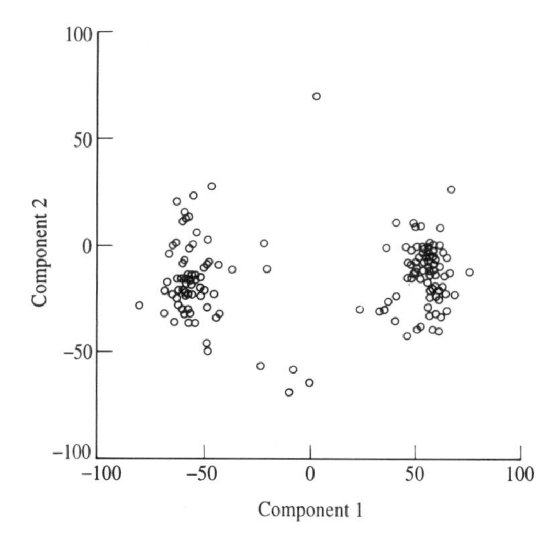

Fig. 8.31 Scatterplot of the first two principal components for the ring torsion angles τ_3–τ_7.

that linear combination of the variables which gives the 'best fit' straight line through the data when they are plotted in the v-dimensional space. More specifically, the first principal component maximises the *variance* in the data so that the data have their greatest 'spread' of data values along the first principal component. This is clear in the two-dimensional example shown in Figure 8.30. The second and subsequent principal components account for the maximum variance in the data not already accounted for by previous principal components. Each principal component corresponds to an axis in a v-dimensional space, and each principal component is orthogonal to all the other principal components. There can clearly be as many principal components as there are dimensions in the original data, and indeed in order to explain all of the variation in the data it is usually necessary to include all the principal components. However, in many cases only a few principal components may be required to explain a significant proportion of the variation in the data. If only one or two principal components can explain most of the data then a graphical representation is possible.

The principal components are calculated using standard matrix techniques [Chatfield and Collins 1980]. The first step is to calculate the variance–covariance matrix. If there are s observations, each of which contains v values, then the data set can be represented as a matrix $\mathbf{D}$ with v rows and s columns. The variance–covariance matrix $\mathbf{Z}$ is:

$$\mathbf{Z} = \mathbf{D}^{\mathrm{T}}\mathbf{D} \tag{8.40}$$

The eigenvectors of $\mathbf{Z}$ are the coefficients of the principal components. As $\mathbf{Z}$ is a square symmetric matrix its eigenvectors will be orthogonal (provided there are no degenerate eigenvalues). The eigenvalues and their associated eigenvectors can be obtained by solving the secular equation $|\mathbf{Z} - \lambda\mathbf{I}| = 0$ or by matrix diagonalisation. The first principal component corresponds to the largest eigenvalue, the second principal component to the second largest eigenvalue, and so on. The ith principal component accounts for a proportion $\lambda_i/\sum_{j=1}^{v} \lambda_j$ of the total variance in the data. The first m principal components therefore account for $\sum_{j=1}^{m} \lambda_j / \sum_{j=1}^{v} \lambda_j$ of the total variation in the data.

As an example of the application of principal components analysis, we shall consider the conformations adopted by the five-membered ribose ring in our set of conformations extracted from the Cambridge Structural Database. The conformation of a five-membered ring can be described in terms of five torsion angles ($\tau_3 \ldots \tau_7$ in Figure 8.28). As we cannot visualise points in a five-dimensional space it would clearly be useful to reduce the dimensionality of the data set. When a principal components analysis is performed on the data, the following results are obtained:

Principal component	*Proportion of variance explained*	$c(\tau_3)$	$c(\tau_4)$	$c(\tau_5)$	$c(\tau_6)$	$c(\tau_7)$
1	85.9%	−0.14	−0.26	0.55	−0.61	0.48
2	14.0%	−0.63	0.59	−0.31	−0.06	0.41
3	0.0002%	−0.19	0.50	0.65	−0.004	−0.53
4	0.0001%	−0.47	−0.38	0.12	0.71	0.19
5	0.0001%	0.58	0.43	0.28	0.35	0.53

It can thus be seen that most of the variation in the data (85.9%) is explained by the first principal component, with all but a fraction being explained by the first two components. These two principal components can be plotted as a scatter graph, as shown in Figure 8.31, suggesting that there does indeed seem to be some clustering of the conformations of the five-membered ring in this particular data set.

8.13 The role of conformational analysis in predicting the structures of peptides and proteins

Peptides and proteins are constructed from sequences of amino acids. There are 20 common naturally occurring amino acids, shown in Figure 8.32. The amino acids are linked together via amide bonds to give a polypeptide chain. All the naturally occurring amino acids have the same relative stereochemistry at the alpha-carbon. The side chains have different sizes, shapes, hydrogen bonding capabilities and charge distributions which enable proteins to display a vast array of functions.

The biological function of a protein or peptide is often intimately dependent upon the conformation(s) that the molecule can adopt. X-ray crystallography and NMR are the two methods used to provide detailed information about protein structures. Unfortunately, the rate at which new protein sequences are determined

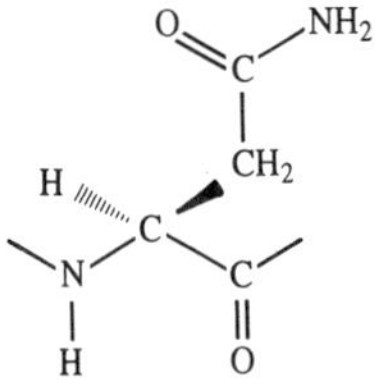

Asparagine
Asn
N

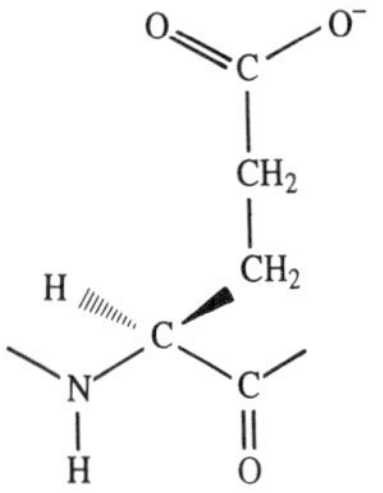

Glutamine
Gln
G

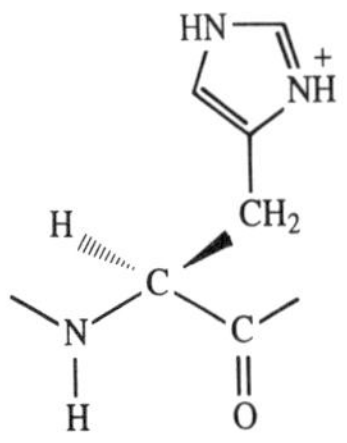

Histidine
His
H

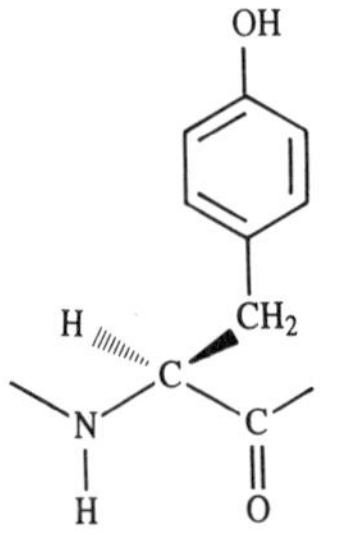

Tyrosine
Tyr
Y

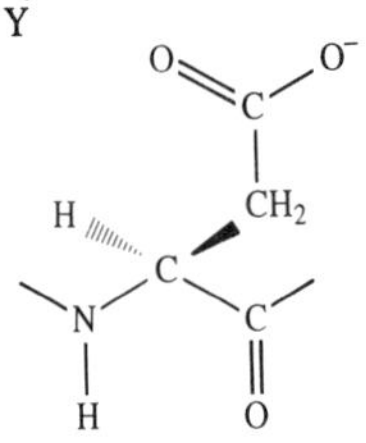

Aspartic
Asp
D

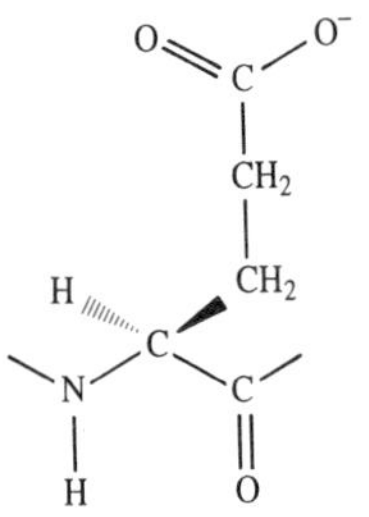

Glutamic acid
Glu
E

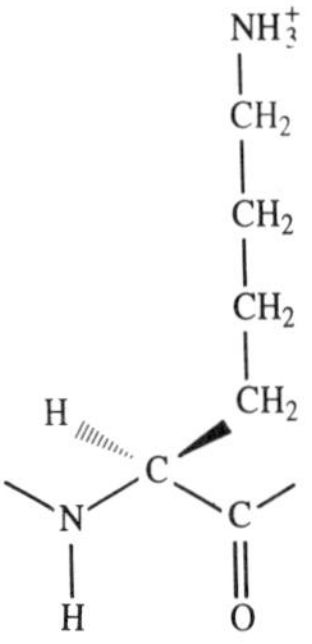

Lysine
Lys
K

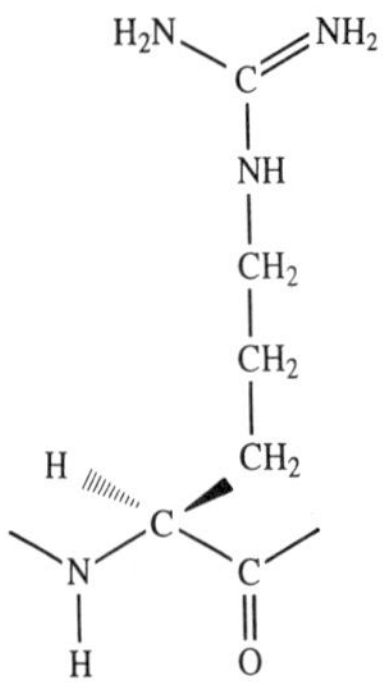

Arginine
Arg
R

Fig. 8.32 The 20 naturally occurring amino acids and their codes.

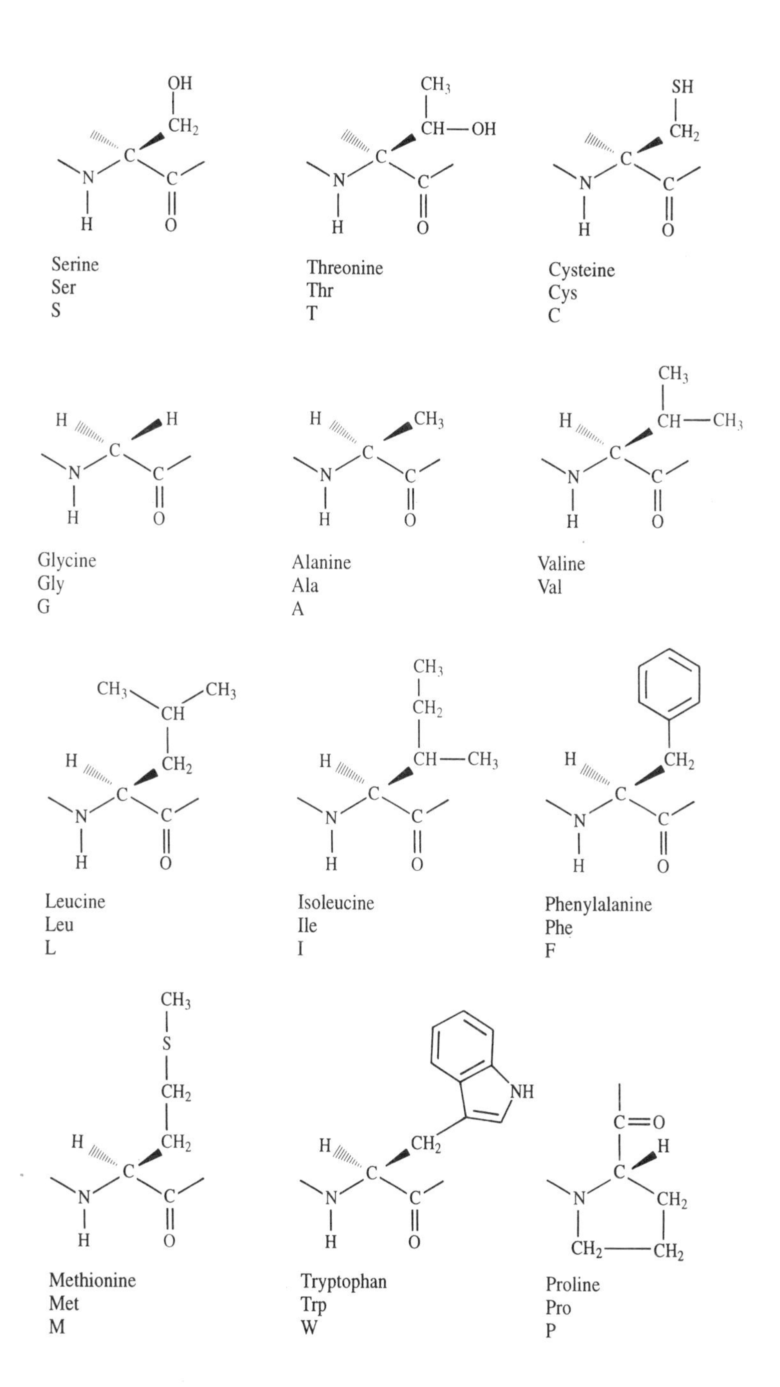

Fig. 8.32 Continued.

far exceeds the rate at which protein structures are determined experimentally. There is thus considerable interest in theoretical methods for predicting the three-dimensional structure of proteins from the amino acid sequence: the *protein folding problem.*

8.13.1 Some basic principles of protein structure

The first X-ray structures revealed that proteins did not adopt regular or symmetrical structures, but appeared to be highly irregular. However, certain structural motifs were observed to occur frequently. The most common motifs are the α-helix and the β-strand, shown in Figure 8.33. Together these constitute the *secondary structure* of a protein. Linus Pauling had predicted that the α-helix would

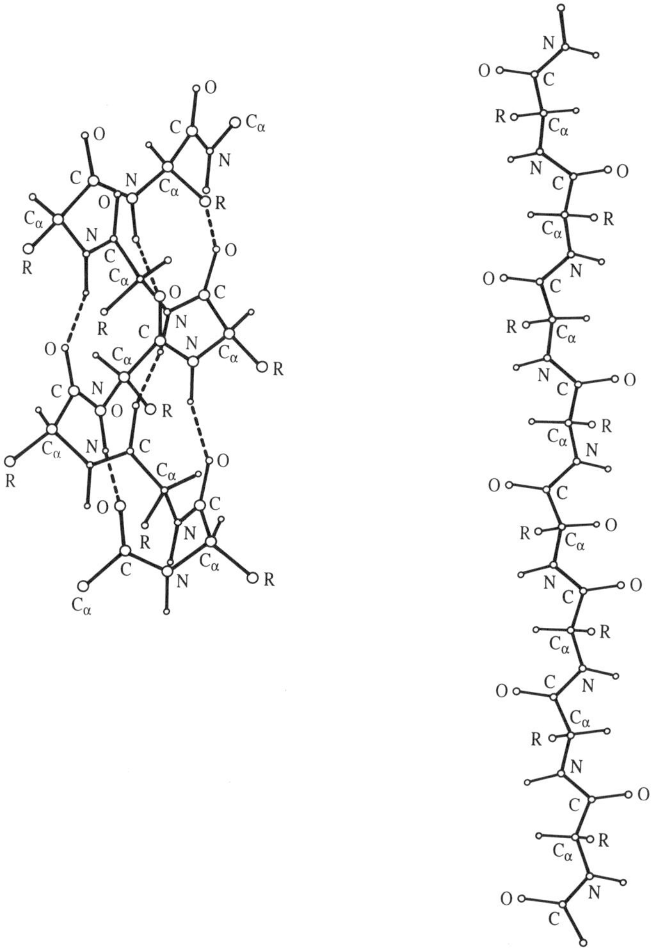

Fig. 8.33 The α-helix and β-strand structures.

be a stable element of polypeptide structure well before the first protein structure was solved [Pauling *et al.* 1951]. His prediction was based upon mechanical models constructed after a careful analysis of the geometry of the peptide unit in crystal structures of small molecules and can be considered a classic example of the predictive power of molecular modelling. Another type of helix, the 3_{10} helix is also found in relatively small amounts. The β-strands often form extended structures called β-sheets in which the strands are hydrogen bonded to each other. In a β-sheet the strands can run in either parallel or anti-parallel directions, as shown in Figure 8.34. Secondary structural elements are connected by regions often referred to as 'loops' that adopt less regular structures. Nevertheless, common conformations can be identified in certain types of loop structure, such

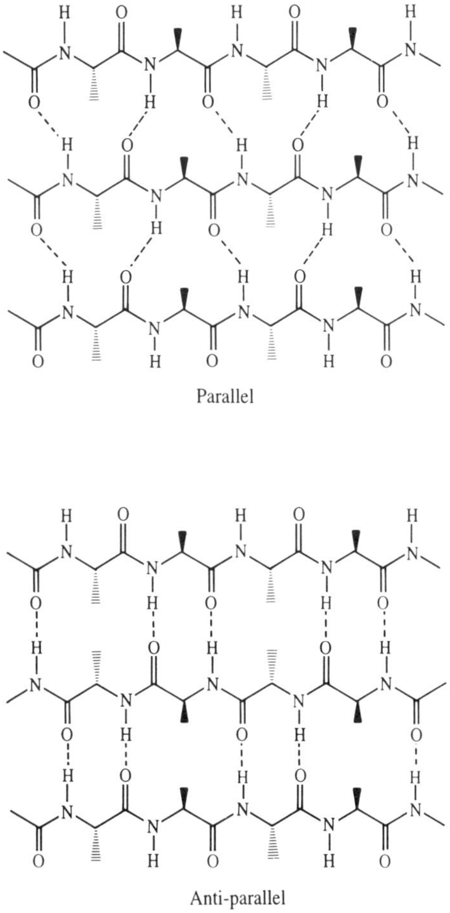

Parallel

Anti-parallel

Fig. 8.34 The formation of parallel and anti-parallel β-sheets.

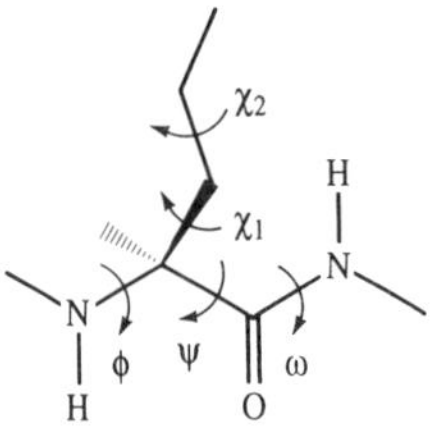

χ_2 χ_1 H N N H ϕ ψ O ω

Fig. 8.35 The torsion angles that define the conformation of an amino acid.

as conformations that are commonly adopted by the 'β-turn' regions between β-strands, Figure 8.35.

If we ignore the small variations in bond angles and bond lengths the conformation of an amino acid residue in a protein or peptide can be classified according to the torsion angles of its rotatable bonds. There are three backbone torsion angles, labelled ϕ, ψ and ω (Figure 8.35). The conformations of the side chains are characterised by the torsion angles χ_1, χ_2, etc. The amide bond has a relatively high energy barrier for rotation away from planarity and so ω rarely deviates significantly from 0° or 180°. Moreover, there is a significant preference for the *trans* ($\omega = 180°$) conformation (except for proline, which typically shows a relatively high proportion of *cis*-peptide linkages). We have already noted the contribution of Ramachandran to our understanding of protein structure. Examination of X-ray structures of proteins shows that most amino acids occupy one of the low-energy regions in the Ramachandran contour map. Indeed, it is now common practice when assessing an X-ray or NMR structure determination to construct its Ramachandran map, and to examine closely any residues which adopt a conformation outside the preferred regions. The side chains also tend to adopt preferred conformations though there were many examples of unusual or higher-energy structures [Ponder and Richards 1987]. Subsequent investigations have revealed that the side-chain conformations are often correlated with the backbone structure; for certain conformations of the backbone only particular side-chain structures are possible [Summers *et al.* 1987].

8.13.2 The hydrophobic effect

Water soluble globular proteins usually have an interior composed almost entirely of non-polar, hydrophobic amino acids such as phenylalanine, tryptophan, valine and leucine with polar amino acids such as lysine and arginine located on the surface of the molecule. This packing of hydrophobic residues is a consequence of the *hydrophobic effect* which is the most important factor that contributes to protein stability. The molecular basis for the hydrophobic effect continues to be the subject of some debate, but is generally considered to be entropic in origin. Moreover, it is the entropy change of the solvent that is important. The contribution to the overall free energy of folding due to the packing of the non-polar amino acids is positive (i.e. unfavourable) on both enthalpic and entropic grounds. The enthalpy change to remove the non-polar amino acids from water is positive due to dipole/induced-dipole electrostatic interactions between the polar water molecules and the hydrocarbon side chains. When packed, there are only (weaker) dispersion interactions between the side chains. The entropy change associated with packing the amino acids is negative because the unfolded state is less ordered than the packed state (for example, more conformational degrees of freedom are accessible). Water molecules form a cage-like structure around a non-polar solute, which has been likened to a local 'iceberg'. The water molecules in this region are locally ordered with most of the hydrogen bonding network of pure water intact. The area of the non-polar interface is much larger for the unfolded protein (Figure 8.36) and so the entropy change of the water when the protein folds is large. The enthalpy change of the water is negative as the disruption to the hydrogen bonding network is less for the folded protein. Of these four contributions, the two enthalpy terms are believed to be small, with the entropy change associated with the ordering of the solvent water molecules being the dominant term.

Not all proteins are water soluble; a very important class is the membrane-bound proteins, which include receptors and ion channels. The arrangement of the amino acids in these proteins is very different in the membrane-spanning regions. The membrane provides a very hydrophobic environment and so hydrophobic residues are often located on the outside, towards the membrane. It is very difficult to obtain X-ray crystal structures of membrane-bound proteins due to the problems of obtaining satisfactory crystals. The crystal structure of the photosynthetic reaction centre, which earned Michel, Deisenhofer and Huber the Nobel prize in 1988, was obtained after much painstaking work in which the protein was crystallised from a detergent solution. Electron microscopy has been used to determine the structures of membrane-bound proteins; in favourable cases the resolution of this technique approaches that of X-ray crystallography but is usually much lower. Henderson and Unwin have pioneered the application of this method to membrane proteins with their determination of the structures of bacteriorhodospin and rhodopsin [Henderson *et al.* 1990; Havelka *et al.* 1995]. Both of these proteins contain seven *trans*-membrane helices which are connected by loops in the extracellular and intracellular regions.

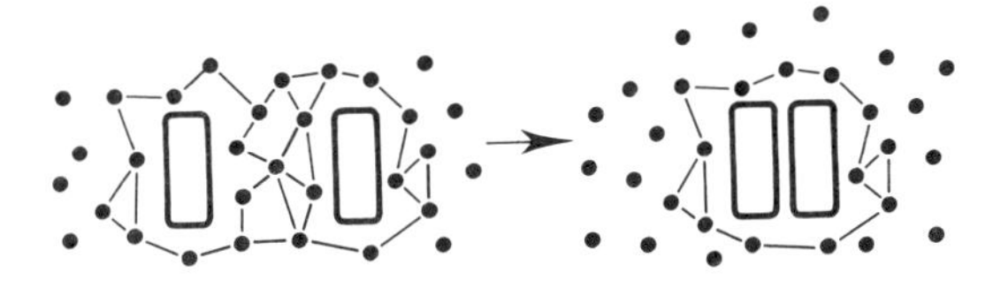

Fig. 8.36 The hydrophobic effect. Water molecules around a non-polar solute form a cage-like structure which reduces the entropy. When two non-polar groups associate, water molecules are liberated increasing the entropy.

No solution has yet been found to the protein folding problem, but a variety of promising approaches have been developed. In the next sections we shall consider some of these methods for predicting the structures of proteins and peptides. Our discussion will first consider methods that attempt to predict the structures of proteins from first principles. We will then discuss methods that use a stepwise approach, in which elements of secondary structure are first identified and then these elements are packed together. Finally, we will consider the prediction of protein structures by homology modelling (sometimes referred to as comparative modelling), where the structure of the unknown protein is based upon the known structure of a related (i.e. homologous) protein.

8.13.3 First-principles methods for predicting protein structure

The most ambitious approaches to the protein folding problem attempt to solve it from first principles (*ab initio*). As such, the problem is to explore the conformational space of the molecule in order to identify the most appropriate structure. The total number of possible conformations is invariably very large and so it is usual to try and find only the very lowest energy structure(s). Some form of empirical force field is usually used, often augmented with a solvation term (see section 9.8). The global minimum in the energy function is assumed to correspond to the naturally occurring structure of the molecule.

All of the conformational search methods that were described in sections 8.2–8.7 have been used at some stage to explore the conformational space of small peptides. Here we will describe some of the methods designed specifically for tackling the problem for peptides and proteins.

H A Scheraga has devised many novel methods with his colleagues for exploring the conformational space of peptides and proteins. Each new method is rigorously tested using a standard test molecule, met-enkephalin (H–Tyr–Gly–Phy–Met–OH). One method is the 'build-up' approach, in which the peptide is constructed from three-dimensional amino acid templates [Gibson and Scheraga 1987]. Each template corresponds to a low-energy region of the Ramachandran map. To explore the conformational space of a peptide, a dipeptide fragment is first constructed by joining together all possible pairs of templates available to the first two amino acids. Each dipeptide fragment is minimised and the lowest energy structures are retained for the next step, in which the third amino acid is connected. The peptide is gradually built up in this way, with energy minimisation and selection of the lowest energy structures at each stage.

The simplest of Scheraga's random search methods is a random dihedral search, in which a single dihedral is selected at each iteration and randomly rotated [Li and Scheraga 1987]. The resulting structure is minimised and then accepted or rejected according to the Metropolis criterion. The electrostatically driven Monte Carlo method [Ripoll and Scheraga 1988; Ripoll and Scheraga 1989] is a more complex random search method which recognises the importance of long-range electrostatic interactions in polypeptides and proteins. It is based upon the observation that the local dipoles of amide units often adopt a favourable alignment in the electrostatic field of the protein. Two different types of move are thus used in the scheme. In the first type of move, an amide unit is randomly selected and the backbone (ϕ, ψ) torsion angles are changed to enable its dipole to be optimally aligned in its local electrostatic field. The resulting conformation is minimised and then accepted or rejected according to the Metropolis criterion. The second type of move involves a random change to a randomly selected dihedral followed by minimisation and acceptance or rejection in the usual way. This approach thus combines moves designed to optimise the long-range electrostatic interactions with moves that have a more local influence on the conformation.

Proteins have many more rotatable bonds than peptides, and so it is common to use some form of simplified model of the protein to make the problem tractable. The energy surface of a model with fewer degrees of freedom should have a smaller number of minima than the energy surface of a more detailed model; it must be assumed that the energy surface of the simplified model reproduces the general features of the more detailed representation, but without the fine structure (Figure 8.37). Various simplified models have been developed for investigating the conformational space of proteins. Many of these models are analogous to the models used to perform Monte Carlo simulations of polymers, such as the lattice and 'bead' models.

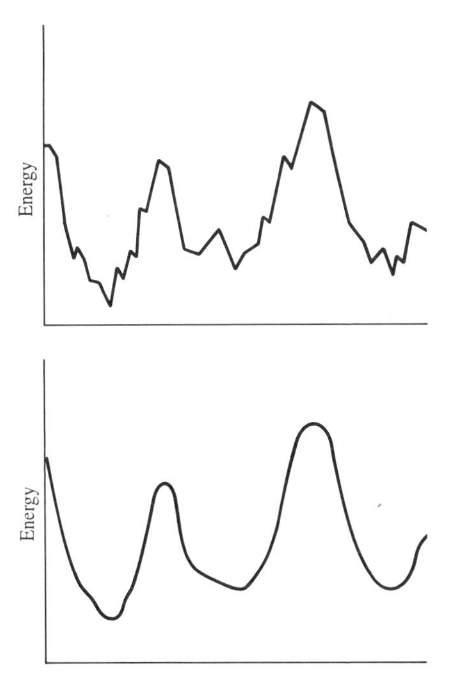

Fig. 8.37 Schematic energy surfaces for 'all-atom' (top) and simplified (bottom) models.

8.13.4 Lattice models for investigating protein structure

An advantage of the simple lattice models is that they can be used to try and answer some of the fundamental questions about protein structure. For example, it may be possible to enumerate all possible conformations for a chain of a given length on the lattice. From this set of states statistical mechanics can be used to derive thermodynamic properties and to investigate the relationship between the structure and the sequence. For example, in the 'HP' model of Chan and Dill [Chan and Dill 1993] a protein is modelled as a sequence of hydrophobic (H) and hydrophilic (P) monomers. The sequence is grown onto a two-dimensional lattice using a self-avoiding walk and the energy of the resulting conformation is calculated by summing interactions between pairs of monomers that occupy adjacent lattice sites but are not covalently bonded (Figure 8.38). In the model of Chan and Dill the interaction between two hydrophobic monomers was favoured by a constant energy increment with all other interaction energies set to zero. Exhaustive enumeration of all the conformations was possible for chains of 30 or so monomers and the global energy minimum for each chain was determined. Several interesting features arose from studies of this model. When the H–H interaction energy is small a large number of conformations are accessible. As the hydrophobic interaction energy increases there is a sharp decrease in the number of compact conformations containing hydrophobic cores. Another interesting feature of this and similar models is that α-helices and β-sheets naturally arise in the compact cores of such models. This suggests that the formation of secondary structure in a protein is not driven by specific hydrogen bonding interactions between amino acids, but rather by the compact nature of the core; conformations other than helices or sheets are not viable.

The simple lattice models are intended to address general questions about protein folding and structure. More sophisticated lattice models have been designed which are used to predict the actual structures of specific proteins. With such methods it is usually not possible to exhaustively explore the conformational space even on the lattice and so methods such as Monte Carlo simulated

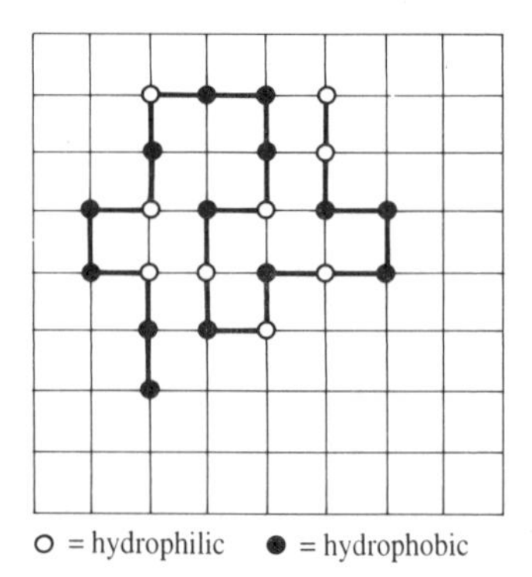

Fig. 8.38 The HP model of Chan and Dill.

annealing are used to generate low-energy structure(s). Skolnick has developed several lattice models, which are used in a three-stage procedure to construct a model of the protein [Godzik *et al.* 1993]. In the first stage a 'coarse' lattice is used, in which five different types of move are permitted, excluded volume effects due to side chain packing are taken into account and the interaction energy model contains a total of seven terms. A set of low-energy structures are obtained using Monte Carlo simulated annealing which are then refined using a finer lattice model. This second model is closer to the actual structures of proteins and uses a more accurate representation of the side chains. The structures obtained with the finer lattice typically show RMS deviations between 2 Å and 4 Å when compared to the experimentally determined structures. The conformations obtained from the finer lattice model may then be converted to a full atomic model for refinement using a technique such as energy minimisation with a standard force field.

Some simplified models of proteins represent each amino acid residue as one or more 'pseudo-atoms', according to the size and chemical nature of the amino acid. These models are analogous to the 'bead' polymer models and full conformational freedom is possible. All of the standard types of calculation can be performed with these simplified models, encompassing techniques such as energy minimisation and molecular dynamics. An empirical model is used to calculate the residue–residue interaction energies. The parameters for these empirical models can be derived in a variety of ways. One option is to parametrise the simple model to reproduce the results of a more detailed, all-atom model. An early attempt to develop such a representation was made by Levitt [Levitt 1976] who used energy minimisation to predict the structures of small proteins. In this model the interaction between each pair of residues is equal to the average of the calculated interaction over all spatial orientations of the two residues. Minimisation of a polypeptide chain from an initial open structure resulted in compact conformations with the same size and shape of experimentally determined protein structures, together with features such as secondary structure and β-turns. Some of Levitt's observations are still very pertinent. In particular, he notes that the 'wrong' structure may still have a lower energy (as predicted by the energy function) than the 'correct' structure; this is also found to be the case even with more complex molecular mechanics potential functions [Novotny *et al.* 1988].

To summarise, first-principles methods have been successfully used to predict the naturally occurring conformations of small peptides, but are not yet sufficiently reliable to predict accurately the structures of proteins, though in some cases the general fold of the molecule is quite similar to the native structure.

8.13.5 Rule-based approaches

Most protein structures contain a significant amount of secondary structure (α-helices and β-strands). An obvious way to tackle the problem of predicting a protein's three-dimensional structure is to first determine which stretches of amino acids should adopt each type of secondary structure, and then pack these secondary structural elements together.

The first step in this procedure requires the secondary structural elements to be predicted. In other words, each amino acid must be assigned to one of three classes: α-helix, β-strand or coil (i.e. neither helix nor strand). Some approaches also predict whether an amino acid is present in a turn structure. One of the most widely used methods for secondary structure prediction was devised by Chou and Fasman [Chou and Fasman 1978]. Theirs is a statistical method, based upon the observed propensity of each of the 20 amino acids to exist as α-helix, β-strand and coil. These propensities were originally determined by analysing 15 protein X-ray structures. The fractional occurrence of each residue in each of these three states was calculated, as was the fractional occurrence of the residue over all 15 structures. The propensity of that residue for a given type of secondary structure then equals the ratio of these two values. Each residue was also classified according to its propensity to act as an 'initiator' or as a 'breaker' for α-helices and β-strands. To predict the secondary structure, the amino acid sequence is searched for potential α-helix or β-strand initiating residues. The helix or strand is then extended so long as the average propensity value for a window of five or six residues exceeded a threshold value. β-turns are also predicted using a statistical measure of the propensity of an amino acid to exist in such structures.

Many other methods have been proposed for predicting the secondary structure of a protein from its sequence, including approaches based upon information theory and neural networks [Ning and Sejnowski 1988]. However, the performance of even the best methods is barely more than 65–70% (a success rate of 33% would be achieved purely by chance, if helix, sheet and coil structures were present in equal amounts). The rather disappointing performance of these methods can be ascribed to the fact that these prediction methods only consider local interactions; they neglect interactions between amino acids that are far apart in the sequence but close in three-dimensional space.

Having predicted the secondary structural elements, it is then necessary to determine how they could pack together in order to achieve a low-energy structure. Cohen, Sternberg and Taylor analysed the packing of α-helices and β-sheets in a number of proteins and deduced a series of rules that could be used to derive favourable packing arrangements [Cohen *et al.* 1982]. For example, from an analysis of 18 protein structures they observed that an α-helix usually packs against a β-sheet in a parallel arrangement involving two rows of non-polar residues on the helix. These rules were then used to pack the α-helices and β-sheets into a stable core structure [Sternberg *et al.* 1982]. The number of possible packing arrangements was usually very large, but this number could be drastically reduced using two simple filters. First, there had to be a sufficient number of residues between sequential elements of secondary structure to span the distance between them, and secondly there should be no unfavourable interactions between the helices and sheets in the packed structure. Having generated one or more approximate structures the model was submitted to an energy refinement.

This rule-based approach to protein structure prediction is obviously heavily reliant on the quality of the initial secondary structure prediction, which is unfortunately not particularly accurate. The method tends to work best if it is known to which structural class the protein belongs, as can sometimes be deduced from experimental techniques such as circular dichroism. For example, some proteins only contain α-helices, which obviously makes the problem of predicting the secondary structure considerably easier. The rule-based methods have been used to predict the structures of various proteins, with varying success.

8.13.6 Homology and comparative modelling methods

There are striking similarities between the three-dimensional structures of some proteins. For example, the three-dimensional structures of trypsin, chymotrypsin and thrombin are shown in Figure 8.39 (colour plate section), from which it is obvious that they adopt very similar conformations. These proteins are all members of the trypsin-like serine protease family of enzymes but it is also possible for biologically unrelated proteins to show significant structural similarity. For example, many proteins have a structure consisting of eight twisted parallel β-strands arranged in a barrel-like structure with the β-strands connected by α-helices (Figure 8.40). This structure is often referred to as a 'TIM barrel' after triosephosphate isomerase which was the first protein with this framework to have its structure determined by X-ray crystallography.

Homology modelling exploits the structural similarities between proteins by constructing a three-dimensional structure using the known structure of another protein as a template. To do this, it is first necessary to decide which protein to use as the template, and then to decide how to match the amino acids in the unknown structure with the amino acids in the known structure.

If the biological function of the protein is known it is often relatively straightforward to decide which protein(s) one might wish to consider as templates from

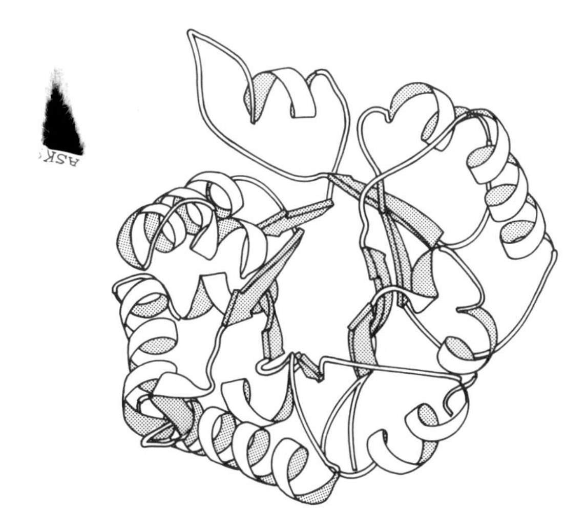

Fig. 8.40 The 'TIM barrel' [Noble *et al.* 1991].

which to build the model. In other cases the function of the protein may not be known, but it may be possible to deduce to which family it belongs by searching a sequence database for the presence of particular combinations of amino acids (called *motifs*) that often imply a particular function or structural feature. In other cases, the template is chosen as the protein whose sequence is the closest match for the unknown protein. For this reason, the term 'comparative' modelling is now preferred to 'homology' modelling; the latter implies some similarity of function between the unknown protein and the template, but this may not necessarily be the case.

8.13.7 Aligning protein sequences

When the three-dimensional structures of trypsin, chymotrypsin and thrombin are overlaid then identical amino acids are found at many positions in space, including the active site serine, histidine and aspartic acid residues, as shown in Figure 8.41 (colour plate section). If we compare the amino acid sequences of trypsin, chymotrypsin and thrombin then we find many striking similarities. A part of this *sequence alignment* is shown in Figure 8.42, written using the one-letter codes for the amino acids.

The objective of a sequence alignment is to position the amino acid sequences so that stretches of amino acids are matched with the expectation that these correspond to common structural features (such as the secondary structure and catalytic residues). Gaps in the aligned sequences correspond to regions where polypeptide loops are deleted or inserted.

If we know the sequences of other proteins in the same family then it is usually preferable to create a *multiple sequence alignment* where more than two sequences are matched. A multiple sequence alignment is often more reliable than a pairwise alignment between the known protein and the template as it is easier to detect any clear trends; with just two sequences it is easy to be misled by some chance correspondence.

Trypsin	SQWVVSAA**H**C	....	YKSGIQVRLG	EDNINVVEGN	E.QFIS[illegible]S
Chymotrypsin	EDWVVTAA**H**C	....	GVTTSDVVVA	GEFDQGLETE	DTQVLKIGKV
Thrombin	DRWVLTAA**H**C	LLYPPWDKNF	TVDDLLVRIG	KHSRTRYERK	VEKISMLDKI
Trypsin	IVHPSYN.SN	TLNN**D**IMLIK	LKSAASLNSR	VASISLP...	TSCA..SAGT
Chymotrypsin	FKNPKFS.IL	TVRN**D**ITLLK	LATPAQFSET	VSAVCLP...	SADEDFPAGM
Thrombin	YIHPRYNWKE	NLDR**D**IALLK	LKRPIELSDY	IHPVCLPDKQ	TAAKLLHAGF
Trypsin	QCLISGWGN.	....TKSSGT	SYPDVLKCLK	APILSDSSCK	SAYPGQITSN
Chymotrypsin	LCATTGWGK.	....TKYNAL	KTPDKLQQAT	LPIVSNTDCR	KYWGSRVTDV
Thrombin	KGRVTGWGNR	RETWTTSVAE	VQPSVLQVVN	LPLVERPVCK	ASTRIRITDN
Trypsin	MFCAGYLEGG	...KDSCQGD	**S**GGPVV..CS	GK....LQGI	VSWGSGCAQK
Chymotrypsin	MICAG..ASG	...VSSCMGD	**S**GGPLV..CQ	KNGAWTLAGI	VSWGSSTCST
Thrombin	MFCAGYKPGE	GKRGDACEGD	**S**GGPFVMKSP	YNNRWYQMGI	VSWGEGCDRD

Fig. 8.42 Sequence alignment of trypsin, chymotrypsin and thrombin (bovine). The active sites histidine, aspartic acid and serine are highlighted.

How is a sequence alignment obtained? We first consider the problem of aligning just two sequences. Any alignment algorithm requires some method for 'scoring' an arbitrary alignment of the two sequences. The objective is to find the alignment that gives the 'best' score. The simplest scoring metric is the *sequence identity* which gives the percentage of amino acids that are the same in the two sequences; identical pairs score 1 and all others score 0. An alternative approach is to recognise that topologically equivalent residues in two structurally homologous proteins may not be identical, but may have very similar shape, electronic, hydrogen bonding and hydrophobic properties. Such 'conservative' substitutions can often be made with little disruption to the three-dimensional structure of the protein and so it is desirable to be able to take this into account in the scoring scheme. For example, in the alignment of the serine proteases position 54 is restricted to either threonine or serine due to the need to form a hydrogen bond to position 43. Dayhoff and co-workers have analysed substitution frequencies in aligned sequences and have published a series of tables which give the probability of mutating one amino acid to another. These probabilities are usually stored as 20×20 matrices known as PAM matrices (PAM stands for percentage of acceptable point mutations per 10^8 years) that can be used as part of a more sophisticated scoring scheme.

A variety of methods have been developed to find the best way to align protein sequences. Manual alignment is generally not feasible except for the simplest cases; nor is it advisable to use an automatic alignment without checking it carefully as the results of an automatic alignment program can often be improved by some manual intervention. An algorithm developed by Needleman and Wunsch is widely used for aligning pairs of sequences; this algorithm guarantees to find the optimal alignment based upon the scoring matrix used [Needleman and Wunsch 1970]. The Needleman–Wunsch algorithm uses *dynamic programming*. A matrix $\mathbf{M}$ is constructed with R rows and C columns, to represent the R amino acids of one protein and the C amino acids of the other protein. Each element M_{ij} of this matrix is initially assigned a value that represents the degree to which the two two amino acids i and j are considered to be similar. Thus if sequence identity is being used as the scoring method then the element M_{ij} will be equal to one if the amino acids i and j are identical, and zero otherwise. The example originally used by Needleman and Wunsch to illustrate their method is shown in Figure 8.43 for two stretches of polypeptide (A and B) which have different numbers of amino acids (there are 12 residues in A and 13 in B). Thus the element $M_{7,8}$ equals 1 because residue 7 in sequence A is a cysteine (C) as is residue 8 from sequence B. An alignment can be obtained by tracing a pathway through this matrix as shown. The objective is to determine the alignment (or alignments) that provides the maximum score.

The dynamic programming method identifies the optimal match for the two sequences by calculating the optimal match for pairs of subsequences. For example, consider the matrix element $M_{10,11}$. We need to determine the maximum score that could be achieved if residue 10 from sequence A is matched with residue 11 from sequence B. We thus consider residues 10–12

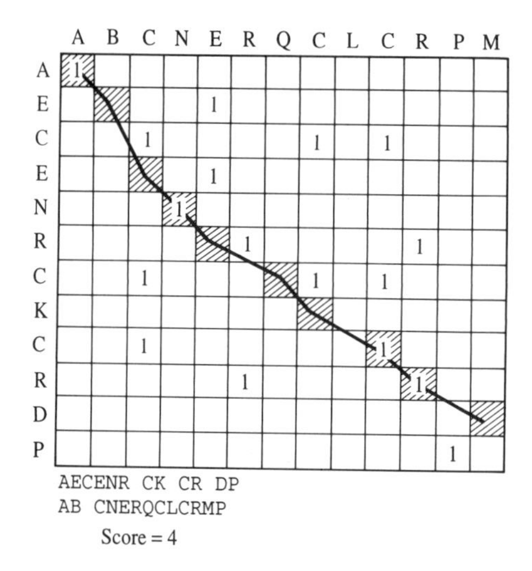

Fig. 8.43 Finding the optimal sequence alignment using dynamic programming and the Needleman–Wunsch algorithm. Sequence A = AECENRCKCRDP; sequence B = ABCNERQCLCRPM. The non-optimal alignment shown has a score of 4. Figure adapted from Needleman S B and C D Wunsch 1970. A General Method Applicable to the Search for Similarities in the Amino Acid Sequences of Two Proteins. *Journal of Molecular Biology* **48**: 443–453.

from A (i.e. the sequence RDP) and residues 11–13 from B (RPM). In this case the two amino acids are the same (R) and it is also possible to match one of the subsequent pairs (residues 12 from the first and 12 from the second are both P). The 'score' for this pair is set to 2 and this value is entered into the matrix. Now let us consider the element $M_{8,6}$. The two sequences that start with these residues are KCRDP and RQCLCRPM. The maximum possible score when aligning these two sequences is 3. The maximal scores for these subsequences can be efficiently calculated by starting at the termini of the two sequences (i.e. i = P and j = M) and working backwards. The optimal alignment is then determined by tracing back down the matrix, taking care to take account of alternative matches, as indicated in Figure 8.44.

The optimal Needleman–Wunsch alignment may contain a large, unrealistic number of gaps; this can be overcome by adding a 'gap' penalty to the scoring function. If the structure of one or both sequences is known then further improvements can be obtained by penalising even more severely gaps that occur in an α-helix or β-strand.

The Needleman–Wunsch algorithm, whilst guaranteeing to produce the optimal alignment, is of limited use for obtaining multiple sequence alignments due to the computational demands; in practice the method is limited to comparisons of three sequences. Alternative methods must be used to perform multiple sequence alignments. A two-stage procedure is commonly used [Barton 1996]. First, all pairwise sequence alignments are generated. The multiple alignment is

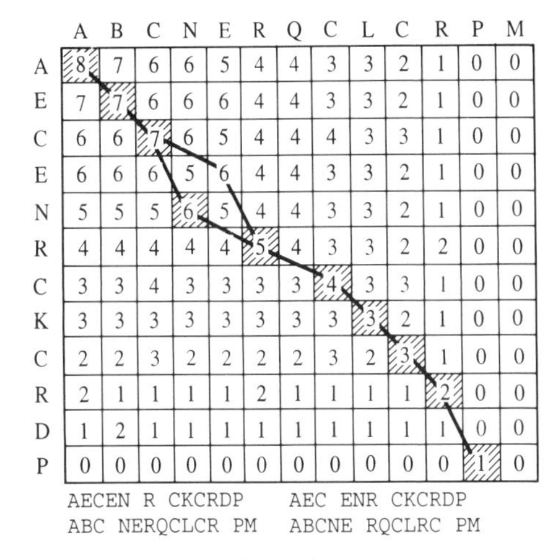

Fig. 8.44 The final Needleman–Wunsch matrix with the two optimal alignments indicated. The value in each matrix element corresponds to the optimal score for the appropriate pair of subsequences. Adapted from Needleman S B and C D Wunsch 1970. A General Method Applicable to the Search for Similarities in the Amino Acid Sequences of Two Proteins. *Journal of Molecular Biology* **48**: 443–453.

then constructed by first aligning the most similar pair of sequences, then the next most similar pair, and so on. Most automatic alignments often benefit from some manual intervention. Computer graphics programs which can display all of the sequences, colour coded by amino acid type or properties can greatly facilitate this process.

The sequence alignment establishes the correspondences between the amino acids in the unknown protein and the template protein from which it will be built. The three-dimensional structures of two or more related proteins are conveniently divided into *structurally conserved regions* (SCRs) and *structurally variable regions* (SVRs). The structurally conserved regions correspond to those stretches of maximum sequence identity or sequence similarity where one expects the conformation of the unknown protein to be very similar to that of the template protein. The structurally conserved regions are often found in the core of the protein. The structurally variable regions usually correspond to polypeptide loops which connect the secondary structure elements together. These loops can show significant differences in sequences and be of completely different lengths. The generation of a three-dimensional structure from the alignment usually proceeds via a three-stage process. First, the amino acid backbone for the structurally conserved regions is generated. This gives the 'core' of the protein to which the loops are then added. Finally, the side chains are placed and the model is validated. The final model may then be subjected to some form of refinement, such as energy minimisation.

8.13.8 Constructing and evaluating a homology model

Construction of the protein core is often relatively straightforward; in some cases the backbone conformation can be simply transferred from the template to the unknown protein. The next task is to determine the conformations of the loop regions. These generally occur on the surface of the molecule. Each loop must obviously adopt a conformation that enables it to properly join together the appropriate parts of the core. The loop conformation should also have a low internal energy and not have any unfavourable interactions with the rest of the molecule. In certain cases the loops may be restricted to a set of *canonical structures*. For example, it has been proposed that the conformations of some antibody loops fall into a small number of classes. Similarly, the loops that connect certain types of secondary structure often show distinct conformations. The β-turns that connect strands of β-sheets, have been classified into a small number of distinct families. In other cases we require an alternative method for predicting the loop conformations. Here we discuss just a few of the many methods that have been proposed for modelling polypeptide loops.

For loops that contain fewer than seven rotatable bonds, an algorithm devised by Go and Scheraga [Go and Sheraga 1970] can be used to calculate possible loop geometries directly. Go and Scheraga showed that it was possible to determine the torsion angles that would permit the end-to-end distance of the loop to achieve the desired value. The original Go and Scheraga method was developed for a model with fixed bond lengths and bond angles; more recent variants permit the bond angles to deviate slightly from their equilibrium values and so have a higher chance of finding an acceptable match [Bruccoleri and Karplus 1985]. In the CONGEN program of Bruccoleri and Karplus a systematic search is used to explore the space of $N - 6$ rotatable bonds (where N is the number of ϕ and ψ torsions in the loop) [Bruccoleri and Karplus 1987]. For each conformation that is generated, the Go and Scheraga chain closure algorithm is used to complete the structure. Purely systematic search methods can also be used to generate loop conformations. One interesting way to try and alleviate the combinatorial explosion is to construct the loop from both ends simultaneously; the half-complete loops are then joined in the middle (Figure 8.45).

Methods based on random algorithms have also been devised for modelling protein loops. One interesting method is the random tweak algorithm [Shenkin *et al.* 1987] which calculates the changes in the backbone ϕ and ψ torsion angles that will enable a randomly generated loop conformation to fit a set of distance constraints. An advantage of the random tweak procedure is that almost every chain can be 'tweaked' so that it satisfies these constraints; it is also extremely fast because it scales with the number of constraints rather than with the length of the chain. However, no information is included about the interactions with the rest of the protein in the calculation and this has to be checked once a loop conformation has been generated.

Loop conformations can be obtained by searching the protein databank for stretches of polypeptide chan that contain the appropriate number of amino acids and also have the correct spatial relationship between the two ends [Jones and

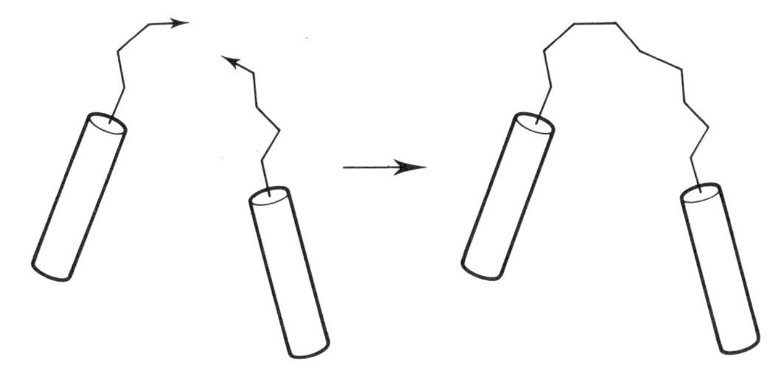

Fig. 8.45 An effective way to construct loops using a systematic search algorithm is to grow the two ends of the chain until they meet.

Thirup 1986]. This procedure can be made very fast by precalculating the necessary geometric information from loops in the protein databank and then using screening methods to identify the loops that can fit. This is similar to the use of polypeptide fragments when solving X-ray structures of proteins.

Once a backbone conformation (or conformations) has been derived for the protein, including the loop regions, it is then necessary to assign conformations to the side chains. In the core region there may be a high degree of sequence identity between the unknown protein and the template, and the side chain conformations can often be transferred directly from the template. Changes in amino acids in the core are often very conservative (e.g., a change from a phenylalanine to a tyrosine) and it is also easy to model the side chain in such cases. Where there is less correspondence between the amino acid sequences (and especially for the loop regions) then the side chains must be added without reference to the template. A variety of systematic and random methods have been used to predict side chain conformations; Monte Carlo, simulated annealing and genetic algorithm methods are particularly common. A popular tactic is to restrict the conformations of the side chains to those that are observed in experimentally determined protein structures [Ponder and Richards 1987].

The initial structures obtained from a homology modelling exercise can often be rather high in energy. Energy minimisation is thus often performed to refine the structure, though one should be careful to ensure that the minimisation does not cause any drastic changes in the structure.

Once a protein model has been constructed, it is important to examine it for flaws. Much of this analysis can be performed automatically using computer programs that examine the structure and report any significant deviations from the norm. A simple test is to generate a Ramachandran map, in order to determine whether the amino acid residues occupy the energetically favourable regions. The conformations of side chains can also be examined to identify any significant deviations from the structures commonly observed in X-ray structures. More sophisticated tests can also be performed. One popular approach is Eisenberg's '3D profiles' method [Bowie *et al.* 1991; Lüthy *et al.* 1992]. This calculates three properties for each amino acid in the proposed structure: the total surface area of the residue that is buried in the protein, the fraction of the side chain area that is covered by polar atoms and the local secondary structure. These three parameters are then used to allocate the residue to one of 18 environment classes. The buried surface area and fraction covered by polar atoms give six classes (Figure 8.46) for each of the three types of secondary structure (α helix, β sheet or coil). Each amino acid is given a score that reflects the compatibility of that amino acid for that environment, based upon a statistical analysis of known protein structures. Specifically, the score for a residue i in an environment j is calculated using:

$$\text{score} = \ln\left(\frac{P(\text{i}:\text{j})}{P_\text{i}}\right) \tag{8.41}$$

$P(\text{i}:\text{j})$ is the probability of finding residue i in environment j and P_i is the overall probability of finding residue i in any environment. For example, $P(\text{i:j})$ is -0.45 for a valine residue in a partially buried environment with a high fraction ($>67\%$) of the surface covered by polar atoms in an α-helix. The negative number indicates that this environment is not favoured for valine. However, this environment is more favoured by arginine, for which the $P(\text{i:j})$ value is 0.50.

The 3D profiles method can be used to calculate an overall score for a protein model. It was found that deliberately misfolded protein models have low scores because they contain residues in environments with which they are not compatible. Such misfolded models often cannot be distinguished from the correct structures using molecular mechanics energies. The 3D profile can also be used to identify whether a generally correct model contains regions of incorrectly assigned residues. This is usually done by plotting the score as a function of the sequence, as shown in Figure 8.47. Any residues for which the score falls signifi-

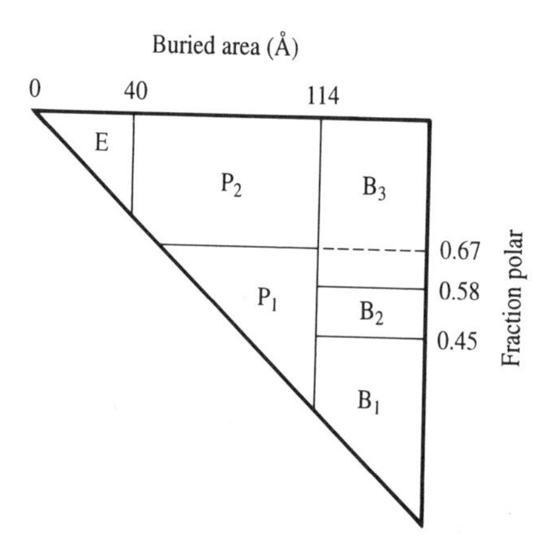

Fig. 8.46 The six environment categories used by the 3D profiles method. After Bowie J U, R. Lüthy and D Eisenberg 1991. A Method to Identify Protein Sequences That Fold into a Known Three-Dimensional Structure. *Science* **253**: 164–170.

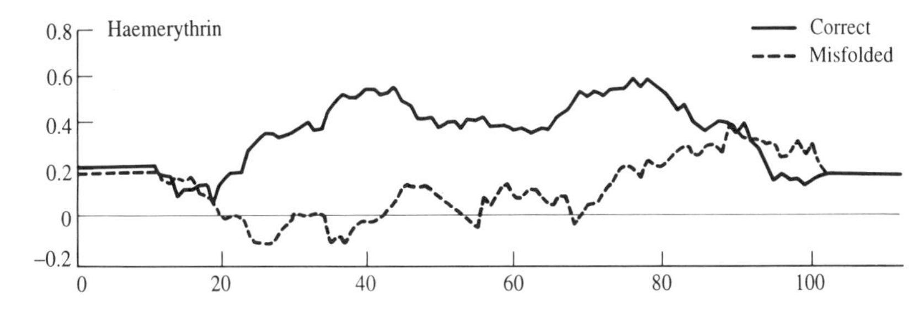

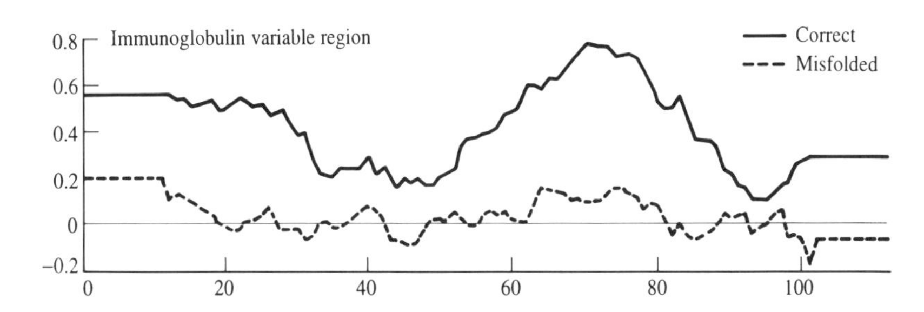

Fig. 8.47 The 3D profiles output for incorrect and partially incorrect protein models compared to the correct structures. The vertical axis gives the average profile score for a 21-residue window. Figure redrawn from Lüthy R, J U Bowie and D Eisenberg 1992. Assessment of Protein Models with Three-Dimensional Profiles. *Nature* **356**: 83–85.

cantly below the average score should be investigated to check whether the model is faulty in that particular region.

8.13.9 Predicting protein structures by 'threading'

Threading is a method that is increasing being used to suggest the structures of proteins [Jones *et al.* 1992; Jones and Thornton 1993]. The basic threading concept is very simple. Suppose we wish to predict the structure of an amino acid sequence, and that we have available a number of three-dimensional protein structures, typically chosen to represent common structural classes. Threading aims to determine which structure is most compatible with the sequence of the unknown protein. This is done by 'threading' the sequence through each protein structure in turn. The amino acids are advanced to occupy the location occupied in the previous iteration by its predecessor. A score is calculated for each structure that is generated. The process is repeated until the sequence has been entirely threaded through the structure. Typically a database containing a large number of protein structures is considered with the result being the structure or structures that correspond to the lowest values of the scoring function. Due to the large number of possibilities threading algorithms use special searching methods such as dynamic programming to efficiently identify the best ways to match the sequence to the structure.

A threading calculation can require a large number of possibilities to be considered, and so the scoring functions are usually quite simple. This also reflects the low resolution nature of the problem, in that one is usually only attempting to predict the basic fold of the protein. Each amino acid is typically treated as a single interaction site. Many of the scoring functions used in threading algorithms are potentials of mean force that provide an estimate of the free energy of interaction between two residues as a function of their separation. These potentials of mean force are calculated from statistical analyses of known protein structures. For example, if one plots the distribution of distances observed in X-ray protein structures between amino acids that are separated by three other residues in the sequence (i.e. between residues i and $i+4$) then a large peak is observed in the interval 5.9–6.5 Å and a broad shoulder from 11.4–13.3 Å. These reflect the presence of α-helices and β-strand conformations. It is from these distribution frequencies that one can determine the residue–residue potentials of mean force. Sippl has provided an example of the use of such potentials [Sippl 1990]. The pentapeptide sequence valine-asparagine-threonine-phenylalanine-valine (VNTFV in the one-letter amino acid code) adopts an α-helical conformation in the protein erythrocruorin but a β-strand conformation in ribonuclease. The potential of a mean force suggests that the β-strand is the more stable conformation for the isolated pentapeptide, but that when it is flanked by aspartic acid at one end and alanine at the other (as in erthrocruorin) the α-helix does indeed become the more stable conformation. For threading algorithms one is particularly interested in the interactions between amino acids that are close in three-dimensional space but far apart in the sequence and the potentials used in such calculations are derived appropriately.

8.13.10 A comparison of comparative modelling strategies

Homology modelling is currently the most popular method for predicting the three-dimensional structure of a protein from its sequence. The utility of a homology model depends upon the use to which it is put. In some cases one is only interested in the general fold that the protein adopts and so a relatively low resolution structure is acceptable. For other applications such as drug design the model must be much more accurate, including the loops and side chains. In such cases, a poor model may often be far worse than no model at all, as it can be seriously misleading.

To evaluate the then available techniques for calculating protein models, a 'competition' was organised in 1994–1995 in which entrants were invited to predict the three-dimensional structures of seven proteins from their amino acid sequences [Mosimann *et al.* 1995]. The structures of these seven proteins were simultaneously solved useing X-ray crystallography but these structures were not made available to the modellers. A total of 43 separate structures were submitted by 13 research groups, each model then being compared with the X-ray structure.

The quality of each model was also assessed using a variety of methods including the calculation of Ramachandran maps and 3D profiles.

Each competitor first had to decide which of the known protein structures they wished to use as the template; they then had to construct a sequence alignment (possibly using other protein sequences from the same family), and finally had to construct the model. The degree of sequence identity for the seven proteins ranged from 22% to 77%. One reassuring conclusion was that in favourable cases very accurate models could be constructed; the 'best' structure had an RMS difference with the X-ray structure of just 0.6 Å. This result was obtained for a prediction of the structure of the protein NM23, which had the highest sequence identity with its template and also the same number of amino acid residues. Overall, the accuracy of the models largely depended upon two factors: the percentage sequence identity and the presence of substantial insertions or deletions between the template and target structures. The need for an accurate sequence alignment was evident; an incorrect alignment almost always resulted in an incorrect structure. In those proteins where there were large insertion loops, these were invariably predicted incorrectly, demonstrating a clear need for new strategies to tackle this problem. In some cases models of the same protein were generated using both 'hands on' approaches (where the modeller directed the construction of the structure) and wholly automatic procedures. In all cases the manual structure was superior to the automatic model.

One disturbing finding was that a significant proportion of the models contained errors including amino acids with the wrong stereochemistry. In addition, significant deviations from planarity of the amide bond were noted in many structures and the torsion angles of the side chains often deviated from the distributions observed in experimental protein structures. Some of the models contained high energy steric interactions between non-bonded atoms and unlikely distributions of amino acids (i.e. hydrophilic residues on the inside and hydrophobic residues on the outside). Most, if not all of these problems can be identified very simply using publically available software and the organisers of the competition suggested that any models submitted for publication should be accompanied by output from a structure verification program to allow an objective assessment of the quality of the model. The use of energy minimisation to refine models was identified as the cause of many of these problems, thereby highlighting the need for alternative protocols for the refinement stage of a homology modelling exercise.

Further reading

Aldenderfer M S and R K Blahfield 1984. *Cluster Analysis*. Newbury, Park, Sage. New York, Garland Publishing.

Branden C and J Tooze 1991. *Introduction to Protein Structure.*

Chatfield C and A J Collins 1980. *Introduction to Multivariate Analysis*. London, Chapman and Hall.

Leach A R 1991. A Survey of Methods for Searching the Conformational Space of Small and Medium-Sized Molecules in Lipkowitz K B and D B Boyd (Editors). *Reviews in Computational Chemistry* vol 2. New York, VCH Publishers: 1–55.

Perutz M 1992. *Protein Structure. New Approaches to Disease And Therapy.* New York, W H Freeman.

Scheraga H A 1993. Searching Conformational Space. In van Gunsteren W F, P K Weiner and A J Wilkinson (Editors) *Computer Simulation of Biomolecular Systems* vol 2. Leiden, ESCOM Science Publishers.

Schulz G E and R H Schirmer 1979. *Principles of Protein Structure*. New York, Springer–Verlag.

References

Allen F H, S A Bellard, M D Brice, B A Cartwright, A Doubleday, H Higgs, T Hummelink, B G Hummelink-Peters, O Kennard, W D S Motherwell, J R Rodgers and D G Watson 1979. The Cambridge Crystallographic Data Centre: Computer-Based Search, Retrieval, Analysis and Display of Information. *Acta Crystallographica* **B35**:2331–2339.

Allen F H, O Kennard, D G Watson, L Brammer, A G Orpen and R Taylor 1987. Tables of Bond Lengths Determined by X-Ray and Neutron Diffraction. 1. Bond Lengths in Organic Compounds. *Journal of the Chemical Society Perkin Transactions* **II**:S1–S19.

Barton D H R 1950. The Conformation of the Steroid Nucleus. *Experientia* **6**:316–320

Barton G J 1996 Protein Sequence Alignment and Database Scanning. In Sternberg M E (Editor) *Protein Structure Prediction—A Practical Approach.* IRL Press, Oxford. Also at http://geoff.biop.ox.ac.uk/papers/rev93_1/rev93_1.html.

Bergerhoff G, R Hundt, R Sievers and I S Brown 1983. The Inorganic Crystal Structure Database. *Journal of Chemical Information and Computer Sciences* **23**:66–69.

Bernstein F C, T F Koetzle, G J B Williams, E Meyer, M D Bryce, J R Rogers, O Kennard, T Shikanouchi and M Tasumi 1977. The Protein Data Bank: A Computer-Based Archival File for Macromolecular Structures. *Journal of Molecular Biology* **112**:535–542.

Birktoft J J and D M Blow 1972. The structure of Crystalline Alpha-Chymotrypsin V. The Atomic Structure of Tosyl-Alpha-Chymotrypsin at 2 Angstroms Resolution. *Journal Of Molecular Biology* **68**:187–240.

Bowie J U, Lüthy and D Eisenberg 1991. A Method to Identify Protein Sequences That Fold into a Known Three-Dimensional Structure: *Science* **253**:164–170.

Bruccoleri R E and M Karplus 1985. Chain Closure with Bond Angle Variations. *Macromolecules* **18**:2767–1773.

Bruccoleri R E and M Karplus 1987. Prediction of the Folding of Short Polypeptide Segments by Uniform Conformational Sampling. *Biopolymers* **26**: 137–168.

Brunger A T, J Kuriyan and M Karplus 1987. Crystallographic R-Factor Refinement by Molecular-Dynamics. *Science* **235**:458–460.

Bryant S H and C E Lawrence 1993. An Empirical Energy Function for Threading Protein Sequence Through the Folding Motif. *Proteins: Structure, Function and Genetics* **16**:92–112.

Chan H S and K A Dill 1993. The Protein Folding Problem. *Physics Today* **Feb**: 24–32.

Chang G, W. C. Guida and W C Still 1989. An Internal Coordinate Monte Carlo Method for Searching Conformational Space. *The Journal of the American Chemical Society* **111**:4379–4386.

Chatfield C and A J Collins 1980. *Introduction to Multivariate Analysis*. London Chapman and Hall.

Chou P Y and G D Fasman 1978. Prediction of the Secondary Structure of Proteins from their Amino Acid Sequence. *Advances in Enzymology* **47**:45–148.

Chung C-W, R M Cooke, A E I Proudfoot and T N C Wells 1995. The Three-Dimensional Structure of RANTES. *Biochemistry* **34**:9307–9314.

Cohen F E, M J E Sternberg and W R Taylor 1982 Analysis and Prediction of the Packing of α-Helices against a β-sheet in the Tertiary Structure of Globular Proteins. *Journal of Molecular Biology* **156**:821–862.

Crippen G M 1981. *Distance Geometry and Conformational Calculations*. Chemometrics Research Studies Series 1. New York, Wiley.

Crippen G M and T F Havel 1988. *Distance Geometry and Molecular Conformation*. Chemometrics Research Studies Series 15. New York, Wiley.

Derome A E 1987. *Modern NMR Techniques for Chemistry Research*. Oxford, Pergamon.

Downs G M, P Willett and W Fisanick 1994. Similarity Searching and Clustering of Chemical Structure Databases using Molecular Property Data. *Journal of Chemical Information and Computer Sciences* **34**:1094–1102.

Ferguson D M and D J Raber 1989. A New Approach to Probing Conformational Space with Molecular Mechanics: Random Incremental Pulse Search. *The Journal of the American Chemical Society* **111**:4371–4378.

Ferro D R and J Hermans 1977. A Different Best Rigid-Body Molecular Fit Routine. *Acta Crystallographica* **A33**:345–347.

Gibson K D and H A Scheraga 1987. Revised Algorithms for the Build-Up Procedure for Predicting Protein Conformations by Energy Minimization. *Journal of Computational Chemistry* **8**:826–834.

Glusker J P 1995. Intermolecular Interactions Around Functional Groups in Crystals: Data for Modeling the Binding of Drugs to Biological Macromolecules. *Acta Crystallographica* **D51**:418–427.

Go N and H A Scheraga 1970. Ring Closure and Local Conformational Deformations of Chain Molecules. *Macromolecules* **3**:178–187.

Godzik A, A Kolinski A and J Skolnick 1993. De Novo and Inverse Folding Predictions of Protein Structure and Dynamics. *Journal of Computer-Aided Molecular Design* **7**:397–438.

Goldberg D E 1989. *Genetic Algorithms in Search, Optimization and Machine Learning*. Reading, Massachusetts, Addison-Wesley.

Greasley S E, H Jhoti, C Teahan, R Solari, A Fensom, G M H Thomas, S Cockroft and B Bax 1995. The Structure of Rat ADP-Ribosylation Factor-1 (ARF-1) Complexed to GDP Determined from Two Different Crystal Forms. *Nature Structural Biology* **2**:797–806.

Havelka W A, R Henderson and D Oesterhelt 1995. 3-Dimensional Structure of Halorhodopsin at 7-Angstrom Resolution. *Journal of Molecular Biology* **247**: 726–738.

Henderson R, J M Baldwin, T A Ceska, F Zemlin, E Beckmann and K H Downing 1990. Model for the Structure of Bacteriorhodopsin based on High-Resolution Electron Cryo-microscopy. *Journal of Molecular Biology* **213**:899–929.

Jack A and M Levitt 1978. Refinement of Large Structures by Simultaneous Minimization of Energy and R factor. *Acta Crystallographica* **A34**:931–935.

Jarvis R A and E A Patrick 1973. Clustering Using a Similarity Measure Based on Shared Near Neighbours. *IEEE Transactions in Computers* **C-22**:1025–1034.

Jones D and J Thornton 1993. Protein Fold Recognition. *Journal of Computer-Aided Molecular Design* **7**:439–456.

Jones D T, W R Taylor and J M Thornton 1992. A New Approach to Protein Fold Recognition. *Nature* **358**:86–89.

Jones T A and S Thirup 1986. Using Known Substructures in Protein Model Building and Crystallography. *EMBO Journal* **5**:819–822.

Judson R S, W P Jaeger, A M Treasurywala and M L Peterson 1993. Conformational Searching Methods For Small Molecules. 2. Genetic Algorithm Approach. *Journal of Computational Chemistry* **14**:1407–1414.

Kabsch W 1978. A Discussion of the Solution for the Best Rotation to Relate Two Sets of Vectors. *Acta Crystallographica* **A34**:827–828.

Kirkpatrick S, C D Gelatt and M P Vecchi 1983. Optimization by Simulated Annealing. *Science* **220**:671–680.

Leach A R, K Prout and D P Dolata. 1988. An Investigation into the Construction of Molecular Models using the Template Joining Method. *Journal of Computer-Aided Molecular Design* **2**:107–123.

Leach A R, D P Dolata and K Prout 1990. Automated Conformational Analysis and Structure Generation: Algorithms for Molecular Perception. *Journal of Chemical Information and Computer Science* **30**:316–324.

Levitt M 1976. A Simplified Representation of Protein Conformations for Rapid Simulation of Protein Folding. *Journal of Molecular Biology* **104**:59–107.

Li Z Q and H A Scheraga 1987. Monte-Carlo-Minimization Approach to the Multiple-Minima Problem in Protein Folding. *Proceedings Of The National Academy Of Sciences USA* **84**:6611–6615.

Lüthy R, J U Bowie and D Eisenberg 1992. Assessment of Protein Models with Three-Dimensional Profiles. *Nature* **356**:83–85.

McGarrah D B and R S Judson 1993. Analysis of the Genetic Algorithm Method of Molecular-Conformation Determination. *Journal Of Computational Chemistry* **14**:1385–1395.

Mosimann S, S Meleshko and M N G Jones 1995. A Critical Assessment of Comparative Molecular Modeling of Tertiary Structures of Proteins. *Proteins: Structure, Function and Genetics* **23**:301–317.

Murray-Rust P M and J P Glusker 1984. Directional Hydrogen Bonding to sp^2 and sp^3-Hybridized Oxygen Atoms and Its Relevance to Ligand–Macromolecule Interactions. *The Journal of the American Chemical Society* **106**:1018–1025.

Needleman S B and C D Wunsch 1970. A General Method Applicable to the Search for Similarities in the Amino Acid Sequences of Two Proteins. *Journal of Molecular Biology* **48**:443–453.

Ning Q and T J Sejnowski 1988. Predicting The Secondary Structure of Globular-Proteins Using Neural Network Models. *Journal of Molecular Biology* **202**:865–888.

Noble M E M, R K Wierenga, A-M Lambeir, F R Opperdoes, A-M W H Thunnissen, K H Kalk, H Groendijk and W G J Hol 1991. The Adaptability of the Active Site of Trypanosomal Triosephosphate Isomerase as Observed in the Crystal Structures of Three Different Complexes. *Proteins: Structure, Function and Genetics* **10**:50–69.

Novotny J, A A Rashin and R E Bruccoleri 1988. Criteria that Discriminate between Native Proteins and Incorrectly Folded Models. *Proteins: Structure, Function And Genetics* **4**:19–30.

Pauling L, R B Corey and H R Bronson 1951. The Structure of Proteins: Two Hydrogen-Bonded Helical Configurations of the Polypeptide Chain. *Proceedings of the National Academy of Sciences USA* **37**:205–211.

Ponder J W and F M Richards 1987. Tertiary Templates for Proteins. Use of Packing Criteria in the Enumeration of Allowed Sequences for Different Structural Classes. *Journal of Molecular Biology* **193**:775–791.

Ramachadran G N, C Ramakrishnan and V Sasiekharan 1963. Stereochemistry of Polypeptide Chain Configurations. *Journal of Molecular Biology* **7**:95-99.

Ripoll D R and H A Scheraga 1988. On the Multiple-Minimum Problem in the Conformational Analysis of Polypeptides. II. An Electrostatistically Driven Monte Carlo Method: Tests on Poly(L-Alanine). *Biopolymers* **27**:1283–1303.

Ripoll D R and H A Scheraga 1989. On the Multiple-Minimum Problem in the Conformational Analysis of Polypeptides. III. An Electrostatically Driven Monte Carlo Method: Tests on met-Enkephalin. *Journal of Protein Chemistry* **8**: 263–287.

Saunders M 1987. Stochastic Exploration of Molecular Mechanics Energy Surface: Hunting for the Global Minimum. *The Journal of the American Chemical Society* **109**:3150–3152.

Saunders M, K N Houk, Y-D Wu, W C Still, M Lipton, G Chang and W C Guida 1990. Conformations of Cycloheptadecane. A Comparison of Methods for Conformational Searching. *The Journal of the American Chemical Society* **112**:1419–1427.

Shenkin P S, D L Yarmusch, R M Fine, H Wang and C Levinthal 1987. Predicting Antibody Hypervariable Loop Conformation. I. Ensembles of Random Conformations for Ringlink Structures. *Biopolymers* **26**:2053–2085.

Sippl M J 1990. Calculation of Conformational Ensembles from Potentials of Mean Force. An Approach to the Knowledge-Based Prediction of Local Structures in Globular Proteins. *Journal of Molecular Biology* **213**:859–883.

Sternberg M J E, F E Cohen and W R Taylor 1982 A Combinatorial Approach to the Prediction of the Tertiary Fold of Globular Proteins. *Biochemical Society Transactions* **10**:299–301.

Summers N L, W D Carlson and M Karplus 1987. Analysis Of Side-Chain Orientations In Homologous Proteins. *Journal of Molecular Biology* **196**:175–198.

Torda A E, R M Scheek and W F van Gunsteren 1990 Time-Averaged Nuclear Overhauser Effect Distance Restraints Applied to Tendamistat. *Journal of Molecular Biology* **214**:223–235.

Turk D, J Sturzebecher and W Bode 1991. Geometry of Binding of the *N*-Alpha-Tosylated Piperidides of *meta*-Amidino-Phenylalanine, Para Amidino-Phenylalanine and *para*-Guanidino-Phenylalanine to Thrombin and Trypsin—X-ray Crystal Structures of Their Trypsin Complexes and Modeling of their Thrombin Complexes. *Febs Letters* **287**:133–138.

Turk D, H W Hoeffken, D Grosse, J Stuerzebecher, P D Martin, B F P Edwards and W Bode 1992. Refined 2.3 Ångstroms X-Ray Crystal Structure of Bovine Thrombin Complexes Formed with the 3 Benzamidine and Arginine-Based Thrombin Inhibitors NAPAP, 4-TAPAP and MQPA: A Starting Point for Improving Antithrombotics. *Journal Of Molecular Biology* **226**:1085–1099.

Weiner S J, P A Kollman, D A Case, U C Singh, C Ghio, G. Alagona, S Profeta and P Weiner 1984. A New Force Field for Molecular Mechanical Simulation of Nucleic Acids and Proteins. *The Journal of the American Chemical Society* **106**:765–784.

9 Three challenges in molecular modelling: free energies, solvation and simulating reactions

In Chapters 1 to 8 we examined many different molecular modelling techniques and saw how these methods can be used to calculate a wide variety of properties. In this chapter we shall consider three important problems in molecular modelling. First we discuss the problem of calculating free energies. We then consider continuum solvent models which enable the effects of the solvent to be incorporated into a calculation without requiring the solvent molecules to be represented explicitly. Finally, we shall consider the simulation of chemical reactions.

9.1 The difficulty of calculating free energies by computer

The free energy is often considered to be the most important quantity in thermodynamics. The free energy is usually expressed as the Helmholtz function, A, or the Gibbs function, G. The Helmholtz free energy is appropriate to a system with constant number of particles, temperature and volume (constant NVT), whereas the Gibbs free energy is appropriate to constant number of particles, temperature and pressure (constant NPT). Most experiments are conducted under conditions of constant temperature and pressure, where the Gibbs function is the appropriate free energy quantity.

Unfortunately, the free energy is a difficult quantity to obtain for systems such as liquids or flexible macromolecules that have many minimum energy configurations separated by low-energy barriers. Associated quantities such as the entropy and the chemical potential are also difficult to calculate. As we showed in section 5.3, the free energy cannot be accurately determined from a 'standard' molecular dynamics or Monte Carlo simulation because such simulations do not adequately sample from those regions of phase space that make important contributions to the free energy. Specifically, we showed that the Helmholtz free energy is given by:

$$A = k_B T \ln\left(\iint d\mathbf{p}^N d\mathbf{r}^N \exp\left(\frac{+\mathscr{H}(\mathbf{p}^N, \mathbf{r}^N)}{k_B T}\right)\rho(\mathbf{p}^N, \mathbf{r}^N)\right) \tag{9.1}$$

The term $\exp[+\mathscr{H}(\mathbf{p}^N, \mathbf{r}^N)/k_B T]$ makes important contributions to the integral but a simulation using either Monte Carlo or molecular dynamics sampling seeks out the lower energy regions of phase space. Such simulations will never adequately sample the important high-energy regions and so to calculate the free energy using a conventional simulation will lead to poorly converged and inaccurate values. The grand canonical and particle insertion methods do provide a route to the free energy but they are not applicable to many of the systems of interest which contain complex molecules at high densities.

9.2 The calculation of free energy differences

Let us consider a closely related, but slightly different problem: the calculation of the free energy difference of two states. As an example, we will consider the problem of calculating the free energy difference between ethanol (CH_3CH_2OH) and ethane thiol (CH_3CH_2SH) in water. As we shall see, this is a problem that can be tackled using methods that use Monte Carlo or molecular dynamics sampling. There are three methods that have been proposed for calculating free energy differences: thermodynamic perturbation, thermodynamic integration and slow growth. We shall consider each of these in turn.

9.2.1 *Thermodynamic perturbation*

Consider two well-defined states X and Y. For example, X could be a system comprising a molecule of ethanol in a periodic box of water and Y could be ethane thiol in water. X contains N particles interacting according to the Hamiltonian $\mathscr{H}_X$. Y contains N particles interacting according to $\mathscr{H}_Y$. The free energy difference (ΔA) between the two states is as follows:

$$\Delta A = A_Y - A_X = -k_B T \ln \frac{Q_Y}{Q_X} \tag{9.2}$$

$$\Delta A = -k_B T \left\{ \frac{\iint \mathrm{d}\mathbf{p}^N \mathrm{d}\mathbf{r}^N \exp[-\mathscr{H}_Y(\mathbf{p}^N, \mathbf{r}^N)/k_B T]}{\iint \mathrm{d}\mathbf{p}^N \mathrm{d}\mathbf{r}^N \exp[-\mathscr{H}_X(\mathbf{p}^N, \mathbf{r}^N)/k_B T]} \right\} \tag{9.3}$$

Substituting 1 in the form $\exp[+\mathscr{H}_X(\mathbf{p}^N, \mathbf{r}^N)/k_B T] \exp[-\mathscr{H}_X(\mathbf{p}^N, \mathbf{r}^N)/k_B T]$ into the numerator gives:

$$\Delta A = -k_B T \left\{ \frac{\begin{array}{c}\iint \mathrm{d}\mathbf{r}^N \mathrm{d}\mathbf{p}^N \exp\left(-\dfrac{\mathscr{H}_Y(\mathbf{r}^N, \mathbf{p}^N)}{k_B T}\right) \\ \times \exp\left(+\dfrac{\mathscr{H}_X(\mathbf{r}^N, \mathbf{p}^N)}{k_B T}\right) \exp\left(-\dfrac{\mathscr{H}_X(\mathbf{r}^N, \mathbf{p}^N)}{k_B\ T}\right)\end{array}}{\iint \mathrm{d}\mathbf{r}^N \mathrm{d}\mathbf{p}^N \exp\left(\dfrac{-\mathscr{H}_X(\mathbf{r}^N, \mathbf{p}^N)}{k_B T}\right)} \right\} \tag{9.4}$$

Equation 9.4 can be written in terms of an ensemble average, as follows:

$$\Delta A = -k_{\mathrm{B}}T\left\{\frac{\iint \mathrm{d}\mathbf{p}^N \mathrm{d}\mathbf{r}^N \exp[-\mathcal{H}_{\mathrm{Y}}(\mathbf{p}^N,\mathbf{r}^N)/k_{\mathrm{B}}T] \times \exp[+\mathcal{H}_{\mathrm{X}}(\mathbf{p}^N,\mathbf{r}^N)/k_{\mathrm{B}}T]\exp[-\mathcal{H}_{\mathrm{X}}(\mathbf{p}^N,\mathbf{r}^N)/k_{\mathrm{B}}T]}{\iint \mathrm{d}\mathbf{p}^N \mathrm{d}\mathbf{r}^N \exp[-\mathcal{H}_{\mathrm{X}}(\mathbf{p}^N,\mathbf{r}^N)/k_{\mathrm{B}}T]}\right\}$$

$$= -k_{\mathrm{B}}T\langle\exp[+(\mathcal{H}_{\mathrm{Y}}(\mathbf{p}^N,\mathbf{r}^N) - \mathcal{H}_{\mathrm{X}}(\mathbf{p}^N,\mathbf{r}^N))/k_{\mathrm{B}}T]\rangle_0 \tag{9.5}$$

The subscript 0 indicates averaging over the ensemble of configurations representative of the initial state X. An analogous expression can also be derived where the averaging is over the ensemble corresponding to the final state Y (indicated by the subscript 1):

$$\Delta A = -k_{\mathrm{B}}T \ln\langle\exp[-(\mathcal{H}_{\mathrm{Y}} - \mathcal{H}_{\mathrm{X}})/k_{\mathrm{B}}T]\rangle_1 \tag{9.6}$$

This approach to the calculation of free energy differences, equation (9.6), is generally attributed to Zwanzig [Zwanzig 1954]. To perform a thermodynamic perturbation calculation we must first define $\mathcal{H}_{\mathrm{Y}}$ and $\mathcal{H}_{\mathrm{X}}$ and then run a simulation at the state X, forming the ensemble average of $\exp[(\mathcal{H}_{\mathrm{Y}} - \mathcal{H}_{\mathrm{X}})/k_{\mathrm{B}}T]$ as we proceed. Analogously, we could run a simulation at the state Y and obtain the ensemble average of $\exp[(\mathcal{H}_{\mathrm{X}} - \mathcal{H}_{\mathrm{Y}})/k_{\mathrm{B}}T]$. Thus, if X corresponds to ethanol and Y to ethane thiol, the free energy difference could thus be obtained from a simulation of ethanol in a periodic box of water as follows. For each configuration we calculate the value of the energy for every instantaneous conformation of ethanol in which the oxygen atom is temporarily assigned the potential energy parameters of sulphur. Alternatively, we could simulate ethane thiol and for each configuration we calculate the energy of the system in which the sulphur is 'mutated' into oxygen.

If X and Y do not overlap in phase space then the value of the free energy difference calculated using equation (9.6) will not be very accurate, because we will not adequately sample the phase space of Y when simulating X. This problem arises when the energy difference between the two states is much larger than $k_{\mathrm{B}}T$: $|\mathcal{H}_{\mathrm{Y}} - \mathcal{H}_{\mathrm{X}}| \gg k_{\mathrm{B}}T$. How then can we obtain accurate estimates of the free energy difference under such circumstances? Consider what happens if we introduce a state that is intermediate between X and Y, with a Hamiltonian $\mathcal{H}_1$ and a free energy $A(1)$:

$$\begin{aligned}\Delta A &= A(Y) - A(X)\\ &= (A(Y) - A(1)) + (A(1) - A(X))\\ &= -k_{\mathrm{B}}T\ln\left[\frac{Q(\mathrm{Y})}{Q(1)}\cdot\frac{Q(1)}{Q(\mathrm{X})}\right]\\ &= -k_{\mathrm{B}}T\ln\langle\exp[-(\mathcal{H}_{\mathrm{Y}} - \mathcal{H}_1)/k_{\mathrm{B}}T]\rangle\\ &\quad - k_{\mathrm{B}}T\ln\langle\exp[-(\mathcal{H}_1 - \mathcal{H}_{\mathrm{X}})/k_{\mathrm{B}}T]\rangle\end{aligned} \tag{9.7}$$

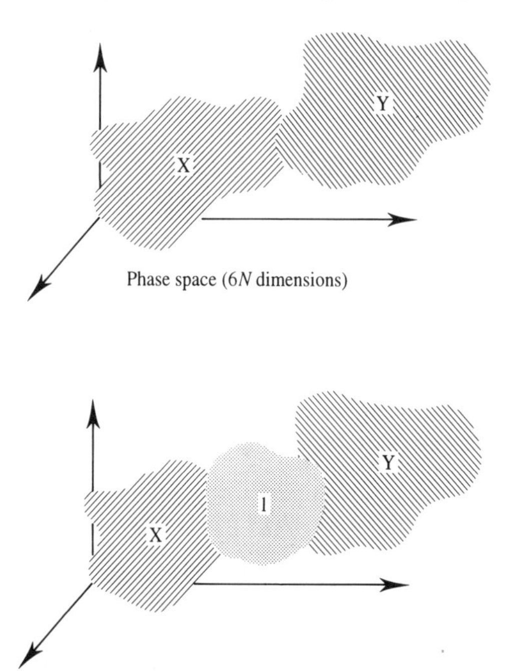

Fig. 9.1 An intermediate state (labelled 1) can improve the degree of overlap in phase space and lead to improved sampling.

If we define region 1 so that it overlaps with X and Y we may improve the sampling and obtain a more reliable value. This is shown in Figure 9.1.

An obvious extension is to use several different intermediate states in progressing from $\mathscr{H}_X$ to $\mathscr{H}_Y$:

$$\begin{aligned}\Delta A &= A(\mathrm{Y}) - A(\mathrm{X}) \\ &= (A(\mathrm{Y}) - A(N)) + (A(N) - A(N-1)) + \ldots \\ &\quad + (A(2) - A(1)) + (A(1) - A(X)) \\ &= -k_B T \ln\left[\frac{Q(\mathrm{Y})}{Q(N)} \cdot \frac{Q(N)}{Q(N-1)} \cdot \frac{Q(N-1)}{Q(N-2)} \cdots \frac{Q(2)}{Q(1)} \frac{Q(1)}{Q(\mathrm{X})}\right] \end{aligned} \tag{9.8}$$

The important point to notice is that the intermediate terms cancel out and so we are free to choose as many intermediate states as are necessary to get good overlaps and thus reliable values of the free energy differences.

9.2.2 Implementation of free energy perturbation

Suppose we are using an empirical energy function such as the following to describe the inter- and intramolecular interactions within our ethanol/ethane thiol system:

$$\mathscr{V}(\mathbf{r}^N) = \sum_{\text{bonds}} \frac{k_i}{2}(l_i - l_{i,0})^2 + \sum_{\text{angles}} \frac{k_i}{2}(\theta_i - \theta_{i,0})^2 + \sum_{\text{torsions}} \frac{V_n}{2}(1+\cos(n\omega - \gamma))$$
$$+ \sum_{i=1}^{N} \sum_{j=i+1}^{N} \left(4\varepsilon_{ij}\left[\left(\frac{\sigma_{ij}}{r_{ij}}\right)^{12} - \left(\frac{\sigma_{ij}}{r_{ij}}\right)^{6}\right] + \frac{q_i q_j}{4\pi\varepsilon_0 r_{ij}} \right) \tag{9.9}$$

The force field model for ethanol contains C–O and O–H bond stretching contributions; in ethane thiol these are replaced by C–S and S–H parameters. Similarly, in ethanol there will be angle bending terms due to C–O–H, C–C–O and H–C–O angles; in ethane thiol these will be C–S–H, C–C–S and H–C–S. The torsional contribution will be modified appropriately, as will the van der Waals and electrostatic interactions (both those within the solute and between the solute and solvent). The partial atomic charges for all of the atoms in ethanol may all be different from those for ethane thiol.

The relationship between the initial, final and intermediate states is usefully described in terms of a *coupling parameter*, λ. As λ is changed from 0 to 1, the Hamiltonian varies from $\mathscr{H}_X$ to $\mathscr{H}_Y$. Each of the terms in the force field for an intermediate state λ can be written as a linear combination of the values for X and Y:

1. Bonds:
$$k_l(\lambda) = \lambda k_l(\text{Y}) + (1 - \lambda)k_l(\text{X}) \tag{9.10}$$
$$l_0(\lambda) = \lambda l_0(\text{Y} + (1 - \lambda)l_0(\text{X}) \tag{9.11}$$
2. Angles:
$$k_\theta(\lambda) = \lambda k_\theta(\text{Y}) + (1 - \lambda)k_\theta(\text{X}) \tag{9.12}$$
$$\theta_0(\lambda) = \lambda\theta_0(\text{Y}) + (1 - \lambda)\theta_0(\text{X}) \tag{9.13}$$
3. Dihedrals:
$$\mathscr{V}_\omega(\lambda) = \lambda\mathscr{V}_\omega(\text{Y}) + (1 - \lambda)\mathscr{V}_\omega(\text{X}) \tag{9.14}$$
4. Electrostatics:
$$q_i(\lambda) = \lambda q_i(\text{Y}) + (1 - \lambda)q_i(\text{X}) \tag{9.15}$$
5. van der Waals:
$$\varepsilon(\lambda) = \lambda\varepsilon(\text{Y}) + (1 - \lambda)\varepsilon(\text{X}) \tag{9.16}$$
$$\sigma(\lambda) = \lambda\sigma(\text{Y}) + (1 - \lambda)\sigma(\text{X}) \tag{9.17}$$

For each value of λ (λ_i) a simulation is performed (using either Monte Carlo or molecular dynamics as appropriate) using the appropriate force field parameters. First, the system is equilibrated using the force field parameters appropriate to λ_i. A production phase is then performed during which the free energy difference $\Delta A(\lambda_i \rightarrow \lambda_{i+1})$ is accumulated as $-k_B T \ln\langle\exp(-\Delta\mathscr{H}_i/k_B T)\rangle$ where $\Delta\mathscr{H}_i = \mathscr{H}_{i+1} - \mathscr{H}_i$. The total free energy change for $\lambda = 0$ to $\lambda = 1$ is then the

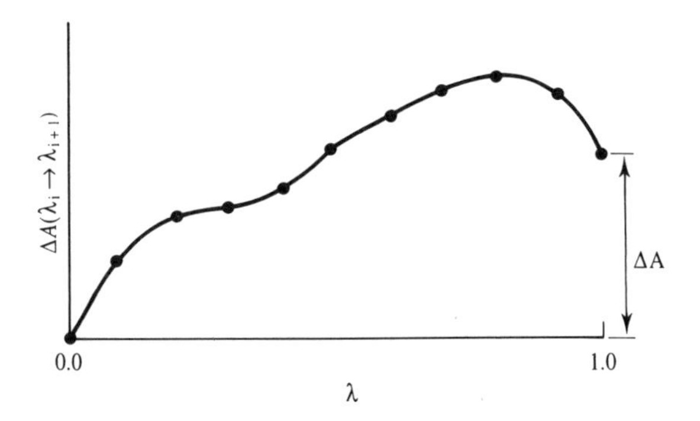

Fig. 9.2 Calculation of the free energy difference using perturbation.

sum of the free energy changes for the various values of λ_i, as shown in Figure 9.2.

The approach that we have described so far is known as *forward sampling* because the free energy is determined for $\lambda_i \rightarrow \lambda_{i+1}$. In *backward sampling*, the free energy difference between λ_i and λ_{i-1} is determined. The coupling parameter λ still increases from 0 to 1; it is just that the free energies are accumulated in a different manner. In *double-wide sampling* the free energy differences for both $\lambda_i \rightarrow \lambda_{i+1}$ as well as $\lambda_i \rightarrow \lambda_{i-1}$ are obtained from a simulation as illustrated in Figure 9.3. Consider point B in Figure 9.3, which corresponds to a coupling parameter λ_i. A simulation performed using λ_i can be used to obtain values for both the free energy difference $\Delta A(\lambda_i \rightarrow \lambda_{i+1})$ and for the free energy difference $\Delta A(\lambda_i \rightarrow \lambda_{i-1})$. This is clearly a more efficient way to obtain the desired free energy as twice as many free energy differences can be obtained from a single simulation.

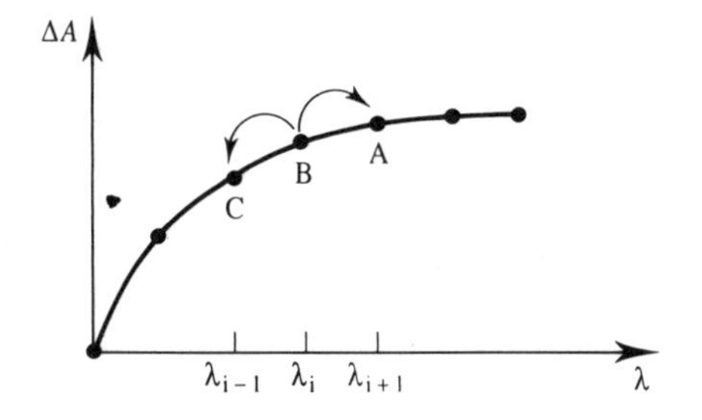

Fig. 9.3 Double wide sampling enables two free energies to be accumulated from a single simulation.

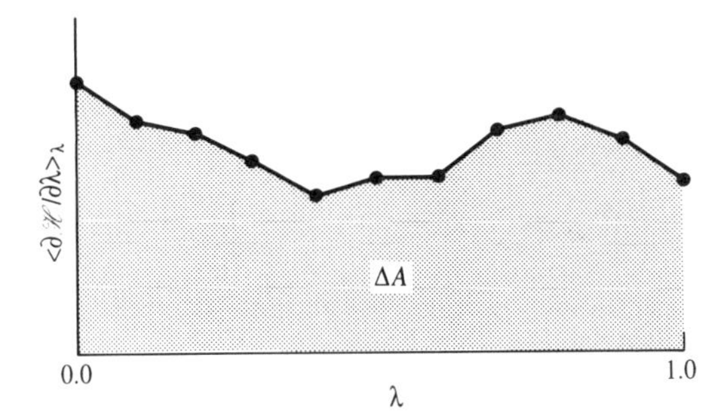

Fig. 9.4 Calculation of free energy differences by thermodynamic integration.

9.2.3 Thermodynamic integration

An alternative way to calculate the free energy difference uses thermodynamic integration. The formula for the free energy difference is derived in Appendix 9.1 and is:

$$\Delta A = \int_{\lambda=0}^{\lambda=1} \left\langle \frac{\partial \mathscr{H}(\mathbf{p}^N, \mathbf{r}^N)}{\partial \lambda} \right\rangle_\lambda \mathrm{d}\lambda \tag{9.18}$$

To calculate a free energy difference using thermodynamic integration we thus need to determine the integral in equation (9.18). In practice this is achieved by performing a series of simulations corresponding to discrete values of λ between 0 and 1. For each value of λ, the average of

$$\left\langle \frac{\partial \mathscr{H}(\mathbf{p}^N, \mathbf{r}^N)}{\partial \lambda} \right\rangle_\lambda$$

is determined. These partial derivatives are calculated analytically in some programs but in others a finite difference approximation is used ($\partial \mathscr{H}/\partial \lambda \approx \Delta \mathscr{H}/\Delta \lambda$). The total free energy difference ΔA is then equal to the area under the graph of

$$\left\langle \frac{\partial \mathscr{H}(\mathbf{p}^N, \mathbf{r}^N)}{\partial \lambda} \right\rangle_\lambda$$

versus λ as illustrated in Figure 9.4.

9.2.4 The 'slow growth' method

A third approach for the calculation of free energy differences from computer simulation is the slow growth method. Here, the Hamiltonian changes by a very small, constant amount at each step of the calculation. This means that at each stage the Hamiltonian $\mathscr{H}(\lambda_{i+1})$ is very nearly equal to $\mathscr{H}(\lambda_i)$. The free energy difference is given by:

$$\Delta A = \sum_{i=1;\lambda=0}^{i=N_{step};\lambda=1} (\mathscr{H}_{i+1} - \mathscr{H}_i) \tag{9.19}$$

This expression is derived in Appendix 9.2.

In principle, all three methods for calculating the free energy difference should give the same result, as the free energy is a state function and so independent of path. However, there may be practical reasons for choosing one method or another, as we shall discuss in section 9.6. The other point to note at this stage is that our formulation of the free energy has been in terms of the partition function, Q and the Hamiltonian $\mathscr{H}(\mathbf{p}^N, \mathbf{r}^N)$, which have contributions from both kinetic and potential energies. When the kinetic energy contributions are integrated out they cancel and so the various equations can be written in terms of the difference between the potential function, $\mathscr{V}(\mathbf{r}^N)$ rather than the Hamiltonian, $\mathscr{H}(\mathbf{p}^N, \mathbf{r}^N)$, Q is then replaced by the configurational integral, Z and the free energy values that are obtained are excess free energies, relative to an ideal gas.

Our discussion so far has considered the calculation of Helmholtz free energies, which are obtained by performing simulations at constant *NVT*. For proper comparison with experimental values we usually require the Gibbs free energy, G. Gibbs free energies are obtained using a simulation at constant *NPT*.

9.3 Applications of methods for calculating free energy differences

9.3.1 Thermodynamic cycles

One of the earliest applications of the free energy perturbation method was the determination of the free energy required to create a cavity in a solvent. Postma, Berendsen and Haak determined the free energy to create a cavity ($\lambda = 1$) in pure water ($\lambda = 0$) using isothermal–isobaric molecular dynamics simulations [Postma *et al.* 1982]. Five different cavity sizes were considered and the results showed that as expected the free energy of cavity formation increased with the size of the cavity, the results being in good agreement with analytical theories. For small cavities (< 1 Å radius) the results were inaccurate due to poor sampling. The calculations provided not only the free energy of cavity formation for the different cavity sizes but also structural and dynamic properties of the water molecules around the cavity. For example, the water structure varied with the cavity size. A cavity of 1.78 Å radius had the most pronounced shell structure with a high first-neighbour peak and a significant second-neighbour peak in the cavity–water pair distribution function.

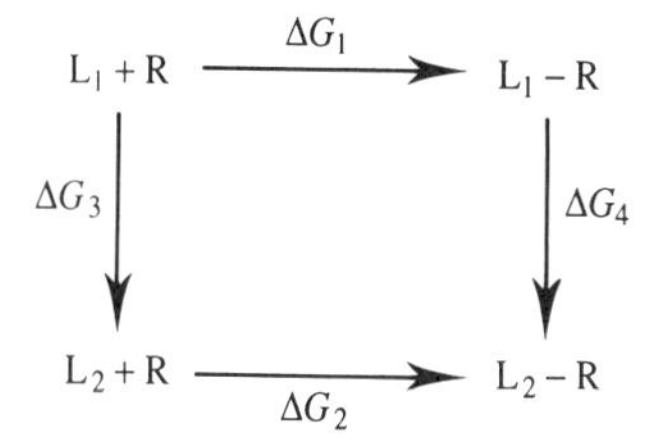

Fig. 9.5 Thermodynamic cycle for binding ligands L_1 and L_2 to receptor R.

Many processes of interest to the molecular modeller involve an equilibrium between molecules that interact via non-covalent forces, the free energy being related to the equilibrium constant by $\Delta G = -RT \ln K$. Let us consider the binding of two different ligands (L_1 and L_2) to a receptor molecule (R). L_1 and L_2 could be putative inhibitors of an enzyme R or two 'guests' for a host R. The thermodynamic cycle for the two binding processes is shown in Figure 9.5. The relative binding affinity of L_1 and L_2 equals $\Delta G_2 - \Delta G_1$ and is commonly written $\Delta\Delta G$. In principle, it would be possible to calculate values of ΔG_1 and ΔG_2 by simulating the actual association process. To do this we would bring the ligand and the receptor together from an initial large separation to gradually form the intermolecular complex. However, in most cases this would involve such a major reorganisation of the receptor, the ligand and the solvent that it would be difficult to ensure adequate sampling of phase space.

The free energy is a state function, and so its value round a thermodynamic cycle must be zero. Thus $\Delta G_2 - \Delta G_1 = \Delta G_4 - \Delta G_3$, Figure 9.5. ΔG_3 corresponds to the free energy difference of the two ligands in solution; ΔG_4 is the free energy difference of the two intermolecular complexes. The changes ΔG_3 and ΔG_4 do not correspond to any transformation that can be performed in the laboratory, but they are quite feasible in the computer. The free energy difference only depends upon the endpoints, and so we are at liberty to change the Hamiltonians in any way we wish. The free energy differences obtained from such non-physical pathways are likely to be much more reliable than the 'physically plausible' processes as they should involve much less reorganisation of the system. This is particularly so if the two ligands L_1 and L_2 have similar structures. To calculate the relative free energy of binding of the two ligands we would therefore 'mutate' L_1 into L_2 in solution and L_1 to L_2 within the receptor. This is the *thermodynamic cycle perturbation approach* to calculating relative free energies.

9.3.2 Applications of the thermodynamic cycle perturbation method

One of the first applications of the thermodynamic cycle perturbation approach to the calculation of relative binding constants was the study by Lybrand, McCammon and Wipff of the synthetic macrocycle SC24, which when proto-

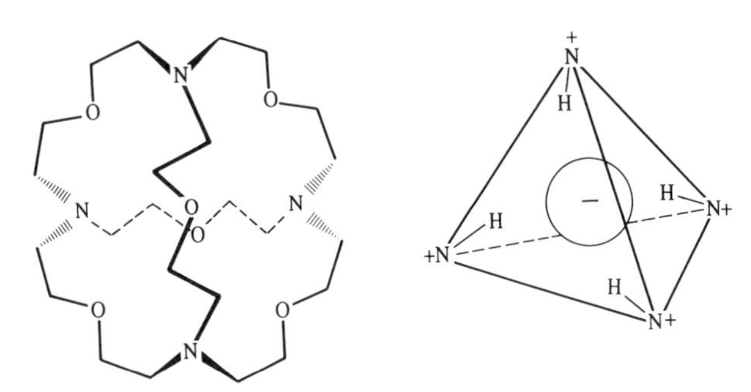

Fig. 9.6 The SC24/halide system. Figure adapted from Lybrand T P, J A McCammon and G Wipff 1986. Theoretical Calculation of Relative Binding Affinity in Host–Guest Systems. *Proceedings of the National Academy of Sciences USA* **83**:833–835.

nated can bind halide ions (Figure 9.6) [Lybrand *et al.* 1986]. SC24 binds Cl^- 4.30 kcal/mol more strongly than Br^-. Two simulations were performed to determine a theoretical value for this relative free energy using the free energy perturbation method with molecular dynamics. First, Cl^- was mutated to Br^- in aqueous solution, giving a free energy difference of 3.35 kcal/mol. The same mutation was then performed within the macrocycle, in a periodic box of waters. The value obtained for this step was 7.50 kcal/mol, giving an overall relative free energy of binding of 4.15 kcal/mol. The experimental value was approximately 4.3 kcal/mol. Thus, although the free energy to desolvate Cl^- is unfavourable compared to Br^-, this is more than compensated by favourable interactions between Cl^- and the host; Br^- is slightly too large to fit comfortably in the inflexible SC24 molecule.

One of the most attractive applications of the free energy techniques is for predicting the relative free energies of binding of inhibitors of biological macromolecules such as proteins or DNA. If we know the binding constant of an inhibitor then we can, in principle at least, calculate the binding constant of a related inhibitor. The free energy cycle used to perform this calculation is analogous to that shown in Figure 9.5: we perform two separate free energy calculations: ligand L_1 is mutated to ligand L_2 in solution and within the binding site. An early calculation of this type was performed by Bash and colleagues who studied two inhibitors of thermolysin (an enzyme which cleaves the amide bonds in peptides and proteins) [Bash *et al.* 1987]. The two inhibitors investigated had the general formula carbobenzoxy-Gly^{P}(X)-L-Leu-L-Leu [Barlett and Marlowe 1987] (Figure 9.7). The experimentally determined binding constants (K_i) of the $X \equiv NH$ and $X \equiv O$ inhibitors were 9.1 nM and 9000 nM, i.e. the former binds 1000 times more strongly. This difference in binding constants is equivalent to 4.1 kcal/mol. X-ray crystallographic analysis showed that the two inhibitors bind in almost identical positions. The calculated free energy difference was determined to be 4.2 ± 0.5 kcal/mol, in good agreement with the experimental result. In the active site of the enzyme the X group of the inhibitor interacts with the backbone carbonyl oxygen of one of the amino acids (Ala 113). The ester

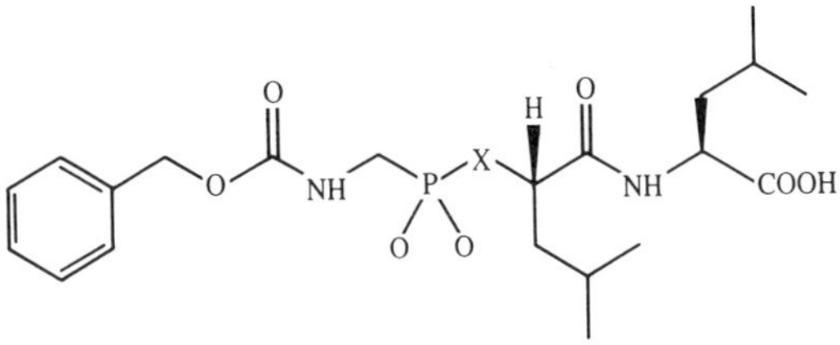

X = NH, X = O

Fig. 9.7 Thermolysin inhibitors [Bartlett and Marlowe 1987].

oxygen interacts unfavourably with this carbonyl but the amide can form a hydrogen bond. The relative free energy of binding of the amide inhibitor to the protein was calculated to be 7.6 kcal/mol lower than the ester but this was counteracted by the difference in the free energies of solvation which was calculated to be 3.4 kcal/mol. The amide inhibitor thus incurs a greater desolvation penalty than the ester.

This study obviously gave very satisfactory agreement with the experimental data. However, a subsequent calculation by Merz and Kollman showed that the results were very sensitive to the charge model used for the inhibitor [Merz and Kollman 1989]. The charges for the inhibitor were obtained by electrostatic potential fitting in each case, though with different basis sets. This second calculation gave a free energy difference of 5.9 kcal/mol. Other studies have also shown that calculated free energies can be very sensitive to the charge model used; we will discuss some of the problems with performing free energy calculations in section 9.6.

As one final example of the application of free energy calculations we will examine the determination of relative partitition coefficients. The partition coefficient (P) is the equilibrium constant for the transfer of a solute between two solvents. The logarithm of the partition coefficient ($\log P$) for transfer between water and a variety of solvents (primarily 1-octanol) is widely used to derive structure–activity relationships (see section 10.7), in which the biological activity of a molecule is correlated with its physicochemical properties. The thermodynamic cycle for the partition of two solutes, A and B, between two solvents is shown in Figure 9.8. If it were possible to calculate the free energy of transfer from one solvent to another (i.e. ΔG_1 or ΔG_2 in Figure 9.8) then this would give the partition coefficient directly. However, such a simulation would require an inordinate amount of time and probably be very inaccurate. A relative partition coefficient

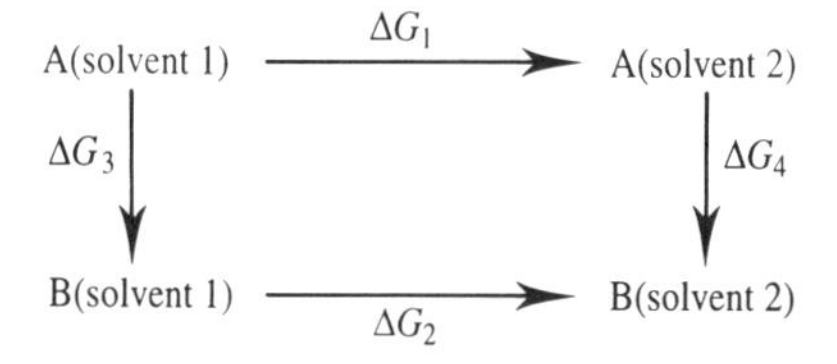

Fig. 9.8 Thermodynamic cycle for calculating relative partition coefficients.

can be determined by mutating one solute into the other in the two separate solvents.

Calculations of relative partition coefficients have been reported using the free energy perturbation method with the molecular dynamics and Monte Carlo simulation methods. For example, Essex, Reynolds and Richards calculated the difference in partition coefficients of methanol and ethanol partitioned between water and carbon tetrachloride with molecular dynamics sampling [Essex *et al.* 1989]. The results agreed remarkably well with experiment (within 0.06 kcal/mol). Jorgensen, Briggs and Contreras used Monte Carlo methods to calculate the relative partition coefficients for eight pairs of solutes (including methanol/thylamine, acetic acid/acetamide and pyrazine/pyridine) between water and chloroform [Jorgensen *et al.* 1990]. For these eight systems good qualitative agreement with experimental data was obtained. However, the results involving acetic acid gave too broad a spread of values. This was traced to the relative free energies of hydration which varied over too wide a range and indicated some areas for improvement in the force field model.

9.3.3 The calculation of absolute free energies

In some cases it is possible to devise thermodynamic cycles which enable the absolute free energy of a change to be determined using free energy perturbation methods [Jorgensen *et al.* 1988]. Figure 9.9 shows a thermodynamic cycle for the association of L and R to give a complex LR in both the gas phase and in solution. ΔG_{ass} is the free energy of association in solution and is given by:

$$\Delta G_{ass} = \Delta G_{gas}(L + R \rightarrow LR) + \Delta G_{sol}(LR) - \Delta G_{sol}(L) - \Delta G_{sol}(R) \tag{9.20}$$

$\Delta G_{sol}(X)$ is the solvation free energy of the species X (the free energy of transfer from the gas phase to solvent). The solvation free energy can be written in terms of perturbations where the species disappear to nothing in the gas phase and in solution, $\Delta G_{sol}(X) = \Delta G_{gas}(X \rightarrow 0) - \Delta G_{sol}(X \rightarrow 0)$. The free energy of association, ΔG_{ass} can then be written:

$$\begin{aligned}\Delta G_{ass} = {} & \Delta G_{gas}(L + R \rightarrow LR) - \Delta G_{gas}(L \rightarrow 0) + \Delta G_{sol}(L \rightarrow 0) \\ & - \Delta G_{gas}(R \rightarrow 0) + \Delta G_{sol}(R \rightarrow 0) + \Delta G_{gas}(LR \rightarrow 0) \\ & - \Delta G_{sol}(LR \rightarrow 0)\end{aligned} \tag{9.21}$$

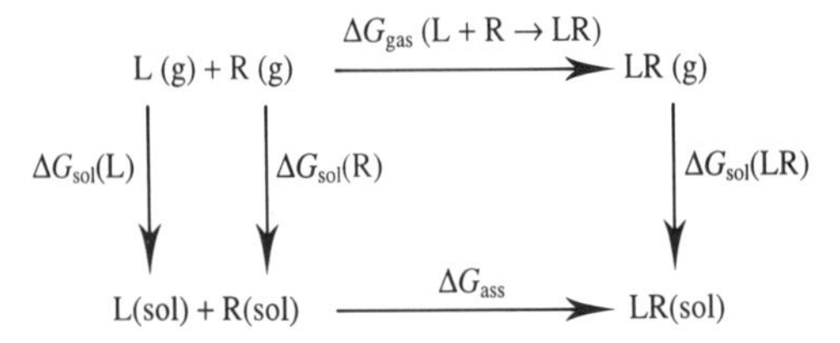

Fig. 9.9 Thermodynamic cycle used to calculate absolute free energies [Jorgensen *et al.* 1988].

The gas-phase terms cancel and $\Delta G_{sol}(\text{LR} \rightarrow 0)$ can be written as the sum of two separate calculations:

$$\Delta G_{sol}(\text{LR} \rightarrow 0) = \Delta G_{sol}(\text{LR} \rightarrow R) + \Delta G_{sol}(\text{R} \rightarrow 0)$$

Thus, the overall free energy change can be written:

$$\Delta G_{ass} = \Delta G_{sol}(\text{L} \rightarrow 0) - \Delta G_{sol}(\text{LR} \rightarrow \text{R}) \qquad (9.22)$$

We thus need only perform two simulations, L to nothing in water and L to nothing in the LR complex. The first application of this approach was to the association of two methane molecules in water, where both species (L and R) are identical. In general, L should be chosen as the smaller component.

9.4 The calculation of enthalpy and entropy differences

Free energy changes can now be routinely calculated with the errors being less than 1 kcal/mol in favourable cases. How does this compare with the error with which the enthalpy or entropy difference can be determined? One way to determine the enthalpy change would be to perform two separate simulations, one of the initial system and one of the final system. For example, the difference in the enthalpy of solvation of ethanol and ethane thiol in water could be determined by simulating the two species separately and then taking the difference in the total enthalpies of the two systems. These total energies are invariably large numbers, with relatively large errors. The error in the calculated enthalpy difference would be comparable in magnitude to the error in the energy of each system. By contrast, the free energy is determined solely in terms of interactions involving the solute. This means that the free energy can be calculated much more accurately. More efficient ways to calculate the enthalpy and entropy change have been proposed for use with both free energy perturbation and thermodynamic integration schemes [Fleischman and Brooks 1987; Yu and Karplus 1988]. The uncertainties in the enthalpies and entropies so calculated are better than would be obtained by subtracting the differences in total energies, but they are still about one order of magnitude larger than the corresponding free energies.

9.5 Partitioning the free energy

The overall free energy can be partitioned into individual contributions if the thermodynamic integration method is used [Boresch *et al.* 1994]. The starting point is the thermodynamic integration formula for the free energy:

$$\Delta A = \int_{\lambda=0}^{\lambda=1} \left\langle \frac{\partial \mathcal{H}(\mathbf{p}^N, \mathbf{r}^N)}{\partial \lambda} \right\rangle_\lambda \mathrm{d}\lambda$$

The Hamiltonian can be written as a sum of contributions from bond stretching, angle bending and so on:

$$\left\langle \frac{\partial \mathcal{H}(\lambda)}{\partial \lambda} \right\rangle_\lambda = \left\langle \frac{\partial \mathcal{H}_{\text{bonds}}(\lambda)}{\partial \lambda} + \frac{\partial \mathcal{H}_{\text{angles}}(\lambda)}{\partial \lambda} + \cdots \right\rangle_\lambda \qquad (9.23)$$

So the free energy is given by:

$$\Delta A = \int_{\lambda=0}^{\lambda=1} \left\langle \frac{\partial \mathcal{H}_{\text{bonds}}(\lambda)}{\partial \lambda} \right\rangle_{\lambda} \mathrm{d}\lambda + \int_{\lambda=0}^{\lambda=1} \left\langle \frac{\partial \mathcal{H}_{\text{angles}}(\lambda)}{\partial \lambda} \right\rangle_{\lambda} \mathrm{d}\lambda + \cdots$$
$$= \Delta A_{\text{bonds}} + \Delta A_{\text{angles}} + \cdots \tag{9.24}$$

We should remember that only the sum of the contributions is truly meaningful as the individual contributions are not state functions. This has led some to criticise any use of such partitioning schemes, though they may be useful for indicating which interactions contribute the most to the overall free energy, and may suggest the source of most of the error in the calculation. It is not possible to perform such a partitioning using thermodynamic perturbation.

An example of this partitioning scheme is the study by Ha and colleagues of the anomeric equilibrium between the α and β anomers of D-glucose [Ha *et al.* 1991]. D-glucose can exist in two tautomeric forms: α-D glucose in which the C_1 hydroxyl group is axial, and β-D glucose in which it is equatorial (Figure 9.10). In the gas phase the axial α isomer is more stable than the equatorial β isomer due to the anomeric effect which is considered to arise from unfavourable dipole–dipole interactions and delocalisation of the lone pair on the ring oxygen into an anti-bonding σ^* orbital. However, the β-D (equatorial) anomer is more stable than the α-D (axial) anomer by 0.3 kcal/mol in aqueous solution. The free energy difference between the two isomers in water was calculated by Ha *et al.* using both free energy perturbation and thermodynamic integration to be -0.3 ± 0.4 kcal/mol for $\beta \rightarrow \alpha$. A partitioning of the free energy showed that this small difference was found to arise from the cancellation of two large terms: the α isomer was predicted to be 3.6 kcal/mol more favourable than the β isomer in the gas phase, due mainly to electrostatic effects. However, the β isomer was favoured over the α isomer in aqueous solution, again due to electrostatic effects, such as the enhanced hydrogen bonding capability of the β isomer with the solvent. We should note that these results were not supported by a comparable calculation performed using semi-empirical quantum mechanics, as discussed in section 9.9.2.

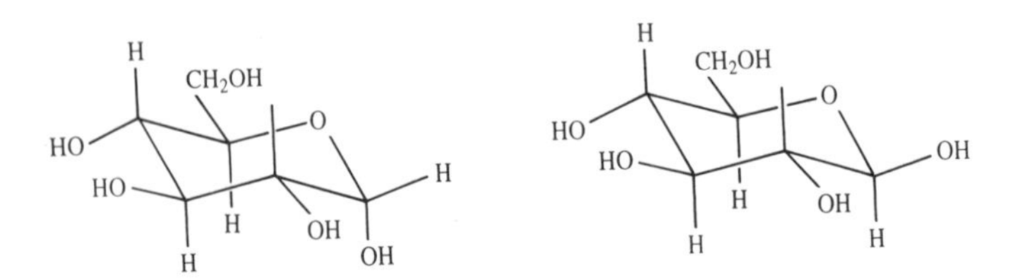

Fig. 9.10 The α and β anomers of D-glucose.

9.6 Potential pitfalls with free energy calculations

There are two major sources of error associated with the calculation of free energies from computer simulations. Errors may arise from inaccuracies in the Hamiltonian, be it the potential model chosen or its implementation (the treatment of long-range forces, etc.). The second source of error arises from an insufficient sampling of phase space.

Unfortunately there is no set recipe that guarantees adequate coverage of phase space and thus reliable free energy values [Mitchell and McCammon 1991]. The errors associated with inadequate sampling may be identified by running the simulation for longer periods of time (molecular dynamics) or for more iterations (Monte Carlo); the perturbation can be performed in both forward and reverse directions; a different scheme could be used to determine the free energy difference (e.g. thermodynamic perturbation and thermodynamic integration). At the very least, the simulation should be run in both directions; the difference in the calculated free energy values (often referred to as the *hysteresis*) gives a lower-bound estimate of the error in the calculation.

One possible pitfall to be aware of when estimating errors is that an excessively short simulation may give an almost zero difference between the forward and reverse directions. If the time of the simulation is much longer than the relaxation time of the system then the change can be performed reversibly. If the simulation time is of the same order of magnitude as the relaxation time then one would expect a significant degree of hysteresis. However, if the simulation is much shorter than the relaxation time then approximately zero hysteresis may result, due to the inability of the system to adjust to the changes. In such a situation the free energies for both forward and reverse directions may be approximately the same, but quite likely incorrect.

9.6.1 Implementation aspects

The allure of methods for calculating free energies and their associated thermodynamic values such as equilibrium constants has resulted in considerable interest in free energy calculations. A number of decisions must be made about the way that the calculation is performed. One obvious choice concerns the simulation method; in principle, either Monte Carlo or molecular dynamics can be used; in practice molecular dynamics is almost always used for systems where there is a significant degree of conformational flexibility whereas Monte Carlo can give very good results for small molecules which are either rigid or have limited conformational freedom.

One must choose from the thermodynamic perturbation, thermodynamic integration and slow growth methods. Each of these methods has been extensively used, but the slow growth method is not now recommended. This method suffers from a phenomenon known as 'Hamiltonian lag'; the system never has time to properly equilibrate for a given value of the coupling parameter because the

$CH_3CHO \rightleftharpoons CH_2{=}CHOH$

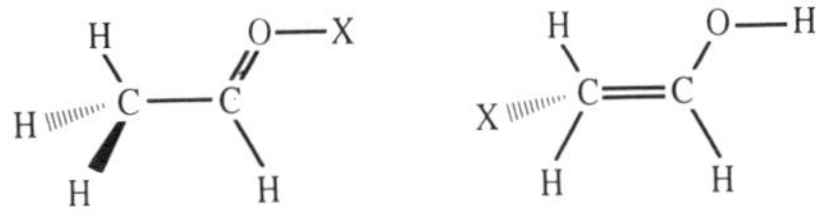

Fig. 9.11 To calculate the free energy difference between the aldehyde and enol forms of acetaldehyde the single topology method uses dummy atoms (X).

potential function changes at every step. An additional advantage of the integration and perturbation approaches is that, should one decide at the end of a simulation that more sampling needs to be done for particular values of λ, or that more λ values are required over a particular range, then this can easily be done without losing information from other parts of the calculation. With slow growth one would have to redo the simulation from scratch.

Prior to the calculation, the increment $\delta\lambda$ in the coupling parameter must be specified. Traditionally, $\delta\lambda$ is set to a constant value before the simulation commences. It is important that there is enough overlap between successive states λ_i and λ_{i+1} so that reliable values can be obtained. An alternative approach is to use small changes in λ when the free energy is changing quickly, and a larger change in λ when the free energy is changing more slowly. This is the basis of a method called *dynamically modified windows*, in which the slope of the free energy versus λ curve is used to determine the value of $\delta\lambda$ to use in the next iteration [Pearlman and Kollman 1989].

As the free energy is a thermodynamic state function the free energy difference between the initial and final states should be independent of the path along which the change is made, so long as it is reversible. It may be possible to proceed from the initial to the final states along more than one pathway. A change that involves high energy barriers will require much smaller increments to be made in the coupling parameter λ to ensure reversibility than a pathway that proceeds via a lower barrier.

Many free energy calculations involve changes in the molecular topologies of the species concerned; there are often different numbers of atoms in the initial and final states and the atoms may be bonded in different ways. For example, suppose we wish to determine the free energy difference between acetaldehyde and its enol, Figure 9.11, in which a hydrogen atom migrates from the methyl carbon atom to the carbonyl oxygen. The system can be represented within the calculation using either a 'single' topology or a 'dual' topology. In the single topology method, the molecular topology at all stages is the union of the initial and final states, using dummy atoms where necessary. A dummy atom does not interact with the other atoms in the system. Thus the hydrogen atom bonded to the oxygen in the enol form would be represented as a dummy atom when the simulation reached the endpoint corresponding to the aldehyde as shown in Figure 9.11.

The alternative to the single topology representation is the dual topology method. Here, both the molecular topologies are maintained during the entire simulation, such that both species 'exist' (in a topological sense) but do not interact with each other. The Hamiltonian that describes the interaction between these groups and the environment can be described in a number of ways, the simplest of which is the linear relationship:

$$\mathscr{H}(\lambda) = \lambda\mathscr{H}_{\mathrm{Y}} + (1 - \lambda)\mathscr{H}_{\mathrm{X}} \tag{9.25}$$

9.7 Potentials of mean force

The free energy changes that we have considered so far correspond to chemical 'mutations'. We may also be interested to know how the free energy changes as a function of some inter- or intramolecular coordinate, such as the distance between two atoms, or the torsion angle of a bond within a molecule. The free energy surface along the chosen coordinate is known as a *potential of mean force* (PMF). When the system is in a solvent, the potential of mean force incorporates solvent effects as well as the intrinsic interaction between the two particles. Potentials of mean force were introduced in our discussion of Langevin dynamics (section 6.8), where we noted that the ratio of *trans* to *gauche* conformers of 1,2-dichloroethane was significantly different in the liquid than for an isolated molecule. Unlike the mutations so common in free energy perturbation calculations which are often along non-physical pathways, the potential of mean force is calculated for a physically achievable process. Consequently, the point of highest energy on the free energy profile that is obtained from a PMF calculation corresponds to the transition state for the process, from which it is possible to derive kinetic quantities such as rate constants.

Various methods have been proposed for calculating potentials of mean force. The simplest type of PMF is the free energy change as the separation (r) between two particles is changed. We might anticipate that we could calculate the potential of mean force from the radial distribution function using the following expression for the Helmholtz free energy:

$$A(r) = -k_{\mathrm{B}}T \ln g(r) + \text{constant} \tag{9.26}$$

The constant is often chosen so that the most probable distribution corresponds to a free energy of zero.

Unfortunately, the potential of mean force may vary by several multiples of $k_{\mathrm{B}}T$ over the relevant range of the parameter r. The logarithmic relationship between the potential of mean force and the radial distribution function means that a relatively small change in the free energy (i.e. a small multiple of $k_{\mathrm{B}}T$) may correspond to $g(r)$ changing by an order of magnitude from its most likely value. Unfortunately, standard Monte Carlo or molecular dynamics simulation methods do not adequately sample regions where the radial distribution function differs drastically from the most likely value, leading to inaccurate values for the potential of mean force. The traditional way to avoid this problem uses a technique called *umbrella sampling*.

9.7.1 Umbrella sampling

Umbrella sampling attempts to overcome the sampling problem by modifying the potential function so that the unfavourable states are sampled sufficiently. The method can be used with both Monte Carlo and molecular dynamics simulations. The modification of the potential function can be written as a perturbation:

$$\mathscr{V}'(\mathbf{r}^N) = \mathscr{V}(\mathbf{r}^N) + W(\mathbf{r}^N) \tag{9.27}$$

$W(\mathbf{r}^N)$ is a weighting function, which often takes a quadratic form:

$$W(\mathbf{r}^N) = k_W(\mathbf{r}^N - \mathbf{r}_0^N)^2 \tag{9.28}$$

For configurations that are far from the equilibrium state $\mathbf{r}_0^N$ the weighting function will be large and so a simulation using the modified energy function $\mathscr{V}'(\mathbf{r}^N)$ will be biased away from the configuration $\mathbf{r}_0^N$. The resulting distribution will, of course, be non-Boltzmann. The corresponding Boltzmann averages can be extracted from the non-Boltzmann distribution using a method introduced by Torrie and Valleau [Torrie and Valleau 1977]. The result is:

$$\langle A \rangle = \frac{\langle A(\mathbf{r}^N)\exp[+W(\mathbf{r}^N)/k_B T]\rangle_W}{\langle \exp[+W(\mathbf{r}^N)/k_B T]\rangle_W} \tag{9.29}$$

The subscript W indicates that the average is based on the probability $P_W(\mathbf{r}^N)$, which in turn is determined by the modified energy function $\mathscr{V}'(\mathbf{r}^N)$. For example, to obtain the potential of mean force via the radial distribution function (equation (9.26)) the distribution function with the forcing potential would be determined and then corrected to give the 'true' radial distribution function from which the free energy can be calculated as a function of the separation. It is usual to perform an umbrella sampling calculation in a series of stages, each of which is characterised by a particular value of the coordinate and an appropriate value of the forcing potential $W(\mathbf{r}^N)$. However, if the forcing potential is too large, the denominator in equation (9.29) is dominated by contributions from only a few configurations with especially large values of $\exp[W(\mathbf{r}^N)]$ and the averages take too long to converge.

To illustrate the use of umbrella sampling let us consider how the technique has been used to determine the potential of mean force for rotation of the central C–C bond of butane in aqueous solution. The barrier between the *trans* and *gauche* conformations of butane is approximately 3.5 kcal/mol which is sufficiently high to give sampling problems in simulations. For example, in the molecular dynamics simulation of Ryckaert and Bellemans the mean time between *gauche–trans* transitions was about 10 ps [Ryckaert and Bellemans 1978]. Jorgensen, Gao and Ravimohan used umbrella sampling with Monte Carlo simulations to calculate the potential of mean force as the central bond in butane is rotated in a periodic box of water molecules, to determine the effect of the solvent on the relative populations of the different conformations [Jorgensen *et al.* 1985]. The results predicted a shift in the expected populations of *trans* and *gauche* isomers from 68% *trans* in the gas phase to 54% in aqueous solution, a change of 14% In addition, the barrier height was reduced in solution. Jorgensen and colleagues performed many calculations on similar systems using umbrella sampling and Monte Carlo simulations; he recommended that to reduce the barriers to a value between 1 kcal/mol and 3 kcal/mol was appropriate. In some cases it is possible to use a barrier height of zero, though the barriers cannot be reduced too severely as this makes the forcing potential too large.

It is also possible to calculate potentials of mean force using the free energy perturbation method using a molecular dynamics or Monte Carlo simulation. As usual, the calculation is broken into a series of steps that are characterised by a coupling parameter λ. With molecular dynamics, holonomic constraint methods are used to fix the desired coordinates without affecting the dynamical motion of the system. This is the essence of the extension of the SHAKE procedure by Tobias and Brooks to cope with general coordinate changes [Tobias and Brooks 1988] (see section 6.5). In a Monte Carlo simulation the required coordinates are simply fixed at the desired value(s). This contrasts with umbrella sampling in which the coordinate(s) of interest would be able to vary over their range of values throughout the simulation, subjected to a potential that has been modified using the forcing function. At each step of the perturbation calculation, the difference in the energy between the configuration and the configuration that corresponds to $\lambda + \delta\lambda$ is determined and the free energy accumulated in the appropriate way.

To compare the perturbation and umbrella sampling methods for calculating potentials of mean force, Jorgensen and Buckner repeated the PMF calculation for butane in water using the perturbation method [Jorgensen and Buckner 1987]. The *gauche* population was calculated to increase by 12.3% using this method, in accordance with the previous umbrella sampling calculations. Jorgensen put forward several arguments in favour of the perturbation approach. A major concern with umbrella sampling is that a proper sampling of the phase space may not be achieved. In some cases the presence of bottlenecks in phase space may be identified if separate simulations starting from different configurations give different results but even this approach is not fail-safe as all simulations may encounter the same problem. Indeed, Jorgensen suggested that just such a bottleneck may have occurred in a previous simulation of pentane in water using umbrella sampling (which involved 5 million Monte Carlo steps). The only real problem with the perturbation method is the need to choose an appropriate value of $\delta\lambda$, so that there is adequate overlap between the configuration corresponding to λ and that corresponding to $\lambda + \delta\lambda$. Jorgensen and Buckner varied the central torsion angle in their simulation of butane using 15° increments.

9.7.2 Calculating the potential of mean force for flexible molecules

To calculate a potential of mean force using free energy perturbation (or indeed umbrella sampling) it is necessary to determine the pathway for the transition of interest. This is trivial for simple problems such as the separation of two particles or the rotation of butane, but can be quite complicated for more detailed changes

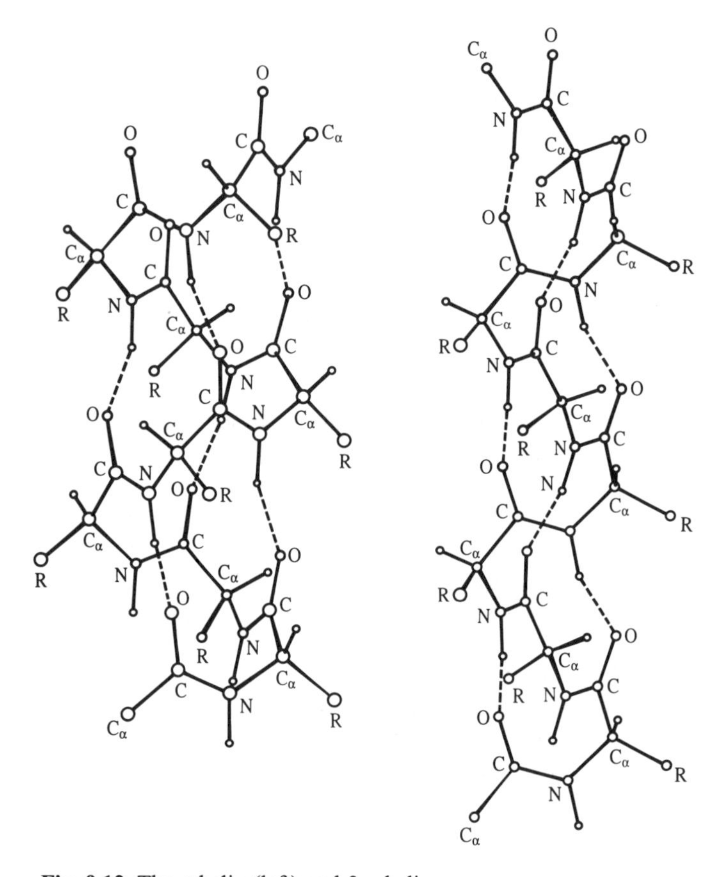

Fig. 9.12 The α-helix (left) and 3_{10}-helix.

such as conformational interconversions. The reaction path methods discussed in section 4.9.3 may be helpful in determining these pathways.

To illustrate the calculation of potentials of mean force for flexible systems we will consider helical conformations of polypeptide chains. We have already met the α-helix that is commonly observed in protein structures (see section 8.13.1). In this conformation hydrogen bonds are formed between residues i and $i + 4$. Polypeptide chains can also form a different type of helix, called a 3_{10}-helix. Here, the hydrogen bonds are formed between residues i and $i + 3$. These two helices are compared in Figure 9.12. The backbone conformations of such helices do not differ significantly; the α-helix has backbone torsion angles ($\phi = -60°$, $\psi = -50°$) and the 3_{10}-helix has ($\phi = -50°$, $\psi = -28°$). The 3_{10}-helix is found to a small extent in protein structures, usually at the ends of α-helices. However, the 3_{10}-helix is much more common in peptides formed from α,α-dialkyl amino acids, which have two alkyl substituents at the α-carbon atom. The prototypical member of this class of amino acids is α-methylalanine (MeA; see Figure 9.13). Peptides containing this amino acid can form both α-helices and

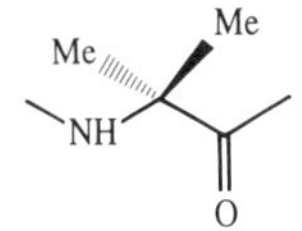

Fig. 9.13 Methylalanine.

3_{10}-helices with the actual conformation present being rather sensitive to the conditions; such peptides are 3_{10}-helical in $CDCl_3$ and α-helical in $(CD)_3SO$.

To calculate the potential of mean force for interconverting the α-helix to the 3_{10}-helix requires an appropriate reaction coordinate to be determined. Here we describe the calculations of three groups who all used different approaches. Smythe, Huston and Marshall studied a decamer of α-methylalanine, CH_3CO-MeA_{10}-NMe using umbrella sampling [Smythe *et al.* 1993; Smythe *et al.* 1995]. They used the self penalty walk method described in section 4.9.3 to determine the transition pathway and observed that the reaction coordinate correlated well with a smooth change in the end-to-end distance from the 3_{10}-helix (19 Å) to the α-helix (13 Å). Their umbrella sampling calculations were performed using molecular dynamics with the end-to-end distance being subjected to a restraining potential. Simulations were performed in various solvents; in water the free energy change for the α-helix $\rightarrow$ 3_{10}-helix transition was calculated to be 7.6 kcal/mol with the value in dichloromethane being 5.8 kcal/mol and *in vacuo* 3.2 kcal/mol. Although a distinct energy barrier was found for the vacuum calculations, no transition barrier was found for either solution calculation.

Zhang and Hermans studied a 10-residue alanine peptide as well as a 10-residue α-methylalanine peptide, *in vacuo* and in water [Zhang and Hermans 1994]. In their calculations the transition from one conformation to the other was performed using a restraining potential that forced the structure to exchange one set of hydrogen bonds to the set of hydrogen bonds appropriate to the other structure. This additional potential function could be used to drive the molecule back and forth between the two conformations by varying a coupling parameter, λ, between 0 and 1. Free energy profiles were determined for the α-helix to 3_{10}-helix transition using molecular dynamics and the slow growth method. The results showed that the alanine peptide had a clear preference for the α-helix both *in vacuo* and in water but that the free energy change for the MeA peptide was approximately zero in water and that the 3_{10}-helix was preferred *in vacuo*. It was proposed by Zhang and Hermans that the discrepancy between these results for the α-methylalanine peptide and those obtained by Smythe, Huston and Marshall was probably due to the different force field models employed; Smythe *et al.* used a united atom model whereas Zhang and Hermans used an all-atom model.

Tirado-Reeves, Maxwell and Jorgensen used yet another approach for calculating the potential of mean force, this time for an undecaalanine peptide in water [Tirado-Reeves *et al.* 1993]. The free energy profile was calculated using the perturbation method and Monte Carlo simulations by gradually varying the ψ backbone torsion angles, keeping the ϕ torsion angles fixed at $-60°$. The free energy difference between the two conformations was calculated to be 10.6 kcal/mol in favour of the α-helix, with a small activation barrier of 2.8 kcal/mol for the 3_{10}-

to α-helical transition. *In vacuo* a larger free energy difference was predicted (13.6 kcal/mol).

These three studies have been described at some length, in part to illustrate the different approaches now available for calculating thermodynamic properties of complex systems but also to emphasise the fact that different methods can given quite different (and sometimes contradictory) results. Such comparative studies serve to highlight the fact that it is necessary to examine critically the methods and models used in a calculation. All three studies were in part prompted by experimental electron spin resonance results that suggested that a 16-residue alanine-based peptide adopted a 3_{10}-helical conformation in water [Miick *et al.* 1992]. These results were contradicted by all the simulations, and indeed prompted Smythe and Marshall to undertake similar experiments on their conformationally constrained peptides, experiments which showed that these peptides were α-helical, in agreement with the calculations.

9.8 Continuum representations of the solvent

Most chemical processes take place in a solvent and so it is clearly important to consider how the solvent affects the behaviour of a system. In some cases, solvent molecules are directly involved, as in ester hydrolysis reactions or in systems where solvent molecules are so tightly bound that they are effectively an integral part of the solute. Such solvent molecules should be modelled explicitly. In other systems the solvent does not directly interact with the solute but it provides an environment that strongly affects the behaviour of the solute. For example, the highly anisotropic environment within a liquid crystal or lipid bilayer strongly influences the conformations of dissolved solutes. Here, it may not be necessary to explicitly model the solvent molecules, though special treatments such as mean field theories (see section 6.10) may be required. In the third case, the solvent merely acts to provide a 'bulk medium' but can still significantly affect solute behaviour with the dielectric properties of the solvent often being particularly important. In this case it would clearly be useful not to have to explicitly include every single solvent molecule in the system, to enable us to concentrate on the behaviour of the solute(s). The solvent acts as a perturbation on the gas-phase behaviour of the system. This is the purpose of the 'continuum' solvent models [Smith and Pettitt 1994]. A considerable variety of such models have been proposed, for use with both quantum mechanics and empirical models. Our discussion will be restricted to a few of the more widely used methods.

9.8.1 Thermodynamic background

The solvation free energy (ΔG_{sol}) is the free energy change to transfer a molecule from vacuum to solvent. The solvation free energy can be considered to have three components:

$$\Delta G_{\text{sol}} = \Delta G_{\text{elec}} + \Delta G_{\text{vdw}} + \Delta G_{\text{cav}} \tag{9.30}$$

ΔG_{elec} is the electrostatic component. This contribution is particularly important for polar and charged solutes due to the polarisation of the solvent, which we will model as a uniform medium of constant dielectric ε. ΔG_{vdw} is the van der Waals interaction between the solute and solvent; this may in turn be divided into a repulsive term, ΔG_{rep}, and an attractive dispersion term, ΔG_{disp}. ΔG_{cav} is the free energy required to form the solute cavity within the solvent. This component is positive and comprises the entropic penalty associated with the reorganisation of the solvent molecules around the solute together with the work done against the solvent pressure in creating the cavity. In addition to the above three components, an explicit hydrogen-bonding term, ΔG_{hb} may be added for those systems where there is localised hydrogen bonding between the solute and solvent. Initially, we will discuss the electrostatic contribution to the free energy of solvation. We will then consider the van der Waals and cavity contributions.

9.9 The electrostatic contribution to the free energy of solvation: the Born and Onsager models

Two important contributions to the study of solvation effects were made by Born (in 1920) and Onsager (in 1936). Born derived the electrostatic component of the free energy of solvation for placing a charge within a spherical solvent cavity [Born 1920] and Onsager extended this to a dipole in a spherical cavity (Figure 9.14) [Onsager 1936]. In the Born model, ΔG_{elec} of an ion is equal to the work done to transfer the ion from vacuum to the medium. This in turn is equal to the difference in the electrostatic work to charge the ion in the two environments. The work to charge an ion in a medium of dielectric constant ε equals $q^2/2\varepsilon a$ where

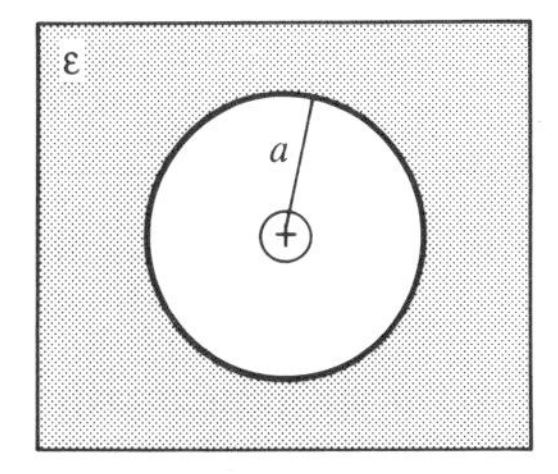

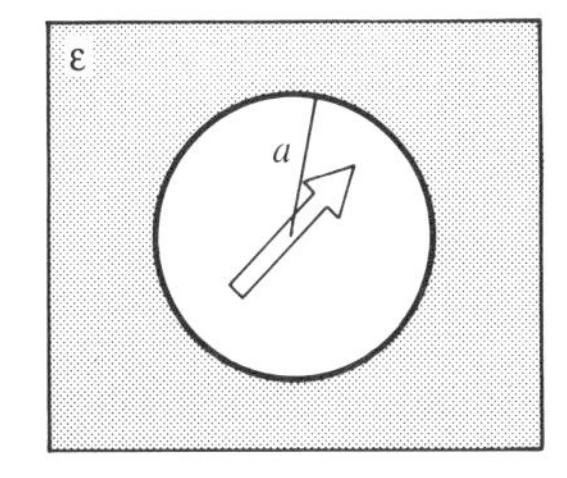

Fig. 9.14 The Born and Onsager models.

q is the charge on the ion and a is the radius of the cavity. The electrostatic contribution to the solvation free energy is thus the difference in the work done in charging the ion in the dielectric and in vacuo:

$$\Delta G_{\mathrm{elec}} = -\frac{q^2}{2a}\left(1 - \frac{1}{\varepsilon}\right) \tag{9.31}$$

Note that in this equation, as throughout our discussion, we have used reduced electrostatic units, in which the factor $4\pi\varepsilon_0$ is ignored. This is common practice in the literature. The Born model is very simple, yet can be quite successful. It is necessary to choose a set of cavity radii. Traditionally, ionic radii from crystal structures are used. However, for the alkali halides it is found that adding 0.1 Å to the radii of anions and 0.85 Å to the radii of cations gives much better agreement with experimental data. Justification for this adjustment was provided by Rashin and Honig who examined electron density distributions in crystals and concluded that the ionic radii are reasonably good indicators of cavity size for anions but that for cations it is more appropriate to use covalent radii [Rashin and Honig 1985]. They subsequently suggested that the optimal agreement with experiment could be obtained by increasing these radii by an empirical factor of 7%.

9.9.1 Calculating the electrostatic contribution via quantum mechanics

The Born model is obviously only appropriate to species with a formal charge. Onsager's dipole model is relevant to many more molecules (in fact, the Onsager model is a special case of the result derived by Kirkwood [Kirkwood 1934] who considered an arbitrary distribution of charges within a spherical cavity). The solute dipole within the cavity induces a dipole in the surrounding medium which in turn induces an electric field within the cavity (the *reaction field*). The reaction field then interacts with the solute dipole, so providing additional stabilisation of the system. The magnitude of the reaction field was determined by Onsager to be:

$$\phi_{\mathrm{RF}} = \frac{2(\varepsilon - 1)}{(2\varepsilon + 1)a^3}\mu \tag{9.32}$$

μ is the dipole moment of the solute; a and ε are the radius of the cavity and the dielectric constant of the medium, as before. The energy of a dipole in an electric field ϕ_{RF} is $-\phi_{\mathrm{RF}}\mu$, but for a polarisable dipole it is necessary to add an additional term which represents the work done assembling the charge distribution within the cavity. This additional term has magnitude $\phi_{\mathrm{RF}}\mu/2$ and so the electrostatic contribution to the free energy of solvation in this model is:

$$\Delta G_{\mathrm{elec}} = -\frac{\phi_{\mathrm{RF}}\mu}{2} \tag{9.33}$$

If the species is charged then an appropriate Born term must also be added. The reaction field model can be incorporated into quantum mechanics, where it is commonly referred to as the *self-consistent reaction field* (SCRF) method, by considering the reaction field to be a perturbation of the Hamiltonian for an isolated molecule. The modified Hamiltonian of the sytem is then given by:

$$\mathscr{H}_{\text{tot}} = \mathscr{H}_0 + \mathscr{H}_{\text{RF}} \tag{9.34}$$

$\mathscr{H}_0$ is the Hamiltonian of the isolated molecule and $\mathscr{H}_{\text{RF}}$ is the perturbation, given by [Tapia and Goscinski, 1975]:

$$\mathscr{H}_{\text{RF}} = -\hat{\mu}^{\text{T}} \frac{2(\varepsilon - 1)}{(2\varepsilon + 1)a^3} \langle \Psi | \hat{\mu} | \Psi \rangle \tag{9.35}$$

$\hat{\mu}$ is the dipole moment operator written in matrix form and $\hat{\mu}^{\text{T}}$ is its transpose. The wavefunction Ψ for the modified Hamiltonian is determined and the electrostatic contribution to the solvation free energy is then given by:

$$\Delta G_{\text{elec}} = \langle \Psi | \mathscr{H}_{\text{tot}} | \Psi \rangle - \langle \Psi_0 | \mathscr{H}_0 | \Psi_0 \rangle + \frac{1}{2} \frac{2(\varepsilon - 1)}{(2\varepsilon + 1)a^3} \mu^2 \tag{9.36}$$

The third term in equation (9.36) is the correction factor corresponding to the work done in creating the charge distribution of the solute within the cavity in the dielectric medium. Ψ_0 is the gas phase wavefunction.

A drawback of the SCRF method is its use of a spherical cavity; molecules are rarely exactly spherical in shape. However, a spherical representation can be a reasonable first approximation to the shape of many molecules. It is also possible to use an ellipsoidal cavity; this may be a more appropriate shape for some molecules. For both the spherical and ellipsoidal cavities analytical expressions for the first and second derivatives of the energy can be derived, so enabling geometry optimisations to be performed efficiently. For these cavities it is necessary to define their size. In the case of a spherical cavity a value for the radius can be calculated from the molecular volume:

$$a^3 = 3V_m / 4\pi N_{\text{A}} \tag{9.37}$$

The molecular volume V_m can in turn be obtained by dividing the molecular weight by the density or from refractivity measurements; N_{A} is Avogadro's number. The cavity radius can also be estimated from the largest interatomic distance within the molecule. A third approach is to calculate the 'volume' of the molecule from a suitable electron density contour. The radii obtained by these procedures are often adjusted by adding an empirical constant to give the 'true' cavity radius. This extra value accounts for the fact that solvent molecules cannot approach right up to the molecule. An additional extension to the simple SCRF procedure is the use of a multipolar expansion to represent the solute [Rinaldi *et al.* 1983]. This overcomes a drawback of the basic model in which a molecule with a zero dipole would have zero solvation energy.

A yet more realistic cavity shape is that obtained from the van der Waals radii of the atoms of the solute. This is the approach taken in the *polarisable continuum* method (PCM) [Miertus *et al.* 1981] which has been implemented in a variety of *ab initio* and semi-empirical quantum mechanical programs. Due to the non-analytical nature of the cavity shapes in the PCM approach, it is necessary to calculate ΔG_{elec} numerically. The cavity surface is divided into a large

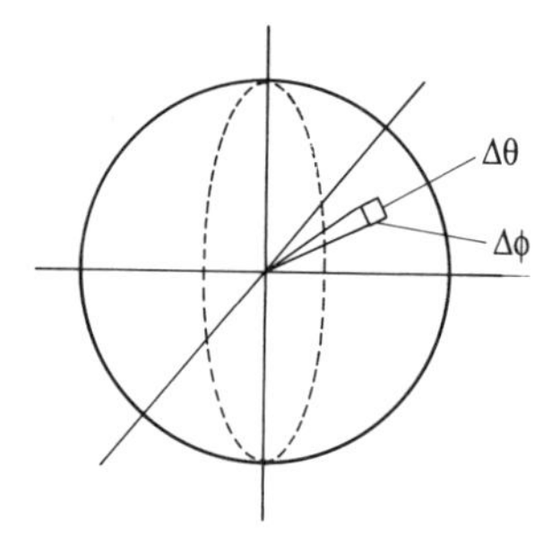

Fig. 9.15 Small surface elements can be created on the van der Waals surface of an atom using constant increments of the polar angles, θ and ϕ.

number of small surface elements. There is a point charge associated with each surface element. This system of point charges represents the polarisation of the solvent, and the magnitude of each surface charge is proportional to the electric field gradient at that point. The total electrostatic potential at each surface element equals the sum of the potential due to the solute and the potential due to the other surface charges:

$$\phi(\mathbf{r}) = \phi_{\rho}(\mathbf{r}) + \phi_{\sigma}(\mathbf{r}) \tag{9.38}$$

$\phi_{\rho}(\mathbf{r})$ is the potential due to the solute and $\phi_{\sigma}(\mathbf{r})$ is the potential due to the surface charges. The PCM algorithm is as follows. First, the cavity surface is determined from the van der Waals radii of the atoms. That fraction of each atom's van der Waals sphere which contributes to the cavity is then divided into a number of small surface elements of calculable surface area. The simplest way to to this is to define a local polar coordinate frame at the centre of each atom's van der Waals sphere and to use fixed increments of $\Delta\theta$ and $\Delta\phi$ to give rectangular surface elements (Figure 9.15). The surface can also be divided using tessellation methods [Paschual-Ahuir *et al.* 1987]. An initial value of the point charge for each surface element is then calculated from the electric field gradient due to the solute alone:

$$q_{\text{i}} = -\left[\frac{\varepsilon - 1}{4\pi\varepsilon}\right]E_i\Delta S \tag{9.39}$$

ε is the dielectric constant of the medium, E_i is the electric field gradient and ΔS is the area of the surface element. The contribution $\phi_{\sigma}(\mathbf{r})$ due to the other point charges can then be calculated using Coulomb's law. These charges are modified iteratively until they are self-consistent. The potential $\phi_{\sigma}(\mathbf{r})$ from the final part of the charge is then added to the solute Hamiltonian ($\mathcal{H} = \mathcal{H}_0 + \phi_{\sigma}(\mathbf{r})$) and the SCF calculation initiated. After each SCF calculation new values of the surface charges are calculated from the current wavefunction to give a new value of $\phi_{\sigma}(\mathbf{r})$ which is used in the next iteration until the solute wavefunction and the surface charges are self-consistent.

To calculate ΔG_{elec} we must take account of the work done in creating the charge distribution within the cavity in the dielectric medium. This is equal to one-half of the electrostatic interaction energy between the solute charge distribution and the polarised dielectric and so:

$$\Delta G_{\text{elec}} = \int \Psi \mathscr{H} \Psi \mathrm{d}\tau - \int \Psi_0 \mathscr{H}_0 \Psi_0 \mathrm{d}\tau - \frac{1}{2} \int \phi(\mathbf{r}) \rho(\mathbf{r}) \mathrm{d}\mathbf{r} \tag{9.40}$$

$\rho(\mathbf{r})$ is the charge distribution of the surface elements.

There are two slight complications with the PCM approach. The first of these arises as a consequence of representing a continuous charge distribution over the cavity surface as a set of single point charges. When calculating the electrostatic potential due to the charges on the surface elements one must exclude the charge for the current surface element. To include it would cause the charges to diverge rather than converge. The contribution of the charge on that surface element is therefore determined separately using the Gauss theorem. The second complication arises because the wavefunction of the solute extends beyond the cavity. Thus the sum of the charges on the surface is not equal and opposite to the charge of the solute. This problem can be easily overcome by scaling the charge distribution on the surface so that it is equal and opposite to the charge of the solute.

The SCRF and PCM models have been used to investigate the effect of solvent upon energetics and equilibria. For example, Wong, Wiberg and Frisch used the SCRF method to investigate the effect of different solvents upon the tautomeric equilibria of 2-pyridone (Figure 9.16) [Wong *et al.* 1992]. Geometry optimisations were performed for the various tautomeric species at high levels of theory and vibrational frequencies were calculated. Results were reported for the gas phase, for a non-polar solvent (cyclohexane, $\varepsilon = 2.0$) and for an aprotic polar solvent (acetonitrile, $\varepsilon = 35.9$). The calculated free energy changes in the gas phase, cyclohexane and acetonitrile were $-0.64\,\text{kcal mol}^{-1}$, $0.36\,\text{kcal mol}^{-1}$ and $2.32\,\text{kcal mol}^{-1}$, which compared favourably with the experimental values of $-0.81\,\text{kcal mol}^{-1}$, $0.33\,\text{kcal mol}^{-1}$ and $2.96\,\text{kcal mol}^{-1}$. The dielectric medium was found to have a much more pronounced effect on the structure, charge distribution and vibrational frequencies of the keto form than of the enol form. This was ascribed to the more polar nature of the keto tautomer.

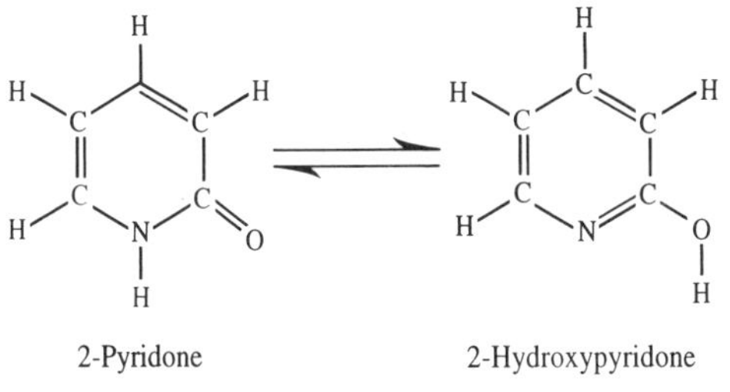

2-Pyridone 2-Hydroxypyridone

Fig. 9.16 Tautomers of 2-pyridone.

9.9.2 *Continuum models for molecular mechanics*

In many cases solvent effects can be incorporated into a force field model using one of the theories that we have just examined. It is possible to study larger systems with the empirical models, in which case it is necessary to take account of the dielectric properties of the solute as well as those of the solvent. Before reading this section it may be useful to revise the definitions of molecular surface and accessible surface given in section 1.5, as these are widely referenced.

The *boundary element method* of Rashin is similar in spirit to the polarisable continuum model, but the surface of the cavity is taken to be the molecular surface of the solute [Rashin and Namboodiri 1987; Rashin 1990]. This cavity surface is divided into small boundary elements. The solute is modelled as a set of atoms with point polarisabilities. This electric field induces a dipole proportional to its polarisability. The electric field at an atom has contributions from dipoles on other atoms in the molecule, from polarisation charges on the boundary, and (where appropriate) from the charges of electrolytes in the solution. The charge density is assumed to be constant within each boundary element but is not reduced to a single point as in the PCM model. A set of linear equations can be set up to describe the electrostatic interactions within the system. The solutions to these equations give the boundary element charge distribution and the induced dipoles, from which thermodynamic quantities can be determined.

The *generalised Born equation* has been widely used to represent the electrostatic contribution to the free energy of solvation [Constanciel and Contreras 1984]. The model comprises a system of particles with radii a_i and charges q_i. The total electrostatic free energy of such a system is given by the sum of the Coulomb energy and the Born free energy of solvation in a medium of relative permittivity ε:

$$G_{\text{elec}} = \sum_{i=1}^{N}\sum_{j=i+1}^{N}\frac{q_i q_j}{\varepsilon r_{ij}} - \frac{1}{2}\left(1-\frac{1}{\varepsilon}\right)\sum_{i=1}^{N}\frac{q_i^2}{a_i} \tag{9.41}$$

The first term in equation (9.41) can be written as the sum of a Coulomb interaction *in vacuo* and a second term in $(1 - 1/\varepsilon)$:

$$\sum_{i=1}^{N}\sum_{j=i+1}^{N}\frac{q_i q_j}{\varepsilon r_{ij}} = \sum_{i=1}^{N}\sum_{j=i+1}^{N}\frac{q_i q_j}{r_{ij}} - \left(1-\frac{1}{\varepsilon}\right)\sum_{i=1}^{N}\sum_{j=i+1}^{N}\frac{q_i q_j}{r_{ij}} \tag{9.42}$$

In the generalised Born approach the total electrostatic energy is written as a sum of three terms, the first of which is the Coulomb interaction between the charges in vacuo:

$$G_{\text{elec}} = \sum_{i=1}^{N}\sum_{j=i+1}^{N}\frac{q_i q_j}{r_{ij}} - \left(1-\frac{1}{\varepsilon}\right)\sum_{i=1}^{N}\sum_{j=i+1}^{N}\frac{q_i q_j}{r_{ij}} - \frac{1}{2}\left(1-\frac{1}{\varepsilon}\right)\sum_{i=1}^{N}\frac{q_i^2}{a_i} \tag{9.43}$$

ΔG_{elec} equals the difference between G_{el} and the Coulomb energy *in vacuo*. This is the generalised Born (GB) equation:

$$\Delta G_{\text{elec}} = -\left(1 - \frac{1}{\varepsilon}\right) \sum_{i=1}^{N} \sum_{j=i+1}^{N} \frac{q_i q_j}{r_{ij}} - \frac{1}{2}\left(1 - \frac{1}{\varepsilon}\right) \sum_{i=1}^{N} \frac{q_i^2}{a_i} \tag{9.44}$$

The generalised Born equation has been incorporated into both molecular mechanics calculations (by Still and co-workers [Still *et al.* 1990]), and into semi-empirical quantum mechanics calculations (by Cramer and Truhlar [Cramer and Truhlar 1992]). In these treatments the two terms in equation (9.44) are combined into a single expression of the following form:

$$\Delta G_{\text{elec}} = -\frac{1}{2}\left(1 - \frac{1}{\varepsilon}\right) \sum_{i=1}^{N} \sum_{j=1}^{N} \frac{q_i q_j}{f(r_{ij}, a_{ij})} \tag{9.45}$$

$f(r_{ij}, a_{ij})$ depends upon the interparticle distances r_{ij} and the Born radii a_i. A variety of forms are possible for the function f; that proposed by Still and colleagues was:

$$f(r_{ij}, a_{ij}) = \sqrt{(r_{ij}^2 + a_{ij}^2 e^{-D})}, \text{ where } a_{ij} = \sqrt{(a_i a_j)} \text{ and } D = r_{ij}^2/(2a_{ij})^2 \tag{9.46}$$

This form of the function f can be justified for the following reasons. When $i = j$, the equation returns the Born expression; for two charges close together (i.e. a dipole, in which r_{ij} is small compared to a_i and a_j) the expression is close to the Onsager result; and for two charges separated by a significant distance ($r_{ij} \gg a_i, a_j$) the result is very close to the sum of the Coulomb and Born expressions. A further advantage of this functional form is that the expression can be differentiated analytically, thereby enabling the solvation term to be included in gradient-based optimisation methods and molecular dynamics simulations.

A rather complex procedure is used to determine the Born radii a_i, values of which are calculated for each atom in the molecule that carries a charge or a partial charge. The Born radius of an atom (more correctly considered to be an 'effective' Born radius) corresponds to the radius that would return the electrostatic energy of the system according to the Born equation if all other atoms in the molecule were uncharged (i.e. if the other atoms only acted to define the dielectric boundary between the solute and the solvent). In Still's force field implementation atomic radii from the OPLS force field are assigned to each atom and amended by an empirically determined offset of -0.09 Å to define the dielectric boundary. In the quantum mechanical approach of Cramer and Truhlar the radius of each atom is a function of the charge on the atom. The dielectric boundary is then taken to be the union of the relevant radii.

The electrostatic energy of an atom i is calculated numerically by constructing a series of spherical shells until the outer shell (shell M) entirely contains the entire van der Waals surface of the molecule, as shown in Figure 9.17. The Born electrostatic energies of the dielectric in these shells are determined using the following equation:

$$\Delta G_{\text{elec}} = -\frac{1}{2}\left(1-\frac{1}{\varepsilon}\right)q_{\text{i}}^{2}\left\{\sum_{k=1}^{M}\frac{A_k}{4\pi r_k^2}\left[\left(\frac{1}{r_k-0.5T_k}\right)-\left(\frac{1}{r_k+0.5T_k}\right)\right] + \frac{1}{r_{M+1}-0.5T_{M+1}}\right\} \tag{9.47}$$

The radius of the kth shell (r_k) is measured at the middle of the shell. A_k is the amount of surface area of a sphere of radius r_k that is not contained within the van der Waals surface of the molecule. The thickness of each shell, T_k, increases with distance from the atom as follows:

$$T_{k+1} = (1+F)T_k \tag{9.48}$$

F (the expansion factor) and T_1 (the radius of the first shell) are parameters, chosen by Still to be 0.5 and 0.1 Å respectively. The shells are constructed from where the dielectric boundary commences (i.e. in Still's case the shells start 0.9 Å inside the van der Waals radii). The final term in equation (9.47) is the contribution due to the dielectric that lies beyond the van der Waals surface of the molecule. The effective Born radius is then given by equating equation (9.47) with the Born equation for the atom, and so:

$$\frac{1}{a_{\text{i}}} = \sum_{k=1}^{M}\frac{A_k}{4\pi r_k^2}\left[\left(\frac{1}{r_k-0.5T_k}\right)-\left(\frac{1}{r_k+0.5T_k}\right)\right] + \frac{1}{r_{M+1}-0.5T_{M+1}} \tag{9.49}$$

The effective Born radii do not change very much and so are recalculated whenever the non-bonded list is updated. The Still formulation of the generalised Born equation requires the surface areas A_k of the spherical shells that are exposed to solvent to be calculated. For this, a fast numerical method devised by

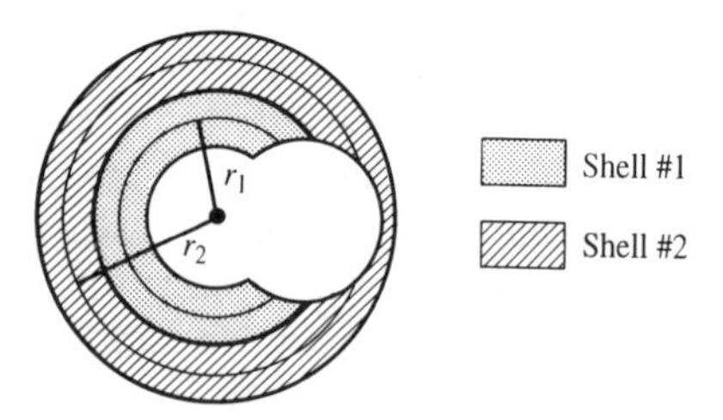

Fig. 9.17 Calculation of the effective Born radius in the generalised Born model. Shells are constructed until they contain the entire molecule. For each shell the amount of exposed surface area is determined for the middle of the shell. Figure adapted from Still W C, A Tempczyrk, R C Hawley and T Hendrickson 1990. Semianalytical Treatment of Solvation for Molecular Mechanics and Dynamics. *The Journal of the American Chemical Society* **112**:6127–6129.

Wodak and Janin [Wodak and Janin 1980] is used in which the accessible surface area is given by:

$$A_i = S_i \prod_j (1.0 - b_{ij}/S_i) \tag{9.50}$$

S_i is the total accessible surface area of an atom i with radius r_i as defined with a solvent probe of radius r_s. b_{ij} is the amount of surface area removed due to overlap with an atom j which is a distance d_{ij} from atom i:

$$S_i = 4\pi(r_i + r_s)^2 \tag{9.51}$$

$$b_{ij} = \pi(r_i + r_s)(r_j + r_i - 2r_s - d_{ij})[1.0 + (r_i - r_j)/d_{ij}] \tag{9.52}$$

The Wodak–Janin method is only approximate for more than two spheres. Exact values of A_i can be calculated, but only with a significant computational effort. A comparison of results obtained with the approximate and exact methods showed that for molecules significantly larger than the probe the approximation was valid (Wodak and Janin's expression was intended to be used to study solvent effects in proteins). Still showed that it was also possible to reduce the b_{ij} term by an empirical constant and obtain accurate results for smaller systems.

The generalised Born model has been incorporated into a number of quantum mechanical and molecular mechanical programs. Still and his group have made extensive use of the model in conformational searching and for calculating relative free energies of binding with free energy perturbation methods. For example, the relative free energies of binding of D- and L-enantiomeric α-amino acid-derived substrates to a podand ionophore (**1**; Figure 9.18) were calculated in good agreement with experiment using a mixed Monte Carlo/dynamics method and the generalised Born model for chloroform [Burger *et al.* 1994]. Similar calculations were then used to predict which of a variety of substituted ionophores (varying the group X in Figure 9.18) would be expected to show the greatest selectivity for the guest (Y = NHMe). The derivative **2** was predicted to show the greatest enantioselectivity. Unfortunately, this particular compound was too insoluble to measure binding affinities, but a related compound, **3**, did show the desired selectivity. Moreover, when the calculations were repeated on **3** the predicted difference in binding affinity was within 0.3 kcal/mol of the experimental result. Such studies clearly illustrate the potential applicability of such calculations, but it should be noted that this system was carefully chosen to minimise any errors associated with the force field parameters (through the use of enantiomeric guests) and sampling (as the host is locked into a single binding conformation). Even so, to achieve accurate results it was usually necessary to perform simulations of the order of 10 ns; such simulations could only be realistically achieved using a continuum model of the solvent.

Cramer and Truhlar used their implementation of the generalised Born model within the AM1 and PM3 semi-empirical Hamiltonians to investigate the anomeric and conformational equilibria of glucuse in aqueous solution [Cramer and Truhlar 1993]. Thermodynamic integration calculations on this system were described in section 9.5. The conclusion from that study was that the small free

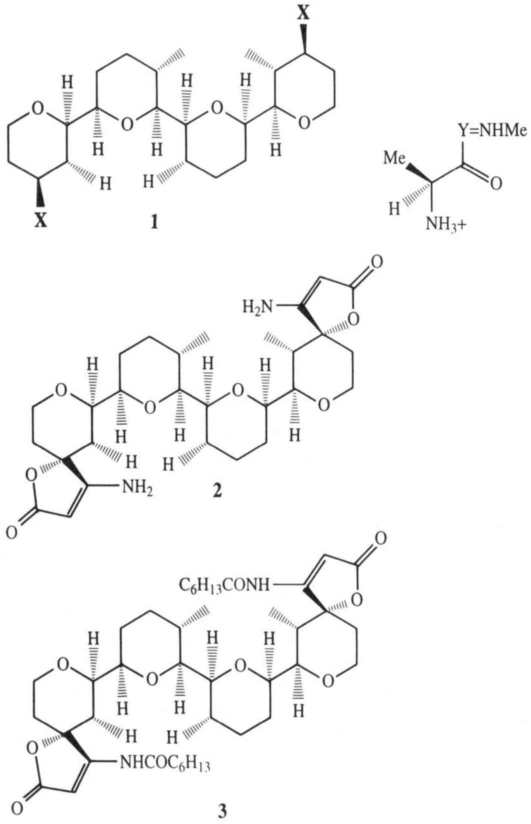

Fig. 9.18 Ionophores that selectively bind amino acids [Burger *et al.* 1994].

energy difference between the free energy of solvation of the α-D and β-D isomers was due to a cancellation of two large contributions. With the semi-empirical calculations, a rather different conclusion was reached; that there was no significant solvation effect on this equilibrium. The actual results did depend upon the level of theory employed (*ab initio* calculations were used to calculate the gas-phase free energy differences); for example, the intramolecular free energy difference (β-D → α-D) was calculated to be −1.7 kcal/mol at the 4-31G level and −0.5 kcal/mol using the 6-31G* basis set. The empirical value of Ha *et al.* was −3.6 kcal/mol. The overall free energy difference was predicted to be −0.5 kcal/mol at the 6-31G* level, leading to the conclusion that solvation effects were negligible. Cramer and Truhlar suggested that the reason for the discrepancy between the two studies could be due to the fact that the force field model used the same charge distribution for both the α and β isomers in the force field model; this was not an assumption made in the quantum mechanics calculations.

9.9.3 *The Langevin dipole model*

The Langevin dipole method of Warshel and Levitt [Warshel and Levitt 1976] is intermediate between a continuum and an explicit solvation model. A three-dimensional grid of rotatable point dipoles is established in the region beyond the boundary (which can be of arbitrary shape; Figure 9.19). For macromolecules the boundary corresponds to the solvent accessible surface. These dipoles represent the molecular dipoles of the solvent molecules in the outer region and the separation between them is chosen accordingly. The electric field $\mathbf{E}_i$ at each dipole has a contribution from the solute and from other solvent dipoles. The size and direction of each dipole is determined using the Langevin equation:

$$\boldsymbol{\mu}_i = \mu_0 \frac{\mathbf{E}_i}{|\mathbf{E}_i|}\left[\frac{\exp[C\mu_0|\mathbf{E}_i|/k_B T] + \exp[-C\mu_0|\mathbf{E}_i|/k_B T]}{\exp[C\mu_0|\mathbf{E}_i|/k_B T] - \exp[-C\mu_0|\mathbf{E}_i|/k_B T]} - \frac{1}{C\mu_0|\mathbf{E}_i|/k_B T}\right] \quad (9.53)$$

μ_0 is the size of the dipole moment of a solvent molecule and C is a parameter that represents the degree to which the dipoles resist reorientation; its value may be obtained from a separate simulation using explicit solvent. Converged values of the dipoles are usually obtained within a few iterations. The free energy of the Langevin dipoles is then given by:

$$\Delta G_{sol} = -\frac{1}{2}\sum_i \boldsymbol{\mu}_i \cdot \mathbf{E}_i^0 \quad (9.54)$$

$\mathbf{E}_i^0$ is the field due to the solute charges alone. The Langevin dipole method has been widely used by Warshel in his studies of enzyme reactions (see section 9.8.3).

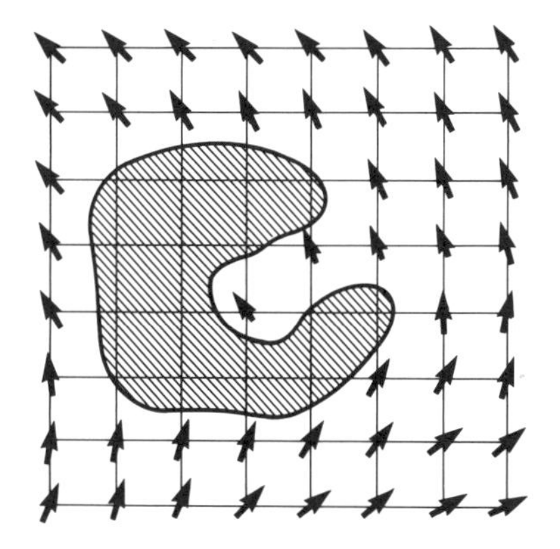

Fig. 9.19 The Langevin dipole model.

9.9.4 Methods based upon the Poisson–Boltzmann equation

The final class of methods that we shall consider for calculating the electrostatic component of the solvation free energy are based upon the Poisson or the Poisson–Boltzmann equations. These methods have been particularly useful for investigating the electrostatic properties of biological macromolecules such as proteins and DNA. The solute is treated as a body of constant low dielectric (usually between 2 and 4) and the solvent is modelled as a continuum of high dielectric. The Poisson equation relates the variation in the potential ϕ within a medium of uniform dielectric constant ε to the charge density ρ:

$$\nabla^2\phi(\mathbf{r}) = -\frac{\rho(\mathbf{r})}{\varepsilon_0\varepsilon} \tag{9.55}$$

In reduced electrostatic units, the factor $4\pi\varepsilon_0$ is eliminated and the Poisson equation becomes:

$$\nabla^2\phi(\mathbf{r}) = -\frac{4\pi\rho(\mathbf{r})}{\varepsilon} \tag{9.56}$$

The charge density is simply the distribution of charge throughout the system and has SI. units of $\mathrm{C\,m^{-3}}$. The Poisson equation is thus a second-order differential equation (∇^2 is the usual abbreviation for $(\partial^2/\partial x^2) + (\partial^2/\partial y^2) + (\partial^2/\partial z^2)$). For a set of point charges in a constant dielectric the Poisson equation reduces to Coulomb's law. However, if the dielectric is not constant, but varies with position, then Coulomb's law is not applicable and the Poisson equation is:

$$\nabla\cdot\varepsilon(\mathbf{r})\nabla\phi(\mathbf{r}) = -4\pi\rho(\mathbf{r}) \tag{9.57}$$

The Poisson equation must be modified when mobile ions are present, to account for their redistribution in the solution in response to the electric potential. The ions are prevented from congregating at the locations of extreme electrostatic potential due to repulsive interactions with other ions and their natural thermal motion. The ion distribution is described by a Boltzmann distribution of the following form:

$$n(\mathbf{r}) = \mathcal{N}\exp(-\mathcal{V}(\mathbf{r})/k_\mathrm{B}T) \tag{9.58}$$

$n(\mathbf{r})$ is the number density of ions at a particular location $\mathbf{r}$, $\mathcal{N}$ is the bulk number density and $\mathcal{V}(\mathbf{r})$ is the energy change to bring the ion from infinity to the position $\mathbf{r}$. When these effects are incorporated into the Poisson equation the result is the *Poisson–Boltzmann equation*:

$$\nabla\cdot\varepsilon(\mathbf{r})\nabla\phi(\mathbf{r}) - \kappa'\sinh[\phi(\mathbf{r})] = -4\pi\rho(\mathbf{r}) \tag{9.59}$$

κ' is related to the Debye–Hückel inverse length, κ, by:

$$\kappa^2 = \frac{\kappa'^2}{\varepsilon} = \frac{8\pi N_\mathrm{A}e^2 I}{1000\varepsilon k_\mathrm{B}T} \tag{9.60}$$

e is the electronic charge, I is the ionic strength of the solution and N_A is Avogadro's number. This is a non-linear differential equation that can be written in an alternative form by expanding the hyperbolic sine function as a Taylor series:

$$\nabla \cdot \varepsilon(\mathbf{r})\nabla\phi(\mathbf{r}) - \kappa'\phi(\mathbf{r})\left[1 + \frac{\phi(\mathbf{r})^2}{6} + \frac{\phi(\mathbf{r})^4}{120} + \cdots\right] = -4\pi\rho(\mathbf{r}) \tag{9.61}$$

The linearised Poisson–Boltzmann equation is obtained by taking only the first term in the expansion, giving:

$$\nabla \cdot \varepsilon(\mathbf{r})\nabla\phi(\mathbf{r}) - \kappa'\phi(\mathbf{r}) = -4\pi\rho(\mathbf{r})$$

How can equation (9.61) be solved? Before computers were available only simple shapes could be considered. For example, proteins were modelled as spheres or ellipses (Tanford–Kirkwood theory); DNA as a uniformly charged cylinder; and membranes as planes (Gouy–Chapman theory). With computers numerical approaches can be used to solve the Poisson–Boltzmann equation. A variety of numerical methods can be employed, including finite element and boundary element methods, but we will restrict our discussion to the finite difference method first introduced for proteins by Warwicker and Watson [Warwicker and Watson 1982].

Several groups have implemented finite difference Poisson–Boltzmann methods; here we concentrate on the work of Honig's group whose DelPhi program has been widely used. A cubic lattice is superimposed onto the solute(s) and the surrounding solvent. Values of the electrostatic potential, charge density, dielectric constant and ionic strength are assigned to each grid point. The atomic charges do not usually coincide with a grid point and so the charge is allocated to the eight surrounding grid points in such a way that the closer the charge to the grid point, the greater the proportion of its total charge that is allocated. The derivatives in the Poisson–Boltzmann equation are then determined by a finite difference formula. Consider the cube of side h surrounding the grid point shown in Figure 9.20. A charge q_0 is associated with the grid point; this is equivalent to a uniform charge density of q_0/h^3 within the cube (i.e. $\rho_0 = q_0/h^3$). The potential at the grid point is given by:

$$\phi_0 = \frac{\sum \varepsilon_i\phi_i + 4\pi\frac{q_0}{h}}{\sum \varepsilon_i + k'^2_0 f(\phi_0)} \tag{9.62}$$

The summations are over the potentials ϕ_i at the six adjoining grid points and the dielectric constants ε_i which are associated with the midpoints of the lines between the grid points. The function $f(\phi_0)$ in the denominator has the value 1 for the linear Poisson–Boltzmann equation and is equivalent to the series expansion $(1 + \phi_0^2/6 + \phi_0^4/120 + \cdots)$ for the non-linear case. κ'^2 is obtained from the ionic strength at the grid point. The crucial feature is that the potential at each grid point influences the potential at the neighbouring grid points, and so by iteratively repeating the calculation converged values will eventually be obtained.

To perform a Poisson–Boltzmann calculation it is necessary to allocate a value for the dielectric constant to each grid point, which requires us to decide which

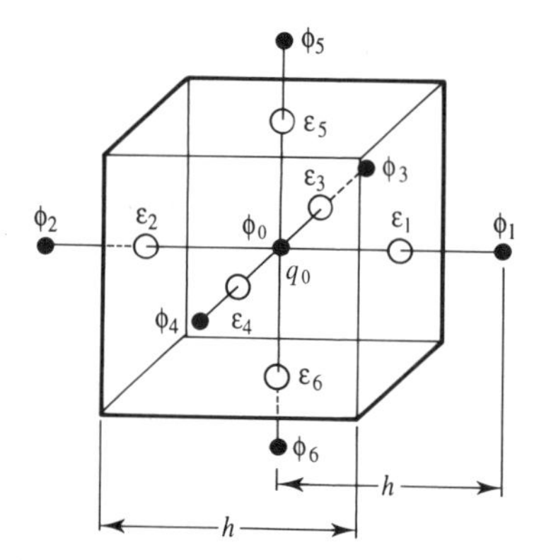

Fig. 9.20 The cube used in the finite difference method for solving the Poisson–Boltzmann equation. Figure adapted from Klapper I, R Hagstrom, R Fine, K Sharp and B Honig 1986. Focusing of Electric Fields in the Active Site of Cu-Zn Superoxide Dismutase: Effects of Ionic Strength and Amino-Acid Substitution. *Proteins: Structure, Function and Genetics* **1**:47–59.

grid points lie within the solute(s) and which are in the solvent. The boundary between the solute and solvent is defined as either the molecular surface or the accessible surface. All grid points outside this surface are assigned a high dielectric constant (80 for water) and an ionic strength value. Grid points within the surface are assigned the dielectric constant of the macromolecule, which is usually considered to lie between 2 and 4. This value of the dielectric constant is justified by the following arguments. The dielectric constant of a material is due to several factors, including its inherent polarisability and its ability to reorientate internal dipoles within a changing electric field. A molecule that is fixed in conformation will not be able to change the orientations of its dipolar groups and so the only contribution to the dielectric constant will be due to polarisation effects. Polarisation effects alone lead to a dielectric constant of about 2 for organic liquids. If the conformation of the molecule can change then the dipolar effects should be taken into account, leading to an increased dielectric of 4. The atomic charges and van der Waals radii are often taken directly from an existing force field, though parameter sets designed specifically for use with the Poisson–Boltzmann method have been developed [Sitkoff *et al.* 1994].

The correct choice of grid size can be crucial to the success of a finite difference Poisson–Boltzmann calculation. The finer the lattice, the more accurate the results, though more computer time will be required. A grid size of 65^3 has been widely used. A technique called *focusing* can help alleviate some of these problems. In this method, a series of calculations are performed with the system occupying an ever greater fraction of the total grid box at each step. The boundary points in each new grid are internal points from its predecessor, as shown in Figure 9.21. Focusing enables better estimates of the potential values at the boundary to be obtained. The results can also depend upon the orientation of the solute(s) within the grid. The error associated with this can be reduced by per-

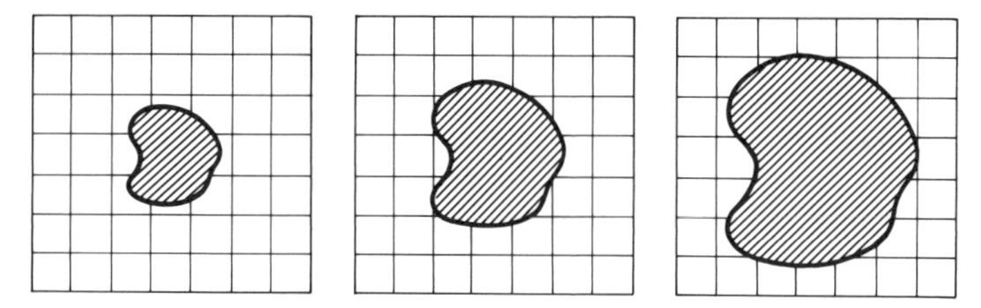

Fig. 9.21 Focussing can improve the accuracy of finite difference Poisson–Boltzmann calculations.

forming a series of calculations on randomly translated and rotated copies of the system and then averaging the results.

9.9.5 Applications of finite difference Poisson–Boltzmann calculations

A wide variety of problems have been studied using the finite difference Poisson–Boltzmann (FDPB) method. In addition to the numerical values that the method can provide, significant insights can often be gleaned by graphical examination of the electrostatic potential around the molecule [Honig and Nicholls 1995]. It is often found that the electrostatic potential around a protein calculated using the FDPB method differs significantly from that obtained with a uniform dielectric model. The location of the charged and polar groups in the protein and the shape of the molecule (which determines the shape of the boundary between the regions of high and low dielectric) significantly influence the shape of the potential. This can be seen in Figure 9.22 (colour plate section), which shows the electrostatic potential around the enzyme trypsin. The activity of this enzyme is regulated *in vivo* by trypsin inhibitor, which is a smaller protein that binds strongly to trypsin. However, both trypsin and trypsin inhibitor have net positive charges. How then do the two molecules associate? If the electrostatic potential around trypsin is calculated assuming a uniform dielectric constant of 80 then, as expected, the potential is positive everywhere. However, when the effects of the dielectric boundary are included then a region of negative electrostatic potential appears in the region where the inhibitor binds. A second example is provided by the enzyme Cu-Zn superoxide dismutase, which catalyses the conversion of the O_2^- radical to O_2 and H_2O_2. The rate constant for the reaction is high, being only about one order of magnitude smaller than the expected collision rate of the substrate with the entire enzyme. However, the active site constitutes a very small proportion of the surface and uniform collisions of the substrate over the protein surface would not explain the observed kinetics. It has therefore been suggested that the substrate is 'steered' into the active site by the electric field of the protein. Figure 9.23 (colour plate section) shows the electrostatic potential around the enzyme (which is a dimer) with the active sites at the top left and bottom right. As can be seen, a concentrated region of positive electrostatic potential extends from the active site into solution [Klapper *et al.* 1986]. The cleft-like nature of the protein around the active site enhances the positive electrostatic potential by focusing electric field lines out into the solvent.

The finite difference Poisson–Boltzmann method can be used to calculate the electrostatic contribution to various processes such as solvation and the formation of intermolecular complexes. The electrostatic component of the solvation free energy equals the change in electrostatic energy for transfer from vacuum to the solvent where the electrostatic energy of a charge q_i in a potential ϕ_i equals $q_i\phi_i$. The solvation free energy is determined by performing two separate calculations using the same grids and the same solute dielectric but exterior dielectrics of 80 (when the solvent is water) and 1 (for the vacuum). Then ΔG_{elec} is given by:

$$\Delta G_{elec} = \frac{1}{2}\sum_i q_i(\phi_i^{80} - \phi_i^1) \quad (9.63)$$

The summation in equation (9.63) is over all charges in the solute.

The change in free energy for the association of two molecules (assumed to have the same internal dielectric constant, ε_m) can be calculated using the finite difference Poisson–Boltzmann method. This problem is usefully discussed by dividing the free energy of association into a series of steps, as shown in Figure 9.24 [Gilson and Honig 1988]. First, the free energy associated with the transfer of the two isolated species from the solvent (dielectric constant ε_s) to a medium of dielectric ε_m is calculated in the same manner as for the solvation free energy (but here the transfer is from the solvent to a medium of dielectric ε_m, not to a vacuum). The free energy to bring the two molecules together is calculated using Coulomb's law in a medium of dielectric ε_m. Finally, the energy to transfer the complex from the medium of dielectric ε_m to the solvent is determined. The same procedure can be applied to other processes, such as the calculation of the free energy difference between two conformations in solution.

9.10 Non-electrostatic contributions to the solvation free energy

So far, we have only considered the electrostatic contribution to the free energy of solvation. Important though this is, there are some additional factors that contribute to the overall free energy of solvation, as shown in equation (9.30). These extra contributions can be especially significant for solutes that are neither charged nor highly polar. The cavity and van der Waals terms are often combined together and represented using an equation of the following form:

$$\Delta G_{cav} + \Delta G_{vdw} = \gamma A + b \quad (9.64)$$

A is the total solvent accessible area and γ and b are constants. This linear dependence upon the area A can be explained as follows. The cavity term equals the work to create the cavity against the solvent pressure and the entropy penalty associated with the reorganisation of solvent molecules around the solute. The solvent molecules most affected by this reorganisation are those in the first solvation shell. The number of solvent molecules in the first solvation shell is approxi-

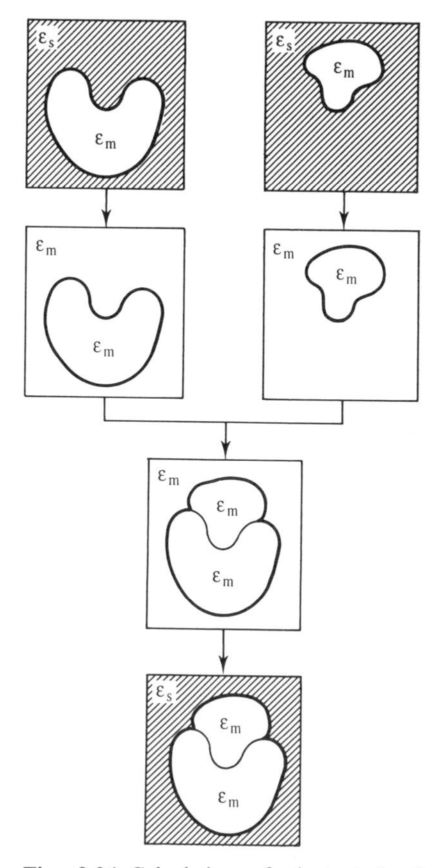

Fig. 9.24 Calculation of electrostatic free energy of association of two molecules. Figure adapted from Gilson M K and B Hong 1988. Calculation of the Total Electrostic Energy of a Macromolecular System: Solvation Energies, Binding Energies and Conformational Analysis. *Proteins*: *Structure*, *Function and Genetics* **4**:7–18.

mately proportional to the accessible surface area of the solute. The solute–solvent van der Waals interaction energy would also be expected to be dependent primarily upon the number of solvent molecules in the first solvent shell as van der Waals interactions fall off rapidly with distance. Hence both the cavity and van der Waals terms should be approximately proportional to the solvent accessible surface area. The parameters γ and b in equation (9.64) are usually taken from experimentally determined free energies for the transfer of alkanes from vacuum to water. The parameter b is commonly set to zero making the cavity plus van der Waals terms directly proportional to the solvent accessible surface area. Still's generalised Born/surface area model (GB/SA) uses the generalised Born approach for the electrostatic contribution together with a cavity and van der Waals surface area term in which the surface area is calculated using a variant of the Wodak and Janin algorithm, the constant γ having the value $7.2\,\text{cal}/(\text{mol}\,\text{Å}^2)$ [Hasel *et al.* 1988]. As we have already stated, analytical first and second derivatives of the surface area with respect to the atomic coordinates can be rapidly determined using the Wodak–Janin method, so enabling the GB/SA method to be incorporated into energy minimisation and molecular dynamics calculations.

The cavity and van der Waals contributions may also be modelled as separate terms. In some implementations an estimate of the cavity term may be obtained using scaled particle theory [Pierotti 1965; Claverie *et al.* 1978], which uses an equation of the form:

$$\Delta G_{\mathrm{cav}} = K_0 + K_1 a_{12} + K_2 a_{12}^2 \tag{9.65}$$

The constants K depend upon the volume of the solvent molecule (assumed to be spherical in shape) and the number density of the solvent. a_{12} is the average of the diameters of a solvent molecule and a spherical solute molecule. This equation may be applied to solutes of a more general shape by calculating the contribution of each atom and then scaling this by the fraction of that atom's surface that is actually exposed to the solvent. The dispersion contribution to the solvation free energy can be modelled as a continuous distribution function that is integrated over the cavity surface [Floris and Tomasi 1989].

9.11 Very simple solvation models

Some particularly simple solvation models include all contributions to the solvation free energy (including the electrostatic contribution) in an equation of the following form:

$$\Delta G_{\mathrm{sol}} = \sum_{\mathrm{i}} a_{\mathrm{i}} S_{\mathrm{i}} \tag{9.66}$$

S_{i} is the exposed solvent accessible surface area of atom i, and the summation is over all atoms in the solute. a_{i} is a parameter that depends upon the nature of atom i. Despite the obvious assumptions inherent in such an approach, it does have the advantage of providing an extremely rapid way to calculate a solvation contribution. Eisenberg and McLachlan developed such a model to study proteins with the parameters a_{i} being derived by considering just five classes of atoms (carbon, neutral oxygen and nitrogen, charged oxygen, charged nitrogen and sulphur) [Eisenberg and McLachlan 1986]. The values themselves were obtained by fitting to experimentally determined free energies of transfer. Eisenberg and McLachlan applied their solvation model to a variety of problems such as the recognition of misfolded protein structures and ligand binding.

9.12 Modelling chemical reactions

It is obviously important to be able to model chemical reactions as these lie at the heart of chemistry and biochemistry. Most reactions of interest do not take place in the gas phase, but in some medium, be it in a solvent, in an enzyme or on the surface of a catalyst. The environment can have a significant impact upon the reaction by speeding it up or slowing it down or even changing the reaction pathway. Good agreement can sometimes be obtained for calculations performed on isolated systems (i.e. in the gas phase) but to properly model the system the environment must be taken into account.

The preferred technique for modelling chemical reactions is usually considered to be quantum mechanics. Unfortunately, if one wishes to explicitly represent the whole system, the large number of atoms that must be considered means that *ab initio* quantum mechanics is rarely practical. Here we will consider three methods that have been used to study chemical reactions involving large systems. One strategy is to use a purely empirical approach. An alternative is to divide the system into two and treat the 'reaction region' using quantum mechanics, with the rest of the system being modelled using molecular mechanics. Thirdly, we shall consider techniques such as the Car–Parrinello method and density functional theory which, when allied to extremely powerful computers, can enable the entire reacting system to be simulated using quantum mechanics.

9.12.1 Empirical approaches to simulating reactions

Despite the often held belief that reactions can only be studied using quantum mechanics, this is by no means the case. Many research groups have developed force field models for studying reactions which can provide very satisfactory results. Such force fields are used to estimate the activation energies of possible transition states to explain and to predict the stereo- and regioselectivity of the reaction. The force field model is usually derived by extending an existing force field to enable the structures and relative energies of transition structures to be determined.

Here we will illustrate the method using a single example. The aldol reaction between an enol boronate and an aldehyde can lead to four possible stereoisomers (Figure 9.25). Many of these reactions proceed with a high degree of diastereoselectivity (i.e. *syn*:*anti*) and/or enantioselectivity (*syn*-I:*syn*-II and *anti*-I:*anti*-II). Bernardi, Capelli, Gennari, Goodman and Paterson studied this reaction using a force field based on MM2 [Bernardi *et al.* 1990]. The force field was parametrised to reproduce the geometries and relative energies of the chair and twist-boat transition structures with unsubstituted reactants, previously determined using *ab initio* methods (see Figure 9.26). It was assumed that the stereoselectivity was determined by the relative energies of the various possible transition structures (i.e. the reaction is assumed to be kinetically controlled).

The force field was then used to predict the results for the addition of the E and Z isomers of the enol boronate of butanone ($R^1 = Me$) to ethanol ($R^2 = Me$). The relevant transition structures are shown in Figure 9.27. A Boltzmann distribution, calculated at the temperature of the reaction (−78°C), predicted that the Z isomer would show almost complete *syn* selectivity (*syn*:*anti* = 99:1) and that the E isomer would be selective for the *anti* product (*anti*:*syn* = 86:14). These results

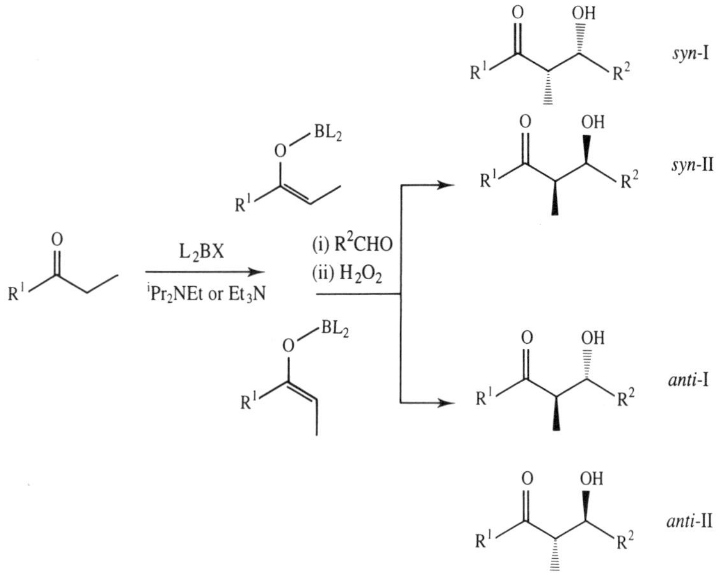

Fig. 9.25 The aldol reaction between an enol boronate and an aldehyde leads to four possible stereoisomers.

were in good agreement with the experimental observations. The major product in each case was obtained from a chair-like transition structure, but the reduced fraction of the *anti* product for the *E*-isomer was due to a significant contribution from the boat pathway which leads to the *syn* product.

9.12.2 The potential of mean force of a reaction

A complete description of a chemical reaction needs to take account of solvent effects. The most realistic way to achieve this is by including explicit solvent molecules. A classic example of how to tackle this problem is Jorgensen's study of the nucleophilic attack of the chloride anion on methyl chloride [Chandrasekhar *et al.* 1985; Chandrasekhar and Jorgensen 1985]. This reaction

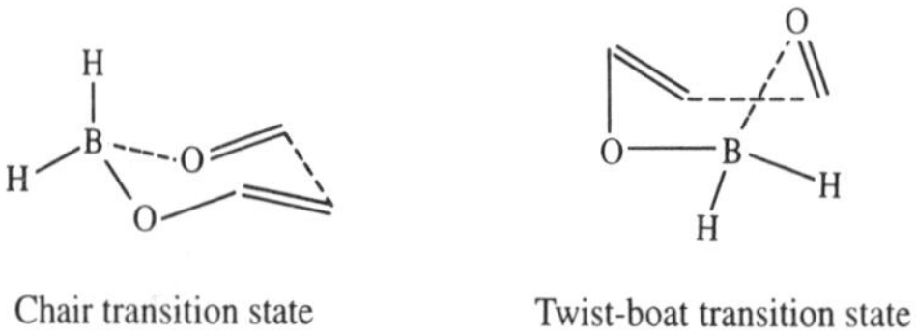

Fig. 9.26 Transition structures for the enol boronoate/aldehyde reaction.

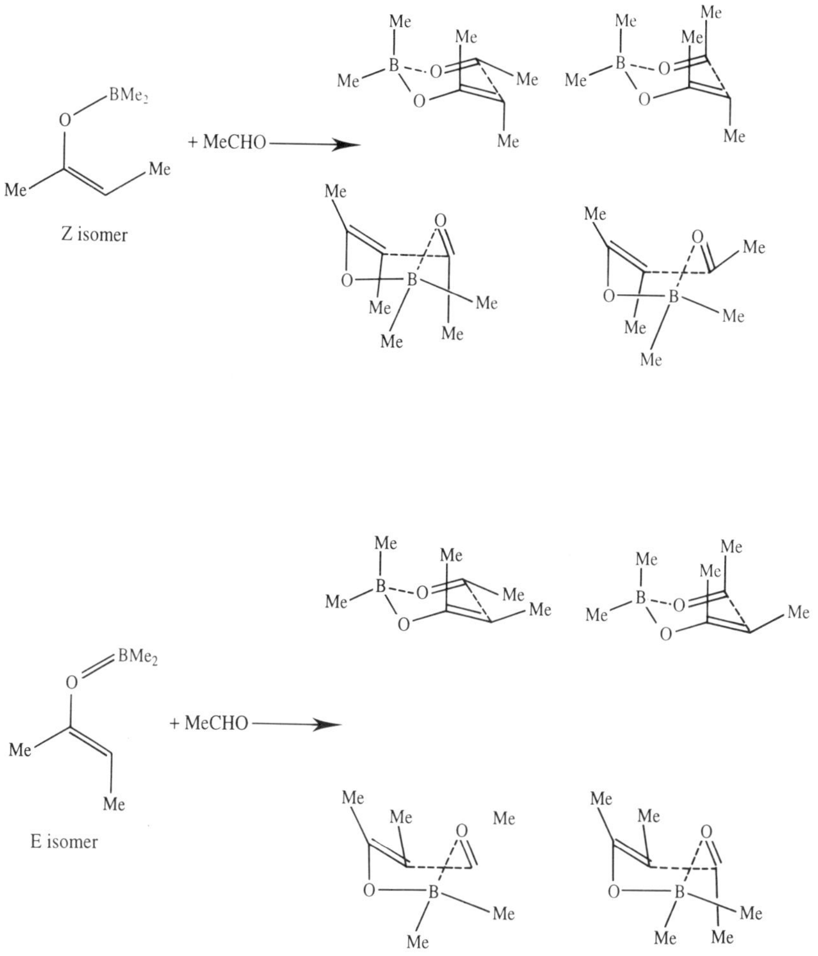

Fig. 9.27 Transition states for aldol reaction between butanone and ethanol.

proceeds via the S_N2 reaction in which the chloride anion approaches along the carbon–chlorine bond of methyl chloride to give a five-coordinate transition state which then collapses to give the products. We considered some aspects of the energy surface for this system in section 4.9, though there we were only interested in the energy change for the gas-phase reaction. The aim of Jorgensen's calculation was to obtain a potential of mean force for the reaction (i.e. the change in the free energy as a function of the reaction coordinate) in a variety of solvents.

The first step was to determine the quantum mechanical reaction pathway; a series of geometries along the path were determined using the path-following method of Gonzalez and Schlegel [Gonzalez and Schlegel 1988]. The solute–solvent interactions were modelled using Lennard–Jones and electrostatic terms in which the parameters smoothly varied with the reaction coordinate. To perform the Monte Carlo simulations, umbrella sampling was employed to constrain the geometry of the solute to a series of windows along the pathway, and thus calculate the potential of mean force. Preferential sampling methods were used, so that solvent molecules near the solute were sampled more often than solvent molecules further away.

The results are summarised in Figure 9.28, which shows how the potential of mean force varies for the reaction in the gas phase, in water and in dimethyl formamide (DMF). The results show a number of interesting features. In the gas phase an ion–dipole complex forms, giving a minimum in the free energy profile. There is then an activation barrier of approximately 13.9 kcal/mol to reach the pentagonal transition state. In aqueous solution, no ion–dipole minimum is observed. This is because any favourable contribution due to the formation of the ion–dipole is compensated by the energy lost in the desolvation of the chloride ion. There is then a large activation free energy barrier of approximately 26.3kcal/mol from the ion–dipole pair to the transition state. This barrier is much larger than in the gas phase because of the poorer solvation of the transition state relative to the ion–dipole complex. In DMF (a solvent with smaller anion solvating ability), the ion–dipole complex is at a minimum in the free energy as less energy is required to desolvate the chloride anion in this solvent.

9.12.3 Combined quantum mechanical/molecular mechanical approaches

One approach to the simulation of chemical reactions in solution is to use a combination of quantum mechanics and molecular mechanics. The 'reacting' parts of the system are treated quantum mechanically with the remainder being modelled using the force field. The total energy E_{TOT} for the system can be written:

$$E_{\mathrm{TOT}} = E_{\mathrm{QM}} + E_{\mathrm{MM}} + E_{\mathrm{QM/MM}} \tag{9.67}$$

E_{QM} is the energy of those parts of the system are treated exclusively with quantum mechanics, and E_{MM} is the energy of the purely molecular mechanical parts of the system. $E_{\mathrm{QM/MM}}$ is the energy of interaction between the quantum mechanical and molecular mechanical parts of the system. This is described by a Hamiltonian $\mathcal{H}_{\mathrm{QM/MM}}$. In some cases $E_{\mathrm{QM/MM}}$ is due entirely to non-bonded interactions between the quantum mechanical and molecular mechanical atoms. An example where this could arise would be if all of the atoms in the reacting species were treated quantum mechanically with molecular mechanics being used exclusively for the solvent. For example, Cl^- and MeCl could be treated using

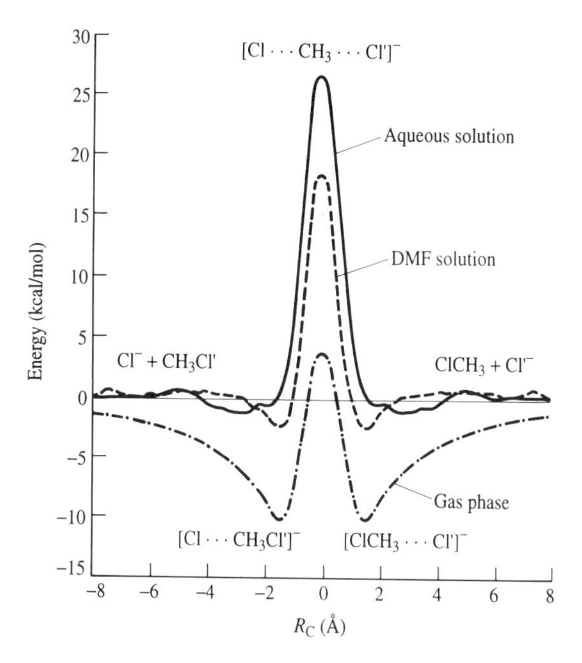

Fig. 9.28 Potential of mean force for the $Cl^- + MeCl$ reaction in various solvents. Figure redrawn from Chandrasekhar J and W L Jorgensen 1985. Energy Profile for a Nonconcerted S_N2 Reaction in Solution. *The Journal of the American Chemical Society* **107**:2974–2975.

quantum mechanics and solvent with molecular mechanics. In this case the Hamiltonian $\mathscr{H}_{QM/MM}$ can be written:

$$\mathscr{H}_{QM/MM} = -\sum_{i}\sum_{M}\frac{q_M}{r_{i,M}} + \sum_{\alpha}\sum_{M}\frac{Z_\alpha q_M}{R_{\alpha,M}} + \sum_{\alpha}\sum_{M}\left(\frac{A_{\alpha,M}}{R^{12}_{\alpha,M}} - \frac{C_{\alpha,M}}{R^{6}_{\alpha,M}}\right) \qquad (9.68)$$

The subscript i in equation (9.68) refers to a quantum mechanical electron and the subscript α to a quantum mechanical nucleus. The subscript M indicates a molecular mechanical nucleus and q_M is its partial atomic charge. There are thus electrostatic interactions between the electrons of the quantum mechanical region and the molecular mechanical nuclei, electrostatic interactions between quantum mechanical and molecular mechanical nuclei and van der Waals interactions between the quantum mechanical and molecular mechanical atoms. The second and third terms in equation (9.68) do not involve electronic coordinates and so can be calculated in a straightforward way (i.e. they are constant for a given nuclear configuration). The first term must be incorporated into the quantum mechanical calculation via one-electron integrals added to the one electron matrix, H^{core}. These one-electron integrals have the form:

$$\int \phi_\mu(1)\frac{1}{r_{1,M}}\phi_\nu(1)\mathrm{d}v(1) \tag{9.69}$$

In some cases the quantum mechanical and molecular mechanical regions are in the same molecule and so there are bonds between atoms from each region. The energy $E_{\text{QM/MM}}$ must now contain terms that describe this interaction. This can be done by adding a molecular mechanical-like energy which contains bond stretching, angle bending and torsional terms for atoms from both the quantum mechanical and molecular mechanical sets. This is illustrated in Figure 9.29 which shows which terms would be included in $E_{\text{QM/MM}}$.

Various combined quantum mechanical/molecular mechanical implementations have been described [Warshel and Levitt 1976; Singh and Kollman 1986; Field *et al.* 1990]. These implementations differ in the quantum mechanical theory that is used (semi-empirical, *ab initio*, valence bond or density functional theory), the molecular mechanical model, and the way in which the solvent is represented (either explicitly or using a simplified model). These methods are not of course restricted to studies of reactions, but can also be used to study association processes and conformational transitions.

The objective of many of the research groups involved in the development of combined quantum mechanical/molecular mechanical models has been the simulation of enzyme reactions. Warshel has reported studies in which the reaction centre is treated using a valence bond model [Warshel 1991; Åqvist and Warshel 1993]. The first part of his strategy is a calibration of the valence bond model for the reference reaction in solution. This model is then used to simulate the enzyme reaction using molecular dynamics and free energy perturbation methods, with solvent effects being treated using the Langevin dipole model. Warshel has exten-

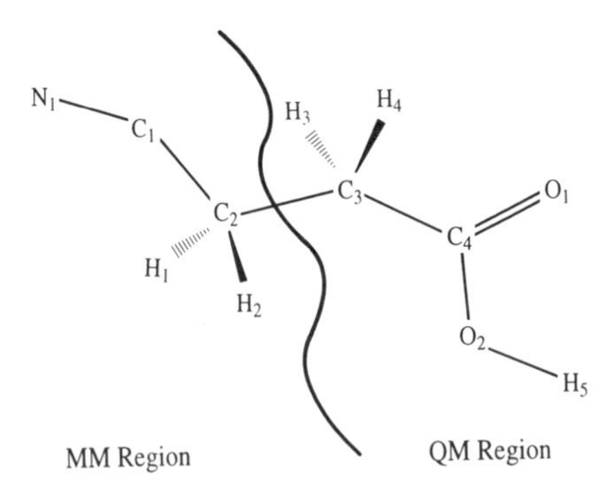

MM bonds: C_2–C_3
MM angles: C_1–C_2–C_3, C_2–C_3–C_4, etc
MM torsions: C_1–C_2–C_3–C_4, H_1–C_2–C_3–C_4, C_2–C_1–C_4–O_1, C_2–C_1–C_4–O_2, C_3–C_2–C_1–N_1, etc.

Fig. 9.29 The division of a molecule into quantum mechanical and molecular mechanical regions with the molecular mechanical contributions as indicated.

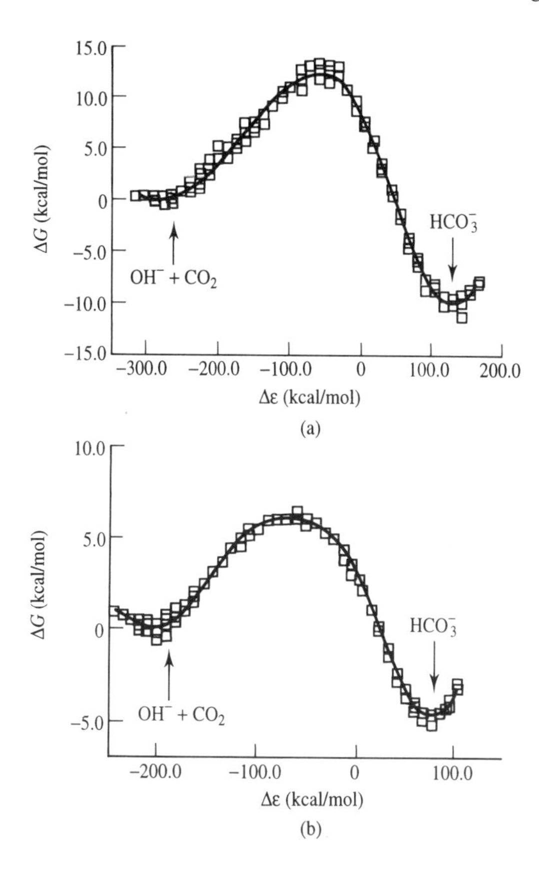

Fig. 9.30 Free energy profile for the nucleophilic attack of water on CO_2 (a) in aqueous solution and (b) in the enzyme carbonic anhydrase. Graphs redrawn from Åqvist J, M Fothergill and A Warshel 1993. Computer Simulation of the CO_2/HCO_3^- Interconversion Step in Human Carbonic Anhydrase I. *The Journal of the American Chemical Society* **115**:631–635.

sively studied a wide range of enzyme systems. One example is his study of the enzyme carbonic anhydrase which is a zinc-containing enzyme that catalyses the reversible hydration of carbon dioxide according to the following mechanism:

$$\text{E-H}_2\text{O} \rightleftharpoons \text{H}^+\text{-E-OH}^- \rightleftharpoons \text{E-OH}^- \tag{9.70}$$

$$\text{E-OH}^- + \text{CO}_2 \rightleftharpoons \text{E-HCO}_3{}^- \rightleftharpoons \text{E-H}_2\text{O} + \text{HCO}_3{}^- \tag{9.71}$$

E represents the enzyme. In the first step a bound water molecule is proteolised and the protein is transferred to solution. This is the rate-determining step of the reaction. In the second step CO_2 is converted to $HCO_3{}^-$. Åqvist, Fothergill and Warshel have examined both steps; here we concentrate on their results for the nucleophilic attack on the carbon dioxide (equation 9.71) [Åqvist *et al.* 1993]. Simulations of the hydration reaction of CO_2 in water were performed to find valence bond parameters which reproduced the experimentally observed value. This valence bond model was then used to simulate the same reaction in the enzyme. The resulting free energy profiles for the reference reaction and the enzyme reaction are shown in Figure 9.30. These results suggest that the enzyme markedly lowers the activation barrier for the reaction and that the reaction is less exothermic in the enzyme than in water ($\Delta G^{\ddagger} = 6.3$ kcal/mol vs 11.9 kcal/mol; $\Delta G^{0} = -4.8$ kcal/mol vs -10.5 kcal/mol). The experimental values were estimated to be $\Delta G^{\ddagger} = 7.1$ kcal/mol and $\Delta G^{0} = -4.1$ kcal/mol. The enzyme thus speeds up the reaction by a factor of about 10^3 compared to aqueous solution. The transition state geometry obtained from the simulation was found to be similar to geometries obtained using gas-phase *ab initio* calculations.

9.13 Density functional theory

Density functional theory is an approach to the electronic structure of atoms and molecules which has enjoyed a dramatic surge of interest since the 1980's [Parr 1983; Wimmer 1991]. The approach is based upon a theory presented by Hohenberg and Kohn in 1964 [Hohenberg and Kohn 1964] which states that all the ground-state properties of a system are functions of the charge density. The Hohenberg–Kohn theorem thus enables us to write the total electronic energy as a function of the electron density ρ:

$$E(\rho) = E_{\mathrm{KE}}(\rho) + E_{\mathrm{C}}(\rho) + E_{\mathrm{H}}(\rho) + E_{\mathrm{XC}}(\rho) \tag{9.72}$$

$E_{\mathrm{KE}}(\rho)$ is the kinetic energy, $E_{\mathrm{C}}(\rho)$ is the electron–nuclear interaction term, $E_{\mathrm{H}}(\rho)$ is the electron–electron Coulombic energy, and $E_{\mathrm{XC}}(\rho)$ contains the exchange and correlation contributions. All of the electron–electron interactions are thus contained within the E_{H} and E_{XC} terms. A crucial conclusion from the Hohenberg–Kohn theorem is that the ground-state properties of a system are determined by the density. An incorrect density gives an energy above the true energy. To perform a density functional calculation it is necessary to write the various terms in equation (9.72) in terms of the density and then optimise the energy with respect to the density, subject to any constraints on the system.

The starting point is a wavefunction that is taken to be an antisymmetrised product (i.e. a Slater determinant) of molecular orbitals which are both real and orthonormal. The charge density at a point **r** can then be written as the sum over occupied molecular orbitals of ψ^2 (see also section 2.14.3):

$$\rho(\mathbf{r}) = \sum_{\mathrm{i}=1}^{N_{\mathrm{occ}}} |\psi_{\mathrm{i}}(\mathbf{r})|^2 \tag{9.73}$$

The various components that contribute to the energy (equation (9.72)) must first be expressed in terms of the density. For the first three contributions in equation (9.72) standard expressions can be used as follows:

$$E(\rho) = 2\sum \int \psi_i\left(-\frac{\nabla^2}{2}\right)\psi_i \mathrm{d}\nu + \int \mathscr{V}_{\text{nuclear}}\rho(\mathbf{r})\mathrm{d}\nu$$

$$+\frac{1}{2}\iint \mathrm{d}\nu\mathrm{d}\nu' \frac{\rho(\mathbf{r})\rho(\mathbf{r}')}{|\mathbf{r}-\mathbf{r}'|} + \mathrm{E}_{\mathrm{XC}}[\rho(\mathbf{r})] \tag{9.74}$$

For the exchange–correlation term $E_{\mathrm{XC}}[\rho(\mathbf{r})]$ some approximations must be made. The most common way to obtain this contribution uses the so-called *local density approximation*, which is based upon a model called the uniform electron gas. In a uniform electron gas the electron density is constant throughout all space. The exchange–correlation energy can be determined for this model. The local density approximation assumes that the charge density varies slowly throughout a molecule so that a localised region of the molecule behaves like a uniform electron gas. If $\varepsilon_{\mathrm{XC}}$ is the exchange–correlation energy per particle in a uniform electron gas then the total exchange–correlation energy, E_{XC} for the system can be obtained by integrating over all space:

$$E_{\mathrm{XC}}[\rho(\mathbf{r})] \cong \int \rho(\mathbf{r})\varepsilon_{\mathrm{XC}}[\rho(\mathbf{r})]\mathrm{d}\mathbf{r} \tag{9.75}$$

Various analytical representations of the exchange energy of the uniform electron gas have been proposed. One is the following power-law relationship [Kohn and Vashista 1983]:

$$\varepsilon_{\text{exch}}(\rho(\mathbf{r})) = -0.785\,587\,70[\rho(\mathbf{r})]^{1/3} \tag{9.76}$$

The correlation contribution can also be represented by a parametric relationship such as the following result of Perdew and Zunger [Perdew and Zunger 1981]:

$$\varepsilon_{\text{corr}}(\rho(\mathbf{r})) = \gamma(1 + \beta_1 r_s^{1/2} + \beta_2 r_s) \tag{9.77}$$

$$r_s^3 = \frac{3}{4\pi\rho(\mathbf{r})};\ \gamma = -0.142\,30,\ \beta_1 = 1.052\,90 \text{ and } \beta_2 = 0.333\,40 \tag{9.78}$$

This result applies when the number of up spins equals the number of down spins and so is not applicable to systems with an odd number of electrons.

In a density functional calculation, the energy is optimised with respect to the density. The one feature still missing from our discussion is the way in which the density $\rho(\mathbf{r})$ is represented. In the commonly used Kohn and Sham implementation the density is represented as if it were derived from a single Slater determinant with orthonormal orbitals [Kohn and Sham 1965]. The use of these 'Kohn–Sham orbitals' enables the energy to be optimised by solving a set of one-electron equations, but with electron correlation included. This is one of the key advantages of the density functional approach. The Kohn–Sham equations are:

$$\left[-\frac{\nabla^2}{2}+\mathcal{V}_{\text{nuclear}}(\mathbf{r})+\int \mathrm{d}v'\frac{\rho(\mathbf{r})}{|\mathbf{r}-\mathbf{r}'|}+\mathcal{V}_{\text{XC}}(\mathbf{r})\right]\psi_{\text{i}}(\mathbf{r})=\varepsilon_{\text{i}}\psi_{\text{i}}(\mathbf{r}) \tag{9.79}$$

The exchange–correlation functional, $\mathcal{V}_{\text{XC}}(\mathbf{r})$ is the derivative of the exchange–correlation energy with respect to the density:

$$\mathcal{V}_{\text{XC}}(\mathbf{r})=\frac{\delta E_{\text{XC}}[\rho(\mathbf{r})]}{\delta\rho(\mathbf{r})} \tag{9.80}$$

$\mathcal{V}_{\text{XC}}(\mathbf{r})$ can be easily obtained from the appropriate analytical expressions for the local density approximation.

The molecular orbitals in density functional calculations are usually written as a linear expansion of atomic orbitals (i.e. basis functions) which can be represented using Gaussian functions, Slater orbitals or as numerical orbitals. One way to construct a numerical basis set is to represent the basis functions as values on a spherical polar grid centred on each atom, with the variation in the function at each grid point being stored as a cubic spline (this enables analytical gradients to be calculated).

9.13.1 *Density functional methods in the study of processes on solids*

Our focus here will be on the use of density functional methods to study reactions on or in solids. Two areas of particular interest are the properties of silicon and germanium (due to their use in computer chip manufacture) and metal oxides (which are used as ceramics and catalysts). The quantum mechanical methods used to study such systems are often somewhat different to those traditionally employed for studies of individual molecules or intermolecular complexes. Solids consist of an infinite set of repeating units, and so it is necessary to ensure that edge effects can be eliminated if possible. One way to deal with this is via the *periodic Hartree–Fock* method where periodic boundary conditions are built in to the self-consistent field equations [Dovesi *et al.* 1981; Dovesi *et al.* 1983]. Such calculations involve infinite sums of the Coulomb and exchange integrals for which exact formulae are generally not available. Moreover, convergence can be difficult to achieve. Nevertheless, methods similar in spirit to the cell multipole method (see section 5.9) have been devised. Perfect crystals are relatively straightforward to deal with using periodic Hartree–Fock methods because the repeating unit (or unit cell) is usually quite small. It is more difficult to study defects on surfaces using these methods because the number of atoms that can be included is limited and there are often difficulties in achieving convergence. A second difficulty with the study of solid systems is that they often contain elements much later in the periodic table than are usually encountered in Hartree–Fock calculations. It is therefore typical to consider explicitly only the valence electrons with the core electrons being subsumed into the nuclear core. *Pseudopotentials* are usually used to represent the wavefunctions of the valence

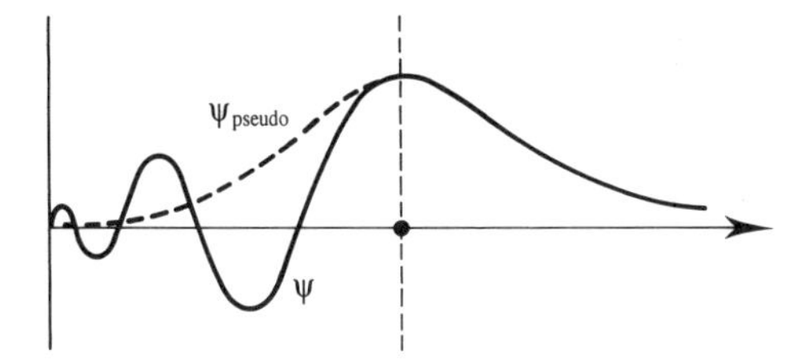

Fig. 9.31 Schematic representation of a pseudopotential. After Payne M C, M P Teter, D C Allan, R A Arias and D J Joannopoulos 1992. Iterative Minimisation Techniques for *Ab Initio* Total-Energy Calculations: Molecular Dynamics and Conjugate Gradients. *Reviews of Modern Physics* **64**:1045–1097.

electrons under the influence of the nucleus and the core electrons [Heine 1970]. A pseudopotential is a potential function that gives wavefunctions with the same shape as the true wavefunction outside the core region but with fewer nodes inside the core region, as illustrated in Figure 9.31. The number of nodes is important because the basis sets used to study solids often consist of a superposition of plane waves (rather than, say, Gaussians). A plane wave is an exponential function of the form exp($i\mathbf{k}.\mathbf{r}$), where $\mathbf{k}$ is related to the momentum of the wave by $\mathbf{p} = h\mathbf{k}/2\pi$. For example, the plane wave exp(ikz) is the wavefunction of a particle travelling in the z direction with momentum $kh/2\pi$. If there are many nodes in the inner core region then many plane waves are required to provide an adequate representation and the amount of calculation required would be overwhelming; the use of pseudopotentials reduces the number of terms in the plane-wave expansion of the wavefunction which in turn drastically reduces the scale of the computational problem.

9.13.2 The Car–Parrinello method

Hartree–Fock calculations are normally solved using iterative matrix diagonalisation techniques as discussed in Chapter 2. Density functional calculations can also be tackled using such methods. However, for systems with many atoms and/or basis functions such calculations can be very time-consuming and it may also be difficult to achieve convergence. Even with pseudopotentials, the number of plane-wave basis functions that can be required for density functional calculations may be very large, and as the number of occupied orbitals is often considerable, it can be a major task to solve the Kohn–Sham equations to determine the energy of a given configuration of atoms. In 1985 Car and Parrinello described a method that brought together a number of key concepts that we have considered in earlier chapters of the book [Car and Parrinello 1985; Remler and Madden 1990]. Car and Parrinello were primarily concerned with the problem of performing *ab initio* simulations involving both the electronic and the nuclear motions ('total energy' simulations or '*ab initio* molecular dynamics'). However, their scheme can be used to perform energy minimisation or simply to determine the basis set coefficients for a fixed atomic configuration.

A key feature of the Car–Parrinello proposal was the use of molecular dynamics and simulated annealing to search for the values of the basis set coefficients that minimise the electronic energy. In this sense their approach provides an alternative to the traditional matrix diagonalisation methods. In the Car–Parrinello scheme 'equations of motion' for the coefficients are set up, and then molecular dynamics is used to move the system through the space of the basis set coefficients. Starting from a random set of coefficients (which correspond to a high energy) the system moves downhill on the energy surface, accumulating 'kinetic energy'. Simulated annealing was proposed as a mechanism for preventing the system becoming trapped in a local minimum. The SHAKE algorithm (see section 6.5) is used to impose constraints on the system to ensure that the orbitals remain orthonormal.

To perform *ab initio* molecular dynamics, Car and Parrinello suggested that the electronic and nuclear dynamics could be performed simultaneously. Somewhat surprisingly, it is found that in the Car–Parrinello scheme it is not necessary for the electronic configuration to be at a minimum in coefficient space for each molecular dynamics time step, even though this gives errors in the forces on the nuclei. It can be shown that the errors in the nuclear forces are cancelled by the associated errors in the electronic motion. This rather strange (and fortuitous!) result can be understood if we consider the motion of atom with a single occupied molecular orbital. If the nucleus starts to move with a constant velocity, then the orbital will initially lag behind the nucleus. The orbital starts to accelerate until it eventually overtakes the nucleus. Having overtaken the nucleus, the orbital starts to slow down, until the nucleus overtakes the orbital, and so on as illustrated in Figure 9.32.

An alternative to the Car–Parrinello method is the following scheme which separates the electronic and nuclear motions:

1. Calculate the forces on the nuclei.
2. Move the nuclei according to the molecular dynamics integration scheme.
3. Optimise the electronic configuration for the new nuclear configuration.
4. Go to step 1.

This algorithm alternates between the electronic structure problem and the nuclear motion. It turns out that to generate an accurate nuclear trajectory using this decoupled algorithm the electrons must be fully relaxed to the ground state at each iteration, in contrast to the Car–Parrinello approach where some error is tolerated. This need for very accurate basis set coefficients means that the minimum in the space of the coefficients must be located very accurately, which can be computationally very expensive. However, it has been found that a conjugate gradients minimisation algorithm is an effective way to derive accurate plane-wave coefficients especially if information from previous steps is incorporated [Payne *et al.* 1992]. This reduces the number of minimisation steps required to locate accurately the best set of basis set coefficients.

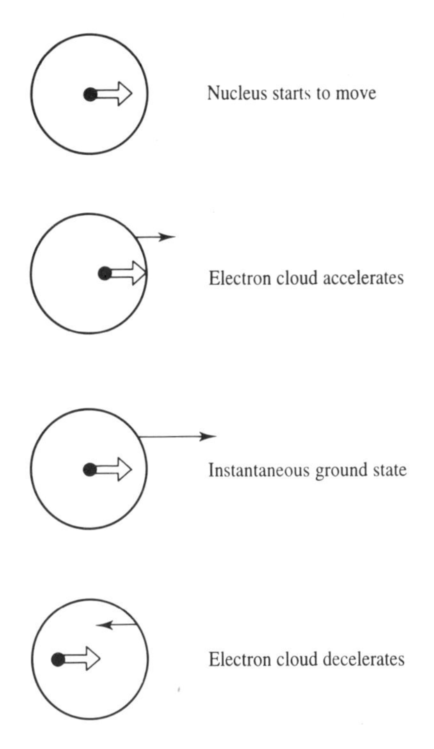

Fig. 9.32 Lag effects in *ab initio* molecular dynamics. Figure redrawn from Payne M C, M P Teter, D C Allan, R A Arias and D J Joannopoulos 1992. Iterative Minimisation Techniques for *Ab Initio* Total-Energy Calculations: Molecular Dynamics and Conjugate Gradients. *Reviews of Modern Physics* **64**:1045–1097.

9.13.3 **Ab initio** ***molecular dynamics applied to a chemisorption process***

An impressive application of the methods that we have discussed in this section is the *ab initio* molecular dynamics simulation of the reaction between a chlorine molecule and a silicon surface [Stich *et al.* 1994]. This reaction is particularly

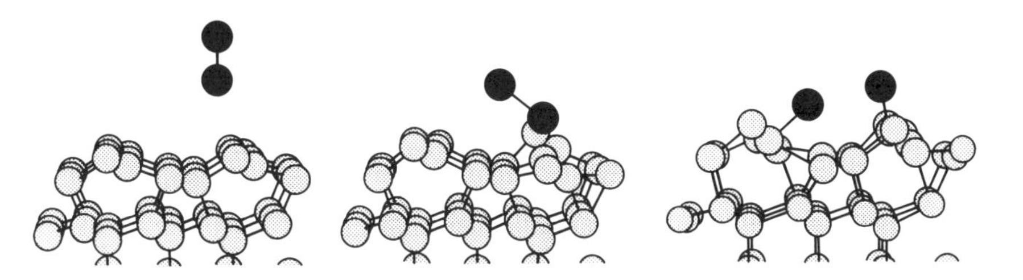

Fig. 9.33 Structural changes observed during the reaction of a chlorine molecule with a silicon surface. Figure reproduced from Stich I, A De Vita, M C Payne, M J Gillan and L J Clarke 1994. Surface Dissociation from First Principles: Dynamics and Chemistry. *Physical Review B* **49**:8076–8085.

important in silicon chip manufacture where the dissociative chemisorption of chlorine (and other halogens) is widely used for processes such as dry etching and surface cleaning. A series of simulations were performed, in each of which a chlorine molecule was 'fired' towards the silicon surface. The subsequent motion and reaction was then determined using the *ab initio* molecular dynamics approach based upon conjugate gradients minimisation. The motions of the nuclei were determined using the Verlet algorithm with a time step of approximately 0.5 fs and each simulation was performed for a total time between 200 fs and 400 fs.

The silicon surface contains chains of atoms that are formally bonded to just three other atoms. These atoms compensate for the lack of a full valence complement of bonds by π-bonding along the chains (Figure 9.33). These π-bonded chains represent regions of high electron density, with the valleys between the chains being of relatively low density. Despite this difference in electron density, the chlorine molecule dissociated when it was directed towards either of these two regions. Bonds formed between the chlorine atoms and π-bonded silicon atoms which in turn causes a change in the local hydridisation from sp^2 to sp^3. This then causes a large local deformation which lifts the silicon atoms involved above the π-bonded chains.

Appendix 9.1 Calculating free energy differences using thermodynamic integration

If the free energy, A, is a continuous function of λ then we can write:

$$\Delta A = \int_0^1 \frac{\partial A(\lambda)}{\partial \lambda} \mathrm{d}\lambda \tag{9.81}$$

Now

$$A(\lambda) = -k_\mathrm{B} T \ln Q(\lambda) \tag{9.82}$$

Thus

$$\Delta A = -k_\mathrm{B} T \int_0^1 \left[\frac{\partial \ln Q(\lambda)}{\partial \lambda} \right] \mathrm{d}\lambda = \int_0^1 \frac{-k_\mathrm{B} T}{Q(\lambda)} \frac{\partial Q(\lambda)}{\partial \lambda} \mathrm{d}\lambda \tag{9.83}$$

From the definition of Q (section 5.1.1):

$$Q_{NVT} = \frac{1}{N!} \frac{1}{h^{3N}} \iint \mathrm{d}\mathbf{p}^N \mathrm{d}\mathbf{r}^N \exp\left[-\frac{\mathscr{H}(\mathbf{p}^N, \mathbf{r}^N)}{k_\mathrm{B} T} \right] \tag{9.84}$$

we can write the following for $\partial Q(\lambda)/\partial \lambda$:

$$\frac{\partial Q(\lambda)}{\partial \lambda} = \frac{1}{N!} \frac{1}{h^{3N}} \iint \mathrm{d}\mathbf{p}^N \mathrm{d}\mathbf{r}^N \frac{\partial}{\partial \lambda} \exp\left[-\frac{\mathscr{H}(\mathbf{p}^N, \mathbf{r}^N)}{k_\mathrm{B} T} \right] \tag{9.85}$$

Applying the chain rule:

$$\frac{\partial Q(\lambda)}{\partial \lambda} = -\frac{1}{N!}\frac{1}{h^{3N}}\frac{1}{k_B T}\iint \mathrm{d}\mathbf{p}^N \mathrm{d}\mathbf{r}^N \frac{\partial \mathscr{H}(\mathbf{p}^N, \mathbf{r}^N)}{\partial \lambda} \exp\left[-\frac{\mathscr{H}(\mathbf{p}^N, \mathbf{r}^N)}{k_B T}\right] \tag{9.86}$$

Substituting back into the expression for $\partial A / \partial \lambda$ gives:

$$\begin{aligned}
\frac{\partial A(\lambda)}{\partial \lambda} &= \frac{1}{N!}\frac{1}{h^{3N}}\frac{1}{Q(\lambda)}\iint \mathrm{d}\mathbf{p}^N \mathrm{d}\mathbf{r}^N \frac{\partial \mathscr{H}(\mathbf{p}^N, \mathbf{r}^N)}{\partial \lambda} \exp\left[-\frac{\mathscr{H}(\mathbf{p}^N, \mathbf{r}^N)}{k_B T}\right] \\
&= \iint \mathrm{d}\mathbf{p}^N \mathrm{d}\mathbf{r}^N \frac{\partial H(\mathbf{p}^N, \mathbf{r}^N)}{\partial \lambda} \left\{ \frac{\exp\left[-\frac{\mathscr{H}(\mathbf{p}^N, \mathbf{r}^N)}{k_B T}\right]}{Q(\lambda)} \right\} \\
&= \left\langle \frac{\partial \mathscr{H}(\mathbf{p}^N, \mathbf{r}^N, \lambda}{\partial \lambda} \right\rangle_\lambda
\end{aligned} \tag{9.87}$$

Thus

$$\Delta A = \int_{\lambda=0}^{\lambda=1} \left\langle \frac{\partial \mathscr{H}(\mathbf{p}^N, \mathbf{r}^N, \lambda)}{\partial \lambda} \right\rangle_\lambda \mathrm{d}\lambda \tag{9.88}$$

Appendix 9.2 Using the slow growth method for calculating free energy differences

The slow growth expression can be derived from the thermodynamic perturbation expression (equation 9.7) if it is written as a Taylor series:

$$\Delta A = -k_B T \sum_{\mathrm{i}=0}^{N_{\text{step}}-1} \ln\langle \exp(-[\mathscr{H}(\lambda_{\mathrm{i}+1}) - \mathscr{H}(\lambda_{\mathrm{i}})]/k_B T)\rangle_{NVT} \tag{9.89}$$

$$\Delta A \approx -k_B T \sum_{\mathrm{i}=0}^{N_{\text{step}}-1} \ln\langle 1 - [\mathscr{H}(\lambda_{\mathrm{i}+1}) - \mathscr{H}(\lambda_{\mathrm{i}})]/k_B T + \cdots\rangle_{NVT} \tag{9.90}$$

$$\Delta A \approx -k_B T \sum_{\mathrm{i}=0}^{N_{\text{step}}-1} \ln\left\{1 - \frac{1}{k_B T}\langle[\mathscr{H}(\lambda_{\mathrm{i}+1}) - \mathscr{H}(\lambda_{\mathrm{i}})]\rangle_{NVT} + \cdots\right\} \tag{9.91}$$

$$\Delta A \approx \sum_{\mathrm{i}=0}^{N_{\text{step}}-1} \langle[\mathscr{H}(\lambda_{\mathrm{i}+1}) - \mathscr{H}(\lambda_{\mathrm{i}})]\rangle_{NVT} \tag{9.92}$$

Further reading

Beveridge D L and F M DiCapua 1989. Free Energy via Molecular Simulation: A Primer in van Gunsteren W F and P K Weiner (Editors). *Computer Simulation of Biomolecular Systems*. Leiden, ESCOM Science Publishers: 1–26.

Jorgensen W L 1983. Theoretical Studies of Medium Effects on Conformationl Equilibria. *Journal of Physical Chemistry* **87**:5304–5314.

King P M 1993. Free Energy via Molecular Simulation: A Primer in van Gunsteren W F, P K Weiner and A J Wilkinson (Editors). *Computer Simulation of Biomolecular Systems*. Volume 2. Leiden, ESCOM Science Publishers: 267–314

Kollman P A 1993. Free Energy Calculations: Applications to Chemical and Biochemical Phenomena. *Chemical Reviews* **93**:2395–2417.

Mark A E and van Gunsteren W F 1995. Free Energy Calculations in Drug Design: A Practical Guide in Dean P M, G Jolles and C G Newton (Editors). *New Perspectives in Drug Design*. London, Academic Press: 185–200.

Mezei M and D L Beveridge 1986. Free Energy Simulations in Beveridge D L and W L Jorgensen (Editors). Computer Simulation of Chemical and Biomolecular Systems. *Annals of the New York Academy of Sciences* **482**:1–23.

van Gunsteren W F 1989. Methods for Calculation of Free Energies and Binding Constants: Successes and Problems in van Gunsteren and P K Weiner (Editors). *Computer Simulation of Biomolecular Systems*. Leiden, ESCOM Science Publishers: 27–59.

References

Åqvist J and A Warshel 1993. Simulation of Enzyme Reactions Using Valence Bond Force Fields and Other Hybrid Quantum/Classical Approaches. *Chemical Reviews* **93**:2523–2544.

Åqvist J, Fothergill M and A Warshel 1993. Computer Simulation of the CO_2/HCO_3^- Interconversion Step in Human Carbonic Anhydrase I. *The Journal of the American Chemical Society* **115**:631–635.

Bartlett P A and C K Marlowe 1987. Evaluation of Intrinsic Binding Energy from a Hydrogen-bonding Group in an Enzyme Inhibitor. *Science* **235**:569–571.

Bash P A, U C Singh, F K Brown, R Langridge and P A Kollman 1987. Calculation of the Relative Change in Binding Free-Energy of a Protein-Inhibitor Complex. *Science* **235**:574–576.

Bernardi A, A M Capelli, A Comotti, C Gannari, J M Goodman and I Paterson 1990. Transition-State Modeling of the Aldol Reaction of Boron Enolates: A Force Field Approach. *Journal of Organic Chemistry* **55**:3576–3581.

Boresch S, G Archontis and M Karplus 1994. Free Energy Simulations: The Meaning of the Individual Contributions From a Component Analysis. *Proteins: Structure, Function and Genetics* **20**:25–33.

Born M 1920. Volumen und Hydratationswärme der Ionen. *Zeitschrift für Physik* **1**:45–48.

Burger M T, A Armstrong, F Guarnieri, D Q McDonald and W C Still 1994. Free Energy Calculations in Molecular Design: Predictions by Theory and Reality by Experiment with Enantioselective Podand Ionophores. *The Journal of the American Chemical Society* **116**:3593–3594.

Car R and M Parrinello 1985. Unified Approach For Molecular-Dynamics and Density-Functional Theory. *Physical Review Letters* **55**:2471–2474.

Chandrasekhar J and W L Jorgensen 1985. Energy Profile for a Nonconcerted S_N2 Reaction in Solution. *The Journal of the American Chemical Society* **107**:2974–2975.

Chandrasekhar J, S F Smith and W L Jorgsensen 1985. Theoretical Examination of the S_N2 Reaction Involving Chloride Ion and Methyl Chloride In the Gas Phase and Aqueous Solution. *The Journal of the American Chemical Society* **107**:154–163.

Claverie P, J P Daudey, J Langlet, B Pullman, D Piazzola and M J Huron 1978. Studies of Solvent Effects. I. Discrete, Continuum and Discrete-Continuum Models and Their Comparison for Some Simple Cases: NH_4^+, CH_3OH and substituted NH_4^+. *Journal of Physical Chemistry* **82**:405–418.

Constanciel R and R Contreras 1984. Self-Consistent Field-Theory of Solvent Effects Representation by Continuum Models—Introduction of Desolvation Contribution. *Theoretica Chimica Acta* **65**:1–11.

Cramer C J and D G Truhlar 1992. AM1-SM2 and PM3-SM3 Parameterized SCF Solvation Models for Free Energies in Aqueous Solution. *Journal of Computer-Aided Molecular Design* **6**:629–666.

Cramer C J and D G Truhlar 1993. Quantum Chemical Conformational Analysis of Glucose in Aqueous Solution. *The Journal of the American Chemical Society* **115**:5745–5753.

Dovesi R, C Pisani, C Roetti and P Dellarole 1981. Exact-Exchange Hartree–Fock Calculations For Periodic-Systems. 4. Ground-State Properties Of Cubic Boron-Nitride. *Physical Review B, Condensed Matter* **24**:4170–4176.

Dovesi R, C Pisani, C Roetti and V R Saunders 1983. Treatment of Coulomb Interactions in Hartree–Fock Calculations of Periodic-Systems. *Physical Review B, Condensed Matter* **28**:5781–5792.

Eisenberg D and A D McLachlan 1986. Solvation Energy in Protein Folding and Binding. *Nature* **319**:199–203.

Essex J W, C A Reynolds and W G Richards 1989. Relative Partition Coefficients from Partition Functions: A Theoretical Approach to Drug Transport. *Journal of the Chemical Society Chemical Communications*: 1152–1154.

Field M J, P A Bash and M Karplus 1990. A Combined Quantum Mechanical and Molecular Mechanical Potential for Molecular Dynamics Simulations. *Journal of Computational Chemistry* **11**:700–733.

Fleischman S H and C L Brooks III 1987. Thermodynamics of Aqueous Solvation—Solution Properties of Alcohols and Alkanes. *The Journal of Chemical Physics* **87**:3029–3037.

Floris F and J Tomasi 1989. Evaluation of the Dispersion Contribution to the

Solvation Energy—A Simple Computational Model in the Continuum Approximation. *Journal of Computational Chemistry* **10**:616–627.

Gilson M K and B Honig 1988. Calculation of the Total Electrostatic Energy of a Macromolecular System: Solvation Energies, Binding Energies and Conformational Analysis. *Proteins: Structure, Function and Genetics* **4**:7–18.

Gonzalez C and H B Schlegel 1988. An Improved Algorithm for Reaction Path Following. *The Journal of Chemical Physics* **90**:2154–2161.

Ha S, J Gao, B Tidor, J W Brady and M Karplus 1991. Solvent Effect on the Anomeric Equilibrium in D-Glucose: A Free Energy Simulation Analysis. *The Journal of the American Chemical Society* **113**:1553–1557.

Hasel W, T F Hendrickson and W C Still 1988. A Rapid Approximation to the Solvent Accessible Surface Areas of Atoms. *Tetrahedron Computer Methodology* **1**:103–116.

Heine V 1970. The Pseudopotential Concept. *Solid State Physics* **24**:1–36.

Hohenberg P and Kohn W 1964. Inhomogeneous Electron Gas. *Physical Review* **136**:B864–871.

Honig B and A Nicholls 1995. Classical Electrostatics in Biology and Chemistry. *Science* **268**:1144–1149.

Jorgensen W L and J K Buckner 1987. Use of Statistical Perturbation Theory for Computing Solvent Effects on Molecular Conformation. Butane in Water. *Journal of Physical Chemistry* **91**:6083–6085.

Jorgensen W L, J Gao and C Ravimohan 1985. Monte Carlo Simulations of Alkanes in Water: Hydration Numbers and the Hydrophobic Effect. *The Journal of Physical Chemistry* **89**:3470–3473.

Jorgensen W L, J K Buckner, S Boudon and J Tirado-Reeves 1988. Efficient Computation of Absolute Free Energies of Binding by Computer Simulations—Applications to the Methane Dimer in Water. *The Journal of Chemical Physics* **89**:3742–3746.

Jorgensen W L, J M Briggs and M L Contreras 1990. Relative Partition Coefficients for Organic Solutes from Fluid Simulations. *Journal of Physical Chemistry.* **94**:1683–1686.

Kirkwood J G 1934. Theory of Solutions of Molecules Containing Widely Separated Charges with Special Application to Zwitterions. *The Journal of Chemical Physics* **2**:351–361.

Klapper I, R Hagstrom, R Fine, K Sharp and B Honig 1986. Focusing of Electric Fields in the Active Site of Cu–Zn Superoxide Dismutase: Effects of Ionic Strength and Amino-Acid Substitution. *Proteins: Structure, Function and Genetics* **1**:47–59.

Kohn W and L J Sham 1965. Self-consistent Equations Including Exchange and Correlation Effects. *Physical Review* **140**:A1133–1138.

Kohn W and P Vashita 1983. General Density Functional Theory. In *Theory of Inhomogeneous Electron Gas*. Lundquist S and N H March (Editors). New York, Plenum: 79–148.

Lybrand T P, J A McCammon and G Wipff 1986. Theoretical Calculation of Relative Binding Affinity in Host–Guest Systems. *Proceedings of the National Academy of Sciences USA* **83**:833–835.

McRee D E, S M Redford, E D Getzoff, J R Lepock, R A Hallewell and J A Tainer 1990. Changes in Crystallographic Structure and Thermostability of a Cu, Zn Superoxide Dismutase Mutant Resulting from the Removal of Buried Cysteine. *Journal of Biological Chemistry* **265**:14234–14241.

Marquart M, J Walter, J Deisenhofer, W Bode, R Huber 1983. The Geometry of the Reactive Site and of the Peptide Groups in Trypsin, Trypsinogen and its Complexes with Inhibitors. *Acta Crystallographica* **B39**:480–490.

Merz K M Jr and P A Kollman 1989. Free Energy Perturbation Simulations of the Inhibition of Thermolysin: Prediction of the Free Energy of Binding of a New Inhibitor. *The Journal of the American Chemical Society* **111**:5649–5658.

Miertus S, E Scrocco and J Tomasi 1981. Electrostatic Interaction of a Solute with a Continuum—A Direct Utilization of *Ab Initio* Molecular Potentials for the Provision of Solvent Effects. *Chemical Physics* **55**:117–129.

Miick S M, G V Martinez, W R Fiori, A P Todd and G L Millhauser 1992. Short Alanine-Based Peptides May Form 3(10)-Helices and not Alpha-Helices in Aqueous-Solution. *Nature* **359**:653–655.

Mitchell M J and J A McCammon 1991. Free Energy Difference Calculations by Thermodynamic Integration: Difficulties in Obtaining a Precise Value. *Journal of Computational Chemistry* **12**:271–275.

Onsager L 1936. Electric Moments of Molecules in Liquids. *The Journal of the American Chemical Society* **58**:1486–1493.

Parr R G 1983. Density Functional Theory. *Annual Review of Physical Chemistry* **34**:631–656.

Paschual-Ahuir J L, E Silla, J Tomasi and R Bonaccorsi 1987. Electrostatic Interaction of a Solute with a Continuum. Improved Description of the Cavity and of the Surface Cavity Bound Charge Distribution. *Journal of Computational Chemistry* **8**:778–787.

Payne M C, M P Teter, D C Allan, R A Arias and D J Joannopoulos 1992. Iterative Minimisation Techniques for *Ab Initio* Total-Energy Calculations: Molecular Dynamics and Conjugate Gradients. *Reviews of Modern Physics* **64**:1045–1097.

Pearlman D A and P A Kollman 1989. A New Method for Carrying Out Free-Energy Perturbation Calculations—Dynamically Modified Windows. *Journal Of Chemical Physics* **90**:2460–2470.

Perdew J P and A Zunger 1981. Self-Interaction Correction to Density-Functional Approximations for Many-Electron Systems. *Physical Review B, Condensed Matter* **23**:5048–5079.

Pierotti R 1965. Aqueous Solutions of Nonpolar Gases. *Journal of Physical Chemistry* **69**:281–288.

Postma J P M, Berendsen H J C and J R Haak 1982. Thermodynamics of Cavity Formation in Water. *Faraday Symposium of the Chemical Society* **17**:55–67.

Rashin A A 1990. Hydration Phenomena, Classical Electrostatics, and the Boundary Element Method. *Journal of Physical Chemistry* **94**:1725–1733.

Rashin A A and B Honig 1985. Reevaluation of the Born Model of Ion Hydration. *Journal of Physical Chemistry* **89**:5588–5593.

Rashin A A and K Namboodiri 1987. A Simple Method for the Calculation of

Hydration Enthalpies of Polar Molecules with Arbitrary Shapes. *Journal of Physical Chemistry* **91**:6003–6012.

Remler D K and P A Madden 1990. Molecular Dynamics without Effective Potentials via the Car–Parrinello Approach. *Molecular Physics* **70**:921–966.

Rinaldi D, M F Ruiz-Lopez and J L Rivail 1983. *Ab Initio* SCF Calculations on Electrostatically Solvated Molecules Using a Deformable Three Axes Ellipsoidal Cavity. *The Journal of Chemical Physics* **78**:834–838.

Ryckaert J-P and A Bellemans 1978. Molecular Dynamics of Liquid Alkanes. **20**:95–106.

Singh U C and P A Kollman 1986. A Combined *Ab Initio* Quantum Mechanical and Molecular Mechanical Method for Carrying out Simulations on Complex Molecular Systems: Applications to the $CH_3Cl + Cl^-$ Exchange Reaction and Gas Phase Protonation of Polyethers. *Journal of Computational Chemistry* **7**:718–730.

Sitkoff D, K A Sharp and B Honig 1994. Accurate Calculation of Hydration Free Energies Using Macroscopic Solvent Models. *Journal of Physical Chemistry* **98**:1978–1988.

Smith P E and B M Pettitt 1994. Modeling Solvent in Biomolecular Systems. *Journal of Physical Chemistry* **98**:9700–9711.

Smythe M L, S E Huston and G R Marshall 1993. Free Energy Profile of a 3_{10} to α-Helical Transition of an Oligopeptide in Various Solvents. *The Journal of the American Chemical Society* **115**:11594–11595.

Smythe M L, S E Huston and G R Marshall 1995. The Molten Helix: Effects of Solvation on the α- to 3_{10}-Helical Transition. *The Journal of the American Chemical Society* **117**:5445–5452.

Stich I, A De Vita, M C Payne, M J Gilland and L J Clarke 1994. Surface Dissociation from First Principles: Dynamics and Chemistry. *Physical Review B* **49**:8076–8085.

Still W C, A Tempczyrk, R C Hawley and T Hendrickson 1990. Semianalytical Treatment of Solvation for Molecular Mechanics and Dynamics. *The Journal of the American Chemical Society* **112**:6127–6129.

Tapia O and O Goscinski 1975 Self-Consistent Reaction Field Theory of Solvent Effects. *Molecular Physics* **29**:1653–1661.

Tirado-Reeves J, D S Maxwell and W L Jorgensen 1993. Molecular Dynamics and Monte Carlo Simulations Favor the α-Helical Form for Alanine-Based Peptides in Water. *The Journal of the American Chemical Society* **115**:11590–11593.

Tobias D J and C L Brooks III 1988. Molecular Dynamics with Internal Coordinate Constraints. *The Journal of Chemical Physics* **89**:5115–5126.

Torrie G M and J P Valleau 1977. Nonphysical Sampling Distributions in Monte Carlo Free-Energy Estimation: Umbrella Sampling. *Journal of Computational Physics* **23**:187–199.

Warshel A 1991. *Computer Modelling of Chemical Reactions in Enzymes and Solutions*. New York, John Wiley.

Warshel A and M Levitt 1976. Theoretical Studies of Enzymic Reactions: Dielectric, Electrostatic and Steric Stabilization of the Carbonium Ion in the Reaction of Lysozyme. *Journal of Molecular Biology* **103**:227–249.

Warwicker J and H C Watson 1982. Calculation of the Electric-Potential in the Active-Site Cleft Due to Alpha-Helix Dipoles. *Journal Of Molecular Biology* **157**:671–679.

Wimmer E. 1991 Density Functional Theory for Solids, Surface and Molecules: from Energy Bands to Molecular Bonds in Labanowski J R and J W Andzelm (Editors). *Density Functional Methods in Chemistry*: 7–31. Berlin, Springer Verlag

Wodak S J and J Janin 1980. Analytical Approximation to the Solvent Accessible Surface Area of Proteins. *Proceedings of the National Academy of Sciences USA* **77**:1736–1740.

Wong M W, K B Wiberg and M J Frisch 1992. Solvent Effects. 3. Tautomeric Equilibria of Formamide and 2-Pyridone in the Gas Phase and Solution. An *Ab Initio* SCRF Study. *The Journal of the American Chemical Society* **114**:1645–1652.

Yu H-A and M Karplus 1988. A Thermodynamic Analysis of Solvation. *The Journal of Chemical Physics* **89**:2366–2379.

Zhang L and J Hermans 1994. 3_{10} Helix versus α-Helix: A Molecular Dynamics Study of Conformational Preferences of Aib and Alanine. *The Journal of the American Chemical Society* **116**:11915–11921.

Zwanzig R W 1954. High-temperature Equation of State by a Perturbation Method. I. Nonpolar Gases. *The Journal of Chemical Physics* **22**: 1420–1426.

10 The use of molecular modelling to discover and design new molecules

Molecular modelling techniques are widely used in the chemical, pharmaceutical and agrochemical industries. Much of this modelling activity employs the tools that we have discussed in earlier chapters, such as energy minimisation, molecular dynamics and Monte Carlo simulations and conformational analysis. In this chapter we will discuss a number of methods that do not fit naturally into these categories. Our discussion will use examples drawn primarily from the pharmaceutical industry, though many of the techniques are applicable to molecular design in other areas.

10.1 Molecular modelling in drug discovery

The development of a new medicine is a long and expensive process. A new compound must not only produce the desired response with minimal side-effects, but must be demonstrably better than existing therapies. The first task in many drug discovery programmes is to identify one or more *lead compounds*. A lead compound is a molecule that is active in an appropriate assay. The lead compound is then modified to enhance its potency and selectivity, to ensure that neither it nor any of its metabolites is toxic, and to provide appropriate transport characteristics to enable it to pass through cell membranes and reach its target.

Many drugs produce their effect by interacting with a biological macrocycle such as an enzyme, DNA, glycoprotein or receptor. The interaction between a ligand and its target* may be due entirely to non-bonded forces, but in some cases a covalent interaction may be involved. Drugs which interact with receptor proteins can be classified as *agonists*, *antagonists* or *inverse agonists*. Agonists produce the same or elevated effect as the natural substrate or effector molecule whereas antagonists inhibit the effect of the natural ligand. Inverse agonists create an effect which appears opposite to that of the agonist. Tight-binding ligands often have a high degree of complementarity with the target. This complementarity can be assessed and measured in various ways. Many ligands show significant shape complementarity with the region of the macromolecule where they bind (the binding site). This can be observed by constructing the molecular surfaces as

*We shall use the generic term 'ligand' to indicate the inhibitor or substrate, and the term 'receptor' to indicate the macromolecule to which it binds, be it an enzyme, a gene or a receptor protein.

illustrated in Figure 10.1 (colour plate section), which shows the molecular surfaces of avidin bound to biotin. The ligand often forms hydrogen bonds with the receptor. Some receptors have hydrophobic 'pockets', formed by groups of non-polar amino acids, into which the ligand can place a hydrophobic group of an appropriate size. It is also crucial to remember that a good drug does more than simply bind tightly to its target. After administration, a drug must get to the site of action. This transport process often requires the drug to pass through cell membranes. A cell membrane is a hydrophobic environment and so the drug must be sufficiently lipophilic (lipid loving) to partition into the membrane. Once inside the cell, the drug can access its target. During this process the molecule may also be removed from the body by metabolism, excretion and other pathways.

Finding novel lead compounds can be a difficult problem, and one where the molecular modeller has much to offer. Serendipity often played an important role, a classic example being the discovery of penicillin by Alexander Fleming. For many years pharmaceutical companies have screened soil and other biological samples to find new leads, but it can be difficult to extract and purify any active ingredients. Modern combinatorial chemistry techniques are now used to generate large numbers of molecules for screening using highly automated, robotic techniques.

In an increasing number of cases a three-dimensional structure of the target macromolecule is available. Such structures are obtained using X-ray crystallography or NMR or from a theoretical method such as homology modelling (see section 8.13). Even when detailed structural information about the target receptor is not available it may be possible to derive an abstract model called a *pharmacophore* that indicates the key features of a series of active molecules. The role of the molecular modeller is to suggest compounds that would be expected to interact favourably with the receptor, or compounds that contain the required functional groups of the pharmacophore.

Once a lead compound has been identified, a programme of chemical modification is undertaken to enhance its properties. The molecular modelling techniques described in earlier chapters can play a significant role in suggesting what modifications to make and in understanding the experimental binding results. Quantum mechanical and molecular mechanical calculations can provide information about the electronic and conformational properties of the moleculates. Molecular dynamics simulations can be used to investigate the dynamical behaviour of the ligand–receptor system and to calculate the relative free energies of binding of alternative compounds.

10.2 Deriving and using three-dimensional pharmacophores

In drug design, the term pharmacophore refers to a set of features that is common to a series of active molecules. Hydrogen bond donors and acceptors, positively and negatively charged groups, and hydrophobic groups are typical features. We will refer to such features as 'pharmacophoric groups'. A *three-dimensional (3D)*

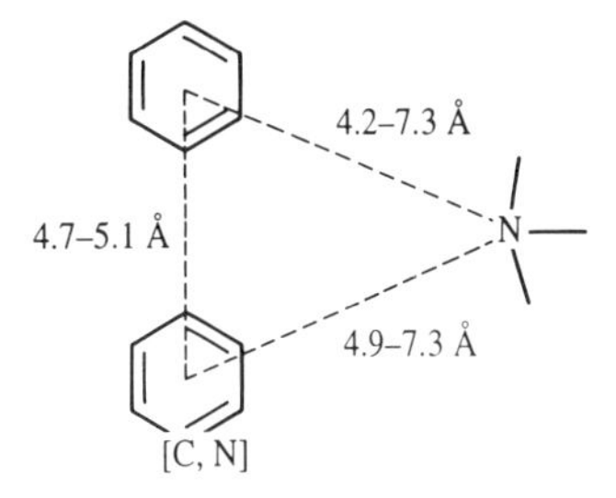

Fig. 10.2 Antihistamine 3D pharmacophore.

pharmacophore specifies the spatial relationships between the groups. These relationships are often expressed as distances or distance ranges but may also include other geometrical measures such as angles and planes. For example, a commonly used 3D pharmacophore for antihistamines contains two aromatic rings and a tertiary nitrogen distributed as shown in Figure 10.2. The development of methods for studying the conformations of ligands has stimulated an interest in the influence of the three-dimensional structures of molecules on their chemical and biological activity. The objective of a procedure known as *pharmacophore mapping* is to determine possible 3D pharmacophores for a series of active compounds. Once a pharmacophore has been developed, it can then be used to find or suggest other active molecules.

There are two problems to consider when calculating 3D pharmacophores. Unless the molecules are all completely rigid, one must take account of their conformational properties. The second problem is to determine those combinations of pharmacophoric groups that are common to the molecules and can be positioned in a similar orientation in space. More than one pharmacophore may be possible; indeed, some algorithms can generate hundreds of possible pharmacophores which must then be evaluated to determine which best fits the data. It is important to realise that all of these approaches to finding 3D pharmacophores assume that all of the molecules bind in a common manner to the macromolecule.

10.2.1 Constrained systematic search

In some cases it is relatively straightforward to deduce which features are required for activity. A well-known example is the pharmacophore for the angiotension-converting enzyme (ACE), which is involved in regulating blood pressure. Four typical ACE inhibitors are shown in Figure 10.3, including captopril which is widely used to treat hypertension. Angiotension-converting enzyme is a zinc metallopeptidase whose X-ray structure has not been solved at the time of writing. Three features within the class of inhibitors such as captopril are required for activity: a terminal carboxyl group (believed to interact with an arginine residue in the enzyme), an amido carbonyl group (which hydrogen bonds to an amide carbonyl in the enzyme), and a zinc-binding group. The problem is to deter-

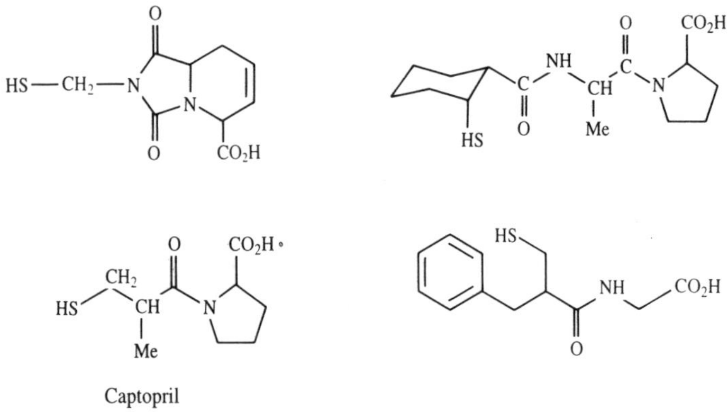

Captopril

Fig. 10.3 Four typical ACE inhibitors.

mine conformations in which the inhibitors can position these three pharmacophoric groups in the same relative position in space.

One of the most widely used methods for tackling this problem is the constrained systematic search method of Dammkoehler, Motoc and Marshall [Dammkoehler *et al.* 1989]. At first sight, it would appear that a systematic search over 20–30 molecules would greatly magnify the combinatorial explosion associated with a systematic conformational analysis. In fact one can significantly reduce the scale of the problem by making use of information about molecules whose conformational space has already been considered. Thus, we are only interested in those conformations that would enable the current molecule's pharmacophoric groups to be positioned in the same locations that have already

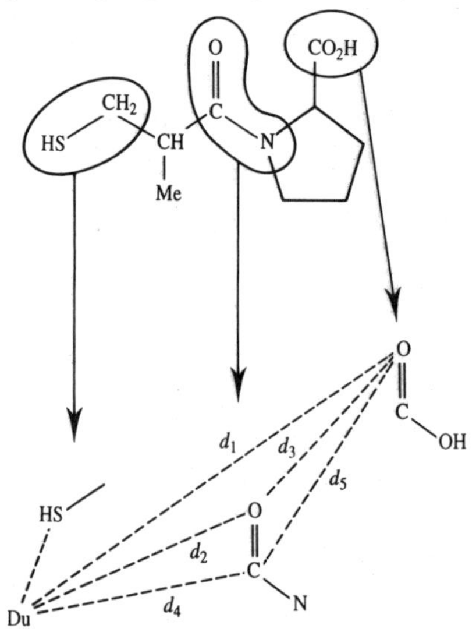

Fig. 10.4 Four points and five distances define the ACE pharmacophore.

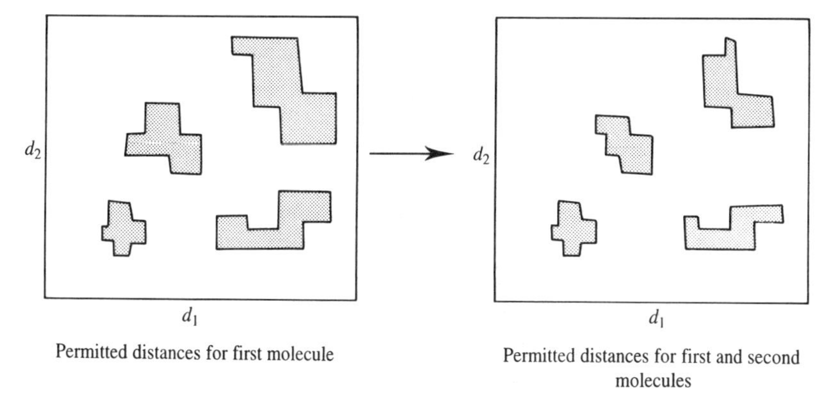

Fig. 10.5 A distance map indicates the distances available to specified groups. As more molecules are considered, the permitted regions get smaller.

been found for previous molecules. Dammkoehler and colleagues showed that it is possible to determine what torsion angles of the rotatable bonds will enable conformations consistent with the previous results to be obtained. It is best to choose the most conformationally restricted molecules first as these will have a reduced conformational space.

To derive an ACE pharmacophore, four points were defined for each molecule. The derivation of these four points for captopril is shown in Figure 10.4. Five distances (also shown in Figure 10.4) were defined between these four points.

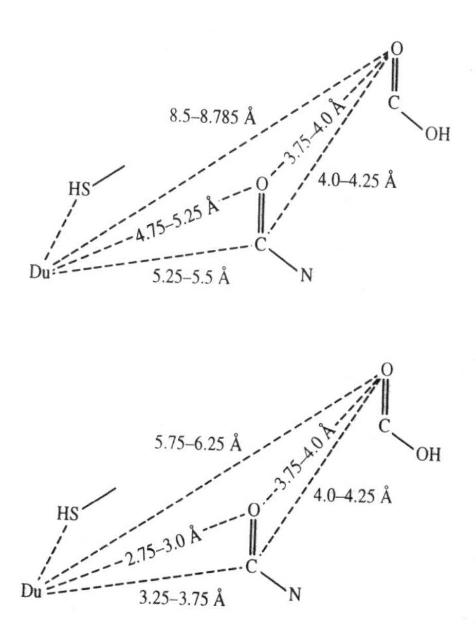

Fig. 10.6 Two ACE pharmacophores identified by the constrained systematic search [Dammkoehler *et al.* 1989].

Note that one of the points corresponds to the presumed location of the enzyme's zinc atom. The number of rotatable bonds in each inhibitor varied between 3 and 9 and the molecules were considered in order of increasing number of rotatable bonds. The entire conformational space was explored for the first (most inflexible) mol-ecule. For each conformation a point was registered in a five-dimensional hyperspace that correspond to that particular combination of the five distances. When the second molecule was considered, only those torsion angles that would enable these distances to be achieved were permitted to the rotatable bonds. As more molecules were examined, so the common regions in the five-dimensional hypersurface were reduced, as illustrated schematically for a two-dimensional example in Figure 10.5. Two distinct 3D pharmacophores were obtained from the search, shown in Figure 10.6. The constrained search could be performed three orders of magnitude faster than the approach involving a separate systematic search on all the molecules.

10.2.2 Ensemble distance geometry and ensemble molecular dynamics

A variant of distance geometry called *ensemble distance geometry* [Sheridan *et al.* 1986] can be used to simultaneously derive a set of conformations with a previously defined set of pharmacophoric groups overlaid. Ensemble distance geometry uses the same steps as standard distance geometry with the special feature that the conformational spaces of all the molecules are considered simultaneously. This is done using much larger bounds and distance matrices, with dimensions equal to the sum of the atoms in all the molecules. In these matrices, elements 1 to N_1 correspond to the N_1 atoms of molecule 1, elements $N_1 + 1$ to $N_1 + N_2$ to the N_2 atoms of molecule 2, and so on (Figure 10.7). Elements (i, j) and (j, i) of the bounds matrix thus represent the upper and lower bounds between atoms i and j (which may or may not be in the same molecule). The upper and

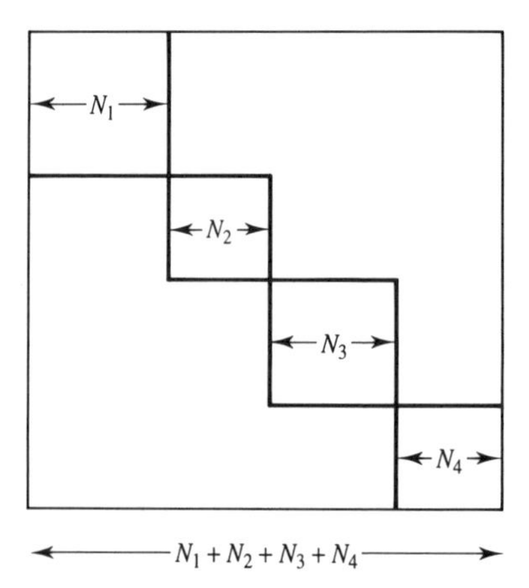

Fig. 10.7 Distance matrix used in ensemble distance geometry. There are N_1 atoms in the first molecule, N_2 in the second and so on.

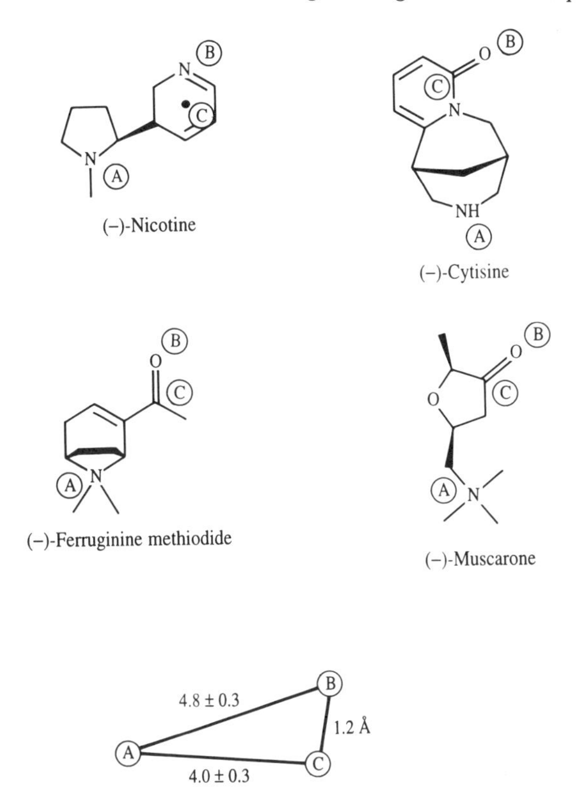

Fig. 10.8 Four molecules used to derive the nicotinic pharmacophore by distance geometry and the pharmacophore obtained.

lower bound distances between two atoms that are in the same molecule are set in the usual way. The lower bounds for atoms that are in different molecules are set to zero. This enables the molecules to be overlaid in three-dimensional space. The upper bounds for pairs of atoms that are in different molecules are set to a large value, except for those atoms that need to be superimposed in the pharmacophore which are set to a small tolerance parameter. Having defined the bounds matrix, the usual distance geometry steps are followed: smoothing, assignment of random distances, and optimisation against the initial bounds.

The first application of ensemble distance geometry was to derive a model of the nicotinic pharmacophore using the four nicotinic agonists shown in Figure 10.8. Three sets of atoms were selected as the pharmacophoric groups labelled A, B and C in Figure 10.8. The ensemble distance geometry algorithm generated several different solutions, but after eliminating those that contained distorted bond lengths or angles or unfavourable van der Waals contacts the remaining solutions corresponded to a single pharmacophore. This pharmacophore can be represented as a triangle (Figure 10.8). Note that the B–C distance is fixed at the length of the C=O bond. Also note that in (−)-nicotine the centroid of the pyridine ring is defined as one of the pharmacophoric points. The pharmacophore was then tested by determining whether low-energy conformations could be generated for other known nicotinic agonists that were consistent with the distance constraints of the pharmacophore.

Ensemble molecular dynamics derives a pharmacophore using restrained molecular dynamics for a collection of molecules. A force field model is set up so that none of the atoms in each molecule 'sees' the atoms in any other molecule. This enables the molecules to be overlaid in space. A restraint term is included in the potential that forces the appropriate atoms or functional groups to be overlaid in space.

10.2.3 Clique detection methods for finding pharmacophores

When there are many pharmacophoric groups present in the molecule it may be very difficult to identify all possible combinations of the functional groups (there may be thousands of possible pharmacophores). To tackle this problem, *clique detection* algorithms can be applied to a set of precalculated conformations of the molecules. To understand what we mean by a clique we need to understand some simple elements of graph theory.

A *graph* contains *nodes* that are connected by *edges*. Two examples are shown in Figure 10.9. In a molecular graph the nodes correspond to the atoms and the edges to the bonds, as shown for acetic acid in Figure 10.10. The locations of the nodes and edges of a graph on the page are irrelevant; only the way in which the nodes are connected together matters. The search trees that we met in section 8.2 are a special kind of graph. A *subgraph* is a subset of the nodes and edges of a graph; thus the graph for CH_3 is a subgraph of the graph of acetic acid. A graph is said to be *completely connected* if there is an edge between all pairs of nodes. Only in rare cases is the molecular graph a completely connected graph, one example being the P_4 form of elemental phosphorus.

A clique is defined as a 'maximal completely connected subgraph'. This definition is best understood by considering a simple example. Consider the graph G in

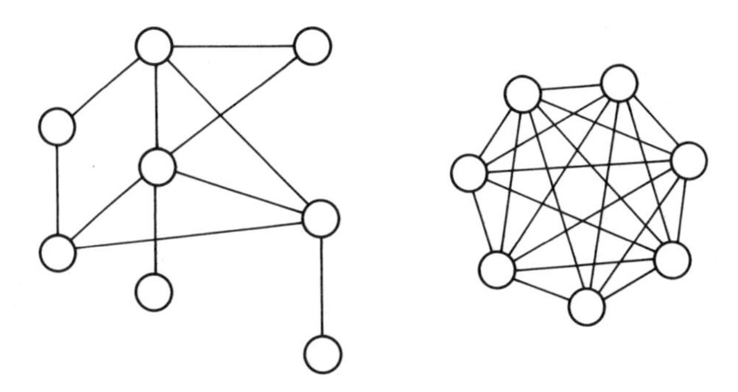

Fig. 10.9 Graphs contain nodes connected by edges. A completely connected graph (right) has an edge between all pairs of nodes.

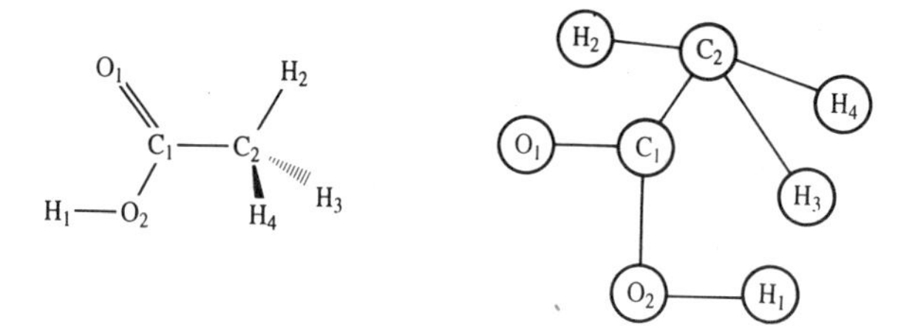

Fig. 10.10 The molecular graph of acetic acid.

Figure 10.11 together with various subgraphs. G is not a completely connected graph because there is not an edge between all the nodes. The subgraph S_1 is not a completely connected subgraph because there is no edge between nodes 1 and 8. The subgraph S_2 is a completely connected subgraph because there are edges between all the nodes. However, S_2 is not a clique because it is not a maximal completely connected subgraph; it is possible to add node 8 in order

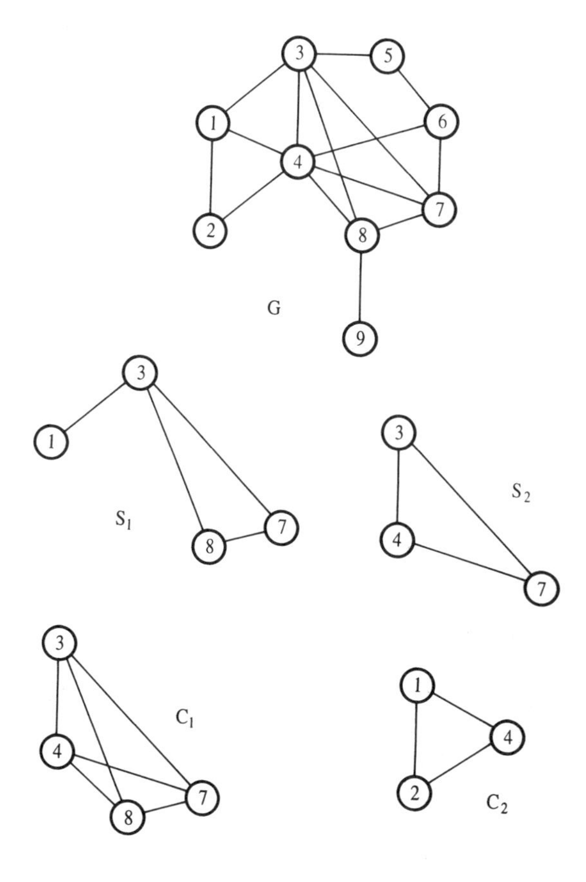

Fig. 10.11 Identifying cliques in a graph.

to obtain the clique C_1. A graph may contain many cliques; thus Figure 10.11 also shows a second clique, C_2. Finding the cliques in a graph belongs to a class of problems that are known as NP-complete. This means that the computational time required to find an exact solution increases in an exponential fashion with the size of the problem. Many algorithms have been devised for finding cliques; the method of Bron and Kerbosch has been found to be suitably efficient for pharmacophore identification [Bron and Kerbôsch 1973].

How is clique detection related to the identification of pharmacophores [Martin *et al.* 1993]? Let us suppose that we are comparing two conformations of two molecules, A and B (Figure 10.12). We construct a graph in which there is a node for every pair of matching pharmacophoric groups in the two structures. The two hydrogen bond acceptors in molecule A (O_1(A) and O_2(A)) and the two in molecule B (O_1(B) and O_2(B)) give rise to four nodes in the joint graph. There is one hydrogen bond donor in molecule A (H(A)) but three in molecule B (H_1(B), H_2(B) and H_3(B)), giving rise to three nodes in the graph. The intramolecular distances between these groups are indicated in Figure 10.12. An edge is drawn between each pair of nodes when the distance between the corresponding groups in the two molecules is the same, within some tolerance. For example, the distance between O_1(A) and O_2(A) is 4.73 Å and the distance between O_1(B) and O_2(B) is 4.19 Å. If the tolerance is at least 0.54 Å then the two distances would be considered equal and so an edge is drawn between the corresponding pairs of nodes in the graph, as shown in Figure 10.13. The full graph is shown in Figure 10.13 where we have assumed a tolerance of 0.6 Å. Clique detection is used to find maximal sets of matching groups for the two molecules; in this simple example there are three cliques, two containing just two overlapping atoms and one containing three matching atoms (Table 10.1).

In the clique detection approach, the first step is to generate a family of low-energy conformations for the molecules. The molecule with the smallest number of conformations is used as the starting point, with each of its conformations being used in turn as the reference structure. Each conformation of every other molecule is then compared with the reference conformations and the cliques identified. The cliques for each molecule are obtained by combining the results for each of its

Table 10.1 Cliques found when matching the molecules in Figure 10.12.

Clique number	Atom from A	Atom from B
1	O_1	O_2
	H	H_2
2	O_1	O_2
	O_2	O_1
3	O_1	O_1
	O_2	O_2
	H	H_3

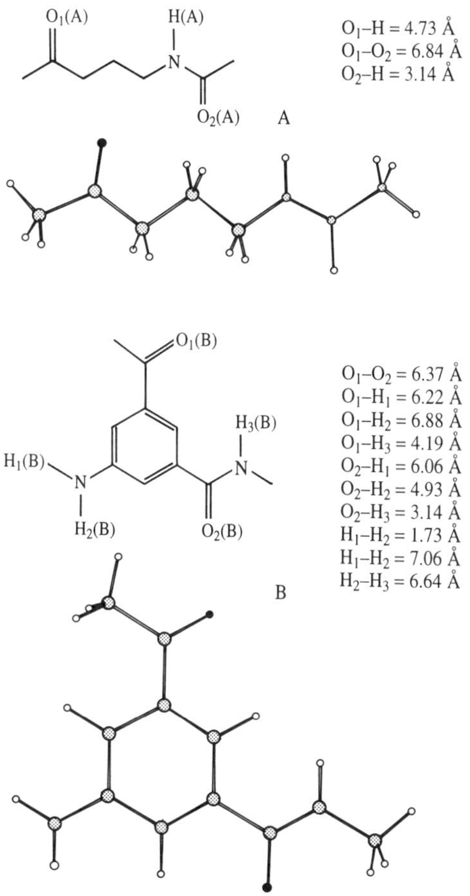

O1(A)
H(A)
N
O2(A)
A
O1–H = 4.73 Å
O1–O2 = 6.84 Å
O2–H = 3.14 Å
O1(B)
H3(B)
H1(B)
N
N
H2(B)
O2(B)
B
O1–O2 = 6.37 Å
O1–H1 = 6.22 Å
O1–H2 = 6.88 Å
O1–H3 = 4.19 Å
O2–H1 = 6.06 Å
O2–H2 = 4.93 Å
O2–H3 = 3.14 Å
H1–H2 = 1.73 Å
H1–H2 = 7.06 Å
H2–H3 = 6.64 Å

Fig. 10.12 Two molecules used to illustrate clique detection.

O1(A)
O2(B)
O1(A)
O1(B)
O2(A)
O1(B)
H(A)
H1(B)
O2(A)
O2(B)
H(A)
H2(B)
H(A)
H3(B)

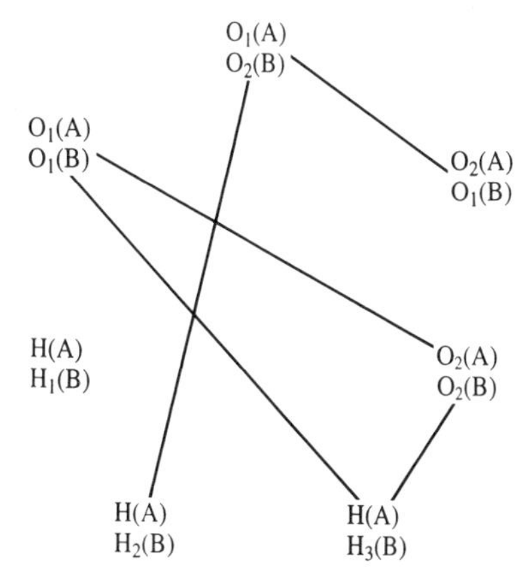

Fig. 10.13 The matching graph for the two molecules in Figure 10.12.

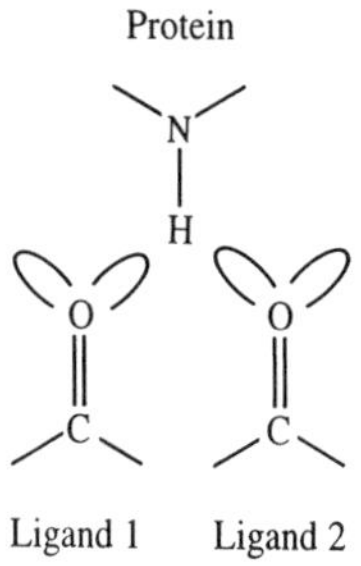

Fig. 10.14 Two ligands may be able to position a hydrogen bond acceptor in different locations in space yet still interact with the same hydrogen bond donor.

conformations. Those cliques that are common to at least one conformation from each molecule can then be combined to give a possible 3D pharmacophore for the entire set.

10.2.4 Incorporating geometric features in a 3D pharmacophore

The features used to define a 3D pharmacophore are most easily derived from the positions of specific atoms within each molecule. It may be more appropriate to consider locations around the molecule where the receptor might position its functional groups. This is especially useful for hydrogen bond donors and acceptors; two ligands may be able to hydrogen bond to the same protein atom with the ligand atoms being in a completely different location in the binding site, as illustrated in Figure 10.14. 3D pharmacophores may also be defined in terms of specific geometrical relationships between the pharmacophoric groups, such as the angle between the planes of two aromatic rings. The 3D pharmacophore may also contain features that are designed to mimic the presence of the receptor. These are commonly represented as *exclusion spheres*, which indicate locations within the 3D pharmacophore where no part of a ligand is permitted to be positioned. Some of these additional features are illustrated in Figure 10.15.

10.3 Molecular docking

In molecular docking we attempt to predict the structure (or structures) of the intermolecular complex formed between two or more molecules. Docking is widely used to suggest binding modes of protein inhibitors. Most docking algorithms are able to generate a large number of possible structures, and so they also require a means to score each structure to identify which are of most interest. The 'docking problem' is thus concerned with the generation and evaluation of plausible structures of intermolecular complexes [Blaney and Dixon 1993].

The docking problem involves many degrees of freedom. There are six degrees of translational and rotational freedom of one molecule relative to the other as well

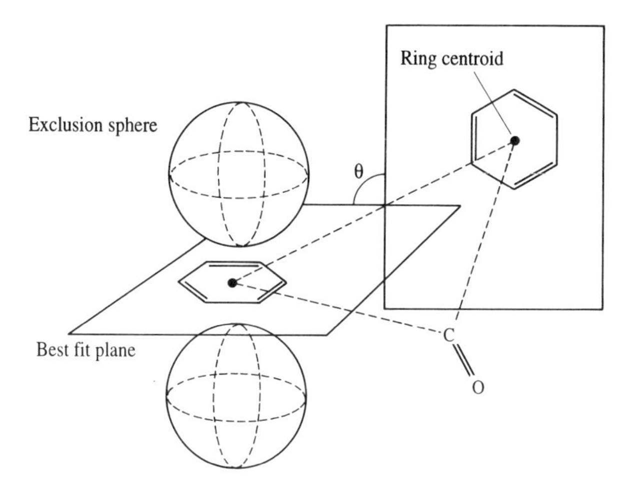

Fig. 10.15 Features that can be incorporated into 3D pharmacophores.

as the conformational degrees of freedom of each molecule. The docking problem can be tackled manually, using interactive computer graphics. This 'hands on' approach can be very effective if we have a good idea of the expected binding mode, for example because we already know the binding mode of a closely related ligand. However, in such cases one must be wary; X-ray crystallographic experiments have revealed that quite different binding modes may be adopted even by very similar inhibitors. Automatic docking algorithms can be less biased than human modellers and usually consider many more possibilities.

Various algorithms have been developed to tackle the docking problem. These algorithms can be characterised according to the number of degrees of freedom that they ignore. Thus, the simplest algorithms treat the two molecules as rigid bodies, and only explore the six degrees of translational and rotational freedom. The earliest algorithms for docking small molecule ligands into the binding sites of proteins and DNA used this approximation. A well-known example of such an algorithm is the DOCK program of Kuntz and co-workers [Kuntz *et al.* 1982]. DOCK is designed to find molecules with a high degree of shape complementarity to the binding site. The program first derives a 'negative image' of the binding site from the molecular surface of the macromolecule. This negative image consists of a collection of overlapping spheres of varying radii that touch the molecular surface at just two points, as shown schematically in Figure 10.16. Ligand atoms are then matched to the sphere centres to find matching sets (cliques) in which all the distances between the ligand atoms in the set are equal to the corresponding sphere centre–sphere centre distances (within some user-specified tolerance). The ligand can then be oriented within the site by performing a least-squares fit of the atoms to the sphere centres, as shown in Figure 10.17. The orientation is checked to ensure there are no unacceptable steric interactions between the ligand and the receptor. If the orientation is acceptable then an inter-

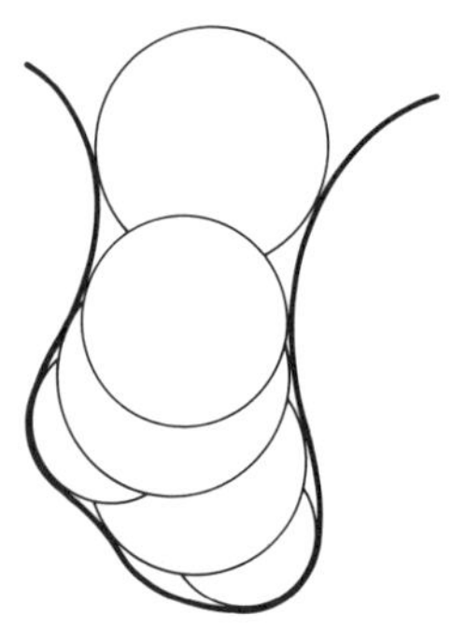

Fig. 10.16 A binding site represented as a collection of overlapping spheres.

action energy is computed to give the 'score' for that binding mode. New orientations are generated by matching different sets of atoms and sphere centres. The top scoring orientations are retained for subsequent analysis.

To perform conformationally flexible docking the conformational degrees of freedom need to be taken into account. Most of the methods that attempt to include the conformational degrees of freedom only consider the conformational space of the ligand; the receptor is invariably assumed to be rigid. All of the common methods for searching conformational space have been incorporated at some stage into a docking algorithm. For example, Monte Carlo methods have been used to perform molecular docking, often in conjunction with simulated annealing [Goodsell and Olson 1990]. At each iteration of the Monte Carlo procedure the internal conformation of the ligand is changed (by rotating about a bond) or the entire molecule is randomly translated or rotated. The energy of the ligand within the binding site is calculated using molecular mechanics and the move is then accepted or rejected using the standard Metropolis criterion. Genetic algorithms can also be used to perform molecular docking [Judson *et al.* 1994; Jones *et al.* 1995; Oshiro *et al.* 1995]. Each chromosome codes not

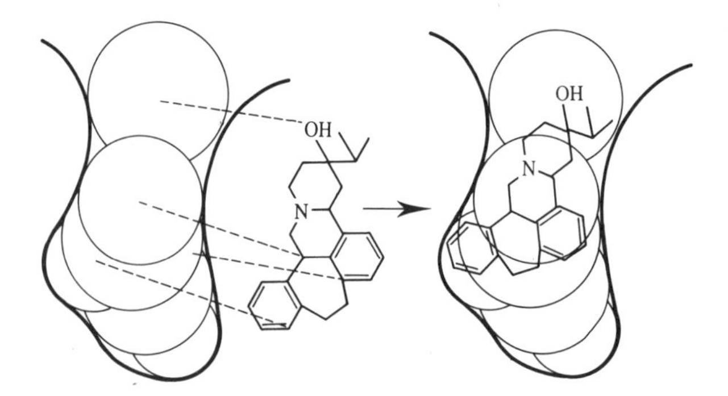

Fig. 10.17 The DOCK algorithm [Kuntz *et al.* 1982]. Atoms are matched to sphere centres and then the molecule is positioned within the binding site.

only for the internal conformation of the ligand as described in section 8.5, but also for the orientation of the ligand within the receptor site. Both the orientation and the internal conformation will thus vary as the populations evolve. The score of each docked structure within the site acts as the fitness function used to select the individuals for the next iteration.

Distance geometry can be used to perform molecular docking. The major problem to be addressed with this method is to find a way to generate conformations of the ligand within the binding site. One way to achieve this is by using a modified penalty function that forces the ligand conformation to remain within the binding site. For example, an additional penalty term can be added which has the effect of forcing the ligand to lie in the DOCK-derived cluster of spheres that represents the binding site.

The ideal docking method would allow both ligand and receptor to explore their conformational degrees of freedom. Perhaps the most ‘natural’ way to incorporate the flexibility of the binding site is via a molecular dynamics simulation of the ligand–receptor complex. However, such calculations are computationally very demanding and are in practice only useful for refining structures produced using other docking methods; molecular dynamics does not explore the range of binding modes very well except for very small, mobile ligands. For many systems, the energy barriers that separate one binding mode from another are often too large to be overcome.

10.4 Structure-based methods to identify lead compounds

If we have a three-dimensional structure of the target or a 3D pharmacophore, then there are two ways in which we can attempt to find new lead compounds. We could try to find a known molecule which would be expected to interact with the receptor. This approach is usually implemented by searching a database. Alternatively, we could attempt to design entirely new molecules ‘from scratch’, an approach commonly referred to as *de novo* ligand design.

10.4.1 Finding lead compounds by searching 3D databases

Many organisations maintain databases of chemical compounds. Some of these databases are publically accessible; others are proprietary. A database may contain an extremely large number of compounds; several hundred thousand is common, and the database maintained by the American Chemicals Society contains more than 10 million compounds. With the advent of three-dimensional information about drug design targets in the form of X-ray structures, homology models and pharmacophores it was recognised that it was important to take account of the conformational properties of possible inhibitors, thus leading to the concept of the *three-dimensional (3D) database*. In a 3D database search one tries to find molecules that satisfy both the chemical and the geometrical requirements of the receptor. There are two flavours of 3D database search. A program such as DOCK can be used to search a database by generating orientations of each molecule within the binding site. Alternatively, one can search the database for compounds that match a 3D pharmacophore.

10.4.2 Sources of 3D data

In a 3D database search one needs to consider the three-dimensional structures of the molecules. Where does such structural information come from? An obvious source is the Cambridge Structural Database, which contains experimental X-ray structures of more than 100 000 compounds. However, for most of the compounds in a typical compound database no crystal structure is available. Structure generation programs are designed to produce one or more low-energy conformations solely from the molecular graph. As the number of compounds may be very large, such programs must be able to operate automatically, rapidly, and with little or no user intervention (i.e. without crashing!). The most widely used structure generator to date is a program called CONCORD [Rusinko *et al.* 1988], which uses a knowledge-based approach combined with energy minimisation to generate three-dimensional structures.

Most structure generation algorithms only produce a single conformation for each molecule. With the possible exception of wholly rigid molecules there is no guaranteee that this structure corresponds to the conformation adopted when the molecule binds. We therefore need some way to take conformational flexibility into account during the 3D database search. The simplest way to do this is to store information about many conformations. To store conformations explicitly would usually require a large amount of disk space and so the information is usually compressed into a more compact form. A common way to do this is to derive a set of distance 'keys' for the pairs of pharmacophoric groups in the molecule. Each key is a binary number in which each bit corresponds to a distance range between the appropriate pair of groups (donor–donor, donor–acceptor, etc.). For example, the first bit could correspond to a distance in the range 2.0–2.5 Å, the second bit to the range 2.5–3.0 Å and so on. In fact, it is more efficient to use smaller bin sizes for the more 'common' distances and larger bins for less common distances because the distribution of distances between such groups in molecules is not uniform. The key initially contains all zeros. As each conformation is generated, the distances between the pharmacophoric groups are calculated and the appropriate bits in the relevant keys are changed to ones (Figure 10.18). To search the database, the keys corresponding to the pharmacophore are calculated. The pharmacophore's keys are then ANDed with each molecular key, so identifying all molecules which could match the pharmacophore. The AND operation is very fast and so the database can be screened very rapidly. Separate formula screens are also used; these contain information about the number of each type of feature (e.g. number of donors). If the molecule does not contain the minimal number of groups in the pharmacophore then it obviously cannot match and so can be discarded before its conformational properties need to be considered.

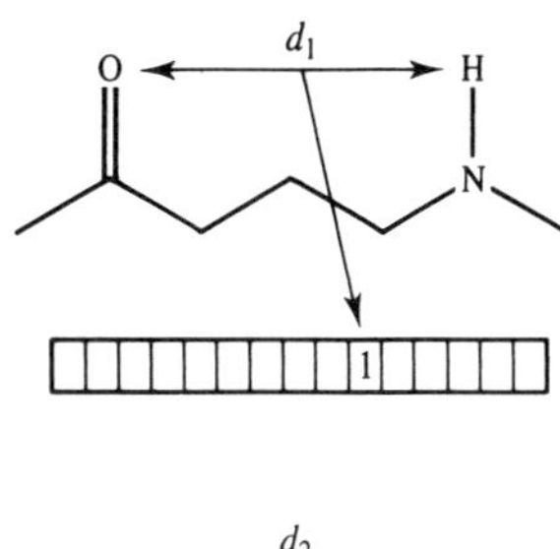

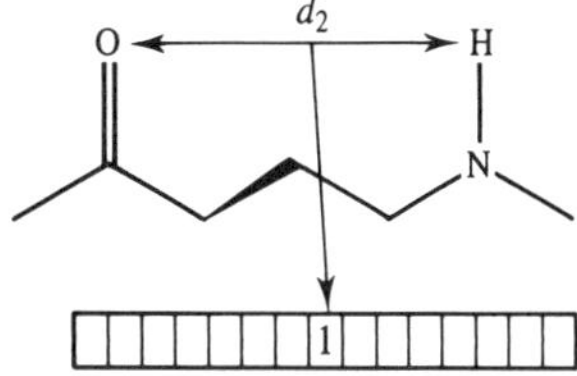

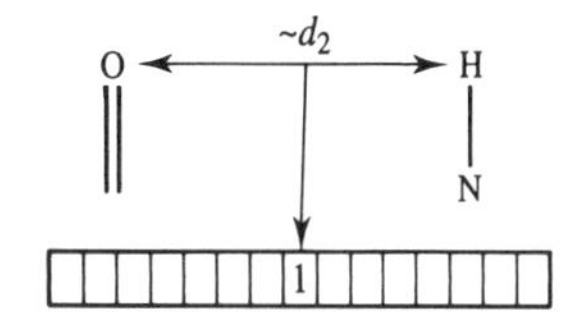

Fig. 10.18 3D database searching. As each conformation is generated an appropriate bit is set in the binary key. At search time the binary key appropriate to the pharmacophore is set up and compared with the keys in the database.

An alternative strategy is to explore the conformational space for each molecule during the database search. Systems which employ such an approach rely heavily upon screens which identify and reject molecules that could not satisfy the requirements of the pharmacophore before their conformational space is explored. These screens can be determined solely from the molecular graph and are typically represented as distance ranges. Triangle smoothing (section 8.6) is one way in which such distance screens can be calculated; it provides the upper and lower bounds on interatomic distances. However, the distance ranges provided by triangle smoothing can be much wider than the actual distances that are observed in real structures. A simple example suffices to illustrate this point: the distance obtained when triangle smoothing is used to calculate the lower bound distance between the amide nitrogen and the carbonyl oxygen of the carboxylic acid group in 4-acetamido benzoic acid (Figure 10.19) is equal to the sum of the van der Waals radii (approximately 3.3 Å depending upon the van der Waals radii used), compared to a distance of about 6.4 Å in all the accessible conformations of this molecule.

Having eliminated those molecules that could not possibly satisfy the geometric and chemical requirements of the pharmacophore, the program must explore the

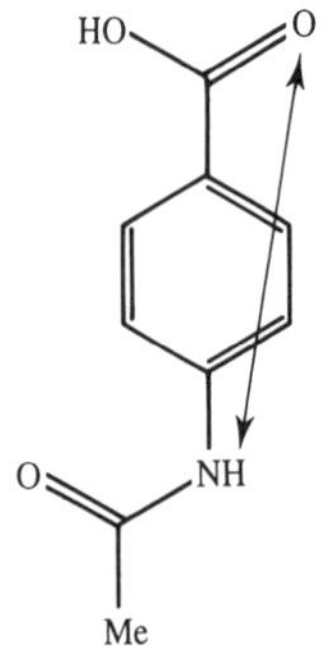

Fig. 10.19 4-Acetamido benzoic acid. Triangle smoothing predicts that the lower bound distance between the amide nitrogen and the cabonyl oxygen is equal to the sum of the van der Waals radii. The actual distance is about 6.4 Å.

conformational degrees of freedom of the molecules that remain. This is done using methods to rapidly identify one or more conformations which satisfy the constraints of the pharmacophore. A natural method to use would be distance geometry, in which the pharmacophoric constraints would be incorporated into the bounds matrix, thereby leading to the generation of conformations that satisfy the constraints. However, distance geometry is rather too slow for this purpose. An alternative strategy is to 'adjust' or 'tweak' the conformation by rotating about single bonds, to force it to fit the pharmacophore. Adjustment is usually performed in torsional space (i.e. only the torsion angles may vary) by minimising an appropriate potential function expressed in terms of distances.

There are now a number of published studies that demonstrate the utility of database searching in drug design. Kuntz's group have used the DOCK program against a number of targets, including the HIV protease, DNA, thymidylate synthase and haemagglutinin [Kuntz 1992; Kuntz *et al.* 1994]. In each case one or more inhibitors of modest potency were discovered. Information about hits from the first generation were then used to perform more exhaustive database searches to identify yet more potent compounds. The structures of some of the 'hits' were determined by X-ray crystallography, revealing that not all of the ligands bound in the same way as predicted by the docking algorithm. A degree of serendipity is still important even with automated docking methods.

10.5 *De novo* ligand design

Database searching is an attractive way to discover new lead compounds; in favourable cases the hits can be tested immediately or the molecule can be synthesised using a published method. However, database searching does not provide molecules that are truly 'novel'. Moreover, many databases are biased towards particular classes of compounds, so limiting the range of structures that can be obtained. In *de novo* design, the three-dimensional structure of the receptor or the 3D pharmacophore is used to design new molecules. There are two basic types of *de novo* design algorithms. The first class of methods have been described as 'outside in' methods [Lewis and Leach 1994]. Here, the binding site is first analysed to determine where specific functional groups might bind tightly. These groups are connected together to give molecular skeletons which are then converted into 'real' molecules. In the 'inside out' approach, molecules are grown within the binding site, under the control of an appropriate search algorithm with each suggestion being evaluated using an energy function. These two approaches are compared in Figure 10.20.

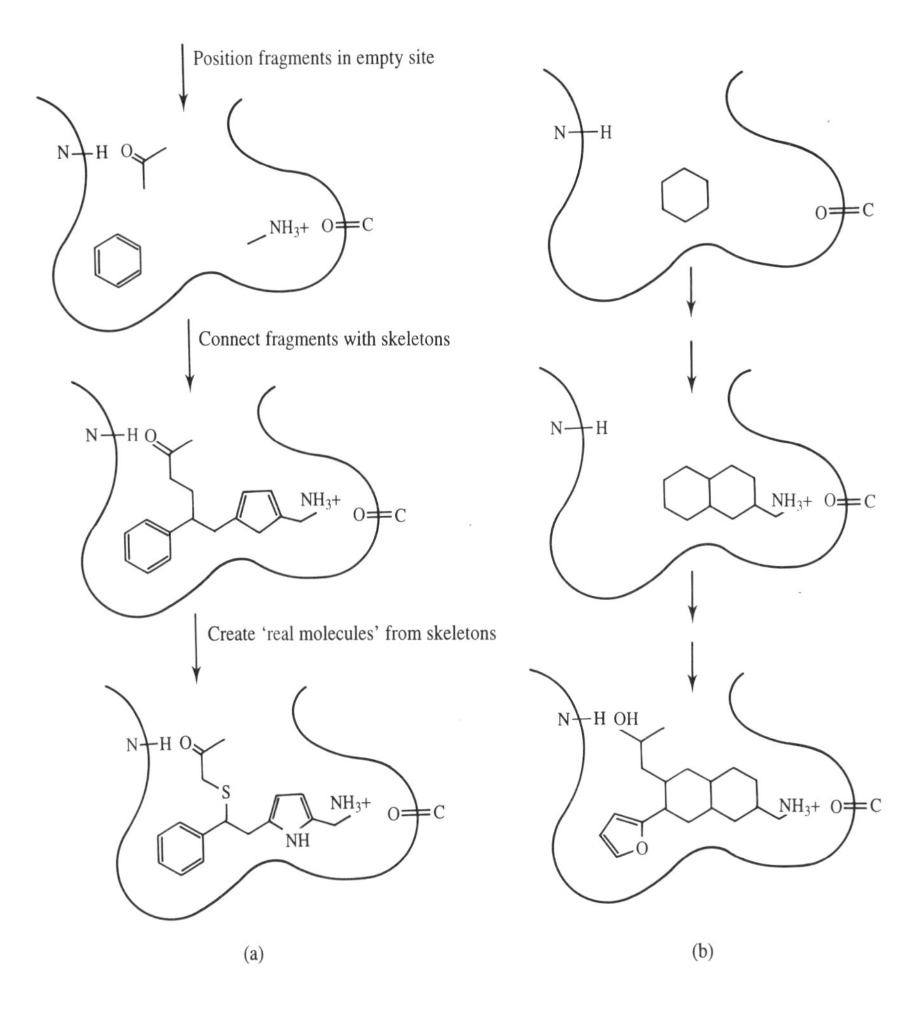

Fig. 10.20 Approaches to *de novo* design: (a) outside in, (b) inside out.

10.5.1 Favourable positions of molecular fragments within a binding site

One of the most widely used tools in structure-based ligand design is the GRID program [Goodford 1985]. A regular grid is superimposed upon the binding site. A probe group is then placed at the vertices of the grid and the interaction energy of the probe with the protein is determined using an empirical energy function. The result is a three-dimensional grid with an energy value at each vertex; this data can then be analysed to find those locations where it might be favourable to position a particular probe. An example of the output produced by GRID is shown in Figure 10.21 (colour plate section) for the binding site of neuraminidase. Parameters for many probes have been developed, covering a variety of small molecules and common functional groups. An alternative to the use of a grid is to permit each fragment to explore the entire binding site using energy minimisation or some form of simulation method. In the multiple-copy simultaneous search (MCSS) approach [Miranker and Karplus 1991] the binding site is initially filled with many copies of the same fragment, distributed randomly. A molecular mechanics energy model is used in which the protein interacts simultaneously with all of the fragments, but there are no interactions between the individual fragments. Energy minimisation is then used to try and find energetically favourable positions. The minimisation is done in a series of stages with the orientations of the fragments being clustered at the end of each stage to remove duplicates.

An alternative to the energy-based methods such as GRID is to suggest possible binding positions using a knowledge-based approach. Analysis of experimentally determined structures of ligand–receptor complexes reveals that they often contain certain types of interactions. For example, many ligands form hydrogen bonds with their receptors. The knowledge-based approaches generate binding modes that contain these commonly observed interactions with the fragments being positioned to reproduce the most commonly observed geometries. For example, in most hydrogen bonds the distance between the donor hydrogen and its acceptor is close to 1.8 Å and the angle subtended at the hydrogen is rarely less than 120°. Information about the preferred geometries of such interactions can be obtained from analyses of X-ray crystallographic databases (described in section 8.9). A program called LUDI has been widely used to dock small molecular fragments within protein binding sites using such an approach [Böhm 1992].

The knowledge-based docking approach to ligand design requires the receptor site to be surveyed to identify possible hydrogen bonding donors and acceptors as well as regions where other groups might favourably be positioned. The results of such an analysis are often converted into a distribution of *site points*. A site point is a location within the binding site where an appropriate ligand atom or group could be placed. For example, a hydrogen bonding analysis would typically result in a series of donor and acceptor site points. When generating the site points one should take account of any preferred geometries for that particular type of interaction; there is usually more than one site point associated with each donor or acceptor atom in the receptor to reflect the fact that a distribution of geometries is found in the crystal structure analyses. The range of preferred geometries can also be represented as a continuous region, as shown in Figure 10.22 (colour plate section). Having surveyed the binding site, each molecular fragment is examined to determine which features it contains and the fragment is positioned in the site by fitting the appropriate atoms to their corresponding site points.

10.5.2 Connecting molecular fragments in a binding site

Having positioned molecular fragments in favourable positions within the binding site using either an energy scheme or a knowledge-based approach, the next stage is to connect the fragments together into 'real' molecules. One way to tackle this problem is by searching a database of molecular connectors. Bartlett's CAVEAT program was one of the first methods for tackling this problem [Lauri and Bartlett 1994]. The relationship between each pair of fragments can be considered in terms of two vectors that represent the bonds they make to the linker. The geometrical relationship between each pair of linking vectors can be described using a distance, two angles and a dihedral angle as shown in Figure 10.23. CAVEAT searches its database to find connectors that also contain two bond vectors with the same geometrical parameters. The data is stored in an efficient form within the database, and so the search is very rapid. The connectors used by the first versions of CAVEAT consisted of ring systems extracted from the Cambridge Structural Database. The four geometric parameters were calculated between all pairs of bond vectors exocyclic to the ring, as shown in Figure 10.23. Connectors can also be generated using structure generation programs.

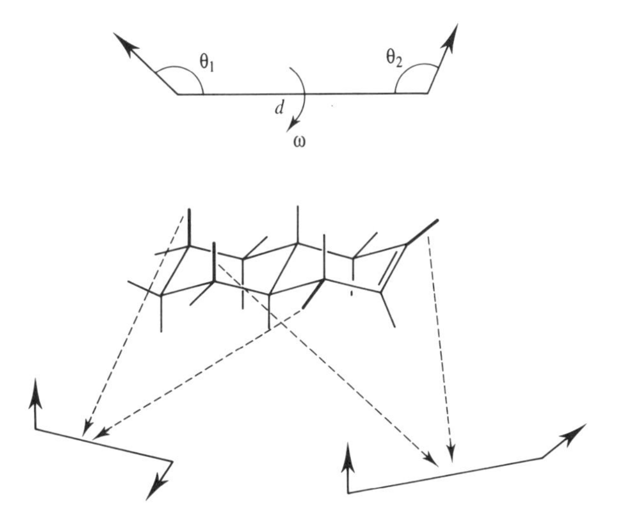

Fig. 10.23 The relationship between two bond vectors can be represented using a distance, two angles and a torsion angle as indicated (top). To derive the data for the database all possible pairs of exocyclic vectors are considered and these four geometric parameters calculated.

Not all problems are amenable to the database searching approach exemplified by CAVEAT; the molecular fragments may be further apart than the longest connector in the database. Moreover, with such an approach one is inherently restricted to the fragments contained within the database. An alternative strategy is to create a skeleton that connects the fragments. The skeleton can be grown one atom at a time or by joining molecular templates. The templates typically comprise rings and acyclic fragments commonly found in drug molecules [Gillett *et al.* 1993].

The final stage in the outside in approach to *de novo* design is to create 'real' molecules from the molecular skeletons that are generated. This is the most difficult part of the procedure, as there are so many ways in which even a small set of atom types can be assigned to a skeleton. The objective is to produce molecules that can be synthesised relatively easily and that also enhance the binding of the ligand to the binding site. It is very difficult to incorporate the concept of 'ease of synthesis' in a computer program, though some encouraging attempts have been made [Myatt 1995].

The outside-in procedure breaks the problem into a number of distinct stages. The inside-out method grows a ligand within the site in one step. Such an approach has been successfully used to generate peptides within protein binding sites [Moon and Howe 1991]. In this case, ligands are constructed from templates that are low-energy conformations of amino acids. Both systematic and random search algorithms can be used to explore the space of possible combinations of templates. In the systematic search, all of the amino acid building blocks are added to the growing ligand in turn. Each new structure is checked to ensure that it does not interact unfavourably with the protein and that it does not contain any high-energy intramolecular interactions. A molecular mechanics energy is then calculated for the structure. It is impractical to keep all of the structures from one stage to the next due to the combinatorial explosion and so only the lowest-energy structures are retained for the next iteration. Alternatively, a Monte Carlo simulated annealing search can be performed in which the Metropolis criterion is used at each stage to decide whether to accept or reject a given structure, based upon its energy and that of its predecessor. Genetic algorithm approaches have also been used to explore the search space [Glen and Payne 1995]. The advantage of applying this method to peptides is that there is a defined way of connecting the building blocks together, and the synthesis of peptides is straightforward. It is more difficult to generate general 'organic' molecules.

Database searching and *de novo* design programs can generate extremely large numbers of candidate structures, often far too many to inspect visually. It may therefore be useful to use cluster analysis to provide representative sets of structures for visual examination. Many structure-based methods provide an estimate of the binding free energy of each ligand which can help to decide which structures are of most interest. Unfortunately, it is very difficult to calculate binding energies accurately and rapidly. The most complete technique uses free energy perturbation (see section 9.2), but this method can only provide the relative binding energies of closely related ligands and the calculations take far too long to be practical for the thousands of suggestions that 3D database searching and *de novo* design methods can generate. It is therefore necessary to use a more approximate method. A common approach is to determine a binding energy using molecular mechanics. Such energies can be calculated rapidly using a potential grid similar to that produced by the GRID program. Suppose we wish to determine the electrostatic interaction energy between the receptor and a ligand. Prior to the search, the electrostatic potential due to the receptor atoms is calculated at each vertex on a grid superimposed on the site. The electrostatic interaction energy with a suggested ligand can then be determined by multiplying the charge on each ligand atom by the electrostatic potential at the atom's coordinates, interpolated from the neighbouring points on the grid. Other contributions to the intermolecular energy can also be determined in this way.

The energy provided by such a calculation takes no account of factors such as the desolvation of the ligand and the loss of conformational freedom. Some of these additional contributions can be significant, and misleading results will be obtained if they are not taken into account. For example, if the receptor site is highly charged then the ligands with the lowest molecular mechanics energies will usually be those that have the largest complementary charge, yet such ligands will also incur the largest desolvation penalty.

10.5.3 Structure-based design methods to design HIV-1 protease inhibitors

An impressive example of the application of structure-based methods was the design of an inhibitor of the HIV protease by a group of scientists at DuPont Merck [Lam *et al.* 1994]. This enzyme is crucial to the replication of the HIV virus and so inhibitors may have therapeutic value as anti-AIDS treatments. The starting point for their work was a series of X-ray crystal structures of the enzyme with a number of inhibitors bound. Their objective was to discover potent, novel leads which were orally available. Many of the previously reported inhibitors of this enzyme possessed substantial peptide character, and so were biologically unstable, poorly absorbed and rapidly metabolised.

The X-ray structures of the HIV protease revealed several key features that were subsequently incorporated into the designed inhibitor. The enzyme is a dimer with C_2 symmetry. It is a member of the aspartyl protease family with the two active-site aspartate residues lying at the bottom of the active site. Many of the crystal structures contained a tetracoordinated water molecule that accepted two hydrogen bonds from the backbone amide hydrogens of two isoleucine residues in the 'flaps' of the enzyme and donated two hydrogen bonds to the carbonyl oxygens of the inhibitor (Figure 10.24, colour plate section).

A flow chart showing the various phases leading to the final compound is reproduced in Figure 10.25. The first step was a 3D database search of a subset of the Cambridge Structural Database. The pharmacophore for this search comprised two hydrophobic groups, and a hydrogen bond donor or acceptor. The hydrophobic groups were intended to bind in two hydrophobic pockets (the S_1 and S_1' pockets) and the hydrogen bond donor or acceptor to bind to the catalytic aspartate residues. The search yielded the hit shown in Figure 10.25. This molecule not only contained the desired elements of the pharmacophore but it also had an oxygen atom that could displace the bound water molecule. Displacement of the water was expected to be energetically favourable due to the increase in entropy. The benzene ring in the original compound was changed to a cyclohexanone, which was able to position the substituents in a more appropriate orientation.

The DuPont Merck group had previously explored a series of peptide-based diols that were potent inhibitors but with poor oral bioavailability. They were keen to retain the diol functionality and so the next step was an expansion of the ring to a seven-membered diol. The ketone was then changed to a cyclic urea, to strengthen the hydrogen bonds to the flaps and to aid the synthesis. Further modelling studies based upon the X-ray structure were performed to predict the optimal stereochemistry and the conformation required for optimal interaction with the enzyme. The results of these studies showed that the 4*R*, 5*S*, 6*S*, 7*R* configuration was most appropriate. Nitrogen substituents were predicted to bind to the S_2 and S_2' pockets of the enzyme, and so various analogues were synthesised in order to enhance the potency whilst maintaining the desired pharmacological properties. The compound eventually chosen for further studies including clinical trials, was a *p*-hydroxymethylbenzyl derivative (Figure 10.25).

10.6 Molecular similarity

The word 'similarity' can mean many things to chemists. Two molecules may be considered 'similar' if they have common functional groups or substructures. This type of similarity may be determined by comparing the molecular graphs. An important concept in ligand design is that of *bioisosteres*, which are atoms or functional groups with similar properties. Some common bioisosteric groups are shown in Figure 10.26. Another type of similarity can be calculated from the three-dimensional structure of the molecule and is measured in terms of those properties of a molecule that are considered to be important in intermolecular interactions, such as the electrostatic potential and the shape. One way to discover new leads is to try and identify molecules that are similar to other compounds which do show the desired activity.

Several measures of the steric and electronic similarity between pairs of molecules have been devised. The *Carbo index* was developed to enable the electron density of two molecules to be compared [Carbo *et al.* 1980]. It uses the following formula:

$$S_{\mathrm{AB}} = \frac{\int \rho_{\mathrm{A}}\rho_{\mathrm{B}}\mathrm{d}v}{\left(\int \rho_{\mathrm{A}\mathrm{d}v}^{2}\right)^{1/2}\left(\int \rho_{\mathrm{B}\mathrm{d}v}^{2}\right)^{1/2}} \tag{10.1}$$

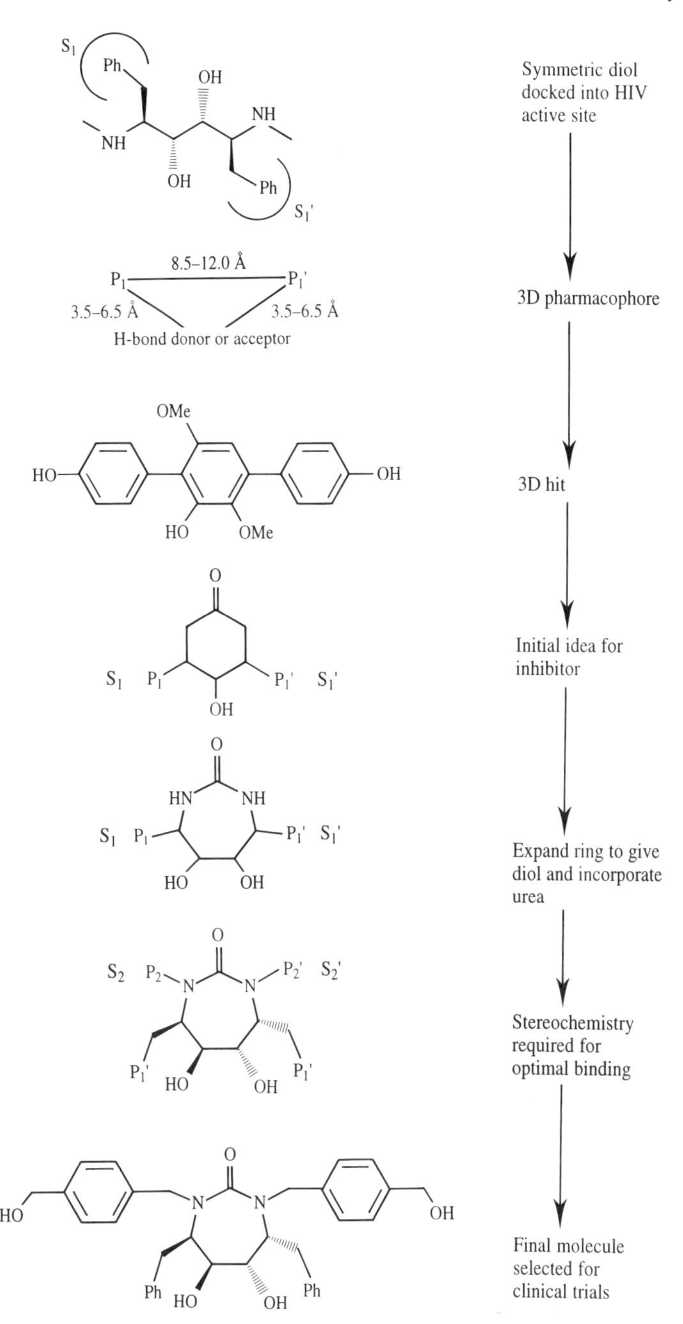

Fig. 10.25 Flow chart showing the design of novel orally active HIV-1 protease inhibitor. Figure adapted from Lam P Y S, P K Jadhav, C E Eyermann, C N Hodge, Y Ru, L T Bacheler, J L Meek, M J Otto, M M Rayner, Y N Wong, C-H Chang, P C Weber, D A Jackson, T R Sharpe and S Erickson-Viitanen 1994. Rational Design of Potent, Bioavailable, Nonpeptide Cyclic Ureas as HIV Protease Inhibitors. *Science* **263**:380–384.

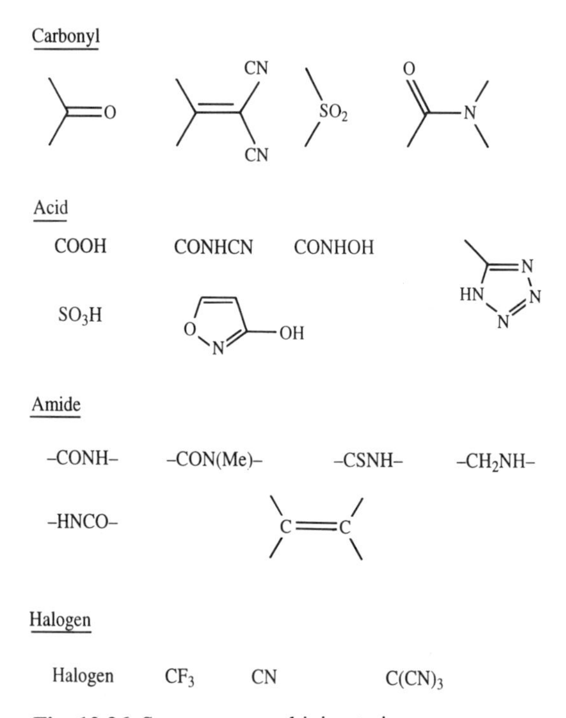

Fig. 10.26 Some common bioisosteric groups.

The electron densities at each point are determined in the usual way from the square of the wavefunction. The value of the Carbo index runs from 0 (no similarity) to 1 (perfect similarity). Unfortunately, the electron density is not an ideal measure of similarity because the density is strongest near the atomic nuclei and so the Carbo formula will be dominated by the extent to which the nuclei overlap. The electrostatic potential is a more appropriate property as it emphasises electronic effects away from the nuclei. Another drawback of the Carbo index is that it does not depend upon the magnitude of the property at a point but just its sign. This means that a location where the potential is positive from one molecule and equal but negative from the other would be weighted the same irrespective of the magnitude of the potential. Hodgkin and Richards suggested the following alternative measure of similarity for use with the electrostatic potential [Hodgkin and Richards 1987]:

$$S_{\mathrm{AB}} = \frac{2\int \phi_{\mathrm{A}}(\mathbf{r})\phi_{\mathrm{B}}(\mathbf{r})\mathrm{d}\mathbf{r}}{\int \phi_{\mathrm{A}}^{2}(\mathbf{r})\mathrm{d}\mathbf{r} + \int \phi_{\mathrm{B}}^{2}(\mathbf{r})\mathrm{d}\mathbf{r}} \tag{10.2}$$

A positive value of the Hodgkin–Richards index is obtained if large charges of the same sign are located in approximately the same regions of space; a negative value is obtained if large charges of opposite sign are located in the same regions of space.

The integrals in the Hodgkin–Richards approach can be evaluated in a number of ways. One approach is to position the molecules within a rectangular grid and to evaluate the electrostatic potential due to each molecule at each grid point. The integrals in equation (10.2) are then determined numerically by summing over the grid points. This can be rather slow, particularly if the molecules are allowed to vary their relative orientations and conformations in order to find the location of maximum similarity. An alternative is to represent the potential using an analytical function. For example, linear combinations of Gaussian functions can be fitted to the potential enabling the similarity measure to be computed much more rapidly [Good *et al.* 1993].

10.7 Quantitative structure–activity relationships

In drug design, it is always important to remember that a molecule may be required to show qualities beyond *in vitro* potency. The most potent enzyme inhibitor is of little use as a drug if it cannot reach its target. The *in vivo* activity of a molecule is often a composite of many factors. A structure–activity study can help to decide which features of a molecule give rise to its activity and help to make modified compounds with enhanced properties. A quantitative structure–activity relationship (QSAR) relates numerical properties of the molecular structure to the activity via a mathematical model. The relationship between these numerical properties and the activity is often described by an equation of the general form:

$$v = f(p) \tag{10.3}$$

v is the activity in question, p are structure-derived properties of the molecule, and f is some function. An early example of a structure–activity relationship was the discovery by Meyer and Overton of a correlation between the potencies of narcotics and the partition coefficient of the compounds between oil and water. Overton's interpretation was that the narcotic effect is due to physical changes caused by the dissolution of the drug in the lipid component of cells.

The first use of QSARs to rationalise biological activity is usually attributed to Hansch [Hansch 1969]. He developed equations which related biological activity to a molecule's electronic characteristics and hydrophobicity. For example:

$$\log(1/C) = k_1 \log P - k_2(\log P)^2 + k_3\sigma + k_4 \tag{10.4}$$

C is the concentration of the compound required to produce a standard response in a given time, log P is the logarithm of the partition coefficient of the compound between 1-octanol and water which was chosen by Hansch as a suitable measure of relative hydrophobicity, σ is the Hammett substituent parameter and k_1 through k_4 are constants.

The hydrophobic component was considered to model the ability of the drug to pass through cell membranes. A crucial contribution was the recognition that there would be an optimal value of the hydrophobicity: too low and the drug would not partition into the cell membrane; too high and the compound would partition into the membrane but tend to remain there rather than proceeding to the actual target. This explains the parabolic dependence of the activity upon log P. An alternative way to express the Hansch equation uses a parameter π. This is the logarithm of the partition coefficient of a compound with substituent X relative to a parent compound in which the substituent is hydrogen:

$$\pi = \log(P_X/P_H) \tag{10.5}$$

Thus

$$\log(1/C) = k_1\pi - k_2\pi^2 + k_3\sigma + k_4$$

The Hammett substituent parameter was used by Hansch as a concise measure of the electronic characteristics of the molecules. Hammett and others (such as Taft) showed that the positions of equilibrium and the reaction rates of series of related compounds such as substituted benzoic acids could be expressed in the following way:

$$\log\left(\frac{k}{k_0}\right) = \rho\sigma \text{ or } \log\left(\frac{K}{K_0}\right) = \rho\sigma \tag{10.6}$$

k_0 and K_0 are the rate constant and equilibrium constant respectively for a 'reference' compound (usually a hydrogen-substituted compound). The substituent parameter σ depends only upon the nature of the substituent and whether it is *meta* or *para* to the carboxyl group. The reaction constant ρ is fixed for a given process under specified experimental conditions. The 'standard' reaction is the dissociation of benzoic acids which have $\rho = 1$. A full discussion of linear free energy relationships can be found in many physical organic textbooks.

The derivation of a QSAR equation involves a number of distinct stages. First, it is obviously necessary to synthesise the compounds and determine their biological activities. QSARs can be derived for very diverse sets of compounds, but it is more common to consider a related series of compounds that differ in just one part of the molecule. These differences can then be expressed as appropriate substituent constants. When planning which compounds to synthesise, it is important to cover the range of properties that may affect the activity. For example, it would be unwise to make a series of compounds with almost identical partition coefficients. The properties chosen for inclusion in the QSAR equation should ideally be uncorrelated with each other. A straightforward way to determine the degree of correlation between two properties is to calculate a correlation coefficient. Pearson's correlation coefficient is given by:

$$r = \frac{\sum_{i=1}^{n} (x_i - \langle x \rangle)(y_i - \langle y \rangle)}{\sqrt{\left[\sum_{i=1}^{n} (x_i - \langle x \rangle)^2\right]\left[\sum_{i=1}^{n} (y_i - \langle y \rangle)^2\right]}} \tag{10.7}$$

$\langle x \rangle$ and $\langle y \rangle$ are the arithmetic means of x and y and n is the number of compounds. A value of 1.0 indicates a perfect positive correlation and the x, y coordinates lie on a straight line with a positive slope. A value of -1.0 indicates a perfect negative correlation with the x, y coordinates on a straight line with a negative slope. A value of 0.0 indicates either no correlation or that the x, y coordinates follow a non-linear scatter. It can often be very useful to plot the values of the parameters in graphical form; this can help to identify any correlations and the presence of 'outliers'. A *Craig plot* is a two-dimensional scatterplot of one parameter against another (e.g. σ against π). Ideally, the substituents chosen should lie in all four quadrants of the Craig plot which should minimise the degree of cross-correlation.

An enormous number of QSAR equations have been reported in the literature, many having a functional form much more complicated than the original Hansch equation. Many different parameters have been used in QSAR equations, most of which are designed to represent the hydrophobic, electronic or steric characteristics of the molecule. Some of these parameters are properties of the entire molecule, such as the partition coefficient. Others such as the Hammett constant σ are 'substituent values', indicating the value relative to a standard substituent (usually hydrogen). Steric effects can be modelled using parameters that are computed from the geometry of each substituent. Molar refractivity is also used to model steric effects. The molar refractivity (MR) is given by:

$$MR = \frac{(n^2 - 1)}{(n^2 + 1)} \frac{MW}{d} \tag{10.8}$$

In equation (10.8) MW is the molecular weight, d is the density and n is the refractive index. The refractive index does not vary much from one organic compound to another and as the molecular weight divided by the density equals the volume, MR indicates the steric bulk of a molecule.

Extensive tables have been published giving the values of common parameters for a wide variety of substituents. Molecular modelling methods can also be used to calculate the values of some of the properties. Many QSAR equations contain parameters that are related to the electronic structure of the molecule such as the dipole moment, the atomic charges and the orbital energies (especially the HOMO and LUMO energies). These parameters must be obtained theoretically using quantum mechanics.

An approach called molecular shape analysis includes descriptors that measure the relative shape of the compounds [Rhyu *et al.* 1995]. A conformational analysis of the compounds is performed to identify the minimum energy structures. These conformations are then overlaid on the reference structure (usually one of the most active compounds in the series). From these overlaid structures it is then possible to calculate the common overlap volume and the non-overlap volume which can then be included in the QSAR equation together with other parameters.

Another type of 'parameter' that often appears in published QSAR equations is an indicator variable. Indicator variables are used to extend a QSAR equation over a variety of different types of molecules and so make the equation more generally applicable. For example, Hansch and colleagues derived the following equation for the binding constants of sulphonamides ($X\text{-}C_6H_4\text{-}SO_2NH_2$) to human carbonic anhydrase [Hansch *et al.* 1985]:

$$\log K = 1.55\sigma + 0.64 \log P - 2.07I_1 - 3.28I_2 + 6.94 \tag{10.9}$$

I_1 takes the value 1 for *meta* substituents (0 for others) and I_2 is 1 for *ortho* substituents (0 for others).

10.7.1 The calculation of partition coefficients

Of all the parameters that may appear in a QSAR equation, the partition coefficient is found most frequently, usually as the logarithm (this converts the value onto a free energy scale). Experimental determination of the partition coefficient can be difficult, particularly for zwitterionic and very lipophilic or polar compounds. Partition coefficients are often measured between octanol and water as originally used by Hansch. In some cases it would be more appropriate to measure the partition coefficient in an alternative system and it is now possible to directly measure partition coefficients between lipid membranes and water.

Various theoretical methods can be used to calculate partition coefficients. The partition coefficient is an equilibrium constant and so is directly related to a free energy change. Partition coefficients can be calculated using free energy perturbation methods, as discussed in section 9.3.2. These methods suffer from the limitations of adequate force field parametrisation and the large amount of computer time that is required to perform such calculations. More widely used are fragment-based approaches in which the partition coefficient is calculated as a sum of individual fragment contributions plus a set of correction factors. Such an approach clearly depends critically upon the definition of the fragments. The widely used CLOGP program of Hansch and Leo [Leo 1993] uses a small number of compounds to accurately define a set of fragment values. CLOGP breaks a molecule into fragments by identifying 'isolating carbons', which are carbon atoms that are not doubly or triply bonded to a heteroatom. These carbon atoms and their attached hydrogens are considered to be hydrophobic fragments with the remaining groups of atoms being the polar fragments. A partition coefficient is calculated by adding together appropriate values for the fragments and the isolating carbons, together with various correction factors. The process is illustrated in Figure 10.27 for two simple molecules, benzyl bromide and *o*-methyl acetanilide. Benzyl bromide contains one aliphatic isolating carbon and six isolating aromatic carbons, together with one bromide fragment. Each of these

Bromide fragment	0.480
1 aliphatic isolating carbon	0.195
6 aromatic isolating carbons	0.780
7 hydrogens on isolating carbons	1.589
1 chain bond	–0.120
Total	2.924

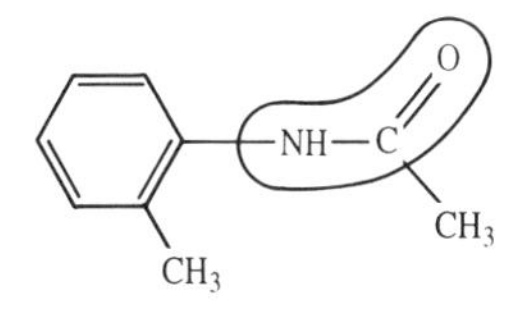

NH-amide fragment	–1.510
2 aliphatic isolating carbons	0.390
6 aromatic isolating carbons	0.780
10 hydrogens on isolating carbons	2.270
1 chain bond	–0.120
1 benzyl bond	–0.150
ortho substituent	–0.760
Total	0.900

Fig. 10.27 CLOGP calculations on benzyl bromide and *o*-methylacetanilide.

fragments contributes a characteristic score to which are added values for the seven hydrogens on the isolating carbons and a contribution from one acyclic bond. *o*-methylacetanilide contains an amide fragment, two aliphatic isolating carbons and six isolating aromatic carbons. In addition there are contributions from the hydrogen atoms, the acyclic bond, a benzyl bond (to the *o*-methyl group) and a factor due to the presence of an *ortho* substituent.

10.7.2 Deriving the QSAR equation

The most widely used technique for deriving QSAR equations is *multiple linear regression*, which uses least-squares fitting to find the 'best' combination of coefficients in the QSAR equation (the technique is also referred to as ordinary least-squares). We can illustrate the least-squares technique using the simple case where the activity is a function of just one property. We therefore want to derive an equation of the form:

$$y = mx + c \tag{10.10}$$

y is known as the dependent variable (the observations) and x is the independent variable (the parameters). For example, y might be the activity and x might be log *P*. The objective of a regression analysis is to find the coefficients m and c that minimises the sum of the deviations of the observations from the fitted equa-

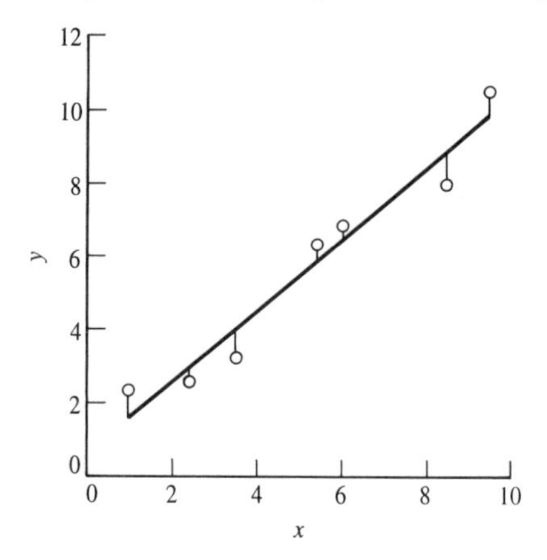

Fig. 10.28 The regression equation is the best-fit line through the data that minimises the sum of the deviations.

tion, as shown in Figure 10.28. The least-squares coefficients m and c in the linear regression equation (10.10) are given by:

$$m = \frac{\sum_{\mathrm{i}=1}^{n} (x_{\mathrm{i}} - \langle x \rangle)(y_{\mathrm{i}} - \langle y \rangle)}{\sum_{\mathrm{i}=1}^{n} (x_{\mathrm{i}} - \langle x \rangle)^2}; \qquad c = \langle y \rangle - m\langle x \rangle \tag{10.11}$$

The regression equation passes through the point $(\langle x \rangle, \langle y \rangle)$, where $\langle x \rangle$ and $\langle y \rangle$ are the means of the dependent and independent variables respectively. The 'quality' of a regression equation is often reported as the R^2 value. This indicates the fraction of the total variation in the dependent variables that is explained by the regression equation. To determine R^2 the sum of squares of the deviations of the observed y values are calculated both from the mean $\langle y \rangle$ and the predicted value from the regression equation, $y_{\mathrm{p,i}}$:

$$SS_{\mathrm{mean}} = \sum_{\mathrm{i}=1}^{n} (y_{\mathrm{i}} - \langle y \rangle)^2 \qquad SS_{\mathrm{error}} = \sum_{\mathrm{i}=1}^{n} (y_{\mathrm{i}} - y_{\mathrm{p,i}})^2 \tag{10.12}$$

$y_{\mathrm{p,i}}$ is obtained by feeding the appropriate x_{i} value into the regression equation. The R^2 is then given by:

$$R^2 = \frac{SS_{\mathrm{mean}} - SS_{\mathrm{error}}}{SS_{\mathrm{mean}}} \tag{10.13}$$

R^2 can adopt values between 0.0 and 1.0; a value of 0.0 indicates that none of the variation in the observations is explained by variation in the independent variables whereas a value of 1.0 indicates that all of the variation in the observations can be explained. A disadvantage of the standard R^2 value is that it is dependent upon the number of independent variables, with higher R^2 values being obtained for larger data sets. More sophisticated statistical measures should ideally be used; for

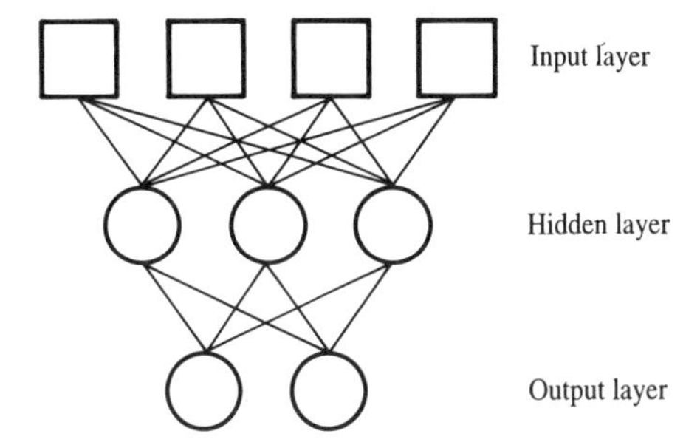

Fig. 10.29 Neural network with four input, three hidden and two output nodes.

example, one can determine whether the addition of a particular descriptor significantly contributes to the model [Montgomery and Peck 1992].

It is straightforward to extend this analysis to more than one independent variable: such calculations are tedious to perform by hand but can be performed using a statistical package. The value of the independent variables are often scaled by subtracting the mean value and then dividing by the standard deviation:

$$x'_i = \frac{x_i - \langle x \rangle}{\sqrt{\frac{1}{n-1}\sum_{i=1}^{n}(x_i - \langle x \rangle)^2}} \tag{10.14}$$

There are some important criteria to consider when using multiple linear regression. To achieve statistically significant results there should be sufficient data; it is often considered that at least five compounds are required for each parameter included in the regression analysis. The selected compounds should give a good spread of values of the parameters, which should be uncorrelated. Compounds which have a value for some parameter that is greatly different from the remainder (i.e. outliers) should be examined very closely.

10.7.3 Non-linear models: neural networks and genetic algorithms

Neural networks have been proposed as an alternative way to generate quantitative structure–activity relationships [Andrea and Kalayeh 1991]. A commonly used type of neural net contains layers of units with connections between all pairs of units in adjacent layers (Figure 10.29). Each unit is in a state represented by a real value between 0 and 1. The state of a unit is determined by the states of the units in the previous layer to which it is connected and the strengths of the weights on these connections. A neural net must first be trained to perform the desired task. To do this the network is presented with a set of sample inputs and outputs. Each input is fed along the connections to the nodes in the next layer where they are operated upon and the results fed into the next layer and so on. During the training period the network adjusts the strengths of the connections until it finds the set of values giving the best agreement between the input and output. Once trained, the net can then be used in a predictive fashion.

In QSAR, the inputs correspond to the value of the various parameters and the network is trained to reproduce the experimentally determined activities. Once trained, the activity of an unknown compound can be predicted by presenting the network with the relevant parameter values. Some encouraging results have been reported with neural networks which have also been applied to a wide range of problems such as predicting the secondary structure of proteins and interpreting NMR spectra. One of their main advantages is the ability to incorporate non-linearity into the model. However, they do present some problems [Manallack *et al.* 1994]; for example, if there are too few data values then the network may simply 'memorise' the data and have no predictive capability. Moreover, it is difficult to assess the importance of the individual terms and the networks can also require a considerable time to train.

Genetic algorithms can also be used to derive QSAR equations [Rogers and Hopfinger 1994]. The genetic algorithm is supplied with the compounds, their activities and information about their properties and other relevant descriptors. From this data the genetic algorithm generates a population of linear regression models, each of which is then evaluated to give the fitness score. A new population of models is then derived using the usual genetic algorithm operators (see section 8.5) with the parameters in the models being selected on the basis of the fitness. Unlike other methods, the genetic algorithm approach provides a family of models from which one can select either the model with the best score or can generate an 'average' model.

10.7.4 Interpreting a QSAR equation

What does one do with a QSAR equation once it has been derived? An obvious use is for predicting the activities of as yet untested (and possibly not yet synthesised) molecules. The predictive ability of a QSAR is generally more accurate for interpolative predictions (i.e. for compounds that have parameter values within the range of those considered in the data set) than for extrapolative predictions (compounds that are outside the range). A QSAR equation may provide insights into the mechanism of the process being studied. As we have already noted, the presence of a parabolic relationship between the activity and the logarithm of the partition coefficient has been interpreted in terms of the transport of a compound to the receptor. An alternative model for transport is the *bilinear model*, in which the activity is related to the partition coefficient by an equation of the following form:

$$\log(1/C) = k_1 \log P - k_2(\log(\beta P + 1)) + k_3 \tag{10.15}$$

The bilinear model enables the ascending and descending parts of the function to have different slopes, whereas the parabolic equation is symmetrical. The parabolic model is generally most applicable to complex *in vivo* systems where a drug must cross several barriers to reach its target whereas the bilinear model often gives the best fit to the data for less complex *in vitro* systems.

Quantitative structure–activity relationships are often interpreted in terms of specific interactions with the macromolecular target. In a number of cases the crystal structure of the ligand–receptor complex was subsequently determined and so it has been possible to use computer molecular graphics to discover whether the parameters in the QSAR equation have any real meaning [Hansch and Klein 1986]. For example, the presence of log P in the QSAR equation for the inhibition of carbonic anhydrase (equation (10.9)) was interpreted as a hydrophobic interaction with the enzyme. The crystal structure of the enzyme revealed the presence of just such a hydrophobic surface along which a *para*-substituted group X could lie. The negative coefficients of the indicator variables for *meta* and *ortho* substitution also had a clear interpretation: such substituents would clash with the enzyme.

The absence of a correlation may also provide useful insights. For example, if one set of parameters gives a better correlation than another then this may indicate one particular mechanism is operating. If there is no correlation with a parameter (e.g. a steric measure) for a series of compounds then this may indicate that the associated property (i.e. steric volume) is of less importance.

10.7.5 Partial least squares

Multiple linear regression cannot deal with data sets where the variables are highly correlated and/or where the number of variables exceeds the number of data values. Two methods are widely used to deal with such situations: principal components regression and partial least squares. In principal components regression the variables are subjected to a principal components analysis (described in section 8.12) and then a regression analysis is performed using the first few principal components. An alternative is to use the technique of *partial least squares* (PLS) [Wold 1982].

The PLS method expresses a dependent variable (y) in terms of linear combinations of the original independent (x) variables as follows:

$$y = b_1 t_1 + b_2 t_2 + b_3 t_3 + \cdots b_m t_m \tag{10.16}$$

where

$$t_1 = c_{11} x_1 + c_{12} x_2 + \cdots c_{1p} x_p \tag{10.17}$$

$$t_2 = c_{21} x_1 + c_{22} x_2 + \cdots c_{2p} x_p \tag{10.18}$$

$$t_m = c_{m1} x_1 + c_{m2} x_2 \cdots c_{mp} x_p \tag{10.19}$$

t_1, t_2, etc. are called *latent variables* (or components) and are constructed in such a way that they form an orthogonal set. The use of orthogonal linear combinations of the x values is very similar to principal components analysis. The major difference is that the latent variables in partial least squares are constructed to explain not

Table 10.2 Data on halogenated hydrocarbons [Dunn *et al.* 1984]

	Compound	y LD_{25}	x_1 MR	x_2 log P	x_3 BP	x_4 H_{vap}	x_5 MW	x_6 d_{20}	x_7 n_{20}^{Na}	x_8 q_C	x_9 q_{Cl}	x_{10} $E\ln C$	x_{11} $E\ln Cl$
1	CH_2Cl_2	0.96	16.56	1.25	40.0	7.57	85	1.326	1.424	0.097	−0.1083	8.88	9.96
2	$CF_3CHBrCl$	1.31	23.54	2.30	50.0	7.11	197	1.484	1.448	0.1883	−0.1001	9.72	10.04
3	$CHCl_3$	1.45	21.43	1.97	61.7	7.50	119	1.483	1.370	0.1805	−0.0870	9.69	10.16
4	CCl_4	1.53	26.30	2.83	76.5	8.27	154	1.589	1.461	0.2662	−0.0666	10.55	10.36
5	$Cl_2C=CHCl$	2.26	26.05	2.29	86.5	8.01	131	1.465	1.456	0.1175	−0.0696	9.90	10.33
6	$Cl_2C=CCl_2$	2.26	30.45	2.60	121.0	9.24	166	1.623	1.506	0.1360	−0.0680	10.08	10.34
7	$CHCl_2CHCl_2$	2.42	30.92	2.66	146.0	9.92	168	1.587	1.494	0.1370	−0.1018	9.27	10.02

The parameters are as follows: LD_{25}: total toxicity measure; MR: molar refractivity; log P: logarithm of the partition coefficient; BP: boiling point; H_{vap}: latent enthalpy of vaporisation; MW: molecular weight; d_{20}: density at 20°C; n_{20}^{Na}: refractive index at 20°C measured using sodium light. q_C, q_{Cl}: charges on the chlorine-bearing carbon and the chlorine atom respectively E ln C; E ln Cl: orbital electronegativities of C, Cl.

only the variation in the x data but also to maximise the degree to which the variation in the observations can be explained.

We will illustrate the partial least squares method using a data set published by Dunn *et al.* which provides the toxicity of a series of halogenated hydrocarbons together with eleven descriptor variables (see Table 10.2).

The first seven variables ($x_1 \ldots x_7$) are standard global molecular descriptors and the final four variables ($x_8 \ldots x_{11}$) are calculated measures of the electronic character of each molecule. Each of the seven compounds is thus described in terms of eleven variables. However, many of these variables are highly correlated with each other. For example, the charge on the chlorine is perfectly correlated with the electronegativity of the chlorine (a consequence of the way in which these two parameters were calculated, using the Gasteiger and Marsili method). Another strong correlation is that between the molar refractivity and the boiling point (correlation coefficient = 0.92).

A partial least squares analysis [Malpass 1994] provides the weightings of the original variables in the latent variables. For example, the weightings for the first three latent variables are given in Table 10.3.

These results suggest that all of the variables contribute to the first component with the higher weightings being due to the molar refractivity (x_1), log P (x_2), boiling point (x_3), latent heat of vaporisation (x_4), d_{20} (x_6) and n_{20}^{Na} (x_7). The first latent variable thus represents a combination of steric, hydrophobic and electronic factors. The highest weightings in the second component are for the charge on the carbon (x_8) and the electronegativity of the carbon (x_{10}) and so this component has a higher contribution from electronic effects.

It is also possible to calculate the degree to which each component explains the variance in each variable, and how far each component explains the variation in the dependent variable (Table 10.4).

The results in Table 10.4 reinforce our earlier conclusion that the first component explains most of the steric and hydrophobic effects with the second component explaining electronic effects. The first component explains 74.1% of the variation in the observed activity, with the first two components explaining a total of 86.9% of the variation.

10.7.6 Partial least squares and molecular field analysis

One of the most popular uses of the partial least squares method in molecular modelling and drug design is comparative molecular field analysis (CoMFA), first described by Cramer and co-workers [Cramer *et al.* 1988]. The starting point for a CoMFA analysis is a set of conformations, one for each molecule in the set. Each conformation should be the presumed active structure of the molecule. The conformations must be overlaid in the suggested binding mode. The molecular fields surrounding each molecule are then calculated by placing appropriate probe groups at points on a regular lattice that encompasses the molecule, in a manner analogous to that used by the GRID program. The results of this analysis

Table 10.3 Weightings of the various parameters in the first three latent variables.

Component	x_1	x_2	x_3	x_4	x_5	x_6	x_7	x_8	x_9	x_{10}	x_{11}
1	0.4320	0.3197	0.4428	0.3875	0.1896	0.3265	0.3271	−0.1038	0.2133	0.1219	0.2105
2	−0.0850	0.2172	−0.2863	−0.1765	0.2833	0.2322	0.0479	0.7111	0.0936	0.4147	0.0982
3	0.1273	0.0985	0.0346	−0.3307	−0.1061	−0.1416	−0.5048	−0.2248	0.4841	0.2352	0.4853

Table 10.4 The degree to which each component explains the variance in each variable, and in the dependent variable (last column).

Component	x_1	x_2	x_3	x_4	x_5	x_6	x_7	x_8	x_9	x_{10}	x_{11}	Total
1	97.2	81.5	78.8	67.7	39.8	86.1	66.6	3.6	32.6	30.2	32.1	74.1
2	0.2	14.6	16.4	21.2	12.8	6.8	5.4	84.5	18.4	57.3	19.2	12.8
3	0.3	1.1	0.5	2.2	22.5	0.9	5.7	3.4	45.9	11.4	45.5	4.7

can be represented as a matrix, **S**, in which each row corresponds to one of the molecules and the columns are the energy values at the grid points (Figure 10.30). If there are N points in the grid and P probe groups are used, then there will be $N \times P$ such columns. The table is completed by adding an additional column that contains the activity of the molecule.

A correlation between the biological activity and the field values is then determined. The general form of the equation that we desire is:

$$\text{activity} = C + \sum_{\mathrm{i}=1}^{N} \sum_{\mathrm{j}=1}^{P} c_{\mathrm{ij}} S_{\mathrm{ij}} \tag{10.20}$$

c_{ij} is the coefficient for the column in the matrix that corresponds to placing probe group j at grid point i. This problem is massively overdetermined as there may be thousands of grid points but often fewer than 30 compounds. However, a successful analysis may often be performed using partial least squares.

The maximum number of latent variables is the smaller of the number of x values or the number of molecules. However, there is an optimum number of latent variables in the model beyond which the predictive ability of the model does not increase. A PLS model is often evaluated according to its ability to predict the activity of compounds not used to derive the model (rather than how well the model reproduces the activity of the compounds actually used to construct the model). A variety of statistical methods can be used to achieve this. One of the most commonly used techniques is *cross-validation*, in which the predictive ability of the model is determined by dividing the data into a number of groups. A series of models are derived by leaving out one of the groups. The omitted data is then used to test the model by calculating the differences between the predicted and actual values. This measures the predictive ability of the model, which is usually reported as the PRESS (predictive residual sum of squares). An alternative way to assess the significance of a model is to randomly reassign the activities, thereby associating the 'wrong' activity with each set of grid values. When this is done then the predictive ability

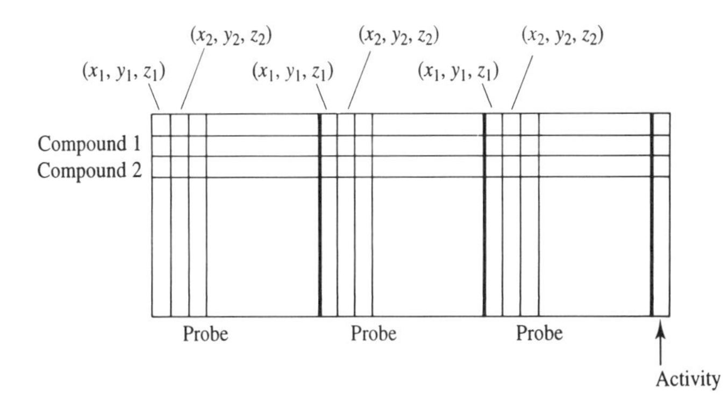

Fig. 10.30 The data structure used in a CoMFA analysis.

of the model should be significantly better for the true data set than for any of the randomised sets.

The CoMFA approach generates a coefficient for each column in the data table. This coefficient indicates the significance of each grid point in explaining the activity. Such data can usefully be represented as a three-dimensional surface that connect points having the same coefficients. These diagrams have been used to identify regions where (for example) changing the steric bulk would increase or decrease binding.

Since its introduction partial least squares has been widely used to calculate 3D QSARs. These studies have demonstrated the validity and usefulness of the approach but have also highlighted the sensitivity of the approach to several factors. These factors include the selection of the active compounds, the different types of probe that can be employed, the force field models to describe the interactions between the probe and each compound, the size and spacing of points in the grid, and indeed the way in which the PLS analysis is performed. One of the main requirements (and indeed limitations) of the CoMFA technique is that it requires the structures of the molecules to be correctly overlaid in what is assumed to be the bioactive conformation (this in turn implies that the compounds have a common binding mode). The first application of CoMFA was to a series of steroid molecules binding to two different targets: human corticosteroid globulins and testosterone binding globulins. In this case the steroid nucleus of each molecule was least-squares fitted to the nucleus of the most active steroid. It can be more difficult to determine the appropriate binding mode in other cases though the pharmacophore identification programs discussed in section 10.2 may help with this problem.

Further reading

Kubinyi H (Editor) 1993. 3D QSAR in *Drug Design, Theory, Methods and Applications*. Leiden, ESCOM Science Publishers.

Kubinyi H 1995. The Quantitative Analysis of Structure–Activity Relationships in Wolff M E (Editor). *Burger's Medicinal Chemistry and Drug Discovery.* 5th Edition, Volume 1. New York, John Wiley: 497–571.

Marshall G R 1955. Molecular Modeling in Drug Design in Wolff M E (Editor). *Burger's Medicinal Chemistry and Drug Discovery.* 5th Edition, Volume 1. New York, John Wiley: 573–659.

Martin Y C. *Quantitative Drug Design. A Critical Introduction.*

Tute M S 1990. History and Objectives of Quantitative Drug Design. In Hansch C, P G Sammes and J B Taylor (Editors). *Comprehensive Medicinal Chemistry Volume 4*, Oxford, Pergamon Press: 1–31.

Waterbeemd H van de 1995. *Chemometric Methods in Molecular Design.* Weinheim, VCH Publishers.

References

Andrea T A and H Kalayeh 1991. Applications of Neural Networks in Quantitative Structure–Activity-Relationships of Dihydrofolate-Reductase Inhibitors. *Journal of Medicinal Chemistry* **34**:2824–2836.

Blaney J M and J S Dixon 1993. A Good Ligand is Hard to Find: Automated Docking Methods. *Perspectives in Drug Discovery and Design* **1**:301–319.

Böhm H J 1992. LUDI—Rule-Based Automatic Design of New Substituents for Enzyme-Inhibitor Leads. *Journal Of Computer-Aided Molecular Design* **6**:593–606.

Brick P, T N Bhat and D M Blow 1989. Structure of Tyrosyl-tRNA Synthetase Refined at 2.3 Å Resolution. Interaction of the Enzyme with the Tyrosyl Adenylate Intermediate. *Journal of Molecular Biology* **208**:83–98.

Bron C and J Kerbosch 1973. Algorithm 475. Finding All Cliques of an Undirected Graph. *Communications of the ACM* **16**:575–577.

Carbo R, L Leyda and M Arnau 1980. An Electron Density Measure of the Similarity Between Two Compounds. *International Journal of Quantum Chemistry* **17**:1185–1189.

Cramer R D III, D E Patterson and J D Bunce 1988. Comparative Molecular Field Analysis (CoMFA). 1. Effect of Shape on Binding of Steroids to Carrier Proteins. *The Journal of the American Chemical Society* **110**:5959–5967.

Dammkoehler R A, S F Karasek, E F B Shands and G R Marshall 1989. Constrained Search of Conformational Hyperspace. *Journal of Computer-Aided Molecular Design* **3**:3–21.

Dunn W J III, S Wold, U Edlund, S Hellberg, J Gasteiger 1984. Multivariate Structure–Activity Relationships Between Data from a Battery of Biological Tests and an Ensemble of Structure Descriptors: The PLS Method. *Quantitative Structure–Activity Relationships* **3**:1313–137.

Gillet V J, A P Johnson, P Mata, S Sik and P Williams 1993. SPROUT—A Program for Structure Generation. *Journal of Computer-Aided Molecular Design* **7**:127–153.

Glen R C and A W R Payne 1995. A Genetic Algorithm for the Automated Generation of Molecules within Constraints. *Journal Of Computer-Aided Molecular Design* **9**:181–202.

Good A C, E E Hodgkin and Richards W G 1993. The Utilisation of Gaussian Functions for the Rapid Evaluation of Molecular Similarity. *Journal of Chemical Information and Computer Science* **32**:188–192.

Goodford P J 1985. A Computational Procedure for Determining Energetically Favorable Binding Sites on Biologically Important Macromolecules. *Journal of Medicinal Chemistry* **28**:849–857.

Goodsell D S and A J Olson 1990. Automated Docking of Substrates to Proteins by Simulated Annealing. *Proteins: Structure, Function and Genetics* **8**:195–202.

Hansch C 1969. A Quantitative Approach to Biochemical Structure–Activity Relationships. *Accounts of Chemical Research* **2**:232–239.

Hansch C and T E Klein 1986. Molecular Graphics and QSAR in the Study of

Enzyme–Ligand Interactions. On the Definition of Bioreceptors. *Accounts of Chemical Research* **19**:392–400.

Hansch C, J McClarin, T Klein and R Langridge 1985. A Quantitative Structure–Activity Relationship and Molecular Graphics Study of Carbonic Anhydrase Inhibitors. *Molecular Pharmacology* **27**:493–498.

Hodgkin E E and W G Richards 1987. Molecular Similarity Based on Electrostatic Potential and Electric Field. *International Journal of Quantum Chemistry. Quantum Biology Symposia* **14**:105–110.

Jones G, P Willett and R C Glen 1995. Molecular Recognition of Receptor-Sites Using a Genetic Algorithm with a Description of Desolvation. *Journal of Molecular Biology* **245**:43–53.

Judson R S, E P Jaeger and A M Treasurywala 1994. A Genetic Algorithm-Based Method for Docking Flexible Molecules. *Journal of Molecular Structure: Theochem* **114**:191–206.

Kuntz I D 1992. Structure-Based Strategies for Drug Design and Discovery. *Science* **257**:1078–1082.

Kuntz I D, J M Blaney, S J Oatley, R Langridge and T E Ferrin 1982. A Geometric Approach to Macromolecule–Ligand Interactions. *Journal of Molecular Biology* **161**:269–288.

Kuntz I D, E C Meng and B K Shoichet 1994. Structure-Based Molecular Design. *Accounts of Chemical Research* **27**:117–123.

Lam P Y S, P K Jadhav, C E Eyermann, C N Hodge, Y Ru, L T Bachelor, J L Meek, M J Otto, M M Rayner, Y N Wong, C-H Chang, P C Weber, D A Jackson, T R Sharpe and S Erickson-Viitanen 1994. Rational Design of Potent, Bioavailable, Nonpeptide Cyclic Ureas as HIV Protease Inhibitors. *Science* **263**:380–384.

Lauri G and P A Bartlett 1994. CAVEAT—A Program to Facilitate the Design of Organic Molecules. *Journal of Computer-Aided Molecular Design* **8**:51–66.

Leo A J 1993. Calculating log P_{oct} from Structures. *Chemical Reviews* **93**:1281–1306.

Lewis R M and A R Leach 1994. Current Methods for Site-Directed Structure Generation. *Journal of Computer-Aided Molecular Design* **8**:467–475.

Malpass J A 1994. *Continuum Regression: Optimised Prediction of Biological Activity.* PhD Thesis, University of Portsmouth (UK).

Manallack D T, D D Ellis and D J Livingstone 1994. Analysis of Linear and Nonlinear QSAR Data Using Neural Networks. *Journal of Computer-Aided Molecular Design* **37**:3758–3767.

Martin Y C, M G Bures, A A Danaher, J DeLazzer, I Lico and P A Pavlik 1993. A Fast New Approach to Pharmacophore Mapping and its Application to Dopaminergic and Benzodiazepine Agonists. *Journal of Computer-Aided Molecular Design* **7**:83–102.

Miranker A and M Karplus 1991. Functionality Maps of Binding-Sites—A Multiple Copy Simultaneous Search Method. *Proteins: Structure, Function and Genetics* **11**:29–34.

Montgomery D C and A A Peck 1992. *Introduction to Linear Regression analysis.* New York, Wiley.

Moon J B and W J Howe 1991. Computer Design of Bioactive Molecules—A Method for Receptor-Based Denovo Ligand Design. *Proteins: Structure, Function And Genetics* **11**:314–328.

Myatt G 1995. Computer-Aided Estimation of Synthetic Accessibility. PhD thesis, University of Leeds.

Oshiro C M, I D Kuntz and J S Dixon 1995. Flexible Ligand Docking Using a Genetic Algorithm. *Journal of Computer-Aided Molecular Design* **9**:113–130.

Priestle J P, A Fassler, J Rosel, M Tintelnot-Blomley, P Strop and M G Gruetter 1995. Comparative Analysis of The X-Ray Structures of HIV-1 and HIV-2 Proteases in Complex with a Novel Pseudosymmetric Inhibitor. *Structure (London)* **3**:381–389

Pugliese L, A Coda, M Malcovati, and M Bolognesi 1993. Three-Dimensional Structure of the Tetrahedral Crystal Form of Egg-White Avidin in its Complex with Biotin at 2.7 Å Resolution. *Journal of Molecular Biology* **231**:698–710.

Rhyu K-B, H C Patel and A J Hopfinger 1995. A 3D-QSAR Study of Anticoccidal Triazines Using Molecular Shape Analysis. *Journal of Chemical Information and Computer Science* **35**:771–778.

Rogers D and A J Hopfinger 1994. Application of Genetic Function Approximation to Quantitative Structure–Activity Relationships and Quantitative Structure–Property Relationships. *Journal of Chemical Information and Computer Science* **34**:854–866.

Rusinko A III, J M Skell, R Balducci, C M McGarity and R S Pearlman 1988 CONCORD: a program for the rapid generation of high quality 3D molecular structures. The University of Texas at Austin and Tripos Associates: St Louis MO.

Sheridan R P, R Nilakantan, J S Dixon and R Venkataraghavan 1986. The Ensemble Approach to Distance Geometry: Application to the Nicotinic Pharmacophore. *Journal of Medicinal Chemistry* **29**:899–906.

Von Itzstein M, W Y Wu, G B Kok, M S Pegg, J C Dyason, B Jin, T V Phan, M L Smythe, H F White, S W Oliver, P M Colman, J N Varghese, D M Ryan, J M Woods, R C Bethell, V J Hotham, J M Cameron and C R Penn 1993. Rational Design of Potent Sialidase-Based Inhibitors of Influenza-Virus Replication. *Nature* **363**:418–423.

Wold H 1982. Soft Modeling. The Basic Design and Some Extensions in K-G Joreskog and H Wold (Editors). *Systems under Indirect Observation*. Volume II. Amsterdam, North Holland.

Index